Advances in Total Knee Arthroplasty

Shuichi Matsuda
Editor

Advances in Total Knee Arthroplasty

JSRA
JAPANESE SOCIETY FOR
REPLACEMENT ARTHROPLASTY

Springer

Editor
Shuichi Matsuda
Department of Orthopaedic Surgery
Kyoto University
Kyoto, Japan

ISBN 978-981-97-4919-5 ISBN 978-981-97-4920-1 (eBook)
https://doi.org/10.1007/978-981-97-4920-1

The translation was completed using an artificial intelligence translation tool. Following this, a human revision was conducted to further refine the work, ensuring that the translation is accurate in terms of content and scientific correctness.

Translation from the Japanese language edition: "Jinkou Hiza Kansetsu Chikan Shujutsu" by Shuichi Matsuda, © The Japanese Society for Replacement Arthroplasty 2022. Published by Nankodo Co., Ltd.. All Rights Reserved.

This Springer imprint is published by the registered company Springer Nature Singapore Pte Ltd.
The registered company address is: 152 Beach Road, #21-01/04 Gateway East, Singapore 189721, Singapore

Preface 1

We are pleased to announce the publication of two academic books, "Advances in Total Hip Arthroplasty" and "Advances in Total Knee Arthroplasty," edited by the Japanese Society for Replacement Arthroplasty (JSRA). Japan has contributed greatly to the progress of joint prostheses and has numerous achievements in basic research on biomaterials and biomechanics as well as various clinical studies. The JSRA plans to publish books with the intention of reconciling arthroplasty research in Japan.

To keep consistency in the structure of the two books, it was decided to follow the same flow: history; basic research, such as biomechanics and biomaterials; surgical techniques; and postoperative results. Hip and knee implants developed in Japan were also introduced.

Since these books are positioned as summaries of arthroplasty research in Japan, the authors were requested to summarize their own research but also additional research conducted in Japan. We thank the authors for their understanding of this objective and for their cooperation during the editing process. It is not possible to produce a logical structure based solely on Japanese research, but we trust that you will agree that Japanese research has made a significant contribution to the progress in arthroplasty surgery.

Many problems associated with arthroplasty remain unresolved. We sincerely hope that this book will motivate orthopedic surgeons, especially young doctors, to attempt to solve these problems through research and encourage further progress in arthroplasty research.

Finally, we would like to express our deepest gratitude to the editorial board members for their hard work and dedication to the editorial process.

President, The Japanese Society for Replacement Arthroplasty — Shuichi Matsuda
Kyoto University
Kyoto, Japan

Preface 2

We are pleased to announce the publication of the book "Advancement in Total Knee Arthroplasty," edited by the Japanese Society for Replacement Arthroplasty (JSRA). As an editorial representative, I would like to offer my greetings. There are many academic books on total knee arthroplasty, but since this was edited by the JSRA, the editorial committee has discussed the book structure in length to ensure that the contents are substantial. Leading Japanese surgeons, who have globally presented their research, were requested to contribute to this book. We would like to thank all the authors for taking the time out of their busy schedules to write chapters for this book.

Here, we introduce some of the content. The history of total knee arthroplasty is introduced first, and the concepts and clinical results of knee prostheses developed in Japan are introduced by developers or doctors familiar with the development of these knee prostheses. In the area of basic research, we have devoted considerable text to the research on anatomical landmarks, kinematics, and biomaterials, research areas in which Japan are global leaders. Regarding the surgical techniques, it was considered important to explain the surgical objectives and rationale prior to describing the specific techniques. Alignment and ligament balance are discussed based on current recommended surgical objectives, centered on biomechanics and clinical outcomes. Experienced surgeons compiled the preoperative planning and various surgical techniques. Computer-assisted surgery has also been covered in detail as this field has made great progress, including robot-assisted surgery. Improvement in range of motion is one of the most important aspects of the Japanese lifestyle and is also covered in detail. Postoperative rehabilitation and functional assessments have made significant progress in recent years, and the latest findings have been included. Finally, several sections on dealing with difficult cases and complications have been compiled by the authors.

We hope that through this volume, readers will deepen their understanding of the history of knee arthroplasty, surgical techniques, and the rationale behind them which will motivate future research for further progress.

Department of Orthopaedic Surgery
Kyoto University
Kyoto, Japan

Shuichi Matsuda

Contents

Part II Anatomy

Part III Biomechanics

Part IV Biomaterials

Part XII Postoperative Rehabilitation

Part XIII Imaging and Functional Assessment

Part XIV Difficult Cases

Part XV Complications

Part I

History of TKA

1 History of Total Knee Arthroplasty Development

Akiho Hoshino

1 World's Oldest Knee Prosthesis

The world's oldest knee prosthesis was performed by Themistocles Gluck in 1890.

Gluck, who was born in Romania and later became a physician in Germany, used an artificial knee made of ivory for surgery performed on a 17-year-old woman with tuberculous arthritis [1] (Fig. 1.1).

For approximately 50 years following Gluck's surgery, no new artificial knee joints were developed. During that time, resection arthroplasty or intermediate insertion membrane surgery served as the primary method for knee arthroplasty.

Intermediate-membrane surgery involved inserting various soft tissues into the damaged joint. Different materials, such as human skin, muscle fascia, chromed porcine bladder, cellophane, and nylon, were used; however, the results were not satisfactory.

In Japan, Jinnaka et al. from Kyushu University used chrome-treated femoral fascia (J-K membrane) for knee surgeries [2].

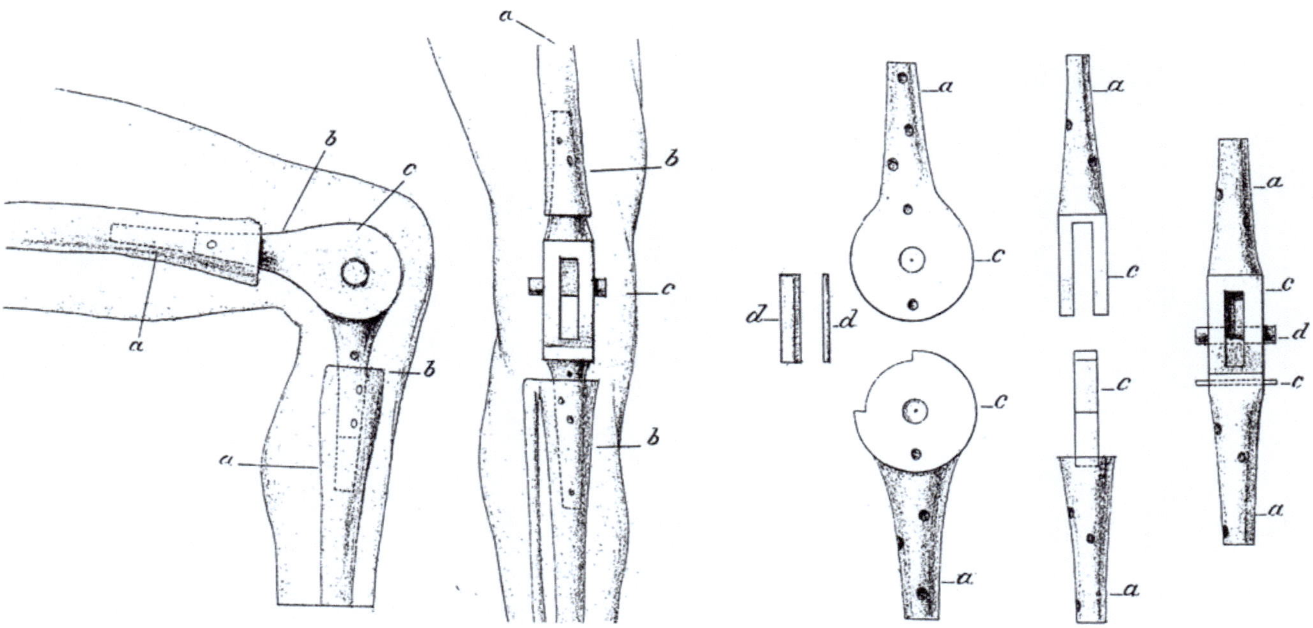

Fig. 1.1 Gluck: Ivory knee prosthesis. (Brand et al. [1])

A. Hoshino (✉)
Kawaguchi Kogyo General Hospital, Kawaguchi, Japan
e-mail: hoshino@kogyohsp.gr.jp

S. Matsuda (ed.), *Advances in Total Knee Arthroplasty*, https://doi.org/10.1007/978-981-97-4920-1_1

2 Dawn of Artificial Knee Joints

Before the introduction of full-scale hinge-type knee prosthetic joints, attempts were made to develop femoral molds and tibial trays as hemi-artificial replacements for either the femoral or tibial sides of the knee joint.

Among the oldest femoral molds was the Campbell (1940) [3], which was inspired by the success of the Smith-Petersen cup. Another significant model was the Jones implant, known as the MGH Femoral Condylar Replacement (1967), which featured a stem [4] (Fig. 1.2).

A tibial tray is essentially a metal plate that replaces only the tibial articular surface. Examples of early tibial trays include the designs by McKeever [5] (Fig. 1.3) and McIntosh [6]. These trays were fixed in place without the use of cement, using driving flanges or screws for stabilization.

While the tibial tray allowed for alignments by changing the thickness of the tray, there were several complications, including deteriorations in the contralateral joints and fixation failures. Consequently, the initially promising short-term results did not prove to be durable in the long run.

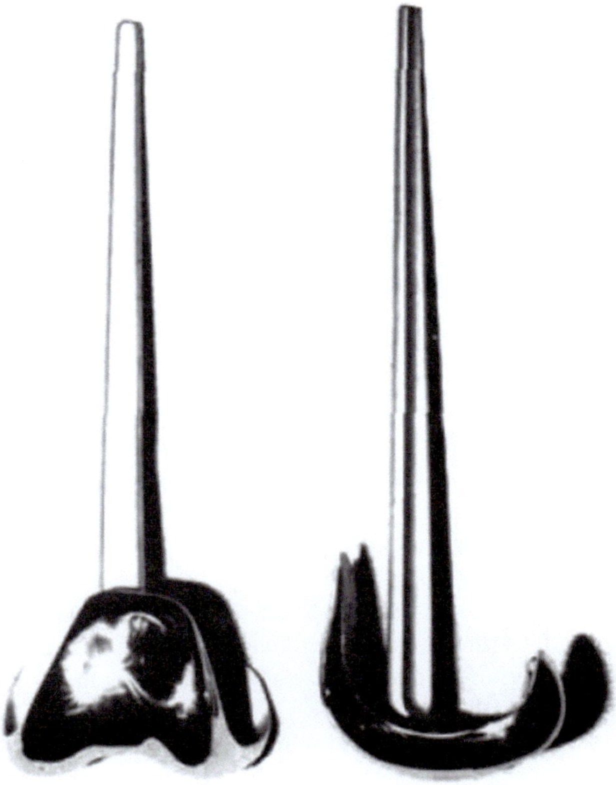

Fig. 1.2 Jones: MGH femoral mold. (Jones et al. [4])

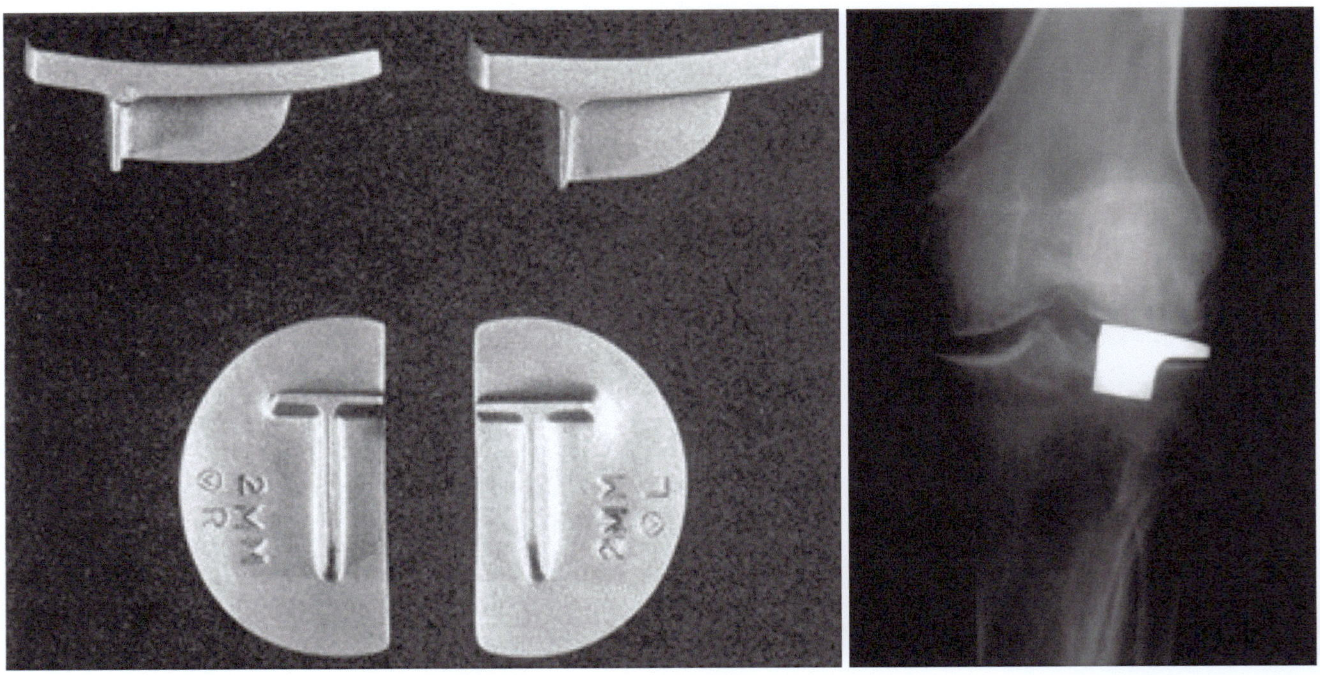

Fig. 1.3 McKeever: tibial tray. (By courtesy of Stryker Japan)

3 Age of Hinge

In the 1960s, full-scale artificial knee joints that replaced both the femoral and tibial articular surfaces were introduced for the first time.

The premise of these appearances was the utilization of stable biomaterials.

One such biomaterial, Vitallium, which is a chromium-cobalt-molybdenum alloy, was initially used by Petersen in 1939 as a cup for the femoral head.

Another significant material is ultra-high-molecular-weight polyethylene, which has been used as a sliding surface material since Charnley introduced it in total hip arthroplasty in 1962.

There are many works and papers that mistakenly refer to polyethylene as high-density polyethylene (HDP), but these statements are incorrect. It seems that this misconception originated from Charnley's own misinterpretation of describing polyethylene as "HDP" in his book.

However, it is important to clarify that HDP has never been used in the history of artificial joints [7].

In addition to Gluck's ivory fabrication, the world's first artificial knee joint was an acrylic hinge (1950) [8] developed by Majnoni d'Intignano. While it has been mentioned that the Judet brothers' (1947) acrylic hinge [9] was the first in the world, it was only manufactured and not used clinically. Therefore, Majnoni d'Intignano's acrylic hinge is considered to be the world's first practical application.

The first commonly used knee prosthesis was Walldius [10]. The first Walldius model was made of acrylic but was later replaced with a cobalt-chromium alloy (1958) (Fig. 1.4). However, the cementless fixation and limited motion (restricted to extension or flexion motion only, with one degree of freedom) often led to loosening. Additionally, the flexion angle was limited to <90° because the center of rotation was located at the intersection of the femoral and tibial axes (in the middle of the knee).

Shiers (1954) made improvements by offsetting the axis of rotation backward and utilizing a cementless long stem for fixation [11]. However, this approach also resulted in stem loosening due to the limited degree of freedom.

The Guepar hinge (1972), developed by an orthopedic group in Paris, marked the pinnacle of the hinge era. It featured a rear-offset axis of rotation, a cement-fixed stem with a valgus angle of 7°, a silicone stop bumper in the extension position, and an anterior flange for patellar gliding [12].

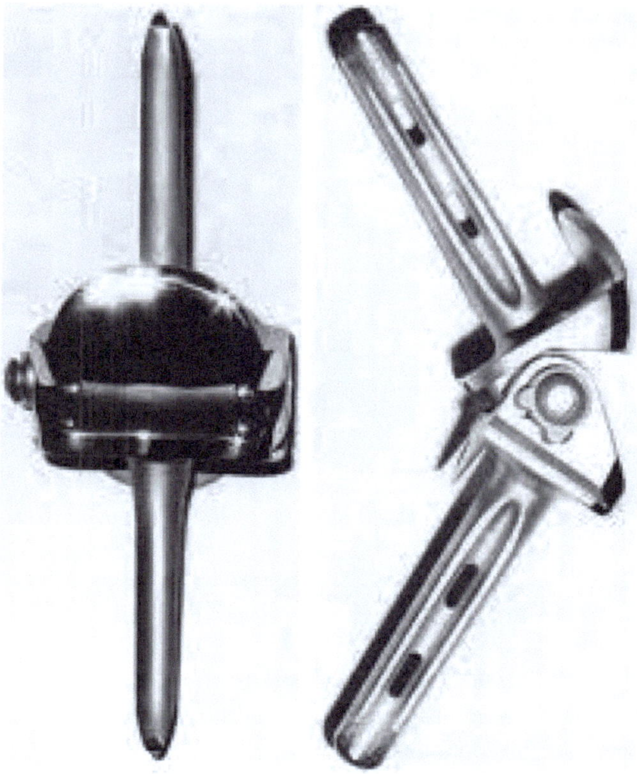

Fig. 1.4 Walldius hinge. (By courtesy of Stryker Japan)

Subsequently, a rotating hinge design, primarily used for revision surgeries, was devised and continues to be used to this day.

4 Link Type

The link type of knee prosthesis offers a distinct design concept separate from these hinge types. It features a complex shape design and serves as a bridge passage between the hinge type and the surface replacement type, as explained further below.

The metal ball protruding from the tibial plane formed a ball-and-socket joint with the femoral cup. All articular surfaces are composed of metal-on-plastic and fixed in place with cement. Unlike the hinge axis, there is no specific axis of rotation, theoretically allowing for six degrees of freedom. However, in practice, it typically provides around three degrees of freedom. However, it allows for progressive varus or valgus motions and internal or external rotations.

Representative designs in this category include the spherocentric knee (1973) [13] (Fig. 1.5) and the Attenborough knee (1978) [14].

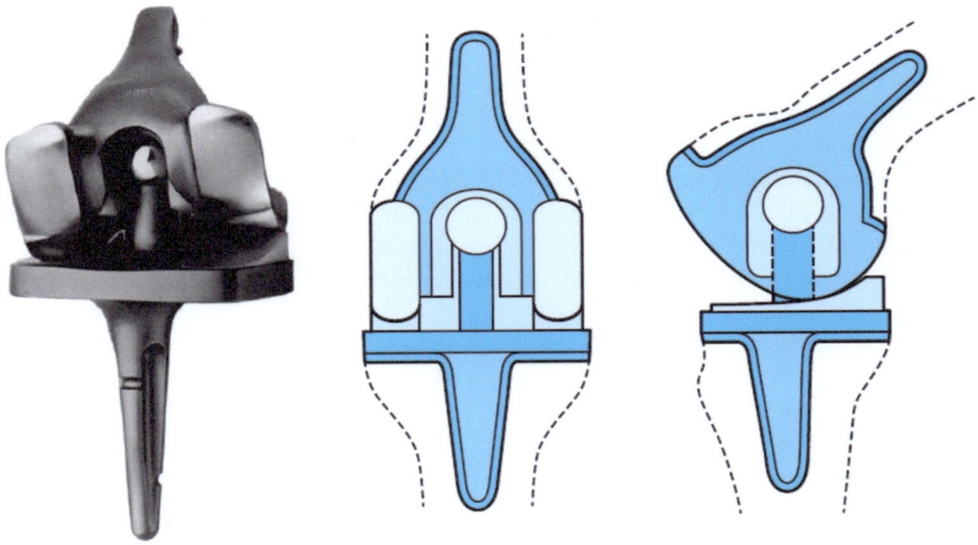

Fig. 1.5 Matthews: Spherocentric knee. (By courtesy of Stryker Japan)

5 Cambrian

During the 1970s to the late 1980s, the Cambrian period of artificial knee joints emerged, and all types of design ideas emerged and continue to this day.

An artificial knee joint with a surface replacement design different from that of the hinge and link was developed.

Gunston's polycentric knee (1971) served as the foundation for this design [15] (Fig. 1.6). Gunston developed the first non-constrained design, drawing from his experience working with Charnley in the United Kingdom and incorporating elements from total hip arthroplasty, such as metal-on-plastic sliding surfaces and cement fixation.

In this design, a wheel-like femoral component known as the runner glides on a polyethylene track on the tibial surface, with both components being cemented in place. While the femoral component is unicentric (single radius), achieving proper positioning for all four components during surgery can be challenging.

In 1971, bone cement, known as polymethyl methacrylate, which had already been in use in Europe, received approval from the United States Food and Drug Administration for its use.

The first surface replacement knee prosthesis, which integrated the femur and tibia to form a single articular surface instead of four separate components, as in polycentric design, was the Geomedic knee (1971) [16] (Fig. 1.7) developed by Coventry of the Mayo Clinic.

The Geomedic knee featured a single-radius femoral articular surface connected by a metal bar, with the tibial joint surface connected anteriorly. This design had high constraints, and maintaining the cruciate ligaments resulted in kinematic conflict and loosening. Subsequently, the restraint of the tibial articular surface was reduced in the later improved Geomedic knee (1973).

The Geomedic design trend further evolved into various designs, including the anametric knee in Howmedica and the

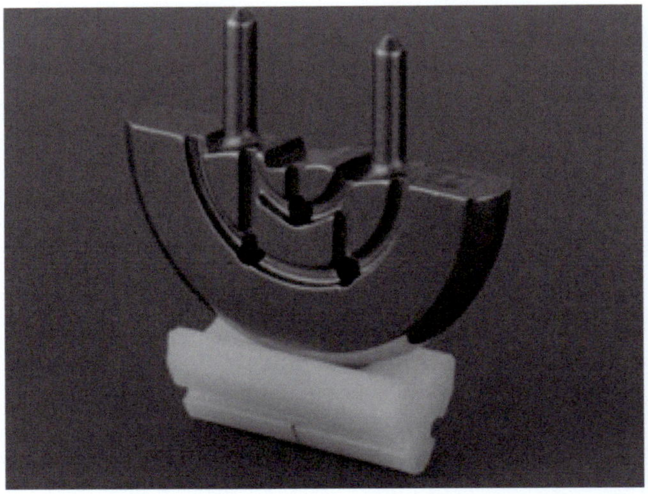

Fig. 1.6 Gunston: Polycentric knee. (By courtesy of Stryker Japan)

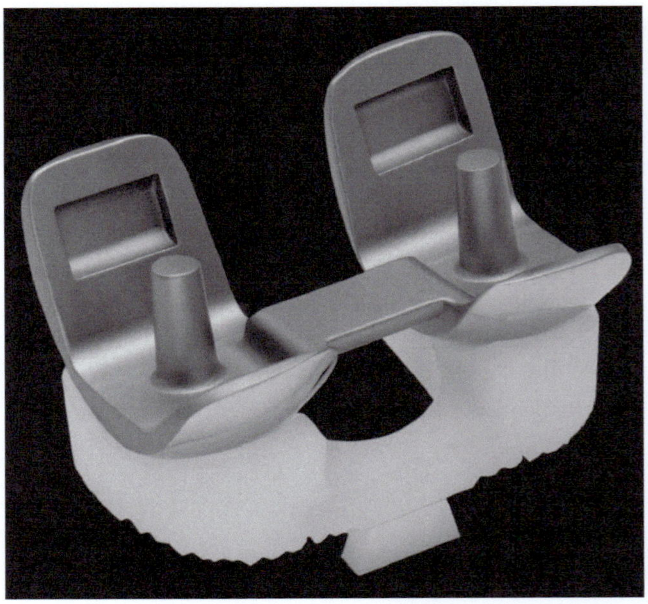

Fig. 1.7 Coventry: Geomedic knee. (By courtesy of Stryker Japan)

Geotibial, Geopatellar, Gipitelar, and Multi-Radius designs in Zimmer.

Subsequently, the integrated femoral and tibial joint surfaces, known as the condylar design, became the mainstream approach for artificial knee joints. The condylar design can be further categorized into two types: anatomical design and functional design.

The anatomical design has an anatomical shape and attempts to obtain stability using ligaments, whereas the functional design primarily reproduces the functional aspects of the knee joint and does not strictly adhere to the anatomical shape.

5.1 Anatomical Design

The oldest anatomical design was the Kodama-Yamamoto MK I (1970) [17] (Fig. 1.8) developed by Sumiki Yamamoto.

The femur was made of a cobalt-phosphorus alloy with a patellar flange, and the unconstrained insert, which preserved the anterior and posterior cruciate ligaments, was cementless and fixed with staples. Yamamoto's design was influenced by Michael Freeman between 1974 and 1975, and in 1985, the MK I design was improved to MK II and MK III, showcasing creative advancements distinct from Freeman's designs.

Another sequential anatomical design was the UCI knee (1972) [18] developed by Ted Waugh et al. It featured femoral components with multiple turning radii and unconstrained horseshoe-shaped inserts, aiming to allow rotation while preserving both cruciate ligaments. This design was later evolved into Genesis I and II by Gustillo.

Fig. 1.8 Kodama-Yamamoto MK I. (By courtesy of Dr. Sumiki Yamamoto)

Townly's Anatomic Knee (1972) [19] included all the aforementioned anatomical aspects and introduced the world's first patellar replacement with the use of a patellar button.

This design had a significant impact on subsequent knee prostheses such as the AMK, Cloutier, AGC, Orthomet, Axiom, and Natural knees.

The Eftekhar MK II (1973) [20] (Fig. 1.9) featured an anatomically shaped femoral component, a metal-backed tibia, and three interchangeable inserts capable of 40° internal or external rotation, as well as varus or valgus motion. The tibia stem was cemented, covering most of the elements found in a modern knee prosthesis.

Walker, who left the hospital for special surgery (HSS) in 1975, developed the kinematic knee system (1978) [21] (Fig. 1.10) with Clement Sledge et al. at Robert B. Brigham Hospital in Boston.

The kinematic system consists of four different designs: The anterior-joined model retained both cruciate ligaments, while the posterior cruciate retention model proved particularly successful. The stabilizer model incorporated a post-cam mechanism, and the rotating hinge model was also part of the system. The posterior cruciate retention model has been superior. It was used worldwide and was later improved to become the Kinemax knee system.

Knee prosthetics that provide both anatomical and functional designs have become commercially available.

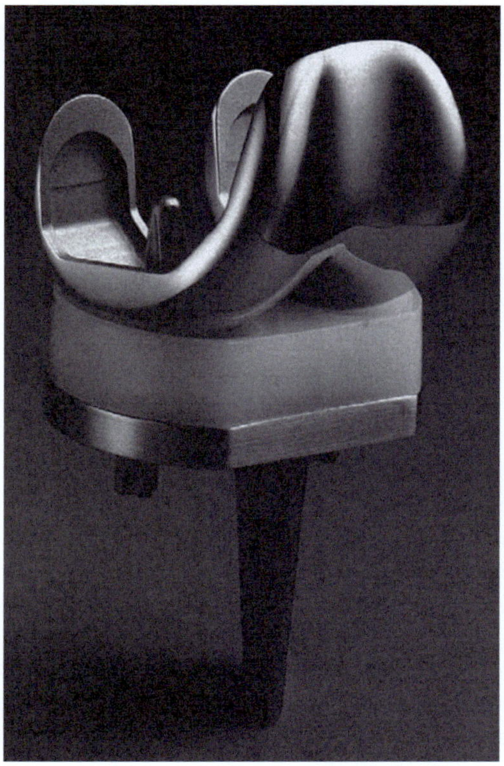

Fig. 1.9 Eftekhar MK II. (By courtesy of Zimmer-Biomet)

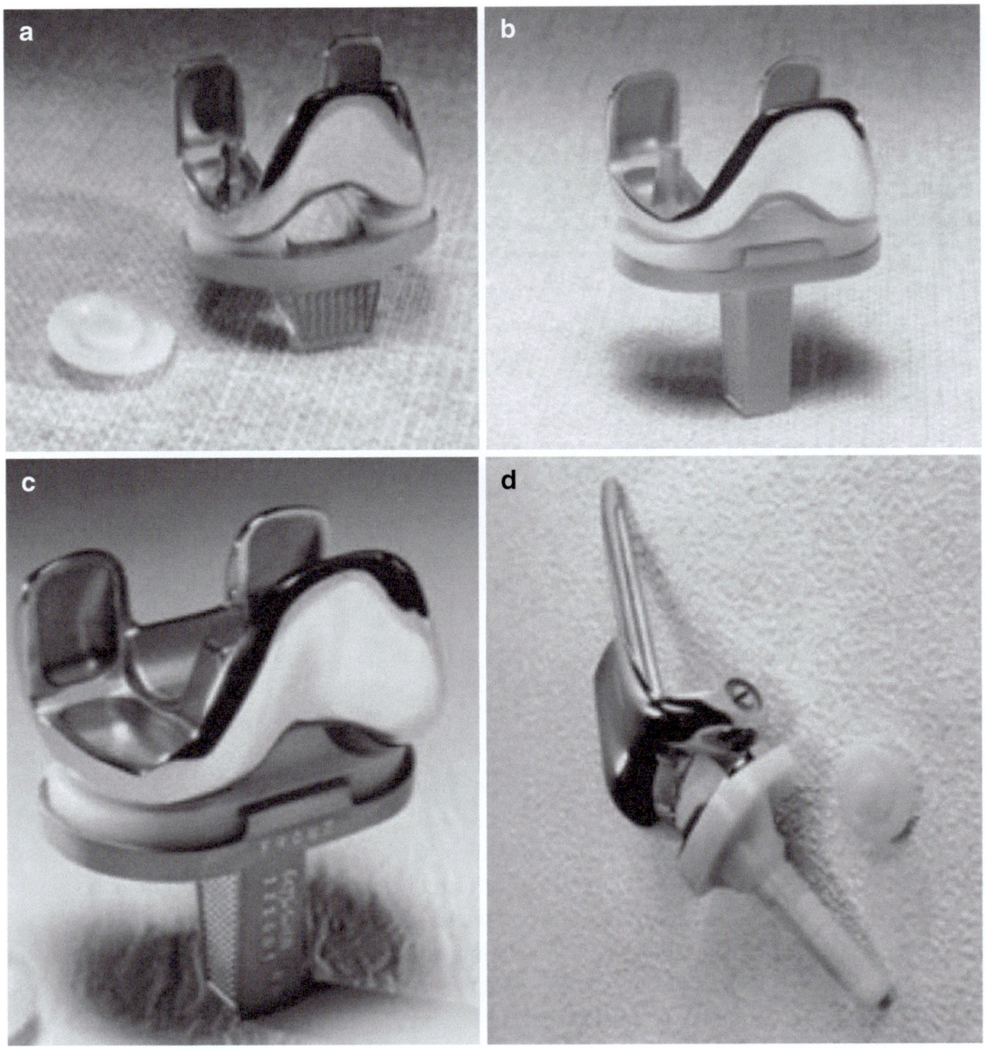

Fig. 1.10 Kinematic knee system. (**a**): Anterior joined. (**b**): Posterior cruciate retention. (**c**) Bottom left: stabilizer. (**d**): Rotating hinge. (By courtesy of Stryker Japan)

5.2 Functional Design

The original functional design was the Freeman-Swanson knee (1970) [22] (Fig. 1.11) by Freeman and Swanson.

This design aimed to reduce load stress on the fixation surface by removing both cruciate ligaments, as they occupied a significant contact area. The goal was to simplify kinematics and achieve a simplified function where the sagittal motion on the "roller-in-trough" shape of the tibial joint surface relative to the femur followed a single radius, mainly relying on rolling rather than sliding.

The femur, with a small peg, and the tibia, with a dovetail-shaped stud, were cemented.

This design has been widely used in the United Kingdom and other regions. However, it had limitations, such as a single size and simple kinematics that did not allow for anterior or posterior movement, varus or valgus motion, or include patellar flanges. Improvements were made to this design, resulting in subsequent knee prostheses such as the

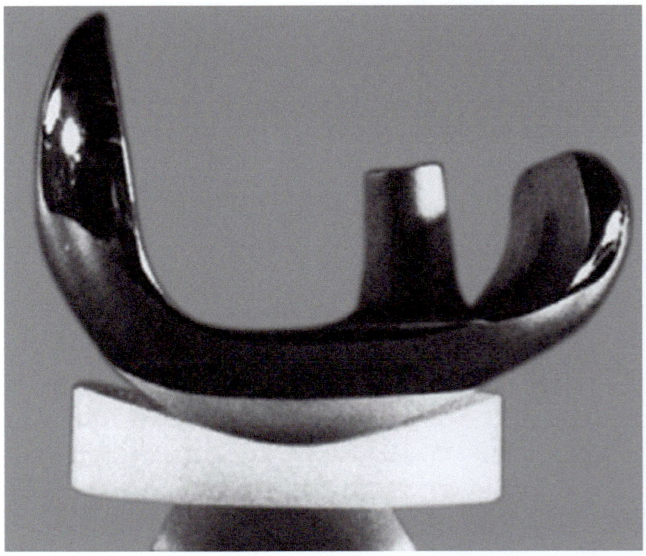

Fig. 1.11 Freeman-Swanson knee (right is front). (By courtesy of Stryker Japan)

Imperial College London Hospital knee (1974) [23], and the Freeman-Swanson knee (1980).

5.3 HSS

In the 1970s, HSS brought together notable individuals such as Insall, Ranawat, Walker, Burstein, and others who laid the foundation for modern knee prostheses.

Insall, a close friend of Freeman at Cambridge University, joined HSS in 1965, and Walker joined Leeds University in 1969. Despite being a surgeon, Insall had excellent engineering experience and immediately began developing a new knee prosthesis.

The Duocondylar and Duopatella (1974) designs were anatomically shaped, and the latter was first used by Ranawat but proved to be inferior to the functionally designed total condylar (TC) (1974) [24] (Fig. 1.12) knee prosthesis, which was being developed concurrently. Ranawat eventually discontinued the pursuit of the anatomical design.

The TC knee prosthesis was greatly influenced by Freeman's work at the Imperial College London Hospital and became the foundation of modern knee prostheses. In this design, both cruciate ligaments are removed, and the articular surface features a double-dished shape with gentle constraint. The loading site at the edge of the insert provides stability due to the so-called "uphill" effect. The TC knee prosthesis replaced all condyles, including a patella button for the patellofemoral joint replacement, and used cement fixation with a tibial central peg.

Although TC knee prostheses have been widely used worldwide with excellent results, two main issues have been identified: the paradoxical motion of the femoral component moving forward during flexion and a limited flexion angle of approximately 90°.

Following Walker's departure in 1975, Burstein took over in 1976 and focused on addressing these problems. The result was the development of the Insall-Burstein Posterior-Stabilized knee (1978) [25]. The most notable feature of this model is the inclusion of a post-cam mechanism on the sliding surface, which induces femur rollback. The later improved Insall-Burstein Posterior-Stabilized II (1988) (Fig. 1.13) is metal-backed and characteristic of its use of direct-molded polyethylene, which has been widely used worldwide.

5.4 Cementless Anatomical Design with Bone Ingrowth

The 1980s witnessed a surge in interest in cementless total knee arthroplasty with bone ingrowth. One notable development during this time was the porous-coated anatomical knee (1980) [26] (Fig. 1.14) developed by Hungerford et al. of Johns Hopkins University. This anatomical design preserved the posterior cruciate ligament and introduced the innovative concept of bone invasion onto fixed surfaces with

Fig. 1.12 Insall: total condylar. (By courtesy of Stryker Japan)

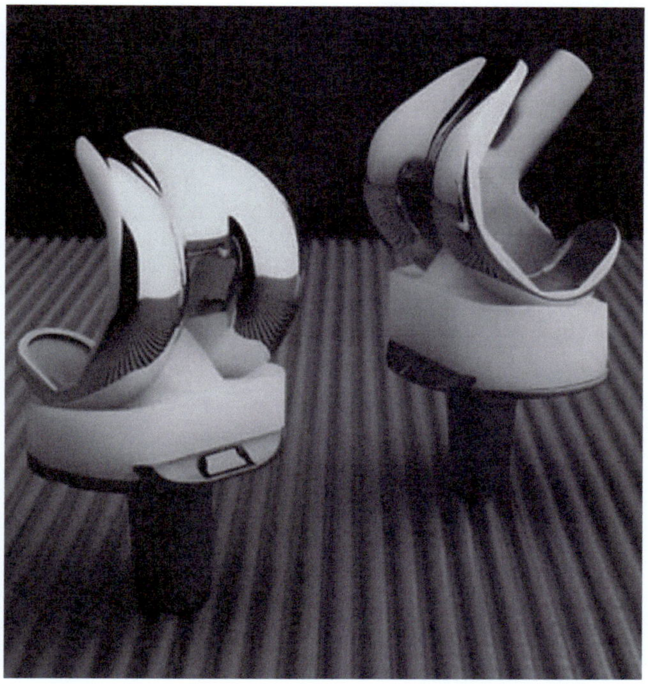

Fig. 1.13 Insall-Burstein

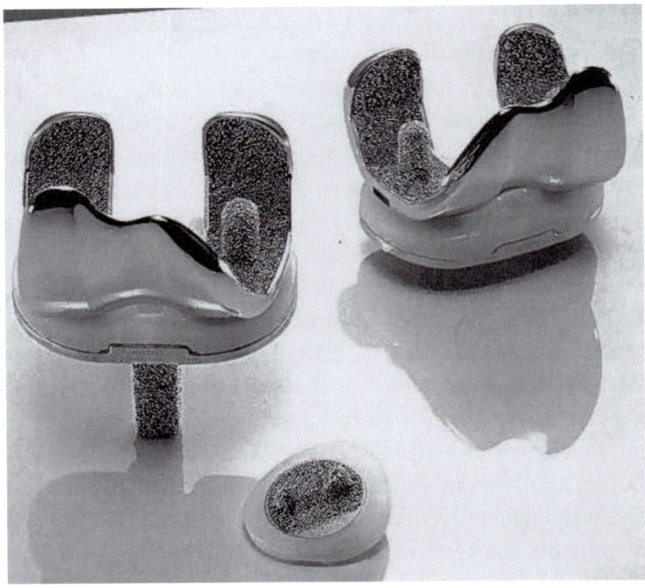

Fig. 1.14 Hungerford: PCA. (By courtesy of Stryker Japan). PCA, posterior cruciate ligament

1.5-mm-diameter sintered metal beads. The bones would adhere to these beads, creating a stable fixation.

However, the porous-coated anatomical knee was eventually abandoned due to premature wear issues with the heat-treated polyethylene, leading to the development of an improved version known as Duracon.

Another prominent design that dominated the era was that of the Miller-Galante (M-G) knee (1985) (Fig. 1.15). This design featured unique flat-on-flat articulated surfaces made of polyethylene reinforced with carbon fibers and bone ingrowth facilitated by a titanium fiber mesh for fixation surfaces. However, challenges such as premature wear of the titanium sliding surface, issues with carbon-fiber-reinforced polyethylene, and patellar dislocation due to incorrect rotation led to the transition to an improved version called the M-G II [27].

While these ingrowth designs demonstrated strength against vertical loads, they were relatively weaker against horizontal shear stresses. Therefore, flat articular surfaces with low constraints were used. In the 1990s, cementless anatomy designs with hydroxyapatite coatings emerged. However, the comparative outcomes of cemented versus cementless approaches are still being evaluated, and conclusive results have yet to be determined.

Fig. 1.15 Miller-Galante I. (By courtesy of Zimmer-Biomet)

Fig. 1.16 Goodfellow: Oxford knee. (By courtesy of Zimmer-Biomet)

5.5 Mobile Bearing Design

The concept of mobile bearings in knee prostheses is not new. The world's first mobile bearing design was the Oxford Knee (1976) [28] (Fig. 1.16) developed by Goodfellow and O'Conner. However, the original purpose of this design was to preserve the anterior and posterior cruciate ligaments rather than specifically reduce wear. The primary objective of this design was to reproduce the movement of a four-bar linkage.

The next one, low contact stress (1977) [29] by Buecel and Pappas, specifically targeted low wear. This design had two variations: the meniscal bearing type, where the medial and lateral inserts moved independently, and the rotating platform type, which integrated the bearings and allowed only rotational mobility. Both types could be fixed with either cement or cementless methods.

Since the introduction of these two models, the United States Food and Drug Administration has classified mobile bearing devices as class III devices, requiring approval for artificial joints. Therefore, no new mobile bearings have been introduced in the United States market for a long time. Consequently, these two models remain virtually the only mobile-bearing models with long-term results, and assessments are divided.

5.6 Since the 1990s

During the Cambrian period of the 1970s and 1980s, numerous ideas emerged in the field of knee prostheses, but many of them have been discarded over time. Despite ongoing developments, no truly innovative designs have emerged thus far. In terms of anatomical design, the preservation of the posterior cruciate ligament has become common practice. Conversely, guided motion by post-cam mechanisms has become the mainstream approach for functional design. Both models are commercially available, providing a selectable range of options for patients.

The only advances have been in materials such as highly cross-linked polyethylene and sterilization methods that prevent oxidation.

References

1. Brand RA, et al. Biographical sketch: Themistocles Gluck (1853-1942). Clin Orthop. 2011;6:1525–7.
2. Matsuno S. History of artificial knee joint development, knee replacement. In: Matsuno S, et al., editors. Basic and clinical. Tokyo: Nankodo; 2005. p. 2–15. [Japanese].
3. Campbell WC. Interposition of Vitallium plates in arthroplasties of the knee: preliminary report. Am J Surg. 1940;47:639–41.
4. Jones WN, et al. Mold arthroplasty of the knee. J Bone Joint Surg Am. 1967;49:1022–5.
5. McKeever DC. The choice of prosthetic materials and evaluation of results. Clin Orthop. 1955;6:17–21.
6. McIntosh DL. Hemiarthroplasty of the knee using a space-occupying prosthesis for painful varus and valgus deformities. J Bone Joint Surg Am. 1958;40:1431–6.
7. Hirakawa K, Phantom HDP. J Joint Surg. 1999;12:92–5. [Japanese]
8. Majnoni d'Intignano JM. Articulation totales en resin acrylique. Rev Orthop Chir Appar Mot. 1950;36:535–7.
9. Judet J, et al. Essais de Prosthese Ostéo-articulaire. Presse Med. 1947;55:302.
10. Walldius B. Arthroplasty of the knee joint using an acrylic prosthesis. Acta Orthop Scand. 1953;23:121–31.
11. Shiers LGP. Arthroplasty of the knee; preliminary report of new method. J Bone Joint Surg Br. 1954;36:553–60.
12. Mazas FB. Guepar total knee prosthesis. Clin Orthop Relat Res. 1973;94:211–21.
13. Matthews LS, et al. The spherocentric knee. Clin Orthop Relat Res. 1973;94:234–41.
14. Attenborough CG. The Attenbough total knee replacement. J Bone Joint Surg Br. 1978;60:320–6.
15. Gunston FH. Polycentric knee arthroplasty, prosthetic simulation of normal knee movement. J Bone Joint Surg Br. 1971;53:272–7.
16. Coventry MB, et al. A new geometric knee for total knee arthroplasty. Clin Orthop Relat Res. 1972;83:157–62.
17. Yamamoto S. Total knee replacement with the Kodama-Yamamoto knee prosthesis. Clin Orthop Relat Res. 1979;145:60–7.
18. Waugh TR, et al. Total knee replacement, operative technic and preliminary results. Clin Orthop Relat Res. 1973;94:196–201.
19. Townley CO. The anatomic total knee resurfacing arthroplasty. Clin Orthop. 1985;192:82–96.
20. Eftekhar NS. Total knee-replacement arthroplasty: results with the intramedullary adjustable total knee prosthesis. J Bone Surg. 1983;65:293–309.
21. Ewald FC, et al. Kinematic total knee replacement. J Bone Joint Surg Am. 1984;66:1032–40.
22. Freeman MAR, et al. Total replacement of the knee using the Freeman-Swanson knee prosthesis. Clin Orthop Relat Res. 1973;94:153–70.
23. Freeman MAR, et al. ICLH arthroplasty of the knee: 1968-1977. J Bone Joint Surg Br. 1978;60:339–44.
24. Insall JN, et al. Total condylar knee replacement: preliminary report. Clin Orthop. 1976;120:149–54.
25. Insall JN, et al. The posterior stabilized condylar prosthesis: a modification of the total condylar design: two to four-year clinical experience. J Bone Joint Surg Am. 1982;64:1317–23.
26. Hungerford DS, et al. The porous-coated anatomic total knee. Orthop Clin North Am. 1982;13:103–22.
27. Berger RA, et al. Long-term follow up of the miller-Galante total knee replacement. Clin Orthop. 2001;388:58–67.
28. Goodfellow J, O'Connor J. The mechanics of the knee and prosthesis design. J Bone Joint Surg Br. 1978;60:358–69.
29. Buechel FF, Pappas MJ. Long-term survivorship analysis of cruciate sparing versus cruciate sacrificing knee prostheses using meniscal bearings. Clin Orthop Relat Res. 1990;260:162–9.

2 TKA Development in Japan: General Remarks

Shuichi Matsuda

The development of TKA in Japan began in the 1970s, and subsequently many implants have been clinically applied. At that time, implants in Europe and the United States were being developed in accordance with Western lifestyles, so acquiring deep flexion was not a vital postoperative goal. Therefore, efforts have been made in Japan to devise a shape that does not compromise durability when the knee joint is moved to achieve deep flexion. In addition, there have been developments to increase size variations and adjust the lateral/anterior components using Japanese human bone shape data, to address the issue of the unsuitable implant sizes for Japanese bone structure. It is also noteworthy that global-leading design improvements have been made to obtain kinematics that closely resemble biological kinematics. This section outlines the development of representative TKA in Japan.

1 Kodama-Yamamoto Type Knee Prosthesis

TKA development in Japan was pioneered by Okayama University in 1970 with the Kodama-Yamamoto knee prosthesis (Kodama-Yamamoto Mark-I). This was relatively early by global standards, when the clinical use of surface-displacement-type TKA began. Subsequently, the improved Mark-II was developed; the surface design allowed for rotational movement in flexion, and surgical instruments were developed simultaneously. Mark-I and II excised the ACLs and PCLs; however, in 1989, the Okayama University PCL-R type was developed to preserve the PCL. This implant improved fixation of the femoral and patellar components, and the design of the patellar groove was modified.

S. Matsuda (✉)
Department of Orthopaedic Surgery, Kyoto University, Kyoto, Japan
e-mail: smat522@kuhp.kyoto-u.ac.jp

2 Improving Range of Motion

The Yoshino-Shoji Total Japan Knee System was developed in 1984 to improve the range of motion of the Yoshino Total Knee, which was developed in 1974. It was one of the earliest TKA systems developed for deep flexion to suit the Japanese lifestyle. It was designed to allow rotational motion during flexion, with a good range of motion. Subsequently, development progressed to the Hy-Flex Total Knee System in 1992, and the Hy-Flex II Total Knee and Ligament Balancing System in 1997.

The Bi-Surface Knee is a uniquely shaped implant that incorporates a ball-and-socket joint posterior to the intercondylar femoral notch to distribute load during deep flexion, and was modified from Type 1 in 1989 to Type 5 in 2011.

The PF joints are key to achieving deep flexion. The Flexible Nichidai Knee (FNK), which was developed in 1995, includes a thin anterior chamber to reduce PF joint pressure. Simultaneously, the posterior femoral condyle is multiradiated to easily achieve deep flexion. The QUEST KNEE, which began clinical use in 2009, was designed to reduce pressure on the PF joint in the deep flexion position by altering the shape of the femoral intercondylar notch where the patella initiates contact in deep flexion, into an anatomical asymmetrical shape allowing the patella to sink into the intercondylar space. The QUEST KNEE also incorporates a shape that avoids edge loading of the posterior femoral condyle.

The ACTIYAS, which was put into clinical use in 2010, adopts a post shape that does not interfere with rotation during flexion. The Vanguard PSRP (2016) has a shape that reduces the risk of POST wear and damage by increasing the contact area on the posterior surface of the tibial post. The Future Knee was designed to easily induce internal rotation of the tibia associated with flexion by eccentrizing the femoral cam inward, and has been clinically used since 2017.

S. Matsuda (ed.), *Advances in Total Knee Arthroplasty*, https://doi.org/10.1007/978-981-97-4920-1_2

3 Kinematics

In 2001, the Fine Knee was the first in the world to adopt an asymmetrical design on the articular surface with a built-in 3° inward inclination between the femoral tibial joints. Smith and Nephew's Journey, which has a similar design concept, began clinical use by its developers in 2004. Furthermore, fluoroscopic analysis after Fine Knee surgery revealed a medial pivot pattern. Physio-Knee, developed in 2005, was designed to reproduce the medial pivot motion with the medial articular surface as a ball-and-socket shape. Initia Knee (2020) curves the sliding surface of the lateral femoral condyle to increase the tolerance of external rotation of the femur.

4 Size Variations

Hi-Tech Knee is an implant who's development began in 1985 and was examined in detail using CT images to fit the knees of Japanese people; Hi-Tech Knee II was put into clinical use in 1992. Other implants developed in Japan also examine the bone morphology of Japanese people in detail and reflect them in implant design and size variations. With an eye on overseas markets, Future Knee examines the bone morphology of not only Japanese but also Thai people in detail, and adjusts the mediolateral and anteroposterior diameters.

5 Ceramic

The widespread use of ceramic implants for femoral components is characteristic of Japanese TKA. KC-1, developed in 1982, is similar in shape to the Mark-II, but the femoral component and tibia base plate are made of alumina ceramic. Bi-Surface Knee also uses ceramics for the femoral component (1989). In 1992, Low Friction Anatomical (LFA)-I was developed, which also included an alumina ceramic femoral component. Subsequently, improvements were made, resulting in the LFA-III and PS types, both made of zirconia ceramics. PhysioKnee, released in 2005, also uses femoral components made of alumina ceramics.

In this chapter, we introduced some of the TKA developments in Japan. However, we have not been able to discuss all the developments. The development of the TKA in Japan has made great progress owing to the cooperation of researchers, including doctors and domestic manufacturers. The challenge is bringing Japanese implants to the global market. This would further require joint development with international doctors, refinement of surgical instruments, and thorough marketing.

TKA Development in Japan: The Kodama, Yamamoto Mark-I/Mark-II/KC-1/PCLR

3

Nobuhiro Abe

Since 1970, condylar knee prostheses have been used to reacquire the motion and stability of the knee joint. The Kodama-Yamamoto knee prosthesis, developed by Okayama University, is the world's first surface replacement type and anatomical design that does not use bone cement. Concurrently, the Freeman-Swanson knee prosthesis developed in the United States adopted the first functional design, a semi-constrained type that excised both anterior and posterior cruciate ligaments and adjusted the extension flexion gap. However, this design used bone cement [1]. Based on these anatomical and functional knee prostheses, numerous studies have been conducted to develop the design and concept of today's knee prostheses.

1 Development of the Kodama-Yamamoto Mark-I (Fig. 3.1)

In 1970, the first surgery was performed using the Kodama-Yamamoto Mark-I knee prosthesis. In the first case, both the femur and tibial sides were made of metal; however, a tibial plate made of polyethylene was also used. When the design was being considered in search of further physiological knee joint surface reconstruction, Dr. Sbarbaro kindly provided us with prototypes of the femoral condyle mold and tibial plateau plate, after which the basic Mark I design was established. It was a surface-replacement design that did not use methacrylate bone cement. It had a femoral condyle mold

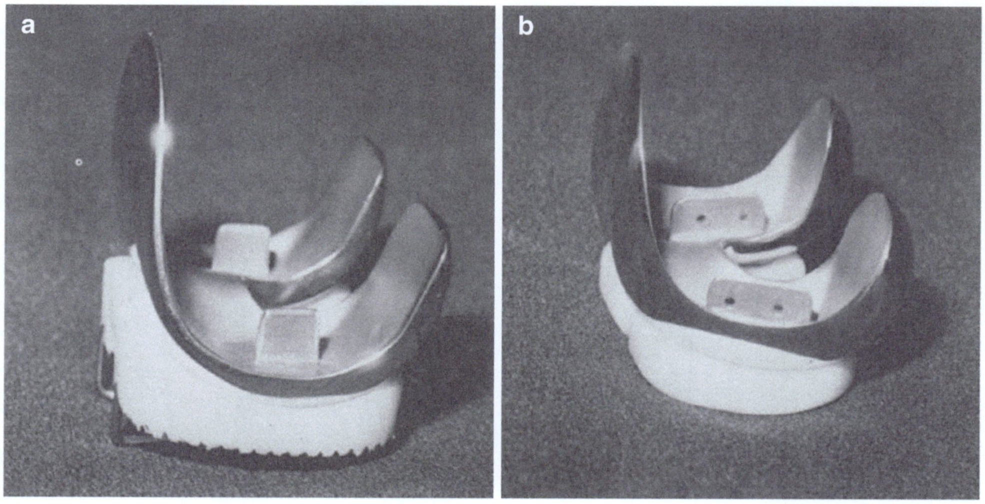

Fig. 3.1 (**a**) The early Mark-I prosthesis. (**b**) The later Mark-I prosthesis (Yamamoto [8])

The Author Nobuhiro Abe died before publication of the English version of this work.

N. Abe (Deceased) (✉)
Department of Orthopaedic Surgery, Kawasaki Medical School General Medical Center, Okayama, Japan
e-mail: smat522@kuhp.kyoto-u.ac.jp

S. Matsuda (ed.), *Advances in Total Knee Arthroplasty*, https://doi.org/10.1007/978-981-97-4920-1_3

(femoral component) made of an alloy COP that was similar to the physiological articular surface, with a slightly convex anterior flange. The upper surface of the tibia was replaced with a relatively flat tibia made of the high-density polyethylene (HDP), with a cut in the middle to preserve both cruciate ligaments. The Mark I series design was later improved from the A-type to the F-type to suit the shape of the Japanese knee joint. In shapes A–C, the anterior wall of the femur mold was changed vertically to reduce pain around the patella, and a slightly smaller implant design was developed to press fit the femoral condylar osteotomized surface due to the increased distal femur osteotomy in the flexion contracture knee. In type D, the femorotibial joint was monocentric (single radius) to improve stability in the early stage [2]. Although this was effective in improving joint stability, it did not provide sufficient flexion owing to the limited femoral rotation and rollback.

2 Trajectory to the Kodama-Yamamoto Mark-II (Fig. 3.2)

The Mark II, which has been improved since 1975, discards the monocentric articular surface used in the D-type and combines spherical surfaces with different circumferences to perform compound movements, such as rotation movements in flexion while providing increased stability during extension, similar to the original knee joint. The implant design and size were determined based on studies of Japanese knee joint morphology, bony construction, and knee rotation mechanisms [3]. Surgical techniques were established and the necessary surgical instruments and devices were developed while the artificial joint design was being developed. Yamamoto reported that it is important to resect both cruciate ligaments from the femoral intercondylar fossa and remove the popliteal tendon approximately 1 cm from the lateral joint. This is because the posterior cruciate ligament, which has degenerated due to arthritic changes, often loses its original properties, such as elasticity. The tibia is restricted in the backward slip movement if the posterior cruciate ligament is preserved during total knee arthroplasty, and the posterior cruciate ligament becomes the fulcrum for hinge movement, resulting in joint flexion limitation. In severe tibial bony defects, the importance of using uneven tibial plates has been clearly stated, as these tibial plates not only fill the defect but also provide ligamentous support. The Mark-II has been continuously used for more than 10 years, and more than 4000 surgeries have been performed in Japan with good results.

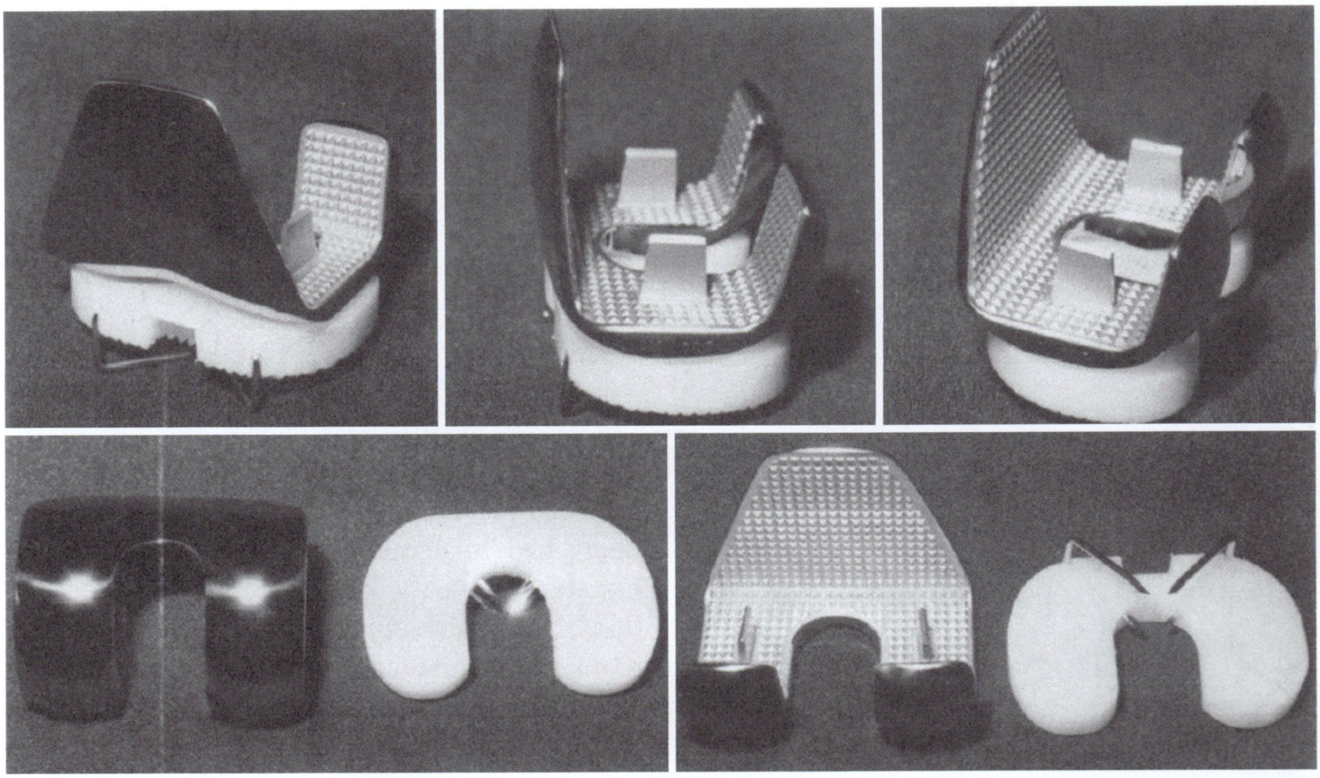

Fig. 3.2 The Kodama-Yamamoto Mark-II. (Yamamoto [8])

3 Introduction of Ceramic KC-1 Artificial Knee Joint (Fig. 3.3)

From 1982 to 1984, the cementless alumina ceramic knee prosthesis (KC-1) was temporarily introduced, and total knee arthroplasties were performed simultaneously with the Kodama-Yamamoto Mark-II [4]. The KC-1 model design is similar to that of the Kodama-Yamamoto Mark-II, but the femoral mold and tibial plates were made with Al2O3 polycrystalline alumina ceramics. The femoral component had a slightly lower posterior wall height than that of the Mark II. Alternatively, the tibial sliding surface is made of the HDP of an elliptical plate and fixed to the ceramic plate with monocrystalline alumina. KC-1 cells were fixed to the tibial osteotomized surface by penetrating internal and external pegs. The clinical outcome did not show a varus inclination of the tibial endoprosthetic plate, which was often seen in the Mark-II follow-up, which is why the KC-1 had a monocoque and a single structure between the HDP and the tibial plate. An increase in the endoprosthesis was not observed in the KC-1 type, and it has been postulated that this may be due to the use of non-deforming ceramic trays that allowed uniform loading.

4 Development of the Okayama PCL-R Prosthesis (Fig. 3.4)

The Mark-II implants loosened at 5 years after surgery, and the number of revision surgeries increased. In particular, the femoral molds were not adequately fixed to the bone. Moreover, bone resorption at the contact area in the distal femur was increased, especially in cases where the anterior femoral mold was not in contact with the cortical bone or where the posterior condyle bony defect was large and fixation of the femoral implant to the posterior condyle was insufficient, [5–7]. To solve problems such as insufficient flexion of the implants, decreased range of motion, and anterior knee pain, including patellar tracking, the Okayama PCL-R was developed and used clinically in 1989. As an improvement on the Mark-II type, the Okayama PCL-R has 3 mm extended from the front wall of the femur mold to widen the contact surface with the bone. Two cylindrical pegs (length 15 mm) were attached to the center of the mold, and wavy irregularities were added to the front and rear inner walls to improve the initial fixation. The front mold part was changed from the

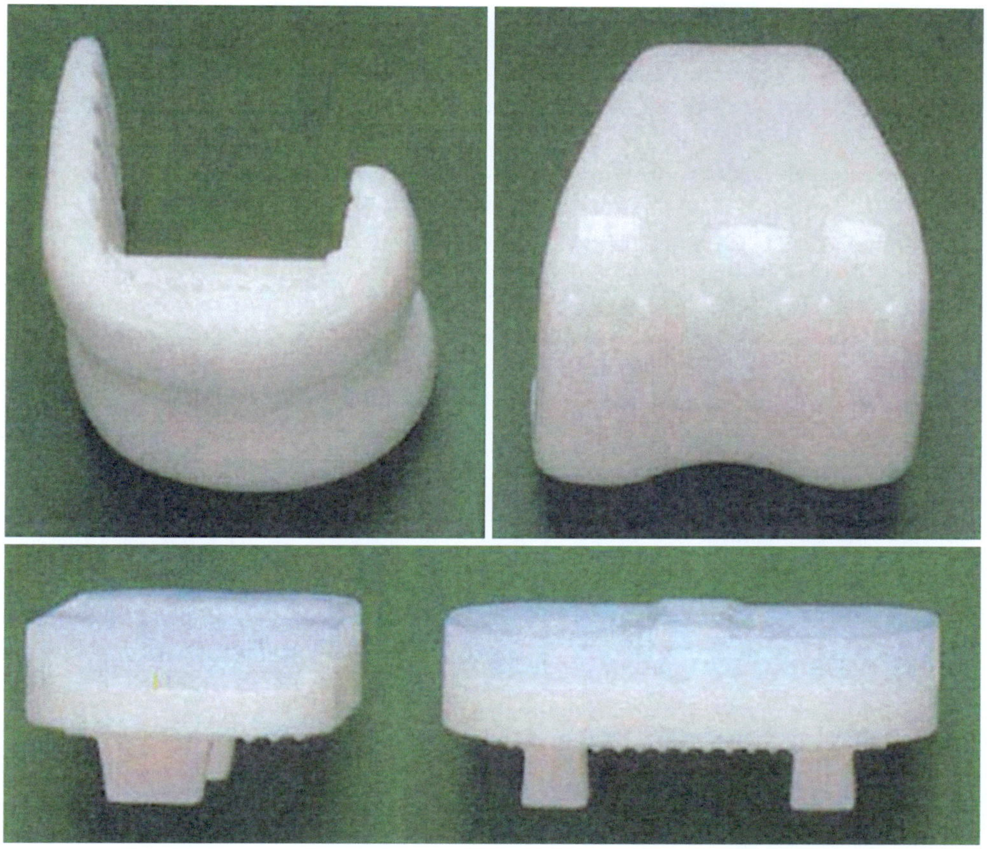

Fig. 3.3 Ceramic KC-1. (By courtesy of Kyocera)

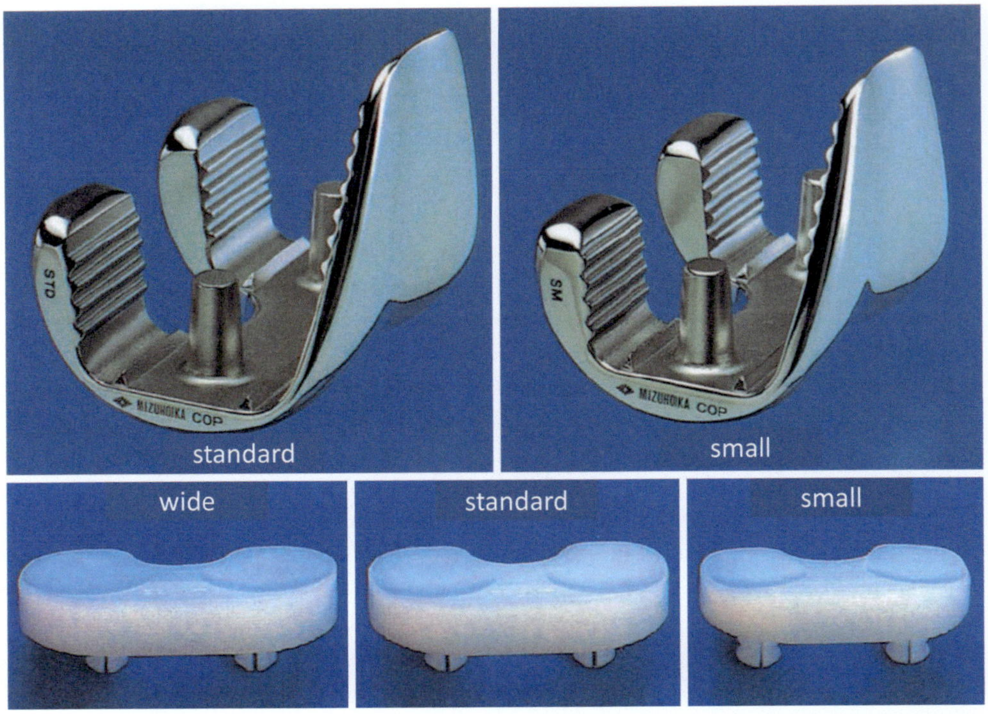

Fig. 3.4 The Okayama PCL-R prosthesis. (By courtesy of Mizuho)

convex surface of the Mark-II to a concave surface to prevent lateral displacement of the patellar dislocation. The tibial side was flattened posteriorly to facilitate rollback when the PCL was preserved, and a greater range was obtained. The tibia was composed of the monocoque HDP. Two pegs (12 mm in length) and a wavy, uneven backside surface were used to ensure good fixation of the trabecular bone. In cases with fragile bone structures, such as in the elderly or patients with rheumatoid arthritis, bone cement can be used for fixation. The clinical outcomes of the PCL-R are generally favorable. Since 2003, we have been developing a new knee joint design with the aim of regaining good joint function and improving the joint range of motion in particular. The development project was discontinued because of the declining demand and supply trends of domestic knee prostheses. Knee prostheses from foreign manufacturers have been used more frequently since then.

References

1. Robinson RP. The early innovators of today's resurfacing condylar knees. J Arthroplast. 2005;20(Suppl 1):2–26.
2. Kodama T, Yamamoto S. Total knee replacement without hinge, total knee replacement. London and New York: Mechanical Engineering Publications; 1975.
3. Yamamoto S, Hachinota M, Yuzuki O, Miyake T. The Kodama-Yamamoto knee prosthesis. Its design and clinical result. Ryumachi. 1981;21:21–8.
4. Kunitomi Y. Follow-up study of alumina ceramic total knee (KC-1 type) replacement. J Okayama Med J. 1987;99:1475–89. (Japanese)
5. Sogawa H. Radiographic follow-up study of cementless total knee replacement (Mark-II, Okayama University type). J Okayama Med Assoc. 1984;96:901–19. (Japanese)
6. Yokoyama Y, Hm I, Ohta Y, Hayashi T, Koura H. Relationship between retention of the posterior cruciate ligament and post-operative flexion in total knee arthroplasty. Acta Med Okayama. 1995;49:295–300.
7. Abe N, InoueH UM, Yukihiro S, Yokoyama Y. Long-term results of TKA (Okayama University Mark II) without bone cement in RA patients: more than 15 years of follow-up. J Jpn Soc Replace Arthroplasty. 1997;27:55–6. (Japanese)
8. Yamamoto S. Total knee replacement with the Kodama-Yamamoto knee prosthesis. Clin Orthop. 1979;145:60–7.

TKA Development in Japan: Hy-Flex II Total Knee and Ligament Balancing System

4

Yasushi Oshima and Tokifumi Majima

Currently, more than 80% of the implants used by TKA in Japan were designed in Europe and the United States. However, several original implants from Japan have been developed thus far. Among them, Hy-Flex II Total Knee and Ligament Balancing System by Professor Emeritus Shinichi Yoshino of Nippon Medical School was the TKA implants designed to achieve deep knee flexion and stable soft tissue balance, with a focus on the Japanese lifestyle.

1 History of Implant Development

Dr. Shinichi Yoshino of Nippon Medical School and his colleagues developed their own TKA implants after the Kodama-Yamamoto TKA, which had been developed in Japan, at a time when total condylar TKA was finally starting to be used overseas. Their initial goal was to remove pain and to obtain the stability of the knee. Later, they tried to achieve a greater knee range of motion (ROM), at least more than 120° in flexion, because they had considered that compared to the Western lifestyle, the Japanese required a larger ROM, including to squat fully to fully flexion on the floor. In addition, since the developers were rheumatoid orthopedic surgeons, most of their cases were secondary knee OA due to RA, and relatively young patients were also included.

2 Development History

The first type was produced as Yoshino Total Knee (Howmedica) in 1974. As this implant was developed with an emphasis on the stability of the knee joint, it was a highly constrained designed. The results in the first 264 patients with an average of 3.2 years follow-up, indicated pain relief and improvement in walking ability, however, knee ROM was not sufficiently regained and the partial pain in PFJ was observed in knee OA patients [1]. In the long-term follow-up of 267 RA patients with an average of 14.3 years, the implant survival rate was 88.6%, which was higher than that of other models of the same period. However, the knee flexion angle was reduced to 76.5° at the end of the follow-up period, from 83.0° immediately after surgery [2].

Therefore, in 1984, the Yoshino-Shoji (Y/S) Total Knee System (Biomet) was developed with the aim of acquiring a larger ROM. They had also considered that TKA was so-called "soft tissue release operation", a ligament tensor was developed that could measure the medial and lateral knee tensions, and the soft tissue was released so that the medial and lateral tensions were balanced. The results with this system showed to obtain well balanced knee and a large flexion angle, an average of 124°, in their RA patients [3]. In addition, 14 patients with 27 knees were able to squat fully in 155 RA patients with 227 knees [4]. In the prognostic study of patients who could squat fully after TKA, out of 327 RA patients with 509 knees, 23 patients with 38 knees were able to squat fully just after TKA, and during the average follow-up of 8.8 years, 7 died of causes unrelated to TKA, and 5 patients with 9 knees of the remaining 16 patients were able to squat fully [5]. To further investigate the impact of patellar replacement, 26 RA patients who underwent bilateral simultaneous TKA were compared with patella replacement on one side and non-replacement on the contralateral side. As a result, patellar replacement was recommended due to the high frequency of pain on standing and stair climbing and tenderness of the PFJ in the non-replacement patients [6].

The tibial insert and metal plate of the Y/S total knee system were integrated as the tibial component, and both the femoral and tibial components were required the cement fixation. Therefore, Hy-Flex Total Knee System (Depuy Co., UK) was developed in 1992 which could be used without cement, had a choice of insert sizes and aimed to achieve deep flexion of more than 120°. Following that, in 1997, Hy-Flex II Total Knee and Ligament Balancing System (Hy-Flex II Total Knee: Depuy Co., UK) was developed [7].

Y. Oshima (✉) · T. Majima
Department of Orthopaedic Surgery, Nippon Medical School, Tokyo, Japan
e-mail: y-oshim@nms.ac.jp

S. Matsuda (ed.), *Advances in Total Knee Arthroplasty*, https://doi.org/10.1007/978-981-97-4920-1_4

3 Features of Hy-Flex II. Total Knee (Fig. 4.1)

3.1 Femoral Component

The articular surface of the femur consisted of four different radii of curvature to achieve a greater flexion angle, in which each radius of curvature was decreased from anterior to posterior. These characteristics could lead to increase the sliding of the femur and enabled to obtain deep knee flexion. The patella groove was 7° valgus with a single radius of curvature of the articular surface to facilitate patellar tracking smoothly. The patella was replaced in all cases to reduce postoperative PFJ pain.

3.2 Tibial Implant

The articular surface of the tibial insert as tilted 4° posteriorly, and a lateral lip of the tibial insert was formed to allow 25° internal and external rotations to induce natural rollback motion.

The PCL-retaining type (CR) insert was applied as much as possible, because they had considered that the PCL function was important for the stability during standing up and climbing stairs. In addition, they continued to use the ligament tensor, which had developed in the Y/S II total knee system to obtain equal tension of medial and lateral soft tissues.

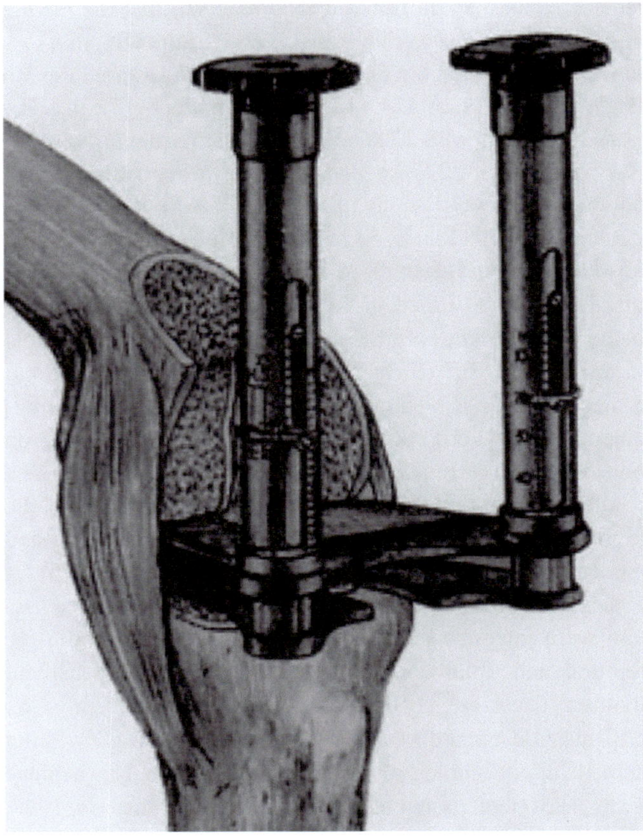

Fig. 4.1 Hy-Flex II total knee implant

For the implant fixations, while the patellar implant was cemented, the femoral and tibial implants were possible to be used both cemented and uncemented, the surfaces of which were porous coating to ensure the bone ingrowth on the implants.

4 Clinical Results of Hy-Flex II Total Knee System

The number of cases was 1271 knees for the Hy-Flex Total Knee System from 1992 to 1997 and 3349 knees for Hy-Flex II Total Knee from 1998 to 2012.

Of these, 84 RA patients with 114 knees (12 males with 17 knees, and 72 females with 97 knees) underwent Hy-Flex II Total Knee from September 1997 to June 1999, and the average age at the surgery was 58.4 ± 11.2 years (23–23 years) [7]. The preoperative evaluation of the Japanese Orthopaedic Association Knee Rating System for Rheumatoid Arthritis (JOA RA knee score) and the Hospital for Special Surgery Knee Scale (HSS score) were 47.3 ± 13.2 and 55.7 ± 12.3 respectively, but significantly improved to 84.5 ± 12.2 and 86.5 ± 10.2, respectively, a year after the surgery.

Knee ROM was 13.6° ± 14.0° in extension and 113.6° ± 20.4° in flexion preoperatively, however, it significantly improved to 0.92° ± 3.7° in extension, and 122.1° ± 15.0° in flexion 1 year after the surgery; 71.9% of patients were able to acquire a flexion angle of greater than 120°, and 13.1% achieved a flexion angle of greater than 140°. In addition, factors associated with flexion angle 1 year after surgery were examined and a correlation was found with preoperative flexion angle, however, not with the joint line elevation, patellar tendon length, patellar thickness or knee flexion angle just after the surgery.

5 A Long-Term Follow-Up Case of Hy-Flex II Total Knee

Since the former Department of Rheumatology of Nippon Medical School was later transferred to the Department of Allergy and Rheumatology and the Department of Orthopaedic Surgery and Rheumatology Surgery, it is not possible to follow up all the patients accurately at present. However, some of them are still being followed after TKA in our department.

The patient was a 79-year-old woman who underwent bilateral THAs and TKAs for RA. The left TKA was

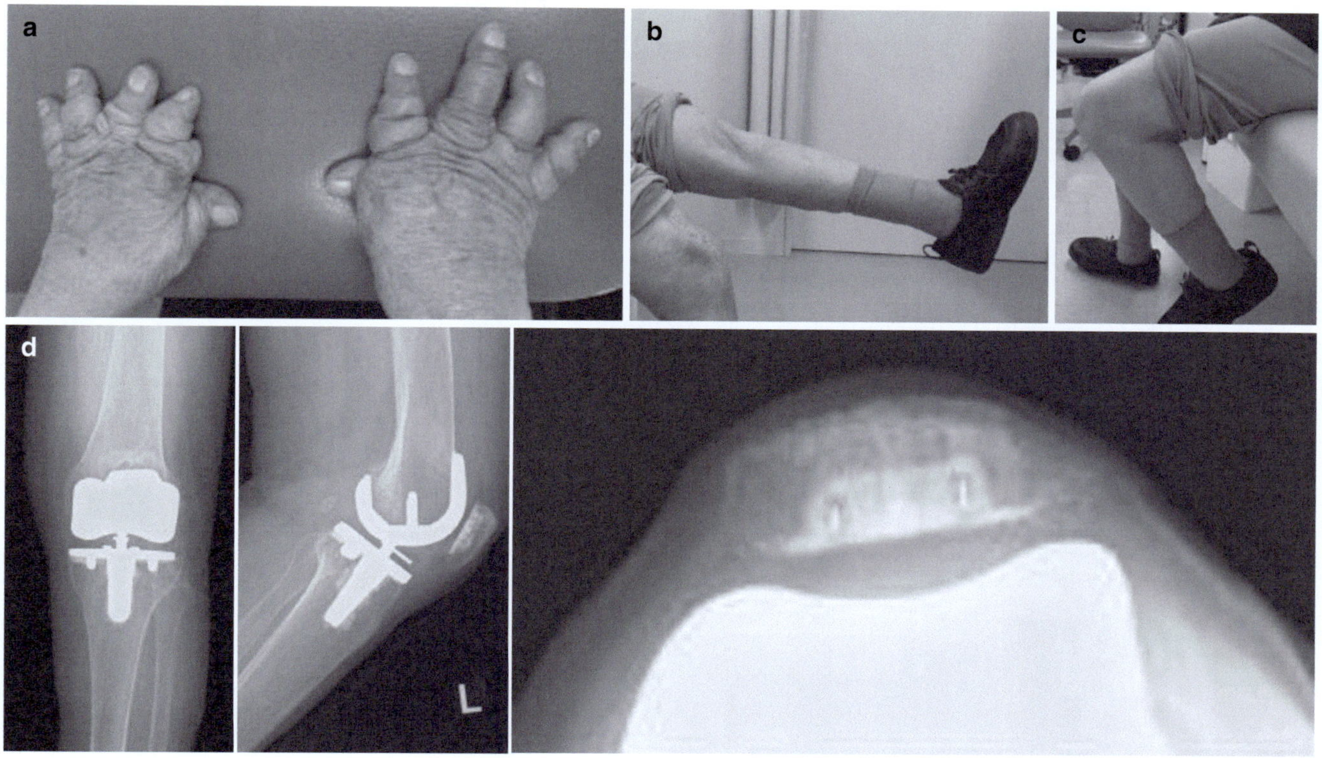

Fig. 4.2 A 79-year-old woman, knee extension and flexion findings and simple X-ray images. (**a**) Mutilans-type finger deformity. (**b**) Knee extension. (**c**) Knee flexion. (**d**) Plain radiograph of the knee joint

performed at the age of 58 and more than 20 years have passed since then. Although all fingers of both hands were severely deformed with arthritis mutilans, the disease activity of RA was well controlled by the medications. The knee ROM was 0° in extension, 120° in flexion, and the extension lag was 0°. Simple X-ray images of the knee still showed no obvious loosening or insert wear (Fig. 4.2).

6 Discussion

In designing the TKA implants, the developers considered the knee pain relief and the Japanese lifestyle, and then, focused on achieving a ROM of greater than 120°, the knee stability from maximum extension to maximum flexion, especially when standing up from maximum flexion, and well-balanced soft tissues in both medial and lateral of the knee.

In addition to designing devices such as implants and tensors, they, as clinicians, also tried pain relief using NSAIDs suppositories prior to rehabilitation, as it was thought to be more effective in improving ROM after TKA if ROM training was performed in a pain-relieved state.

In addition, the purpose of acquiring a large knee ROM was beneficial to Japanese lifestyle, the prevention of peri-implant fractures, as well as accommodating sexual position. This could be attributed to the fact that the developers were rheumatoid orthopedic surgeons who examined and operated not only on the OA of the elderly but also on relatively young women on a daily basis to address sexual inconvenience and pregnancy problems [8].

These clinical experiences and the development and improvement of implants led to the creation of Hy-Flex II Total Knee. The TKA system was based on a number of well-thought-out concepts, including preserving the PCL in RA patients, the use of ligament tensors to balance the medial and lateral soft tissues, the cement fixation of the patella in which all cases were replaced, allowing the femoral and tibial implants to be applied both cement and cementless, and to enable a flexion angle of greater than 120°.

Medications of RA have improved dramatically with the approval of methotrexate since 1999 and biologics since 2003. The results of the Hy-Flex II Total Knee were very successful although the treatments for RA in an era with a few treatment options at that time. At present, this system has no longer been available and cannot be used, however, we would like to carefully observe the remaining patients after the Hy-Flex II Total Knee TKA based on an understanding of the implant characteristics and surgical ingenuity.

Acknowledgments We would like to thank Dr. Hiroshi Nakamura of the Department of Orthopedics and Rheumatology, Junwakai Sanno Hospital for his guidance in preparing this manuscript.

References

1. Yoshino S, Uchida S. Postoperative results of Yoshino total knee prosthesis: a report of 264 arthroplasties performed between 1974 and 1980. Arch Orthop Trauma Surg. 1982;99:239–42.
2. Shiga H, et al. Long-term results of Yoshino total knee arthroplasties in rheumatoid arthritis. Arch Orthop Trauma Surg. 1998;117:15–7.
3. Shoji H, et al. Improved range of motion with the Y/S total knee arthroplasty system. Clin Orthop Relat Res. 1987;218:150–63.
4. Yoshino S, et al. Full flexion after total knee replacement in rheumatoid arthritis. Int Orthop. 1990;14:13–6.
5. Yoshino S, et al. Recovery of full flexion after total knee replacement in rheumatoid arthritis: a follow-up study. Int Orthop. 1997;21:98–100.
6. Kajino A, et al. Comparison of the results of bilateral total knee arthroplasty with and without patellar replacement for rheumatoid arthritis: a follow-up note. J Bone Joint Surg Am. 1997;79:570–4.
7. Yamazaki J, et al. Hy-flex II total knee system and range of motion. Arch Orthop Trauma Surg. 2002;122:156–60.
8. Yoshino S, Uchida S. Sexual problems of women with rheumatoid arthritis. Arch Phys Med Rehabil. 1981;62:122–3.

Total Knee Arthroplasty Development in Japan: Hi-Tech Knee II

5

Takahisa Sasho

1 Concept

During the 1970s and 1980s, the knee prostheses used in Japan were mainly manufactured overseas and did not match the Japanese physique. In addition, the polyethylene insert delaminated relatively early. The manufacturing method of some models was unclear. Therefore, the development of indigenous artificial joint prostheses was necessary. In 1985, the Department of Orthopedic Surgery at Chiba University began collaborating with the High-Tech Research Institute at Kawasaki Steel Chiba Plant to produce a domestic artificial knee prostheses. In 1992, the development was transferred to Nakashima Propeller Co., Ltd. (now Teijin Nakashima Medical Co., Ltd.), resulting in the production of the Hi-Tech Knee II (Fig. 5.1). The development concept was based on the following six points:

1. Shape that fits the Japanese knee

   Computed tomography (CT) images of 200 knees from 100 cadavers were used to determine the size that fits Japanese knees [1].
2. Use of titanium

   A titanium alloy (Ti-4Al6 V) was used to produce the prostheses due to its high bone affinity. The contact surface with the bone had a uniform network structure with a ten-layer fiber mesh. Bone ingrowth into mesh has been demonstrated in animal models and clinically via radiography studies.
3. Cementless fixation

   Cementless fixation was possible due to the use of the titanium fiber mesh, anchor screws, and anchor stem bolts with different screw pitches. These materials increased the initial fixation force of the tibial component, resulting in the strong compression of the tibial component onto the tibial osteotomy surface to achieve a strong initial fixation.
4. Flat surface polyethylene

   By flattening the shape of the polyethylene insert, the load between the bone and the component was reduced. This increases the load on the polyethylene insert, necessitating the development of a highly wear-resistant polyethylene. The flat surface also eliminates defects caused by a rotational mismatch between the femoral and tibial implants. Therefore, the kinematics are highly dependent on ligament balance. A three-dimensional study by Tamai et al. revealed that 50% of knees had a medial-pivot pattern and half had a lateral-pivot pattern [2].
5. Use of a polyethylene patella

   As metallosis is caused by metal-backed patellar polyethylene, an all-polyethylene patella was used to create the prostheses. This also included a cementless fixation in which two cylindrical grooves of varying depths were dug into the patella and press-fitted. Although this is a less common method of patellar fixation, dislocation of the patella is rare.
6. Development of highly wear-resistant polyethylene

   A direct compression molding process was used, and sterilization with ethylene oxide provided high abrasion resistance. In the 1990s, the number of internal cracks was reduced and the oxidative process within polyethylene was suppressed by mixing vitamin E with ultra-high molecular weight polyethylene (UHMWPE) powder, as described by Tomita et al. [3, 4] Following several basic experiments, clinical trials using polyethylene supplemented with vitamin E were conducted on 65 knees from 2004 to 2005 [5]. Consequently, Nakashima Medical Co., Ltd. became the first company in the world to sell this compound under the trade name BLEND-E in 2010.

T. Sasho (✉)
Center for Preventive Medicine, Musculoskeletal Disease, & Department of Orthopaedic Surgery, School of Medicine, Chiba University, Chiba, Japan
e-mail: sasho@faculty.chiba-u.jp

S. Matsuda (ed.), *Advances in Total Knee Arthroplasty*, https://doi.org/10.1007/978-981-97-4920-1_5

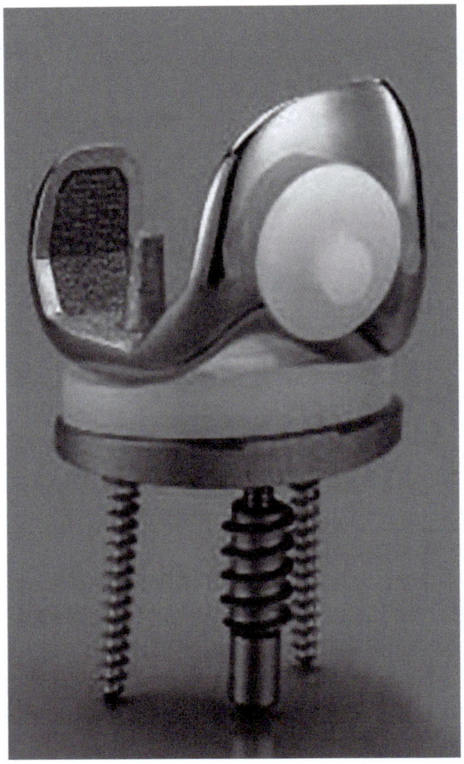

Fig. 5.1 Hi-Tech Knee II

2 When to Use Hi-Tech Knee II

The clinical use of the Japan-based knee prosthesis began in 1994. Initially, only the cruciate-retaining (CR) was used, though the use of posterior-stabilized (PS) began in 1999. The prostheses were used clinically for approximately 20 years, from 1994 to 2016. A total of 1918 surgeries were performed, mainly at Chiba University and its affiliated facilities, with more than 100 surgeries performed annually between 2002 and 2009. However, the postoperative range of motion of prosthetic knee joints has been reported to be inferior to subsequent prosthetic models [6, 7].

References

1. Tsuchida T, et al. Development of a new knee prosthesis in Japan: development of hi-tech knee II. Clin Orthop Surg. 1999;34:127–34.
2. Tamai H, et al. Knee kinematics do not influence the long-term results in posterior cruciate retaining total knee arthroplasties. Chiba Med J. 2008;84:269–76.
3. Teramura S, et al. Reduction of wear volume from ultrahigh molecular weight polyethylene knee components by the addition of vitamin E. J Orthop Res. 2008;26:460–4.
4. Teramura S, et al. Reduction of wear volume from accelerated aged UHMWPE knee components by the addition of vitamin E. J Biomed Mater Res. 2009;4:589–96.
5. Suzuki M. Materials for artificial joints – durability of polyethylene. Orthopaedics. 2011;24:14–21.
6. Kobayashi T, et al. Lateral laxity in flexion increases the postoperative flexion angle in cruciate-retaining total knee arthroplasty. J Arthroplast. 2012;27:260–5.
7. Nakamura J, et al. A comparative study of flat surface design and medial pivot design in posterior cruciate-retaining total knee arthroplasty: a matched pair cohort study of two years. BMC Musculoskelet Disord. 2018;19:234.

6 TKA Development in Japan: LFA

Kazunori Yasuda and Nobuto Kitamura

1 History of Development

The amount of ultra-high molecular weight polyethylene (UHMWPE) wear with a ceramic femoral component is 20% of that with a cobalt-chrome (Co-Cr) alloy component [1]. Therefore, in 1992, the authors developed LFA (low friction, anatomical)-I total knee prosthesis that included an alumina ceramic femoral component. The clinical results of total knee arthroplasty (TKA) with the LFA-I prosthesis were equivalent to those previously reported for common metal prostheses [2]. However, there were some problems with the LFA-I prosthesis. The alumina ceramic component was relatively weak against impact loads, and it was impossible to manufacture an anatomically shaped component using alumina ceramic. To solve these problems, we developed the LFA-III total knee system (Fig. 6.1a–d) in 1999. In this system, the anatomically designed femoral component was made of zirconia ceramic, which has high toughness and low wear properties. In this system, cruciate-retention and posterior-stabilizing types were clinically available in 1999 and 2002, respectively.

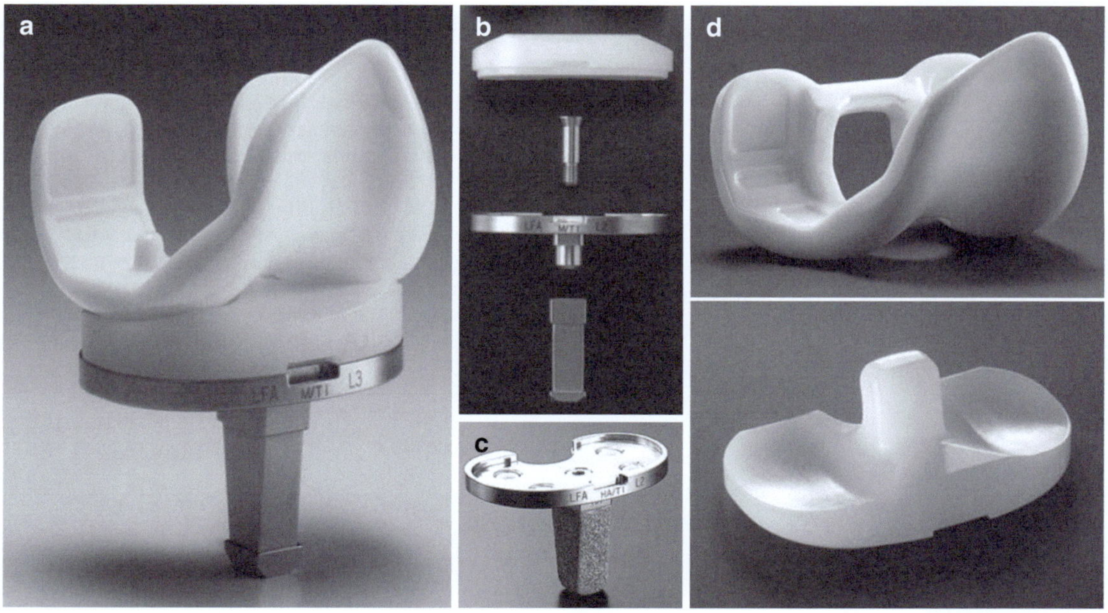

Fig. 6.1 LFA-III total knee system. (**a**): Cruciate-retention type (for cement fixation). (**b**): Modular structure of the tibial component. (**c**): Tibial component for cementless fixation. (**d**): Posterior-stabilizing type

K. Yasuda (✉)
Hokkaido University Graduate School of Medicine, Sapporo, Japan
e-mail: kazu-yasuda@rouge.plala.or.jp

N. Kitamura
St. Luke's International Hospital, Tokyo, Japan

S. Matsuda (ed.), *Advances in Total Knee Arthroplasty*, https://doi.org/10.1007/978-981-97-4920-1_6

2 Concepts Related to the Development of LFA-III

In the development of the LFA-III total knee system, we first attempted to improve the survival rate by reducing the insert wear, using a zirconia femoral component. Second, we aimed to simultaneously achieve a wide range of knee motion and good stability of the knee postoperatively by creating a femoral component with a shape that approximated the shape of the Japanese femur (Fig. 6.1a). Third, because LFA-III is composed of ceramic and titanium alloy components, we expected that it could be used even in patients with metal allergies. In fixation of the components of this system, the zirconia femoral component was fixed with polymethylmethacrylate cement. The adhesion strength of the bone cement to the dry surface of zirconia surface is comparable to that of Co-Cr alloy surface [3]. The tibial component comprises a titanium alloy tray with a parallelepiped stem and a UHMWPE insert. The tray and stem surfaces had two options: cemented or cementless fixation (Fig. 6.1a–c). The parallelepiped stem can be separated from the tray by removing a screw that fixes the stem to the tray (Fig. 6.1b). This modular structure is useful for removing the tibial component from the tibia without serious bone loss in the unlikely event of revision TKA.

3 Clinical Results

In the 5–9 year follow-up evaluation of LFA-III, the cumulative survival rate was 97.9%, and the Japanese Orthopaedic Association (JOA) score and the Hospital for Special Surgery (HSS) score averaged 84 and 91 points, respectively. The postoperative range of knee motion averaged from 0 to 125 degrees of flexion. No cases of revision were caused by polyethylene wear, ceramic breakage, or postoperative metal allergies. LFA-III was used in 11 knees of 11 patients diagnosed with metal allergies before surgery. Postoperatively, the patients did not complain of any allergic symptoms during the 1- to 10-year follow-up periods [3, 4]. Additionally, in magnetic resonance imaging (MRI) examinations of the knee replaced with LFA-III, clear images were obtained without any halation. This feature of the zirconia ceramic component is useful in diagnosing loosening of the femoral component.

4 Discussion

The cumulative survival rate and postoperative knee function of LFA-III, which has the above-described unique features, were comparable to those of various previously reported total knee prostheses. The fact that the LFA-III could be used in patients with metal allergy and that postoperative MRI examinations of the knee could be performed after replacement showed the absolute advantages of the LFA-III over the previously reported metal prostheses. Future research should be conducted to determine how the advantages of zirconia ceramics over metal alloys, as demonstrated by the LFA-III, can be incorporated into the advancement of TKA.

References

1. Yasuda K, et al. Low friction total knee arthroplasty with the alumina ceramic condylar prosthesis. Bull Hosp Joint Dis Orthop Inst. 1993;53:15–21.
2. Majima T, et al. Clinical results of posterior cruciate ligament retaining TKA with alumina ceramic condylar prosthesis: comparison to Co-Cr alloy prosthesis. Knee Surg Sports Traumatol Arthrosc. 2008;16:152–6.
3. Kumahashi N, et al. Biomechanical comparison of the strength of adhesion of polymethylmethacrylate cement to zirconia ceramic and cobalt-chromium alloy components in a total knee arthroplasty. J Orthop Sci. 2014;19:940–7.
4. Kamishima T, et al. Experimental MR imaging of zirconia ceramic joint implants at 1.5 and 3 T. Clin Radiol. 2010;65:387–90.

TKA Development in Japan: FNK

7

Keinosuke Ryu

The Flexible Nichidai Knee (FNK) total knee system was developed through a joint research project collaboration with the Joint Group of the Department of Orthopaedic Surgery, Nihon University (Tokyo, JAPAN), and Nakashima Propeller Co., Ltd. (Okayama, JAPAN). The conventional imported total knee arthroplasty (TKA) implants used in Japan were designed to suit the bone shape and lifestyle of Westerners, resulting in problems such as size, weight, and low flexion angle after TKA in Japanese individuals. The FNK was developed based on the basic concept of "best suited to the bone shape and lifestyle of the Japanese." The cruciate-retaining (CR) type was launched in 1995, followed by the posterior-stabilized (PS) type (Fig. 7.1) in 1998; since then, over 9000 cases have been performed in Japan.

1 Design Concept

The design concept of FNK is to obtain a good range of motion (ROM) and preserve bone stock [1]. To obtain a good ROM, the patellar groove of the femur component was made deeper, and the anterior flange was made thinner to reduce the pressure on the patello-femoral joint and enable smooth knee extension to flexion movement. The condylar geometry

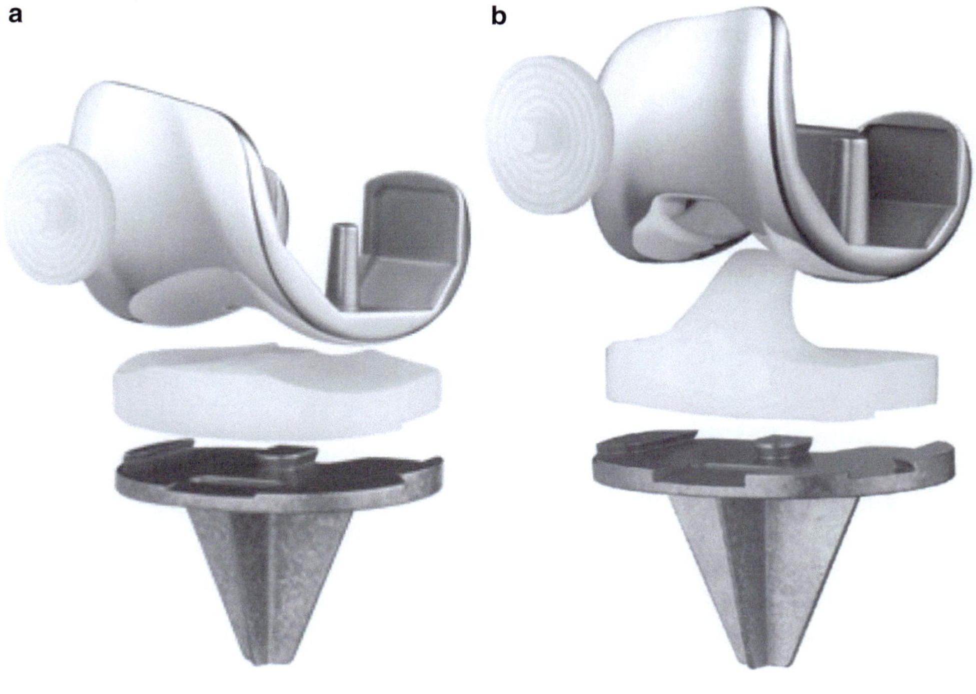

Fig. 7.1 FNK (Flexible Nichidai Knee). (**a**): CR. (**b**): PS. (By courtesy of Nakashima Medical)

K. Ryu (✉)
Department of Orthopaedic Surgery, Fukushima Medical University School of Medicine, Fukushima, Japan
e-mail: ryuk@fmu.ac.jp

S. Matsuda (ed.), *Advances in Total Knee Arthroplasty*, https://doi.org/10.1007/978-981-97-4920-1_7

in the sagittal plane shows a multiradius design based on anatomical geometry, such as the small radius of the posterior condyle, allowing good femoral rollback and deep knee flexion of up to 135°. To preserve bone stock, the femur component used is made of Co-Cr-Mo alloy and designed as thin as possible, while the tibial component used is made of Ti-6Al-4 V alloy and designed thinner with keel shape of the cross to improve the stability of the tibial component. For the polyethylene insert and the patellar component, ultrahigh molecular weight polyethylene (UHMWPe) was molded using the direct compression method. Furthermore, vitamin E–blended UHMWPe has been adopted since 2010 because of its long-term durability.

As a feature of the TKA implant design, the geometry in the coronal plane of the femorotibial (F-T) joint surface adopts a round-on-round design, which is called the modern TKA design. The F-T joint surface in the sagittal plane adopts a flat design posteriorly on the tibial surface, allowing anteroposterior and rotational knee mobility. The PS type adopts a closed PS box design with minimal bone resection in the femoral component. Compared to other PS TKA implants, the post in the tibial insert is significantly higher and wider, enhancing anteroposterior stability, rotation, and varus-valgus during the knee extension to flexion movement. This PS type design is used to treat severely deformed and highly unstable cases in TKA.

2 Clinical Performance

2.1 Long-Term Service Life

It was reported that high survival rates of 99.4% at 5 years and 96.2% at 12.5 years were achieved in 1055 knees of 595 patients with primary cemented TKA using FNK [2]. Compared with the results from registries in other countries, of other TKA implants, our report showed excellent long-term clinical results.

2.2 Clinical Results

Good postoperative knee scores (KS) and function scores (FS) were reported at 8.3 years of follow-up, in addition to good improvement in ROM after TKA for the FNK CR and PS types [2]. Moreover, 34 knees with FNK PS type and 41 knees with bicruciate stabilized (BCS) TKA were matched for preoperative age, sex, osteoarthritis stage, and femur–tibia angle, while cases of TKA using the FNK PS type reportedly resulted in good postoperative KS, FS, and ROM and a good degree of quad and hamstring recovery [3].

In addition, a kinematic analysis after TKA in 19 knees (15 patients) of the FNK PS type showed that engaging the PS post and cam in the early stage of knee extension and flexion movement reproduced a mild medial pivot and smooth rollback [4]. In a study of stair climbing and descending ability with a handrail after FNK PS type versus BCS TKA, as with BCS, a decreasing frequency of handrail use improved knee stability during descending down stair activity [5]. The postoperative infection rate was 0.8% (17 knees) in 1146 patients (2022 knees) in primary TKA performed between 1996 and 2006 (mean age, 72 years; mean follow-up, 3.5 years); furthermore, male sex, obesity, a history of open reduction and internal fixation (ORIF), and residual internal fixation were risk factors for infection [6]. A survey of sports activities after TKA using FNK reported that 21.1% of patients returned to low-impact sports with no complications; moreover, good clinical results were maintained in 109 patients (156 knees) with a mean age of 74.8 years and a mean follow-up period of 2.5 years [7, 8]. FNK demonstrated excellent postoperative and good long-term clinical outcomes.

References

1. Ryu J, et al. Simultaneous bilateral total knee arthroplasty. J Orthop Sci. 1996;1:351–5.
2. Hosaka K, et al. Asian-specific total knee system: 5-14 year follow-up study. BMC Musculoskelet Disord. 2011;12:251.
3. Suzuki G, et al. Previous fracture surgery is a major risk factor for infection after total knee arthroplasty. Knee Surg Sports Traumatol Arthrosc. 2011;19:2040–4.
4. Sumino T, et al. Semiconstrained posterior stabilized total knee arthroplasty reproduces natural deep knee bending kinematics. BMC Musculoskelet Disord. 2020;21:107.
5. Iriuchishima T, et al. Bicruciate substituting total knee arthroplasty improves stair climbing ability when compared with cruciate-retain or posterior stabilizing total knee arthroplasty. Indian J Orthop. 2019;53:641–5.
6. Takubo A, et al. Comparison of muscle recovery following bicruciate substituting versus posterior stabilized total knee arthroplasty in the Asian population. J Knee Surg. 2017;30:725–9.
7. Keinosuke R, et al. Golf and joint surgery after TKA for knee osteoarthritis. J Sports Med. 2012;31:32–8.
8. Keinosuke R, et al. Sports activities after knee replacement surgery for osteoarthritis of the knee joint. J Eastern Japan Assoc Orthopaed Traumatol. 2011;23:281–4.

TKA Development in Japan: Bi-Surface Knee

8

Shinichiro Nakamura

1 Background of Bi-Surface Knee Development

The original design of recent knee prosthesis was developed in the 1960s and 1970s. While durability gradually improved, ROM did not improve. Postoperative ROM is required in Japan and other Asian and Middle Eastern countries because standing and sitting from the floor or ground are necessary for ADLs. Therefore, a new knee prosthesis design is required to achieve the conflicting goals of postoperative ROM and long-term durability to meet Japanese lifestyle needs.

2 History of Bi-Surface Knee Development

The Bi-Surface Knee used a conventional design to enhance flexion function to accommodate the Japanese lifestyle and achieve stable long-term results. This model has two material and design features.

Ceramic, which was used as the femoral component, features low friction, and wear tests demonstrated that its wear is significantly reduced because of its smooth surface [1]. Thus, the ceramic femoral component is considered less likely to cause osteolysis and other problems caused by polyethylene wear particles.

In terms of design, the flexion mechanism and weight-bearing mechanism were separated. In addition to the normal articular surfaces of the medial and lateral condyles, a small joint with excellent rotational function was incorporated into the rearranged posterior part of the joint (Fig. 8.1). The articular surface of the medial and lateral condyles mainly carries the load in extension and mid-flexion, while the ball and socket joint accommodates deep flexion, inducing femoral rollback and sharing the load with the medial and lateral articular surfaces. The femur rotates externally with the knee flexed; however, its rotation is not restricted [2].

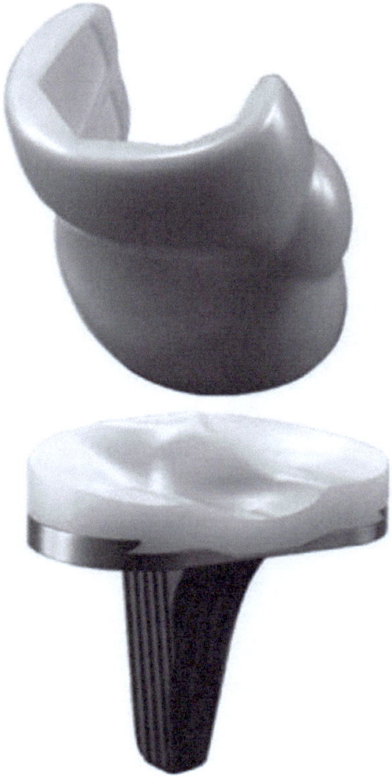

Fig. 8.1 Schematic of Bi-Surface Knee. The bi-surface knee has a ball joint in the posterior central portion of the femur and a corresponding socket attached to the tibial insert. (By courtesy of Kyocera)

3 Evolution of 3 Bi-Surface Knee

Bi-Surface Knee has improved in terms of materials and design from the initial model (Type 1) in 1989 to the present model (Type 5) (Fig. 8.2).

S. Nakamura (✉)
Department of Orthopaedic Surgery, Kyoto University, Kyoto, Japan
e-mail: shnk@kuhp.kyoto-u.ac.jp

S. Matsuda (ed.), *Advances in Total Knee Arthroplasty*, https://doi.org/10.1007/978-981-97-4920-1_8

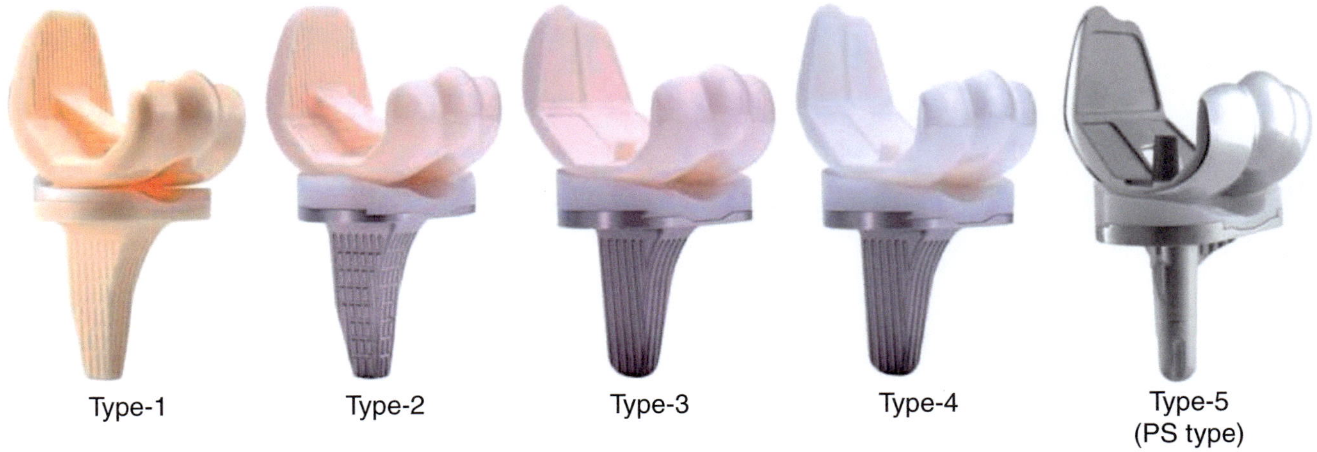

Fig. 8.2 History of Bi-Surface Knee. From left to right, Type 1 to 5 (posterior-stabilized type). (By courtesy of Kyocera)

1. Type 1 to Type 2 (1991)

   The tibial component was changed from ceramic to titanium alloy.

   - The medial and lateral walls were fabricated of polyethylene to improve conformity.
2. Type 2 to Type 3 (1997)
   - Since the ceramic strength had been improved, the strut in the middle of the femur was no longer needed.
   - The femur was designed differently on the left and right sides.
3. Type 3 to Type 4 (2001)

   The material on the femoral side was changed from alumina to zirconia, and the strength was increased.

   - A dish-type insert was added to improve conformity.
4. Type 4 to Type 4+ (2006)

   A smaller mediolateral femoral width was added to increase the choice of size.

   The shape of the posterior condyle of the femur was changed to accommodate deep flexion.

   The box's inner surface and the peg height were changed to increase adhesion to the cement.
5. Type 4+ to Type 5 (2011)

   The insert on the posterolateral surface was flattened, while those on the medial and lateral condyles were made asymmetrical.

   - A metal implant with a post was introduced to improve posterior tibial stability.

The material advantages of ceramic, a major feature of this model since its development in 1989, and the design characteristics of the ball and socket joints were integrated into the current model, for which good clinical results have been reported [3].

References

1. Oonishi H, et al. Ceramic versus cobalt-chrome femoral components: wear of polyethylene insert in total knee prosthesis. J Arthroplast. 2009;24:374–82.
2. Nakamura S, et al. Fluoroscopic and computed tomographic analysis of knee kinematics during very deep flexion after total knee arthroplasty. J Arthroplast. 2010;25:486–91.
3. Nakamura S, et al. Long-term durability of ceramic tri-condylar knee implants: a minimum 15-year follow-up. J Arthroplast. 2017;32:1874–9.

9 TKA Development in Japan: FINE Knee

Toru Suguro

1 Features of FINE Knee

The most significant feature of FINE Knee is that globally, it is the first to adopt an asymmetrical design for the tibial joint surface of the femur, similar to that of a biological knee. Following the basic bone resection theory of TKA, this implant aims to reproduce biological knee joint function by using mechanical alignment for stability and a built-in 3° medial tilt to the F-T joint surface (Fig. 9.1a). This results in a fusion of anatomical and mechanical alignment, leading to kinematic alignment in terms of function, thus achieving ideal reconstruction in terms of ligament function.

1.1 Key Features of FINE Knee

I. Shape of the femoral component
   1. The joint line is tilted 3°
   2. Contact with the tibial joint surface is at a 40% posterior position while standing
   3. In the flexed position, the medial condyle has contact with 1/3 of the posterior position
   4. The medial condyle is round, and the lateral condyle is flat (large radius)
   5. Toe-in is the lowest-point trajectory of the medial condyle
   6. Lower medial surface of the PF articular surface
   7. Patera Groove radius is smaller
   8. Provides a V-shaped overhang to deepen the patellar groove
   9. The main peg is moved anteriorly (reduce stress shielding)
   10. Sub-pegs installed posteriorly

II. Shape of tibial polyethylene
   1. The joint line is tilted by 3° (lower on the medial side)
   2. The medial condyle of the articular surface is round, and the lateral condyle is flat (large radius).
   3. Toe-in is the lowest-point trajectory of the medial condyle
   4. A pivot point is provided at the medial condyle contact point in flexion to induce rotational movement
   5. The posterior aspect of the lateral condyle is flat (large radius)
   6. A jumping gap is provided anteriorly

T. Suguro (✉)
Japan Research Institute of Artifical Joint, Chiba, Japan
e-mail: smat522@kuhp.kyoto-u.ac.jp

S. Matsuda (ed.), *Advances in Total Knee Arthroplasty*, https://doi.org/10.1007/978-981-97-4920-1_9

Fig. 9.1 FINE Knee (Teijin Nakashima Medical Co., Ltd.). (**a**): Coronal surface design. (**b**): Sagittal design and articular shape

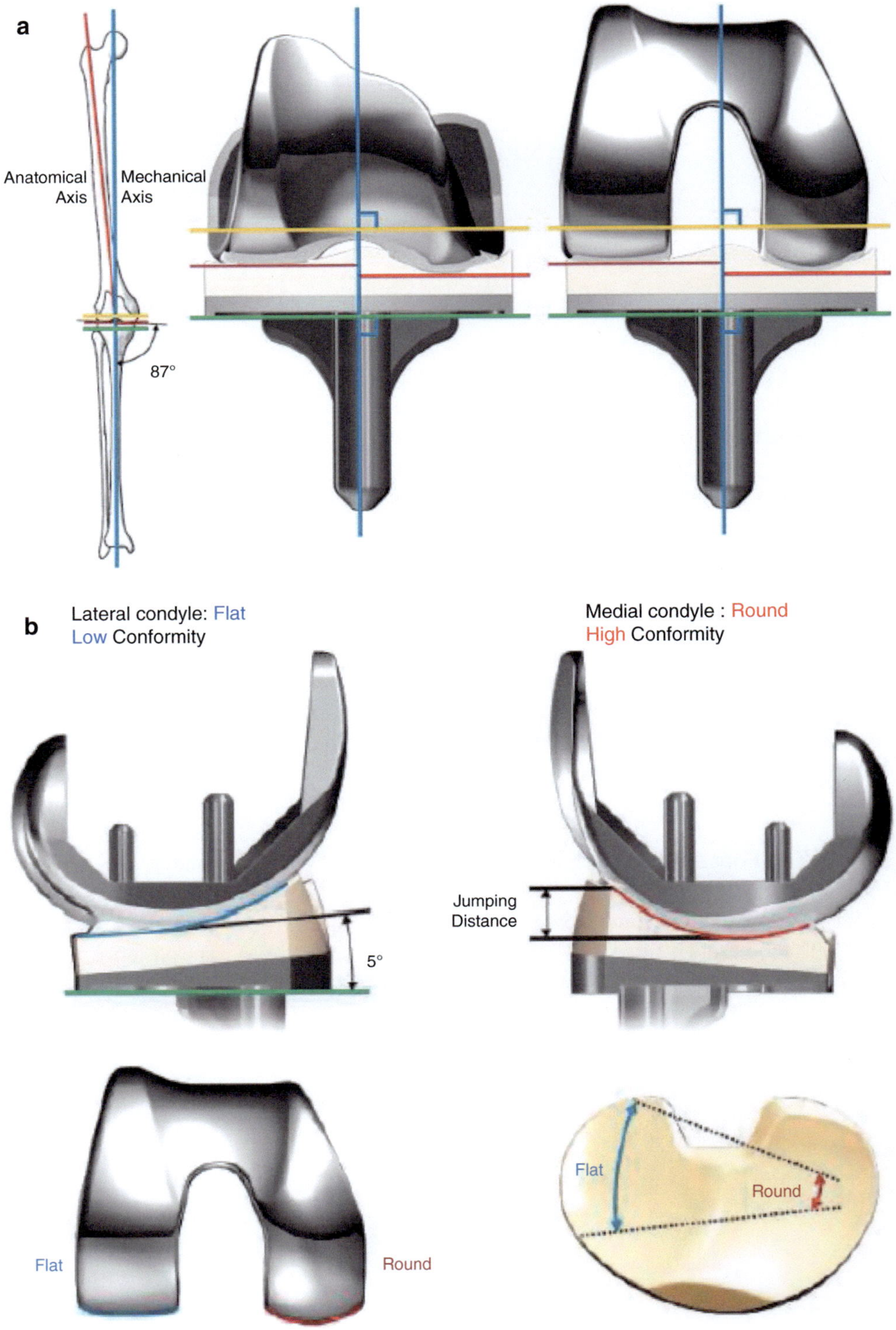

2 The Origin of FINE Knee's Development

To optimize soft tissue function around the knee joint, it is important to reconstruct an articular surface shape that matches the anatomical shape. In pursuit of the ultimate resurfacing TKA, the design and size variation has been optimized by evaluating the bone morphology and joint surface shape of Japanese people in detail, providing an anteroposterior knee motion constraint, and allowing the knee to follow the physiological dynamics, with internal rotation of the lower leg.

Based on the results of bone morphometry measurements, we sought to optimize the femoral articular surface geometry and set a difference in the curvature of the posterior femoral condylar joint surface, which functions in knee flexion, in relation to the distal femoral joint surface, which functions in knee extension, and set progressively smaller curvatures for the posterior condylar surface. Simultaneously, the design was constructed such that the curvature of the medial articular surface was larger than that of the lateral articular surface in the sagittal plane to build a step equivalent to 3° on the F-T joint surface.

In particular, the anterior and posterior lips of the intratibial condylar joint surface increase anteroposterior restraint in the knee extension area, while a change in the curvature of the femoral component articular surface geometry provides anteroposterior mobility in the knee flexion range. On the other hand, by providing only the anterior lip on the lateral condylar joint surface of the tibia and adopting a flat shape that corresponds to the rollback with the rotation of the femur on the posterior surface, a design was constructed that does not inhibit the internal rotation of the lower leg with the medial condylar surface of the tibia as the center of rotation (Fig. 9.1b). Furthermore, due to the decrease in bone density of the anterior femoral condyle after TKA, anterior and posterior femoral condylar pegs are used to prevent bone density loss and improve bone support after TKA surgery.

Clinical application began with CR type in 2001, followed by PS type with shifting post and cam to support internal rotation of the lower leg in 2004. This was followed by the PS type femoral component, introduced in 2016, to which augmentation and stem can be added. Initially, we selected cobalt-chromium alloy and titanium alloy for the femoral components in anticipation of metal allergy cases. The cementless variation uses a laminated titanium fiber mesh. The biggest advancement has been including vitamin E (dl-α-tocopherol) as an antioxidant in the tibial polyethylene plate material before ultrahigh molecular weight polyethylene molding with BLEND-E where direct pressure molding is performed using the Direct Compression Mold method. Clinical application started in 2010, and long-term wear resistance is expected due to the excellent wear resistance properties.

Presently, we have a clinical track record of more than 23,000 cases since commencing clinical use, and simultaneously, good postoperative ROM has been achieved due to the ingenuity of surgical techniques. Furthermore, excellent clinical results have been reported in patient-reported outcome measures in aspects such as pain and ADL [1, 2]. Postoperative knee dynamic evaluation showed natural knee joint movements such as internal rotation of the lower leg, including a medial pivot pattern and knee flexion mobility with stability of the internal condylar surface [3, 4].

References

1. Nakane K. To gain a good range of motion with TKA. Jpn J Rehabil Med. 2018;55:1017–25.
2. Nakajima A, et al. Three-year clinical and radiological results of a cruciate-retaining type of the knee prosthesis with anatomical geometry developed in Japan. BMC Musculoskelet Disord. 2021;22:241.
3. Miyazaki Y, et al. Analysis of the kinematics of total knee prostheses with a medial pivot design. J Arthroplast. 2011;26:1038–44.
4. Fujito T, et al. Influence of posterior tibial slope on kinematics after cruciate-retaining total knee arthroplasty. J Arthroplast. 2018;33:3778–82.

10 TKA Development in Japan: Physio-Knee

Akio Kobayashi and Yoshinori Kadoya

The development of condylar surface-replacement TKA dates back to the 1970s, when two main concepts emerged [1]. One is an anatomic approach in which the PCL is preserved, performing osteotomy while preserving the anatomical bone surface, and adjusting the ligament balance by soft tissue release as the final step. This later led to CR-type TKA, which employed a measured resection technique. The second approach was termed the functional approach in which the PCL is resected and operated by a gap balancing technique using a device with a high-conforming articulating surface, that addresses low wear from an engineering perspective. This later evolved into a PS-type TKA.

The Freeman-Swanson TKA with a roller-in-trough shape on the articulating surface marked the origin of the functional approach [1]. The FS1000 TKA, launched in 1994 as an improved version, was the world's first medial pivot TKA (MPK) with a ball-in-socket shape on the medial tibiofemoral articular surface.

1 Development of Physio-Knee

From 1992 to 1996, Kakoya, Kobayashi, and others studied the concept of MPK with Mr. MAR Freeman in London, England. After returning to Japan, they began developing MPK using ceramics for femoral components in collaboration with Kyocera in 1997. It was designed specifically for the size and shape of the Japanese population, and a special surgical instrument named GPS (Gap, Position, Size) was produced for the modified gap technique. Clinical usage of MPK began in July 2001. However, an improved version called Physio-knee was launched in 2005 after making certain modifications. These modifications included adjustments to the AP-ML ratio in the femur and tibial components as well as alterations to the center position of the spherical shape of the femoral condyle.

2 Features of Physio-Knee

1. The medial tibiofemoral joint surface has a ball-in-socket shape, and the medial femoral condyle is part of a sphere with a radius of 24 mm and maintains a spherical shape up to 105° posteriorly, with the most distal position at 0° (Fig. 10.1).
2. The femoral component is made of alumina ceramics, which have less polyethylene wear than metals, based on the results of simulator tests [2] and wear particle analysis in joint fluid [3]. Additionally, ceramics, which are more wear-resistant than metals, have advantages regarding polyethylene wear because bone fragments, cement powder, and precipitated calcium compounds can penetrate the sliding surface and worsen the surface roughness of the femoral component.
3. Alumina ceramics possess high X-ray transparency and nonmagnetic properties that enable them to take clear images of postoperative X-rays, CT, and MRI, thus enabling the evaluation of the bone-material interface, including the surrounding soft tissues such as ligaments. Moreover, recesses were made where tantalum ball markers can be placed at four points on the anterior flange and posterior condyles of the femoral component. Sharp spikes are installed in the tibial component (titanium alloy) to improve the accuracy of image evaluation, such as in postoperative kinematic analysis (Fig. 10.1).

A. Kobayashi (✉)
Department Orthopedic Surgery, Shiraniwa Hospital, Ikoma, Nara, Japan
e-mail: Akobayashi@omu.ac.jp

Y. Kadoya
Department of Orthopedic Surgery, Hanwa Joint Reconstruction Center, Sakai, Osaka, Japan

S. Matsuda (ed.), *Advances in Total Knee Arthroplasty*, https://doi.org/10.1007/978-981-97-4920-1_10

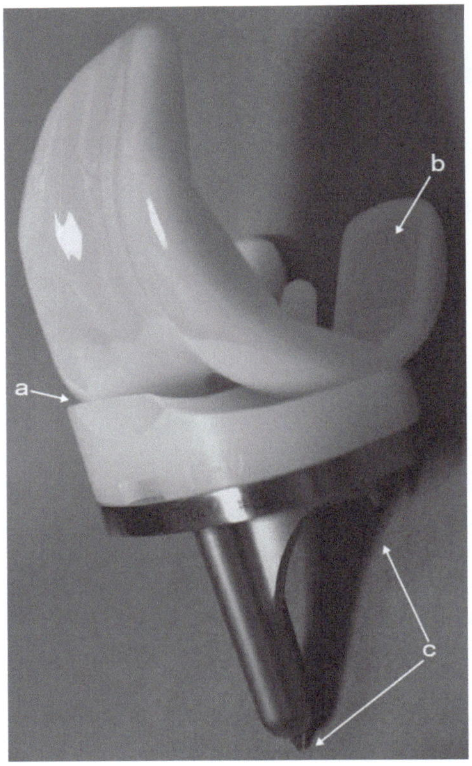

Fig. 10.1 Physio-Knee. (**a**): The medial part of the femoral component is designed as part of the sphere, contributing to the ball-in-socket geometry of the articulating surface. (**b**): Recess for marker placement on the inner surface of the lateral posterior condyle of the femoral component. (**c**): Spikes on the lower surface of the tibial component

3 Clinical Evaluation of Physio-Knee

The first version of the ceramic MPK was utilized in 122 knees in approximately 2 years starting from July 2001; the improved Physio-knee was released in June 2005 and was used in 2690 cases by the end of 2021.

In a report of 107 knees in 80 patients with ceramic MPK operated from 2001 to 2003 with at least 10 years of follow-up (mean 11.8 years), the mean age at surgery was 72 years, the average Knee Society Knee Score at the time of the final follow-up was 89 points (14 points preoperatively), the mean maximum extension angle was −1° (preoperatively −8°), and the average maximum flexion angle was 117° (110° preoperatively). Only one case of revision occurred owing to a fracture around the tibial component caused by a fall 2 years after surgery, and the 10-year survival rate using revision as an endpoint was 99.1%. A concern that postoperative flexion would worsen because of the ball-in-socket shape of the medial region and the lack of rollback of the femur existed; however, this was not recognized in this study [4].

The physio-knee group showed similar clinical results, but debonding of the cement-implant interface of the femoral component was reported in 12 patients (0.4%). Although the robustness of the cement-implant interface is affected by the surface morphology of the implant, it is generally accepted that ceramics adhere less well to cement than metals. The application of cement to a ceramic surface, while it is still in liquid form, is vital to prevent debonding.

4 Wear Particle Analysis and Kinematic Study

The concepts of physio-knee include low polyethylene wear and reproduction of the medial pivot motion. Analysis of polyethylene wear particles in the joint fluid using an electron microscope showed that the number of wear particles was lower than that of the PS type and the MPK with a metal femoral component [3]. In addition, the results of motion analysis using fluoroscopy showed that the medial tibiofemoral movement was small and the lateral condyle moved backward with flexion, resulting in medial pivot movement [5]. Thus, the Physio-Knee provides stability in the midflexion position and low wear with high conforming surface geometry, which improves patient satisfaction [6] and is suitable for high-intensity activities such as sports.

References

1. Robinson RP. The early innovators of today's resurfacing condylar knees. J Arthroplast. 2005;20(Suppl 1):2–26.
2. Oonishi H, et al. Ceramic versus cobalt-chrome femoral components; wear of polyethylene inserts in total knee prosthesis. J Arthroplast. 2009;24:374–82.
3. Minoda Y, et al. In vivo analysis of polyethylene wear particles after total knee arthroplasty: the influence of improved materials and designs. J Bone Joint Surg Am. 2009;91(Suppl 6):67–73.
4. Nakamura S, et al. Clinical results of alumina medial pivot total knee arthroplasty at a minimum follow-up of 10years. Knee. 2017;24:434–8.
5. Iwakiri K, et al. Difference between lowest points and contact points as a reference in kinematic analysis of medial pivot total knee arthroplasty. Trans Orthop Res Soc. 2007:1826.
6. Pritchett JW. Patients prefer a bicruciate-retaining or medial pivot total knee prosthesis. J Arthroplast. 2011;26:224–8.

TKA Development in Japan: ACTIYAS

11

Toshifumi Watanabe

1 Development Concept and Outline

The ACTIYAS is a PS prosthesis developed by Tokyo Medical and Dental University and Kyocera Corporation that aims to enable a comfortable Japanese lifestyle (Fig. 11.1). The ACTIYAS was designed for Japanese people with the aim of enabling smooth deep flexion and ensuring stability, durability, and strength. The PS prosthesis is more versatile than the CR prosthesis and can be widely used in primary TKA. Surgical instruments are sophisticated and easy to use, and simultaneous bilateral surgery can be easily performed [1]. The name ACTIYAS is coined from ACT (activities) and IYASU (healing), which expresses the hope that knee joint disease will be cured and an active lifestyle restored. The first TKA using this prosthesis was performed in November 2010, and more than 3000 cases have been performed to date.

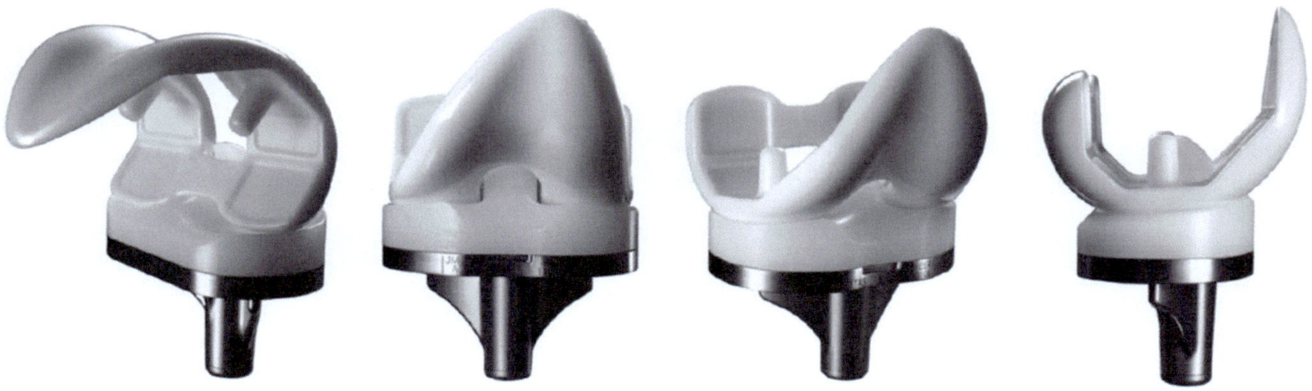

Fig. 11.1 ACTIYAS suitable for the Japanese population. (By courtesy of Kyocera)

T. Watanabe (✉)
Dokkyo Medical University Saitama Medical Center,
Saitama, Japan
e-mail: toshi-w@dokkyomed.ac.jp

S. Matsuda (ed.), *Advances in Total Knee Arthroplasty*, https://doi.org/10.1007/978-981-97-4920-1_11

2 Design Features and Materials

To achieve deep flexion, the design avoids high tension of the anterior soft tissue, the post/cam shape does not interfere with the rotation during flexion, and the size variations are sufficient to maintain the posterior condylar offset. Moreover, the component design was based on CT images of Japanese osteoarthritic knees and is well-suited to Asian populations [2], while the bone resection was minimized for bone preservation. Given that the femoral component is made of zirconia ceramics, low wear of the polyethylene insert is expected in terms of durability. Furthermore, ceramics have excellent biocompatibility and can be used in patients with metal allergies, in contrast to cobalt-chromium alloys, which are normally used on the femoral side.

3 Points to Note

Anchoring with cement would be the prime concern for ceramic implants. When the cement is ready, it is applied to the femoral implant at an early stage, and the osteotomy surface is firmly coated with cement before the implants are fixed so that strong fixation can be achieved. The issue of size variability for large patients is the next concern. There may not be a size that fits large men. Larger foreign-made models should be selected, if implant size is an issue during preoperative planning. Finally, because this model was designed for normal primary TKA, it is not indicated for patients with large bone defects or medial or lateral ligament instability. Metal blocks, wedges, and extension stems are available only for the tibial side.

4 Clinical Performance

The ACTIYAS has shown good clinical results in accordance with its design concept. It demonstrated a comparable ROM to that of NexGen LPS-Flex (Zimmer), a high-flexion PS model that is extensively used worldwide [3]. Similar to other PS implants, anterior post-impingement may occur during stair climbing [4]; however, the rounded design of the intercondylar notch of the femoral component and the anterior aspect of the tibial post maintains low stress on the post and functions like a BCS implant [5]. Coronal and sagittal laxity is adequately maintained in knees with good postoperative function [6], and better patient-reported outcomes have been achieved in patients with consistent femoral rotation during knee flexion [7]. The examination of Japanese-style sitting activities on the floor revealed reasonable knee kinematics adapted to each activity [8]. The survival rate of the implants at 9 years was 95.5% [9].

References

1. Ariga A, et al. Simultaneous bilateral total knee arthroplasty is safe and effective for patients in different ages. J Orthop Sci. 2021;28:173. https://doi.org/10.1016/j.jos.2021.08.011.
2. Yanagisawa K, et al. Do the distal femur and the proximal tibia have narrower aspect ratios in smaller knees?: a morphological analysis of osteoarthritic knees in the Japanese population using computed tomography. Knee. 2021;33:84–92.
3. Watanabe T, et al. In vivo kinematics of high-flex posterior-stabilized total knee prosthesis designed for Asian populations. Int Orthop. 2016;40:2295–302.
4. Watanabe T, et al. Anterior tibial post impingement during stair climbing: a kinematic analysis and clinical outcomes. J Arthroplast. 2019;34:379–84.
5. Watanabe T, et al. Postcam design and contact stress on tibial posts in posterior-stabilized total knee prostheses: comparison between a rounded and a squared design. J Arthroplast. 2017;32:3757–62.
6. Watanabe T, et al. Coronal and sagittal laxity affects clinical outcomes in posterior-stabilized total knee arthroplasty: assessment of well-functioning knees. Knee Surg Sports Traumatol Arthrosc. 2020;28:1400–9.
7. Watanabe T, et al. Consistent femoral external rotation during weight-bearing knee flexion is associated with better patient-reported pain and mediolateral balance after total knee arthroplasty. Clin Biomech (Bristol, Avon). 2021;88:105438.
8. Watanabe T, et al. Characteristic kinematics of floor-sitting activities after posterior-stabilized total knee arthroplasty determined using model-based shape-matching techniques. Knee. 2021;29:571–9.
9. Nakagawa Y, et al. Mid-term clinical outcomes of a posterior stabilized total knee prosthesis for Japanese patients: a minimum follow-up of 5 years. J Orthop Sci. 2022;

TKA Development in Japan: Quest Knee

12

Tatsuhiko Kutsuna

1 Implant Design Concept

The MERA Quest Knee System (Senko Medical Instrument Mfg. CO., LTD., Tokyo, Japan) is a total knee arthroplasty (TKA) implant developed based on the concept of adapting to the anatomical shape of the Japanese knee joint, supporting deep knee flexion, and acquiring long-term durability owing to low polyethylene wear (Fig. 12.1). Prioritization of the development of a cruciate retaining (CR) design and its clinical application started in October 2009, and Quest Knee has been used in approximately 1600 cases in Japan. In this section, we describe its design features and clinical results at Ehime University Hospital.

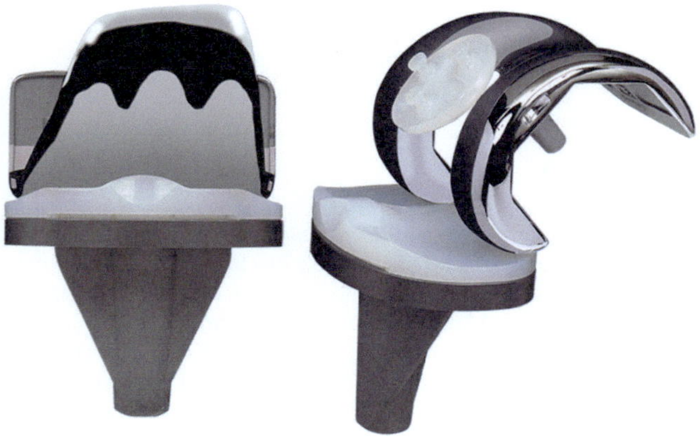

Fig. 12.1 MERA Quest Knee System. (Provided by Senko Medical Instrument Mfg. Co., Ltd., Japan)

2 Design Features

2.1 Its Design Conforms to the Anatomical Features of the Knee Joint of Japanese

Asians, who are known to have a smaller physique, such as smaller proximal tibias and distal femurs than Westerners. Therefore, in Quest Knee, the size variation was shifted to a smaller size, and the size pitch was set narrow to fit the small skeletal shape of Japanese. In addition, as the ratio of the anteroposterior diameter to the mediolateral diameter of the medial femoral condyle at the level of the distal femoral bone cut surface in Japanese is smaller and flatter than that in Westerners, the femoral component of Quest Knee was designed to flatten as the size increased.

2.2 Intercondylar Fossa Shape Considering Deep Knee Flexion

In the normal knee, the shape of the intercondylar fossa, in which the patella contacts with deep knee flexion, has a steep curve at the lateral condyle with a large space toward the lateral side. Accordingly, the patella sinks deeply into the intercondylar fossa [1]. Quest Knee incorporates this joint shape in the biological knee and has a shape in which the lateral condyle is deeply engraved. Therefore, the pressure of the patellofemoral joint in deep knee flexion will reduce, thus achieving better flexion range of motion (ROM) (Fig. 12.2) [2].

T. Kutsuna (✉)
Department of Bone and Joint Surgery, Ehime University Graduate School of Medicine, Ehime, Japan
e-mail: kutsuna@m.ehime-u.ac.jp

S. Matsuda (ed.), *Advances in Total Knee Arthroplasty*, https://doi.org/10.1007/978-981-97-4920-1_12

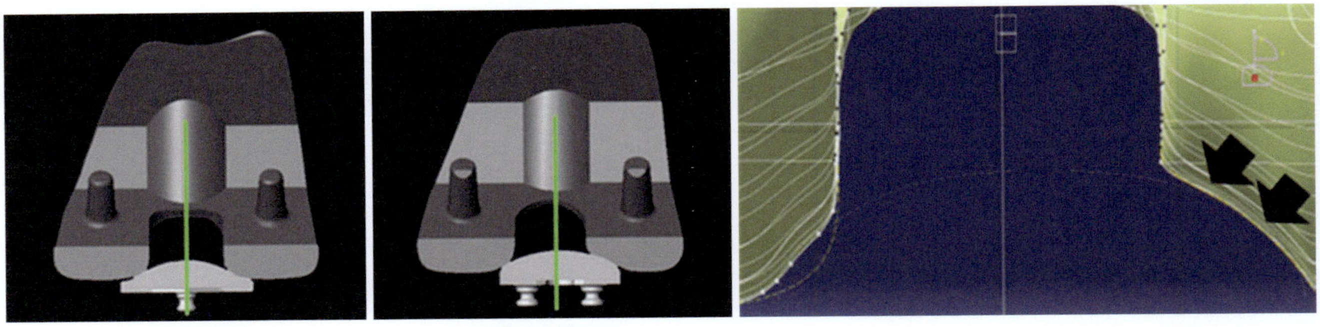

Fig. 12.2 Intercondylar fossa shape considering deep knee flexion. Quest Knee has a shape in which the lateral condyle is deeply engraved, thus reducing the pressure of the patellofemoral joint in deep knee flexion

2.3 Acquisition of Long-Term Durability Through Stability and Low Polyethylene Wear

Quest Knee aims to improve stability through extension to deep knee flexion by gradually reducing the sagittal radius of the femoral component. Moreover, to avoid point contact between the edge of the posterior condyle of the femoral component and the insert during deep knee flexion, the thickness of the posterior flange of the femoral component is increased. Furthermore, gradually increasing the radius of the terminal of the posterior condyle of the femur enables 155° of knee flexion.

3 Clinical Results

Of the 204 knees that underwent TKA using Quest Knee at Ehime University Hospital from April 2010 to December 2016, we investigated the clinical results of 117 knees with medial knee osteoarthritis (OA). TKA was conducted by a single surgeon and the follow-up period was at least 1 year (follow-up rate: 85%, mean follow-up period: 4 years). Knee score and ROM significantly improved from 40.4 points and 122° preoperatively to 95.7 points and 125° postoperatively, respectively. Furthermore, deep knee flexion of >130° was obtained in 50% of the cases in this series.

We performed CR-TKA on 298 knees in our institute using Quest Knee. Only one knee required revision surgery, which was a debridement, antibiotics, and implant retention (DAIR) with polyethylene insert exchange because of suspected periprosthetic joint infection (PJI). No case of revision due to aseptic loosening for mechanical failure or excessive polyethylene wear has been experienced, with preferable mid-term clinical outcome [3].

Compatibility with the Japanese lifestyle and accomplishment of sufficient function and long-term durability are considered the ultimate goals in TKA. Hence, confirming postoperative implant kinematics and clinical outcomes and providing feedback are important for developing future design and improving surgical techniques.

References

1. Moro-oka T, et al. Patellar tracking and patellofemoral geometry in deep knee flexion. Clin Orthop. 2002;394:161–8.
2. Ishimaru M, et al. Three-dimensional motion analysis of the patellar component in total knee arthroplasty by the image matching methods using image correlations. J Orthop Res. 2014;32:619–26.
3. Onishi Y, et al. The influence of tibial resection on the PCL in PCL-retaining total knee arthroplasty: a clinical and cadaveric study. J Orthop Sci. 2016;21:798–803.

TKA Development in Japan: Future Knee

13

Hiroshi Tsumura

1 History of Artificial Joint Research and Development

The "project to produce artificial joints tailored to individual patients" by Nakashima Propeller Co., Ltd. was certified in 2008 as a Super Special Zone (a Special Zone for Advanced Medical Development) by the Cabinet Office. The main focus of this project was the development of a series of technologies for designing and manufacturing joint prostheses optimized for each individual based on data on the patient's joint morphology, which resulted in the development of materials and three-dimensional (3D) modeling technologies. In 2011, it was certified as a "Personalized Joint Prostheses Function and Safety Evaluation Project" as a part of the Asian Standards Certification Promotion Project by the Ministry of Economy, Trade and Industry (METI). The objectives of this project were: (1) standardization of technologies related to bone morphological extraction; (2) analysis of bone morphology of the Asian people and construction of a morphological database; (3) development of evaluation methods using surgical support technology; and (4) formulation of standardization proposals for functional and safety evaluation standards for personalized joint prostheses. This study was conducted in collaboration with research institutions in five countries: Japan, Thailand, Hong Kong, Taiwan, and Singapore. We measured and evaluated the 3D morphology of the lower limb using CT in groups of different races, sexes, and age groups, mainly with the development of artificial knee joints in mind. We concluded that there are clear racial and sex-based differences in knee joint morphology and that it is appropriate to use the standard postosteotomy surface as a measurement method on the premise of artificial joint replacement. Finally, we decided on the ISO 19233-1:2017 standard by Japan. The New Energy and Industrial Technology Development Organization (NEDO) certified the "NEDO International Demonstration: R&D and Demonstration of Construction of Artificial Joint and Surgical Support System" Project, which led to the development of "Future Knee."

H. Tsumura (✉)
Oita University, Oita, Japan
e-mail: htsumura@oita-u.ac.jp

2 Outline of the "NEDO International Demonstration: R&D and Demonstration of Construction of Artificial Joint and Surgical Support System" Project

This project aims to analyze the bone morphology of Asians based on medical images to design artificial joints adapted to their lifestyles and conduct functional and safety tests, dynamic analysis, and cadaveric tests. The project targeted the Kingdom of Thailand, aiming to enter overseas markets, particularly in Asia, by providing support for the development of joint prostheses that are compatible with bone morphology and lifestyles and the establishment of an evaluation system for joint prostheses. This was a joint research project with the National Metal and Materials Technology Center (MTEC) in Thailand. With funding from NEDO, a simulator capable of performing artificial joint durability tests was installed at MTEC, and 5 million wear tests of Future Knee were conducted.

3 Future Knee Concept and Features

Future Knee is a PS-type implant. Figure 13.1 shows a graph of the ML and MAP sizes of the distal femur end of women in data collected by NEDO. The distribution of Thais with large circles and Japanese individuals shown as small circles shows that the ML–MAP ratios are similar, although Thai individuals are slightly smaller. Based on this ratio, the implant was designed to optimize bone morphology, and the rectangular boxes indicate the area where femoral components of each size can be adequately covered [1] with the smallest being specifically for Thai people.

S. Matsuda (ed.), *Advances in Total Knee Arthroplasty*, https://doi.org/10.1007/978-981-97-4920-1_13

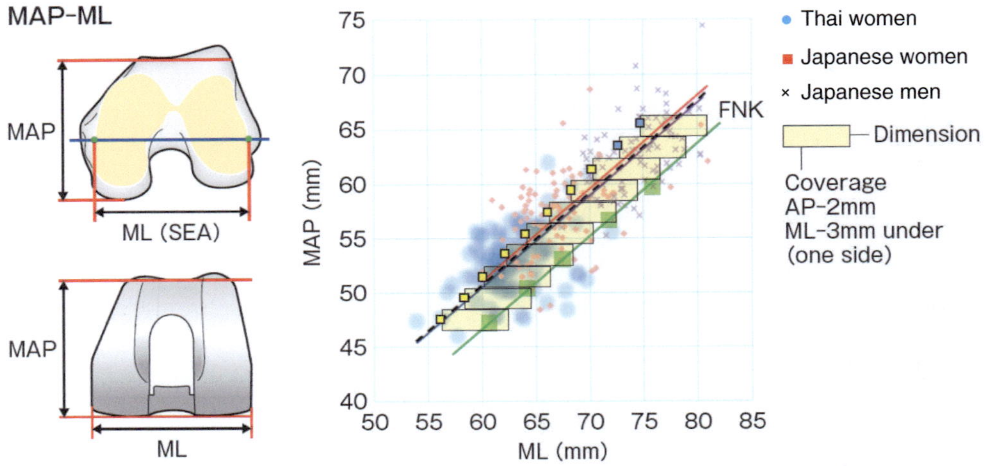

Fig. 13.1 Measurement results of the distal femur. Vertical and horizontal sizes of the distal femur in Thai and Japanese women. The aspect ratio is similar in the two groups, but Thai women have a smaller aspect ratio than Japanese women, as indicated by the slightly larger circle. (By courtesy of Teijin Nakashima Medical Co., Ltd.)

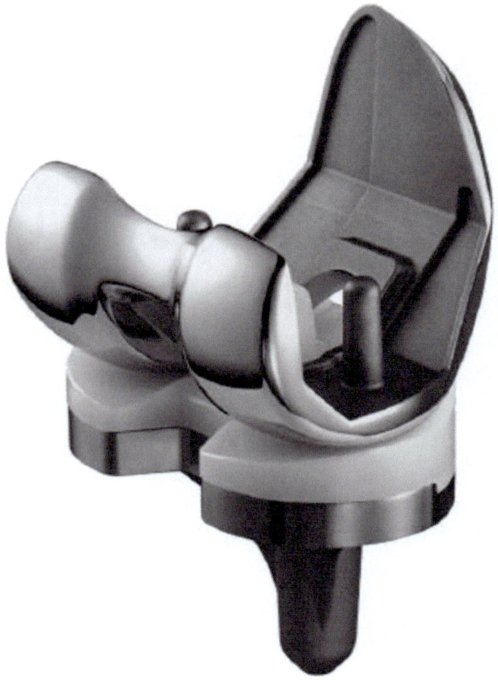

Fig. 13.2 Future Knee and postoperative X-ray images. It is a PS-type implant designed based on the morphometric results of Thai and Japanese people and has a high convolution tolerance. Postoperative radiography revealed a sharp anterior flange. (By courtesy of Teijin Nakashima Medical Co., Ltd.)

First, to minimize midflexion instability and optimize MCL function, the anatomical shape of the condylar articular surface in the femur is reproduced, and the rotational position of the surgical instrument can be set while maintaining a constant medial gap relative to the medial articular surface. Second, rotational tolerance is increased by ±10° for extension and ± 30° for flexion, and the femoral cam is eccentric to the medial side to facilitate the induction of internal rotation of the lower leg in flexion. Additionally, the anterior flange and lateral posterior condyle are narrowed to prevent impingement at each site. Figure 13.2 shows a photograph of the rear of the implant and a postoperative radiograph. The narrow anterior flange is impressive.

Clinical use began in October 2017 at a limited number of institutions. As of December 2021, there have been more than 1300 cases nationwide, and good feedback has been reported regarding ROM and size matching [2–4].

References

1. Nagamine R, et al. Evaluation of kinematics of the new PS system future knee using cadaveric knees. J West-Jpn Soc Orthop Trauma. 2018;67(Suppl 2):150. [Japanese]
2. Ryu K, et al. Knee prostheses from Asia. Future knee: Asian specific posterior stabilized total knee arthroplasty. JOSKAS. 2018;43:474.
3. Matsuda S, et al. Short-term clinical results of future knee TKA. JOSKAS. 2020;45:428–9. [Japanese]
4. Noji H, et al. Short-term clinical evaluation of future knee. J East Jpn Assoc Orthop Trauma. 2021;33:445. [Japanese]

14 TKA Development in Japan: Initia Knee

Shuichi Matsuda

Initia Knee is a TKA system developed in collaboration with Kyoto University and Kyocera Corporation, together with the support of A-Step of Japan Science and Technology Agency (Fig. 14.1).

The basic concepts of Initia Knee are as follows:

1. Maintain MCL tension throughout the range of motion

   When the femur shape is regarded as a single radius from 0° to 90° combined with the shape of the tibial joint surface, the position of the medial epicondyle does not change, even in the rollback, with up to 1/3 of the posterior portion observed in a biological knee (Fig. 14.2) [1]. The position of the medial epicondyle does not elevate even in the presence of an LIP posterior to the tibial articular surface, by making the radius of the femoral curvature smaller at 120° of flexion.
2. Reduce forward movement of the femur

   The lip anterior to the tibia is slightly raised to reduce the anterior movement of the femur, the so-called paradoxical motion, during flexion. Although rotational stability is also a concern, ligament tension does not increase during rotation because the entire articular surface is spherical [2].
3. Allowing external rotation of the femur

   The articular surface of the lateral femoral condyle is curved to increase the tolerance of external femoral rotation (Fig. 14.3).
4. Reduce PFJ contact pressure

   The shape of the intercondylar femoral component is deepened on both the medial and lateral sides, reducing the load on the PFJ.
5. Easy-to-use surgical instruments

   The posteromedial reference of the femur allows fine adjustment of the anteroposterior position even after anterior osteotomy. The tibial guide can also be adjusted in 1° increments on the posterior slope of the tibia. The instruments were designed to be visually appealing in terms of design and color.

Fig. 14.1 Appearance of Initia Knee

Clinical use started in 2020.

S. Matsuda (✉)
Department of Orthopaedic Surgery, Kyoto University, Kyoto, Japan
e-mail: smat522@kuhp.kyoto-u.ac.jp

S. Matsuda (ed.), *Advances in Total Knee Arthroplasty*, https://doi.org/10.1007/978-981-97-4920-1_14

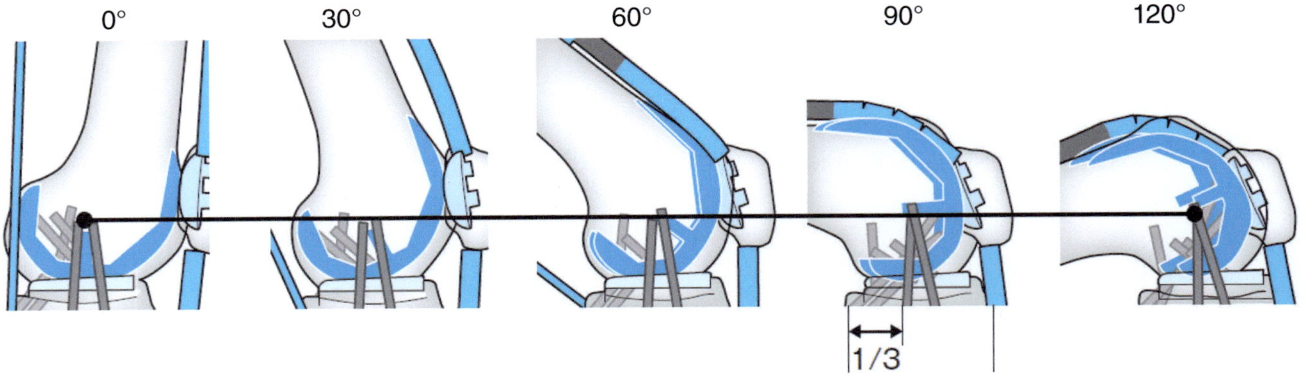

Fig. 14.2 Position of the medial epicondyle at 0–120° flexion

Fig. 14.3 Shape of the posterior femoral condyle

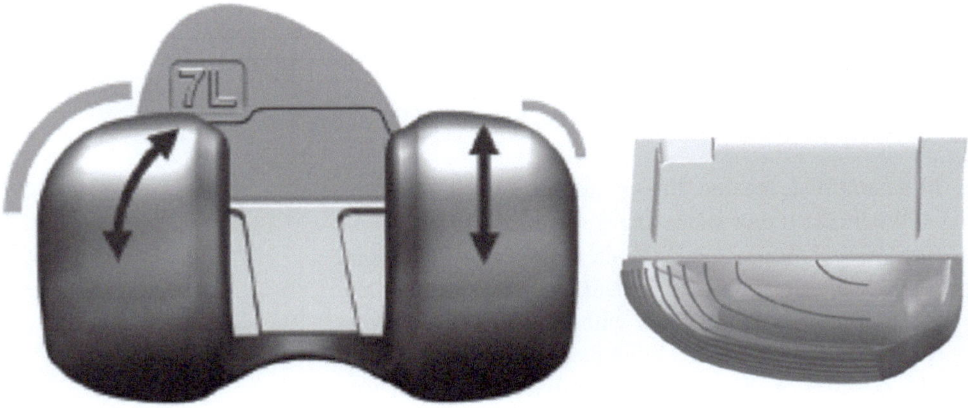

References

1. Fukagawa S, et al. Posterior displacement of the tibia increases in deep flexion of the knee. Clin Orthop Relat Res. 2010;468:1107–14.
2. Kuriyama S, et al. Malrotated tibial component increases medial collateral ligament tension in total knee arthroplasty. J Orthop Res. 2014;32:1658–66.

TKA Development in Japan: Vanguard PSRP

15

Yoshinori Kadoya

1 Overview

Vanguard PSRP was introduced in 2016 as a successor to the Vanguard RP High-Flex (used since 2006), with an improved box ratio [1], inner and outer width, and outer posterior condyle shape, as well as with additional size variations. The design aimed to safely achieve deep flexion by combining the PS design and mobile mechanism (Fig. 15.1); currently, more than 28,000 knees are being used with good results.

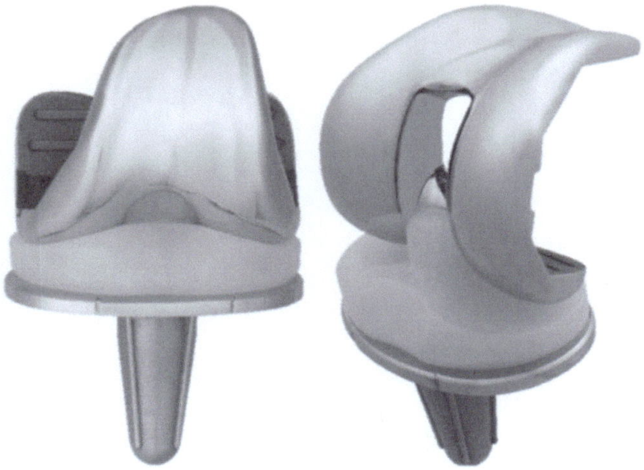

Fig. 15.1 Appearance of Vanguard PSRP. (By courtesy of Zimmer-Biomet)

2 Design Concept

2.1 Large Contact Area and Stability

By using the posterior surface of the tibial post as an articular surface, a large contact surface is ensured in all ROMs. The corresponding femoral side also has a saddle shape and provides high lateral stability (Fig. 15.2). The contact area is 1148 mm^2, maintaining 450 ~ 700 mm^2 at 30–120° in the extended position. The design even maintains a large contact area of 298 mm^2 at a deep flexion of 150° (Fig. 15.3). Cam is designed to engage early, initially acting at 30° for stability during ADLs. The mechanical strength of the post must be maintained when the posterior surface is used as a joint surface. To this end, we optimized the shape, front and rear, and inner and outer diameters by using the finite element analysis. The design ensures 10° hyperextension tolerance and maintains surface contact even when an anterior impingement occurs. Low-wear ArCom is used for polyethylene. High wear resistance is achieved with the combination of these elements.

2.2 Physiological Movement and Wear Resistance

A mobile mechanism was introduced, to achieve both physiological movements, such as rotation and a wide contact area and stability,

Y. Kadoya (✉)
Hanwa Joint Reconstruction Center, Osaka, Japan
e-mail: kadoya.yoshinori@kinshukai.or.jp

S. Matsuda (ed.), *Advances in Total Knee Arthroplasty*, https://doi.org/10.1007/978-981-97-4920-1_15

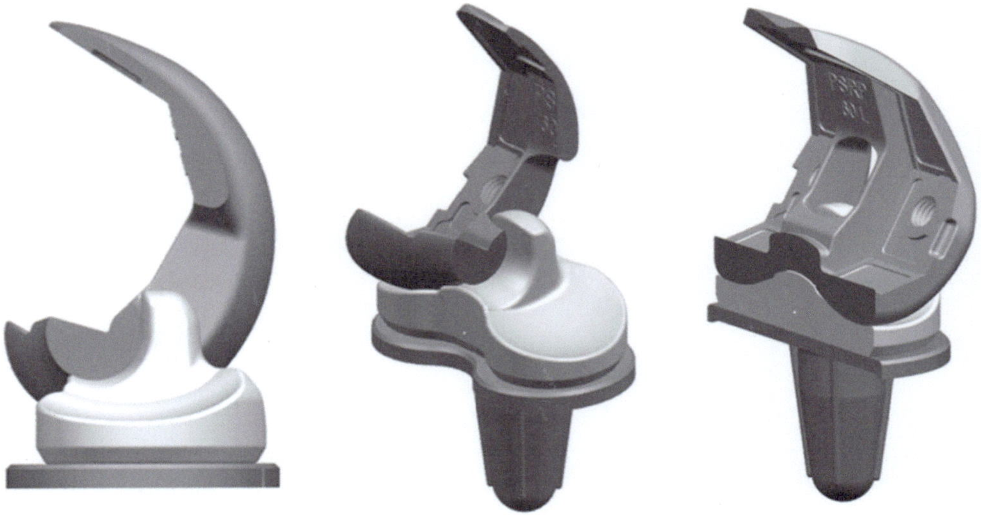

Fig. 15.2 Saddle shape of the articular surface and posterior condyle of the posterior surface of the post. (By courtesy of Zimmer-Biomet)

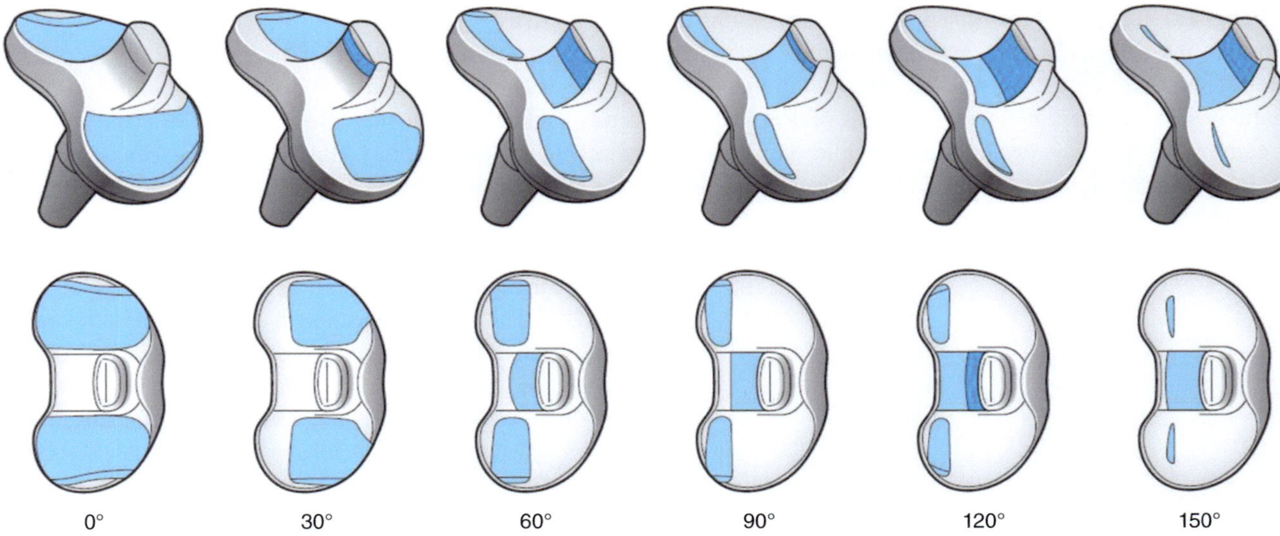

Fig. 15.3 Change in contact surface with various flexion angles. (By courtesy of Zimmer-Biomet)

The advantages of the mobile mechanism are:

1. Stress reduction in the deep flexion position [2] (Safety valve mechanism); and
2. Tolerance of rotational mismatch in the deformed knee [3] (self-aligning mechanism).

It is expected to reproduce physiological movements and safely achieve deep flexion.

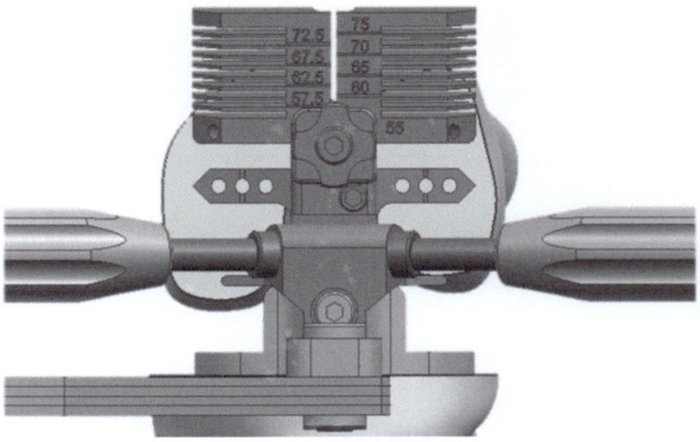

Fig. 15.4 Pro-Flex G: a special instrument for the gap balancing technique. (By courtesy of Zimmer-Biomet)

2.3 Development of Specialized Surgical Instruments

To perform surgery with good reproducibility, we developed a special instrument (Pro-Flex G) for the gap balancing technique. The basic function is to guide the anterior-posterior and rotational positions of the femoral cutting block (Fig. 15.4), making it easy and reliable to adjust the appropriate flexion gap [4].

With the above characteristics, it has been reported that high stability and physiological movement patterns [5–7] can be obtained, and this model is considered to provide good clinical results in safely achieving deep flexion.

References

1. Fukunaga K, et al. The incidence of the patellar clunk syndrome in a recently designed mobile-bearing posteriorly stabilized total knee replacement. J Bone Joint Surg Br. 2009;91:463–8.
2. Nakagawa S, et al. Tibiofemoral movement 3: full flexion in the living knee study by MRI. J Bone Joint Surg Br. 2000;82:1199–200.
3. Matsui Y, et al. Rotational deformity in varus osteoarthritis of the knee: analysis with computed tomography. Clin Orthop. 2005;433:147–51.
4. Kadoya Y. Hanwa joint reconstruction Center TKA manual -basic course. Tokyo: Medical View Co., Ltd.; 2019.
5. Minoda Y, et al. Mobile-bearing TKA improved the anteroposterior joint stability in mid-flexion range comparing to fixed-bearing TKA. Arch Orthop Trauma Surg. 2016;136:1601–6.
6. Kurita M, et al. In vivo kinematics of high-flex mobile-bearing total knee arthroplasty, with a new post-cam design, in deep knee bending motion. Int Orthop. 2012;36:2465–71.
7. Kage T, et al. In vivo kinematics of a newly updated posterior-stabilised mobile-bearing total knee arthroplasty in weight-bearing and non-weight-bearing high-flexion activities. Knee. 2021;29:183–9.

Part II

Anatomy

16 Functional Anatomy of the Knee Joint

Yasushi Oshima

1 Joint Structure

The knee joint consists of tibiofemoral and patellofemoral joints.

As the tibiofemoral joint is a hinge joint, which does not provide bony stability, the stabilization is maintained by the static stabilizers such as ligaments and meniscus, and the dynamic stabilizers such as muscles and tendons. Therefore, it is indispensable to preserve knee stability with the appropriate soft tissue balancing as well as to reproduce accurate lower limb alignment in TKA.

2 Nerve and Vascular Innervation (Fig. 16.1)

The anterior surface of the knee is innervated by the medial cutaneous nerve, the intermediate cutaneous nerve, and the saphenous subpatellar branch, which are branches of the femoral cutaneous nerve. Therefore, postoperative neuropathy often occurs lateral to the skin incision.

In addition, the skin on the front of the knee is nourished with the descending knee artery from the saphenous artery by penetrating through the deep fascia, however, these vas-

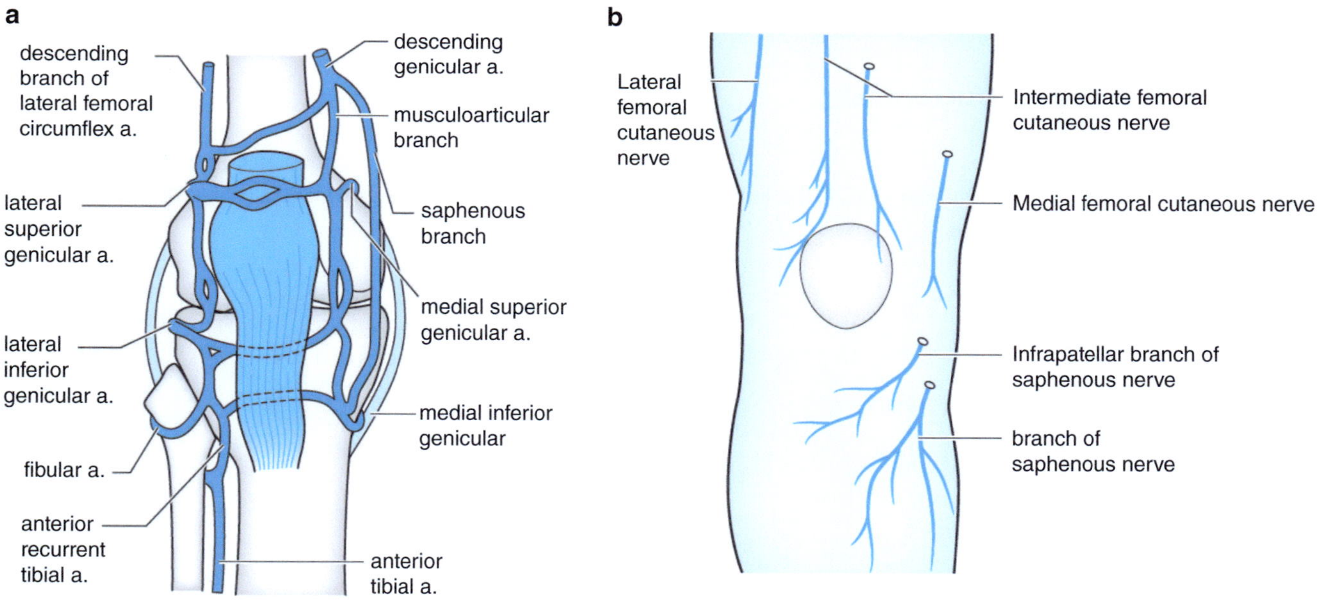

Fig. 16.1 Nerve and vascular anatomy on the front of the knee. (**a**): Blood circulation on the front of the knee. (**b**): Nerve tracts of the skin on front of the knee. (Made based on Artificial joint replacement: Procedure and issues, Igaku-Shoin Ltd., p. 2, Fig. 16.1)

Y. Oshima (✉)
Department of Orthopaedic Surgery, Nippon Medical School, Tokyo, Japan
e-mail: y-oshim@nms.ac.jp

S. Matsuda (ed.), *Advances in Total Knee Arthroplasty*, https://doi.org/10.1007/978-981-97-4920-1_16

cular branches do not infiltrate to the surrounding subcutaneous tissues. Moreover, because most of the branches, which are penetrating the deep fascia, are often located in the medial of the knee, postoperative skin disorders also often occur lateral to the skin incision. Therefore, the skin incision should not be located too medial, subcutaneous detachment should be minimized, and the superficial fascial layer needs to be preserved. In addition, to prevent from the operative wound dehiscence and the pain and paresthesia during kneeling motion, the distal end of the incision should be avoided just above the tibial tuberosity surface. These neurovascular disorders are also considered desirable to be explained to the patients preoperatively as one of the postoperative complications.

3 Extension-Flexion Mechanism

The knee extension mechanism consists of the quadriceps, which is the largest muscles in the human body, patella, and patellar tendon, as dynamic stabilizers in front of the knee. The quadriceps muscle efficiently transfers extension force to the lower leg by increasing the moment arm with the thickness of the patella. Posteriorly, there are the biceps, semimembranous muscles, and semitendinosus muscles, which act on the extension of the hip joint and flexion of the knee joint and are called hamstrings.

4 The Soft Tissue Structures at the Medial Knee

The soft tissue structures on the media side of the knee can be divided into three anatomic layers.

The first layer is the superficial layer, where the medial patellar retinaculum extends to cover the entire medial side of the knee, including the popliteal fascia and fascia lata. Proximally, it extends over the MCL and vastus medialis, and distally, it reaches the tibia, covering the sartorius and other muscles.

The second layer consists of the superficial MCL (sMCL), MPFL, and posterior oblique ligament (POL), which connects to the third layer, involving the deep MCL (dMCL) and the joint capsule [1, 2] (Fig. 16.2).

Robinson et al. classified the medial soft tissue structures into three sites: anterior, middle, and posterior, with the anterior part from the medial border of the patellar tendon to the anterior border of the longitudinal fibers of the sMCL, the middle part from the anterior border to the posterior border of the sMCL. The posterior part contains the posteromedial corner of the knee and extends from the posterior border of the longitudinal fibers of the sMCL to the medial edge of the medial head of the gastrocnemius muscle. The posteromedial corner consists of five components, the POL, semimembranosus tendon, oblique popliteal ligament, posteromedial joint capsule, and posterior horn of the MM, and plays an important role to maintain the rotational stability of the knee joint [3, 4]. On the other hand, it has been reported that the sMCL and dMCL were added to the posteromedial corner, and these sMCL, dMCL, and POL were responsible for the main stabilizers of the posteromedial of the knee [5].

4.1 MCL

The MCL consists of two components, the sMCL as the second layer of the medial side of the knee and the dMCL as the third layer.

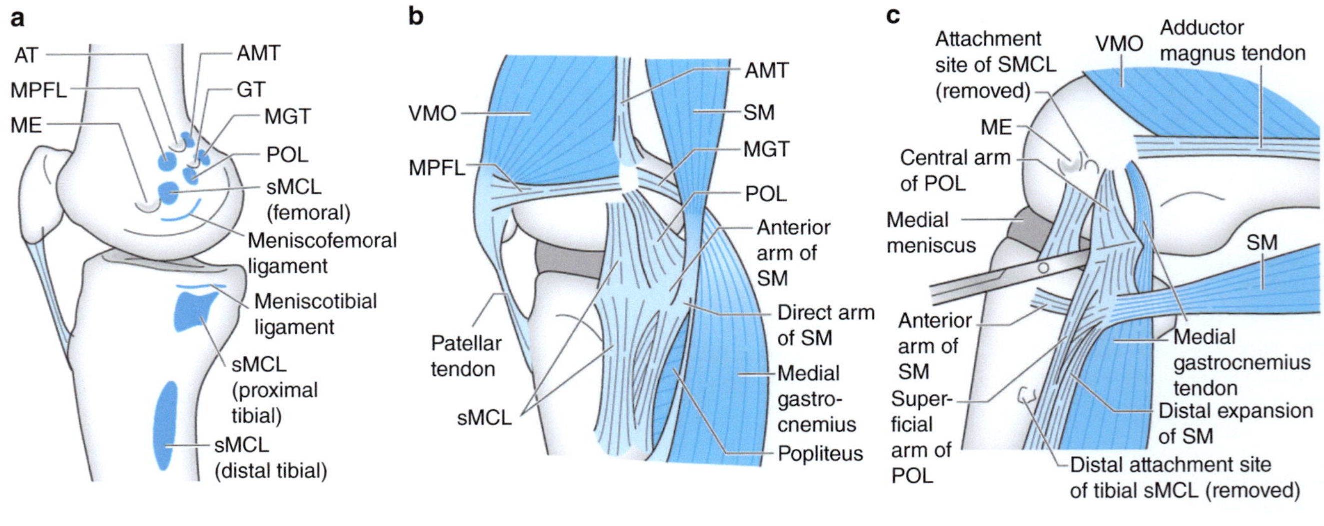

Fig. 16.2 Attachment and structure of the medial soft tissues of the knee. (**a**): Bone landmarks and soft tissue attachment on the medial side of the knee. (**b**): Soft tissues on the medial side of the knee. (**c**): POL and dMCL structure, sMCL removed. (Made based on LaPrade et al. [7])

1. sMCL

According to Liu et al., the sMCL was a thin, anteroposteriorly broad, and triangular in shape with the proximal, meddle, and distal widths of 10.9 ± 1.2 mm, 17.7 ± 2.1 mm, and 10.7 ± 1.8 mm, respectively [6]. The insertion area on the femur was 79.7 ± 17.6 mm^2, and the centroid of the femoral insertion was 31.1 ± 4.6 mm from the femoral joint line, 2.9 ± 0.8 mm from the medial epicondyle, 14.7 ± 4.5 mm from the medial gastrocnemius tubercle, 16.5 ± 1.6 mm from the adductor tubercle, and was slightly proximal to the medial epicondyle [6]. LaPrade et al. found that the femoral attachment of the sMCL was round to slightly oval in shape and was located in the depression that was an average of 3.2 mm proximal and 4.8 mm posterior to the medial epicondyle [7]. The tibial insertion from the tibial joint line was 62.4 ± 5.5 mm, and the overall length from the femoral insertion to the centroid of the tibial insertion was 100.7 ± 9.5 mm. It was also reported that the sMCL had two separate tibial attachments: proximal and distal [7]. Although the proximal site was stated to be an average of 12.2 mm from the joint line, most of it blended to the anterior aspect of the tibial attachment of the semimembranous muscle, but did not attach firmly [6]. The distal attachment on the tibia was broad, and was located just anterior to the posteromedial crest of the tibia, forming the bursa together with the pes anserinus, and the insertion area was 348.6 ± 42.8 mm^2. The posterior border of the sMCL combined with the semimembranosus tendon.

2. dMCL (Fig. 16.3)

The dMCL is the capsular ligament that forms part of the joint capsule in contact with the MM, and is divided into the proximal part as meniscofemoral ligament (MFL) and the distal part as meniscotibial ligament (MTL). The femoral insertion of the MFL was the anteroposterior width of 9.9 ± 3.2 mm, the length of 9.4 ± 3.3 mm, and the area of 71.9 ± 14.8 mm^2, and the centroid of the attachment was located 20.5 ± 4.2 mm from the femoral joint line. The tibial insertion of the MFL was the anteroposterior width of 18.0 ± 4.0 mm, the length of 5.1 ± 1.8 mm, and the area of 63.6 ± 13.4 mm^2, and the centroid of the insertion was located 6.5 ± 1.3 mm from the tibial joint line. The anteroposterior widths of the MFL and the MTL were 10.6 ± 4.5 mm and 17.9 ± 2.7 mm, and the lengths were 26.2 ± 5.6 mm and 9.2 ± 1.8 mm, respectively [6].

As the posterior fibers of sMCL and dMCL were firmly attached to the MM, during MM resection in TKA, care should be taken not to damage the sMCL and dMCL. In addition, the MTL attached at a length of 5 mm, approximately 4 mm to 9 mm from the tibial joint line. Since the osteophyte on the medial tibial condyle in knee OA is formed proximal to the tibial insertion of the MTL, osteophyte resection can be performed while protecting the MTL. However, in cases of large bone defects on the medial of the tibia, the MTL may be degenerated or disappeared.

The anterior edge of the dMCL is parallel to the anterior edge of the sMCL, and can be distinguished by its location in

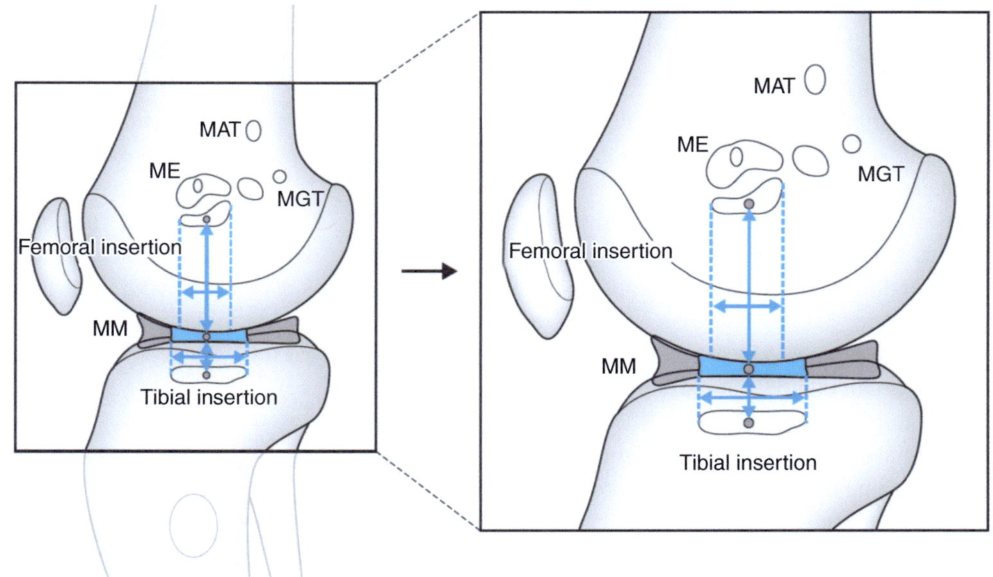

Fig. 16.3 Structure of dMCL. (Made based on Liu et al. [6])

the thinning part of the joint capsule from the medial patella. When performing the arthrotomy through the medial approach, the anterior margin of the dMCL should be identified to preserve the dMCL by not releasing it posteriorly.

4.2 POL

The POL was previously thought to be an oblique fiber of the sMCL. Recently, the POL has been reported to be located posteriorly to the sMCL at the distal end of the semimembranous tendon and to consist of the superficial, central, and capsule arms [8]. The femoral insertion was an average of 7.7 mm distal and 6.4 mm posterior to the adductor muscle tubercle, was 1.4 mm distal and 2.9 mm anterior to the gastrocnemius tubercle, and was proximal and posterior to the femoral attachment of the sMCL [7].

The superficial arm of the POL consisted of a thin fascial expansion. It blended into the central arm proximally, coursed medial to the anterior arm of the semimembranous tendon, and was parallel to the posterior border of the sMCL distally until it blended into the distal tibial expansion of the semimembranous tendon. The central arm was the largest and thickest portion of the POL, coursed from the distal aspect of the main semimembranous tendon, was a thick fascial reinforcement of the posteromedial capsule, and had a stout attachment to the MM. The capsular arm consisted of a thin fascial expansion off the anterior aspect of the distal part of the semimembranous tendon proximally, and did not attach to the tibia.

For the knee valgus load, the sMCL was the first restraint, the dMCL was the second restraint, and the PCL compensated for the MCL function. Therefore, sMCL injury produced valgus instability in all ROM, while dMCL injury alone did not [9]. Besides the valgus stabilizer, the sMCL and dMCL were the external rotation stabilizers of the tibia. Conversely, the POL was the internal rotation stabilizer of the tibia in the knee extension [5]. Thus, the sMCL injury, as well as the dMCL and POL injuries caused rotational instability, and the dMCL and POL injuries caused valgus instability [10].

In medial knee OA, the marginal osteophytes of the medial tibial condyle are often formed extensively from medial to posterior medial. Insufficient exposure of the posteromedial knee can cause the residual MCL tightness due to the remaining osteophytes, or the rotational instability due to the damage to the posteromedial corner including the POL. Thus, it is important to expose the posteromedial of the knee sufficiently by the so-called Ransall maneuver with knee flexion, foot external rotation, and the anterior drawer of the tibia, to resect the posteromedial osteophyte, and to protect the posteromedial corner [10, 11].

5 MPFL

The MPFL is the first restraint against the lateral patellar dislocation, which is located in the second layer of the medial side of the knee. The proximal edge attached to the distal end of the vastus medialis oblique, and the distal edge was discernible from the proximal-medial edge of the patella to the femoral attachment. It attached to the superomedial aspect of the medial border of the patella, and the midpoint of the attachment was located 41.4% of the length from the proximal tip of the patella. The femoral attachment was an average of 10.6 mm proximal and 8.8 mm posterior to the medial epicondyle and 1.9 mm anterior and 3.8 mm distal to the adductor tubercle. The average length was 65.2 mm, and the width was approximately 10–20 mm [7, 12].

The MPFL was most important in the range of 20–90° of flexion to act as the stabilizer for patella [13]. The patella was in contact with the trochlear groove of femur when knee flexion exceeds 30° and stabilized when flexion exceeds 60–70°.

In the medial parapatellar approach of TKA, the MPLF, medial patellar retinaculum and joint capsule are dissected at the beginning, and then are sutured after the implantation of the prostheses. As an inappropriate suture of these tissues causes patellar instability and dislocation, PFJ pain, and the limitation of knee flexion postoperatively, an incision should be made leaving sufficient seam allowance in knee arthrotomy, and these tissues should be sutured appropriately with confirming the patella tracking and medial soft tissue tightness.

6 Semimembranous Tendon

The tibial attachment consisted of the anterior and direct arms. The anterior arm was in an oval shape, and attached deep to the proximal tibial attachment of the sMCL and to the tibial joint line. The direct arm attached to the posteromedial part of the tibia, distal to the posteromedial joint line. It is a constituent of hamstrings, and has the effects of knee flexion, tibial internal rotation, and hip extension and internal rotation.

7 Pes Anserine Tendon

The pes anserine tibial attachment consisted of the sartorius, gracilis, and semitendinosus tendons; the sartorius tendon fascia was located in the first layer of the medial side of the knee, and the gracilis and semitendinosus tendons were located between the first and second layers [2]. It attached to the anteromedial aspect of the proximal tibia in the order of sartorius, gracilis, and semitendinosus tendons, and the average tendon widths at the tibial attachment sites were 8.0 mm for the sartorius, 8.4 mm for the gracilis, and 11.3 mm for the semitendinosus. The midpoint of the lateral attachment of the gracilis on the tibia averaged 8.2 mm proximal and 13.4 mm anterior to the distal osseous attachment of the sMCL. Compared to the sMCL, dMCL, and POL, the function of the pes anserine for knee stability was less important [5].

8 Intra-Articular Structures

There involves ACL, PCL, MM, and LM. Recently, bicruciate retaining TKA that preserves ACL and PCL has also been attempted, however, the number of these cases has been still smaller than the conventional CR and PS TKAs [14]. At least, the PCL function is necessary to understand in TKA because CR-TKA, in which the ACL is excised and the PCL is preserved, and PS-TKA, in which the ACL and PCL are excised and the PCL is replaced by post-cam mechanisms.

The PCL is the largest and strongest intra-articular ligament in the knee, consists of an anterolateral bundle (ALB) and a posteromedial bundle (PMB), and the ALB is larger. The origin is the anterior part of the intercondylar fossa of the medial femoral condyle, and the distal end attaches to the middle of the posterior margin of the tibia. The most important function is the first restraint to the posterior drawer of the tibia relative to the femur in the entire ROM. Both bundles act in the entire ROM, but ALB functions particularly in flexion from 70° to 105°, and PMB in flexion from 0° to 15°. It is responsible for the first restraint for internal rotation at 90° or more and part of the restraint for external rotation. In addition, it is the secondly restraint for internal rotation and valgus load [15, 16] (Fig. 16.4).

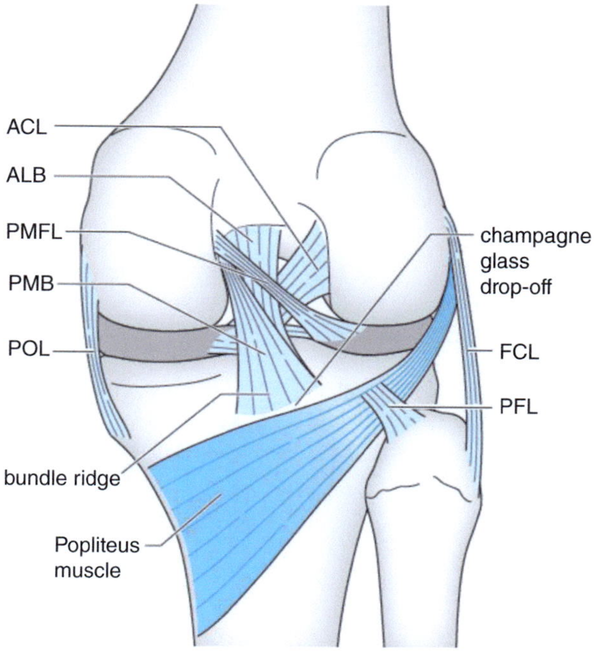

Fig. 16.4 Structure of the PCL and posterolateral corner of the knee. (Made based on Pache et al. [15])

Until now, the optimal tibiofemoral joint gap in TKA has been considered as equal and rectangular in extension and flexion and in the medial and lateral of the knee. To obtain the optimal gap, soft tissue release has been performed in TKA. During this process, the PCL resection has been reported to significantly increase the flexion gap compared with the extension gap [17, 18]. Recently, it has been examined that the flexion gap was wider than the extension gap, and the lateral was wider than the medial in the normal knee [19]. The excessive release of the medial soft tissue has also been reported to be a result of postoperative valgus and rotational instability. For these reasons, the recent TKA technique has been emphasized to preserve medial stability with minimal soft tissue release, and has been allowed the trapezoidal gap with slightly wider in flexion gap than extension gap. When the medial soft tissues are preserved, the MCL is the primary restraint and the PCL is the secondary restraint against valgus loads. Therefore, the PCL resection with the existence of the MCL has been reported not to increase the flexion gap [20]. In performing TKAs, it is important to understand that the effects of the PCL resections on the tibiofemoral gaps are varied with the state of the medial soft tissues.

9 Posterolateral Corner (PLC)

Posterolateral stability is maintained by static stabilizers, including the LCL, popliteus tendon, popliteofibular ligament, fabellofibular ligament, arcuate ligament, and posterolateral capsule, and dynamic stabilizers, such as the biceps tendon, iliotibial tract, lateral head of gastrocnemius, and popliteus muscle-tendon complex [21] (Fig. 16.5). The main tissues are the LCL, popliteus tendon, and popliteofibular ligament [22].

9.1 LCL

The femoral attachment was on average of 1.4 mm proximal and 3.1 mm posterior to the lateral epicondyle of the femur, and some fibers covered the lateral epicondyle. The average femoral attachment area was 0.48 cm^2, which was distal and posterior to the femoral attachment of the MCL [23]. The LCL distally attached 8.2 mm posteriorly to the anterior edge of the fibular head and 28.4 mm distal to the tip of the fibular styloid process, and the fibular attachment area was

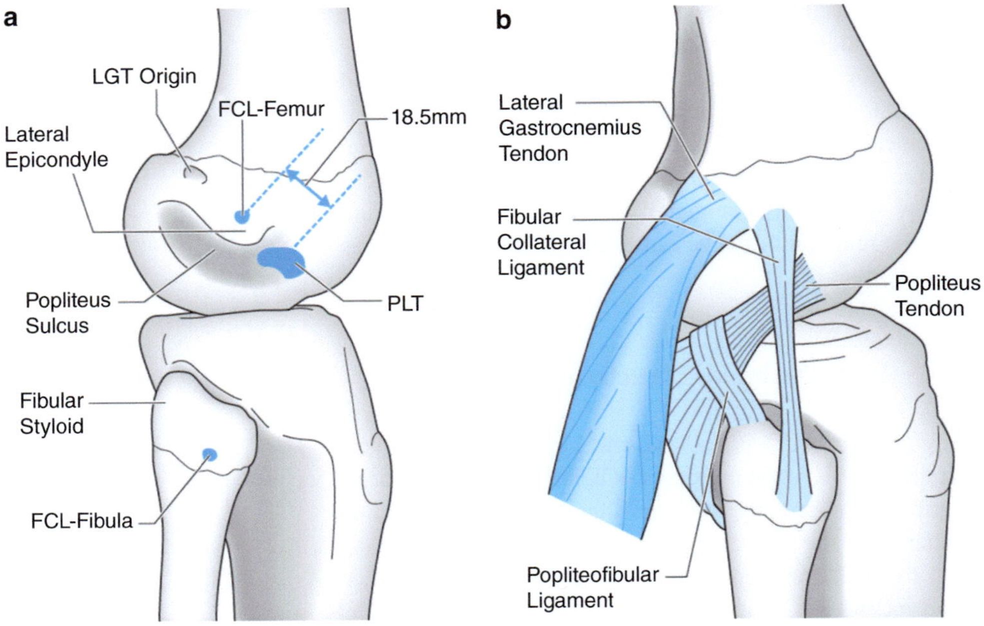

Fig. 16.5 Attachment and structure of lateral soft tissues of the knee. (**a**): The lateral bony landmark of the knee and the attachment of the posterolateral corner. (**b**): Posterolateral corner. (Made based on LaPrade et al. [24])

on average of 0.43 cm^2. The average total length of the LCL between its attachments was 69.6 mm [22, 24].

9.2 Popliteus Tendon

The popliteus muscle attached to the posteromedial tibia, coursed proximally and laterally, gave rise to the popliteal tendon the lateral one-third of the popliteal fossa, coursed intraarticular, and ran medial to the LCL. The attachment was at the most anterior fifth and the proximal half of the popliteal sulcus, and the attachment area was 0.59 cm2. The attachment on the femur was always anterior to the LCL attachment, and the average length from its femoral attachment to its musculotendinous junction was 54.5 mm [24].

9.3 Popliteofibular Ligament

The popliteofibular ligament originated at the musculotendinousn junction of the popliteus muscle and was divided into anterior and posterior components. The average fibular attachment of the anterior division was 2.8 mm distal to the tip of the fibular styloid process on its anteromedial downslope, and the average attachment of the posterior division was 1.6 mm distal to the tip of the fibular styloid process. The average width of the anterior division's attachment was 2.6 mm, and the average width of the posterior component was 5.8 mm [24].

PLC was the first restraint for varus load and external rotation of the tibia and was the secondary restraint for the anterior-posterior translation of the tibia. Among these, the LCL was the primary restraint for internal rotation, and the other PLC tissues were the secondary restraints. For external rotation of the tibia at 30 ~ 40°flexion, the LCL and popliteal tendon complex were the first restraints, and the PCL was the second restraint. The popliteus tendon, along with the arcuate ligament and other PLCs, were the stabilizes for the tibial external rotation, with the anterior fibers tensioned and the posterior fibers relaxed in knee flexion, and this phenomenon was enhanced with the increase of the tibial external rotation [25]. In the full extension of the knee, it is responsible for the secondary restraint against the anteroposterior movement of the tibia, which was important for maintaining posture in the standing position [26]. The popliteofibular ligament compensated the functions of varus restraint of the LCL and external rotational restraint of the popliteus tendon [22].

Compared to the medial approach, the lateral approach for the severe lateral knee OA is not often performed, it requires to understand the iliotibial ligament and PLC anatomy preoperatively.

References

1. Warren LF, Marshall JL. The supporting structures and layers on the medial side of the knee:an anatomical analysis. J Bone Joint Surg Am. 1979;61:56–62.
2. Curtis BR, et al. Pes anserinus:anatomy and pathology of native and harvested tendons. AJR Am J Roentgenol. 2019;213:1107–16.
3. Robinson JR, et al. The posteromedial corner revisited. An anatomical description of the passive restraining structures of the medial aspect of the human knee. J Bone Joint Surg Br. 2004;86:674–81.
4. Dold AP, et al. The posteromedial corner of the knee:anatomy, pathology, and management strategies. J Am Acad Orthop Surg. 2017;25:752–61.

5. Chahla J, et al. The posteromedial corner of the knee:an international expert consensus statement on diagnosis, classification, treatment, and rehabilitation. Knee Surg Sports Traumatol Arthrosc. 2021;29:2976–86.
6. Liu F, et al. Morphology of the medial collateral ligament of the knee. J Orthop Surg Res. 2010;16:69.
7. LaPrade RF, et al. The anatomy of the medial part of the knee. J Bone Joint Surg Am. 2007;89:2000–10.
8. Hughston JC, Eilers AF. The role of the posterior oblique ligament in repairs of acute medial (collateral) ligament tears of the knee. J Bone Joint Surg Am. 1973;55:923–40.
9. Griffith CJ, et al. Medial knee injury:part 1, static function of the individual components of the main medial knee structures. Am J Sports Med. 2009;37:1762–70.
10. Iizawa N, et al. Influence of the medial knee structures on valgus and rotatory stability in total knee arthroplasty. J Arthroplast. 2016;31:688–93.
11. Rossi R, et al. Total knee arthroplasty in the varus knee:tips and tricks. Int Orthop. 2019;43:151–8.
12. Aframian A, et al. Origin and insertion of the medial patellofemoral ligament:a systematic review of anatomy. Knee Surg Sports Traumatol Arthrosc. 2017;25:3755–72.
13. Nomura E, et al. A mid-term follow-up of medial patellofemoral ligament reconstruction using an artificial ligament for recurrent patellar dislocation. Knee. 2000;7:211–5.
14. Christensen JC, et al. Higher frequency of reoperation with a new bicruciate-retaining total knee arthroplasty. Clin Orthop Relat Res. 2017;475:62–9.
15. Pache S, et al. Posterior cruciate ligament:current concepts review. Arch Bone Jt Surg. 2018;6:8–18.
16. Saeki K, et al. Stability after medial collateral ligament release in total knee arthroplasty. Clin Orthop Relat Res. 2001;392:184–9.
17. Kadoya Y, et al. Effects of posterior cruciate ligament resection on the tibiofemoral joint gap. Clin Orthop Relat Res. 2001;391:210–7.
18. Mihalko WM, Krackow KA. Posterior cruciate ligament effects on the flexion space in total knee arthroplasty. Clin Orthop Relat Res. 1999;360:243–50.
19. Okazaki K, et al. Asymmetry of mediolateral laxity of the normal knee. J Orthop Sci. 2006;11:264–6.
20. Oshima Y, et al. The influence of posterior cruciate ligament resection on tibiofemoral joint gap in varus osteoarthritic knees. J Knee Surg. 2022;35:323–30.
21. Chun YM, et al. Evaluation of the mechanical properties of posterolateral structures and supporting posterolateral instability of the knee. J Orthop Res. 2008;26:1371–6.
22. Chahla J, et al. Posterolateral corner of the knee:current concepts. Arch Bone Jt Surg. 2016;4:97–103.
23. Otake N, et al. Morphologic study of the lateral and medial collateral ligaments of the human knee. Okajimas Folia Anat Jpn. 2007;83:115–22.
24. LaPrade RF, et al. The posterolateral attachments of the knee:a qualitative and quantitative morphologic analysis of the fibular collateral ligament, popliteus tendon, popliteofibular ligament, and lateral gastrocnemius tendon. Am J Sports Med. 2003;31:854–60.
25. Watanabe Y, et al. Functional anatomy of the posterolateral structures of the knee. Arthroscopy. 1993;9:57–62.
26. Frosch KH, et al. Arthroscopic reconstruction of the popliteus complex:accuracy and reproducibility of a new surgical technique. Knee Surg Sports Traumatol Arthrosc. 2015;23:3114–20.

Anatomical Landmarks Required for TKA

17

Masao Akagi

1 Rotational Landmarks for the Femoral Component

The rotational position of the femoral component has implications for patellar tracking, medial and lateral ligament balance in knee flexion, rotational alignment against the tibial component, and foot orientation in extension. Therefore, determining the correct rotational position of the implant is necessary. Figure 17.1 illustrates commonly used anatomical landmarks for determining femoral rotation are. Patellar tracking deteriorates when the femur is internally rotated beyond the posterior condylar axis (PCA) [1]. However, the medial gap in the flexion position loosens when the femur is externally rotated beyond the clinical epicondylar axis (CEA). In addition, excessive external rotation of the femoral component results in external rotation of the lower leg and foot during knee extension. Conversely, excessive internal rotation results in internal rotation of the lower legs and feet. Before cementing the implants, it is essential to check foot orientation and rotational discrepancies between the femoral components and tibial inserts. The CEA or PCA is rotated externally or internally by approximately 3° relative to the surgical epicondylar axis (SEA) [2–5]. When using mechanical alignment in TKA and positioning the femoral component rotationally between the CEA and PCA, it appears not to cause any major clinical problems mentioned above.

The CEA, SEA, and PCA can be used to determine the rotational position of the femur using CT- or MRI-based 3D preoperative planning software. When mechanical alignment was selected for TKA, the femoral component was set parallel to the SEA. However, when the kinematic alignment was selected, it was planned to be set parallel to the PCA.

When surgeons use the gap technique, the femoral rotational position is determined based on the medial and lateral

M. Akagi (✉)
Department of Orthopaedic Surgery, Kashimoto Hospital, Osaka-Sayama, Osaka, Japan
e-mail: makagi@kashimoto.or.jp

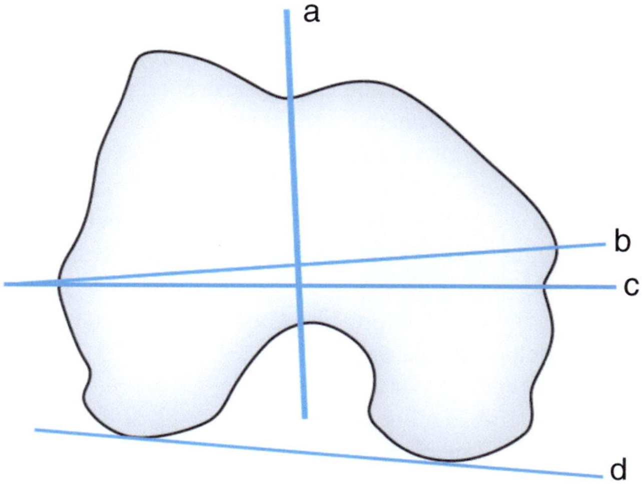

Fig. 17.1 Rotational landmarks for the femoral component. (**a**): Whiteside line, (**b**): clinical epicondylar axis (CEA), (**c**): surgical epicondylar axis (SEA), (**d**): posterior condylar axis (PCA)

ligament balance in a knee flexion position of 90°. However, determining the femoral rotational position using only the medial and lateral ligament balance may be dangerous. This is because the lateral soft tissue of the knee is looser than the medial soft tissue in the flexion position in physiological knees [6], and the lateral soft tissue is often looser than the physiological state in medial knee OA [7, 8]. Lateral soft tissue laxity in the flexion position often persists even after adjusting the medial and lateral balance in the extension position using the medial releasing technique. This can result in a significant variation in femoral rotational position, from −4° to 13° [9]. Applying the same tension to the medial and lateral sides in the knee flexion position, could lead to excessive internal rotation of the femur relative to the functional axis of the tibia. Consequently, the femoral component may rotate excessively externally concerning the rotational axis of the femur (SEA) when aiming for a parallel flexion gap. As a result, many surgeons employing the gap balancing technique in posterior stabilized TKA

S. Matsuda (ed.), *Advances in Total Knee Arthroplasty*, https://doi.org/10.1007/978-981-97-4920-1_17

allow some lateral laxity (approximately 2 mm between medial and lateral soft tissue) and adjust the rotational position of the femoral component with reference to anatomical landmarks [10, 11]. Surgeons employing measured resection techniques frequently emphasize the anatomical landmarks for femoral rotation and the need for appropriate tension in the medial soft tissue throughout the entire range of knee movement [12]. This emphasizes their understanding of the interplay between the frontal and rotational alignment of the femoral component and the balance of the medial and lateral ligaments in the knee.

Individual anatomical landmarks of the femur are described below.

1. Whiteside line

   Arima and Whiteside defined the anteroposterior line of the femur as a line connecting the deepest points of the patellar groove from the anterior part of the groove to the middle of the intercondylar notch [13, 14]. This anatomical landmark can be easily identified in the operative field. It is especially useful when the PCA cannot be used for valgus knee OA or RA with a small lateral femoral condyle [3, 13]. However, it may not be used when the patellar groove is deformed due to OA, RA, or congenital morphological abnormalities.
2. Clinical epicondylar axis (CEA)

   The CEA is a line connecting the apex of the medial epicondyle (attachment of the MCL superficial layer) to the apex of the lateral epicondyle (attachment of the LCL) [15]. The CEA is reported to be rotated externally by 5–6° with respect to the PCA [2, 3, 16, 17]. The apex of the medial epicondyle can be palpated as a bony prominence.
3. Surgical epicondylar axis (SEA)

   It connects the recess of the medial epicondyle (attachment of the MCL deep layer) to the apex of the lateral epicondyle [15]. In a normal knee, the tibia rotates around this axis [5, 16–19]. However, in some cases, the medial epicondyle recess may not be identified on CT or MRI [3]. Therefore, it appears challenging for surgeons to identify SEA intraoperatively in such cases [20, 21].
4. Posterior condylar axis (PCA)

   It is a line connecting the most posterior point of the medial and lateral femoral condyles, which can be identified preoperatively by CT or MRI [3, 22–24]. Given that the PCA is reported to internally rotate by an average of 3° in relation to the SEA, the measured resection technique determines the rotational position of the femoral component by applying a 3° external rotation from the PCA. Intraoperatively, the PCA is identified by using the paddle portion of the posterior referencing instrument on the medial and lateral posterior condyles. However, due to significant variations in the angle between the PCA and SEA (surgical epicondylar axis), it is advisable to measure this angle preoperatively using CT or MRI [1, 3]. Caution is needed in cases of valgus knee OA and RA as the posterior part of the lateral femoral condyle may be hypoplastic or destroyed [3, 5, 13]. In the kinematic alignment method, the femoral rotational position is set parallel to the PCA. In the gap balancing technique, the femoral rotational position is determined with a tensor, which is placed between the proximal tibial cut surface and the medial and lateral posterior condyles during knee flexion. In this case, the PCA was used as an anatomical landmark for femoral rotation. However, when the PCA is used, precautions must be taken as the femoral component may be placed in an external rotated position of a few degrees owing to the cartilage wear of the medial posterior condyle [24, 25].

2 Rotational Landmarks for the Tibial Component

The rotational position of the tibial component affects patellar tracking [26–29], polyethylene wear [30], coverage of the osteotomized tibial surface [31], femorotibial rotational alignment [32], postoperative knee contracture [33], orientation of the lower leg and foot during knee extension, and postoperative pain and discomfort [34, 35]. When the proximal tibial cut has a posterior slope, the anteroposterior direction of the tibial cut affects varus and valgus alignment in the frontal plane [26, 28, 29]. As the primary source of rotational discrepancy in the femorotibial joint is attributed to the tibial side [32], it is essential to accurately determine the anteroposterior direction of the tibial joint surface, considering both femoral and tibial anatomical landmarks. Figure 17.2 illustrates commonly used anatomical landmarks associated with determining tibial rotation, as found in textbooks and literature. Caution is needed when relaying on anatomical landmarks, given the presence of various rotational contractures in the femorotibial joint in varus knee OA [36].

2.1 Medial 1/3 of the Tibial Tubercle (Fig. 17.2a)

The method of determining rotational position using the medial 1/3 of the tibial tubercle as the anterior landmark has been used since the development of TKA [26]. However, to the best of our knowledge, the designer and developer of this anatomical landmark and its underlying rationale remain

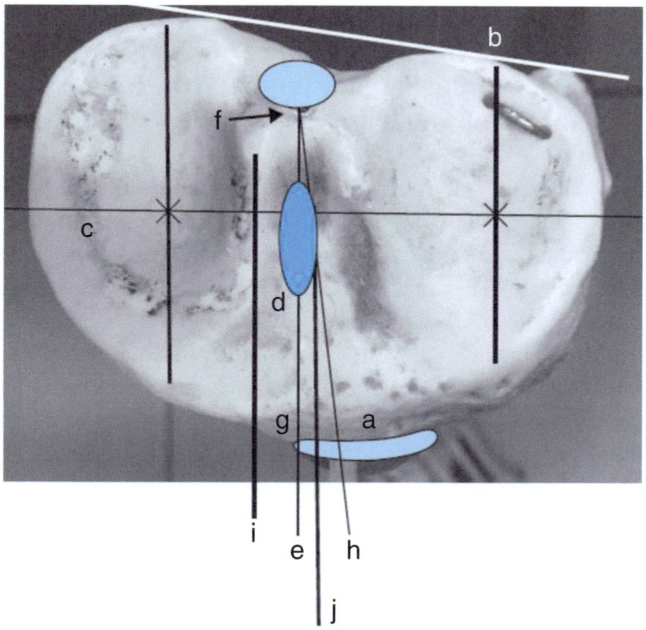

Fig. 17.2 Rotational landmarks for the tibial component. (**a**): Medial 1/3 of the tibial tubercle, (**b**): tibial posterior condylar axis, (**c**): transcondylar axis of the tibia, (**d**): mid-sulcus line, (**e**): Akagi's anteroposterior line, (**f**): PCL center, (**g**): medial border of the patellar tendon attachment to the tibial tubercle, (**h**): a line connecting the center of the PCL and the medial 1/3 of the tibial tubercle, (**i**): a line connecting the medial tibial intercondylar spine and the medial edge of the patellar tendon at the tibial plateau level, (**j**):a line connecting the center of the tibial cut surface and the medial 1/6 of the patellar tendon attachment to the tibia. The lines **d**, **e**, **j**, and **i** are approximately parallel

unknown. Furthermore, the anteroposterior direction could not be determined using a single anterior landmark. It is recommended to select a trial that conforms to the tibial cut surface and directs the anterior center of the tibial component toward the medial 1/3 of the tibial tubercle [29]. However, it is challenging for inexperienced surgeons to determine the tibial rotational position with only the medial 1/3 of the tibial tubercle, and a certain percentage of cases may have poor postoperative results owing to incorrect rotational placement of the tibial component.

2.2 Posterior Condylar Axis of the Tibia (Fig. 17.2b)

This is the tangential line between the medial and lateral posterior condyles of the tibia. In most cases, a line obtained by 10° external rotation of the posterior condylar axis of the tibia indicates an appropriate tibial transverse axis [37]. However, its accuracy is undeniably inferior owing to large individual differences, deformities, and osteophyte formation on the tibial joint surface. In addition, it is difficult to use this axis in revision TKA.

2.3 Transcondylar Axis of the Tibia (Fig. 17.2c)

It is a line connecting the center points of the medial and lateral tibial plateaus. Eckhoff et al. reported that the posterior condylar line (PCL) of the femur and the transverse axis of the tibial condyle were parallel in a healthy knee [38]. However, owing to deformity and osteophyte formation in the tibial plateau, the accuracy is not high. Furthermore, there exists a logical contradiction in that determining the anteroposterior direction of the tibia is a prerequisite for establishing the medial and lateral center points of the tibial plateau.

2.4 Mid-Sulcus Line Between the Medial and Lateral Intercondylar Spines (Fig. 17.2d)

Dalury et al. reported that a line connecting the center between the medial and lateral intercondylar spines and a point 1 mm medial to the medial border of the tibial tubercle could be used to determine the anteroposterior orientation of the tibia [39]. This line is called the "mid-sulcus line." However, when tibial intercondylar ridge deformity is severe, the line becomes unclear.

2.5 Akagi's Anteroposterior Line (Fig. 17.2e)

Akagi proposed the anteroposterior axis of the tibia to determine the rotational alignment of the tibial component. They used the middle of the PCL tibial attachment as a posterior landmark (Fig. 17.2f) and the medial edge of the patellar tendon at the level of the tibial attachment as an anterior landmark (Fig. 17.2g) [40]. These anatomical landmarks were determined using axial CT images of the extended healthy knee so that the line connecting these landmarks was perpendicular to the SEA (Fig. 17.2). During CT imaging, a thin pillow was placed in the popliteal region to avoid hyperextension of the knee; therefore, terminal external rotation of the tibia was not considered. This line was conceived based on the idea that the rotational position of the femoral component should be set parallel to the SEA using the mechanical alignment method. These anatomical landmarks can be easily identified in the surgical field of TKA and 3D images, such as CT and MRI. Aglietti et al. introduced this line in their paper as "Akagi's anteroposterior line" [41]. This line evidenced less variation than conventional extra-articular landmarks, including the transmalleolar axis of the ankle and the second metatarsal axis [42]. Kawahara et al. measured the angle between Akagi's line and the SEA in knees with OA and stated that Akagi's line is slightly internally rotated in relation to the SEA [43]. The line connecting the middle of the PCL attachment to the mid-sulcus is very similar to Akagi's line [44].

2.6 Line Connecting the Center of the PCL Tibial Attachment and the Medial 1/3 of the Tibial Tubercle (Fig. 17.2h)

Jeffcote et al. proposed a method to identify the medial edge of the tibial tubercle in which a finger is applied to the medial side of the tibial tubercle. They stated that the proximal tibia can be accurately cut perpendicular to the tibial functional axis when the medial border of the rod of the tibial extramedullary cut guide is set to the medial edge of the tibial tubercle [45, 46]. The medial edge of the tibial tubercle identified using this method was almost identical to the medial edge of the patellar tendon attached to the tibial tubercle [40]. Akagi et al. used CT images to report that a line connecting the center of the PCL tibial attachment and the medial 1/3 of the patellar tendon attachment is rotated externally by 9–10° with respect to Akagi's line [40].

2.7 The Medial Wall of the Intercondylar Notch

It is an anatomical landmark proposed as a reference axis to determine the rotational position of the tibial component of medial UKA [46]. When osteophytes form in the medial wall of the inter-condylar notch, they must be removed. In addition, precautions must be taken in the rotational relationship between the femur and tibia because the femoral anatomy is used to determine the anteroposterior direction of the tibia.

2.8 Line Connecting the Medial Tibial Intercondylar Spine and Medial Edge of the Patellar Tendon at the Tibial Plateau Level (Fig. 17.2i)

It is a landmark proposed as the anteroposterior orientation of the tibia during medial UKA and it is parallel to Akagi's line [47]. It can be identified using CT and MRI and can be used for 3D preoperative planning.

2.9 A Line Connecting the Center of the Tibial Cut Surface and the Medial 1/6 of the Patellar Tendon Attachment to the Tibia (Fig. 17.2j)

Kawahara et al. reported that a line passing through the geometric center (GC) and parallel to Akagi's line passes through the medial 1/6 of the patellar tendon attachment to the tibial tubercle [48]. This anatomical reference line was devised as a modification of Akagi's line because the PCL attachment to the tibia is unclear after osteotomy of the proximal end of the tibia. As this line passes through the GC, it coincides with the anteroposterior line of the symmetrical tibial component. However, it is challenging to identify the GC of the tibial cut surface during surgery.

2.10 A Shape of the Tibial Cut Surface

This method uses the shape of the tibial cut surface as an anatomical landmark to determine the rotational position of the tibial component. Components with tibial anatomical geometry can be used in this method, and their component size and rotational position can be adjusted to maximize the coverage of the tibial cut surface [49, 50]. However, in the actual operative field, it may be necessary to combine this method with other reference axes, such as the posterior condylar line of the tibia [51].

3 Coronal Landmarks for the Tibial Component

Determination of coronal alignment involves finding proximal and distal landmarks and connecting them. Therefore, these can be divided into landmarks on the proximal tibia, tibial shaft, and ankle. The basic axes in the coronal plane were the tibial anatomical (shaft) axis (Fig. 17.3A-a), tibial functional axis (Fig. 17.3A-b) (the axis connecting the middle of the ankle joint [Fig. 17.3A-c]), and the intercondylar groove (the center of the tibial plateau) (Fig. 17.3A-d). In cases without tibia vara, both axes coincide; however, in cases with tibia vara, there is a divergence between the two axes. The selection of anatomical landmarks in the preoperative plan or operative field varies based on the axis along which the proximal tibia is cut perpendicularly. Moreover, there are two types of proximal tibial cut guides: extramedullary and

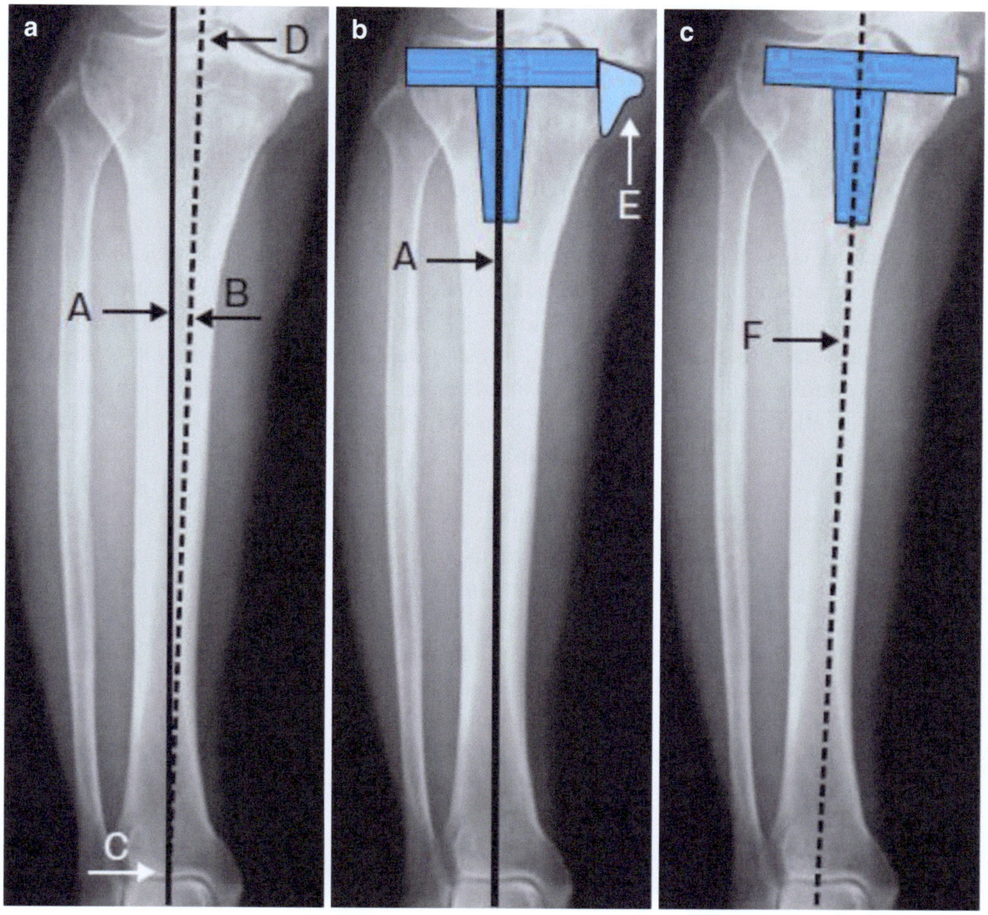

Fig. 17.3 Planning on the tibial shaft axis or the tibial functional axis. (**a**): (A) Tibial shaft axis and (B) tibial functional axis. The tibial shaft axis (A) connects the center of the tibial shaft. The tibial functional axis (B) connects the ankle joint center (C) and the intercondylar groove (the center of the tibial plateau) (D). In a knee with tibia vara, the tibial shaft axis (A) passes near the lateral tibial intercondylar spine. In a straight tibia, both axes coincide; however, in the tibia vara, both axes differ. (**b**): When placed perpendicular to the tibial shaft axis (A), the component is sized downward and lateralized, and the uncovered medial condyle (E) is resected (medial reduction osteotomy). (**c**): When placed perpendicular to the postoperative functional axis (F), a larger component can be used; however, the tibia vara remains

intramedullary. The choice of landmarks depends on the guide selected. Hence, a practical approach between the two methods–the tibial functional axis and tibial anatomical (shaft) axis methods–would involve aligning the tibial component perpendicular to the postoperative functional axis while minimizing tibia vara as much as possible.

3.1 The Position of the Tibia Perpendicular to the Tibial Anatomical (Shaft) Axis (Fig. 17.3b)

In cases of a severe tibia vara, size reduction and component lateralization should be performed to align the stem axis of the tibial component with the tibial anatomical (shaft) axis (Fig. 17.3B-a). Because the medial condyle of the tibia remains uncovered by the component, medial resection osteotomy of the uncovered bone is performed (Fig. 17.3B-e) [52, 53]. Bone resection relieves tension in the MCL and prevents excessive MCL release. In this method, the postoperative tibial functional axis and tibial anatomical axis coincide, and the tibia vara disappears. If the proximal tibial cut is performed perpendicular to the tibial anatomical axis, and the maximum component covering the tibial cut surface is selected for a knee with tibia vara, the tibial component might be placed in a valgus orientation concerning the postoperative tibial functional axis, frequently necessitating the release of the MCL in many cases [54].

3.2 When the Tibia Is Positioned Perpendicularly to the Postoperative Tibial Functional Axis (Fig. 17.3c)

During templating, select the tibial component with a maximum size covering the proximal cut surface of the tibia, and plan the placement so that the component stem axis points to the middle of the ankle joint, resulting in the tibial component placed perpendicular to the postoperative tibial functional axis (Fig. 17.3C-f). Although the tibia vara remains after the surgery, the hip-knee-ankle (HKA) angle of 0° can be achieved with little release of the MCL. This method is often used when applying the extramedullary tibial cutting guides [54].

The anatomical landmarks for determining tibial coronal alignment are divided by site: the proximal tibia, tibial shaft, and ankle joint.

1. Landmarks in the proximal tibia landmarks
   (a) Mid-sulcus line between the tibial intercondylar spines: Because the ACL attachment seems to be located at the center of the knee, the intercondylar groove (sulcus) is used as the proximal point of the tibial functional axis. It can be well recognized in the frontal view of plain tibial radiographs used in preoperative planning and in the operative field. As previously mentioned, the mid-sulcus line indicates the anteroposterior direction of the tibia; therefore, if the spike of the proximal fixation bar of the extramedullary cutting guide is driven on the intercondylar sulcus parallel to the mid-sulcus line, the coronal and rotational alignment of the tibia can be set correctly simultaneously.
   (b) Lateral tibial intercondylar spine: If the tibia vara is severe, the tibial shaft axis often extends proximally near the apex of the lateral tibial intercondylar spine. The lateral tibial intercondylar spine is used as an anatomical landmark for the proximal tibial cut when an entry hole is created for the intramedullary guide.
   (c) The medial edge of the tibial tubercle (medial edge of the patellar tendon attachment): Jeffcote et al. proposed a method to identify the medial edge of the tibial tubercle with a belly of the index finger and to align the proximal medial edge of the vertical rod of the extramedullary cutting guide to the edge [45]. As an extension of the mid-sulcus line emerges at the medial edge of the tibial tubercle, this landmark is used when an extramedullary guide is employed.
   (d) A line connecting the medial 1/3 of the tibial tubercle and the distal 1/4 of the anterior margin of the tibia: By using CT data, it has been found that the line connecting the medial 1/3 of the tubercle of the tibia and the distal 1/4 of the anterior margin of the tibia is parallel to the tibial functional axis [55].
2. Landmarks on the tibial shaft
   (a) Anterior border of the tibia: As a landmark used when it is difficult to identify the middle of the ankle joint, it has been reported that the line connecting the proximal 1/3 and distal 1/3 of the anterior tibia is parallel to the tibial functional axis [56].
   (b) Tibial medullary cavity: When an intramedullary guide is used, the tibial isthmus is an anatomical landmark. It is used when an osteotomy is performed perpendicular to the tibial shaft and the entry hole of the intramedullary rod at the tibial plateau is determined during preoperative planning.
3. Landmarks around the ankle joint

   When using an extramedullary tibial guide, the distal end of the vertical rod must be positioned at the center of the ankle joint. Alignment checks were also performed after osteotomy by positioning the distal end of the vertical rod at the center of the ankle joint. However, it is difficult to identify the middle of the ankle joint accurately, and various methods have been proposed. In practice, it is preferable to combine several familiar landmarks.
   (a) Lower leg proximal to the ankle joint: The distal end of the lower leg is grasped by a supra-malleolar clamp, and the center of the ankle joint is located 5 mm medial to the clamp center [45, 57].
   (b) Medial and lateral ankle malleolus: The ankle joint center is thought to be located 3–5 mm medial to the midpoint of the medial and lateral ankle malleolus [26, 57–59]. In the navigation system, it is necessary to find the three-dimensional ankle center, and the medial and lateral ankle malleolus are used as landmarks. The midpoint of these landmarks is approximately 4.5 mm lateral and 2.7 mm posterior to the ankle center [60]. Furthermore, it has been reported that the ankle joint center is 57% lateral to a line connecting the medial and lateral malleolus [61]. A method has been proposed in which the anteroposterior direction of the extramedullary guide is aligned with Akagi's line, and the apex of the medial and lateral malleolus is captured with a flat paddle, with the midpoint of the paddle at the center of the ankle joint [62, 63].
   (c) Second metatarsal bone: When the ankle joint is held in a neutral position, the second metatarsal bone aligns with the middle of the ankle joint [64]. The distal end of the extramedullary cutting guide rod is aligned with the second metatarsal bone. Accuracy may be reduced by variations in ankle joint positioning and deformities and contractures.
   (d) Extensor hallucis longus tendon, anterior tibialis tendon, Achilles tendon, and dorsal artery: Reports provide insights into these landmarks. For instance, the lateral edge of the extensor hallucis longus tendon nearly intersects the middle of the talus [65], the lateral edge of the tibialis anterior tendon is located approximately 1 cm medial to the center of the talus [66], a line connecting the Achilles tendon and the extensor hallucis longus tendon passes through the middle of the ankle joint [67], and the dorsal artery traverses the center of the ankle joint [68].

References

1. Akagi M, et al. Effect of rotational alignment on patellar tracking in total knee arthroplasty. Clin Orthop Relat Res. 1999;366:155–63.
2. Nagamine R, et al. Reliability of the anteroposterior axis and the posterior condylar axis for determining rotational alignment of the femoral component in total knee arthroplasty. J Orthop Sci. 1998;3:194–8.

3. Akagi M, et al. Relationship between frontal knee alignment and reference axes in the distal femur. Clin Orthop Relat Res. 2001;388:147–56.
4. Yoshino N, et al. Computed tomography measurement of the surgical and clinical transepicondylar axis of the distal femur in osteoarthritic knees. J Arthroplast. 2001;16:493–7.
5. Matsuda S, et al. Anatomical analysis of the femoral condyle in normal and osteoarthritic knees. J Orthop Res. 2004;22:104–9.
6. Okazaki K, et al. Asymmetry of mediolateral laxity of the normal knee. J Orthop Sci. 2006;11:264–6.
7. Bellemans J, et al. The influence of coronal plane deformity on mediolateral ligament status: an observational study in varus knees. Knee Surg Sports Traumatol Arthrosc. 2010;18:152–6.
8. Okamoto S, et al. Lateral soft tissue laxity increases but medial laxity does not contract with varus deformity in total knee arthroplasty. Clin Orthop Relat Res. 2013;471:1334–42.
9. Heesterbeek PJ, et al. Effects of the balanced gap technique on femoral component rotation in TKA. Clin Orthop Relat Res. 2009;467:1015–22.
10. Nagai K, et al. The influence of joint distraction force on the soft-tissue balance using modified gap-balancing technique in posterior-stabilized total knee arthroplasty. J Arthroplast. 2017;32:2995–9.
11. Nagai K, et al. Intraoperative soft tissue balance using novel medial preserving gap technique in posterior-stabilized total knee arthroplasty: comparison to measured resection technique. Knee Surg Sports Traumatol Arthrosc. 2018;26:3474–81.
12. Matsuda S, et al. Ligament balancing in total knee arthroplasty-medial stabilizing technique. Pac J Sports Med Arthrosc Rehabil Technol. 2015;2:108–13.
13. Arima J, et al. Femoral rotational alignment, based on the anteroposterior axis, in total knee arthroplasty in a valgus knee: a technical note. J Bone Joint Surg Am. 1995;77:1331–4.
14. Whiteside LA, et al. The anteroposterior axis for femoral rotational alignment in valgus total knee arthroplasty. Clin Orthop Relat Res. 1995;321:168–72.
15. Berger RA, et al. Determining the rotational alignment of the femoral component in total knee arthroplasty using the epicondylar axis. Clin Orthop Relat Res. 1993;286:40–7.
16. Kurosawa H, et al. Geometry and motion of the knee for implant and orthotic design. J Biomech. 1985;18:493–9.
17. Yoshioka Y, et al. The anatomy and functional axes of the femur. J Bone Joint Surg Am. 1987;69:873–80.
18. Asano T, et al. In vivo three-dimensional knee kinematics using a biplanar image-matching technique. Clin Orthop Relat Res. 2001;388:157–66.
19. Asano T, et al. The functional flexion-extension axis of the knee corresponds to the surgical epicondylar axis: in vivo analysis using a biplanar image-matching technique. J Arthroplast. 2005;20:1060–7.
20. Kinzel V, et al. Can the epicondylar axis be defined accurately in total knee arthroplasty? Knee. 2005;12:293–6.
21. Jenny JY, et al. Low reproducibility of the intra-operative measurement of the transepicondylar axis during total knee replacement. Acta Orthop Scand. 2004;75:74–7.
22. Moon YW, et al. Variability in femoral component rotation reference axes measured during navigation-assisted total knee arthroplasty using gap technique. J Arthroplast. 2010;25:238–43.
23. Asada S, et al. Effects of cartilage remnants of the posterior femoral condyles on femoral component rotation in varus knee osteoarthritis. Knee. 2012;19:185–9.
24. Tashiro Y, et al. Articular cartilage of the posterior condyle can affect rotational alignment in total knee arthroplasty. Knee Surg Sports Traumatol Arthrosc. 2012;20:1463–9.
25. Miyasaka T, et al. Impact of posterior femoral condylar cartilage and posterior intercondylar distance on rotation of femoral component in total knee arthroplasty. BMC Musculoskelet Disord. 2020;21:498.
26. Insall JN. Surgical techniques and instrumentation in total knee arthroplasty. In: Insall JN, et al., editors. Surgery of the knee. 2nd ed. New York: Churchill Livingstone; 1993. p. 739–804.
27. Nagamine R, et al. Patellar tracking after total knee arthroplasty. The effect of tibial tray malrotation and articular surface configuration. Clin Orthop Relat Res. 1994;304:262–71.
28. Insall JN, et al. Surgical techniques and instrumentation in total knee arthroplasty. In: Insall JN, et al., editors. Surgery of the knee, vol. 2. 3rd ed. New York: Churchill-Livingstone; 2001. p. 1553–620.
29. Vail TP, et al. Surgical techniques and instrumentation in total knee arthroplasty. In: Scott WN, editor. Insall & Scott surgery of the knee. 6th ed. Philadelphia: Elsevier; 2018. p. 1665–720.
30. Lewis P, et al. Posteromedial tibial polyethylene failure in total knee replacements. Clin Orthop Relat Res. 1994;299:11–7.
31. Martin S, et al. Maximizing tibial coverage is detrimental to proper rotational alignment. Clin Orthop Relat Res. 2014;472:121–5.
32. Watanabe S, et al. Change in tibiofemoral rotational alignment during total knee arthroplasty. J Orthop Sci. 2014;19:571–8.
33. Bédard M, et al. Internal rotation of the tibial component is frequent in stiff total knee arthroplasty. Clin Orthop Relat Res. 2011;469:2346–55.
34. Nicoll D, et al. Internal rotational error of the tibial component is a major cause of pain after total knee replacement. J Bone Joint Surg Br. 2010;92:1238–44.
35. Bell SW, et al. Component rotational alignment in unexplained painful primary total knee arthroplasty. Knee. 2014;21:272–7.
36. Uehara K, et al. Bone anatomy and rotational alignment in total knee arthroplasty. Clin Orthop Relat Res. 2002;402:196–201.
37. Bradley P, et al. Rotational references for total knee arthroplasty tibial components change with level of resection. Clin Orthop Relat Res. 2010;468:2734–8.
38. Eckhoff DG, et al. Version of the osteoarthritic knee. J Arthroplast. 1994;9:73–9.
39. Dalury DF. Observations of the proximal tibia in total knee arthroplasty. Clin Orthop Relat Res. 2001;389:150–5.
40. Akagi M, et al. An anteroposterior axis of the tibia for total knee arthroplasty. Clin Orthop. 2004;420:213–9.
41. Aglietti P, et al. Rotational position of femoral and tibial components in TKA using the femoral transepicondylar axis. Clin Orthop Relat Res. 2008;466:2751–5.
42. Akagi M, et al. Variability of extraarticular tibial rotation references for total knee arthroplasty. Clin Orthop Relat Res. 2005;436:172–6.
43. Kawahara S, et al. Relationship between the tibial anteroposterior axis and the surgical epicondylar axis in varus and valgus knees. Knee Surg Sports Traumatol Arthrosc. 2012;20:2073–7.
44. Sahin N, et al. Centre of the posterior cruciate ligament and the sulcus between tubercle spines are reliable landmarks for tibial component placement. Knee Surg Sports Traumatol Arthrosc. 2013;10:2384–91.
45. Jeffcote B, et al. Varus/valgus alignment of the tibial component in total knee arthroplasty. Knee. 2003;10:243–7.
46. Kawahara S, et al. Is the medial wall of the intercondylar notch useful for tibial rotational reference in unicompartmental knee arthroplasty? Clin Orthop Relat Res. 2012;470:1177–84.
47. Tsukamoto I, et al. Anteroposterior rotational references of the tibia for medial unicompartmental knee arthroplasty in Japanese patients. J Arthroplast. 2017;32:3169–75.
48. Kawahara S, et al. Medial sixth of the patellar tendon at the tibial attachment is useful for the anterior reference in rotational alignment of the tibial component. Knee Surg Sports Traumatol Arthrosc. 2014;22:1070–5.
49. Clary C, et al. Tibial base design and patient morphology affecting tibial coverage and rotational alignment after total knee arthroplasty. Knee Surg Sports Traumatol Arthrosc. 2014;22:3012–8.

50. Stulberg SD, et al. Which tibial tray design achieves maximum coverage and ideal rotation: anatomic, symmetric, or asymmetric? An MRI-based study. J Arthroplast. 2015;30:1839–41.
51. Okazaki Y, et al. The use of an asymmetrical tibial tray in TKA optimizes tibial rotation when fitted to the posterior tibial plateau border. Knee Surg Sports Traumatol Arthrosc. 2020;28:3821–6.
52. Mullaji A, et al. Correcting varus deformity. J Arthroplast. 2007;22:15–9.
53. Mori S, et al. Tibia vara affects the aspect ratio of tibial resected surface in female Japanese patients undergoing TKA. Clin Orthop Relat Res. 2013;471:1465–71.
54. Matsuda S, et al. Tibial shaft axis does not always serve as a correct coronal landmark in total knee arthroplasty for varus knees. J Arthroplast. 2003;18:56–62.
55. Fukagawa S, et al. Anterior border of the tibia as a landmark for extramedullary alignment guide in total knee arthroplasty for varus knees. JOR. 2011;29:919–24.
56. Nishikawa K, et al. Accuracy of proximal tibial bone cut using anterior border of tibia as bony landmark in total knee arthroplasty. J Arthroplast. 2015;30:2121–4.
57. Dennis DA, et al. Intramedullary versus extramedullary tibial alignment systems in total knee arthroplasty. J Arthroplast. 1993;8:43–7.
58. Simmons ED Jr, et al. The accuracy of tibial intramedullary alignment devices in total knee arthroplasty. J Arthroplast. 1991;6:45–50.
59. Fang DM, et al. Coronal alignment in total knee arthroplasty: just how important is it? J Arthroplast. 2009;24:39–43.
60. Siston RA, et al. Evaluation of methods that locate the center of the ankle for computer-assisted total knee arthroplasty. Clin Orthop Relat Res. 2005;439:129–35.
61. Marchant DC, et al. An algorithm for locating the center of the ankle joint in knee navigation surgery. Comput Aided Surg. 2005;10:45–9.
62. Akagi M, et al. Estimation of frontal alignment error of the extramedullary tibial guide on the bi-malleolar technique: a simulation study with magnetic resonance imaging. Knee. 2012;19:836–42.
63. Asada S, et al. Location of the ankle center for total knee arthroplasty. Knee. 2017;24:121–7.
64. Tew M, et al. Tibiofemoral alignment and the results of knee replacement. J Bone Joint Surg Br. 1985;67:551–6.
65. Schneider M, et al. Use of palpable tendons for extramedullary tibial alignment in total knee arthroplasty. J Arthroplast. 2007;22:219–26.
66. Rajadhyaksha AD, et al. Use of tibialis anterior tendon as distal landmark for extramedullary tibial alignment in total knee arthroplasty: an anatomical study. Am J Orthop. 2009;38:68–70.
67. Tiftikçi U, et al. Can Achilles tendon be used as a new distal landmark for coronal tibial component alignment in total knee replacement surgery? An observational MRI study. Ther Clin Risk Manag. 2017;13:81–6.
68. Sugimura N, et al. The dorsal pedis artery as a new distal landmark for extramedullary tibial alignment in total knee arthroplasty. Knee Surg Sports Traumatol Arthrosc. 2014;22:2618–22.

18 Anatomical Variations

Ryuji Nagamine

The Japanese population has bony anatomical variations that are rare in Western populations, mainly in the coronal plane, [1–3] and anatomical variations common with Western populations, mainly in the sagittal plane [4, 5]. In this section, we summarize the contents of each variation, the evaluation method, and the relationship between each osseous variation commonly seen in knees with medial osteoarthritis. Please refer to another article on lateral-type arthropathy [6].

1 Contents of Various Variations and Indicators for Evaluation

1.1 Coronal Plane

Lateral bowing of the femoral diaphysis [2, 7], constitutional varus with a medial shift of the articular surface [2, 8, 9], exacerbation of the standing FTA (lateral angle of the knee), and age-related changes in the tibial articular surface.

1.2 Sagittal Plane

Anterior bowing of the femoral diaphysis [4], flexion of the femoral metaphysis and extension of the epiphysis (Fig. 18.1a), posterior slope of the articular surface created by constitutional posterior rotation of the tibial condyle (Fig. 18.1b) [5, 10, 11], rhomboid transformation of the tibial condyle (Fig. 18.1b) [5, 10], rhomboid transformation of the patella with longitudinal axis extension (Fig. 18.2a) [10], and posterior shift of the ankle joint (Fig. 18.3e).

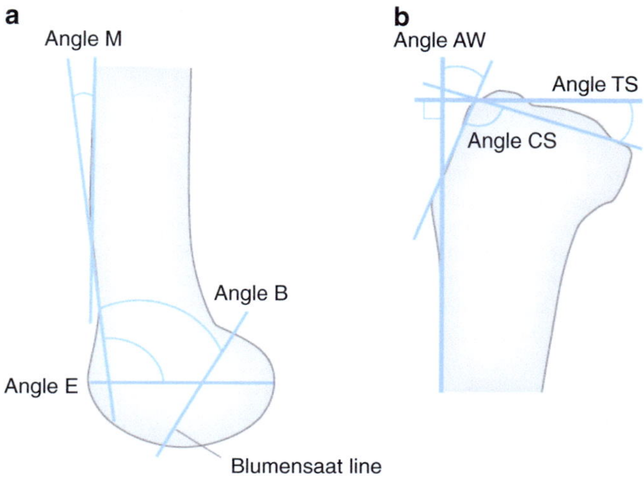

Fig. 18.1 (**a**): An index indicating the angle of the diaphysis, metaphysis, and epiphysis of the distal femur in the sagittal plane. Angle M: Angle of the anterior metaphysis cortex relative to the anterior diaphysis cortex. Angle B: The angle of the Blumensaat line relative to the anterior metaphysis cortex. Angle E: Angle of the AP diameter line of the epiphysis relative to the anterior metaphytic cortex. (**b**): An index indicating the shape of the proximal tibia in the sagittal plane. Angle AW: Angle of the anterior wall of the condyle relative to the anterior cortex of the tibial diaphysis, indicating the degree of the tibial condyle posterior rotation. Angle TS: Angle of posterior slope of the articular surface relative to the perpendicular of the anterior cortex of the tibial diaphysis. Angle CS: Angle of the articular surface relative to the anterior wall of the tibial condyle, indicating the degree of rhomboid transformation

1.3 Transverse Plane

Changes in the position of the central ridge of the patella (change in Wiberg index) (Fig. 18.2b) and narrowing of the medial articular surface of the trochlea (change in trochlear length ratio) (Fig. 18.2b).

R. Nagamine (✉)
Nagamine Rheumatology & Orthopedics Clinic, Fukuoka, Japan
e-mail: takaakane@do6.enjoy.ne.jp

S. Matsuda (ed.), *Advances in Total Knee Arthroplasty*, https://doi.org/10.1007/978-981-97-4920-1_18

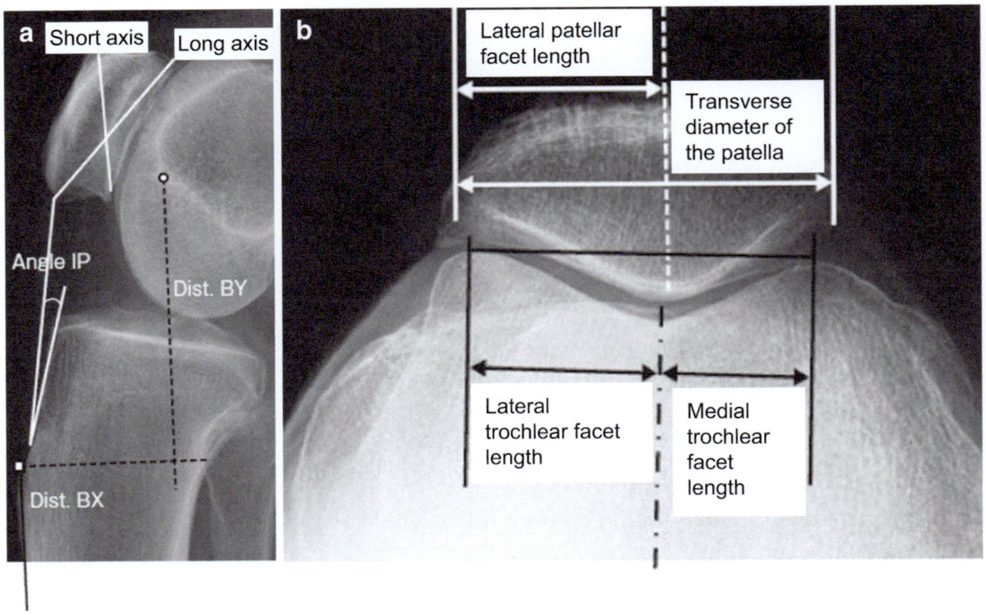

Fig. 18.2 (**a**): An index to evaluate the position of the tibial and femoral condyle in a lateral view of the knee taken at 90° flexion. Angle IP: The angle between the patellar tendon and the anterior wall of the tibial condyle indicating the infrapatellar space. Dist. BX: Horizontal distance from the tibial tuberosity to the anterior point of the Blumensaat line, showing posterior shift of the femoral condyle. Dist. BY: Vertical distance from the the tibial tuberosity to the anterior point of the Blumensaat line showing a distal shift of the femoral condyle. (**b**): Index indicating the shape of the patella and femoral trochlea in the axial view. Wiberg index: Lateral patellar joint distance divided by patellar width, indicating the degree of deviation of the central ridge. Trochlea Length ratio: lateral articular surface distance divided by medial articular surface distance, indicating the shape of the femoral trochlea

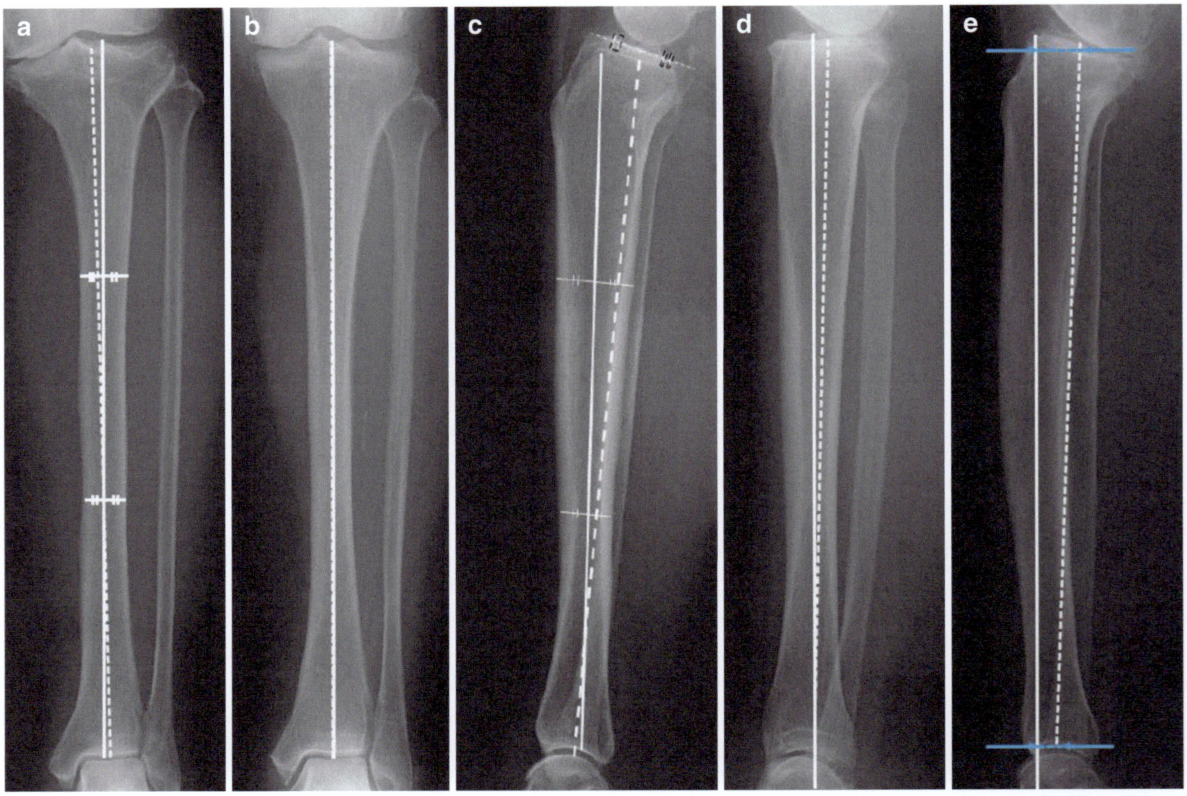

Fig. 18.3 Change in location and angle of the mechanical axis relative to the anatomical axis due to the shape of the tibial condyle. (**a**, **b**): If the varus of the tibial joint surface is large, the discrepancy between the anatomical and mechanical axes is large. (**c**, **d**): If the posterior rotation of the tibial condyle is large, the discrepancy between the anatomical and mechanical axes is also large. (**e**): In cases where the ankle joint shifts posteriorly, the angular discrepancy between the anatomical and mechanical axes is small, but the distance discrepancy is large

2 Constitutional and with Age

The shape of the tibial condyle, including the articular surface, has been reported to be formed by the retardation of bone growth at the metaphysis and is constitutional [5, 8–10].

According to the Hueter–Volkmann law, growth at the metaphysis is retarded by increased compression in the medial and posterior parts, whereas reduced loading accelerates growth in the lateral and anterior parts of the proximal tibia. Therefore, the tibial condyle is tilted medially (constitutional varus), and the articular surface shifts medially in the coronal plane. In the sagittal plane, the tibial condyle is tilted posteriorly (constitutional posterior rotation). Additionally, the tibial condyle changes shape from rectangular to rhomboid (rhomboid transformation) due to the shear stress of body weight loading, and the articular surface shifts posteriorly [5, 10]. In some cases, the medial torsion of the tibia may also be involved [3]. The shape of the patella is not correlated with age or physique and is constitutional.

In contrast, the distance from the center of the tibial articular surface to the center of the intercondylar notch was 0.7 mm on the outside, which positively correlated with age. The shape of the tibial articular surface changes with age. In addition, the angle of lateral bowing in the femoral diaphysis correlates with age, height, %YAM value on DEXA, and TRACP-5b levels. The degree of lateral bowing of the diaphysis increases with age. The angle of flexion at the femoral metaphysis relative to the diaphysis also correlates and changes with age.

While the femur has an overall arc-like shape with bowing in the diaphysis, the middle 1/3 diaphysis of the tibia is almost linear, and the distal 1/3 of the diaphysis tends to be varus and posteriorly shifted in smaller patients (Fig. 18.3c–e).

3 Relation Among Variations and Influence of Each Variation

The coronal and sagittal plane indices weakly correlate between anterior and lateral bowing in the femoral diaphysis, but all other parameters of the femur, tibia, and patellar geometry showed no correlation at all. Asian lifestyles, including seiza (sitting straight), cause bony variations in the coronal plane [2]. However, anterior bowing in the femoral diaphysis and posterior slope of the tibial articular surface in the sagittal plane have been reported worldwide. Varying loads are assumed to be applied to the knee in the coronal and sagittal planes.

In the sagittal plane, the degree of posterior rotation of the tibial condyle is expressed by the Angle AW formed by the line along the anterior diaphysis cortex and the line along the anterior wall of the tibial condyle. The angle AW is an important parameter in determining the sagittal plane alignment of the knee [5, 10]. Upon evaluation of knees using X-ray images taken at 90° knee flexion, the tibial articular surface shifts posteriorly and distally by the posterior rotation of the tibial condyle, resulting in a posterior and distal shift of the femoral condyle, and a posterior shift of the patella, thereby increasing the flexion angle. This phenomenon was confirmed based on the distance from the tibial tuberosity to the anterior point of the Blumensaat line of the femur (Fig. 18.2a). Angle IP (Fig. 18.2a), defined as the angle between the anterior wall of the condyle and the patellar tendon at 90° knee flexion, shows the size of the infrapatellar space. In patients with a small Angle AW, the Angle IP is also smaller, and the patellar tendon and tibia come into contact early during knee flexion. Furthermore, the greater the posterior rotation of the tibial condyle, the longer the long axis of the patella; consequently, the Insall–Salvati ratio is smaller [10]. In our study, the short axis length was also measured (Fig. 18.2a), and the patellar axis ratio (long axis length/short axis length) was calculated. A positive correlation was observed between the long axis length and patellar axis ratio, which proved that the long axis of the patella became longer due to rhomboid transformation [10]. Our other study showed that the Angle AW and posterior shift of the tibial articular surface of the knee correlate with the posterior shift of the ankle joint, and the correlation becomes stronger in small people.

In the sagittal plane, the femoral metaphysis flexed relative to the diaphysis by an average of 3.2°. In our study, the angle of flexion of the epiphysis relative to the metaphysis was evaluated using two parameters (the angle between the anterior-posterior diameter line of the epiphysis and the anterior metaphysis cortex and the angle between the Blumensaat line and the anterior metaphysis cortex) (Fig. 18.1a). The results showed that the more metaphysis flexes relative to the diaphysis, the more epiphysis extends relative to the metaphysis.

In the transverse section, two parameters were used to evaluate the position of the central ridge of the patella and femoral trochlea: Wiberg index (lateral articular surface distance of the patella/width of the patella) (Fig. 18.2b) and trochlea length ratio (lateral trochlea joint distance/medial trochlea joint distance) (Fig. 18.2b). The Wiberg index correlated with the trochlear length ratio (Fig. 18.2b) and varus angle of the tibial articular surface (Fig. 18.2b). In other words, the stronger the varus of the proximal tibia, the greater the medial shift of the central ridge of the patella. The deepest part of the femoral patellar groove is located approximately at the center of the epicondylar line [12], and the smaller the medial trochlear joint distance, the greater the medial shift of the central ridge of the patella. However, no correlation was observed between the varus of the proximal tibia and the trochlea length ratio. The transverse-sectional shape of the patella is influenced by both the femoral trochlear shape and the proximal tibia vara.

4 Evaluation of the Alignment of Each Variation and Its Influence on Component Installation

In the coronal plane, the average angle of lateral bowing of the femoral diaphysis is 1.6°, and the discrepancy between the mechanical and anatomical axes of the femur increases with age [7]. The average angle and distance discrepancy between the mechanical and anatomical tibial axes of the tibia were 0.4° and 3.1 mm on the tibial articular surface. The stronger the varus, the more the articular surface shifts medially, and greater the discrepancy between the two axes (Fig. 18.3a, b), The hip-knee-ankle (HKA) angle underestimates the degree of varus [9]. The anatomical axis drawn in the image of the proximal 1/3 of the tibia (assuming an AP view knee X-ray image) is on average 1.8° more varus than that of the entire tibia. Stronger varus, the greater the tibial condyle rotates medially, and anatomical axis on the AP view radiograph of the knee tilt varus from the anatomical axis of the entire tibia.

In the sagittal plane, the femoral diaphysis shows an average anterior bowing of 10.6°. There are large individual differ-

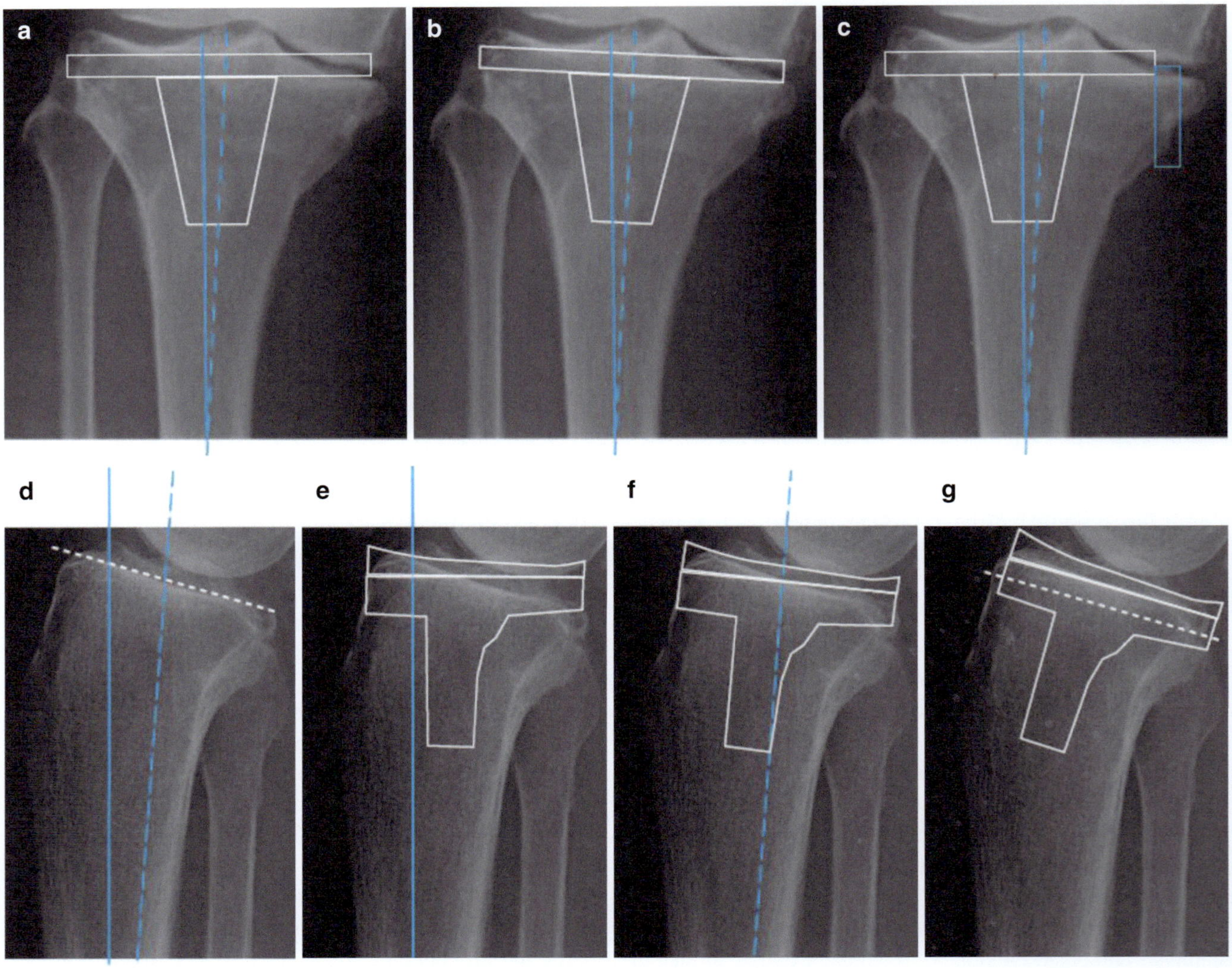

Fig. 18.4 Effects of proximal tibia vara and the tibial condyle posterior rotation on alignment of the tibial component. (**a**): Installed perpendicular to the tibial anatomical axis. Alignment by FTA is neutral but expressed as valgus in the HKA angles. (**b**): Installed perpendicular to the tibial mechanical axis. Alignment by the HKA angle is neutral, but expressed as varus in the FTA. (**c**). Reduction osteotomy, which resects the medial bone of the condyle that is not covered by the components, is a procedure that matches the anatomical axis with the mechanical axis. (**d**): The articular surface shifts posteriorly due to posterior rotation of the tibial condyle, and the discrepancy between the anatomical and mechanical axes increases in the sagittal plane. (**e**–**g**): perpendicular to the tibial anatomical axis, perpendicular to the mechanical axis, and parallel to the articular surface

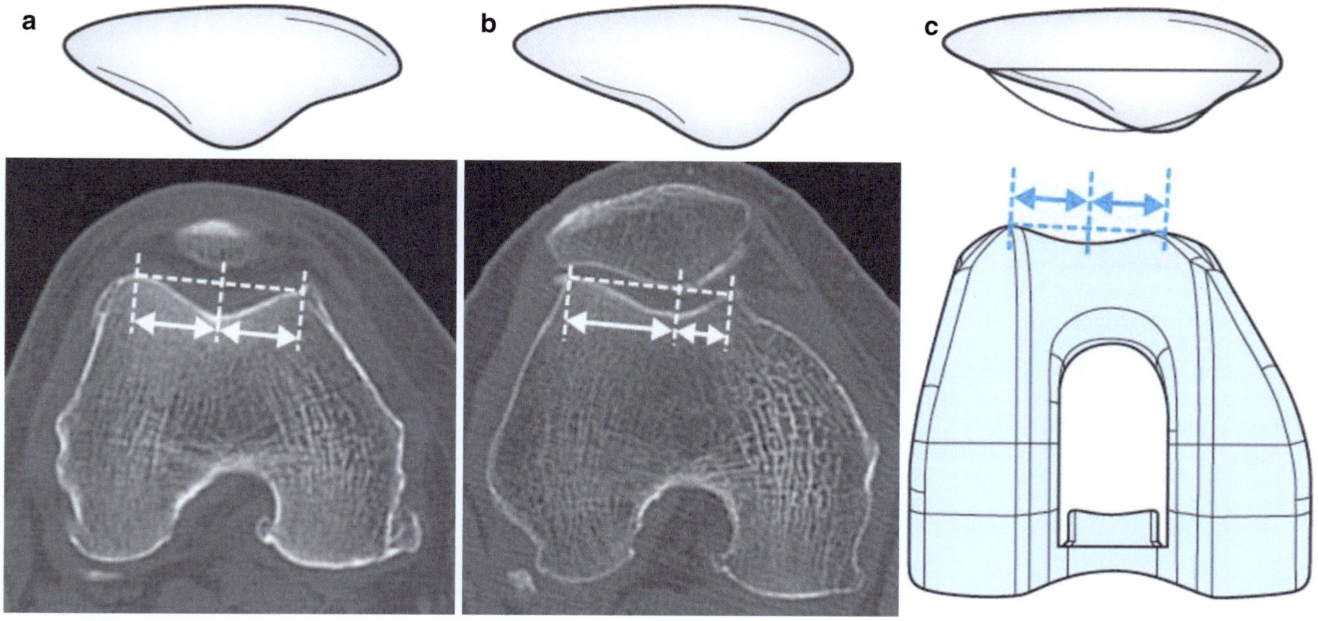

Fig. 18.5 Differences in femoral trochlea and patella shape and component configuration by case. (**a**): The medial trochlear articular surface is wide, and the trochlear length ratio is 1.04. The central ridge was closer to the center. (**b**): Medial trochlear articular surface is small and trochlear length ratio is 1.95. The central ridge has shifted medially. (**c**): The position of the patellar component that fits the configuration of the femoral component trochlea with a TLR of 1.00 is closer to the center of the patella

ences. The flexion angle of the metaphysis relative to the diaphysis changes with age; therefore, the discrepancy between the mechanical and anatomical axes also increases with age. In the tibia, the average discrepancy between the mechanical and anatomical axes is 2.8°, and the discrepancy between the two axes on the knee articular surface is 15.0 mm on average. The larger the posterior rotation of the condyle, the more the articular surface shifts posteriorly, and the discrepancy between the two axes increases (Fig. 18.3c, d). Because of the posterior rotation of the tibial condyle, the anatomical axis on the image of the proximal third of the tibia (assuming a lateral-view knee X-ray image) is tilted posteriorly by an average of 1.1° compared to the anatomical axis of the entire tibia. The stronger the posterior rotation of the condyle, the more tilted the anatomical axis on the lateral view of the knee radiograph posteriorly from the anatomical axis the entire tibia.

In cases of strong femoral anterior and lateral bowing, strong tibial varus, and tibial condyle posterior rotation, simple osteotomy of the distal femur and proximal tibia perpendicular to the mechanical axis may cause various problems [2, 4, 7] (Fig. 18.4). Analysis results at 90° flexion lateral-view radiographs of the knee showed that the sagittal plane alignment, in which the femoral condyle is shifted posteriorly by the posterior rotation of the tibial condyle, which cannot be corrected by TKA. The optimal preparation against anterior and lateral femoral bowing and physiological varus of the tibia and the optimal posterior slope angle of the tibial component remains a matter of debate. Accurate evaluation of the variation of the knee joint by using the angle index alone is impossible. Distance index is also necessary to accurately assess anatomical variations in the three planes. Considering that the proximal tibia vara is associated with a medial shift of the articular surface and that the tibial condyle posterior rotation is associated with a posterior shift of the articular surface, each variation in each case should be accurately evaluated, and the degree of correction should be considered. In Asia, two concepts "replacement" and "arthroplasty" are required. "Replacement" is for cases with no variation, in which each part of the bone is cut with the same thickness of the component. "Arthroplasty" is for cases with strong variation in which the alignment is corrected and the joint is rebuilt.

Regarding resurfacing, because the sagittal plane and transverse-sectional shape of the patella are related to the shape of the tibial and femoral condyles, it is necessary to determine the resurfacing or non-resurfacing of the patella by considering the degree of posterior rotation of the tibial condyle and the difference between the shape of the femoral and tibial condyles and the shape of the femoral and tibial components (Fig. 18.5a–c).

References

1. Kim YH, et al. Clinical faceoff: do we need special strategies for Asian patients with TKA? Clin Orthop Relat Res. 2016;474:1102–7.
2. Nagamine R, et al. Anatomic variations should be considered in total knee arthroplasty. J Orthop Sci. 2000;5:232–7.
3. Nagamine R, et al. Medial torsion of the tibia in Japanese patients with osteoarthritis of the knee. Clin Orthop Relat Res. 2003;408:218–24.
4. Ko JH, et al. Femur bowing could be a risk factor for implant flexion in conventional total knee arthroplasty and notching in navigated total knee arthroplasty. Knee Surg Sports Traumatol Arthrosc. 2016;24:2476–82.
5. Nagamine R, et al. The posterior tibial slope is mainly created by the posterior rotation of the tibial condyles. J Orthop Surg (Hong Kong). 2020;28:2309499020975580.
6. Matsuda S, et al. Anatomical analysis of the femoral condyle in normal and osteoarthritic knees. J Orthop Res. 2004;22:104–9.
7. Nagamine R, et al. Distal femoral cut perpendicular to the mechanical axis may induce varus instability in flexion in medial osteoarthritic knees with varus deformity in total knee arthroplasty: a pitfall of the navigation system. J Orthop Sci. 2004;9:555–9.
8. Bellemans J, et al. The Chitranjan Ranawat award:is neutral mechanical alignment normal for all patients? The concept of constitutional varus. Clin Orthop Relat Res. 2012;70:45–53.
9. Kawasaki M, et al. Proximal tibia vara involves the medial shift of the tibial articular surface. J Orthop Surg (Hong Kong). 2020;28:2309499020902592.
10. Nagamine R, et al. Assessment of influences of posterior rotation of the tibial condyles on the Insall-Salvati ratio. Sci Rep Sci Rep. 2022;12(1):21850.
11. Matsuda S, et al. Posterior tibial slope in the normal and varus knee. Am J Knee Surg. 1999;12:165–8.
12. Nagamine R, et al. Reliability of the anteroposterior axis and the posterior condylar axis for determining rotational alignment of the femoral component in total knee arthroplasty. J Orthop Sci. 1998;3:194–8.

19 Laxity in Normal and Osteoarthritic Knees

Umito Kuwashima

1 Knee Joint Laxity

Laxity of the knee joint can be broadly divided into three forms: anteroposterior, internal valgus, and rotation. Compared to normal knees, OA and RA knees both show changes in laxity, which varies with age and sex. In general, laxity of the knee joint increases with age [1] in normal knees, and joint laxity tends to be greater in women than in men [2].

2 Anteroposterior Laxity in Normal/Medial Compartmental OA Knees (Medial OA)

Anterior laxity in normal knees has been reported to be 6.1 ± 1.8 mm at 63 Nm at 90° knee flexion in a stress test [3]. In medial OA, although the anteroposterior laxity increases significantly compared to the normal knee at 10.0 ± 6.3 mm in grade I in the Kellgren-Lawrence classification (K-L), it gradually decreases as the K-L grade progresses, becoming close to that of the normal knee. In addition, no relationship between the degree of intraoperative ACL/PCL degeneration and the degree of intraoperative ACL/PCL degeneration was found in the medial OA knee, while a previous study reported that posterior laxity was significantly increased in patients with ACL loss in the RA knee, and that anteroposterior laxity was significantly increased in subjects with PCL degeneration.

Anterior and posterior stress X-rays showed laxities of 4.9 ± 2.0 mm at 20° flexion and 4.6 ± 2.7 mm posterior stress at 70° flexion in healthy knees in young people [4]. The same study showed that anteroposterior laxity was significantly greater when generalized laxity was observed than when it was not, although the left-right difference was within 3 mm in both groups. This suggests that ACL/PCL dysfunction should be suspected if the difference between the affected and unaffected sides is >3 mm. In addition, because manual stress imaging may cause variations in evaluation, it is desirable to use an instrument that can apply a certain amount of external force. When evaluating ACL function, the anteroposterior gravity view" (Fig. 19.1) [5] for 15° knee flexion in the prone position has been reported to be a highly reproducible imaging method. While evaluating the PCL function, the "Gravity sag view" (Fig. 19.2) [6] with 90° knee flexion in the supine position and 45° hip flexion has also been reported as a highly reproducible imaging method.

Dynamic analysis of the knee joint using image-matching methods has also been used to analyze ligament balance during a variety of movements. In normal knee flexion, the ACL is tense during knee extension and gradually relaxes with knee flexion, while the degree of laxity tends to decrease at a deep flexion of 120° or more [7]. This tendency was observed

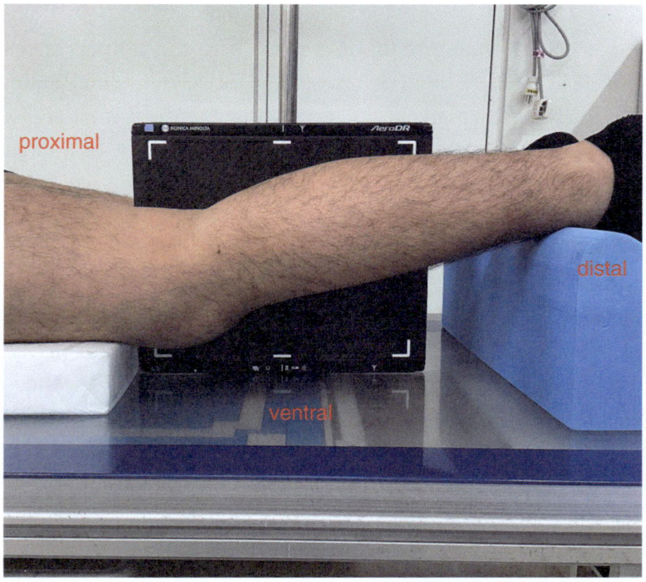

Fig. 19.1 The anterior gravity view

U. Kuwashima (✉)
Department of Orthopaedic Surgery, Tokyo Women's Medical University, Tokyo, Japan
e-mail: kuwashima.umito@twmu.ac.jp

S. Matsuda (ed.), *Advances in Total Knee Arthroplasty*, https://doi.org/10.1007/978-981-97-4920-1_19

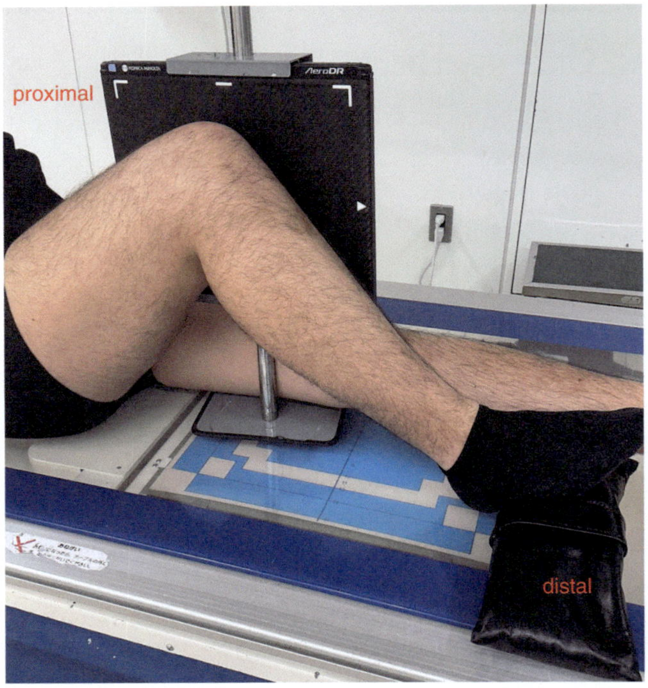

Fig. 19.2 The gravity sag view

both with and without load; however, the tendency to relax was weaker with load. In contrast, the the PCL tends to gradually tense with knee flexion, and because both ACL and PCL functions tend to attenuate in the middle flexion position (knee flexion: 30–90°), stability in the middle flexion position is important in TKA [8].

3 Varus/Valgus Laxity in Normal/Medial OA Knees

A prior study reported that the varus laxity in normal knees was 12.0° ± 3.8° in an 8 Nm stress test at 20° knee flexion [3]. Further, he valgus laxity tended to increase as medial OA progressed, but the difference was not significant. In a large cohort study, OA knees with K-L grade II or higher were found to have higher varus-valgus laxity than normal knees on manual examination, with values that were significantly higher in females than males [9]. In addition, factors such as female sex, radiographic OA changes, and varus deformity have been reported to be associated with varus-valgus laxity. It has also been reported that the valgus laxity in OA knees was significantly correlated with the presence of pain.

Varus-valgus stress radiography is often performed to evaluate knee joint laxity. In normal knees, the joint line convergence angle (JLCA) at varus and valgus stress in the knee extension position (10° flexion) was 4.9° ± 2.0° and 2.4° ± 1.6°, respectively, while the same values were 4.8° ± 3.2°and 1.7° ± 1.4° in knee flexion (80° flexion), respectively [10]. This indicates that lateral laxity is greater than medial laxity in both knee extension and flexion positions. In another study, MRI imaging in the varus-valgus stress limb in the knee flexed position showed that the medial opening in the valgus stress was 2.1 ± 1.1 mm, whereas the lateral opening in the varus stress was 6.7 ± 1.9 mm, a mean difference of 4.6 mm [11]. Even in normal knees, the medial and lateral laxities are not the same in the extension and flexion positions, while the lateral laxity is large. This result is important when considering the balance of varus-valgus laxity after TKA.

One previous study used fluoroscopic analysis to analyze the MCL/LCL in the normal knee. Herein, the MCL was classified as deep (dMCL) or superficial (sMCL), and the dynamics of each were analyzed; the results showed that the anterior and central fibers of the dMCL and the anterior fibers of the sMCL tended to be tense due to knee flexion, especially in middle flexion [7]. In the LCL, the anterior fibers tended to gradually tense with knee flexion; however, the standard deviation was larger than that in the MCL, suggesting the existence of large individual differences. Surgeons should consider these findings when dissecting soft tissues.

The median JLCA of valgus stress in OA knees in extension varus-valgus stress radiographs has been reported to be 0°, which is significantly smaller than that of normal knees (median: 2°), whereas the medial JLCA of varus stress in OA knees has been reported to be 8°, which is significantly larger than that in normal knees (median: 4°). A weak correlation between the FTA and varus-valgus JLCA has also been reported [12]. Conversely, when the gap was evaluated using a tenser during TKA surgery in three groups, mild (<10°), moderate (10°~20°), and severe (>20°) varus deformity, the gap increased significantly in the severer varus deformity on the lateral side [13]. While the gap tended to increase as the varus deformity became more severe on the medial side, a significant difference was observed only between the mild and severe group [13]. Even in patients in whom the preoperative varus deformity was severe, the medial side was not as tight, whereas the lateral side tended to be loose. These results suggest that it is necessary to consider the effects of cartilage wear and patient-specific femur and tibial joint line obliquity on the JLCA when evaluating varus-valgus stress X-rays (Fig. 19.3).

In one study on OA knees, a significant difference was observed between the JLCA for the varus and valgus when measured with 15 kg of stress on X-rays (7.9° ± 1.2° and −1.5° ± 2.2°); however, it should be noted that when the JLCA in the supine anteroposterior X-ray (3.2° ± 2.0°) was used as the reference for evaluation of the JLCA change, the varus JLCA was 4.8° ± 2.1° and valgus JLCA was −4.7° ± 1.8°, which were equivalent in laxity with no significant differences [14]. In another cohort, when the JLCA

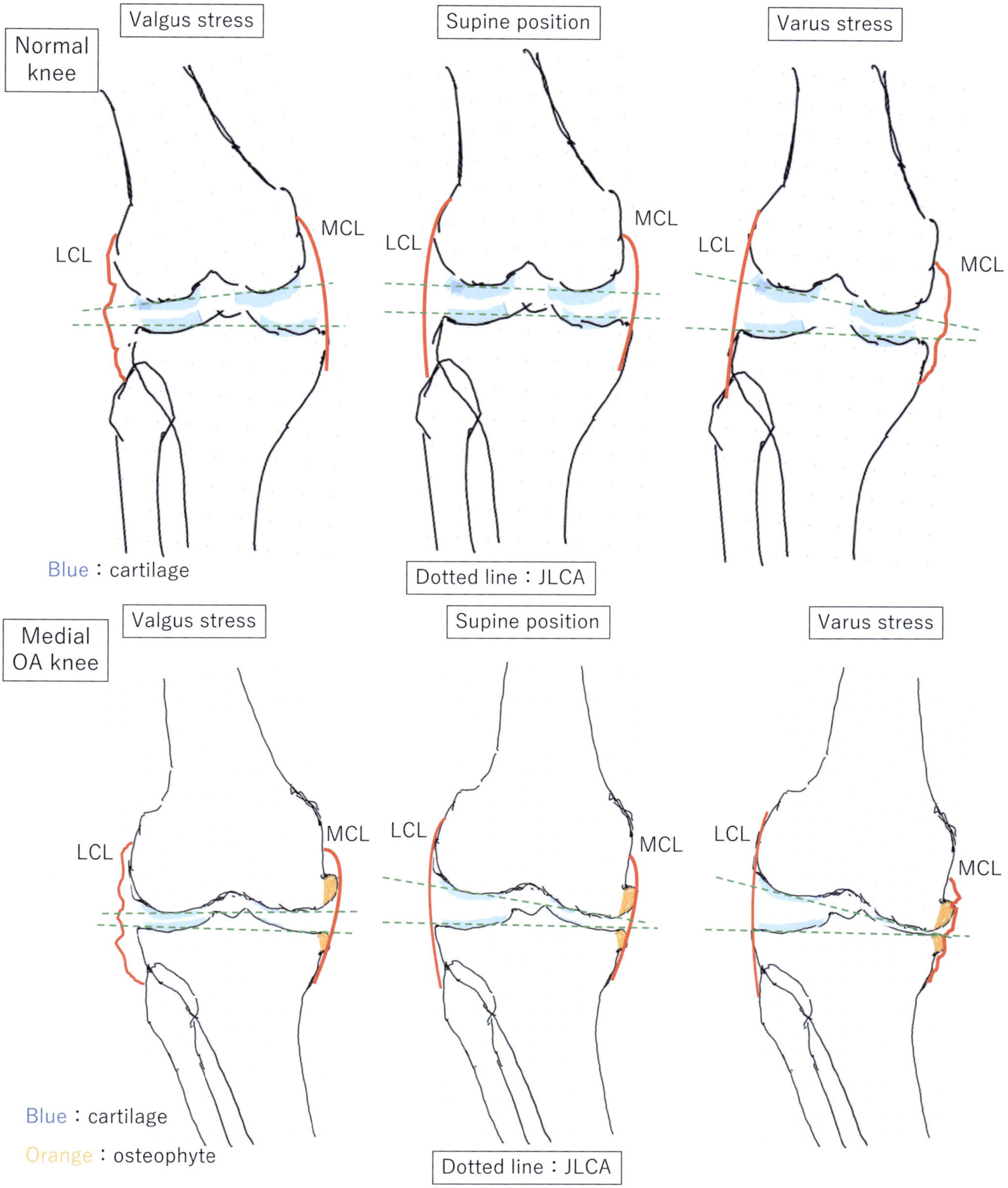

Fig. 19.3 The varus-valgus stresses of normal and medial compartmental OA knees (the angle formed by the dotted line indicates the JLCA on X-ray). In the normal knee, the JLCA was 0° ± 1°. Although the outer laxity was slightly larger, the cartilage thickness was maintained; therefore, on X-rays, the slope was almost the same because of the varus-valgus stress. In the medial OA knee, the cartilage on the medial articular surface is thinned and osteophytes have formed; therefore, the JLCA is inclined medially even in the supine position. As lateral laxity tends to increase with the progression of varus deformity, the JLCA appears to be greatly inclined medially by varus stress on radiography; however, the amount of change in the JLCA under varus or valgus stress is almost the same as that of a normal knee when supine radiography is used as a reference. When evaluating medial contracture and lateral laxity, it is necessary to measure the change with reference to the front of the recumbent position

change from the criterion of varus/valgus stress was evaluated based on the JLCA in the supine anteroposterior radiograph, no significant differences were found in medial and lateral laxity in the advanced OA knee of K-L grade III/IV, and 90% of cases had medial and lateral laxity of 3° [15].

In addition to preoperative lower limb alignment osteophytes are also considered an influencing factor for varus-valgus laxity in medial OA. In a prior study, the FTA and HKA were reported to have moderate correlations with the JLCA on varus stress X-rays [14]. In contrast, FTA shows only a weak correlation with medial laxity [15]. Regarding osteophytes, severe varus deformity tends to increase the number of osteophytes [16]; however, no significant correlation was observed between osteophyte size and medial and lateral laxity [14, 16]. In addition, a relationship between lower limb alignment and quadriceps and hamstring muscle strength has also been reported [17]. Hence, the factors influencing varus/valgus laxity may vary. Therefore, a detailed preoperative evaluation of varus-valgus laxity and an accurate intraoperative technique and assessment are necessary when performing TKA, and the surgeon must visualize the target ligament balance in each case.

4 Intraoperative Data of Tenser Device

There are several types of tenser devices which may be used in the clinic, for which numerous reports have outlined the appropriate tension during intraoperative measurements (see another chapter) Table 19.1 [18] shows the post-osteotomy tension at PS-TKA in OA knees, along with the changes in the center gap and varus angle. The greater the tension of the device, the greater the gap and varus inclination. When the gap was measured from knee extension to 135° of flexion with the femoral trial implanted, the laxity was greater on the lateral side than on the medial side at all flexion angles.

Regarding the extension and flexion gaps, the effect of PCL resection has also been reported, with reports indicating that the change in the flexion gap is larger than that of the extension gap [19–22]. Once study showed that the mean change in the extension gap was approximately 1 mm, whereas the mean change in the flexion gap varied by approximately 3 mm depending on the report. Furthermore, the difference between cases was also large, making prediction difficult [23]. If the PCL is resected, attention should be paid to changes in the gap. It should also be noted that in devices that measure the center gap, a stronger preoperative varus deformity results in a larger extension gap, which tends to appear as lateral laxity increases [24]. Studies investigating difference between extension and flexion gaps have reported that the force required to open the gap by 1 mm when equal tension is applied is greater in extension than in flexion [25]. In other words, the possibility that the flexion gap is overestimated compared to the extension gap at equivalent tension must be considered.

Table 19.1 Changes in tension, center gap, and varus angle in medial compartmental OA knees

	Joint distraction force(mean ± SEM)		
	20 lbs	40 lbs	60 lbs
Joint center gap(mm)			
Extension	23.2 ± 0.2	26.2 ± 0.3	28.3 ± 0.2
Flexion	22.6 ± 0.3	25.7 ± 0.3	27.7 ± 0.3
Varys ligament balance(°)(+varus)			
Extension	4.3 ± 0.3	6.0 ± 0.4	6.9 ± 0.4
Flexion	1.9 ± 0.4	3.9 ± 0.4	4.7 ± 0.5

Source: Nagai et al. [18]

5 Rotational Laxity of Normal/Medial OA Knees

The rotational laxity in normal knees has been reported to be 48.4° ± 5.9° in a 5.4 Nm stress test at 80° knee flexion [3], and the rotational laxity was shown to decrease as degeneration progressed. In addition, a rotational laxity at a knee flexion angle of 20° in a normal knee has a significantly greater effect on external rotation than internal rotation, while rotational laxity decreases with loading [26]. Furthermore, it has been reported that rotational laxity is significantly correlated with general and anteroposterior laxity [2].

6 Joint Laxity of the Lateral Compartmental Knee OA (Lateral OA)

Lateral OA is rarer than medial OA, and there are relatively few reports of the associated joint laxity. Regarding the varus-valgus laxity of lateral OA, in most cases where the valgus deformity is relatively mild, alignment can be corrected by varus stress in knee extension (0–20°) [27]. However, some patients experience medial laxity or lateral contractures during knee extension. Thus, preoperative evaluation using stress radiography is important. In addition, prior research has shown that, like medial OA knees, the lateral side of lateral OA knees is significantly looser than the medial side at 90° flexion [27]. Care must therefore be taken to balance laxity between extension and flexion when performing TKA for lateral OA.

Regarding the rotation in lateral OA, measurement using intraoperative navigation showed that valgus deformity rotates internally with flexion in cases with FTA ≧160°, while in cases with severe valgus deformity (FTA < 160°), the tibia rotates externally with flexion [28]. Attention should therefore also be paid to ligament balance and implant selection during lateral OA surgery.

7 RA Knee Joint Laxity in RA Knees

The joint laxity of knees with RA is known to be slightly different from that of knees with OA. Compared to healthy knees, anterior laxity is greater in knees with RA, with laxity increasing as the Larsen classification grade progresses [3]. Regarding the degree of intraoperative ACL/PCL degeneration, ACL-missing knees showed significantly increased anteroposterior laxity, while PCL-degenerated knees also showed significantly increased anteroposterior laxity. Another study showed that while varus–valgus laxity tended to increase as degeneration progressed in the RA knee, no significant difference was observed [3]. It was further shown that rotational laxity tends to decrease significantly as degeneration progresses.

References

1. Zyroul R, et al. Knee laxity of Malaysian adults: gender differentials, and association with age and anthropometric measures. Knee. 2014;21:557–62.
2. Shultz SJ, et al. Measurement of varus-valgus and internal-external rotational knee laxities in vivo – Part II: relationship with anterior-posterior and general joint laxity in males and females. J Orthop Res. 2007;25:989–96.
3. Wada M. Knee laxity in patients with osteoarthritis and rheumatoid arthritis. Br J Rheumatol. 1996;35:560–3.
4. Lim HC, et al. Anterior and posterior knee laxity in a young adult Korean population. Knee Surg Sports Traumatol Arthrosc. 2011;19:1890–4.
5. Mae T, et al. Anterior laxity of the knee assessed with gravity stress radiograph. Skeletal Radiol. 2018;47:1349–55.
6. Shino K, et al. The gravity sag view: a simple radiographic technique to show posterior laxity of the knee. Arthroscopy. 2000;16:670–2.
7. Kono K, et al. In vivo length change of ligaments of normal knees during dynamic high flexion. BMC Musculoskelet Disord. 2020;21:552.
8. Rao Z, et al. In vivo kinematics and ligamentous function of the knee during weight-bearing flexion: an investigation on mid-range flexion of the knee. Knee Surg Sports Traumatol Arthrosc. 2020;28:797–805.
9. Miura H, et al. Varus-valgus laxity correlates with pain in osteoarthritis of the knee. Knee. 2009;16:30–2.
10. Okazaki K, et al. Asymmetry of mediolateral laxity of the normal knee. J Orthop Sci. 2006;11:264–6.
11. Tokuhara Y. The flexion gap in normal knees. An MRI study. J Bone Joint Surg Br. 2004;86:1133–6.
12. Ishii Y, et al. Preoperative laxity in osteoarthritis patients undergoing total knee arthroplasty. Int Orthop. 2009;33:105–9.
13. Okamoto S, et al. Lateral soft tissue laxity increases but medial laxity does not contract with varus deformity in total knee arthroplasty. Clin Orthop Relat Res. 2013;471:1334–42.
14. Ushio T, et al. Medial soft tissue contracture does not always exist in varus osteoarthritis knees in total knee arthroplasty. Knee Surg Sports Traumatol Arthrosc. 2019;27:1642–50.
15. Ishii Y, et al. Medial and lateral laxity in knees with advanced medial osteoarthritis. Osteoarthr Cartil. 2018;26:666–70.
16. Ishii Y, et al. Size of medial knee osteophytes correlates with knee alignment but not with coronal laxity in patients with medial knee osteoarthritis. J Orthop Res. 2020;38:639–44.
17. Nakagawa K, et al. Associations of knee muscle force, bone malalignment, and knee-joint laxity with osteoarthritis in elderly people. J Phys Ther Sci. 2017;29:461–4.
18. Nagai K, et al. Soft tissue balance changes depending on joint distraction force in total knee arthroplasty. J Arthroplast. 2014;29:520–4.
19. Kadoya Y, et al. Effects of posterior cruciate ligament resection on the tibiofemoral joint gap. Clin Orthop Relat Res. 2001;391:210–7.
20. Baldini A, et al. Flexion-extension gap changes during total knee arthroplasty: effect of posterior cruciate ligament and posterior osteophytes removal. J Knee Surg. 2004;17:69–72.
21. Park SJ, et al. Effect of PCL on flexion-extension gaps and femoral component decision in TKA. Orthopedics. 2009;32:22–5.
22. Kayani B, et al. Posterior cruciate ligament resection in total knee arthroplasty: the effect on flexion-extension gaps, mediolateral laxity, and fixed flexion deformity. Bone Joint J. 2019;101:1230–7.
23. Warth LC, et al. Posterior cruciate ligament resection does not consistently increase the flexion space in contemporary total knee arthroplasty. J Arthroplast. 2021;36:963–9.
24. Wakama H, et al. Preoperative factors associated with extension gap in cruciate-retaining total knee arthroplasty: a retrospective study on continuous determination of distraction force. J Orthop Sci. 2020;25:1035–9.
25. Asano H, et al. Stiffness of soft tissue complex in total knee arthroplasty. Knee Surg Sports Traumatol Arthrosc. 2008;16:51–5.
26. Shultz SJ, et al. Measurement of varus-valgus and internal-external rotational knee laxities in vivo – Part I: assessment of measurement reliability and bilateral asymmetry. J Orthop Res. 2007;25:981–8.
27. McAuliffe MJ, et al. Coronal plane laxity of valgus osteoarthritic knee. J Orthop Surg. 2019;27:2309499019833058.
28. Kawaguchi K, et al. Rotational kinematics differ between mild and severe valgus knees in total knee arthroplasty. Knee. 2021;28:81–8.

Part III

Biomechanics

Biomechanics of the Knee Joint

20

Hiroshi Tsumura

1 Three-Dimensional (3D) Gait Analysis

Among the analysis methods for joint movement during bodily motion, one method identifies and tracks the position by attaching infrared reflection markers to landmarks of the human body and photographing them synchronously with multiple infrared cameras. This method is represented by VICON®, in which a precise analysis is performed based on position accuracy and high measurement density. The method can also synchronize the floor-reaction force and measure it as a 3D vector in real time. Furthermore, if myoelectric measurements are performed, the relative activity of the functioning muscles can be estimated. Gait analysis using this method is described below.

1.1 Normal Gait

Walking is a cyclic movement of both lower limbs, and one gait cycle is defined as the period from the grounding of the lower limb on one side to just before the next grounding, with the stance phase in which the plantar is grounded and the swing phase in which the sole is not grounded. The standing phase is divided into (1) initial contact phase (heel grounding), (2) loading response phase, (3) midstance phase, (4) terminal stance phase, and (5) preswing phase, while the swing phase is divided into (6) initial swing phase, (7) midswing phase, and (8) terminal swing phase (Fig. 20.1). During walking, when both legs are grounded, the standing phase is longer than the swinging phase. The distance from the heel contact point of one foot to that of the opposite foot is called the stride. Walking speed is determined by the number of steps per unit time (cadence) and stride length [1].

The angles of the lower limb joints during the walking cycle are shown in Fig. 20.2. The hip joint extends from the maximum flexion position during the initial contact, reaches the maximum extension position during the preswing phase, and flexes when entering the swing phase. In contrast, the knee joint is extended during initial contact, slightly flexed during the loading response phase, and extended again from the midstance to the terminal stance phases. Flexion begins in the preswing phase and the maximum flexion position is attained in the initial swing phase. Thus, this movement of the knee joint flexing twice in one walking cycle is called "double-knee action." The ankle joint remains grounded in a mild plantar flexion position in the initial contact phase, becomes dorsiflexed toward the midstance phase, and reaches maximum plantar flexion from the terminal stance to the preswing phase.

The floor-reaction force naturally occurs in the stance phase; however, it is a biphasic curve that is largest during the loading response phase and then decreases and increases again from the terminal stance phase to the preswing phase (Fig. 20.3). Looking at each component, the vertical component is the largest owing to the reaction force to the main body weight, followed by the component in the anteroposterior direction. From the initial contact to the loading response period, the heel touches forward; thus, a backward reaction force occurs. From the terminal stance phase to the preswing phase, a forward reaction force is generated against the forward propulsion force (kicking the floor back). The components in the inner and outer directions are considerably smaller than the previous two components. The tergites are turned slightly outward only at the initial contact, but thereafter, they are turned inward.

When viewed from the front, the floor reaction force passes between the inner and outer sides of the knee joint in normal lower limb alignment (Fig. 20.4). This indicates that in normal lower limb alignment, the moment of adduction or abduction is extremely small. In the anteroposterior direction, it does not pass inside the knee joint; however, it can be dynamically balanced by exerting a muscular force in the direction of flexion and extension of the knee joint. Thus, the floor reaction force acts as an external force but is

H. Tsumura (✉)
Oita University, Oita, Japan
e-mail: htsumura@oita-u.ac.jp

S. Matsuda (ed.), *Advances in Total Knee Arthroplasty*, https://doi.org/10.1007/978-981-97-4920-1_20

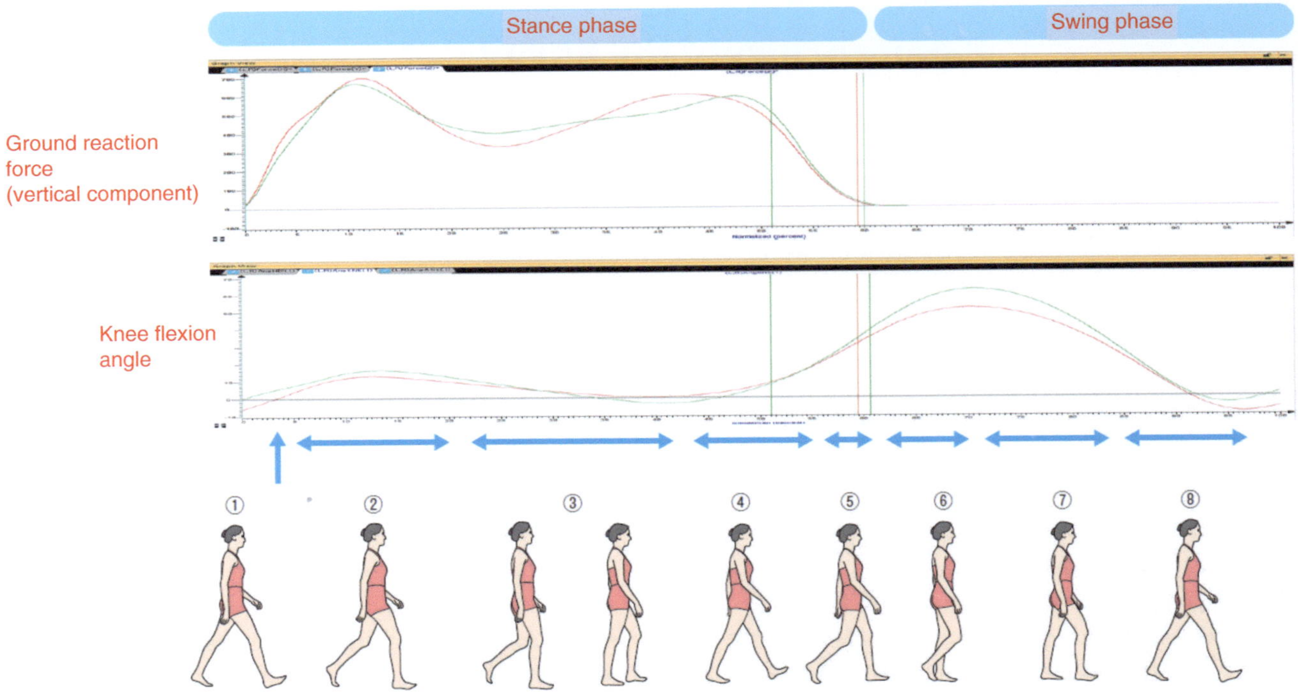

Fig. 20.1 Walking cycle. The upper graph shows the vertical component of the floor-reaction force, and the lower graph shows the knee joint flexion angle. See the text for the names of the phases. Gait analysis data were obtained with the cooperation of Masaya Anan of the Faculty of Welfare and Health Sciences at Oita University

not the only one acting on the knee joint. The force exerted by the muscles around the knee joint acts as an internal force, and the combined force of the floor reaction force (although it is necessary to remove the weight of the lower leg) and the muscular force is the load applied to the knee joint.

Muscle activity in the walking cycle, measured by electromyography, is shown in Fig. 20.5, in which the muscle activities of the quadriceps, hamstrings, and triceps muscles can be observed. At the time of initial contact, both the quadriceps and hamstrings are active simultaneously; however, the muscle activity of the hamstrings rapidly weakens, and the activity of the quadriceps dominates. It increases moderately again in the terminal stance phase. From the midstance to the terminal stance phase, the activity of the quadriceps increases when the knee joint is extended and reaches its maximum from the initial contact to the loading response phase. Muscle activity increases when the knee joint is flexed during the swing phase; however, this is mild, and the highest muscle activity is achieved when the knee joint is extended toward the initial contact. This indicates that the hamstrings increase the rigidity of the knee joint and reduce swaying during eccentric contractions. Muscle activity of the triceps muscle of the calf is simple, indicating that it is active from the midstance to the preswing phase.

1.2 Walking in OA Example

In OA, based on changes in the lower limb alignment, a mechanistic understanding of the pathophysiology is essential. Figure 20.6 shows the floor reaction force during the loading response phase of medial OA, which is the most common case. When viewed from the front, the floor-reaction force passes inward from the knee joint. This acts as an adduction moment for the knee joints. Figure 20.7 shows the transition of the adduction moments in the knee joint in normal and medial OA cases. In normal cases, it varies slightly from the baseline, but in osteoarthritis cases, a large adduction moment is generated. A study examining the trajectory of the load axis movement relative to the proximal tibial articular surface (Fig. 20.8) showed that in the normal knee, it is concentrated in the central portion; however, in mild OA,

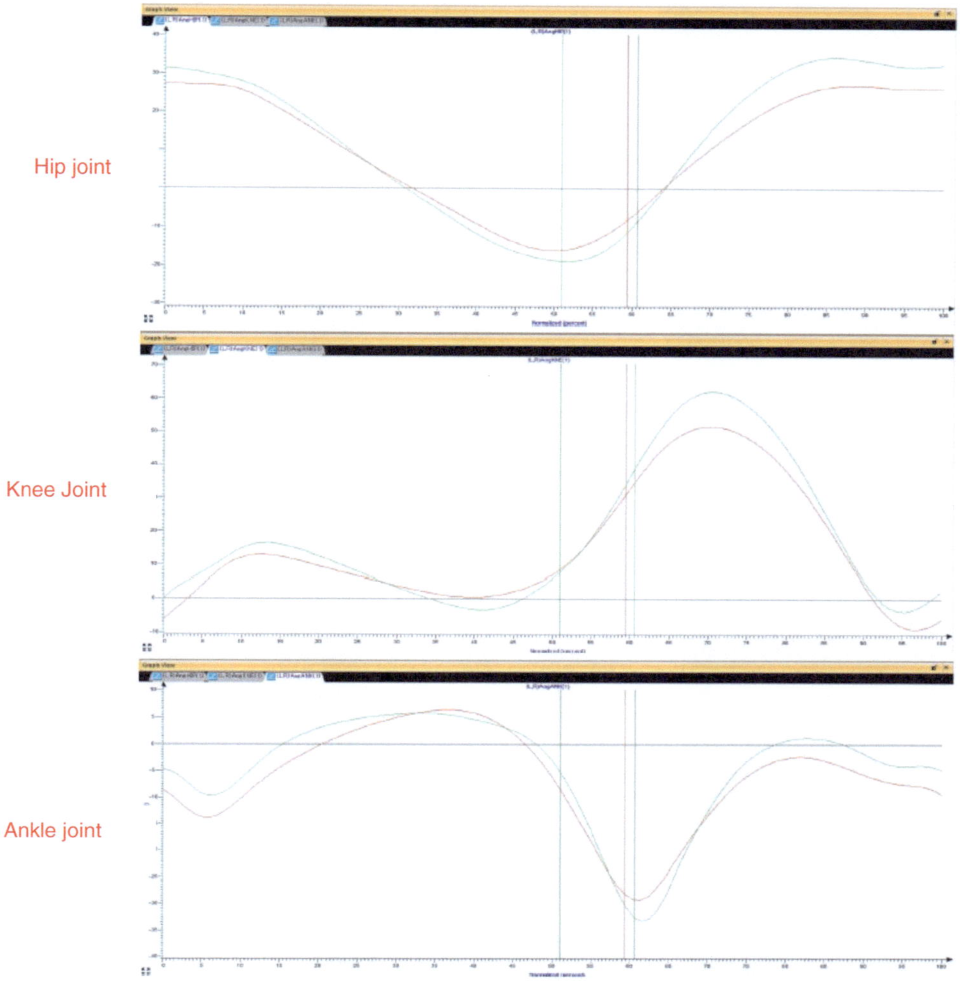

Fig. 20.2 Angle of lower limb joints in the gait cycle. The upper row shows the hip joint, the middle row shows the knee joint, and the lower row shows the ankle joint. The pattern in which the knee joint is flexed twice during the step cycle is called the "double-knee action"

it is concentrated in the medial articular surface, and in severe OA cases, it is more medially away from the articular surface [2].

In contrast, studies that observed upper-body shaking during walking reported that upper-body tilt toward the stance side occurred in the OA group during the stance period compared with the control group. This is presumed to be a compensatory pattern for bringing the load axis closer to the knee joint [3].

1.3 Gait Analysis After TKA

Post-TKA biomechanics have been explained in another chapter, but there is a study on adduction moments after TKA regarding the adduction moments that occur in the knee joint. Even after correcting the lower limb alignment with TKA, there are cases in which a large adduction moment remains, and it has been reported that the greater toe-out angle is large in such cases [4].

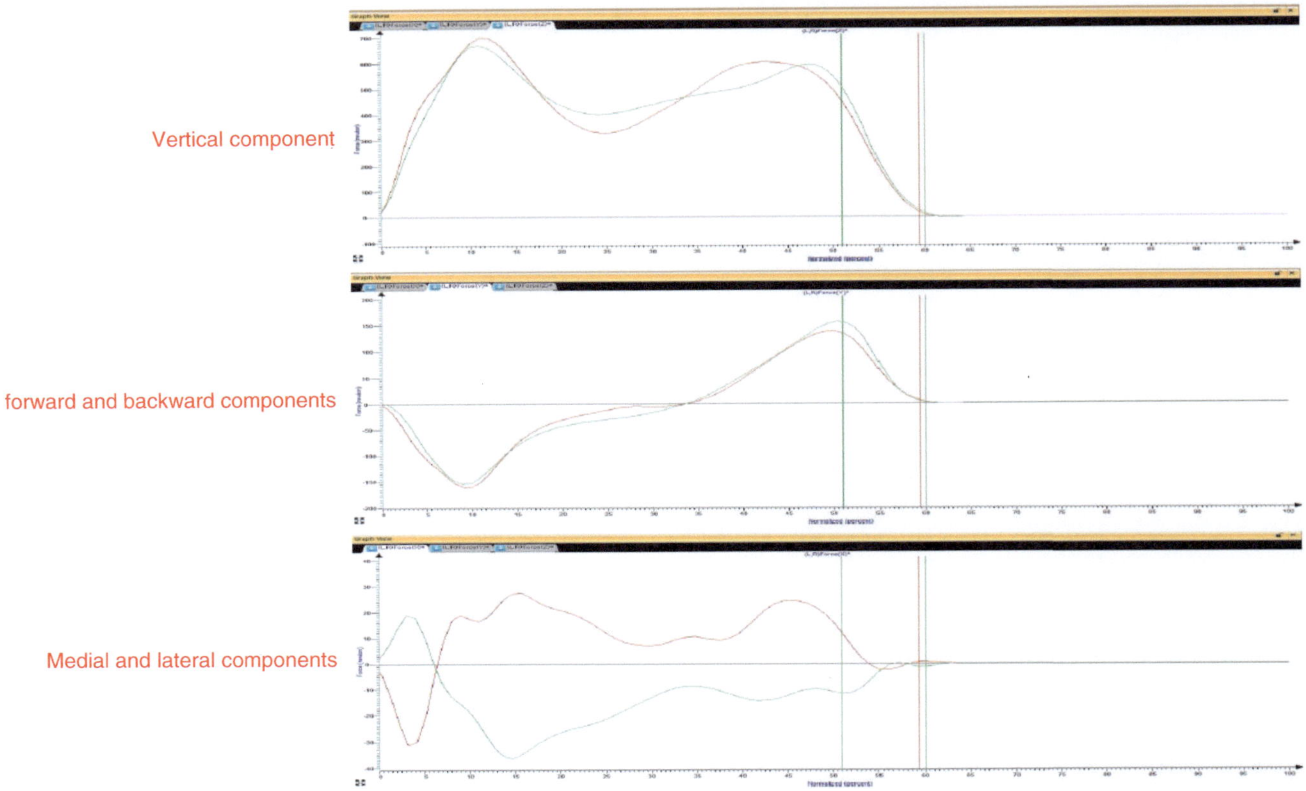

Fig. 20.3 Floor-reaction force in the walking cycle. The upper row has a vertical component (maximum value of approximately 700 N), the middle row has forward and backward components (maximum value of approximately 150 N), and the lower row has medial and lateral components (maximum value of approximately 30 N). The vertical component is the largest

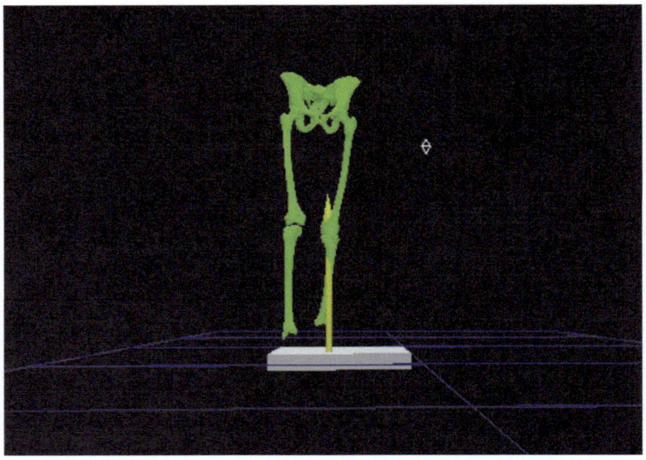

Fig. 20.4 Floor-reaction force at the normal knee joint. When viewed from the front, the floor-reaction force passes between the medial and lateral sides of the knee joint

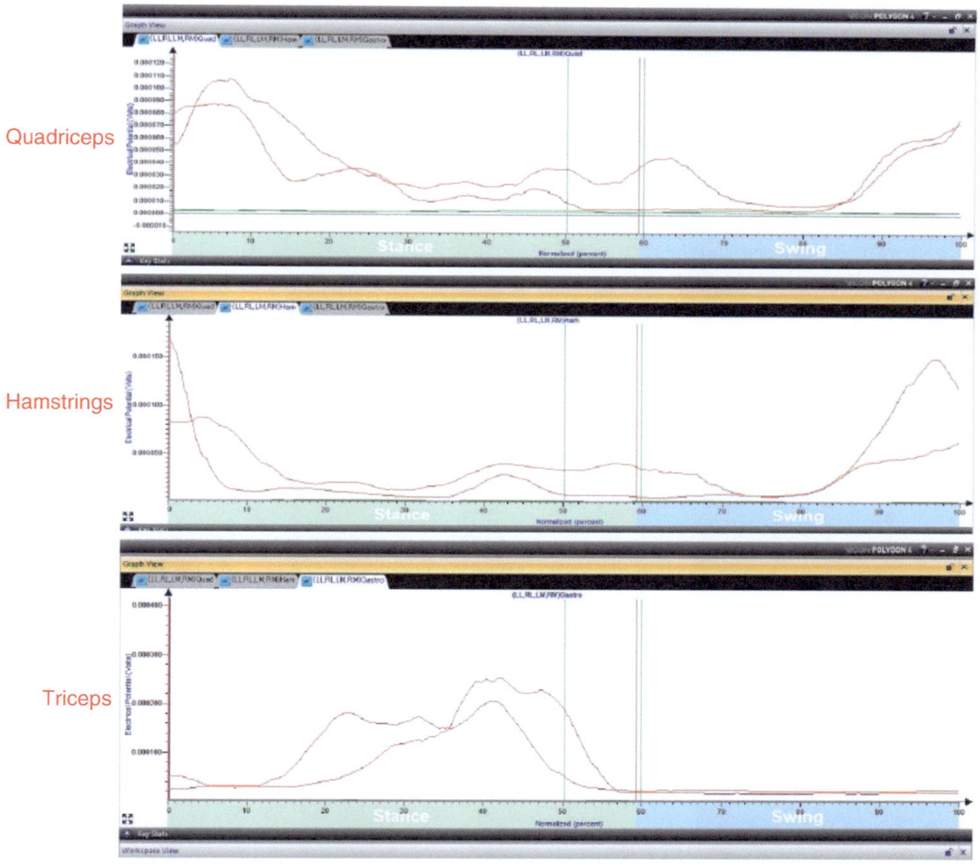

Fig. 20.5 Electromyography of major lower limb muscles in the walking cycle. The upper row shows the quadriceps, the lower row shows the hamstrings, and the lower row shows the triceps muscles of the calf. Simultaneous contraction of the quadriceps and hamstrings is observed from the initial ground contact phase to the loading response phase

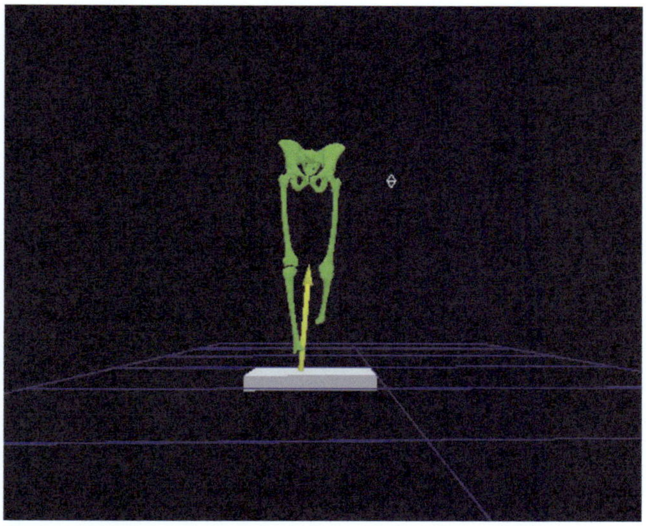

Fig. 20.6 Floor-reaction force in medial OA. The floor reaction force passes inside the knee joint, and the pathology can be clearly observed by comparing it with that shown in Fig. 20.4

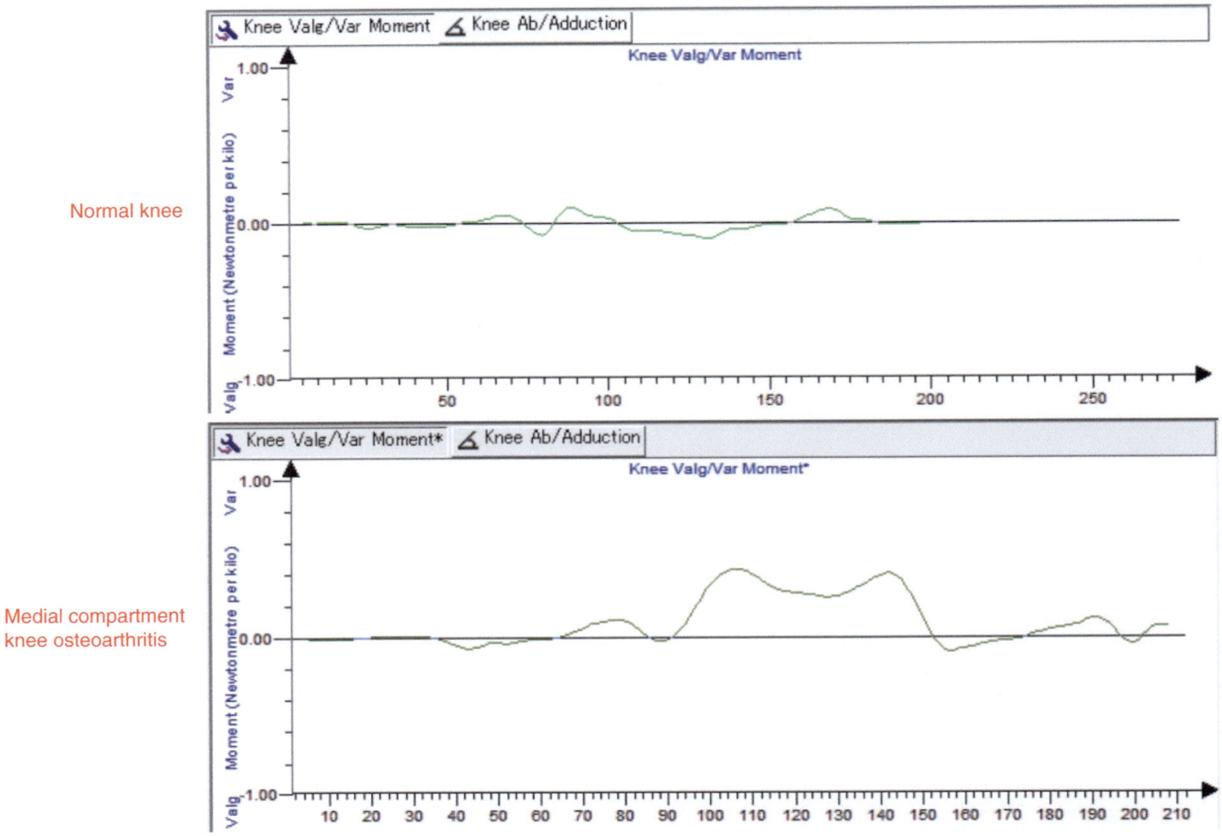

Fig. 20.7 Adduction moment in the knee joint during walking. The upper row shows the normal knee shown in Fig. 20.4, and the lower row shows the medial OA knee shown in Fig. 20.6

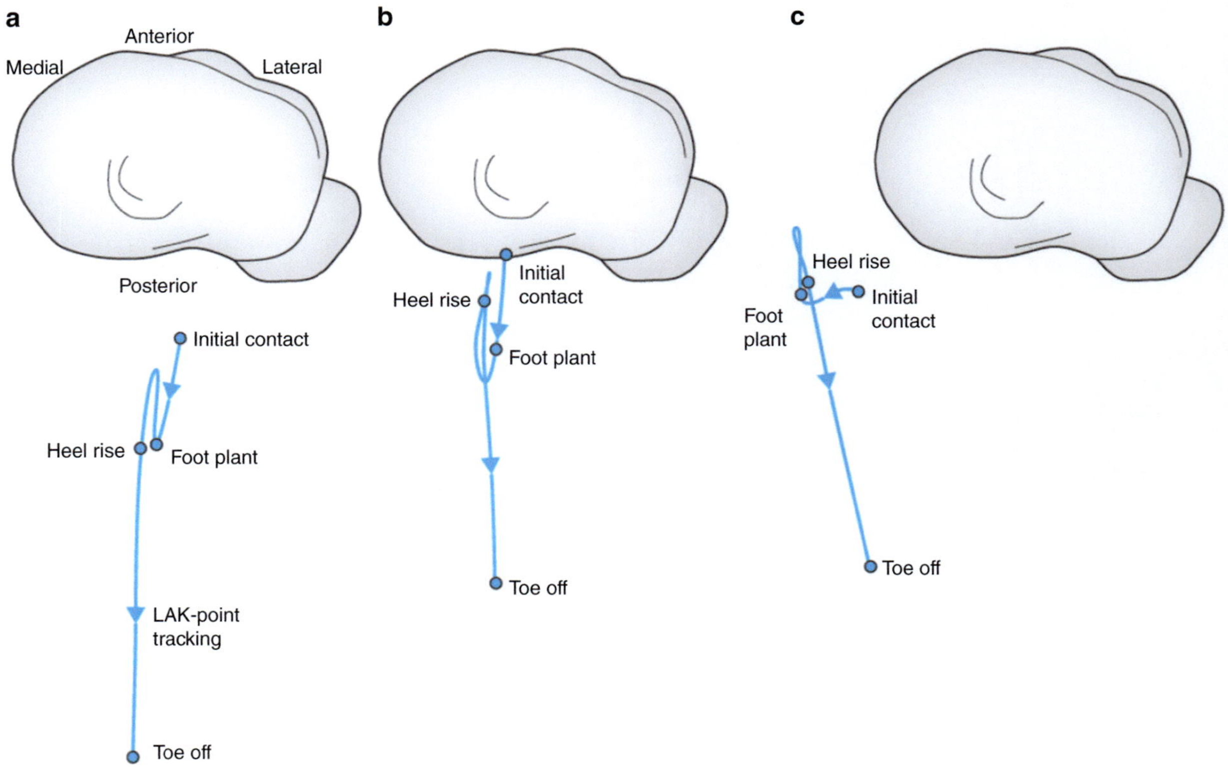

Fig. 20.8 Trajectory of the load axis during walking. (**a**): Normal. (**b**): Mild OA. (**c**): As severe OA symptoms progress, the load axis shifts inward. (Cited from Nishino et al. [2])

2 Force and Contact Pressure Distribution Acting on the Femoral-Tibial Joint

2.1 Estimation of the Force Acting on the Tibial Articular Surface of the Femur Using the Calculation Method

As mentioned previously, the load applied to the knee joint is the sum of the external force expressed in the floor reaction force and the force generated by the knee muscles. As the muscle force cannot be measured, some mathematical and physical models (inverse dynamics) can be estimated from the observed reaction force and angular acceleration; however, these models are not accurate in either case. During joint motion, there is simultaneous activity of the antagonist and the main active muscle. Although the difference between the two muscles causes the motion to be performed, the forces of both are added to the load. This is also evident in the gait analysis described previously, in which not only the quadriceps but also the hamstrings show muscle activity when the knee joint is extended. However, as it is expedient to know the calculation method and compare it with the actual measured value described later, an example is shown with reference to Yoshikazu et al. [5].

The force applied to the knee joint is calculated using the exercise of sitting in a chair and extending the knee joint. To simplify, the extension motion of the lower leg centered on the knee joint (c) is a circular motion. The model consists of a single rod driven only by the quadriceps muscle (Fig. 20.9). From the 3D motion analysis, the angular acceleration (α) and knee flexion angle (α) associated with the movement of the lower leg can be measured. The weight of the lower leg (W) and the distance from c to the center of gravity of the lower leg (b) are calculated based on the literature by Walker [6]. The direction of the quadriceps muscle force (direction of the patellar tendon) (φ) and the distance from c to the point of force (a) are measured using plain radiographs of the lower leg. First, calculate the quadriceps muscle force, and then calculate the force (FJ) acting on the joint. Figures 20.10 and 20.11 show the equations, and Table 20.1 shows an actual example. The equation of motion is "F = ma," but because it is a rotational motion, multiplying the two sides by the radius (r) yields "rF = rma." Furthermore, if acceleration (a) is multiplied by angular acceleration (α) and radius (r), we get "rF = mr^{2}α." "rF" is the first moment of force, denoted as "M," which in this case has the same meaning as torque. "mr^2" on the right side is called the moment of mass inertia and is often written as "I." Therefore, "M = Iα" is the equation of motion of rotation, which is similar to "F = ma." The force acting on the joint generates no moment because the radius is 0. As "M = Iα" is not involved, it is only necessary to consider quadriceps muscle strength and lower leg weight (mass × gravitational acceleration [g]), and quadriceps muscle strength can be obtained using the measured angular acceleration (Fig. 20.10).

The force acting on a joint can be calculated using d'Alembert's theorem, which states that multiple forces acting on a single object are balanced, even at the moment of movement. For an object at rest with multiple forces acting on it, the equilibrium equation is "ΣFi = 0." However, the equation of an object in motion is "ΣFi = ma," and moving the right side to the left gives "ΣFi − ma = 0." The system can be considered balanced [7]. The forces acting on the lower leg are gravity, quadriceps muscle force, force due to angular

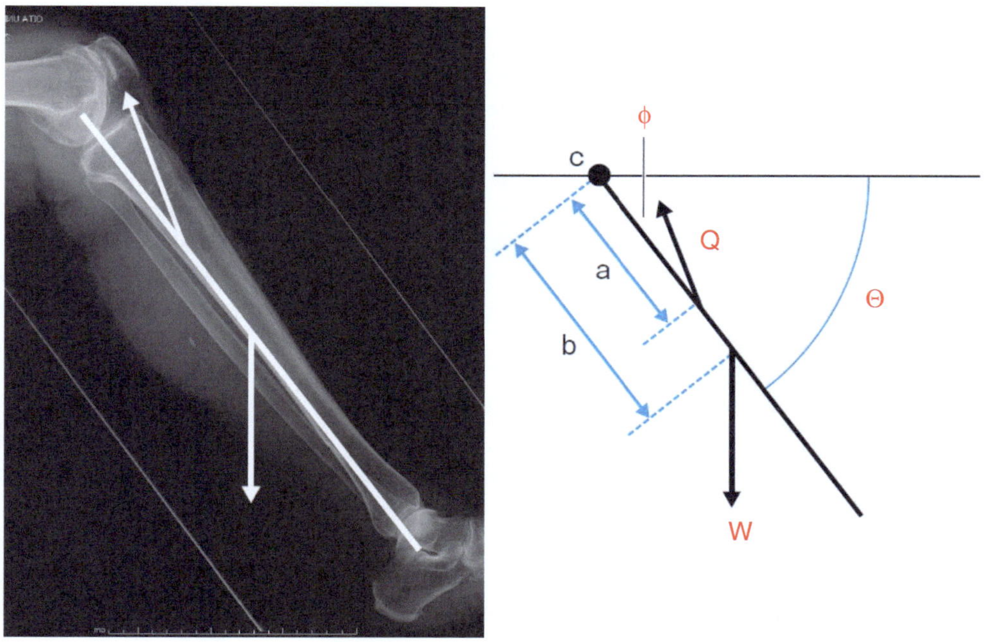

Fig. 20.9 Analysis of the force generated on the quadriceps muscle and knee joint when extending the knee joint in a sitting position. The flexion angle of the knee joint is θ, the angle of the quadriceps muscle relative to the lower leg axis is φ, the distance to the attachment of the quadriceps to the lower leg is a, the distance to the center of gravity of the lower leg is b, and the lower leg weight is W

Quadriceps force and lower leg weight are divided into the lower leg axis (n) and the direction perpendicular to it (t).

- $Q_n = Q\cos\phi$, $Q_t = Q\sin\phi$
- $W_n = Wg\sin\theta$, $W_t = Wg\cos\theta$

Moment of rotation

- $M = aQ_t - bW_t = aQ\sin\phi - bWg\cos\theta$

Angular velocity of rotational motion(α)

Mass moment of inertia (I) = Wb^2

- $aQ\sin\phi - bWg\cos\theta = Wb^2\alpha$
- $Q = \dfrac{Wb^2a + bWg\cos\theta}{a\sin\phi}$

Fig. 20.10 Calculation of quadriceps muscle strength

Fig. 20.11 Calculation of force applied to the knee joint

- According to d'Alembert's theorem, when a force in the opposite direction of the force due to acceleration (-Wbα) is added, the total force in motion is zero.
- Forces occurring at the joint (Fj) are assumed to be those occurring at the fulcrum and divided into the lower leg axis direction and the vertical direction.
- Centrifugal force is $Wb\omega^2$. (ω: angular velocity)
  - $F_{jn} = Q\cos\phi - Wb\omega^2 - Wg\sin\theta$
  - $F_{jt} = Q\sin\phi - Wb\alpha - Wg\cos\theta$
  - $F_j = \sqrt{(F_{jn}^2 + F_{jt}^2)}$

acceleration, force due to angular velocity (ω) (centrifugal force), and the reaction force generated at the joint. Assuming that these are balanced, the reaction force generated at the joint (equivalent to the force acting on the joint) can be obtained. The aforementioned calculation is easier to perform if it develops on the axis of the lower leg and is perpendicular to it than on the X- and Y-axes. This is because the forces due to angular acceleration and angular velocity can be separated (Fig. 20.11).

When the knee joint is extended from 90° to 0° flexion, angular acceleration first occurs in the extension direction. After obtaining the maximum angular velocity, reverse angular acceleration occurs to stop the extension at 0°. The reverse angular acceleration cannot be calculated by this model because it is assumed that the hamstrings are involved instead of the quadriceps. Figure 20.10 shows the calculation while the quadriceps muscle is acting after the beginning of extension. The quadriceps muscle force is approximately 32% of the body weight, and the force acting on the articular surface is 25% of the body weight. Additionally, when the angular acceleration is 0° and the angular velocity is at its maximum, the force acting on the articular surface is 14% of the body weight (Table 20.1). However, the equation shows that the load is larger when moving quickly (angular acceleration is large) than when moving slowly.

2.2 Measurement of the Force Acting on the Tibial Joint Surface of the Femur Using a Sensored Artificial Joint

Previous studies have directly measured the force acting on the knee joint [8]. A pressure sensor, transducer for transmitting signals, and coil for power generation are incorporated

Table 20.1 Calculation based on actual examples

Distance to the attachment of the quadriceps to the lower leg (a): 17 cm
Distance to the center of gravity of the lower leg(b):21:84 cm (52% of length of the lower leg(42 cm)
Patellar tendon angle relative to the lower leg axis (φ)
Lower leg weight:3.84 kg (6.4% of body weight (60 kg))
1. Angular acceleration(α):23.4 rad/s^2
Angular velocity(ω):2.3 rad/s
Knee flexion angle(θ): 60 degrees
Q = 190.3 N(32%BW) Fj = 147.2 N(25%BW)
2. Angular acceleration(α):20.0 rad/s^2
Angular velocity(ω):4.0 rad/s
Knee flexion angle(θ): 45 degrees
Q = 131.8 N(22%BW) Fj = 87.7 N(14%BW)

At the beginning of the movement, a force of 25% of the body weight is generated, and when the angular acceleration becomes 0, a force of approximately 14% of the body weight is generated

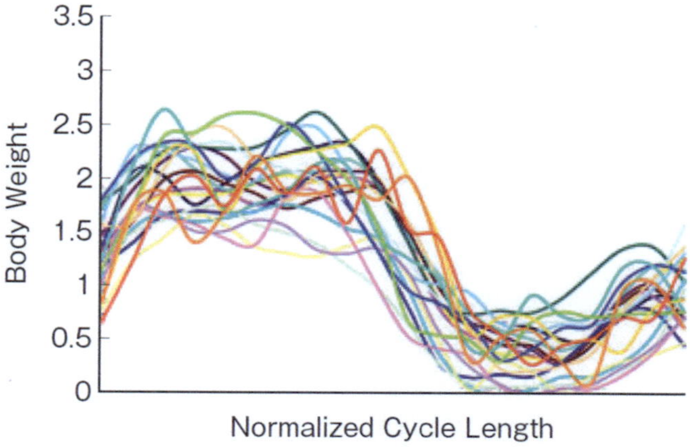

Fig. 20.12 Results of measuring the load during walking with a TKA with a sensor. The maximum load of approximately 2.5 times the body weight is generated. (D'Lima et al. [8])

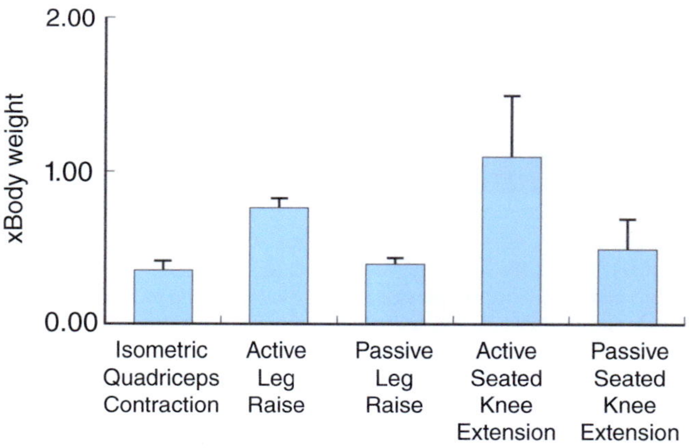

Fig. 20.13 Results of measuring loads during various operations with TKA with sensors. It should be noted that extending the knee joint in the second sitting position from the right generates a load that is more than the body weight, which is larger than the calculated value in Table 20.1. (Cited from D'Lima et al. [8])

into the tibial plate and installed in vivo as an implant that can withstand normal use. The load during ADL and rehabilitation, such as walking, is measured and reported. The results during walking are shown in Fig. 20.12, and the results for the other movements are shown in Fig. 20.13. When walking, a load of 2–2.5 times the body weight is applied at the time of initial contact, and the maximum load value increases as the walking speed increases. A load of more than four times the body weight has been reported for a 5-mile jog. The automatic exercise of sitting in a chair and extending the knees requires a force more than the body weight, and the passive exercise requires approximately half the body weight. In the case of automatic exercise, it is larger than the value calculated from the earlier motion analysis, which is a simultaneous contraction of the hamstrings, an antagonist muscle. This indicates that it is difficult to estimate the exact load value by using the calculation method alone.

2.3 Contact Pressure Distribution on the Tibial Joint Surface of the Femur

Articular surfaces can transmit compressive force but not tension (traction). This is because the two bones that form the joint are easily separated when tension is applied. Therefore, analyzing a model using ordinary elements of the finite element method is impossible, and it is necessary to devise an analysis that includes gap elements. The rigid-body spring model (RBSM) [9] developed for bedrock analysis can be used to analyze such nonlinearities easily. Two-dimensional dissolution has been used to analyze joint contact pressure distribution [10, 11] because the calculation time is extremely short; therefore, it is developed as a simulator for osteotomy around the knee [12]. Figure 20.14 shows the contact pressure distribution between normal alignment (a) and varus alignment (b). In (a), uniform pressure is distributed on the medial and lateral femoral–tibial joint surfaces, whereas in (b), pressure is concentrated on the medial articular surface, tension is generated on the lateral soft tissue, and the lateral articular surface is dissected.

Figure 20.15 shows an example of the RBSM analysis in three dimensions [13]. When a load twice the body weight of 60 kg is applied, the maximum pressure is 1.19 MPa. It is estimated that approximately 1 MPa is the pressure generated at the knee joint.

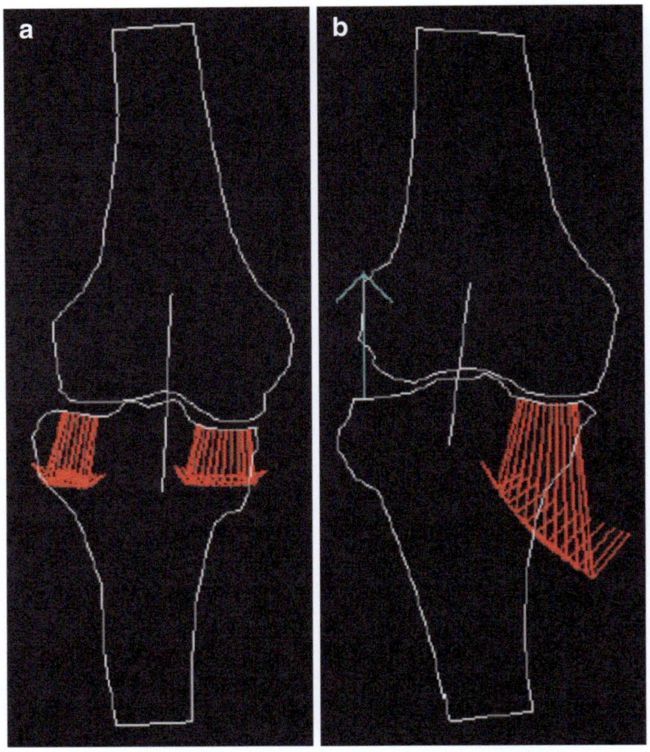

Fig. 20.14 Two-dimensional contact pressure distribution of the knee joint by RBSM. (**a**): Normal alignment. (**b**): Varus alignment. In the right knee joint, uniform pressure is distributed on the medial and outer tibial joint surfaces in A, whereas in B, the pressure is concentrated on the medial articular surface, and tension is generated in the outer soft tissue

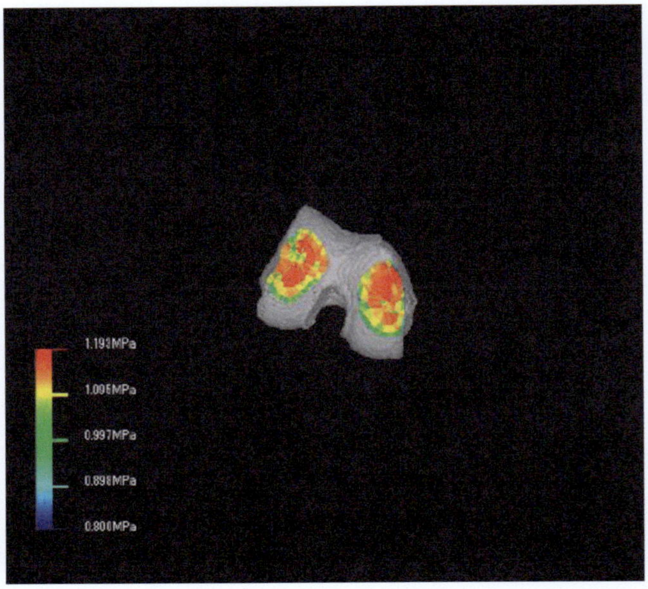

Fig. 20.15 Analysis of contact pressure distribution by 3D RBSM. The maximum pressure was 1.19 MPa when the load was 60 kg

References

1. Götz-Neumann K. Gait analysis by observation (Keiichi Tsukishiro et al., Trans.). Tokyo: Igaku-Shoin Ltd; 2005. p. 5–22 [Japanese].
2. Nishino K, et al. Three-dimensional dynamic analysis of knee joint during gait in medial knee osteoarthritis using loading axis of knee. Gait Posture. 2015;42:127–32.
3. Mine T, et al. Gait oscillation analysis during gait and stair-stepping in elder patients with knee osteoarthritis. J Orthop Surg Res. 2019;14:21.
4. Nagura T, et al. Analysis of the factors that correlate with increased knee adduction moment during gait in the early postoperative period following total knee arthroplasty. Knee. 2017;24:250–7.
5. Ozkaya N. et al. Biomechanics: principles and applications of biomechanics (Yusho Kazu et al., Trans.). Tokyo: NTS; 2001. p. 361–364 [Japanese].
6. Walker PS. Human joints and their artificial replacements. Springfield: Charles C Thomas · Publisher; 1977. p. 55.
7. Hashimoto J. Notes on analytical mechanics. Tokyo: Kodansha Ltd; 2009. p. 32–41. [Japanese]
8. D'Lima DD, et al. 'Lab'-in-a-knee in vivo knee forces, kinematics, and contact analysis. Clin Orthop Relat Res. 2011;469:2953–70.
9. Kawai T, et al. A new element in discrete analysis of plane strain problems. Seisan Kenkyu. 1977;29:204–7.
10. An KN, et al. Pressure distribution on articular surfaces: application to joint stability evaluation. J Biomech. 1990;23:1013–20.
11. Hsu RWW, et al. Normal axial alignment of the lower extremity and load-bearing distribution at the knee. Clin Orthop Relat Res. 1990;255:215–27.
12. Tsumura H, et al. Mechanical analysis of high tibial osteotomy. Joint Surg. 1990;9:635–41. [Japanese]
13. Takeuchi N. Computer extreme analysis series 4: discrete extreme analysis in geodynamics. Tokyo: Baifukan Co., Ltd; 1991. p. 187–2. [Japanese]

Normal and Osteoarthritic Knee Kinematics

21

Satoshi Hamai

1 Normal Knee Kinematics

In a normal knee, the tibia rotates internally with the medial articular surface as the pivot along with knee flexion owing to different articular surface morphologies on the medial and lateral sides of the femur and tibia, cruciate ligament, collateral ligament tension, and physiological lateral laxity. In addition, screw-home movement, a steep change in rotation that occurs near the extension position, has also been reported. Kinematic analysis of normal knees is the key to objectively evaluating pathological and reconstructed knee joints and validating surgery aimed at physiological knee function.

1.1 Analysis of Cadaveric Knee

In a study of the cadaveric knee, Iwaki et al., in collaboration with Freeman et al., reported [1] on sagittal MRI and anatomical examinations that the medial femoral condyle hardly moved outside the tibial articular surface to 110° flexion, while the lateral femoral condyle moved posteriorly by as much as 19 mm from full extension to 120° flexion. Overall, the femur showed a medial pivot motion with the medial articular surface as the pivot with respect to the tibia. Wada et al. used a navigation system in fresh-frozen cadaveric knees to show that the femur rotates externally in the early stage of flexion, internally in the intermediate flexion range, and slowly externally in the late stage of flexion [2, 3]. The center of the femur moved forward in the early stage of flexion and backward in the late stage of flexion, and the surgical epicondylar axis (SEA) showed a medial pivot motion.

S. Hamai (✉)
Department of Orthopedic Surgery, Graduate School of Medical Sciences, Kyushu University, Fukuoka, Japan
e-mail: hamai.satoshi.075@m.kyushu-u.ac.jp

1.2 Analysis Using Open MRI

Following Iwaki et al.'s report [1], Nakagawa et al. reported motion analysis *in vivo* using open MRI in two deep knee flexion positions (from 90° to 133° active flexion and 133° to 162° passive flexion) [4]. The movement of the flexion facet center in the medial condyle of the femur moved 2 mm backward from 90° to 133° flexion, further moved backward by 4.5 mm to the regular sitting position at 162° flexion, and lift-off occurred in which the posterior horn of the medial meniscus impinged on the posterior surface of the femur and separated the articular surfaces. The lateral condyle of the femur moved 28 mm backward from 90° to 162° and maintained only slight contact with the posterior edge of the tibial articular surface, which could be a subluxation position, resulting in internal tibial rotation of 28°.

1.3 Dynamic Analysis Using X-Ray Images

Japanese researchers have reported the dynamic analysis of normal knees using X-ray images. Asano et al. were the first in Japan to report: *in vivo* 3D normal knee dynamics under load conditions using a biplane image-matching technique in 2001 [5]. The coordinate system was the geometric center axis (GCA) of the posterior condyle on the femoral side. Using the anteroposterior axis of the tibial side, the medial condyle of the femur showed forward movement (3.8 mm), the lateral condyle showed backward movement (17.8 mm) in the range of 120° from the full extension, and a rotational motion of 29° was observed. At less than 30°, unlike the dynamics of GCA, posterior femoral rollback was observed at both medial and lateral sides.

Moro-oka et al., in collaboration with Banks et al., found that bone models generated from MR images could be used for single-plane 3D shape registration [6]. CT and MRI imaging were performed on the normal knee, and when each bone model was first prepared and the difference in model surface distance was measured, a difference of 0.18 mm was

S. Matsuda (ed.), *Advances in Total Knee Arthroplasty*, https://doi.org/10.1007/978-981-97-4920-1_21

observed in the femur model and 0.12 mm in the tibia model. Owing to the influence of the distortion of MR images, the accuracy of matching MRI bone models tends to be low; however, they can be used for knee dynamic analysis. Morooka et al. also discussed squats, stair climbing, and kneeling. Three different types of knee flexion movements were analyzed (Fig. 21.1) [7]. The anteroposterior and rotational patterns between the femur and tibia were analyzed by superimposing bone models generated from CT on continuous X-ray images during movements. In addition, the movement of the contact area was analyzed using bone and cartilage models prepared using MRI. The anteroposterior and rotational movements between the femur and tibia differed from the movement of the contact area, and the changes in the contact area decreased. In addition, the anteroposterior and rotational patterns between the femur and tibia were different for each movement, especially with two different maximum flexion movements (squats and kneeling): 20° of maximum rotational difference at the same flexion angle. In other words, it is necessary to recognize that physiological knee joint rotation is not forcibly induced with flexion as a single movement but shows flexible patterns according to the limb position of the foot and the posture of the trunk. Yamazaki et al. also reported 2D/3D registration technique using X-ray fluoroscopy and polygonal 3D bone models generated from CT images [8]. The mean squared error in the plane was within 1 mm and 1° for the femur, tibia, and fibula sides, respectively; however, the error on the patellar side was relatively large. The standing motion from the chair showed internal tibial rotation of 21°, from 2.3° hyperextension to 103° flexion.

In collaboration with Banks and Freeman et al., Mu et al. found that the dynamics during unloaded flexion and extension by stop-motion using open MRI and the dynamics during dynamic loaded squats using 3D-to-2D model-to-image registration methods were shown to be in the same coordinate system [9]. For joint motion analysis, optical tracking systems, including point-cluster methods, are used in addition to MRI and 3D-to-2D model-to-image registration techniques, each of which has its own characteristics in terms of accuracy, load, speed of motion, field of view (FOV), measurability in TKA, and soft tissue delineation. Therefore, selecting the most appropriate method depending on the purpose is preferable. Using the same method, the authors reported the 3D dynamics during deep knee bending [10]. Changes in the contact position of the medial and lateral compartments and internal/external rotation with knee flexion during the lunge motion from 85° to 150° were analyzed (Fig. 21.2). Consistent with the report by Nakagawa et al. using open MRI [4], the contact position of the medial compartment remained in the center even during deep flexion, and the medial meniscus was leveraged to lift the femur, causing a separation between the medial articular surfaces. The contact position of the lateral compartment moved posteriorly and was recognized at the posterior edge of the articular surface. At 150° maximum flexion, the femur exhibited 30° rotation and 5° valgus relative to the tibia. Murakami et al. compared the 3D dynamics during squat and golf swings using image-matching techniques with digitally reconstructed radiography (DRR) generated from CT images (Fig. 21.3) [11]. The rotation of the knee joint during the squat is a medial pivot called the screw home movement near the extension, bicondylar rollback up to 120° with flexion, and a medial pivot after 120°. The rotation of the knee joint during the golf swing (22°) was similar to that during the squat, even in the mild flexion range (15–35°), because of the anterior and posterior movements of the medial and lateral femoral condyles. The center of rotation was centered from the setup to the top, laterally from the top to the impact, and medially from the impact to the finish. It is suggested that the admissibility of rotation in TKA is important not only in the deep flexion range but also in the mild flexion range.

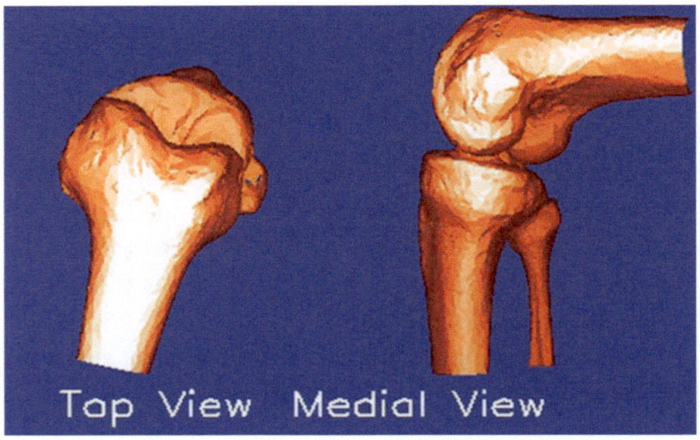

Fig. 21.1 Dynamic analysis of the normal knee during squat, stair climbing, and kneeling motions using the 3D-to-2D model-to-image registration method

Regarding the dynamics of the patella, the authors reported 2D measurements of continuous lateral X-ray images of the knee joint during flexion captured using a flat panel detector (FPD) (Fig. 21.4) [12]. The angle of the patella with respect to the "proximal tibial axis" and the "distal femoral axis" showed a consistent change throughout the ROM with respect to the knee flexion angle during the squat and showed a tendency that the knee flexion angle was always distributed 2:1 to each, respectively. Ishimaru et al. investigated the 3D measurement accuracy in patellar kinematic analysis for normal knees and TKA using image-matching technique with DRR. The DRR used for image correlation is generated by integrating the distance that X-rays pass through the 3D bone model by the voxel value: α, which is an information value such as the bone density. The mean squared error has been reported by Ishimaru et al. to be 0.2 mm and 0.2° with high accuracy [13].

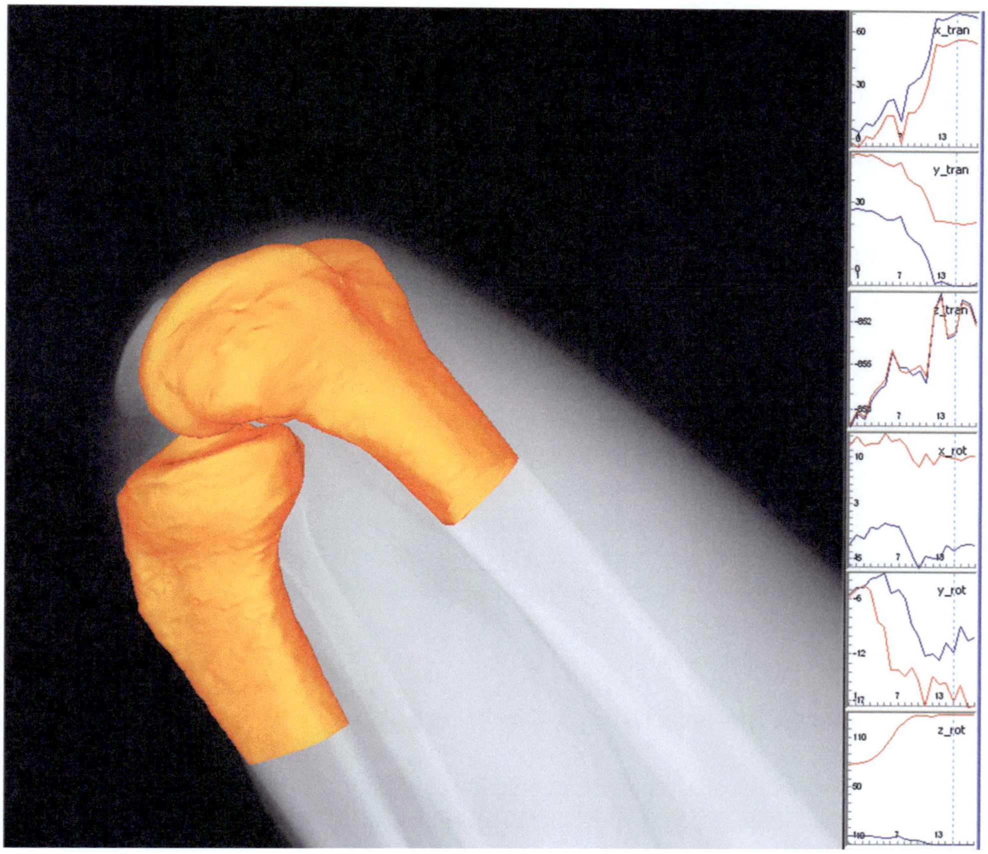

Fig. 21.2 Dynamic analysis of deep flexion of the normal knee using the 3D-to-2D model-to-image registration technique

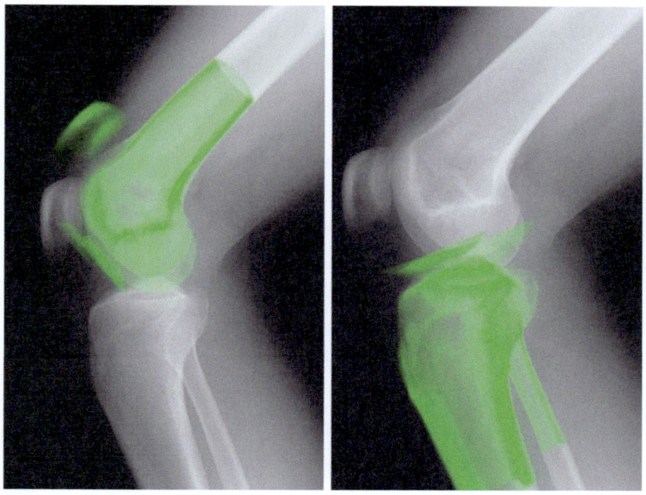

Fig. 21.3 3D dynamic analysis of the normal knee during squat and golf swing motions using digitally reconstructed radiography (DRR) generated from CT images

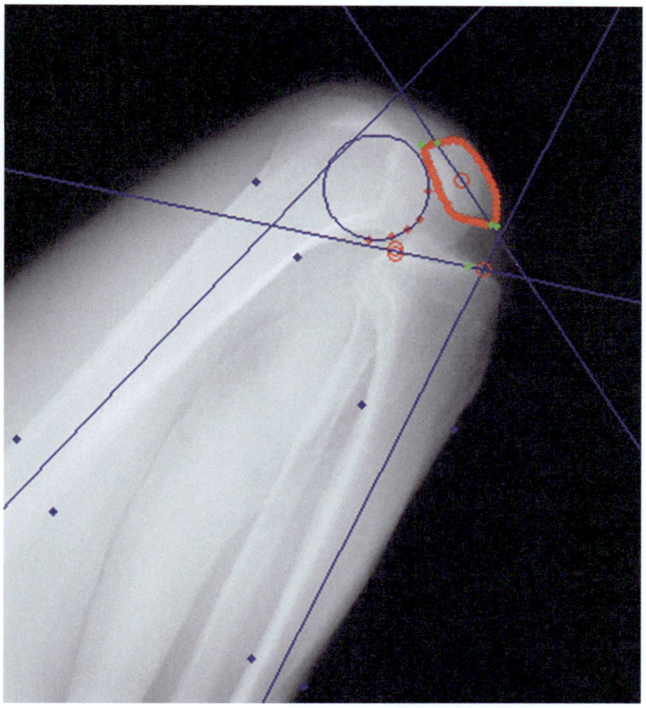

Fig. 21.4 2D dynamic analysis of the patella during knee flexion using a flat panel detector

1.4 Dynamic Analysis Using an Optical Tracking System

Using an optical tracking system, Fukagawa et al. reported the effect of age on the dynamics during deep flexion movement [14]. Elderly people required a significant amount of time for deep flexion movement; the flexion and rotation angles of the knee joint tended to be smaller, and they reported that it was more of an adduction. In addition, hip flexion increased because of the forward tilt of the trunk. When evaluating the function, it is desirable to compare it with an age-matched control group. Kawaji et al. reported that external knee flexion moments in the sagittal plane increased the load on the knee joint in healthy volunteers [15]. They stated that these data are useful when considering effective interventions to improve excessive loading in knee osteoarthritis.

2 Osteoarthritic Knee Kinematics

3D dynamic analysis of osteoarthritic (OA) knees is less frequently reported than that of normal knees and TKA. If one of the goals of TKA is to obtain knee joint function close to that of a normal knee, it is important to understand the dynamics of the OA knee in detail. In the OA knee, the disappearance of screw home movements, a decrease in anteroposterior movement and rotation angle, and a decrease in rotational diversity between movements were observed. Soft tissue contracture and ligament dysfunction have been reported as the causes.

2.1 Dynamic Analysis Using X-Ray Images

Similar to Moro-oka et al.'s report on normal knees [7], the authors analyzed three different movements of squats, stair climbing, and kneeling in knees with OA immediately before TKA (Fig. 21.5) [16]. The extension and flexion angles during squats, stair climbing, and kneeling were 22° → 93°, 82° → 17°, and 92° → 121°, respectively. In the squat, the external rotation of the femur associated with flexion (20° → 80°) was 7° (external rotation 4° → 11°), backward movement was 0 mm medial compartment/1 mm lateral compartment. The internal rotation of the femur associated with extension (80° → 20°) was 7° (9° → 2°) for stair climbing, and forward movement was 0 mm medial compartment/2 mm lateral compartment. Compared with the previously reported dynamics of the normal knee (20–80° flexion) [7], the amount of rotational change was similar in the OA knee; however, the angle of rotation and anteroposterior movement were reduced. Screw home movements in squats were not observed because of flexion contracture. Kneeling with flexion (100–120°) was accompanied by 5° external rotation of the femur (external rotation 10–15°), and the posterior translation was −1 mm medial compartment/5 mm lateral compartment, showing medial pivot motion similar to that of a normal knee. Unlike the normal knee, the flexion and extension angles during movement decreased, the screw home movement disappeared, the rotation angle and amount of anteroposterior movement decreased, and the diversity between the three movements decreased. This is considered to be caused by pathological degeneration of the articulating surfaces and ligaments.

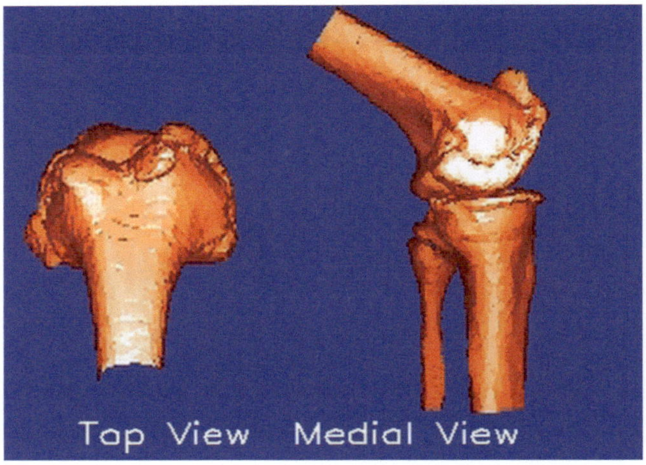

Fig. 21.5 3D dynamic analysis of the OA knee during squat, stair climbing, and kneeling motions using the 3D-to-2D model-to-image registration technique

In an analysis that included mild and moderate OA knees, Ikuta et al. analyzed sitting positions and squat movements in OA knees with Kellgren-Lawrence (KL) classification grades 0-IV using the 3D-to-2D model-to-image registration method [17]. The external rotation and posterior translation of the tibia relative to the femur tended to increase as the disease progressed. The prevention of joint dynamics difference from those of a normal knee might lead to the prevention of stage progression. Matsuki et al. used a similar method to analyze squat and pivot movements in normal knees and knees with KL classification I and II OA knees [18]. Compared to the normal knee, the lateral contact area shifted forward in both squats and pivoting movements in the OA knee. The change in the medial contact area was similar; however, the amount of anteropoterior movement was smaller in knees with OA. These changes in joint dynamics may affect the onset or staging of knee OA. In the active rotational motion of the knee joint, Kiyohara et al. reported twisting behavior in normal knees and KL classification III and IV OA knees by image-matching using DRR [19]. Internal and external rotations of normal knees during twisting (front → ipsilateral twisting → contralateral twisting) were 2.7° external rotation → 11° external rotation → 9.5° internal rotation, and internal and external rotations of the OA knee were 2.2° external rotation → 4.4° external rotation → 2.7° internal rotation. The average rotation was 20° in the normal knee and 7° in the OA knee, respectively. In the normal knees, both internal and external rotations were 11°; however, in the OA knees, the rotations were 4°. Using a similar method, the authors reported the stepping motion of normal knees and knees with KL classification III and IV OA [20]. In the normal knee, the inclination of the tibial functional axis with respect to the ground is 2.3° from the end-of-swinging phase to the stance phase, considering the physiological articular surface inclination (2° to 3°) with respect to the tibial functional axis. The lateral pivot motion reported during walking in the past was not observed as constant pivot motion in either normal or OA knees, but small rotational change was observed. Compared to normal knees, OA knees had (1) decreased extension angle, (2) varus thrust, (3) decreased anteroposterior movement, and (4) a medial deviation of the femur relative to the tibia.

2.2 Dynamic Analysis Using Other Methods

Kawashima et al. used a marker-less volume-based registration method to analyze KL classification III and IV OA knees [21]. Owing to the posterior movement of the medial femoral condyle during knee flexion, no external rotation of the femur was observed. From maximum extension to 90° flexion, the femur showed internal rotation with respect to the tibia. Nagano et al. analyzed the rotational dynamics of normal and OA knees in the standing position using joint ultrasonography [22]. Although a screw home movement of 8.3° was observed in normal knees, the rotation angle decreased, and the flexion and adduction angles increased with disease progression in the OA knee, as reported in the past. Sonoo et al. conducted a systematic review and meta-analysis of reports of knee OA during chair standing motion using motion capture systems, floor reaction force meters, and electromyography [23]. It was reported that the changes in the OA knee were mainly in the sagittal plane, with the trunk of the body showing an anterior tilt and tilting to the healthy side, decreasing the external knee joint flexion moment.

References

1. Iwaki H, et al. Tibiofemoral movement 1: the shapes and relative movements of the femur and tibia in the unloaded cadaver knee. J Bone Joint Surg Br. 2000;82:1189–95.
2. Wada K, et al. Rotational and varus-valgus laxity affects kinematics of the normal knee: a cadaveric study. J Orthop Surg (Hong Kong). 2019;27:2309499019873726.
3. Wada K, et al. Intraoperative analysis of the kinematics of the native knee including two-dimensional translation of the femur using a navigation system: a cadaveric study. J Med Investig. 2019;66:367–71.
4. Nakagawa S, et al. Tibiofemoral movement 3: full flexion in the living knee studied by MRI. J Bone Joint Surg Br. 2000;82:1199–200.
5. Asano T, et al. In vivo three-dimensional knee kinematics using a biplanar image-matching technique. Clin Orthop Relat Res. 2001;388:157–66.
6. Moro-Oka TA, et al. Can magnetic resonance imaging-derived bone models be used for accurate motion measurement with single-plane three-dimensional shape registration? J Orthop Res. 2007;25:867–72.
7. Moro-Oka TA, et al. Dynamic activity dependence of in vivo normal knee kinematics. J Orthop Res. 2008;26:428–34.
8. Yamazaki T, et al. 3D kinematics of normal knee using X-ray fluoroscopy and CT images. IFMBE Proc. 2007;14:2793–6.
9. Mu S, et al. Comparison of static and dynamic knee kinematics during squatting. Clin Biomech (Bristol, Avon). 2011;26:106–8.
10. Hamai S, et al. In vivo healthy knee kinematics during dynamic full flexion. Biomed Res Int. 2013;2013:717546.
11. Murakami K, et al. In vivo kinematics of healthy male knees during squat and golf swing using image-matching techniques. Knee. 2016;23:221–6.
12. Hamai S, et al. Physiological sagittal plane patellar kinematics during dynamic deep knee flexion. Int Orthop. 2013;37:1477–82.
13. Ishimaru M, et al. Three-dimensional motion analysis of the patellar component in total knee arthroplasty by the image matching method using image correlations. J Orthop Res. 2014;32:619–26.
14. Fukagawa S, et al. Age-related changes in kinematics of the knee joint during deep squat. Knee. 2012;19:208–12.
15. Kawaji H, Kojima S. Effect of altered sagittal-plane knee kinematics on loading during the early stance phase of gait. Gait Posture. 2019;74:108–13.
16. Hamai S, et al. Knee kinematics in medial osteoarthritis during in vivo weight-bearing activities. J Orthop Res. 2009;27:1555–61.
17. Ikuta F, et al. Knee kinematics of severe medial knee osteoarthritis showed tibial posterior translation and external rotation: a cross-sectional study. Aging Clin Exp Res. 2020;32:1767–75.
18. Matsuki K, et al. In vivo kinematics of early-stage osteoarthritic knees during pivot and squat activities. Gait Posture. 2017;58:214–9.
19. Kiyohara M, et al. Tibiofemoral kinematics in healthy and osteoarthritic knees during twisting. J Orthop. 2020;21:213–7.
20. Hamai S, et al. In vivo kinematics of healthy and osteoarthritic knees during stepping using density-based image-matching techniques. J Appl Biomech. 2016;32:586–92.
21. Kawashima K, et al. In vivo three-dimensional motion analysis of osteoarthritic knees. Mod Rheumatol. 2013;23:646–52.
22. Nagano Y, et al. Association between in vivo knee kinematics during gait and the severity of knee osteoarthritis. Knee. 2012;19:628–32.
23. Sonoo M, et al. Altered sagittal plane kinematics and kinetics during sit-to-stand in individuals with knee osteoarthritis: a systematic review and meta-analysis. J Biomech. 2019;96:142–6.

22 Kinematics After PS- and CR-Type TKAs

Teruya Ishibashi and Tetsuya Tomita

While the promising long-term results of total knee arthroplasty (TKA) have been well-established, satisfaction is only ~70–80%, which is lower than that of total hip arthroplasty (THA), indicating a need for improvements. Cruciate retaining (CR)-type TKA preserves the posterior cruciate ligament (PCL), while posterior stabilized (PS)-type TKA sacrifices both cruciate ligaments and substitutes for PCL function using a post/cam mechanism. There are two primary types of TKA: Fixed-type TKA, in which the polyethylene insert is fixed to the tibial tray, and Mobile-type TKA, in which wear is reduced by the higher conformity between the femoral component and the polyethylene insert, allowing movement with the polyethylene insert on the tibial tray. Furthermore, several models have been developed to reproduce physiological articular surfaces. In contrast, other models are intended to function with a single radius of femoral sagittal plane alignment. However, it is unclear whether they their actual in vivo function matches these concepts. In the 1990s, a method using X-ray fluoroscopy was developed to evaluate in vivo knee kinematics after TKA [1]. The authors reported a shape-matching method to estimate the three-dimensional position of components by superimposing the contours of the components on captured images and computer-aided design (CAD) models [2–4] (Fig. 22.1). We report the in vivo kinematics after CR-type- and PS-type TKA, with a review of the literature.

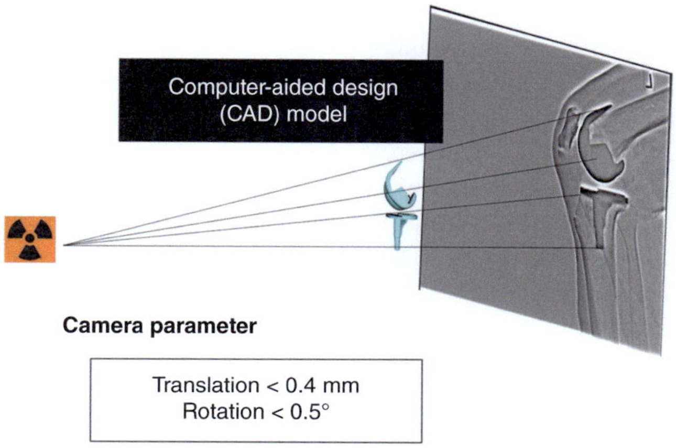

Fig. 22.1 2D/3D registration technique

1 Postoperative Kinematics of TKA

Using X-ray fluoroscopy, if knee flexion in the weight-bearing condition was set as the target motion, the patients were asked to perform the squat motion from maximum extension to maximum flexion. The stair-climbing motion consists of repetitive weight-bearing and non-weight-bearing with knee flexion and extension. Four phases were analyzed: foot strike (non-weight-bearing), stance phase before the cross leg (weight-bearing), stance phase after the cross leg (weight-bearing), and foot off (non-weight-bearing).

The three-dimensional (3D) position of the TKA components can be estimated (2D/3D registration technique) using camera parameters, CAD models of the TKA components, and fluoroscopic images (2D) obtained during the squatting motion (Fig. 22.1).

The relative posture-estimation accuracy of this analysis system is 0.5° rotation and 0.4 mm translation, which is a highly accurate method to evaluate intra-articular kinematics. The relative motion of each component is expressed by six degrees of freedom, comprising rotations and translations around the three axes [5]. These include flexion/extension,

T. Ishibashi (✉)
Department of Orthopaedic Biomaterial Science, Osaka University Graduate School of Medicine, Osaka, Japan

T. Tomita
Graduate School of Health Sciences, Morinomiya University of Medical Sciences, Osaka, Japan
e-mail: tetsuya_tomita@morinomiya-u.ac.jp

S. Matsuda (ed.), *Advances in Total Knee Arthroplasty*, https://doi.org/10.1007/978-981-97-4920-1_22

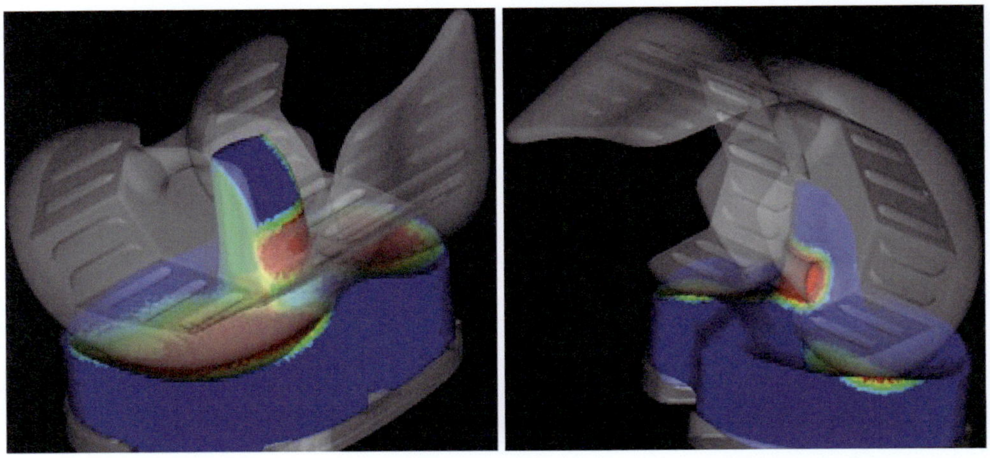

Fig. 22.2 Post/cam engagement (PCE). Left: Anterior PCE. Right: Posterior PCE

internal/external rotation, and varus/valgus in rotation, and anterior/posterior, medial/lateral, and proximal/distal in translation. In general, joint movement is described as proximal-to-distal, that is, the movement of the tibia relative to the femur. Most orthopedic surgeons view TKA as a movement of the femoral component relative to the tibial polyethylene insert. Therefore, the in vivo kinematics of TKA are expressed as the motion of the femoral component relative to the tibial component. Although the polyethylene insert is not extracted radiographically, it is possible to represent the contact point between the femoral component and the polyethylene insert using a known CAD model, as if it were fixed on the tibial tray in Fixed TKA (in Mobile TKA, special procedures are required, such as the installation of tantalum beads in the insert). In our studies, we visualized post-cam engagement by showing a color map of the proximity points on the polyethylene insert [4] (Fig. 22.2).

2 CR-Type TKA

The CR-type TKA involves resection of the ACL, allowing preservation of the PCL, and inducing posterior stability and posterior movement of both femoral condyles during knee flexion (bicondylar rollback). CR-type TKA has further been reported to have better long-term durability than PS-type TKA because of the reduction in stress at the interface between the bone and the implant by sharing the load transfer of the PCL. However, the function of the preserved PCL depends on the conditions of the PCL, such as the flexion/extension gap and implant shape. Therefore, abnormal kinematics such as paradoxical anterior motion [6] and reversed rotation [7] which differ from normal knee kinematics, have been reported. According to an in silico simulation [8], the residual rate of the PCL tibial attachment decreased as the posterior slope increased in proximal tibial osteotomy. To preserve the PCL, additional procedures, such as the island technique, should be performed.

2.1 Deep Knee Bending on Weight-Bearing

The Scorpio NRG CR (Stryker), a relatively unrestricted implant with rotational tolerance for the single radius-shaped femoral component, exhibits external rotation of the femoral component with knee flexion; however, the amount of external rotation is less than 10°. The medial femoral condyle translates anteriorly from maximum extension to 70° and posteriorly from 70° to maximum flexion. The lateral femoral condyle is translated posteriorly from maximum extension to maximum knee flexion. The kinematic pattern of this implant exhibited a central pivot from maximum extension to 70° and bicondylar rollback from 70° to maximum flexion (Fig. 22.3). In contrast, the FINE CR (Teijin Nakashima Medical), which mimics the articular surface shape and has an asymmetrical femur and tibial insert shape, also exhibits external rotation of the femur, although the amount of rotation is less than 10°. The medial femoral condyle translated posteriorly from maximum extension to 40°, anteriorly from 40° to 100°, and posteriorly from 100° to maximum flexion. The lateral femoral condyle translates posteriorly from the maximum extension to 40°, remains steady, and translates posteriorly after 100 °. The kinematic pattern of this implant exhibited a medial pivot, paradoxical anterior motion, and bicondylar rollback. The two CR-type TKA exhibited similar degrees of external rotation (approximately 10°). The kinematics were different from maximum flexion to mid-flexion because of the influence of the implant design, such as articular surface shape and restraint. However, during deep knee flexion, both implants demonstrated bicondylar rollback, suggesting a good PCL function.

2.2 Deep Knee Bending on Non-weight-Bearing

The Scorpio NRG CR (Styker) system allows external rotation of the femoral component in the same manner as under

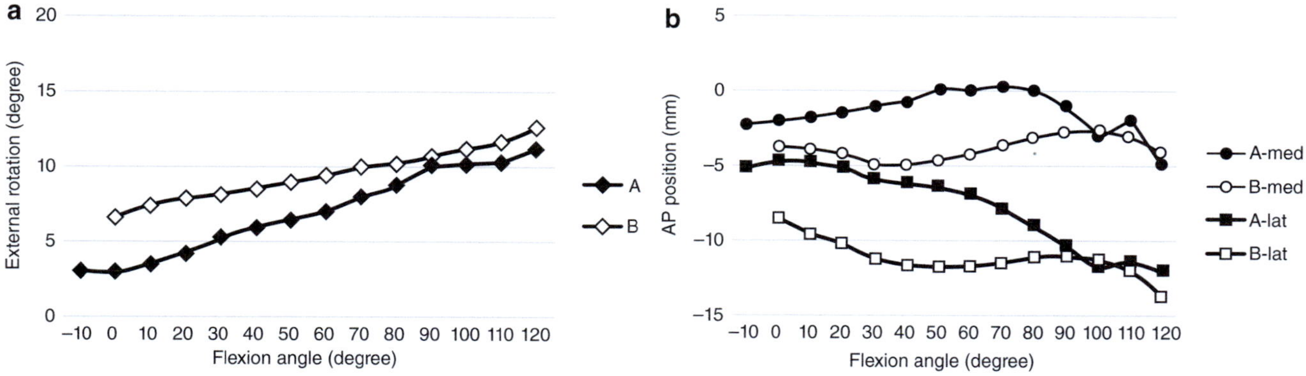

Fig. 22.3 Kinematics of CR-type TKA during weight-bearing conditions. (**a**): Stryker Scorpio NRG CR. (**b**): Teijin Nakashima Medical FINE CR

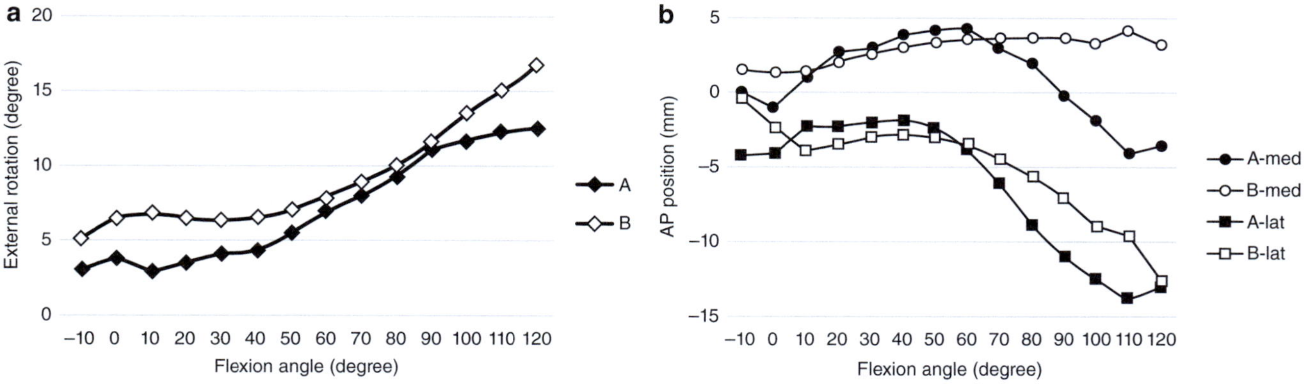

Fig. 22.4 Kinematics of CR-type TKA under non-weight-bearing conditions. (**a**): Stryker Scorpio NRG CR. (**b**): Smith & Nephew JOURNEY II CR

weight-bearing conditions, but with a rotation amount of 10° or less. The medial femoral condyle translated anteriorly to 60° from the maximum extension and then posteriorly. The lateral femoral condyle translated anteriorly to 40° from the maximum extension and then posteriorly. Paradoxical anterior motion was observed from maximum extension to 70°, and after 40°, a bicondylar rollback pattern was observed (Fig. 22.4). The external rotation of the femoral component was approximately 10°in Journey II CR (Smith and Nephew). The medial femoral condyle translated slightly forward from maximum extension to maximum flexion, whereas the lateral femoral condyle moved posteriorly from maximum extension to maximum flexion. The kinematic pattern of this implant exhibited a medial pivot, paradoxical anterior motion, and bicondylar rollback (Fig. 22.4). External rotation of the femur associated with flexion was observed even under non-weight-bearing conditions. Compared to the weight-bearing condition, the anteroposterior translation was larger in mid-flexion in models with low constraints, while the medial femoral condyle exhibited a smaller translation in models with relatively constrained articular surfaces. However, both models exhibited paradoxical anterior motion at mid-flexion, and subsequently exhibited bicondylar rollback.

2.3 Stair Climbing

In FINE CR (Teijin Nakashima Medical), the ascending motion ranging from 62° to 12° knee flexion demonstrated a posterior motion of 4.3 mm. The descending motion ranged from 9° to 36° with almost no anteroposterior shift of the medial and lateral proximal points (Fig. 22.5).

3 PS-Type TKA (Including BCS-Type TKA)

PS-type TKA is commonly used; this technique is designed to resect both cruciate ligaments and induce bicondylar rollback in which the femoral component moves backward with flexion by posteriorly contacting the post on the tibial insert with the cam of the femoral component during post-cam engagement. Post-cam engagement is highly reproducible, and the induction of bicondylar rollback enables deep flexion. This is believed to contribute to the larger maximum flexion angle than in PS-type than CR-type TKA.

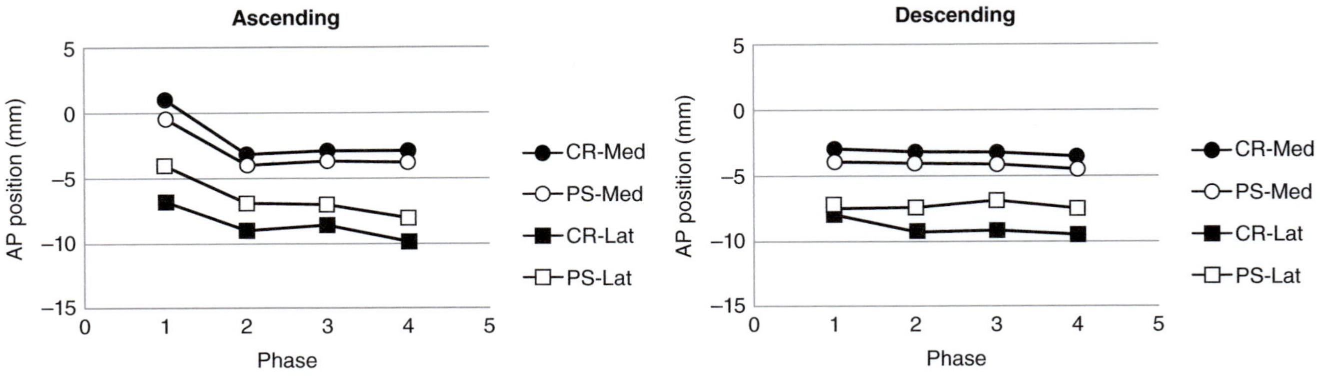

Fig. 22.5 Kinematics of CR-type and PS-type TKA during stair climbing. Left: Stair ascending. Right: Stair descending

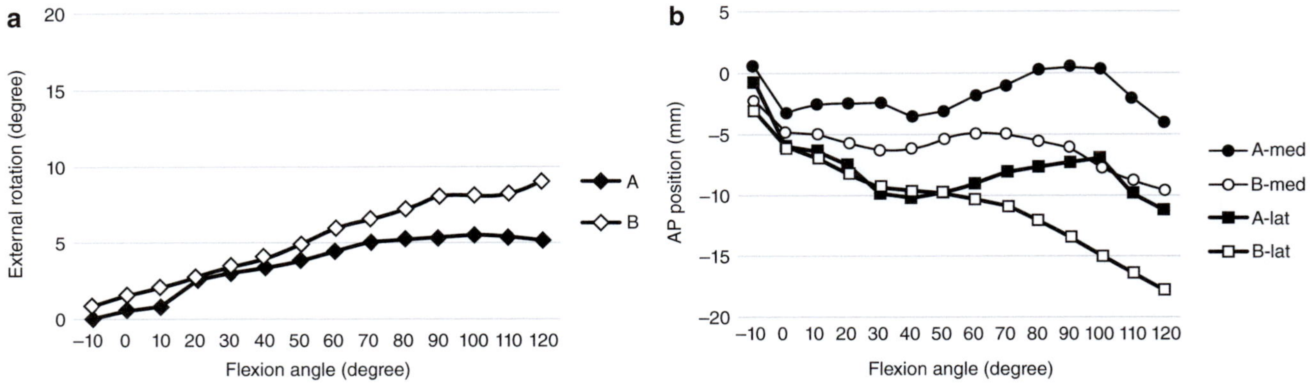

Fig. 22.6 Kinematics of PS-type TKA under weight-bearing conditions. (**a**):Zimmer-Biomet Nexgen LPS-flex. (**b**) Smith & Nephew LEGION PS

3.1 Deep Knee Bending on Weight-Bearing

The Nexgen (Zimmer Biomet) is an implant with a relatively unrestricted tibial insert on a multiradius-shaped femoral component. In LPS-flex, external rotation of the femoral component is allowed with flexion; however, the rotation amount is approximately 5°. The medial femoral condyle moves posteriorly from maximum extension to 40°, anteriorly to 90°, and then posteriorly again. The lateral femoral condyle moves posteriorly from the maximum extension to 40° and then anteriorly to 90°. These movements included bicondylar rollback from maximum extension to 40°, paradoxical anterior motion from 40° to 90°, and bicondylar rollback from 90° (Fig. 22.6). In contrast, LEGION PS (Smith & Nephew), a model with an asymmetrical femur and a tibial insert shape, also exhibits external rotation of the femur, but with a rotational amount of less than 10°. In this system, the lateral femoral condyle moved posteriorly from maximum extension to maximum flexion. The above results show a medial pivot pattern up to 40°, paradoxical anterior motion from 40° to 60°, and then a bicondylar rollback (Fig. 22.6). Similar to CR TKA, the kinematics at maximum extension to mid-flexion differ in implant design and exhibit slight anterior motion; however, once engaged, they exhibit a highly reproducible bicondylar rollback. The starting angle of the rollback differs depending on the assumed angle of engagement.

3.2 Deep Knee Bending on Non-weight-Bearing

The Nexgen LPS-flex (Zimmer Biomet) allows external rotation of the femoral component in the same manner as under weight-bearing conditions, but with a rotation of 10° or less. The medial femoral condyle moves anteriorly from the maximum extension to 60°, and then posteriorly. The lateral femoral condyle moves posteriorly from the maximum extension to 10°, anteriorly to 60°, and then posteriorly. The medial pivot pattern was present from maximum extension to 10°, paradoxical anterior motion up to 60°, and the bicondylar rollback pattern after 60°(Fig. 22.7). In contrast, JOURNEY II BCS (Smith & Nephew), which is designed for anterior post and cam engagement, also has an external rotation of the femoral component of less than 10°. The medial femoral condyle moves slightly anteriorly from max-

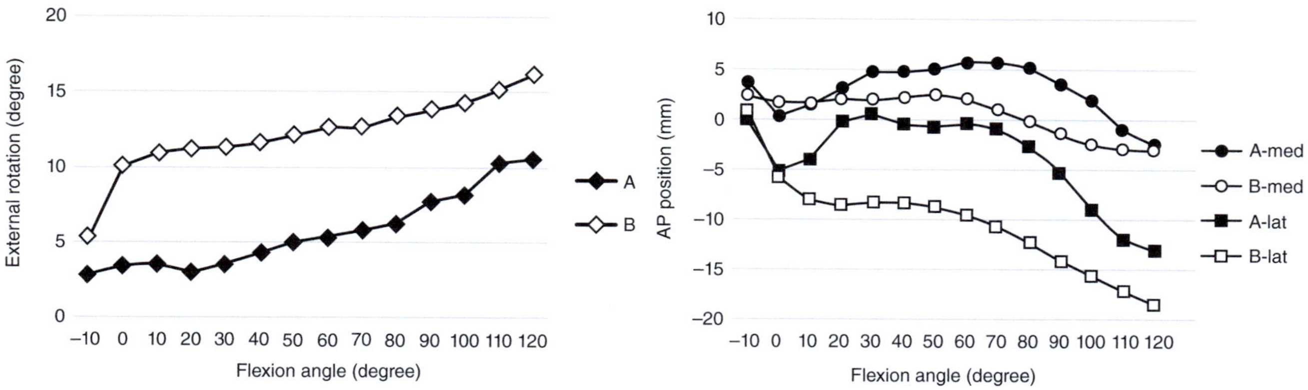

Fig. 22.7 Kinematics of PS-type TKA under non-weight-bearing conditions. (**a**): Zimmer-Biomet Nexgen LPS-flex. (**b**):Smith & Nephew JOURNEY II BCS

imum flexion to 60°, and then posteriorly. The lateral femoral condyle moves posteriorly with the knees flexed. It exhibited a medial pivot pattern at 60° from the maximum extension and a bicondylar rollback at 60° (Fig. 22.7). Similar to the CR type, the external rotations of the PS and BCS types are approximately 10°. Anterior movement in mid-flexion was observed under the non-weight-bearing condition, but changed to bicondylar rollback owing to the subsequent occurrence of post/cam engagement.

3.3 Stair Climbing

In the FINE PS (Teijin Nakashima Medical), the stair-ascending motion ranged from 61° to 11°, the femoral condyle translated posteriorly by 3.5 mm, with a transition from weight-bearing from non-weight-bearing, and no movement was observed thereafter.

4 Comparison Between PS-Type and CR-Type TKA

Although the external rotation of the femoral component with knee flexion is observed in both PS- and CR-type TKA, the rotation amount is approximately 10°, indicating that the external rotation following TKA is smaller than that of a normal knee by 20° or more [9]. In a normal knee, screw home movement occurs in the early stages of flexion [10], but this external rotation movement is not observed in PS-type- and CR-type TKA, which are considered to decrease the rotations. A common feature of both PS- and CR-type TKA is that the anterior cruciate ligament (ACL) is resected, suggesting that the ACL may induce screw home movement. BCS-type TKA is based on the concept of substituting the ACL function with anterior post/cam engagement. However, compared with the conventional PS-type TKA [11] and CR-type with the same joint surface shape [12], no screw home movement was observed, and there was no difference in the amount of rotation. In other words, the rotational pattern may not function as a concept. In the anteroposterior position, the femoral component is positioned posterior to the origin of the tibial tray in knee extension after both PS-type and CR-type TKA, and this phenomenon is also thought to be an effect of ACL deficiency. In BCS-type TKA, anterior post–cam engagement resulted in a significantly higher anterior position than in conventional PS-type TKA [11]. At the mid-flexion angle, both PS- and CR-type TKA exhibits variety of patterns, such as medial pivot, lateral pivot, and central pivot, which is dependent on the radius of curvature (single or multiple radii) in the femoral sagittal plane rather than on both cruciate functions. It has further been reported that the shape and conformity of the polyethylene insert between the femoral component and the tibial insert vary greatly depending on the weight bearing conditions [13]. During deep knee flexion, PS-type TKA frequently induces bicondylar rollback via a post/cam mechanism. In CR-type TKA, the femur may roll back in some cases and exhibit paradoxical anterior motion in other cases [14]. In other words, the surgical technique, in addition to the PCL condition, is greatly affected. If the flexion gap is large, it is more affected under non-weight-bearing conditions, and rollback may be delayed or not observed in some cases. Similarly, when the posterior tibial slope increases, the femoral component is already located posteriorly in mid-flexion and moves anteriorly in flexion [12]. However, even if the posterior tibial slope is increased, an effective rollback can be observed in models with high insert conformity, in which the maximum flexion angle increases [15], and it can be said that it is possible to acquire rollback in deep knee flexion even in CR-type TKA if appropriate surgical techniques and model selection are performed.

5 Deep Knee Flexion Kinematics

In Japan, there is a tendency for patients to engage in deep knee flexion activities, such as sitting on the floor (seiza), even following TKA. Thus, to improve patient satisfaction after TKA, it is important to perform surgery that increases the postoperative flexion angle [16]. Important factors related to the postoperative flexion angle include alignment, ligament balance, accurate osteotomies, implant selection, such as CR-type- and PS-type TKA, and even implant design [17]. In particular, designs that mimic physiological joint surface geometry have been developed in recent years and are expected to achieve deep knee flexion by reproducing normal knee kinematics, such as the anterior femoral position in extension, screw home movement in extension, medial pivot, and bicondylar rollback. JOURNEY II (Smith & Nephew) has a built-in internal inclination of 3° in the coronal section, and both the femur and tibia have an asymmetrical form; the insert shape is concave, with the deepest part slightly forward from the center on the medial side, and the lateral shape slightly convex. CR- and BCS-type TKA are both currently available. When comparing the kinematics of these squat movements, they have the same physiological articular surface shape, but their kinematics are quite different [12]. Even in cases where the CR-type TKA was able to flex more than 120°, the average maximum flexion angle of the CR type was 107.4°, whereas the BCS-type TKA had a large difference of 124.1°. This was correlated with the rollback quantity; that is, under weight-bearing conditions, even models with physiological articular surface shapes could not necessarily reproduce the kinematics as designed in CR-type- and PS-type TKA.

The postoperative kinematics of TKA were reviewed as previously described. Among both the PS and CR types, postoperative kinematics differ significantly owing to differences in implant shape, and the kinematics do not necessarily match the concept. Currently, the goal of TKA is to "reproduce the kinematics of the normal knee", but most of the patient on whom we perform TKA already have ACL-deficient and/or PCL-deficient knees with bone morphological deformity. In the future, definition of the kinematics of individual patients may also need to be considered.

References

1. Banks SA, et al. Accurate measurement of three-dimensional knee replacement kinematics using single-plane fluoroscopy. IEEE Trans Biomed Eng. 1996;43:638–49.
2. Yamazaki T, et al. Improvement of depth position in 2-D/3-D registration of knee implants using single-plane fluoroscopy. IEEE Trans Med Imaging. 2004;23:602–12.
3. Watanabe T, et al. In vivo kinematics of mobile-bearing knee arthroplasty in deep knee bending motion. J Orthop Res. 2004;22:1044–9.
4. Yamazaki T, et al. Visualization of femorotibial contact in total knee arthroplasty using X-ray fluoroscopy. Eur J Radiol. 2005;53:84–9.
5. Grood ES, et al. A joint coordinate system for the clinical description of three-dimensional motions: application to the knee. J Biomech Eng. 1983;105:136–44.
6. Stiehl JB, et al. Fluoroscopic analysis of kinematics after posterior-cruciate-retaining knee arthroplasty. J Bone Joint Surg Br. 1995;77:884–9.
7. Dennis DA, et al. A Multicenter analysis of axial femorotibial rotation after total knee arthroplasty. Clin Orthop Relat Res. 2004;428:180–9.
8. Inou Y, et al. What kinds of posterior cruciate ligament bundles are preserved in cruciate-retaining total knee arthroplasty? A three-dimensional morphology study. J Knee Surg. 2019;32:989–94.
9. Kono K, et al. In vivo three-dimensional kinematics of normal knees during different high-flexion activities. Bone Joint J. 2018;100:50–5.
10. Moglo KE, et al. Cruciate coupling and screw-home mechanism in passive knee joint during extension-flexion. J Biomech. 2005;38:1075–83.
11. Ishibashi T, et al. Kinematics of bicruciate and posterior stabilized total knee arthroplasty during deep knee flexion and stair climbing. J Orthop Res. 2021;39:1262–70.
12. Ishibashi T, et al. Kinematics of bicruciate stabilized and cruciate retaining total knee arthroplasty. J Orthop Res. 2022;40:1547–54.
13. Horiuchi H, et al. In vivo kinematic analysis of cruciate-retaining total knee arthroplasty during weight-bearing and non-weight-bearing deep knee bending. J Arthroplast. 2012;27:1196–202.
14. Bertin KC, et al. In vivo determination of posterior femoral rollback for subjects having a Nexgen posterior cruciate-retaining total knee arthroplasty. J Arthroplast. 2002;17:1040–8.
15. Fujito T, et al. Influence of posterior tibial slope on kinematics after cruciate-retaining total knee arthroplasty. J Arthroplast. 2018;33:3778–82.
16. Matsuda S, et al. Postoperative alignment and ROM affect patient satisfaction after TKA. Clin Orthop Relat Res. 2013;471:127–33.
17. Victor J, et al. Kinematics of posterior cruciate ligament-retaining and -substituting total knee arthroplasty: a prospective randomised outcome study. J Bone Joint Surg Br. 2005;87:646–55.

Kinematics After TKA -BCR

23

Keizo Wada

1 BCR-TKA

BCR-TKA is an operation that aims to acquire stability and kinematics close to a healthy knee by preserving both cruciate ligaments in addition to the MCL and LCL. In this section, summarizes the kinematics studies focusing on the following two BCR implants which have been used in recent years (Fig. 23.1): Vanguard XP (Zimmer-Biomet) with relatively flat inserts and medial and lateral symmetrical designs and Journey II XR (Smith & Nephew), which employs anatomical inserts and asymmetrical designs.

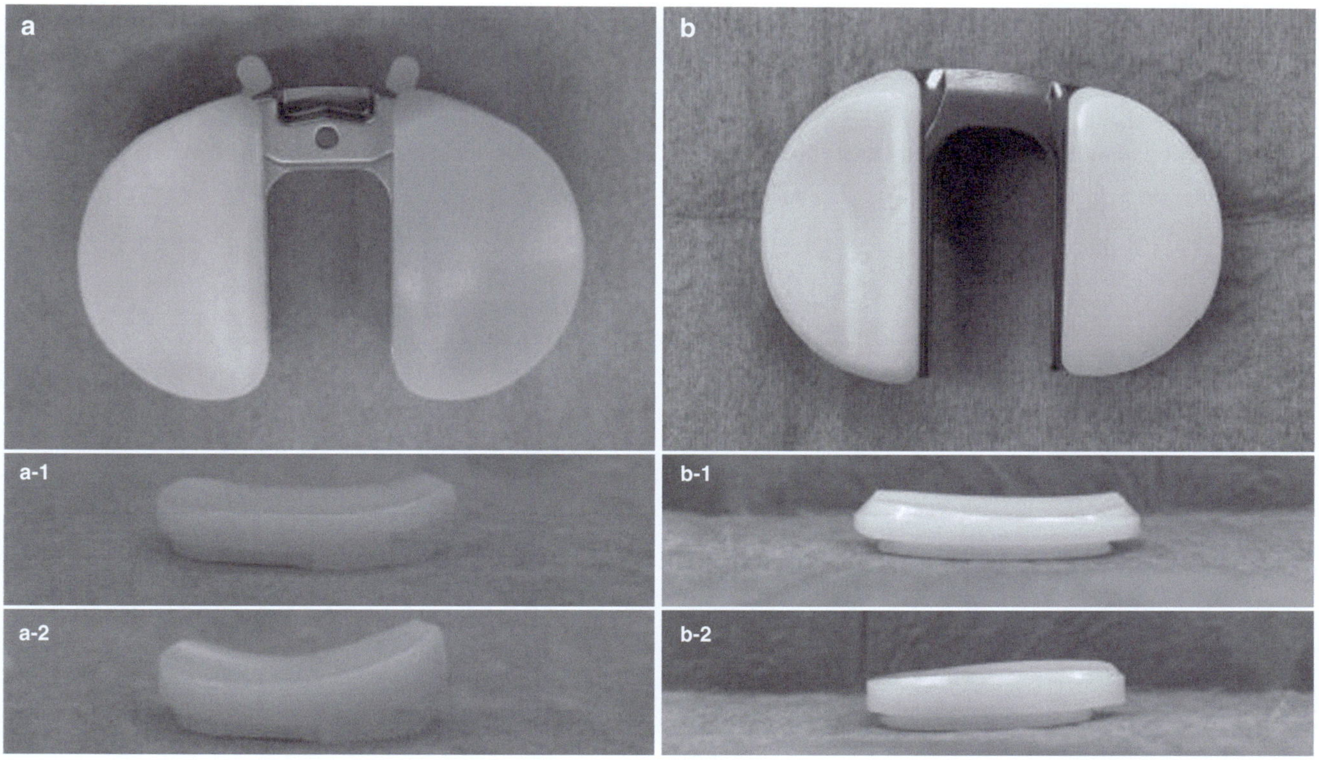

Fig. 23.1 Tibial insert formation of BCR-TKA. (**a**): Vanguard XP, a-1: standard insert, a-2: AS (anterior stabilized) insert. (**b**): Journey II XR, b-1: medial insert, b-2: lateral insert

K. Wada (✉)
Department of Orthopedics, Tokushima University, Tokushima, Japan
e-mail: wada.keizou@tokushima-u.ac.jp

S. Matsuda (ed.), *Advances in Total Knee Arthroplasty*, https://doi.org/10.1007/978-981-97-4920-1_23

2 Kinematics Analysis of BCR-TKA

2.1 In Vitro Study

Heyse et al. evaluated kinematics before and after BCR (Vanguard XP) using the cadaveric knees with a device with 6 degrees of freedom under unloading conditions, as well as under squatting conditions with simulated quadriceps and hamstring forces. As a result, the BCR showed similar kinematics to the healthy knee under unloaded conditions, and a screw-home motion in which the tibia rotates internally against the femur was observed at the beginning of flexion. However, under loading conditions, the BCR had less tibial internal rotation in mid-flexion, slightly more valgus alignment, and abnormal femoral rollback, in which the femoral condyle normally moves posteriorly with flexion. Although the overall difference in kinematics was small, the authors discussed that the difference under load was attributed to the loss of anatomy of the meniscus and tibial articular surfaces [1]. The same research group also reported that the thickness of the insert used had little effect on the kinematics of the BCR [2].

In recent years, Okada et al. used a robotic system with six degrees of freedom to BCR (Vanguard XP) and the state in which the ACL was removed after BCR [3]. As a result, the BCR showed the same femoral rollback and femoral external rotation motion as in the healthy knee, but the femur was located more forward, and more externally rotated than before surgery. In addition, the femur became more posterior in the early stage of flexion after ACL resection, and the femoral external rotation were shown during flexion. Furthermore, the anterior-posterior movement of the joint and the ACL in-situ force were evaluated. As a result, the anteroposterior laxity was not different between a healthy knee and a BCR, but it doubled after ACL resection. The anteroposterior laxities of the BCR and intact knees were comparable at all flexion angles and increased two-fold or more after ACL resection. The ACL in situ force in the BCR knees was two-fold to six-fold higher than that in the intact knees during knee flexion and equivalent to that in the intact knees under anterior loading.

On the other hand, our research group performed kinematics analysis under unloading with the whole-body cadaveric knees using a navigation system and compared the BCR (Vanguard XP) with healthy knees. When we analyzed the results of each surgical protocol from a normal knee to meniscectomy, replacement of only the femoral side, BCR, BCR with ACL resected, and CR, we reported that the kinematics of the normal knee changed little until only the femoral side was replaced, but when the tibial side was replaced, screw-home motion was lost in the early stages of knee flexion [4] (Fig. 23.2). From these results, it was considered that the articular surface shape of the tibia is important for maintaining kinematics. When the medial

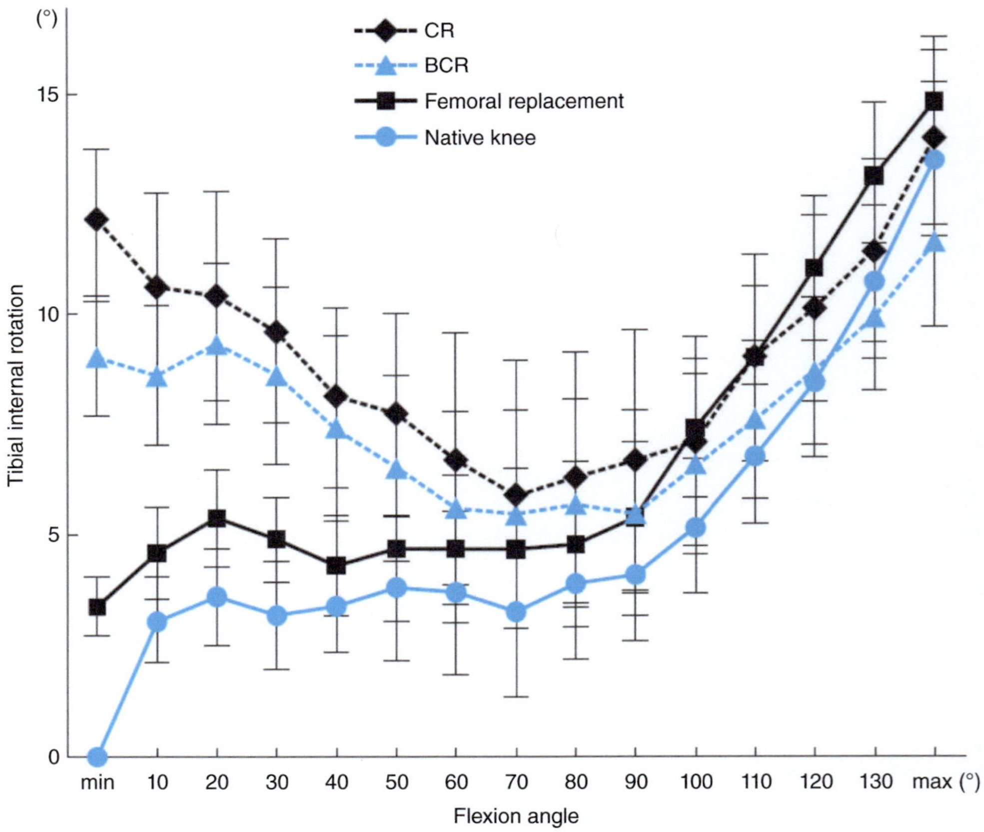

Fig. 23.2 Kinematics changes in TKA. Relative tibial internal rotation angle of the native knee and knees in which femoral replacement, BCR TKA (Vanguard XP), or CR TKA (Vanguard CR) was performed. To maintain consistency of the positional relationship between the femur and tibia, the angle between the surgical epicondylar axis and the line perpendicular to the tibial AP axis at maximum extension of the native knee was defined as the zero position. A positive value indicates relative tibial internal rotation towards the femur. (Hamada et al. [4]. Reproduce with permission)

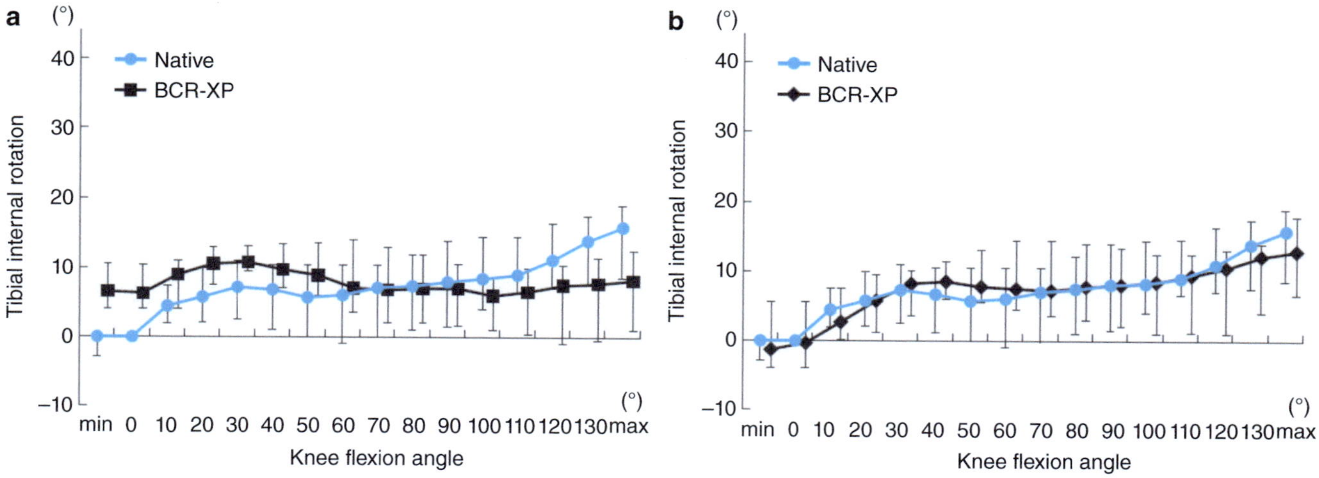

Fig. 23.3 Effect of tibial insert formation for kinematics in BCR-TKA (Vanguard XP). (**a**): Internal rotation of the tibia during knee flexion in the native knee and BCR-XP. The angle of the tibia at extension in the normal knee was set at 0°. (**b**): Internal rotation of the tibia during knee flexion in the native knee and BCR-AS. The angle of the tibia at extension in the normal knee was set at 0°. (Wada et al. [5]. Reproduce with permission)

insert was changed to a constrained (AS; anterior stabilized) one, tibial rotation showed similar to that of a healthy knee [5] (Fig. 23.3). Furthermore, kinematics and joint laxities were examined in BCR and healthy knees. In BCR, we reported that tibial internal rotation during flexion was decreased in cases with less lateral laxity in 90° flexion. In these BCRs with low lateral laxity, changing the lateral insert to a 3° posterior sloped increased the tibial internal rotation, and we considered that a certain degree of lateral laxity is also an important factor for the reproduction of kinematics in healthy knees [6].

However, there are still few studies on Journey II. XR, Arnout et al. conducted studies using fresh frozen cadaveric knees and a device with six degrees of freedom and reported that BCR showed joint laxity almost same as that of a healthy knee, and from this result, it is expected to maintain healthy knee joint function [7].

2.2 In Vivo Study

Simon et al. used motion capture technology and gait analysis using a floor reaction force meter to compare CR (Vanguard CR) to BCR (Vanguard XP). At the same time, electromyography of the lower limb muscles was measured, and although there was no difference between models during flat walking, it was reported that muscle activity of the vastus medialis and rectus femoris muscles in BCR was significantly reduced during downhill walking, and it was considered that the preserved ACL contributed to the stability of the knee joint [8].

Arauz et al. used image-matching technique to analyze the treadmill gait of patients who underwent unilateral BCR (Vanguard XP) and found significantly greater flexion and internal rotation in the stance phase and significantly less adduction compared to the healthy side. Therefore, it was concluded that unilateral BCR cases had an asymmetric gait [9].

The group also conducted an analysis focusing on the contact point of the tibial joint surface compared to the healthy knee and reported that BCR had a smaller rotational range throughout the walking cycle than the healthy knees, that the contact point of the medial condyle of the femur showed a longer anteroposterior movement than the lateral condyle, that lateral pivot rotational motion was observed during the standing phase and that the healthy knee kinematics were not reproduced [10]. It was also reported that the BCR of the lateral femur was located significantly anteriorly compared to the nonoperative side and that the femoral rollback was less, and the external rotation movement of the femur was less than that of the healthy side. In this study, it was concluded that BCR showed kinematics different from those of healthy knees [11, 12].

Smith et al. analyzed kinematics of both BCR (Journey II. XR) and CR (JourneyII. CR Smith & Nephew) during deep knee flexion using image matching technology and compared it with a healthy control group. As a result, kinematics of BCR were more similar to healthy knees than CR in the early stage of flexion, and various patterns in mid-flexion, and no difference in kinematics was observed between BCR, CR, and healthy knee during deep flexion. The authors stated that function of ACL was affected in the extension position, the function of PCL was affected in the deep flexion position, and both the change in articular surface shape and the ACL in the intermediate flexion position were considered to affect kinematics [13].

Kono et al. analyzed the squat and cross-legged sitting motion of BCR (Journey II. XR) using image matching technique and found that BCR exhibits medial-pivot motion from maximum extension to 10° flexion in squat motion and 60° flexion in cross-legged motion. The authors reported that similar rotational motion was shown at the maximum flexion position and concluded that BCR incorporating anatomical features relatively reproduced the kinematics of healthy knee joints [14]. However, the same study group found that UKA (Oxford medial: Zimmer-Bimet) was reproduced healthy knee kinematics better than BCR. In the study, comparing the squat movements of the normal control, the external rotation angle of the femur from the extension position to 10° was larger in the BCR group than in the UKA and healthy control groups, the medial condyle was located forward at the same flexion angle, and the lateral condyle was located forward at flexion angles from 30° to 120°. Compared to UKA, the group reported that BCR did not reproduce kinematics in the healthy knee [15]. In addition, the group calculated the in-situ force of the ACL and PCL from the position information of the femoral tibia in squats and cross-leg movements before and after the BCR. As a result, the kinematics of BCR showed changes from the healthy knee. BCR showed a movement of the femoral condyle moving posteriorly using the medial pivot motion, and there was a significant correlation between the in-situ force and the knee kinematics. The results of this study concluded that it is useful for discussing ligament balance and implant design in BCR [16].

2.3 Computer Simulation Study

Watanabe et al. compared BCR (Vanguard XP) with mechanical alignment and kinematic alignment. Although the kinematics of the knee joint in kinematic alignment was closer to that of the normal knee joint, kinematics could not be completely reproduced and pointed out the usefulness of reproducing anatomical features even in alignment [17].

Koh et al. analyzed kinematics in a healthy knee and BCR with patient-specific articular surfaces, BCR (Vanguard XP), CR (Genesis II. Smith & Nephew). BCR showed kinematics closer to a healthy knee than CR, but the difference was small in BCR that reproduced the patient's unique articular surface [18]. Zumbrunn et al. also compared gait and sitting and deep flexion in a BCR designed from kinematics and joint surface geometry data from in vivo analysis of 40 healthy knee joints and a symmetric design BCR (TKO: Biopro), CR (NexGen: Zimmer-Biomet), and bilateral UKA (MCK: MAKO Surgical). The result was that kinematics of the healthy knee was reproduced in BCR with the design of the articular surface, and pointed out that the reproduction of the articular surface may have contributed to the maintenance of kinematics by reproducing the function of the ACL [19].

3 Future Challenges

Although BCR reproduced some features of kinematics in healthy knee joints, none of the studies have been able to fully reproduce kinematics, especially under high-load exercise conditions. Although the bi-cruciate ligaments preserved by BCR contribute to joint stability, it has been noted that it is necessary to reproduce ligament function more in kinematics, which suggests that reproducing the articular surface shape of a healthy knee joint may be a factor. Changes in the shape of the articular surface change the mechanical environment around the implant, which may lead to implant failure, so further studies are necessary. Evaluation should also be conducted to determine whether the reproduction of kinematics in healthy knees leads to high patient satisfaction and postoperative outcomes in BCR.

There are still few studies on BCR kinematics compared with CR and PS, and it is necessary to proceed with more multifaceted research and discussion, and it is our future challenges.

References

1. Heyse TJ, et al. Kinematics of a bicruciate-retaining total knee arthroplasty. Knee Surg Sports Traumatol Arthrosc. 2017;25:1784–91.
2. Peersman G, et al. The influence of polyethylene bearing thickness on the tibiofemoral kinematics of a bicruciate retaining total knee arthroplasty. Knee. 2017;24:751–60.
3. Okada Y, et al. ACL function in Bicruciate-retaining total knee arthroplasty. J Bone Joint Surg Am. 2018;100:e114.
4. Hamada D, et al. Native rotational knee kinematics are lost in bicruciate-retaining total knee arthroplasty when the tibial component is replaced. Knee Surg Sports Traumatol Arthrosc. 2018;26:3249–56.
5. Wada K, et al. The medial constrained insert restores native knee rotational kinematics after bicruciate-retaining total knee arthroplasty. Knee Surg Sports Traumatol Arthrosc. 2019;27:1621–7.
6. Takasago T, et al. Insufficient lateral joint laxity after bicruciate-retaining total knee arthroplasty potentially influences kinematics during flexion: a biomechanical cadaveric study. Knee. 2021;28:311–8.
7. Arnout N, et al. Knee joint laxity is restored in a bicruciate retaining TKA-design. Knee Surg Sports Traumatol Arthrosc. 2020;28:2863–71.
8. Simon JC, et al. Level and downhill walking to assess implant functionality in bicruciate- and posterior cruciate-retaining total knee arthroplasty. J Arthroplast. 2018;33:2884–9.
9. Arauz P, et al. Knee motion symmetry was not restored in patients with unilateral bicruciate retaining total knee arthroplasty-in vivo three-dimensional kinematic analysis. Int Orthop. 2018;42:2817–23.
10. Tsai TY, et al. Bicruciate retaining total knee arthroplasty does not restore native tibiofemoral articular contact kinematics during gait. J Orthop Res. 2019;37:1929–37.
11. Arauz P, et al. Stair climbing and high knee flexion activities in bicruciate retaining total knee arthroplasty: in vivo kinematics and articular contact analysis. J Arthroplast. 2019;34:570–6.

12. Kwon YM, et al. In vivo kinematics of deep lunges and sit-to-stand activities in patients with bicruciate-retaining total knee arthroplasty. Bone Joint J. 2020;102(6 Suppl):59–65.
13. Smith LA, et al. In vivo knee kinematics:how important are the roles of femoral geometry and the cruciate ligaments? J Arthroplast. 2021;36:1445–54.
14. Kono K, et al. In vivo kinematics of bicruciate-retaining total knee arthroplasty with anatomical articular surface under high-flexion conditions. J Knee Surg. 2021;34:452–9.
15. Kono K, et al. Bicruciate-retaining total knee arthroplasty reproduces in vivo kinematics of normal knees to a lower extent than unicompartmental knee arthroplasty. Knee Surg Sports Traumatol Arthrosc. 2020;28:3007–15.
16. Kono K, et al. In vivo kinematics and cruciate ligament forces in bicruciate-retaining total knee arthroplasty. Sci Rep. 2021;11:5645.
17. Watanabe M, et al. Abnormal knee kinematics caused by mechanical alignment in symmetric bicruciate-retaining total knee arthroplasty are alleviated by kinematic alignment. Knee. 2020;27:1385–95.
18. Koh YG, et al. Preservation of kinematics with posterior cruciate-, bicruciate- and patient-specific bicruciate-retaining prostheses in total knee arthroplasty by using computational simulation with normal knee model. Bone Joint Res. 2017;6:557–65.
19. Zumbrunn T, et al. Regaining native knee kinematics following joint arthroplasty: a novel biomimetic design with ACL and PCL preservation. J Arthroplast. 2015;30:2143–8.

24 Effects of Alignment on Kinematics

Shinichiro Nakamura

1 Coronal Alignment and Kinematics

1.1 Normal Knee Coronal Alignment

Coronal alignment is an important indicator of the even distribution of stress on the medial and lateral compartments of the knee joint. When the lower limb is varus-aligned, the stress is concentrated in the medial compartment, whereas when the lower limb is valgus-aligned, the stress is concentrated in the lateral compartment. This may also be related to the lift-off of the components.

1.2 Effect of Coronal Alignment on Lift-off in TKA

Hamai et al. analyzed the relationship between medial/lateral joint laxity and lift-off during walking [1]. Liftoff was evaluated based on the distance between the femoral component and tibial insert using a 3D model-fitting approach. The results showed that medial/lateral joint laxity did not correlate with lift-off. Further, lift-off was rarely observed in patients with near-neutral postoperative alignment.

Kuriyama et al. analyzed the effects of alignment and lateral laxity on liftoff using computer simulations [2]. In this study, models with LCL laxity and varus femoral components were combined to analyze the effect of ligament laxity or alignment on liftoff. The outer laxity is defined as a lift-off of ≤2 mm when the varus alignment is ≤3°, whereas when the varus alignment is ≥3°, lift-off occurs even in the absence of lateral laxity. To prevent lift-off, it is important to perform surgeries for neutral alignment.

1.3 Effect of Coronal Alignment on Axial Rotation in TKA

Sekiguchi et al. analyzed the effect of coronal alignment on kinematics during the squatting motion [3]. In this study, a posterior stabilized implant was used, and alignment was measured using radiography to determine the femorotibial angle (FTA) and femoral and tibial component alignments according to the Knee Society evaluation methods. The kinematics were evaluated using a 3D model-fitting approach with fluoroscopy.

The femoral component rotated externally by an average of 5.4° from full extension to maximum flexion. The β angle had a negative correlation with the rotation of the femur after 30°, and the FTA was positively correlated after 30°. The rotation angle from full extension to maximum flexion was negatively, positively, and not correlated with the β angle, FTA, and αangle, respectively (Fig. 24.1).

However, all patients in this study underwent mechanical alignment and had an average FTA of 173.3°, with a standard deviation of 2.6°, and an average β angle of 89.0°, with a standard deviation of 2.3°. Therefore, the dynamics of the kinematic alignment method and the tolerability of the varus remain unclear.

S. Nakamura (✉)
Department of Orthopaedic Surgery, Kyoto University, Kyoto, Japan
e-mail: shnk@kuhp.kyoto-u.ac.jp

S. Matsuda (ed.), *Advances in Total Knee Arthroplasty*, https://doi.org/10.1007/978-981-97-4920-1_24

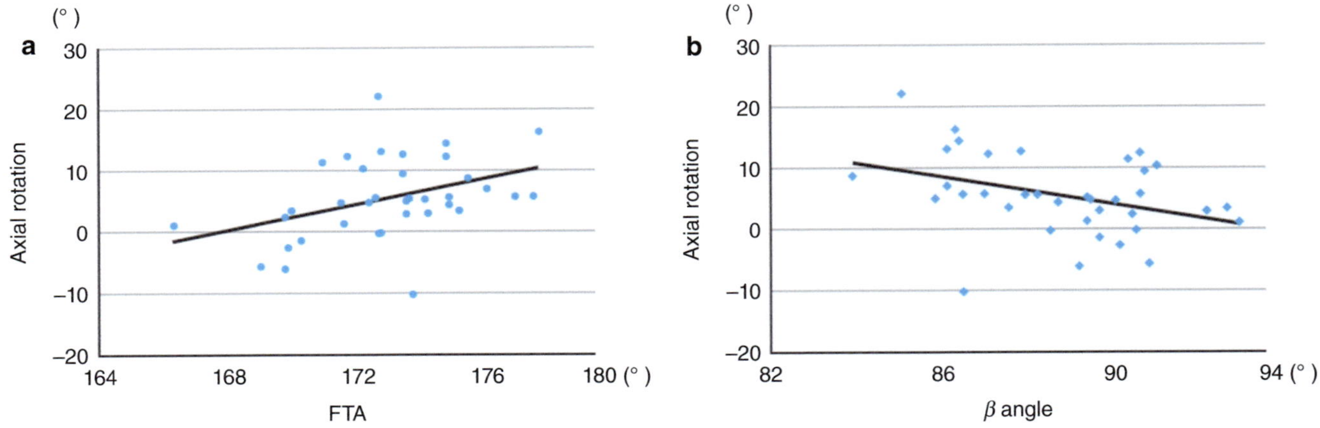

Fig. 24.1 Relative position of implants was analyzed using perspective images and a 3D model fitting approach

2 Sagittal Alignment と Kinematics

2.1 Sagittal Alignment of Normal Knee

The sagittal alignment of the normal knee is defined as anterior bowing on the femoral side and posterior slope on the tibial side. Anterior bowing on the femoral side is greater in women (2.9°) than in men (2.3°) [4], but the clinical significance of anterior bowing is unclear. In contrast, an increase in the posterior slope on the tibial side is a risk factor for ACL injury [5] and is associated with posterior horn injury of the medial meniscus [6]. There is still a debate regarding how sagittal alignment should be determined in TKA.

2.2 Sagittal Alignment in TKA

The sagittal alignment of the tibial component differs between the cruciate-retained (CR) and posterior-stabilized (PS) types. In the CR type, matching the posterior slope of the original articular surface is desirable; however, in the PS type, the posterior slope is slightly reduced. The reason for reducing the posterior slope in the PS type is to avoid anterior impingement in the post and cam mechanisms [7].

2.3 Impact of Sagittal Alignment on Kinematics

The kinematics of flexion of the femoral components have not been evaluated. Fujimoto et al. evaluated the effect of the posterior tibial slope on the postoperative kinematics of a CR-type implant using a 3D model-fitting approach [8] The analysis was conducted during a squatting motion under loading and unloading conditions. The posterior tibial slope was divided into large and small groups at 8°, and the kinematics were analyzed at the nearest points of the medial and lateral femur and tibial components. Overall, the nearest medial point was in almost the same position, whereas the nearest lateral point moved backward owing to flexion. In the unloaded position, the nearest medial point moved forward during mid-flexion, whereas the nearest lateral point moved backward. In addition, external rotation of the femoral component was observed with flexion in both loading and unloading positions.

In the comparison of the posterior tibial slope, the nearest medial point was located backward in the group with a large posterior tibial slope from 10° to 90° of flexion in the loading position. There was no effect of the posterior tibial slope on the lateral compartment. Meanwhile, the posterior tibial slope had almost no influence in both the medial and lateral positions in the non-loading position (Fig. 24.2).

These results suggest that the femoral component falls backward due to gravity when the posterior tibial slope is large and the contact point is located backward (Fig. 24.3). Further studies are required to determine the optimal posterior tibial slope.

Okamoto et al. analyzed the effect of the posterior tibial slope on the PS type using computer simulation [7]. In this study, the posterior tibial slope was changed from 0° to 15°, and the kinematics of the stair-climbing motion were evaluated. When the posterior tibial slope exceeded 10°, the forward movement of the tibia occurred at approximately 65° flexion. In addition, if the posterior tibial slope exceeds 10°, an anterior impingement in the post and cam occurs during extension. Therefore, the posterior tibial slope of the PS type should be ≤5°.

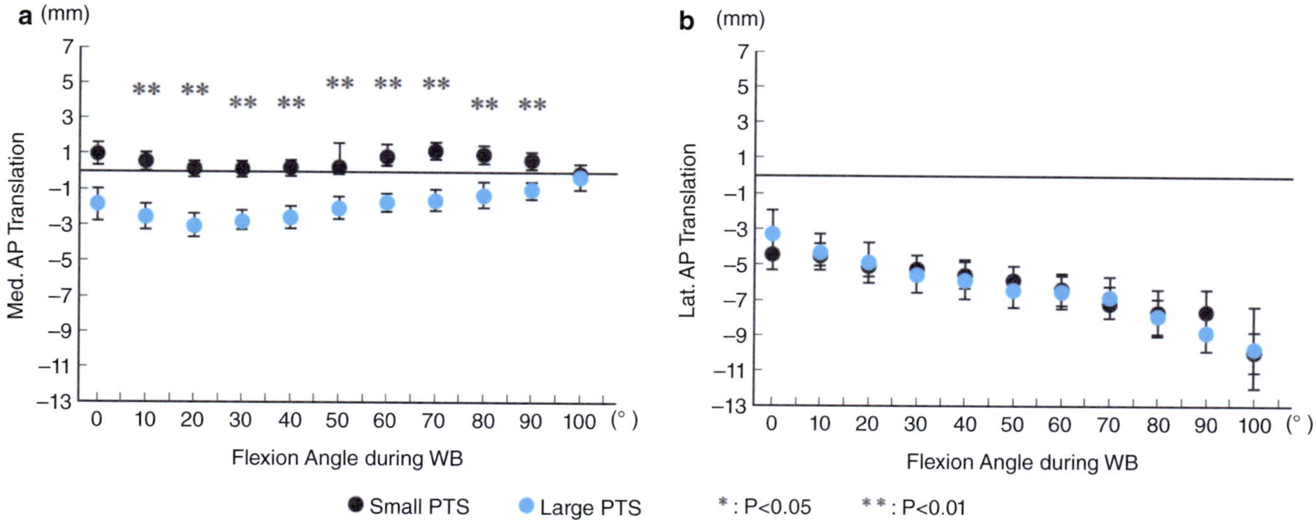

Fig. 24.2 Recent contacts on the inside and outside of the small PTS and large PTS groups with low back slope

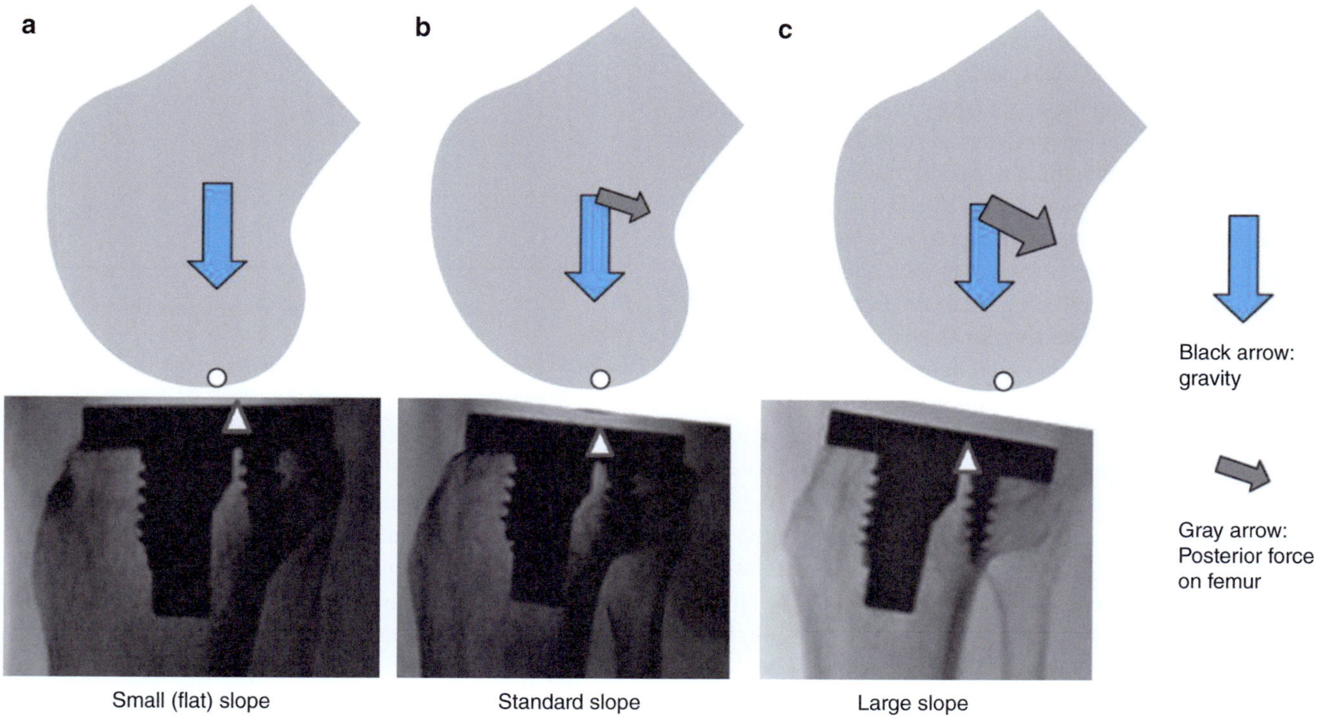

Fig. 24.3 Schematic diagram of the posterior movement of the femur in a group with many backward slopes

3 Axial Alignment と Kinematics

3.1 Axial Alignment of Normal Knee

In a normal knee, axial alignment is defined by the condylar twisting angle of the femur and tibial torsion. In particular, the condylar twisting angle of the femur influences patellar tracking. The condylar twisting angle increases with lateral femoral condyle hypoplasia in valgus knees. Chang et al. reported an association among coronal alignment, femoral condylar twisting angle, and tibial torsion in the OA knee [9]. The condylar twisting angle of the femur is defined as the angle between the line connecting the posterior condyle and the line connecting the medial and lateral epicondyles and the tibial torsion is defined as the angle between the line connecting the proximal posterior tibial condyle and the line connecting the medial and lateral malleolus. It has been reported that as the lower limb alignment changes from varus to valgus, the condylar twisting angle of the femur and tibial torsion increase.

3.2 Effects of Axial Alignment on Clinical Outcomes

The internal rotation of the femoral component causes lateral patellar translation. Internal rotation of the tibial component can also cause maltracking. It has been reported that maltracking of the patella causes postoperative pain, knee joint contracture, and PFJ instability, leading to poor postoperative clinical outcomes [10]. Particularly on the tibial side, malrotation is likely to occur because of the lack of bony landmarks, and there have been cases of rotation failure of ≥15° with conventional methods; therefore, component placement should be carefully performed [11].

3.3 Effects of Axial Alignment on Kinematics

Nakahara et al. analyzed the effect of the rotational alignment of the femur and tibial components on kinematics [12]. Postoperative CT was performed to evaluate the rotational alignment, and the angle between the femoral component and the surgical epicondylar axis and the angle between the tibial component and the Akagi line were measured. Kinematics was evaluated during the stair climbing motion, and the correlation between the rotation of the femoral component relative to the tibial component during extension, 30° and 60° flexion, and rotation of the bones during extension was evaluated. The rotational position of the femoral component did not show a significant correlation with any of the factors, whereas the rotational position of the tibial component was correlated with the rotational position between the components in extension (Fig. 24.4) and between bones (Fig. 24.5). Internal rotation of the tibial component caused a rotational mismatch, in which the tibial component acquired an internal rotation position relative to the femoral component, and interosseous evaluation showed that the tibia was in an external rotation position relative to the femur.

Nakamura et al. also focused on the tibial tubercle-trochlear groove (TT-TG) distance after TKA and analyzed its effects on patellar tilt and patellar shift [13]. The correlation between patellar tilt and patellar shift in extension and skyline views during flexion was evaluated. The TT-TG distance correlated with patellar tilt and patellar shift in extension and flexion, whereas the rotational position of the femoral components was not correlated with patellar tilt and patellar shift, and the rotational position of the tibial components correlated with patellar tilt and patellar shift in the flexion position (Fig. 24.6). Because an increase in the TT-TG distance causes maltracking of the patella, it is necessary to avoid internal rotation of the tibial component, especially factors affecting the TT-TG distance.

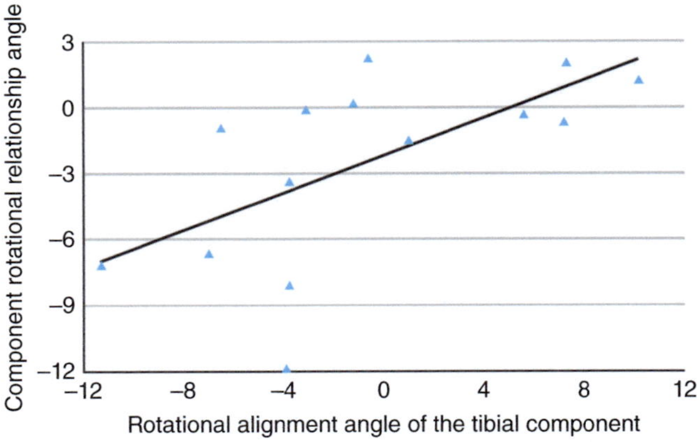

Fig. 24.4 Installation of the rotation of the tibial component (X-axis) and the relative rotation position of the femoral tibial component (Y-axis)

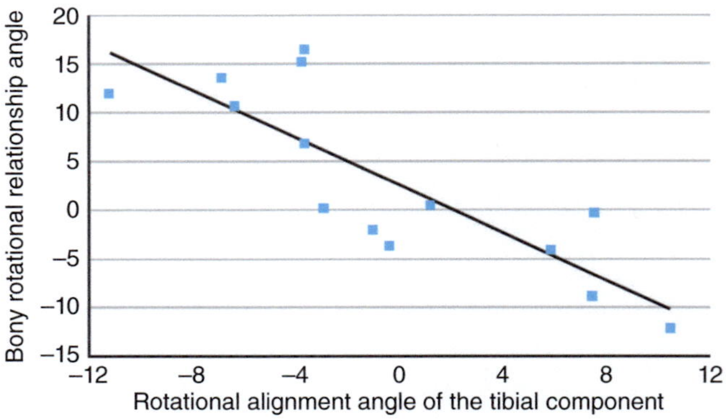

Fig. 24.5 Rotational installation of tibial components (X-axis) and relative rotational position between the bones of the femoral tibia (Y-axis)

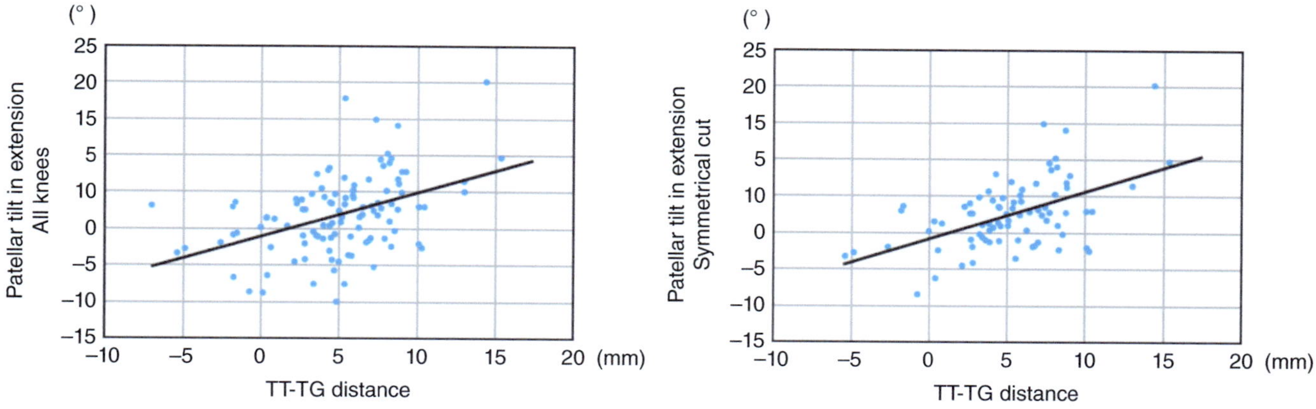

Fig. 24.6 Correlation between the TT-TG distance, femoral component, and patellar tilt in the rotational and extension positions of the tibial component

References

1. Hamai S, et al. No influence of coronal laxity and alignment on lift-off after well-balanced and aligned total knee arthroplasty. Knee Surg Sports Traumatol Arthrosc. 2014;22:1799–804.
2. Kuriyama S, et al. No condylar lift-off occurs because of excessive lateral soft tissue laxity in neutrally aligned total knee arthroplasty: a computer simulation study. Knee Surg Sports Traumatol Arthrosc. 2016;24:2517–24.
3. Sekiguchi K, et al. Varus alignment after total knee arthroplasty results in greater axial rotation during deep knee bend activity. Clin Biomech (Bristol, Avon). 2020;77:105051.
4. Nam JH, et al. Gender difference in bowing of the sagittal femoral morphology measurement using magnetic resonance imaging. Surg Radiol Anat. 2020;42:1231–6.
5. Terauchi M, et al. Sagittal alignment of the knee and its relationship to noncontact anterior cruciate ligament injuries. Am J Sports Med. 2011;39:1090–4.
6. Moon HS, et al. Medial meniscal posterior horn tears are associated with increased posterior tibial slope: a case–control study. Am J Sports Med. 2020;48:1702–10.
7. Okamoto S, et al. Effect of tibial posterior slope on knee kinematics, quadriceps force, and patellofemoral contact force after posterior-stabilized total knee arthroplasty. J Arthroplast. 2015;30:1439–43.
8. Fujimoto E, et al. Significant effect of the posterior tibial slope on the weight-bearing, midflexion in vivo kinematics after cruciate-retaining total knee arthroplasty. J Arthroplast. 2014;29:2324–30.
9. Chang MJ, et al. Relationship between coronal alignment and rotational profile of lower extremity in patients with knee osteoarthritis. J Arthroplast. 2018;33:3773–7.
10. Bedard M, et al. Internal rotation of the tibial component is frequent in stiff total knee arthroplasty. Clin Orthop Relat Res. 2011;469:2346–55.
11. Kuriyama S, et al. Tibial rotational alignment was significantly improved by use of a ct-navigated control device in total knee arthroplasty. J Arthroplast. 2014;29:2352–6.
12. Nakahara H, et al. Rotational alignment of the tibial component affects the kinematic rotation of a weight-bearing knee after total knee arthroplasty. Knee. 2015;22:201–5.
13. Nakamura S, et al. Tibial tubercle-trochlear groove distance influences patellar tilt after total knee arthroplasty. J Arthroplast. 2019;34:3080–7.

25 Effects of Ligament Balance on Kinematics

Toshifumi Watanabe

1 Evaluation of Ligament Balance and Kinematics

Just as normal knee kinematics are based on physiological ligament balance, knee kinematics following TKA are thought to be greatly influenced by ligament balance. However, caution should be exercised when interpreting kinematics after TKA as a wide variety of factors may influence these kinematics, including leg alignment, implant models, knee motions to analyze kinematics, and analysis methods. There are currently several approaches for the evaluation of ligament balance, including intraoperative extension-flexion gap balance, mediolateral gap balance, postoperative mediolateral laxity, and anteroposterior laxity.

2 Evaluation Using Navigation System

The navigation system is a simple method to investigate the relationship between ligament balance and kinematics, which allows evaluation of the intraoperative ligament balance and kinematics of passive knee motion. Although the introduction of a navigation system is necessary for this evaluation, ligament balance and kinematics can be quantified for all ROMs of the knee joint. However, it is difficult to quantify the applied force when evaluating intraoperative ligament balance, and reproducibility is questionable. In addition, the kinematics measured intraoperatively were passive motions under anesthesia. Measurements using a navigation system showed that rotational soft tissue balance was highly correlated with rotational kinematics following TKA (both PS and CR) [1]. In other words, if soft tissues have room for internal rotation of the tibia, the tibia is more likely to rotate internally.

3 Evaluation Using a Tenser and Navigation System

In another method, the intraoperative ligament balance is measured using a special instrument such as a tensor, and the kinematics of passive knee motion are evaluated using a navigation system. This method can be used to evaluate intraoperative ligament balance more accurately. However, knee kinematics still involve passive motion under anesthesia. Such passive non-weight-bearing motion under anesthesia has been correlated with postoperative active knee kinematics to some extent [2].

Several studies have been conducted on various implant types using tensors and navigation systems. In CR TKA, it has been reported that lateral laxity to the medial side at 60° and 90° of flexion correlates with tibial internal rotation [3]. In modified kinematically aligned TKA using a CR implant, significantly greater tibial internal rotation from 60° to 120° and preserved lateral laxity at 90° and 120° compared to mechanically aligned TKA have been reported [4]. Together, these studies indicated that lateral laxity in flexion is desirable in CR TKA to obtain kinematics similar to those of a normal knee.

In contrast, a prior study showed that in PS TKA, the medial gap at 60° and 90° flexion was negatively correlated with tibial internal rotation, while tibial rotation was correlated with the knee flexion angle at 2 years after surgery [5]. Furthermore, one recent study showed that in BCS TKA, tibial internal rotation during knee flexion is significantly smaller when the medial gap in knee flexion was tighter than that in knee extension [6]. These studies indicate that a tight medial gap in knee flexion should be avoided in PS and BCS TKA to obtain kinematics similar to those of a normal knee.

T. Watanabe (✉)
Saitama Medical Center, Dokkyo Medical University, Saitama, Japan
e-mail: toshi-w@dokkyomed.ac.jp

S. Matsuda (ed.), *Advances in Total Knee Arthroplasty*, https://doi.org/10.1007/978-981-97-4920-1_25

4 Evaluation Using a Tenser and the Shape-Matching Technique

The measurement of intraoperative ligament balance using instruments such as a tenser, and the evaluation of postoperative knee kinematics during active motion, such as squatting and kneeling, with the shape-matching technique are time-consuming, but nevertheless provide important information for surgeons [7] (Fig. 25.1). Specifically, many researchers wish to identify which target intraoperative ligament balance can lead to ideal postoperative kinematics. Because it is desirable to evaluate the postoperative knee kinematics of active weight-bearing motion, which is commonly performed in daily life, evaluation using the shape-matching technique is optimal.

In a previous study using a PS implant, the authors found that the gap difference between 135° and 0° was correlated with the total posterior translation of the lateral femoral condyle, femoral external rotation during squatting, and maximum knee flexion [8]. This study indicated that a tight flexion gap should be avoided in PS TKA to achieve kinematics similar to those of normal knees. The authors further noted that well-balanced knees exhibited stable knee kinematics. In a study using a CR implant, it was reported that the anterior displacement of the medial femoral condyle was greater in the group with a gap difference of more than 3 mm between 90° and 0° than in the group with a gap difference of less than 3 mm [9]. The situation in CR TKA is slightly different from that in PS TKA, indicating that a loose flexion gap leads to PCL dysfunction.

Furthermore, studies on PS TKA reported that intraoperative mediolateral laxity was strongly correlated with postoperative varus-valgus kinematics [10]. Although it is natural that mediolateral laxity is directly related to varus-valgus kinematics, excessive imbalance in mediolateral laxity should be avoided.

5 Evaluation Using Stress X-Ray/Arthrometer and the Shape-Matching Technique

The combination of tensor measurements of intraoperative ligament balance and postoperative motion analysis is a useful evaluation method. However, the maintenance of the intraoperative ligament balance after surgery remains a concern. While some studies have reported no change in mediolateral laxity after surgery [11], others reported a decrease in lateral laxity with correction of the preoperative varus knee alignment [12]. By evaluating ligament balance and knee kinematics simultaneously after surgery (Figs. 25.2 and 25.3), the relationship between the two can be investigated by eliminating the deviation of the time axis.

The authors evaluated well-functioning knees after PS TKA and found that the medial laxity was consistent in extension and flexion, and was significantly smaller than the lateral laxity [13]. We further analyzed knee kinematics during squatting activity in the same cohort, and compared the group that consistently showed femoral external rotation with knee flexion to the group that did not. We found that lateral laxity was greater than medial laxity in extension and flexion in both groups; however, the consistent group had significantly smaller lateral laxity in extension than that in the inconsistent group [14]. Overall, these results suggest that medial stability in extension and flexion is important, and that while lateral laxity is acceptable, mediolateral balance in extension should also be considered to some extent.

In CR TKA, coronal laxity and alignment did not affect lift-off under dynamic weight-bearing after well-balanced and aligned total knee arthroplasty [15]. In a study using CS TKA, during the deep knee bend activity, a significant correlation was found between the medial flexion gap and the medial contact point, as well as the axial rotation at full extension. By increasing the medial flexion gap, both contact

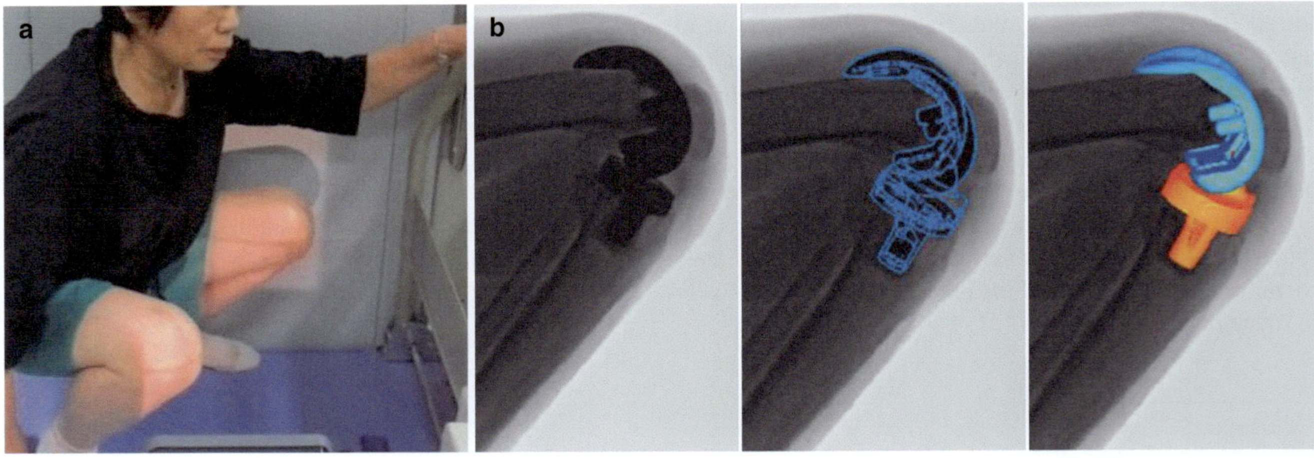

Fig. 25.1 Kinematic analysis of squatting using the shape-matching technique. (Modified from Watanabe et al. [14])

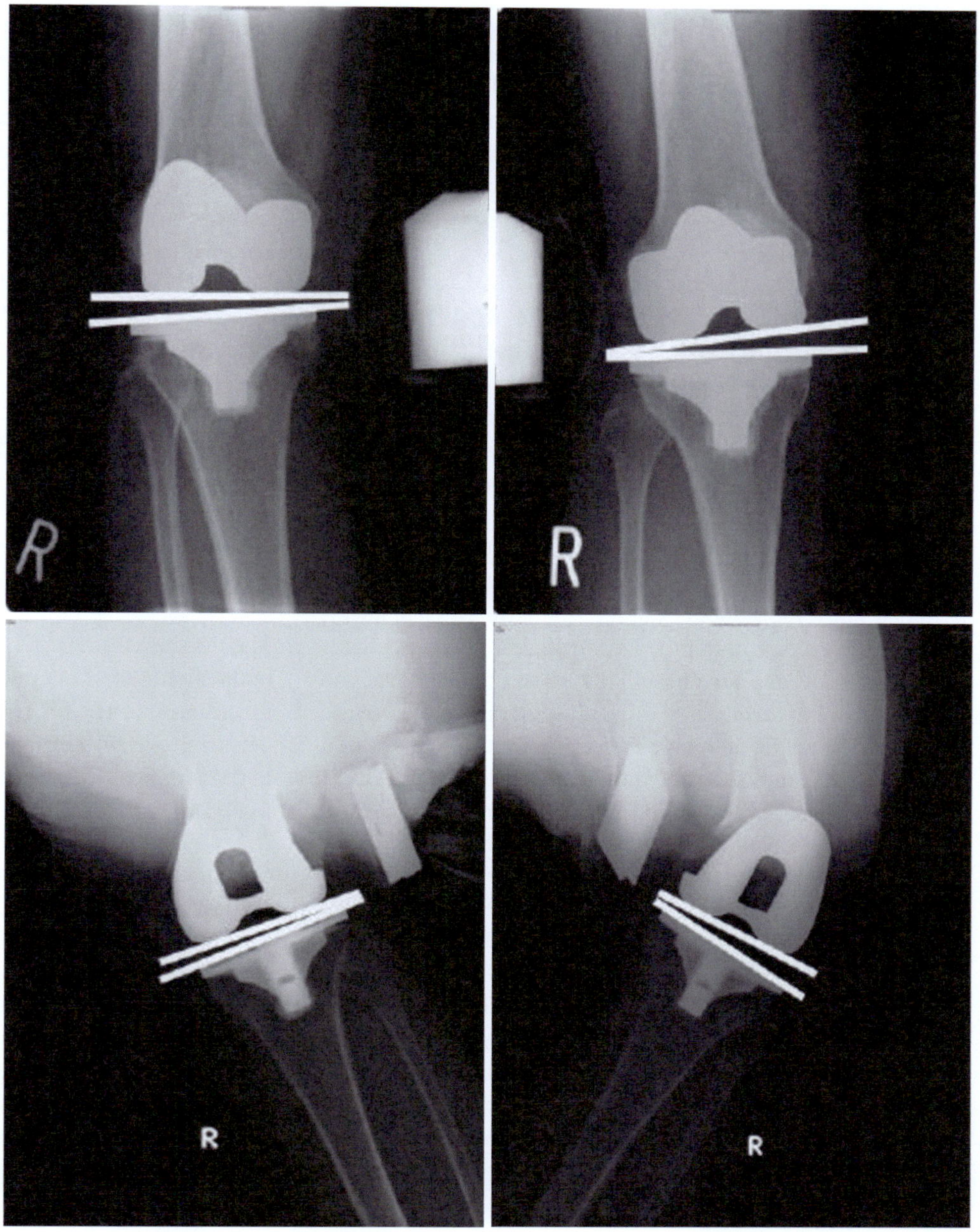

Fig. 25.2 Evaluation of mediolateral laxity in stress X-rays in extension and flexion using the Telos device and hand-held dynamometer. (Modified from Watanabe et al. [13])

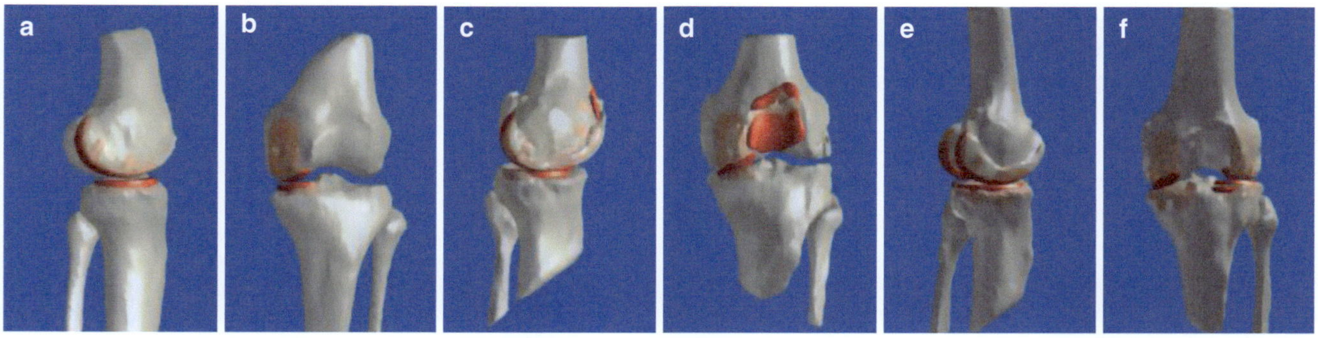

Fig. 25.3 Three-dimensional models of medial UKA, medial UKA + PF arthroplasty, and bicondylar UKA performed by robotic-assisted surgeries

points were positioned anteriorly at complete contact during the kneeling activity, but the effect of the lateral flexion gap was less [16]. This study suggests that excessive medial laxity should be avoided, as it can cause abnormal kinematics. In contrast, another study adapting CS TKA reported that a consistent relationship between laxity and kinematics could not be predicted [17].

6 Evaluation Using the Shape-Matching Technique Under Various Conditions

Prior studies have investigated how the presence or absence of ligaments, degree of joint deformity, mobility of inserts, and the compartment to be replaced may affect postoperative knee kinematics. However, these studies did not directly investigate the relationship between ligament balance and kinematics.

In one study, the authors examined the differences in knee kinematics with or without the PCL using an implant called the ACL-substituting prosthesis [18], which has a ball-and-socket shape that stabilizes the lateral femoral condyle during extension while allowing posterior translation of the lateral femoral condyle during flexion, regardless of the PCL status. Kinematic analysis during squatting showed that the PCL-sacrificing knees had a significantly more anterior location of the medial and lateral femoral condyles in knee flexion of 95° or more compared with the PCL-retaining knees. This may have been due to PCL dysfunction. In contrast, the maximum knee flexion during kneeling was greater in the PCL-sacrificing knees. This study suggests the importance of PCL tensioning when preserving the contracted PCL.

In another study, a different PS model was applied, and knee kinematics were analyzed in three groups: fixed-bearing or mobile-bearing prostheses for moderately deformed knees and fixed-bearing prostheses for severely deformed knees [19].

Although there were differences in the details of kinematics, the total amount of femoral external rotation and condylar anteroposterior translation showed no significant difference between the three groups. This suggests that the PS model can be expected to have postoperative kinematics equivalent to those of moderately deformed knees even when applied to severely deformed knees with more imbalanced ligaments. A mobile-bearing prosthesis may be useful for maintaining rotational balance, but not for increasing the amount of rotation.

Furthermore, the authors compared the in vivo kinematics of knees with various compartmental replacements using robotic-assisted surgeries [20] (Fig. 25.3). All the knees showed significant femoral external rotation and posterior condylar translation with flexion. However, knees with medial UKA and medial UKA + PF arthroplasty showed a greater femoral external rotation and posterior translation, whereas knees with bicondylar UKA showed the lowest level of femoral external rotation and posterior translation. This study showed that kinematics close to those of a normal knee can be obtained by preserving the ACL and PCL. However, the femoral external rotation decreased when the lateral tibiofemoral compartment was replaced. This may be due to the changes in geometry caused by lateral compartment replacement and poor ligament balance in knees with more severe deformities.

7 How Ligament Balance Affects Kinematics

The above results suggest the following about ligament balance from the viewpoint of kinematics:

1. The tibia tends to rotate internally if the soft tissue is lax in the direction of internal rotation of the tibia.
2. In various types of TKA, medial stability in extension and flexion is extremely important to obtain kinematics close to those of healthy knees.
3. In PS and BCS TKA, tight flexion gaps should be avoided.
4. In PS TKA, lateral laxity is allowed to some extent, but mediolateral balance should also be considered.
5. PS TKA can be expected to have the same postoperative kinematics as moderately deformed knees, even in severely deformed knees with greater ligament imbalance, whereas mobile-bearing prostheses do not contribute to an increase in rotation.
6. In CR TKA, an excessive flexion gap should be avoided as it leads to PCL dysfunction. Thus, tensioning the retained PCL is also important.
7. In CR and BCR TKA, lateral laxity during knee flexion is preferable.
8. In partial surface replacement with preservation of both cruciate ligaments, the kinematics worsen with the addition of lateral condylar replacement.

References

1. Kinoshita T, et al. Rotational soft-tissue balance is highly correlated with rotational kinematics in total knee arthroplasty. J Knee Surg. 2021; https://doi.org/10.1055/s-0041-1729619.
2. Wada K, et al. Can intraoperative kinematic analysis predict postoperative kinematics following total knee arthroplasty? A preliminary. J Med Investig. 2017;65:21–6.
3. Matsuzaki T, et al. Tibial internal rotation is affected by lateral laxity in cruciate-retaining total knee arthroplasty: an intraoperative

kinematic study using a navigation system and offset-type tensor. Knee Surg Sports Traumatol Arthrosc. 2014;22:615–20.
4. Matsumoto T, et al. Intraoperative soft tissue balance/kinematics and clinical evaluation of modified kinematically versus mechanically aligned total knee arthroplasty. J Knee Surg. 2020;33:777–84.
5. Kamenaga T, et al. Medial knee stability at flexion increases tibial internal rotation and knee flexion angle after posterior-stabilized total knee arthroplasty. Clin Biomech (Bristol, Avon). 2019;68:16–22.
6. Inui H, et al. The relationship between soft-tissue balance and intraoperative kinematics of guided motion total knee arthroplasty. J Knee Surg. 2019;32:91–6.
7. Banks SA, Hodge WA. Accurate measurement of three-dimensional knee replacement kinematics using single-plane fluoroscopy. IEEE Trans Biomed Eng. 1996;43:638–49.
8. Watanabe T, et al. Intraoperative joint gaps and mediolateral balance affect postoperative knee kinematics in posterior-stabilized total knee arthroplasty. Knee. 2015;22:527–34.
9. Fujimoto E, et al. Intraoperative gaps affect outcome and postoperative kinematics in vivo following cruciate-retaining total knee arthroplasty. Int Orthop. 2016;40:41–9.
10. Moro-oka TA, et al. Modified gap-balancing technique in total knee arthroplasty: evaluation of the postoperative coronal laxity. Knee Surg Sports Traumatol Arthrosc. 2010;18:375–80.
11. Ishii Y, et al. Effect of soft tissue tension on measurements of coronal laxity in mobile-bearing total knee arthroplasty. J Orthop Sci. 2005;10:496–500.
12. Sekiya H, et al. Postoperative lateral ligamentous laxity diminishes with time after TKA in the varus knee. Clin Orthop Relat Res. 2009;467:1582–6.
13. Watanabe T, et al. Coronal and sagittal laxity affects clinical outcomes in posterior-stabilized total knee arthroplasty: assessment of well-functioning knees. Knee Surg Sports Traumatol Arthrosc. 2020;28:1400–9.
14. Watanabe T, et al. Consistent femoral external rotation during weight-bearing knee flexion is associated with better patient-reported pain and mediolateral balance after total knee arthroplasty. Clin Biomech (Bristol, Avon). 2021;88:105438.
15. Hamai S, et al. No influence of coronal laxity and alignment on lift-off after well-balanced and aligned total knee arthroplasty. Knee Surg Sports Traumatol Arthrosc. 2014;22:1799–804.
16. Nakamura S, et al. Analysis of the flexion gap on in vivo knee kinematics using fluoroscopy. J Arthroplast. 2015;30:1237–42.
17. Ishii Y, et al. Effect of knee laxity on in vivo kinematics of meniscal-bearing knee prostheses. Knee. 2007;14:269–74.
18. Watanabe T, et al. Knee kinematics in anterior cruciate ligament-substituting arthroplasty with or without the posterior cruciate ligament. J Arthroplast. 2013;28:548–52.
19. Watanabe T, et al. Matched comparison of kinematics in knees with mild and severe varus deformity using fixed- and mobile-bearing total knee arthroplasty. Clin Biomech (Bristol, Avon). 2012;27:924–8.
20. Watanabe T, et al. In vivo kinematics of a robot-assisted uni- and multicompartmental knee arthroplasty. J Orthop Sci. 2014;19:552–7.

Kinematics After TKA: Kinematics and Postoperative Results

26

Kenichi Kono

To obtain high-precision alignment, the application of total knee arthroplasty (TKA) using computer navigation and intraoperative kinematic evaluation is gradually increasing in Japan. In addition, postoperative kinematics have been analyzed using a variety of devices, such as plain radiography, CT, MRI, body surface markers, and fluoroscopy. Kinematics are evaluated using six dimensions of freedom [rotation (internal and external rotation, varus and valgus, flexion and extension) + movement (anteroposterior, medial, lateral, superior, and inferior)].

Patient-reported outcome measures (PROMs) have become the mainstream index for assessing postoperative clinical outcomes of TKA. There have been reports indicating an association between ROM and patient satisfaction [1]. Therefore, achieving sufficient ROM is considered to improve clinical outcomes.

In recent years, the relationship between kinematics and postoperative outcomes (such as ROMs and PROMs) has been evaluated.

1 Intraoperative Kinematics and Postoperative Outcomes

Regarding the relationship between intraoperative rotational kinematics and postoperative ROM, some studies have reported that external femoral rotation relative to the tibia (internal rotation of the tibia relative to the femur) is correlated with postoperative ROM. 2 4).In relation to intraoperative kinematics and flexion angle in cruciate-retaining (CR) TKA, there was a positive correlation between external femoral rotation during CR-TKA and the maximum 1-year postoperative flexion angle from 60°to the maximum flexion of 124.2° ± 1.3° [2]. In addition, the relationship between intraoperative rotational stress in posterior-stabilized (PS) TKA and the maximum flexion angle at 1 month after surgery was 90° flexion. Intraoperative femoral external rotation was positively correlated with the maximum postoperative flexion angle [3]. Furthermore, a study on intraoperative femur rotation and ROM in PS-TKA showed a positive correlation between intraoperative femoral external rotation and postoperative ROM from extension to 90° flexion and maximum flexion of 115.6° ± 15.5° from extension to postoperative ROM [4].

Regarding the relationship between intraoperative varus/valgus kinematics and postoperative ROM, there was a negative correlation between the intraoperative valgus angle at the maximum extension position (at navigation registration) and the maximum flexion angle at 1 month after surgery [5]. This suggests that the more severe the varus deformity, the more adverse the effect on postoperative ROM.

According to a study evaluating the relationship between intraoperative anteroposterior translation and postoperative ROM at 90° flexion in posterior cruciate ligament (PCL)-retaining CS-TKA, the postoperative flexion angle tends to be inadequate in knees with posterior movement of the tibia during surgery [6]. In knees with non-functioning PCL, such as in posterior sagging, there is no rollback of the femur with flexion, making it difficult to obtain adequate ROM. If the tibia moves posteriorly in the intraoperative flexion position (if the femur does not move backward), switching to a model that induces rollback, such as PS-TKA, can be considered.

According to a study investigating the relationship between the following intraoperative rotational kinematics [six parameters: rotational angular differences (RAD0), maximum flexion and flexion 30° rotational (RAD30), maximum flexion and flexion 60° rotational (RAD60), maximum flexion and flexion 90° rotational (RAD90), rotational angles at maximum flexion, and postoperative clinical outcomes (maximum flexion angle and Knee Injury and Osteoarthritis Outcome Score (KOOS)) at maximum flexion angles], PS-TKA showed a positive correlation with RAD30 at the maximum flexion angle. Bicruciate substituting (BCS)-TKA showed a positive correlation between RAD60 and

K. Kono (✉)
Faculty of Medicine, Department of Orthopaedic Surgery, The University of Tokyo, Tokyo, Japan

S. Matsuda (ed.), *Advances in Total Knee Arthroplasty*, https://doi.org/10.1007/978-981-97-4920-1_26

improvement in the scores of KOOS subscales (pain, symptoms, activities of daily living (ADL), and sports); in addition, RAD90, RAD30, and KOOS-ADL were positively correlated [7]. Thus, PS-TKA and BCS-TKA showed correlations between intraoperative rotational kinematics and postoperative clinical outcomes; however, the relevant parameters differed. Considering that external rotation of the femur was observed from 30° flexion to maximum flexion in PS-TKA and that there was a marked correlation between the rotational amount (external femoral rotation) of maximum flexion from 60° flexion and the degree of improvement of KOOS subscale scores in BCS-TKA, the results suggest that external rotation of the femur from mid to maximum flexion is associated with satisfactory postoperative clinical outcomes regarding intraoperative kinematics. A study investigating the relationship between intraoperative kinematic patterns and postoperative outcomes in PS-TKA suggested that the medial pivot group, which rotated at the medial fulcrum from 0° to 90° flexion, led to significantly better postoperative clinical outcomes in terms of patient satisfaction and functional activities in Knee Society Score 2011 (KSS2011) [8]. This suggests that intraoperative external femoral rotation movement in the medial pivot pattern leads to good postoperative outcomes, at least in PS-TKA.

2 Postoperative Kinematics and Postoperative Outcomes

The relationship between postoperative kinematics and clinical outcomes was evaluated using X-ray fluoroscopy.

2.1 Postoperative Kinematics and ROM

A study examined the correlation between the anteroposterior movement of the lowest point of the femoral component to the tibial component measured from four lateral radiographs (maximum extension, flexion 45°, flexion 90°, and maximum flexion) and the maximum flexion angle measured by goniometer at an average of 17 months after PS-TKA (30 fixed-bearing knees, 26 mobile-bearing knees). For both fixed-bearing and mobile-bearing knees, there was a significant correlation between the anteroposterior position at maximum flexion and the maximum flexion angle [9]; the more posterior the femoral component was at maximum flexion, the greater the maximum flexion angle was. This suggests that rollback leads to good ROM acquisition.

2.2 Association Between Postoperative Kinematics and PROMs

In a study evaluating the relationship between 3D kinematics (using 2D/3D registration after imaging the squat motion from the lateral direction using an X-ray fluoroscope) and PROMs (using KOOS) in patients more than 6 months after BCS-TKA (Figs. 26.1 and 26.2) [10], it was easier to lead a comfortable daily life with less medial anterior movement in the extended-to-mild flexion position. From the viewpoint of reducing medial anterior-posterior movement, surgical techniques such as minimizing medial collateral ligament release and adjusting the medial gap between extension and flexion may be important for obtaining good postoperative ADL.

2.3 Grouping of PROMs and Kinematics

Thirty patients who underwent three different types of PS-TKA were divided into good (high PROM group) and bad (low PROM group) PROM groups (KOOS and KSS2011) by hierarchical cluster analysis, and the kinematics of open kinetic chain (OKC) and closed kinetic chain (CKC) movements at least 6 months postoperatively were analyzed using

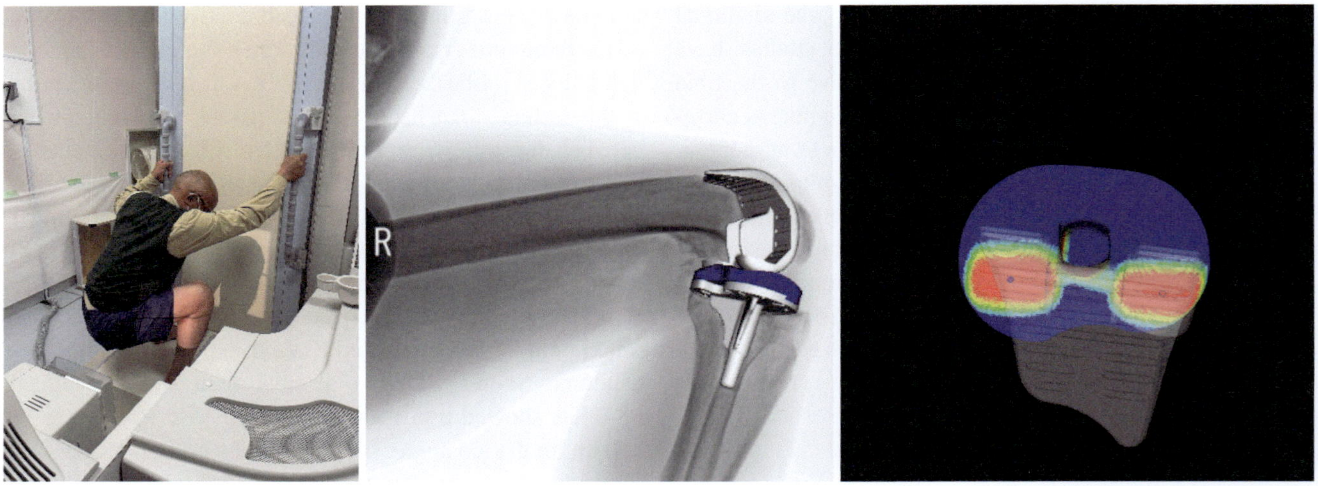

Fig. 26.1 2D/3D registration method

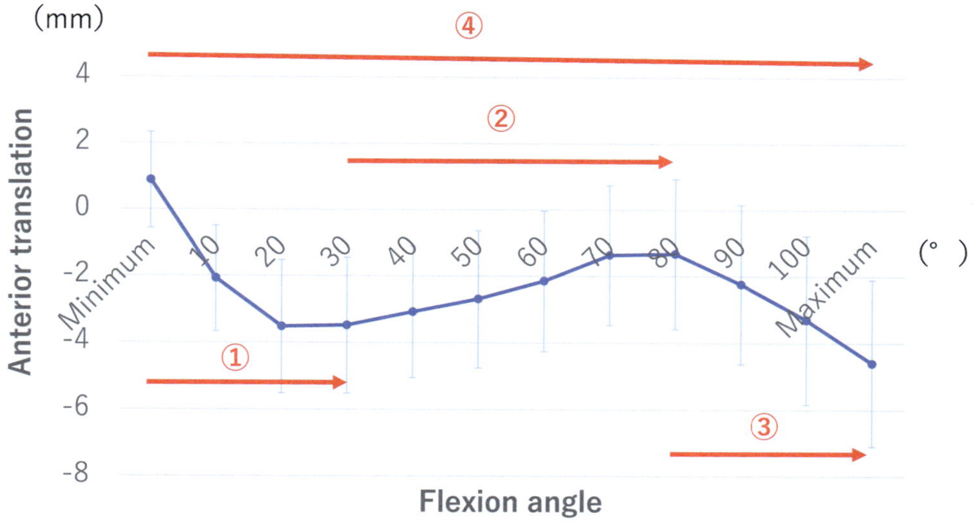

Flexion angle	Postoperative (KOOS)			Improvement (KOOS)		
	Pain	Symptom	ADL	Pain	Symptom	ADL
①Minimum~30°	-0.16	-0.13	-0.33	-0.20	0.11	-0.03
②30° ~80°	0.09	0.02	-0.01	-0.05	-0.04	0.03
③80° ~Maximum	0.01	-0.04	0.00	0.08	0.15	0.03
④Minimum~Maximum	0.18	-0.01	-0.09	-0.17	0.12	-0.06

Fig. 26.2 Relationship between medial anteroposterior translation and KOOS. There was a negative correlation between the degree of movement in (1) and KOOS ADL (numerical value represents the correlation coefficient)

2D/3D registration methods to evaluate whether there were differences. The medial side in the low PROM group moved more anteriorly than that in the high PROM group at 30° extension to flexion in the CKC motion [11]. Furthermore, at 30–60° flexion, there was minimal medial movement in the high PROM group, whereas posterior movement was observed in the low PROM group. Movements suggestive of medial instability, such as the paradoxical anterior motion of the femur at extension and the accompanying abrupt backward movement in the intermediate flexion region, are thought to contribute to the decline in postoperative performance. In addition, the high PROM group showed a significantly greater posterior shift of the lateral femur from 60°to 90°flexion during the CKC motion. Even in a normal knee, posterior translation of the femur is observed from mid-flexion to deep flexion [12, 13]. Kinematics similar to those of a normal knee may improve the lever arm of the quadriceps muscle in deep flexion and lead to the recovery of knee joint function. However, there was no clear difference between PROMs in OKC movements. As most ADLs are in the load position, CKC motion is more likely to affect PROMs.

Some studies conducted evaluation using a single model. Thirty-two patients who underwent BCS-TKA were divided into high and low PROM groups based on hierarchical cluster analysis results, and the squat movements performed at least 6 months after surgery (mean 11.2 ± 3.2 months) using the 2D/3D registration method to compare kinematics were analyzed [14]. There was no significant difference in the angle of rotation associated with flexion; however, the total external rotation of the femoral components was significantly greater in the low PROM group (Fig. 26.3). There was no significant difference in the anteroposterior translation of the medial femur; however, the lateral femur was more anterior in the low PROM group at 0°and 10°of flexion, and the amount of posterior translation was greater in the low PROM group from 0°to 110°of flexion (Fig. 26.4). In the kinematic pathway, which represents the difference between the medial and lateral sides, the high PROM group exhibited a medial pivot pattern from 0°to 20°of flexion and a bicondylar rollback from 70°of flexion. In contrast, the low PROM group exhibited a medial pivot pattern from 0°to 70°of flexion and bicondylar rollback from 70° of flexion. BCS-TKA is characterized by a post-cam mechanism working posteriorly and anteriorly; however, there was no significant difference between the high and low PROM groups. The most important finding of this study was that the external rotation and lateral posterior movement of the femur were greater in the

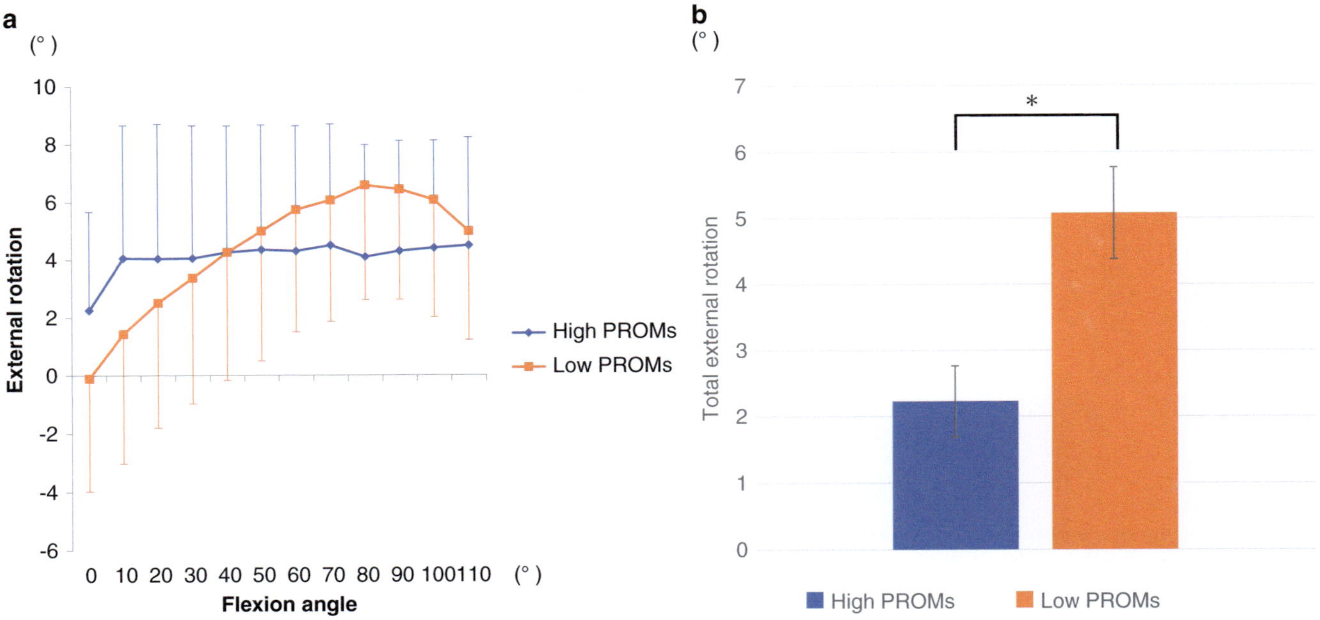

Fig. 26.3 Rotation angle of the femoral component relative to the tibial component. (**a**): Flexion angle and external rotation of the femoral components. (**b**): Total external rotation of the femoral components

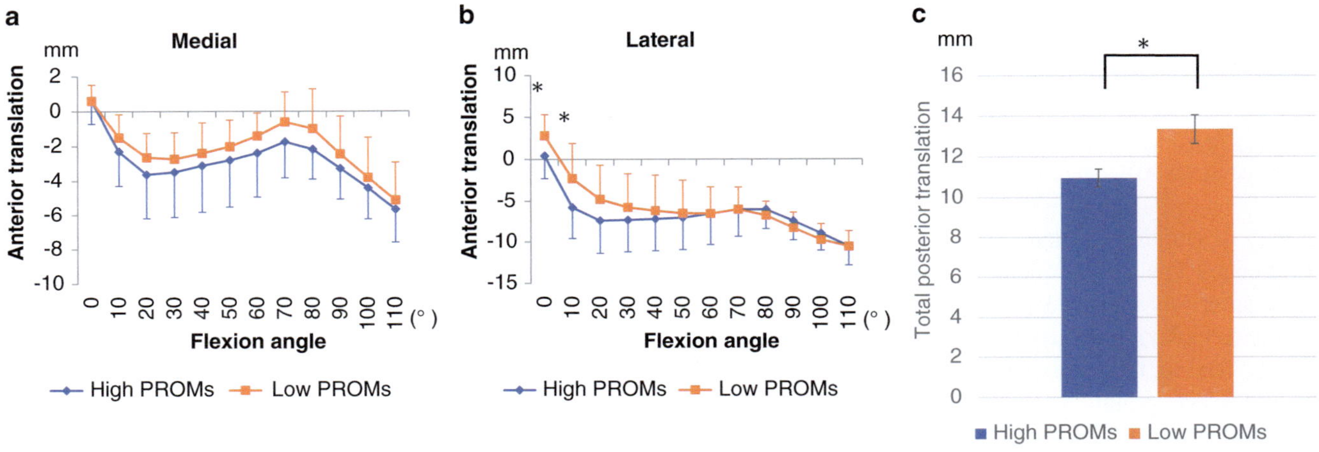

Fig. 26.4 Anteroposterior translation of contact points. (**a**): Flexion angle and medial anteroposterior translation. (**b**): Flexion angle and lateral anteroposterior translation. (**c**): Lateral posterior translation

low PROM group for BCS-TKA. Excessive external rotation of the femur at the medial fulcrum during load flexion is expected to lead to decreased postoperative performance. It was originally noted that iliotibial band traction syndrome might be caused by the rapid induction of external femoral rotation movement by the medial pivot in first-generation BCS-TKA [15]. It has been reported that rollback of ≥6 mm can reduce the KOOS pain score [16]. Internal rotation of the femur due to load flexion [10] and immobility of the lateral side [11] can result in a decrease in postoperative performance; however, in guided-motion models such as BCS-TKA, care should be taken to avoid inducing excessive external femoral rotation or lateral posterior translation.

3 General Review and Future Issues

For satisfactory postoperative outcomes, it is important to maintain medial stability (avoiding excessive anterior femoral movement during flexion) and ensure moderate rollback centered on the lateral side, resulting in external femoral rotation of the medial pivot intraoperatively and postoperatively. These kinematics are similar to those of a normal knee. However, there are reports that post-TKA kinematics have been unable to completely reproduce normal knee kinematics [17], and the extent to which normal knee kinematics can be reproduced or whether it should be reproduced will continue be a topic of discussion. Whether intraoperative kinematics correlate with postoperative kinematics is also a controversial topic [18, 19]. In addition, most evaluations of kinematics and postoperative outcomes have been performed for simple flexion movements. Kinematics differ only by different ADLs in normal and TKA knees [20]. Further evaluation of the relationship between numerous activities (including ADL and sports recreational activities) and postoperative outcomes is expected.

References

1. Matsuda S, et al. Postoperative alignment and ROM affect patient satisfaction after TKA knee. Clin Orthop Relat Res. 2013;471:127–33.
2. Matsuzaki T, et al. Kinematic factors affecting postoperative knee flexion after cruciate-retaining total knee arthroplasty. Int Orthop. 2013;37:803–8.
3. Ishida K. Navigation-based tibial rotation at 90° of flexion is associated with better range of motion in navigated total knee arthroplasty. Knee Surg Sports Traumatol Arthrosc. 2016;24:2447–52.
4. Hanada M, et al. Association of knee flexion angle after posterior-stabilized total knee arthroplasty with postoperative tibial external position relative to the femur and the extent of tibial internal rotation from knee extension to flexion. Knee. 2020;27:1467–75.
5. Ishida K, et al. Factors affecting intraoperative kinematic patterns and flexion angles in navigated total knee arthroplasty. Knee Surg Sports Traumatol Arthrosc. 2015;23:1741–7.
6. Inokuchi T, et al. Intraoperative posterior movement of the tibia at 90° of flexion predicts worse postoperative flexion angles in cruciate-substituting total knee arthroplasty. Knee Surg Sports Traumatol Arthrosc. 2020;28:2816–22.
7. Inui H, et al. Comparison of intraoperative kinematics and their influence on the clinical outcomes between posterior stabilized total knee arthroplasty and bicruciate stabilized total knee arthroplasty. Knee. 2020;27:1263–70.
8. Nishio Y, et al. Intraoperative medial pivot affects deep knee flexion angle and patient-reported outcomes after total knee arthroplasty. J Arthroplast. 2014;29:702–6.
9. Shi K, et al. Kinematic comparison between mobile-bearing and fixed-bearing inserts in NexGen legacy posterior stabilized flex total knee arthroplasty. J Arthroplast. 2008;23:164–9.
10. Kage T, et al. The association between in vivo knee kinematics and patient-reported outcomes during squatting in bicruciate-stabilized total knee arthroplasty. J Knee Surg. 2022;35:1342–8.
11. Van Onsem S, et al. Are TKA kinematics during closed kinetic chain exercises associated with patient-reported outcomes? A preliminary analysis. Clin Orthop Relat Res. 2020;478:255–63.
12. Hill PF, et al. Tibiofemoral movement 2: the loaded and unloaded living knee studied by MRI. J Bone Joint Surg Br. 2000;82:1196–8.
13. Kono K, et al. In vivo three-dimensional kinematics of normal knees during different high-flexion activities. Bone Joint J. 2018;100:50–5.
14. Kono K, et al. The higher patient-reported outcome measure group had smaller external rotation of the femur in bicruciate-stabilized total knee arthroplasty. Knee Surg Sports Traumatol Arthrosc. 2022;30:1292–9.
15. Luycks L, et al. Iliotibial band traction syndrome in guided motion TKA: a new clinical entity after TKA. Acta Orthop Belg. 2010;76:507–12.
16. Twiggs JG, et al. Patient-specific simulated dynamics after total knee arthroplasty correlate with patient-reported outcomes. J Arthroplast. 2018;33:2843–50.
17. Angerame MR, et al. Total knee arthroplasty kinematics. J Arthroplast. 2019;34:2502–10.
18. Belvedere C, et al. Three-dimensional motion analysis of the human knee joint: comparison between intra- and post-operative measurements. Knee Surg Sports Traumatol Arthrosc. 2013;21:2375–83.
19. Wada K, et al. Can intraoperative kinematic analysis predict postoperative kinematics following total knee arthroplasty? A preliminary study. J Med Investig. 2017;65:21–6.
20. Meng F, et al. How prosthetic design influences knee kinematics: a narrative review of tibiofemoral kinematics of healthy and joint-replaced knees. Expert Rev Med Devices. 2019;16:1–15.

Biomechanics of the Patellofemoral Joint (PFJ)

27

Yukio Akasaki

1 PFJ Geometry

The PFJ consists of the patellar articular surface and the femoral trochlear plane, and its conformity changes depending on the angle of knee flexion. Moro-oka et al. [1] acquired MRI images of healthy volunteers at 0°, 45°, 90°, and 135°of knee flexion and analyzed changes in the shape of the PFJ plane of the normal knee. As the knee is flexed from 0° onward, the patella gradually shifts outward (shift) and rotates inward (tilt), sinking into the femoral intercondylar space (Fig. 27.1). The shape of the articular surface at 135° flexion is characteristic of the PFJ, and the inclination of the femoral lateral condyle matches the shape of the articular surface of the patella; thus, a wide contact surface is accessed even during deep flexion.

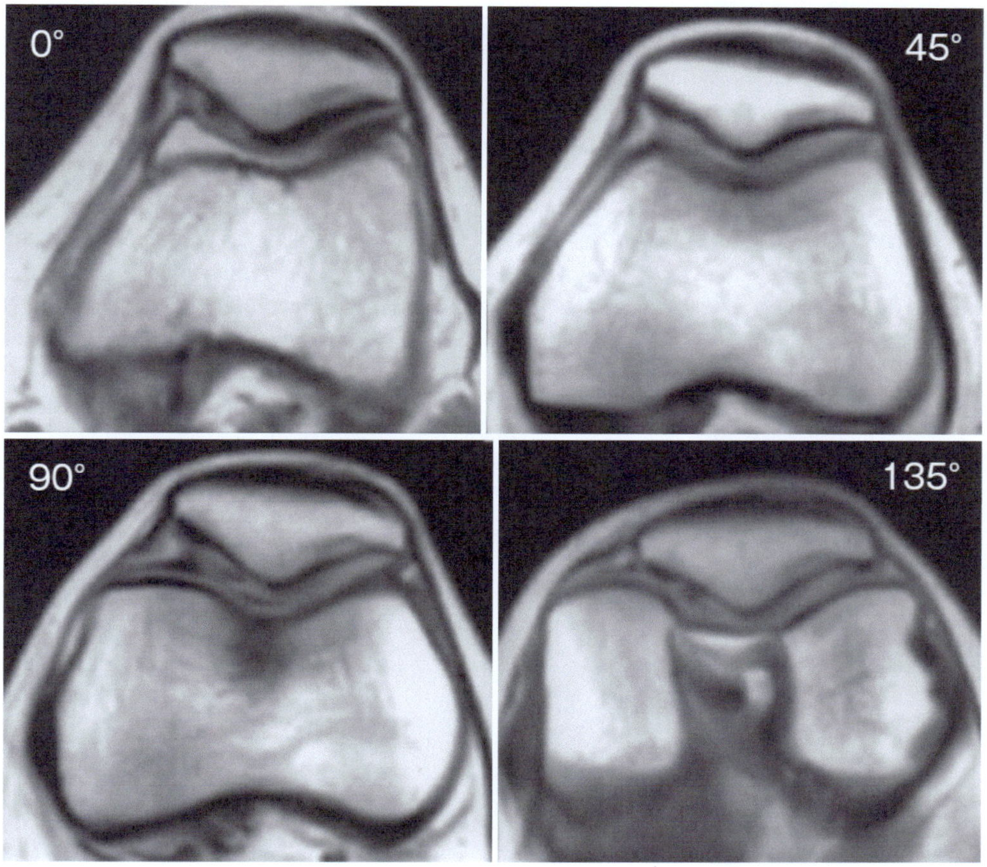

Fig. 27.1 Changes in PFJ geometry. (Reproduced from Moro-oka et al. [1])

Y. Akasaki (✉)
Department of Orthopedic Surgery, Graduate School of Medical Sciences, Kyushu University, Fukuoka, Japan
e-mail: akasaki.yukio.443@m.kyushu-u.ac.jp

S. Matsuda (ed.), *Advances in Total Knee Arthroplasty*, https://doi.org/10.1007/978-981-97-4920-1_27

2 Patellar Tracking

Patellar tracking is defined as the movement of the patella relative to the femur during knee flexion. When using magnetic resonance imaging (MRI) for patellar tracking, the patella displays a six-degree-of-freedom knee joint motion in three dimensions: three translations (movements) in the anterior/posterior, up/down, and inward/outward directions and three rotations in the internal/external, adduction/abduction, and extension/flexion directions. The amount of change can be compared in the normal knee or before and after TKA (Fig. 27.2).

In addition to the aforementioned MRI analyses, studies using fresh cadaver knees are useful for analyzing patellar tracking in normal knees, and the effects of lower leg rotation and quadriceps muscle strength can also be examined. Nagamine et al. [2] installed a fresh cadaver knee without osteoarthritis (OA) on a knee kinematics testing device and found that from 0° to 90°of knee flexion, the patella shifts slightly inward in the initial stage of knee flexion (15°), gradually moving outward as 90° flexion is achieved. The patella tilts slightly inward from 0° to 45° of knee flexion, and then tilts outward during flexion from 45°to 90°. During internal and external rotation, the patella gradually rotates internally as the flexion increases. When the lower leg is rotated internally and externally, the patella shifts inward and outward, and tilts inward and outward, respectively, up to 45° of knee

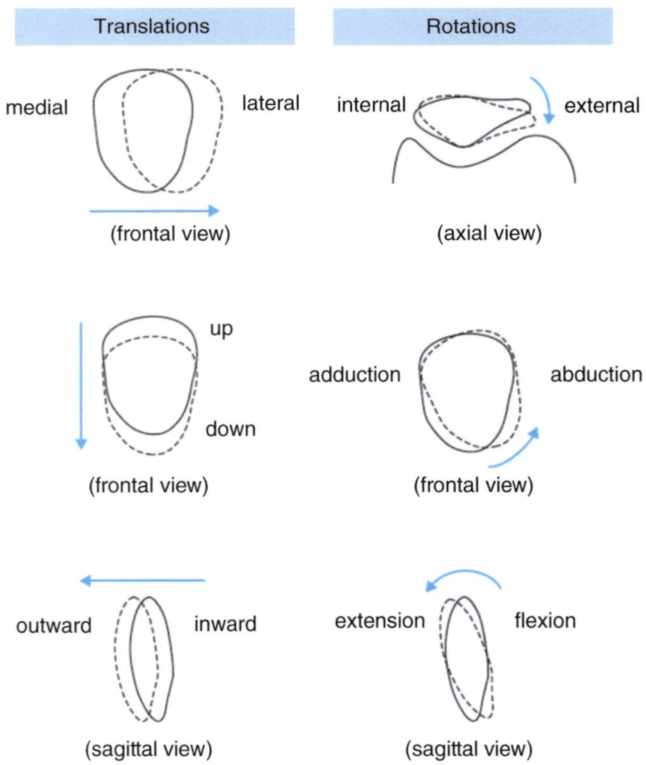

Fig. 27.2 Three-dimensional dynamics of the patella

flexion. The influence of lower leg rotation decreases as the angle of flexion increases.

3 PFJ Contact Pressure by TKA Increases

PFJ pathology is an important issue that has not yet been fully resolved using TKA. This may be one of the reasons for the decline in patient satisfaction caused by residual anterior knee pain after TKA. The degree of PFJ degeneration before surgery, changes in PFJ contact pressure and patellar tracking, and implant design are possible influencing factors. However, it is unclear what preoperative factors may predict the appearance of postoperative PFJ symptoms. It is controversial whether replacement or nonreplacement of patellar components is a dominant factor, as illustrated by the approximately equal proportion of replacement and nonreplacement issues reported in the 2017 report of the Japanese Society for Replacement Arthroplasty.

In vitro experiments using fresh cadaver knees have examined changes in PFJ contact area and pressure due to the replacement or nonreplacement of patellar components during TKA [3–5]. Matsuda et al. [3] compared the nonreplacement, dome-type, and conforming components to the preoperative knee and reported that the PFJ contact pressure of the nonreplacement joint was the same as that of the preoperative knee; however, the contact area of the two substitutions was significantly reduced and the contact pressure increased. On further examination, they concluded that the shape of the femoral implant is greatly affected in the case of nonreplacement, and the contact pressure of the PFJ increased in most shapes [4]. Similarly, Omori et al. [5] reported that the PFJ contact area decreased and the contact pressure increased during knee flexion, even with nonreplacement. A study by Tanikawa et al. using implant designs with enhanced PFJ conformity reported an increase in PFJ pressure after cruciate-retaining (CR), central sensitization (CS), and posterior cruciate-substituting (PS) types of TKA [6]. In other words, the PFJ contact pressure is likely to increase after TKA for both replacement and nonreplacement PFJs.

4 Factors Affecting Patellar Tracking Post-TKA

Placement errors in the patellar components, changes in the femoral and tibial component rotations, and varus/valgus alignment errors lead to poor patellar tracking. The presence or absence of a tourniquet and the surgical approach (parapatellar, midvastus, or subvastus) also affect intraoperative patellar tracking and lateral retinacular tension [7, 8].

4.1 Effects of the Position, Inclination and Thickness of the Patellar Components

Kawano et al. [9] analyzed the postoperative radiographs of 62 TKA cases and found that the lateral tilt and displacement of the patella at 60° of flexion were minimal if the patellar component was placed medially (Fig. 27.3). Similarly, studies using cadaver knees have reported that the combination of medial patellar components and deep femoral trochlear grooves is best for patellar tracking [10].

Fukagawa et al. [11] analyzed postoperative radiographs of 56 TKA patients and reported that the lateral tilt of the patella at 60° of flexion increased in cases where the medial part of the patella was cut more than the lateral part, and that postoperative varus alignment also affected the lateral tilt of the patella.

Tanikawa et al. [12] clarified that the PFJ contact pressure increased by approximately 20% for each +2 mm increase in patellar thickness in cadaver knees, and the effect was greatest at 90° of knee flexion.

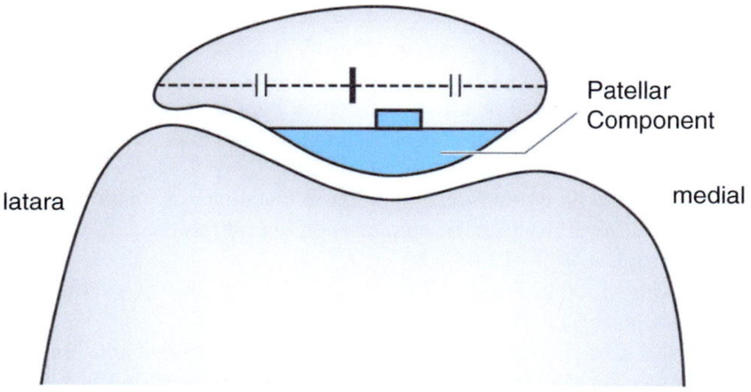

Fig. 27.3 Inner installation of patellar components

Nakamura et al. [13] examined the contact pressure and quadriceps load due to differences in the upper and lower positions of the patellar components using the musculoskeletal model "KneeSIM" and reported that the magnitude of these parameters increased only for the lower installation (Fig. 27.4).

Since the changes reported in each study were small, the overall conclusion to be drawn from these studies is that the patellar component should be placed slightly medial and not downward to avoid an increase in patellar thickness.

4.2 Rotation of Femoral Components and the Influence of Anteroposterior Diameter

Akagi et al. [14] divided 65 TKA cases into a neutral group (32 cases) and a 3–5° external rotation group (33 cases) on the posterior condylar femoral axis according to differences in the rotation of the femoral component. In the external rotation group, only two patients underwent lateral retinacular dissection (compared to the 11 patients in the neutral group), and patellar tracking was reportedly better when the femoral component was positioned for external rotation.

Kawahara et al. [15] measured the PFJ contact pressure before and after an increase in femoral component size during surgery and clarified that an increase in the anteroposterior diameter of the femoral component leads to an increase in PFJ contact pressure at knee flexion of ≥120°.

Fig. 27.4 Differences in the upper and lower placement of patellar components

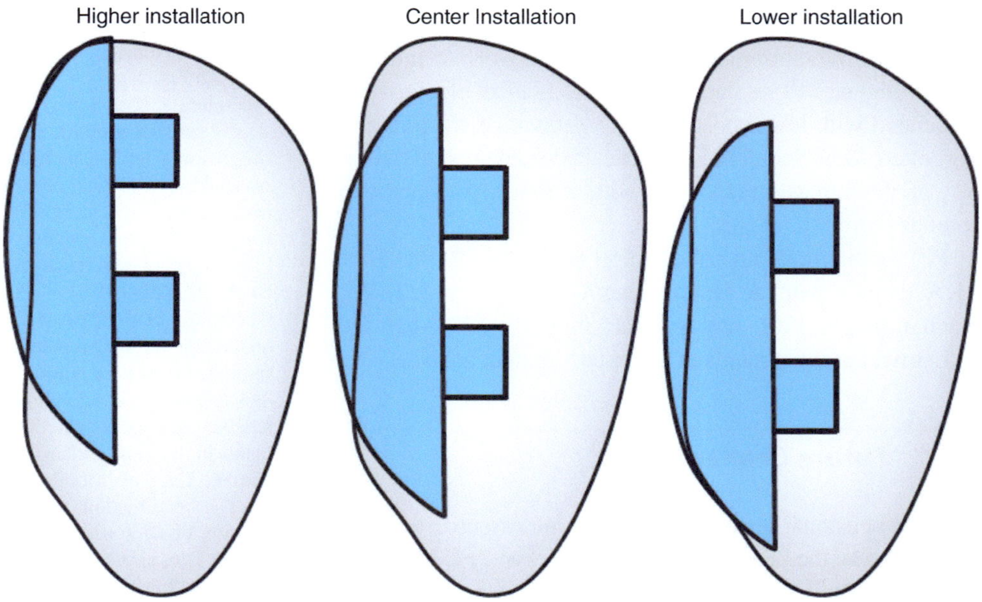

4.3 Influence of Tibial Component Posterior Tilt and Mobile-Bearing-Type Insert

Okamoto et al. [16] used KneeSIM to investigate the effect of a posterior tilt of the tibial components on PFJ contact pressure and quadriceps strength and clarified that PFJ contact pressure and maximum quadriceps muscular strength decrease as the posterior tilt increases. However, they concluded that the posterior tilt should be <5° because an excessive posterior tilt causes anterior instability of the tibia and anterior impingement of the postcam mechanism.

Tibial component rotation is an important factor in PFJ biomechanics. Therefore, many studies have investigated the superiority of mobile bearing prostheses that can tolerate internal and external rotation during patellar tracking. Mobile bearing inserts display a lower PFJ contact pressure in intraoperative measurements than fixed inserts [17, 18]. Intraoperative PFJ contact pressure was lower in patients who were able to reproduce the medial pivot motion that is a characteristic of the kinematics of normal knees [19]. The medial/lateral soft tissue balance correlates with the change in contact pressure between the medial and lateral PFJ during flexion; the lower the PFJ contact pressure during flexion, the deeper the flexion achieved [20].

5 Impact on Clinical Outcome

As mentioned previously, it has been reported that the PFJ contact pressure increases after TKA and that patellar tracking changes compared to the pre-TKA state. Kaneko et al. [21] prospectively examined the relationship between the intraoperative evaluation of PFJ contact pressure and the postoperative clinical outcomes, especially patient-reported outcome assessments. Although none of the 42 TKA patients complained of anterior knee pain postoperatively, an intraoperative higher PFJ contact pressure at 140° of flexion was associated with lower postoperative patient satisfaction and Forgotten Joint Score-12. A higher PFJ contact pressure at 60° of flexion was associated with a lower postoperative patellar score.

PFJ factors that can affect range of motion (ROM) after TKA include patellar tendon length and tension [22–25], patellar height [26], and patellar mobility [27] that are extraarticular components of the extension mechanism.

6 Future Challenges

Three-dimensional imaging techniques are essential to accurately analyze the biomechanical properties of PFJ. However, since the patellar component of TKA is an ultra high-molecular-weight polyethylene that is transparent in radiographs, there is a lack of an accurate and reproducible method for measuring patellar tracking after TKA *in vivo*. Ishimaru et al. [28] have reported a computer-aided image matching method that can analyze the kinematics of normal and post-TKA knees by applying an image window-based analysis method to continuous X-ray scanning. This method is expected to help elucidate the biomechanics of PFJ in the future.

References

1. Moro-oka T, et al. Patellar tracking and patellofemoral geometry in deep knee flexion. Clin Orthop Relat Res. 2002;394:161–8.
2. Nagamine R, et al. Patellar tracking measurement in the normal knee. J Orthop Res. 1995;13:115–22.
3. Matsuda S, et al. Patellofemoral joint after total knee arthroplasty: effect on contact area and contact stress. J Arthroplast. 1997;12:790–7.
4. Matsuda S, et al. Contact stresses with an unresurfaced patella in total knee arthroplasty: the effect of femoral component design. Orthopedics. 2000;23:213–8.
5. Omori G, et al. Contact pressure and three-dimensional tracking of unresurfaced patella in total knee arthroplasty. Knee. 1997:15–21.
6. Tanikawa H, et al. Influence of total knee arthroplasty on patellar kinematics and patellofemoral pressure. J Arthroplast. 2017;32:280–5.
7. Komatsu T, et al. The effect of surgical approaches and tourniquet application on patellofemoral tracking in total knee arthroplasty. J Arthroplast. 2003;18:308–12.
8. Ogata K, et al. Evaluation of patellar retinacular tension during total knee arthroplasty: special emphasis on lateral retinacular release. J Arthroplast. 1997;12:651–6.
9. Kawano T, et al. Factors affecting patellar tracking after total knee arthroplasty. J Arthroplast. 2002;17:942–7.
10. Yoshii I, et al. The effect of patellar button placement and femoral component design on patellar tracking in total knee arthroplasty. Clin Orthop Relat Res. 1992;275:211–9.
11. Fukagawa S, et al. Changes in patellar alignment after total knee arthroplasty. Knee Surg Sports Traumatol Arthrosc. 2011;19:99–104.
12. Tanikawa H, et al. Influence of patella thickness on patellofemoral pressure in total knee arthroplasty. BMC Musculoskelet Disord. 2021;22:298.
13. Nakamura S, et al. Superior-inferior position of patellar component affects patellofemoral kinematics and contact forces in computer simulation. Clin Biomech (Bristol, Avon). 2017;45:19–24.
14. Akagi M, et al. Effect of rotational alignment on patellar tracking in total knee arthroplasty. Clin Orthop Relat Res. 1999;366:155–63.
15. Kawahara S, et al. Upsizing the femoral component increases patellofemoral contact force in total knee replacement. J Bone Joint Surg Br. 2012;94:56–61.
16. Okamoto S, et al. Effect of tibial posterior slope on knee kinematics, quadriceps force, and patellofemoral contact force after posterior-stabilized total knee arthroplasty. J Arthroplast. 2015;30:1439–43.
17. Sawaguchi N, et al. Mobile-bearing total knee arthroplasty improves patellar tracking and patellofemoral contact stress: in vivo measurements in the same patients. J Arthroplast. 2010;25:920–5.
18. Kuroda Y, et al. Mobile-bearing insert reduced patellar contact force at knee flexion during posterior stabilized total knee arthroplasty. Clin Biomech (Bristol, Avon). 2020;76:105022.
19. Konno T, et al. Correlation between knee kinematics and patellofemoral contact pressure in total knee arthroplasty. J Arthroplast. 2014;29:2305–8.

20. Matsumoto T, et al. The influence of intraoperative soft tissue balance on patellar pressure in posterior-stabilized total knee arthroplasty. Knee. 2016;23:540–4.
21. Kaneko T, et al. The influence of compressive forces across the patellofemoral joint on patient-reported outcome after bicruciate stabilized total knee arthroplasty. Bone Joint J. 2018;100:1585–91.
22. Gejo R, et al. Intraoperative patellar tendon strain: predicting the range of knee flexion after total knee arthroplasty. J Orthop Sci. 2009;14:51–5.
23. Ishii Y, et al. Impact of knee flexion on patella length in osteoarthritic patients undergoing total knee arthroplasty. J Orthop Sci. 2013;18:547–51.
24. Ishii Y, et al. Patellar tendon length during knee flexion of meniscal-bearing and rotating total knee arthroplasty implants. Knee Surg Sports Traumatol Arthrosc. 2017;25:3773–8.
25. Sato Y, et al. Quantitative analysis of patellar tendon after total knee arthroplasty using echo intensity: a nonrandomized controlled trial of alpine skiing. J Arthroplast. 2020;35:2858–64.
26. Katagiri H, et al. Increase of patellofemoral height has decreased maximum knee flexion after total knee arthroplasty of posterior cruciate-substituting prosthesis in a clinical series. J Orthop Sci. 2016;21:458–62.
27. Ota S, et al. Is latero-medial patellar mobility related to the range of motion of the knee joint after total knee arthroplasty? Man Ther. 2010;15:574–8.
28. Ishimaru M, et al. Three-dimensional motion analysis of the patellar component in total knee arthroplasty by the image matching method using image correlations. J Orthop Res. 2014;32:619–26.

28 Computer Simulation of the Knee Joint

Shinichi Kuriyama

1 Significance

Total knee arthroplasty (TKA) is an indispensable surgery for end-stage knee osteoarthritis (OA), but approximately 20% of patients in the patient-reported outcome measures have postoperative dissatisfaction. However, the cause of residual knee joint pain and instability is often not identifiable by clinical or imaging findings. The development of new implant designs [1] and surgical techniques [2–4] will require long-term observations to confirm the efficacy. Therefore, simulation analyses can be used to elucidate these issues. Currently, simulators using cadaver knees representing the Oxford Knee rig are established methods to analyze the knee joint biomechanics. In the setup, a cadaver knee with TKA is fixed to the simulator that mimics the lower limb, and the knee motion such as squatting or walking is performed to investigate knee kinematics and kinetics. One problem in using fresh-frozen cadaver knees is that the rigidity is lower than that of in vivo knees, and deterioration may occur during repeated testing. In Japan, formalin-fixed, rigid cadaver knees are readily available, while the facilities that use fresh-frozen, flexible cadaver knees are limited. Thus, a computer simulation analysis is developed as an alternative to cadaver knee biomechanics analysis, such as the finite element method (FEM) [5, 6] and musculoskeletal models [7, 8].

2 FEM Analysis

The analysis predicts the overall behavior by simulating the deformation of individual elements and it achieves this by applying forces to an object, approximating a complex shape and properties, and dividing it into smaller, simpler parts (elements). Moreover, FEM is also applied in numerical analysis and mathematical models [9]. Here, each element consists of a wireframe of tetrahedrons and hexahedrons (Fig. 28.1), and the mesh-like aggregates constitute an object. Tetrahedrons have the advantage to reproduce complex shapes, while hexahedrons have high analysis accuracy. Subdividing elements improves accuracy but increases the amount of computation. Thus, the number of elements is determined according to the analysis task. Additionally, FEM analysis is often used in the stress distribution of elastic polyethylene (e.g., post-cam mechanism of PS-TKA inserts [10–13] or bone [14] rather than rigid metal implants [6]. Furthermore, FEM has been used for TKA design analysis since 1981 in Japan and is now indispensable for determining stresses in various computer analyses.

S. Kuriyama (✉)
Department of Orthopaedic Surgery, Graduate School of Medicine, Kyoto University, Kyoto, Japan
e-mail: kuriyama@kuhp.kyoto-u.ac.jp

S. Matsuda (ed.), *Advances in Total Knee Arthroplasty*, https://doi.org/10.1007/978-981-97-4920-1_28

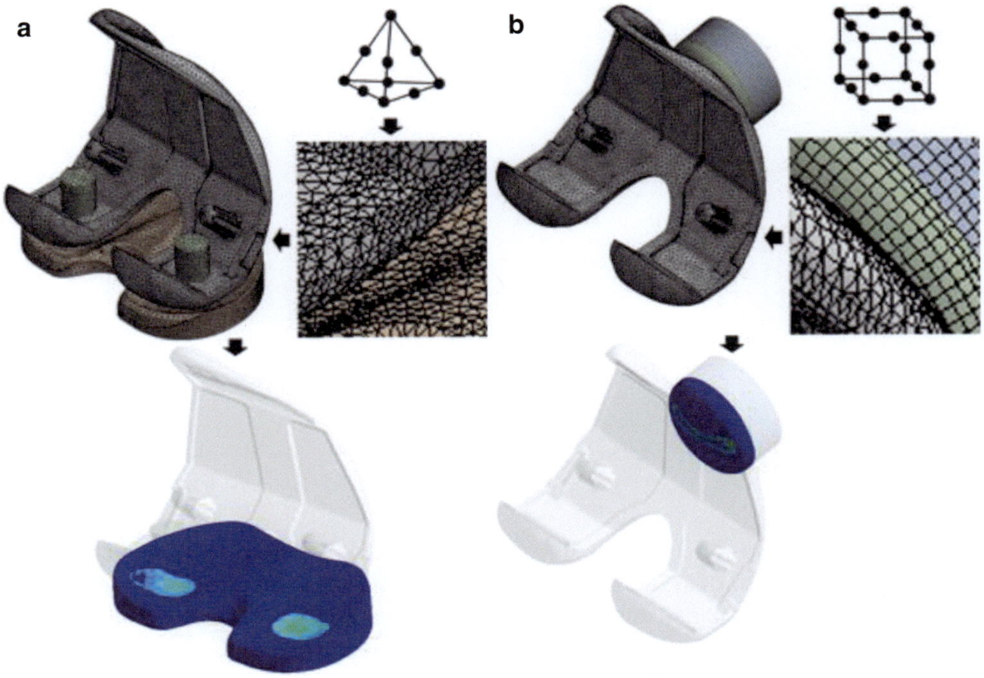

Fig. 28.1 Polyethylene stress analysis using FEM. (**a**): Femoral and tibial inserts with tetrahedral elements. (**b**): Patellar implants with hexahedral elements. (Created from Kuriyama et al. [8])

3 Musculoskeletal Computer Simulation System

3.1 Musculoskeletal Computer Model

Many musculoskeletal computer simulation models are based on KneeSIM (LifeModeler Inc.) as illustrated in Fig. 28.2 developed in the United States [7, 8]. KneeSIM is a computerized cadaver knee simulator that is developed by applying the theory of muscle mathematical models and rigid spring models. The system is a plug-in module specialized in TKA analysis using the ADAMS physics engine. The analysis method consists of a statistical method of design of experiments, which aims to obtain optimal analysis results using efficient experimental methods that includes several factors. The shapes of the femur, tibia, and patella captured from computed tomography data and various implant designs were modeled as stereolithography data. First, the analysis algorithm obtains the minimum force information required to achieve various knee movements using inverse dynamics. Next, weight-bearing loads are applied to the hip joint of the computerized Oxford Knee rig, and squatting and walking are performed based on forward dynamics. Finally, the motion of the knee joint and the distribution of joint forces under different testing conditions are determined in the coordinate axis of the six degrees of freedom.

3.2 Verification Between In Vivo and Computer-Simulated Knees

Consistency between the calculated data of the musculoskeletal knee models and the data obtained from the in vivo knees is important. Various previous studies have verified the reliability of KneeSIM. For example, Mizu-uchi et al. [15] reported almost identical values between the weight-bearing load in patients with tibial implant integrated a load-cell and tibiofemoral contact force calculated by KneeSIM during a squat. The simulator also revealed that the posterior bone spurs protruding from the femoral component and the posterior lip of the tibial insert were impinged, resulting in limitations of the full knee flexion angle [7, 16]. In addition, the patient's actual postoperative flexion angle was similar to the simulator's prediction. Kuriyama et al. [8] demonstrated similarities between the length pattern of the Lateral Collateral Ligament (LCL), Posterior Cruciate Ligament (PCL), and Medial Collateral Ligament (MCL) in cadaver knees after TKA and their respective patterns in KneeSIM. Tanaka et al. [17] compared the anterior-posterior translation of the medial and lateral femoral condyles in three TKA patients who underwent a postoperative image matching procedure [18] and the knee kinematics of the KneeSIM during the squat. They identified almost equivalent in vivo knee kinematics and KneeSIM motion patterns (Fig. 28.3). Similarly, Okamoto et al. [19] and Hada et al. [20] compared the image-matching

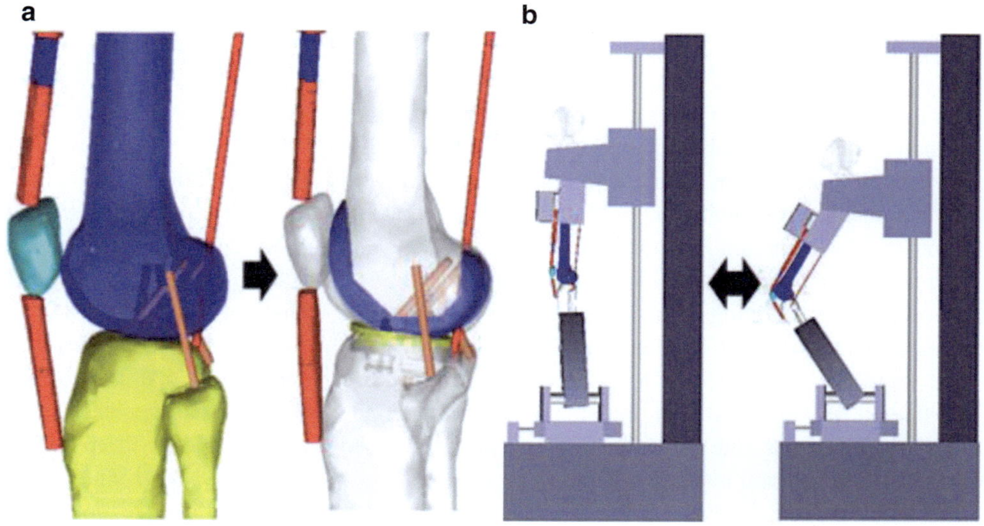

Fig. 28.2 KneeSIM modeling of the Oxford Knee rig. (**a**): Modeling before and after TKA. (**b**): Analysis during knee flexion. (Created from Watanabe et al. [1])

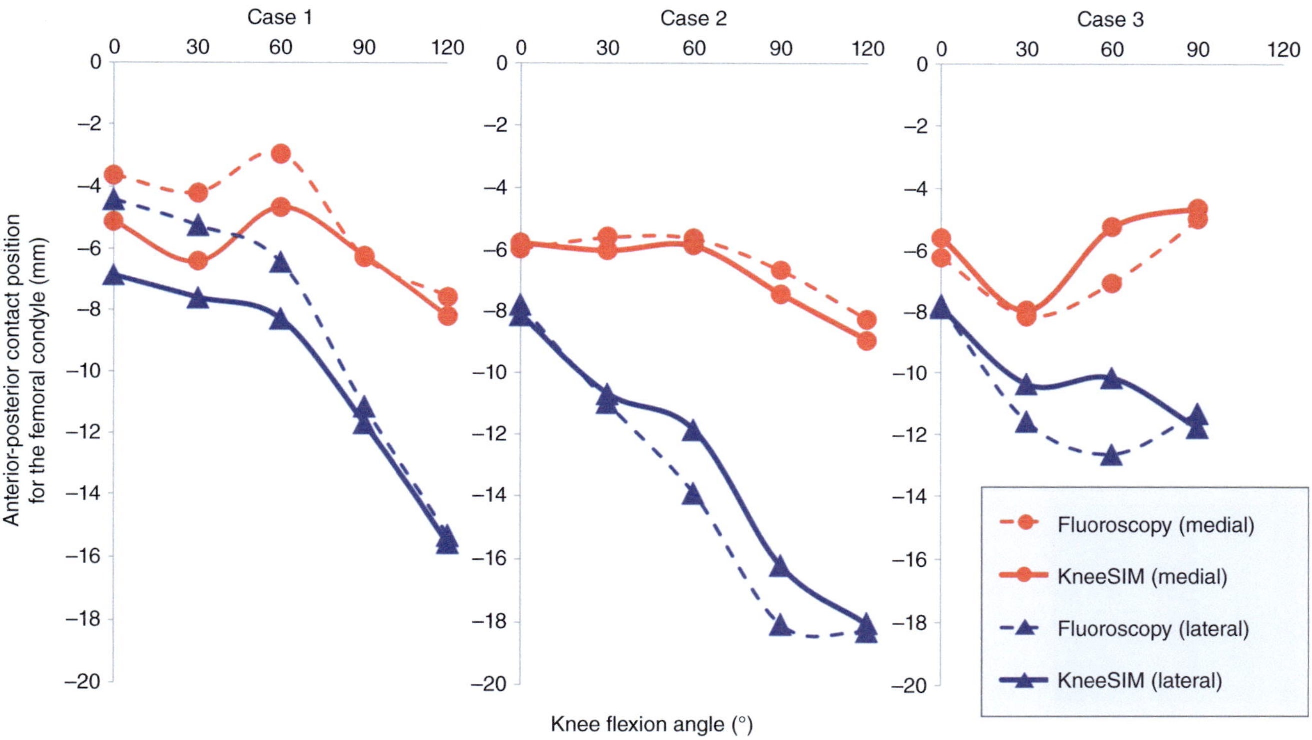

Fig. 28.3 Comparisons of motion pattern during squat between three postoperative TKA patients analyzed by image-matching method using fluoroscopy and corresponding KneeSIM models. (Created from Tanaka et al. [17])

method with the KneeSIM model and reported highly accurate reproducibility.

3.3 Example of Computer Simulator Analysis

KneeSIM is capable of measuring the positional relationships and contact forces of the medial and lateral femoral condyle relative to the tibia or the patella relative to the femur. The tension and length of each ligament of the knee and the force applied to the extensor and flexor mechanisms can also be calculated using KneeSIM. Additionally, the KneeSIM allows for the analysis of knee movements, including closed kinetic chain exercises such as squatting and open kinetic chain exercises such as walking and stair climbing. The system can quantitatively evaluate the effects of changing the angle of implant placement or ligament tension with the knee in a perfectly constant condition. For example, coronal plane analyses have demonstrated that even slight LCL laxity results in a lift-off motion in the lateral tibiofemoral joint in the mid-stance phase of gait when either the femoral component [21] or the tibial component [22] is placed in a varus position. Moreover, when the femoral component with a valgus position is combined with MCL laxity, a medial tibiofemoral lift-off motion is produced [23].

In the sagittal plane, when the posterior tibial slope was reduced, PCL retaining-TKA resulted in an excessive increase in PCL tension in deep knee flexion [24]. In contrast, posterior stabilized TKA reduced anterior-posterior mid-flexion instability by reducing the posterior tibial slope [19], whereas joint line elevation significantly increased anterior-posterior joint instability [25].

In the axial plane, it has been reported that internal rotation of the tibial implant can potentially cause postoperative knee pain, although it is difficult to analyze it clinically. Therefore, KneeSIM analysis is performed during squatting on a model where the tibial implant was adjusted from 15° of internal rotation to 15° of external rotation, in increments of 5° [8]. The results exhibited a sharp increase in MCL tension due to the internally rotational placement of the tibial implant. As the tibial implant rotates internally, the femoral implant also rotates internally due to the constraint of the tibial insert. Consequently, while the tibial MCL attachment remains in the same position, the femoral MCL attachment moves posteriorly along with the internal rotation of the femoral implant. The analysis identified that the distance between the MCL bony attachments was extended, resulting in excessive MCL tension.

3.4 Development of Computer Simulations in Japan

Musculoskeletal computer simulation models centered around the knee have been established in Japan [26]. Recently, Shu et al. [27] developed a combined simulation of the knee model using FEM and a lower limb using a musculoskeletal model. The limitations of FEM analysis up to this point include utilizing a limited analysis range to the knee joint, and an inefficient capability for dynamic analysis. The constraint of KneeSIM, which is based on a musculoskeletal model, was that a detailed analysis was not possible due to simplified components of the knee joint, such as ligaments and muscles. Moreover, FEM analysis could not be performed on KneeSIM alone. Instead, information from one point during knee motion from KneeSIM had to be extracted and performed with another FEM analysis software. On the other hand, the model of Shu et al. [27] proposed a hybrid format that uses a fast-computing musculoskeletal muscle model for the lower limb except for the knee and uses the FEM method only for the knee joint due to the required Central Processing Unit (CPU) power for detailed and extensive analysis. Another system capable of analyzing the hip joint to the ankle joint only by FEM is currently under development at Kyoto University (Fig. 28.4). With further improvement of CPU power, it is expected that a more realistic computer simulation analysis will be achievable in the near future.

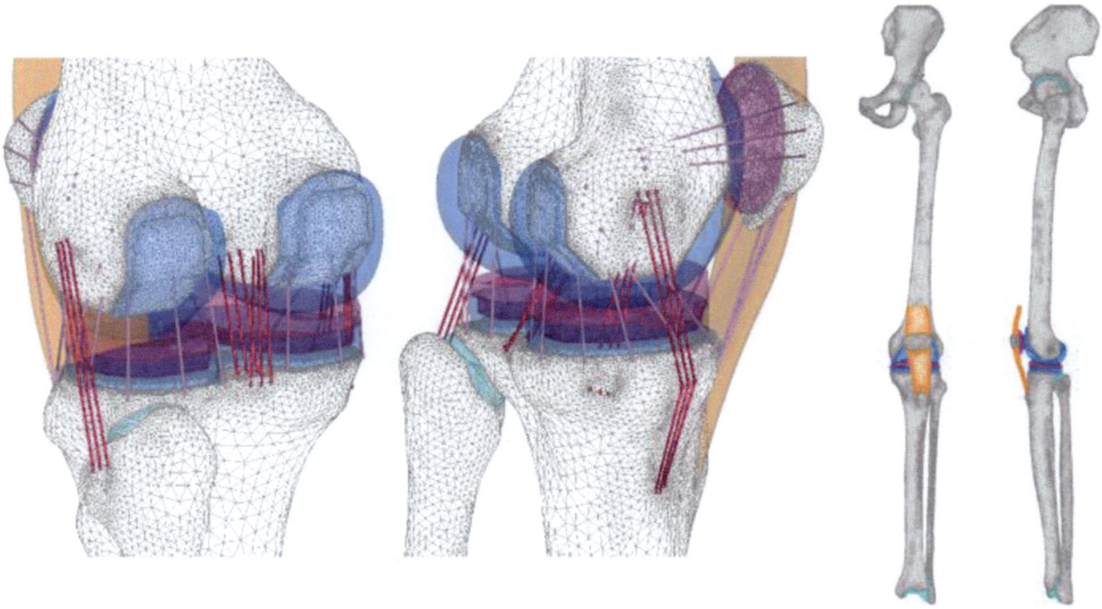

Fig. 28.4 New musculoskeletal computer simulator capable of FEM analysis from the hip joint to the ankle joint

References

1. Watanabe M, et al. Abnormal knee kinematics caused by mechanical alignment in symmetric bicruciate-retaining total knee arthroplasty are alleviated by kinematic alignment. Knee. 2020;27:1385–95.
2. Ishikawa M, et al. Kinematic alignment produces near-normal knee motion but increases contact stress after total knee arthroplasty: a case study on a single implant design. Knee. 2015;22:206–12.
3. Nakamura S, et al. The effects of kinematically aligned total knee arthroplasty on stress at the medial tibia: a case study for varus knee. Bone Joint Res. 2017;6:43–51.
4. Tanaka Y, et al. Medial tilting of the joint line in posterior stabilized total knee arthroplasty increases contact force and stress. Clin Biomech (Bristol, Avon). 2018;53:54–9.
5. Murase K. A finite element analysis of tibial component design in total knee arthroplasty. Fukushima J Med Sci. 1981;27:117–32.
6. Miyoshi S, et al. Analysis of the shape of the tibial tray in total knee arthroplasty using a three-dimension finite element model. Clin Biomech (Bristol, Avon). 2002;17:521–5.
7. Mizu-Uchi H, et al. The importance of bony impingement in restricting flexion after total knee arthroplasty: computer simulation model with clinical correlation. J Arthroplast. 2012;27:1710–6.
8. Kuriyama S, et al. Malrotated tibial component increases medial collateral ligament tension in total knee arthroplasty. J Orthop Res. 2014;32:1658–66.
9. Miura H, et al. Prediction of total knee arthroplasty polyethylene wear using the wear index. J Arthroplast. 2002;17:760–6.
10. Anuar MAM, et al. Mechanics of posterior stabilized total knee arthroplasty during daily activities in relation to prosthesis condylar and post cam design: a finite element study. J Mech Eng. 2017;4:211–28.
11. Osano K, et al. The effect of malrotation of the tibial component of total knee arthroplasty on tibial insert during high flexion using a finite element analysis. Sci World J. 2014;2014:695028.
12. Sakaguchi J, et al. Comparative study for kinetics and kinematics between the CR and PS type artificial knee joints from 3D model analyses. In: World congress on medical physics and biomedical engineering 2006, FMBE proceedings, vol. 14; 2007. p. 3182–3185.
13. Toda M, et al. Computational analysis of stress concentration and wear for tibial insert of PS type knee prosthesis under deep flexion motion. In: 13th international conference on biomedical engineering, vol. 23; 2009. p 1559–1563.
14. Totoribe K, et al. Finite element analysis of the tibial bone graft in cementless total knee arthroplasty. J Orthop Surg Res. 2018;13:113.
15. Mizu-uchi H, et al. Patient-specific computer model of dynamic squatting after total knee arthroplasty. J Arthroplast. 2015;30:870–4.
16. Mizu-uchi H, et al. Effect of total knee arthroplasty implant position on flexion angle before implant-bone impingement. J Arthroplast. 2011;26:721–7.
17. Tanaka Y, et al. How exactly can computer simulation predict the kinematics and contact status after TKA? Examination in individualized models. Clin Biomech (Bristol, Avon). 2016;39:65–70.
18. Sekiguchi K, et al. Varus alignment after total knee arthroplasty results in greater axial rotation during deep knee bend activity. Clin Biomech (Bristol, Avon). 2020;77:105051.
19. Okamoto S, et al. Effect of tibial posterior slope on knee kinematics, quadriceps force, and patellofemoral contact force after posterior-stabilized total knee arthroplasty. J Arthroplast. 2015;30:1439–43.
20. Hada M, et al. Bicruciate stabilized total knee arthroplasty can reduce the risk of knee instability associated with posterior tibial slope. Knee Surg Sports Traumatol Arthrosc. 2018;26:1709–16.
21. Kuriyama S, et al. No condylar lift-off occurs because of excessive lateral soft tissue laxity in neutrally aligned total knee arthroplasty: a computer simulation study. Knee Surg Sports Traumatol Arthrosc. 2016;24:2517–24.
22. Watanabe M, et al. Varus femoral and tibial coronal alignments result in different kinematics and kinetics after total knee arthroplasty. Knee Surg Sports Traumatol Arthrosc. 2017;25:3459–66.
23. Nishitani K, et al. Valgus position of the femoral component causes abnormal kinematics in the presence of medial looseness in total knee arthroplasty: a computer simulation model of TKA for valgus knee osteoarthritis. Knee Surg Sports Traumatol Arthrosc. 2019;27:2051–60.
24. Kuriyama S, et al. Posterior tibial slope and femoral sizing affect posterior cruciate ligament tension in posterior cruciate-retaining total knee arthroplasty. Clin Biomech (Bristol, Avon). 2015;30:676–81.
25. Watanabe M, et al. Impact of intraoperative adjustment method for increased flexion gap on knee kinematics after posterior cruciate ligament-sacrificing total knee arthroplasty. Clin Biomech (Bristol, Avon). 2019;63:85–94.
26. Nikkhoo M, et al. Biomechanical role of the posterior cruciate ligament in total knee arthroplasty: a finite element analysis. Comput Methods Prog Biomed. 2020;183:105109.
27. Shu L, et al. A subject-specific finite element musculoskeletal framework for mechanics analysis of a total knee replacement. J Biomech. 2018;77:146–54.

Part IV

Biomaterials

Polyethylene

29

Naohide Tomita

1 Mechanical Environment of Artificial Knee Joints

The knee joint is in a particularly severe mechanical environment in the human body. Figure 29.1 shows a schematic and symbolic diagram of the structure of the artificial hip and knee joints. The artificial hip joint shown in the upper row is upside-down. In the artificial knee joint shown in the lower part of Fig. 29.1, the contact of the articular surface becomes convex, and the contact area narrows. This structure is required to enable deep knee flexion. Moreover, because extension and in/abduction of the knee joint are strongly restricted, tension applied to the ligament as a limiting force during exercise is added to the load of the sliding contact;

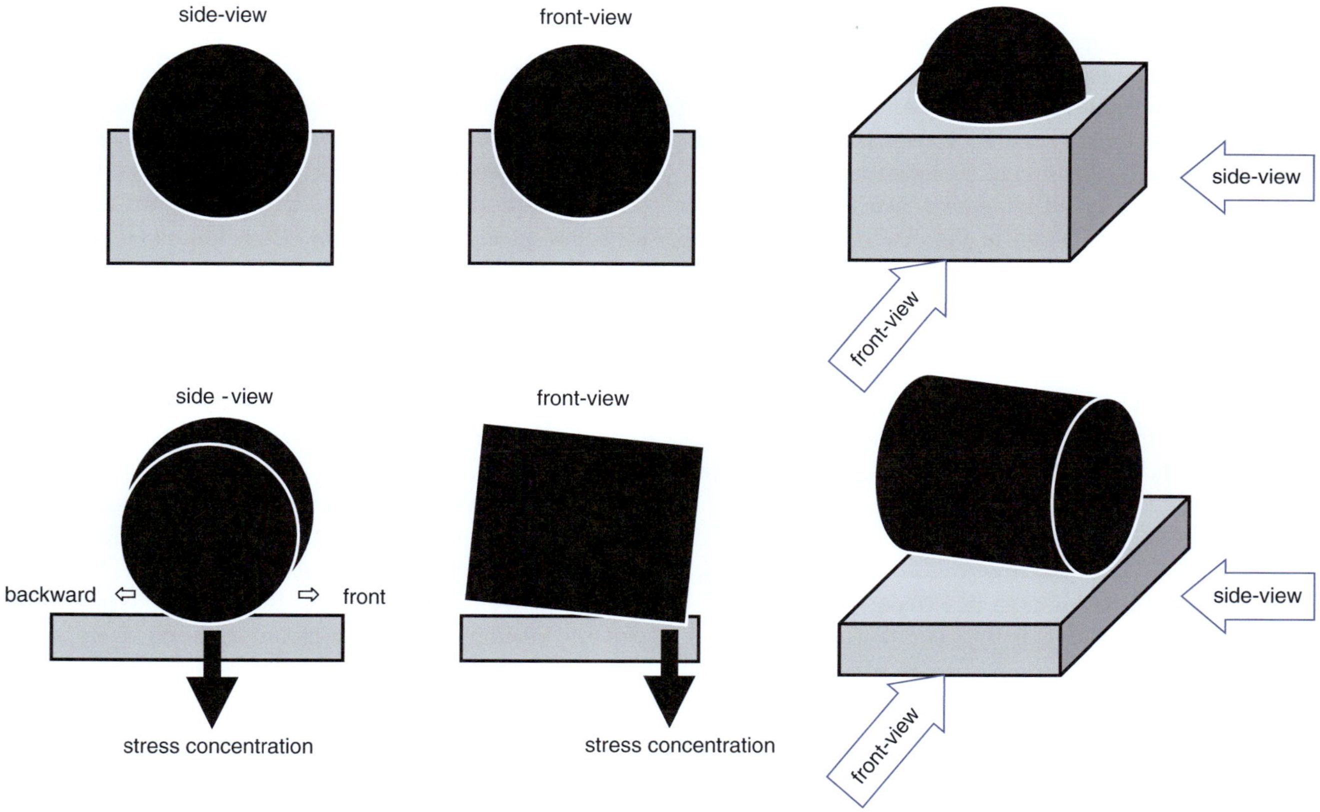

Fig. 29.1 Schematic of the artificial hip and knee joints. The artificial hip joint has a ball-socket structure with a wide contact surface, whereas the artificial knee joint has a convex contact surface

N. Tomita (✉)
Kyoto City University of Arts, Kyoto, Japan
e-mail: n_tomita@kcua.ac.jp

S. Matsuda (ed.), *Advances in Total Knee Arthroplasty*, https://doi.org/10.1007/978-981-97-4920-1_29

therefore, excessive stress is applied to the sliding contact. Thus, the knee joint has a dilemma in which increasing the contact area of the sliding part in the knee joint causes narrowing the range of motion. The applied load during rapid movement causes stress concentration in the sliding material. The reason why knee movement is strongly restricted has not yet been clarified, but it is suggested that this structure is advantageous for fast and sharp movement on various types of surfaces.

2 Joint Design and Material Improvement

Solutions to the problem of stress concentration on the sliding surface of an artificial knee joint can be divided into two methods: improving the joint structure and improving the material. Various efforts have been made to increase the contact area by simulating the meniscus structure and designing joints that reproduce natural knee motions and their installation methods. The former method of devising joint structures includes techniques that do not change the material. One such method prevents stress concentration by reducing the rigidity of the structure by increasing the thickness of polyethylene as much as possible. However, as the space for installing the knee joint is generally limited, the insertion of thick polyethylene may cause movement of the sliding surface and increase the tension of the ligament.

Material improvement is another difficult solution. For example, when the strength of the material increases, the elastic modulus also increases. Because of this increase in the elastic modulus, the stress concentration of the convex contact part becomes more pronounced. In ball-socket structures not in convex contact, such as hip joints, the strengthening method with an increased elastic modulus effectively improves wear resistance; however, care must be taken when using materials with a high elastic modulus in knee joint models with a narrow contact area.

3 Basic Knowledge of Polyethylene

Polymeric materials are composed of chain-like covalent bonds, and various properties can be extracted by changing the higher-order structures of the chain-like molecules. As the glass transition temperature of polyethylene is approximately −120 °C, the amorphous part of polyethylene is in a liquid state at room temperature. In polymer materials, such as polyethylene, the basic mechanical properties are determined by the proportion of crystal parts (crystallinity) with regularly folded arrangements. However, ultra-high molecular weight polyethylene (UHMWPE) has an extremely high viscosity; therefore, even if it is a liquid, the time required for its flow is extremely long. This flow is affected by the fluidity of the non-crystalline part, the cross-linking in which neighboring radicals are chemically bonded in the non-crystalline part, and the tie molecules that connect the crystalline parts.

The molecular weight of polyethylene varies depending on the measurement method; however, polyethylene with a molecular weight of approximately 2 million or more is collectively referred to as UHMWPE. UHMWPE is characterized by excellent mechanical properties such as wear resistance, but it is difficult to mold and process. As mentioned above, the non-crystalline part of polyethylene is in a liquid state at room temperature. However, as viscosity of UHMWPE is extremely high, molding the powder by applying pressure and heat is difficult. Moreover, because the thermal conductivity of polyethylene is low, it is challenging to set conditions such as temperature, time, and deformation speed during molding.

The primary molecular structure of UHMWPE is simple, where carbon and hydrogen are arranged in a regular pattern. However, as described above, there is heterogeneity of 1–10 mm owing to the distribution of heat and pressure during molding and the deformation rate. Another heterogeneity on the order of 10–100 μm exists related to the fusion between powders. Polyethylene is a complex and difficult material to handle, with structures on the order of nanometers, 10–100 μm, and 1–10 mm. The heterogeneity of the nanometer order derived from the structure of molecular chains has been studied by polymer researchers; however, few studies have described the larger structures that affect the actual quality of the product. Thus, the complexity of the fusion and formability of powder interfaces is one of the reasons why research on UHMWPE is difficult.

4 History of Polyethylene Material Development and Osteolysis

Ohnishi et al. irradiated polyethylene with a large dose of 100 Mrad γ in air and used it for artificial hip joints, and reported excellent clinical results over 30 years [1–3]. The cross-linking effect of the non-crystalline part of polyethylene by γ-ray irradiation is thought to be the reason for these positive outcomes. Based on these results, various cross-linked UHMWPE materials have been developed and are now used in many hip prostheses. However, because there has been no engineering experience with large-dose irradiation treatment of 100 Mrad, cross-linked polyethylene with a lower irradiation dose has been put into practical use. Ohnishi's results showed the wear resistance of 100 Mrad-irradiated polyethylene was inferior in pin-on-disk wear tests. However, Ohnishi et al. decided to use 100 Mrad irradiation for clinical use, with an emphasis on the simulator test results that resemble clinical conditions.

On the other hand, The UHMWPE wear powder is phagocytosed or surrounded by macrophages and foreign giant cells in the body, and the released cytokines promote osteoclast differentiation and activity, resulting in osteolysis. As mentioned above, the contact area of the sliding part is relatively small in an artificial knee joint. Therefore, various debates exist regarding applying cross-linked polyethylene to artificial knee joints. For example, because the sliding part is in a convex contact state, the higher elastic modulus of cross-linked polyethylene may increase the stress concentration at the contact part. Fisher et al. noted that using cross-linked polyethylene for artificial knee joints generates relatively high bioactive fine wear particles [4]. In contrast, the National Joint Replacement Registry reported that cross-linked polyethylene reduces osteolysis in artificial knee joints [5]. Recently, the idea that microinfections in the body and tiny amounts of endotoxins produced by bacteria adsorb to the surface of polyethylene to stimulate reactions has been suggested [6]. Therefore, the results of the National Joint Replacement Registry were possibly affected by other factors, such as the use of antibacterial and anti-inflammatory agents, osteoporosis drugs, proton pump inhibitors, and the effect of endotoxin removal from bone cement [7].

5 Development of Polyethylene for Artificial Knee Joints

"Strengthening" in materials science is aimed at increasing strength, fatigue strength, elastic modulus, etc., and in many cases, it makes its fracture more brittle and prone to stress concentration at the contact surface. As mentioned above, when the movement is complicated, and convex contact of the sliding interface is unavoidable, such as in an artificial knee joint, a relatively soft and tough material is required.

The authors mixed dl-α-tocopherol, a type of vitamin E, before molding it into UHMEPE. This material has been reported to prevent oxidative fatigue fractures (delamination fractures) in the polyethylene used for artificial knee joints [8]. UHMWPE has a high viscosity and is produced by pressurization and heat molding from powder. The interface between these powders is easily oxidized, and because it is relatively hard, cracks are likely to occur at this interface because of repeated multidirectional compressive stress [9]. In addition, calcium stearate is removed from medically used polyethylene, suggesting that the powder interface may become more fragile depending on the molding conditions. Considering that dl-α-tocopherol improves the formability of UHMEPE by acting as a lubricant with antioxidant effects, we developed polyethylene with vitamin E for use in artificial knee joints [7]. This material was developed under the guidance of Mr. Yasuo Takuma, who is involved in molding UHMWPE at a polyethylene processing plant. Mr. Takuma continued to provide the authors with prototypes of this material while repeatedly clashing with the president of the company, who prohibited him from being involved in the development of biomaterials.

It was confirmed that the addition of dl-α-tocopherol prevented the formation of cracks at the interface of the powder interface using several simulated mechanical environments of the artificial knee joint. A bidirectional slip fatigue tester was fabricated and a network structure of surface irregularities on the order of 100 μ was observed just before the delamination fracture. This irregularity matched the size of the UHMWPE powder, suggesting the onset of microscopic cracks in the powder-fusion part. The addition of dl-α-tocopherol also suppresses the delamination fractures [10]. In other words, dl-α-tocopherol may have acted as an antioxidant and molding lubricant, helping the powder fuse and preventing internal cracks. Yamamoto et al. reported that adding vitamin E increased the molecular motility of polyethylene molecules [11, 12]. Although it has not yet been clarified, adding tocopherol may change the molecular motility of amorphous polyethylene.

6 Features of Vitamin E-Added Polyethylene for Artificial Knee Joints

Polyethylene with vitamin E for artificial knee joints was developed in Japan to improve the formability of UHMWPE and the fusion of powders to prevent oxidative fatigue destruction at the raw powder interface [7]. Dl-α-tocopherol mixed into the powder is thought to act as a lubricant with antioxidant effects. Incidentally, other substances are better in terms of antioxidant stability and function as lubricants, so why was vitamin E chosen, especially dl-α-tocopherol? Selecting an additive that can ensure long-term safety, durability, and peace of mind while excluding unknown adverse reactions was essential.

Furthermore, Teramura, a graduate student in the author's laboratory in collaboration with the University of Leeds in the United Kingdom, proved through in vitro experiments that the content of dl-α-tocopherol reduces the bioreactivity of wear particles in polyethylene [13, 14]. The results of the mononuclear phagocytosis assay showed that. Cells(human peripheral blood mononuclear cells) contact with polyethylene mixed with vitamin E produced almost no TNF-α, interleukin (IL)-1β, IL-6, and IL-8 14). Although the mechanism underlying this effect has not yet been clarified, Alaa et al. conducted a comprehensive bioinformatics analysis, and reported that dl-α-tocopherol-mixed polyethylene may exhibit anti-osteolytic effects via IL-27 [15]. If this is due to the direct reaction of eluted trace amounts of vitamin E or the suppression of bioreactivity by adsorption on the surface of the wear powder, then it may be temporary effect. However,

If the generation of radicals in polyethylene and the continuous oxidation of polyethylene by radicals are suppressed, a long-term osteolysis inhibitory effect can be expected in vivo. No direct evidence exists regarding whether the inclusion of dl-α-tocopherol suppresses osteolysis in clinic.

Vitamin E mixed with polyethylene has been used in clinical practice since 2010, with manufacturing and marketing approval in Japan. Even the most carefully developed technology is not 100% safe. However, it can be concluded from the results of subsequent basic studies that the possibility of reducing oxidative fatigue fractures (delamination) of artificial knee joints is considerably high [16].

7 Next-Generation Knee Prosthesis Materials

In recent years, the term "human augmentation" has been used in technologies that assist or enhance human functions. Most of them are used in combination with robotics, augmented reality, virtual reality, etc., to expand the field of "can do" functions of the body and the senses. Artificial joints are also a human augmentation technology in a broad sense, and there is a need to expand the range of "can do" such as employment and sports.

As mentioned earlier, if the next generation of artificial joints is required to have high durability to cope with intense dynamic movements, such as work and sports, simulation technology is required to determine the type of load applied to the joints by exercise. In addition, polymeric materials, such as polyethylene, are viscoelastic, meaning their elastic moduli increase under high loading rates. The viscoelastic design of polyethylene materials, which exhibit soft and strong properties even with rapid movement, will also become important in the future. The "durability" of an artificial joint cannot be evaluated by simple measurements, such as abrasion resistance and fatigue resistance. Although the analysis and measurement methods historically constructed in the engineering field guarantee high reproducibility and rigor of meaning, tests that attempt to reproduce clinical situations have low reproducibility and are difficult to interpret because multiple elements are intricately involved. Although it is extremely important to consider the complex interplay of various factors in the development process, various arbitrary interpretations can easily enter the process.

In the industrial world, artistic viewpoints such as "attachment" and "ANSHIN" which are different from "can do" are becoming more important. For example, it is becoming important to seriously consider not only engineering data representing functions, but also how these data affect people's lives. To achieve safety and quality of life, clinicians must acquire basic engineering knowledge, engineers must learn about diversity in clinical practice, and both must have a place for sincere discussions with patients.

References

1. Oonishi H, Kadoya Y. Wear of high-dose gamma irradiated polyethylene in total joint replacement. Long term radiological evaluation, transactions of the 44th annual meeting. Orthop Res Soc. 1998;23:97.
2. Oonishi H, et al. Wear resistance of gamma-ray irradiated UHMW polyethylene socket in total hip prostheses: wear test and long term clinical results. J Biomech. 1990;23:382.
3. Onishi H, et al. Development of ultralow wear CROSS-link polyethylene over 30 years and its transition. Jpn J Rheum Joint Surg. 2003;21:205–13. [Japanese]
4. Fisher J, et al. Wear, debris, and biologic activity of cross-linked polyethylene in the knee, benefits and potential concerns. J Clin Orthop. 2004;428:114–9.
5. Australian Orthopedic Association National Joint Replacement Registry. Annual Report. Adelaide: AOA; 2016. https://aoanjrr.sahmri.com/documents/10180/275066/Hip%2C%20Knee%20%26%20Shoulder%20Arthroplasty
6. Xing Z, et al. Accumulation of LPS by polyethylene particles decreases bone attachment to implants. J Orthop Res. 2006;24:959–66.
7. Brooks RA, et al. Endotoxin contamination of particles produces misleading inflammatory cytokine responses from macrophages in vitro. J Bone Joint Surg Br. 2002;84:295–9.
8. Tomita N, et al. Prevention of fatigue cracks in UHMWPE joint components by addition of vitamin E. J Biomed Res. 1999;48:474.
9. Shibata N, et al. Numerical simulations on fatigue destruction of ultrahigh molecular weight polyethylene using discrete element analyses. J Biomed Mater Res A. 2003;64:570–82.
10. Shibata N, Tomita N. The antioxidative properties of a-tocopherol in g-irradiated UHMWPE with respect to fatigue and oxidation resistance. Biomaterials. 2005;26:5755–62.
11. Takahashi Y, et al. Effects of vitamin E blending on plastic deformation mechanisms of highly crosslinked ultrahigh molecular weight polyethylene (HXL-UHMWPE) in total hip arthroplasty. Acta Biomater. 2015;15:227–36.
12. Yamamoto K, et al. Vitamin E-stabilized highly crosslinked polyethylenes: the role and effectiveness in total hip arthroplasty. J Orthop Sci. 2017;22:384–90.
13. Teramura S, et al. Reduced biological response to wear particles from UHMWPE containing vitamin E. In: 55th annual meeting of the orthopedic research society, Abstract; 2009.
14. Bladen CL, et al. Analysis of wear, wear particles, and reduced inflammatory potential of vitamin E ultrahigh-molecular-weight polyethylene for use in total joint replacement. J Biomed Mater Res Part B. 2013;101:458–66.
15. Terkawi MA, et al. Identification of IL-27 as potent regulator of inflammatory osteolysis associated with vitamin E-blended ultra-high molecular weight polyethylene debris of orthopedic implants. Acta Biomater. 2019;89:242–51.
16. Sakoda H, et al. Evaluating the durability of UHMWPE biomaterials used for articulating surfaces of joint arthroplasty using delamination tests. J Biomed Mater Res B Appl Biomater. 2019;107:65–72.

30 Femoral Component

Masao Akagi

Since femoral components have a sliding surface with polyethylene, the material has (1) high stability in vivo (bioinert), resistance to corrosion (high corrosion resistance), low toxicity (biocompatibility), (2) high hardness (wear resistance), (3) high fatigue strength against repetitive stress (toughness), (4) hard and nondeformable, (5) resistant to impact, (6) light as much as possible, and (7) easy to process and form. These properties have a direct impact on the durability of femoral components, the wear of polyethylene and the occurrence of metal allergies. Materials with high hardness are advantageous as sliding surface materials for artificial joints because the surface can be finished extremely smoothly (mirror surface) by polishing. Further, materials with high hardness have good wear resistance against third bodies. For example, cobalt-chromium alloy is a hard alloy and is often used as a material for femoral components, but surface properties of the retrieved components are coarser than those of ceramic components, and it is known that the amount of polyethylene wear increases when simulator tests are performed using retrieved components made from cobalt-chromium alloy compared to new ones [1, 2]. In addition, adhesion property of the material to bone cement in the case of cement fixation and initial fixation ability and affinity (promotion of bone ingrowth) for bone in the case of cementless fixation can affect the clinical results.

Including materials that are no longer used at present, the characteristics of cobalt chromium alloys (Co-Cr-Mo alloy), titanium alloys (Ti-6Al-4 V alloy), and ceramics (alumina and zirconia) are summarized in Table 30.1. Surface treatment technologies (diffusion injection and coating of ions and molecules) recently introduced are also summarized in Table 30.2.

Long-term observation related to the femoral component materials is necessary to prove their clinical effectiveness. And, further improvements are expected in component materials considering that artificial joint surgeries including TKA tend to be indicated to younger patients.

1 Each Material and Characteristics of the Femoral Component

1.1 Co-Cr Alloy

Cobalt-chromium alloy has been used as a material for osteosynthesis materials (plates and screws) as Vitalium since before the World War II. Vitallium contains Cr (27–30%), Mo (5–7%), and Ni (2.5%). This alloy has high hardness, excellent toughness, high yield strength, high fatigue resistance, and excellent corrosion resistance. In 1956, McKee and Farrar developed a hip prosthesis with a metal-on-metal sliding surface using a cobalt-chromium alloy. In 1960, Müller developed a polyethylene-to-cobalt-chromium alloy hip prosthesis. For TKA, Polycentric Knee [3] was developed by Gunstone in 1969, and Geomedic Knee was developed by Coventry et al., in 1970 [4]. Since then, cobalt-chromium alloy has been used as a material for many femoral components. However, since it contains cobalt-chromium-nickel, it may cause tissue damage and metal allergy due to ionic toxicity. The high specific gravity (heaviness) is another reason for the development of other materials.

Using a cobalt-chromium alloy as the base material, titanium niobium nitride (TiNbN) was applied to the surface by

M. Akagi (✉)
Department of Orthopaedic Surgery, Kashimoto Hospital, Osaka-Sayama, Osaka, Japan
e-mail: makagi@kashimoto.or.jp

S. Matsuda (ed.), *Advances in Total Knee Arthroplasty*, https://doi.org/10.1007/978-981-97-4920-1_30

Table 30.1 Material properties of femoral components (without surface treatment)

	Metal			Ceramics			Composit ceramics		
Material	Pure Titanium (Ti)	Titanium alloy (Ti-6Al-4 V)	Cobalt-chromium alloy (Co-Cr-Mo)	Highly purified alumina (Al2O3)	Yttria-stabilized zirconia(Y-TZP)		Alumina(82%)/ Zirconia (17%) composit ceramics (ZrO2/ Al2O3)	Alumina (75%)/ Zirconia (24%) composit ceramics (ZrO2/ Al2O3)[a]	Alumina (79%)/ Zirconia (18%) composit ceramics (ZrO2/ Al2O3)[a]
Company	–	–	Many	Kyocera	Kyocea	Encore Orthopedics	Ceramtech BIOLOX		Kyocera
Implant	–	–	Many	KU3, physio	KU4, BS5,LFA, ACTIYAS	Foundation knee	Ceramic multigen plus knee	BIOLOX delta femoral ball	ZAUL femoral ball
Specific gravity	4.51	4.42	8.3 ~ 10	3.98	6.05				4.24
Tensile strength(Mpa)		918	1288						
Yielding strength (Mpa)		867	931	600	1450				1400
Melting point	1668	1540–1650							
Young's modulus(Gpa)	106	113	210 ~ 230	400	210				350
Hardness(HV)	100–150	300–322	410–877	1900	1400		1975		1740
Ra (μm)		0.078	0.015(cast) 0.007(wrought)	0.02	0.02			0.002	0.01
Rz		0.48	0.48						

Hardness (HV):HV of diamond is 10,000
TZP tetragonal zirconia polycrystal, *Y-TZP* Yttria (Y2O3)-TZP, *Ce-TZP* Ceria (CeO2)-TZP
[a]Femoral components with these materials are not commercially available at present

Table 30.2 Base materials and technologies of the surface treatment and coating

	Surface treatment			Coating		
Company	S&N	Zimmer		Biomet	DepuySynthes	Implantcast
Products	Lesion, J II	Persona, NexGen LPS-Flex	NexGen CR/PS/ LPS	Vangard	LCS	ACS MB system
Base materials	Zirconium alloy (Zr–2.5%Nb)	Titanium alloy (Ti-6Al-4 V, Tivanium)	Cobalt-chromium alloy(Co-Cr-Mo)	Cobalt-chromium alloy(Co-Cr-Mo)	Titanium alloy (Ti-6Al-4 V)	Cobalt-chromium alloy(Co-Cr-Mo)
Technologies	O2 diffusion	Nitrogen diffusion hardened	Nitrogen diffusion hardened	TiNbN PVD	TiN PVD	
Sliding surface	Zirconia(ZrO2)	Titinium nitride (TiN)	Cobalt-chromium nitiride (Co-CrN)[a]	Titinium nitride niobium (TiNbN)	Titinium nitride(TiN)	
Hardness(HV)	More than twice Co-Cr	929	905	2500	2400–3000	
Ra(μm)	0.082			<0.05	0.035–0.040	0.079
Rz	0.585				0.692	0.627
Thickness of the treatment layer(μm)	5 ~ 7	–	–	3 ~ 6	8	5

[a]Implants with this sliding surface was not commercially available since highly crosslinked polyethylene inserts appeared

physical vapor deposition (PVD). The coated metal is golden, the hardness is four times that of the cobalt-chromium alloy, and the wettability is improved. Therefore, it is expected that this coating technology improve wear characteristics and prevent the elution of metal ions. In a knee simulator test, Malikian et al. [5] reported that the coating reduced the progression of surface roughness but did not improve polyethylene wear. Schroeder et al. [6] reported that TiNbN coatings had slightly higher wear but less elution of metal ions. In a clinical study, Thienpont et al. [7] reported that TiNbN had no difference in clinical results after coating, and there was no revision surgery due to metal allergy when used in patients with a history of metal allergy. Similarly, components coated with titanium nitride (TiN) on sliding surfaces have been developed. Fabry et al. observed that the sliding surface of the component retrieved after several years of use and reported that the TiN coating showed no change over time and that the wear resistance was maintained [8]. In clinical results 5 or 10 years after surgery, TiN on the other hand, Łapaj et al. found no difference between coated and uncoated cobalt chrome alloy components [9, 10]. They said that further improvement of coating technology and quality control would be necessary [11].

1.2 Titanium Alloy

Since titanium alloys have relatively high strength, excellent heat resistance and corrosion resistance, they were developed as military materials during and after the World War II. It was also around the same time that titanium began to be evaluated as a biomaterial. Titanium alloys have a specific gravity of approximately half that of cobalt-chromium alloys and do not contain nickel or cobalt, which are the main causes of metal allergies. In 1940, Bothe et al. reported that titanium was as biocompatible as Vitalium and stainless steel [12] in animal experiments, and in 1953, Clarke and Hickman reported its stability in biological fluids [13]. Similar reports followed, and titanium was tried for clinical use in a variety of medical fields. In the 1960s, Lund University in Sweden reported excellent clinical results of Moore-type artificial hip prosthesis made of titanium. Subsequently, it was shown that the aircraft material Ti-6Al-4 V alloy (6 alumina 4 vanadium titanium alloy) is suitable for orthopedic components, although titanium has lower hardness and strength than cobalt-chromium alloys. In recent years, titanium alloys with high fatigue resistance and titanium alloys containing no vanadium or copper and as little aluminum and other potentially toxic elements as possible have been developed [14].

For TKA, femoral components using titanium alloys have been clinically used since the 1980s. Miller/Galante I is a typical example, but unlike the components currently in use, the sliding surface was not surface treated. The titanium fiber mesh structure was applied to the interface with the bone, and cementless fixation was possible. The patellar component was similarly a metal-backed patella design that allows cementless fixation [15, 16]. However, this model had problems with the design of the PF joint and metallosis generated between the femoral component and the metal-back of the patella due to poor tracking [17]. Titanium alloys have lower hardness and lower wear resistance than cobalt chromium alloys, so there are concerns about third-body wear and metal ion elution. Femoral components made of titanium alloys that have not yet been surface treated are not currently commercially available. In the Miller/Galante II model, which is the successor to the above model, nitrogen diffusion hardened surface treatment was applied to diffuse injection of nitrogen ions on the surface of the base metal. This process forms a thin layer of titanium nitride (TiN) on the surface of the base metal, and the hardness is said to be equivalent to that of cobalt-chromium alloys [18, 19]. Law et al. also found that there were no problems after 5 years of use in patients with self-reported metal allergies [20].

In addition, femoral components (golden) coated with titanium nitride ceramic (TiN) by physical vapor deposition (PVD) using titanium alloy as the base metal are clinically used. TiN is very hard ceramic material (Hardness 2400 ~ 3000). However, the surface roughness of the coating was slightly larger, and there was no superiority in polyethylene wear in the simulation test [21]. In the knee simulator test using carbon fiber reinforced polyethylene (Poly II) as the counter side, TiN coated titanium alloy did not show any scratches in the surface [22]. In addition, it has been reported that this surface-treated component was used in patients with a history of metal allergy, the 10-year survival rate was 95.1%, and there were no clinical problems [23]. However, retrieval observation of femoral head of THA with a TiN PVD-coated titanium alloy showed peeling of the coating that seems to be caused by the third body, and it has been pointed out that it may cause an increase in polyethylene wear [24, 25].

1.3 Ceramics [Alumina (Aluminum Oxide, Al2O3) and Zirconia (Zirconium Oxide, ZrO2), etc.]

Ceramics are extremely stable in vivo (bioinert) and highly corrosion resistant. In addition, the toxicity is extremely low and does not cause allergic reactions due to metal ion elution (biocompatible). The hardness of alumina is the same as that of sapphire, extremely high, and it is resistant to damage. It has many advantages as a material for sliding surfaces, such as very small surface roughness and high wettability [26, 27]. However, firing components with large and complex shapes, such as femoral components, requires advanced technology. After firing, surface resolution formation with a diamond grinding wheel is performed, and then mirror finish is required by polishing using diamond abrasive grains. In THA, femoral heads made of alumina ceramic have been clinically used since the 1970s [28, 29]. Regarding TKA, Kyocera (Kyoto, Japan) developed a cementless alumina ceramic femoral component, and its clinical use began in 1982 by Oonishi et al. [30]. Furthermore, cement alumina ceramic femoral components have been used clinically in Japan since around 1990 [31–33]. Since ceramics generally have low impact resistance (high brittleness), they may be damaged by hammering when fixing with cement. Therefore, the prosthesis has a disadvantage of inferior bone preservation due to increase in thickness of the component [31]. Kyocera has developed high-purity alumina (average grain size: 1.3 μm), which has been used as a material for femoral components since 1998 [27]. However, breakage during fixation of the femoral component with cement cannot be completely avoided. In addition, cases of femoral debonding between the component and cement have been reported. Recently, CerramTec of Germany developed BIOLOX delta, a composite material of alumina (82% by volume) and zirconia (17%), to increase the fracture strength and fracture toughness of alumina. This material is used for femoral components of TKA [34, 35].

To compensate for the above weaknesses of alumina ceramics, Kyocera has been using zirconia ceramics (zirconium oxide, ZrO2, femoral components with an average particle size of 0.3 μm) for the femoral component that are commercially available [36]. Zirconia undergoes a phase transition from tetragonal to monoclinic over time, resulting in an increase in surface roughness due to volume changes. Therefore, zirconia is supplemented with a stabilizer yttria (yttrium oxide, Y2O3) to suppress phase transitions. In addition, zirconia has many properties as a metal including a large specific gravity. So, its strength as a ceramic is somewhat weaker compared to alumina. Except for the Kyocera's, other femoral components made of zirconia have been clinically used (Foundation knee, Encore Orthopedics Inc., Austin, Tx) [37].

1.4 Zirconium Alloy (Zr-2.5Nb)

Zirconium is a titanium-group metal that has excellent biocompatibility and is less likely to cause metal allergies. A femoral component made of zirconium alloy (Zr-2.5Nb) as the base material of the component with a sliding surface of ceramic zirconia (ZrO2) by oxidation treatment (heat treatment at 500 °C in air) is clinically used. Oxygen diffusion causes approximately 5 μm of the component surface to become zirconia ceramic, and the degree of oxidation gradually decreases from the surface to the deeper layer. Unlike coatings, therefore, the zirconia layer on the surface does not peel off even if the main body of zirconium alloy is bend or changes in volume due to heat. Furthermore, since the zirconia ceramic layer formed by this technology is in monoclinic phase, which is the lowest energy state, the surface roughness does not increase over time due to the phase transition [38, 39]. Since the zirconium, the base material of the component, is metal, it is easy to form complex shapes, and it has excellent toughness and strength. The zirconium with the heat-treated sliding surface combines the advantages of metal and ceramic and is a material that compensates for its weaknesses, which has been in clinical use since 1997 for a femoral component [40]. Observations of the retrieved components showed that zirconia ceramic and corresponding polyethylene damage was significantly less than that of cobalt-chromium alloys [41]. However, at present, no superiority in clinical results or survival rate has been found compared to cobalt-chromium alloy femoral components [42, 43].

2 Knee Joint Simulator Test Results

The amount of polyethylene wear (volume or weight wear rate) in the knee simulator test varies greatly depending not only on the material of the femoral component but also on the simulator equipment used, lubricating fluid, exercise and loading conditions, quality of polyethylene on the counter side, etc. In addition, since polyethylene increases in weight due to water absorption during the test, which control is selected for correction of the weight loss (static soak or dynamic soak) can affect simulation results. Although it is difficult to directly compare the results of different tests, the effects of femoral component materials on polyethylene wear are summarized in Table 30.3, [38, 44–46]. These results of simulation tests indicate that amount of polyethylene wear is significantly reduced in ceramic materials with high hardness and wettability, and the amount of wear is further reduced by cross-linking polyethylene.

Table 30.3 Component materials and simulation results

Component materials	Co-Cr	OxZr	Co-Cr	Co-Cr	Co-Cr	Y-TZP	Co-Cr	Alumina	Y-TZP	Co-Cr	Co-Cr	Co-Cr	Ti-Ndium Ti alloy	Co-Cr
Wear rates (mm^3/Mcycle)	4.68 ± 2.30	0.69 ± 0.53	11.1 ± 1.0	6.0 ± 0.7	4.5	Unmeasurable[a]	3.35 ± 1.33	0.15 ± 0.08	0.57 ± 0.18	9.04 ± 1.36	9.12 ± 0.55	0.33(mg/Mcycle)	0.15(mg/Mcycle)	47.7(mg/Mcycle)
PE	Conv.		Conv.	3.5 Mrad XLPE	3.5 Mrad XLPE		Conv.			Conv.		VE HXLPE		Conv.
Knee simulator	AMTI		Shore-Western 6-channel		Shore-Western 6-channel		AMTI			Stallforth & Ungethuem		AMTI		
Authors	Spector BM [38]		Akagi et al. [44]		Tsukamoto et al. [45]		Akagi et al. [26]			Utzschneiders et al. [46]		Zimmer ZRR_WA_2537_12		

Source: Spector et al. [38], Akagi et al. [44], Tsukamoto et al. [45], Akagi et al. [26], Utzschneiders et al. [46]

Co-Cr Cobalt chromium alloy, *OxZr* Oxidized zirconium, *Y-TZP* Yttria-stabilized tetragonal zirconia polycrystal, *PE* Polyethylene, *XLPE* Crosslinked polyethylene, *VE HXLPE* Vitamine E impregnated highly crosslinked polyethylene, Ti-Ndium Ti alloy: Nitriding titanium alloy

[a]Unable to detect weight loss of the polyethylene inserts even if correction of the water absorption during the test was done

References

1. Muratoglu OK, et al. Knee simulator wear of polyethylene tibias articulating against explanted rough femoral components. Clin Orthop Relat Res. 2004;428:108–13.
2. Dowson D, et al. The role of counterface imperfections in the wear of polyethylene. Wear. 1987;119:277–93.
3. Gunston F. Polycentric knee arthroplasty: prosthetic simulation of normal knee movement. J Bone Joint Surg Br. 1971;53:272–7.
4. Coventry MB, et al. Geometric total knee arthroplasty: I. Conception, design, indication, and surgical technique. Clin Orthop Relat Res. 1973;94:171–84.
5. Malikian R, et al. Four station knee simulator wear testing comparing titanium niobium nitride with cobalt chrome. J Bioeng Biomed Sci. 2013;3:3.
6. Schroeder S, et al. Polyethylene wear and metal release of TiNbN-coated knee components. Wear. 2020;458–459:203426.
7. Thienpont E. Titanium niobium nitride knee components are not inferior to chrome cobalt components for primary total knee arthroplasty. Arch Orthop Trauma Surg. 2015;135:1749–54.
8. Fabry C, et al. High wear resistance of femoral components coated with titanium nitride: a retrieval analysis. Knee Surg Sports Traumatol Arthrosc. 2018;26:2630–9.
9. van Hove RP, et al. No clinical benefit of titanium nitride coating in cementless mobile-bearing total knee arthroplasty. Knee Surg Sports Traumatol Arthrosc. 2015;23:1833–40.
10. Louwerens JKG, et al. No clinical difference between TiN-coated versus uncoated cementless CoCrMo mobile-bearing total knee arthroplasty: 10-year follow-up of a randomized controlled trial. Knee Surg Sports Traumatol Arthrosc. 2021;29:750–6.
11. Łapaj Ł, et al. Retrieval analysis of TiN (titanium nitride) coated knee replacements: coating wear and degradation in vivo. J Biomed Mater Res B Appl Biomater. 2020;108:1251–61.
12. Bothe TR, et al. Reaction of bone to multiple metallic components. Sur Gynecol Obst. 1940;71:598–602.
13. Clarke EGC, et al. An investigation into the correlation between the electric potential of metals and their behavior in biological fluids. J Bone Joint Surg Br. 1953;35:467–73.
14. Guillemot F. Recent advances in the design of titanium alloys for orthopedic applications. Expert Rev Med Devices. 2005;2:741–8.
15. Bayley JC, et al. Failure of the metal-backed patellar component after total knee replacement. J Bone Joint Surg Am. 1988;70:668–74.
16. Rader C, et al. The results of total knee arthroplasty with a metal-backed patellar component: a 6-year follow-up study. J Arthroplast. 1996;11:923–30.
17. Effenberger H, et al. Miller-Galante total knee arthroplasty: the importance of material and design on the revision rate. Int Orthop (SICOT). 2001;25:378–81.
18. Shetty RH. Mechanical and corrosion properties of nitrogen diffusion hardened Ti-6Al-4 V alloy. In: Brown SA, et al., editors. Medical applications of titanium and its alloys: the material and biological issues. Philadelphia: ASTM STP 1272, American Society for Testing and Materials; 1996. p. 240–51.
19. Venugopalan R, et al. The effect of nitrogen diffusion hardening on the surface chemistry and scratch resistance of Ti-6Al-4 V alloy. Biomaterials. 2000;21:1669–77.
20. Law JI, et al. Early outcomes of an alternative bearing surface in primary total knee arthroplasty in patients with self-reported metal allergy. Arthroplast Today. 2020;6:639–43.
21. Jones VC, et al. New materials for mobile bearing knee prosthesis-titanium nitride counterface coatings for reduction of polyethylene wear. In: Hamelynck KJ, et al., editors. Mobile bearing knee arthroplasty. LCS®; 2002. p. 349–51.
22. Peterson CD, et al. Component wear of total knee prostheses using Ti-6A1-4 V, titanium nitride coated Ti-6A1-4 V, and cobalt-chromium-molybdenum femoral components. J Biomed Mater Res. 1988;22:887–903.
23. Rossi SMP, et al. Ten-year outcomes of a nitrided Ti-6Al-4 V titanium alloy fixed-bearing total knee replacement with a highly crosslinked polyethylene-bearing in patients with metal allergy. Knee. 2020;27:1519–24.
24. Harman MK, et al. Wear analysis of a retrieved hip component with titanium nitride coating. J Arthroplast. 1997;12:938–45.
25. Raimondi MT, et al. The in vivo wear performance of prosthetic femoral heads with titanium nitride coating. Biomaterials. 2000;21:907–13.
26. Akagi M, et al. Effectiveness of ceramic femoral components on reduction in tibial polyethylene wear: a simulator study. Techniq Knee Surg. 2007;6:206–12.
27. Oonishi H, et al. Ceramic versus cobalt-chrome femoral components; Wear of polyethylene insert in total knee prosthesis. J Arthroplast. 2009;24:374–82.
28. Boutin P. Total hip replacement by sintered alumina prosthesis. Experimental study and first clinical applications. Rev Chir Orthop. 1972;58:229–46.
29. Shikita T, Oonishi H, Hashimoto Y, et al. Wear resistance of irradiated UHMW polyethylenes to Al2O3 ceramics in total hip prostheses. In: Transactions of the 3rd annual meeting of the society for biomaterials, vol. 3; 1977. p. 118.
30. Oonishi H, et al. Alumina versus polyethylene in total knee arthroplasty. Clin Orthop. 1992;282:95–104.
31. Akagi M, et al. The Bisurface total knee replacement: a unique design for flexion. Four- to nine-year follow-up study. J Bone Joint Surg Am. 2000;82:1626–33.
32. Koshino T, et al. Cemented ceramic YMCK total knee arthroplasty in patients with severe rheumatoid arthritis. J Arthroplast. 2002;17:1009–15.
33. Majima T, et al. Clinical results of posterior cruciate ligament retaining TKA with alumina ceramic condylar prosthesis: comparison to Co-Cr alloy prosthesis. Knee Surg Sports Traumatol Arthrosc. 2008;16:152–6.
34. Nguyen TM, et al. Zirconia phase transformation in zirconia-toughened alumina ceramic femoral heads: an component retrieval analysis. J Arthroplast. 2019;34:3094–8.
35. Bergschmidt P, et al. Ceramic femoral components in total knee arthroplasty: two-year follow-up results of an international prospective multicenterentre study. Open Orthop J. 2012;6:172–8.
36. Nakamura T, et al. A new knee prosthesis with Bisurface femoral component made of zirconia-ceramic (report 2). In: Key engineering materials. Trans Tech Publications, Ltd; 2003. p. 607–10.
37. Bal BS, et al. Primary TKA with a zirconia ceramic femoral component. J Knee Surg. 2006;19:89–93.
38. Spector BM, et al. Wear performance of ultrahigh molecular weight polyethylene on oxidized zirconium total knee femoral components. J Bone Joint Surg Am. 2001;83:80–6.
39. Hunter G, et al. Oxidized zirconium. In: Reis MD, et al., editors. Total knee arthroplasty. Berlin, Heidelberg: Springer; 2005. p. 370–7.
40. Innocenti M, et al. Oxidized zirconium femoral component for TKA: a follow-up note of a previous report at a minimum of 10 years. Knee. 2014;21:858.
41. Heyse TJ, et al. Matched-pair total knee arthroplasty retrieval analysis: oxidized zirconium vs. CoCrMo. Knee. 2011;18:448–52.
42. Kim YH, et al. Comparison of genesis II total knee replacement with oxidized zirconium and cobalt-chromium femoral components in the same patients: a prospective, double-blind, randamised controlled study. J Bone Joint Surg Br. 2012;94:1221–7.
43. Vertullo C, et al. Twelve-year outcomes of an Oxinium total knee replacement compared with the same cobalt-chrome design. J Bone Joint Surg Am. 2017;99:275.
44. Akagi M, et al. Wear and toughness of crosslinked polyethylene for total knee replacements: a study using a simulator and small-punch testing. J Orthop Res. 2006;24:2021–7.
45. Tsukamoto R, et al. Y-TZP zirconia run against highly cross-linked UHMWPE tibial inserts: knee simulator wear and phase-transformation studies. J Biomed Mater Res B Appl Biomater. 2008;86:145–53.
46. Utzschneider S, et al. Wear of contemporary total knee replacements: a knee simulator study of six current designs. Clin Biomech. 2009;24:583–8.

Clinical Studies on Polyethylene Wear

31

Yukihide Minoda

Polyethylene wear is one of the causes of failure and revision of TKA. Various materials and designs have been developed to reduce polyethylene wear. This chapter describes the influence of polyethylene wear on the clinical outcomes of TKA.

1 Pathology of Polyethylene Wear in TKA

The following are the problems associated with polyethylene wear in TKA:

1. Submicron-sized polyethylene wear particles formed between the polyethylene and the component (Fig. 31.1) stimulate macrophage reaction. Macrophage releases chemical mediators and finial induces osteoclast, and results in osteolysis and loosening [1]. These particles may also cause edema and cysts around the joints [1–3]. In addition, polyethylene wear occurs not only on the upper surface of the insert but also on the lower surface (Fig. 31.2).
2. Oxidation of polyethylene causes delamination wear, resulting in polyethylene fracture and instability (Fig. 31.3).

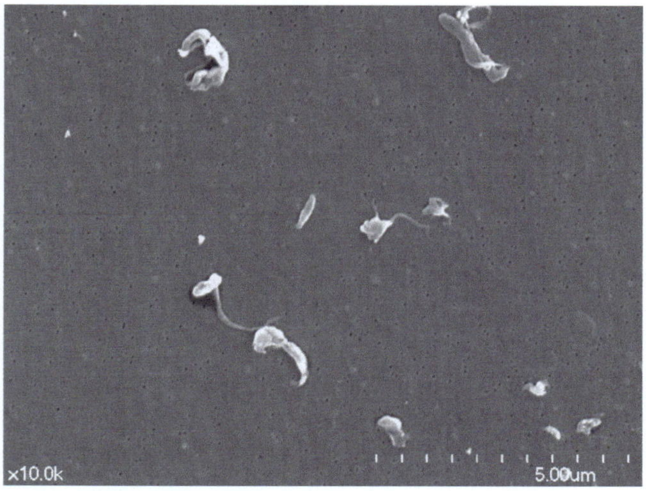

Fig. 31.1 Electron microscope image of polyethylene wear particles produced from the TKA polyethylene insert (×10,000)

2 History of Polyethylene in TKA

In the 1960s, the utilization of polyethylene in artificial joints was introduced. Polyethylene with carbon fibers and heat treatment was also introduced; however, they were discontinued owing to the increase in polyethylene wear [4, 5]. Until the 1990s, polyethylene was mainly sterilized by 25–40 kG gamma irradiation in air. When gamma irradiation was performed in the presence of oxygen, the free radicals in the carbon chain formed cross-links and combined with oxygen, resulting in the oxidation of polyethylene. Mechanical stress in vivo led to a phenomenon called "white band" in the lower layer of the articular surface, resulting in cracking of the polyethylene layer caused by delamination wear owing to reduced mechanical properties (Fig. 31.2). To prevent oxidation during sterilization, gamma irradiation was performed in the presence argon, nitrogen, ethylene oxide, or gas plasma. This sterilization method could reduce but not eliminate the in vivo oxidation of polyethylene. Oxidation of polyethylene within the packaging after sterilization was found to be one of the causes. Therefore, packaging that is less permeable to oxygen were introduced.

Y. Minoda (✉)
Department of Orthopaedic Surgery, Osaka Metropolitan University, Osaka, Japan
e-mail: yminoda@omu.ac.jp

S. Matsuda (ed.), *Advances in Total Knee Arthroplasty*, https://doi.org/10.1007/978-981-97-4920-1_31

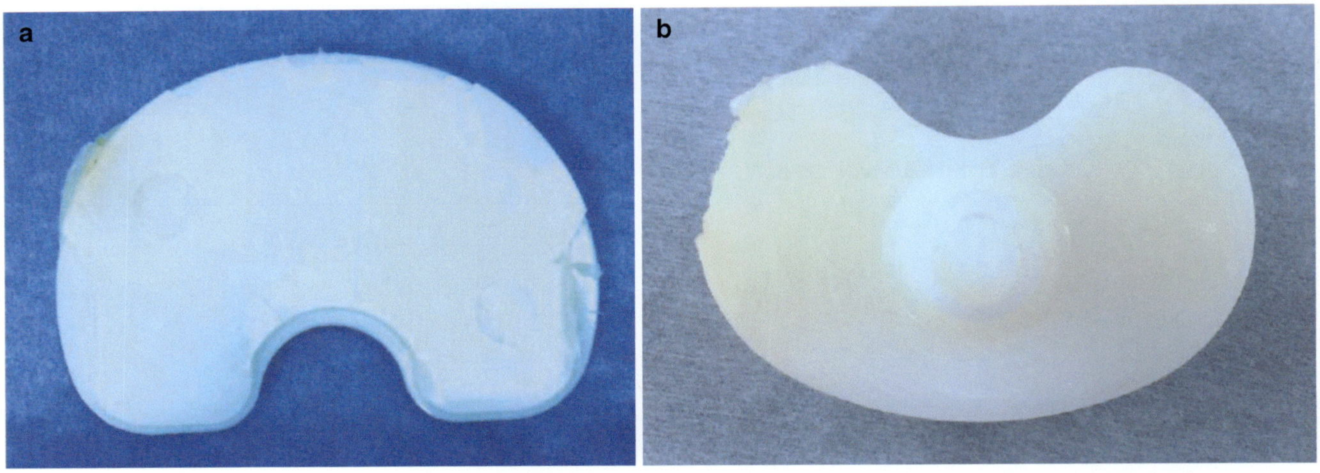

Fig. 31.2 Underside of Fixed-Bearing and Mobile-Bearing inserts. Polyethylene wear also occurred on the underside of the insert. In mobile bearings, wear is observed in concentric circles

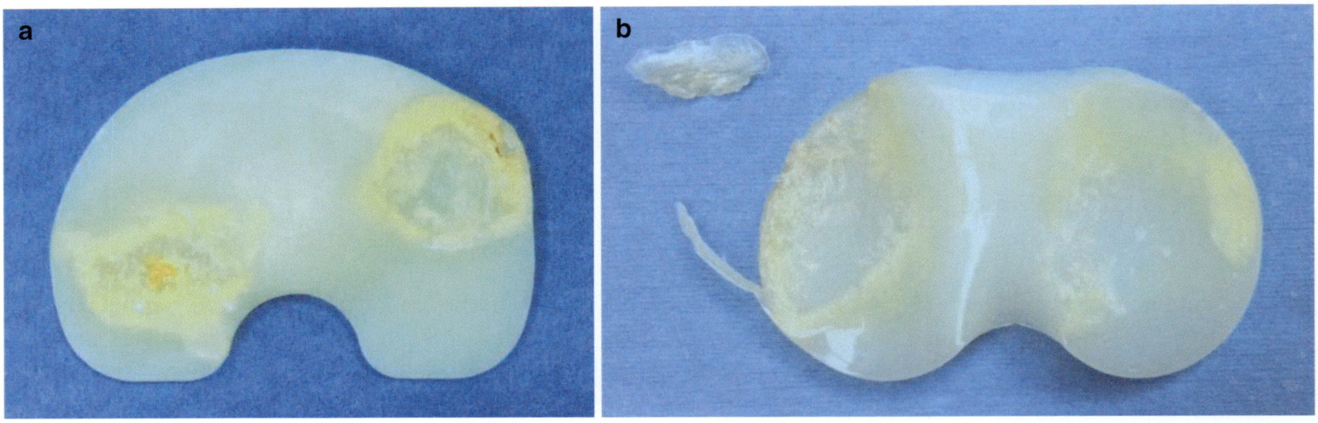

Fig. 31.3 Polyethylene insert fracture owing to delamination wear

In the late 1990s, highly cross-linked polyethylene (HXLPE) was introduced to reduce the biological reactions caused by polyethylene wear particles. Gamma irradiation (50–100 kG) was used to cut the polyethylene carbon chains. While some polyethylene carbon chains underwent cross-linking, others did not, leaving behind free radicals. These residual free radicals combined with oxygen in vivo, resulting in the oxidation of polyethylene. Thermal treatment was performed to eliminate the residual free radicals in the cross-linked polyethylene. Two thermal treatment methods were employed. Remelting, which involved applying heat above the melting point, removed the residual free radicals of the crystalline part but also decreased the degree of crystallinity and fatigue resistance. Annealing, which involved applying heat below the melting point, retained fatigue resistance but did not eliminate free radicals in the crystalline part.

Sequential annealing was introduced to reduce the residual free radicals of the annealed HXLPE, in which low-dose 30 kG gamma irradiation and annealing were repeated three times for a total of 90 kG gamma irradiation.

From the latter half of the 2000s, antioxidants such as vitamin E and Covernox have been used to avoid the decrease in fatigue resistance owing to heat treatment after cross-linking. Because heat treatment is not performed after cross-linking, a decrease in fatigue resistance can be prevented, and residual free radicals can use antioxidants to prevent oxidation. Two methods are available for combining these antioxidants: one involves blending antioxidants during polyethylene formation before cross-linking (blended), while the other involves immersing antioxidants after cross-linking (doped). Therefore, the antioxidant-blended cross-linked polyethylene prevents the production of free radicals during irradiation, making it difficult to predict the percentage of cross-linking. On the other hand, antioxidant-doped cross-linked polyethylene could be reliably crosslinked; however, additional heat treatment is required after cross-linking to infuse antioxidants into the polyethylene.

3 Articulating Surface Design and Polyethylene Wear

The larger the contact surface area between the femoral component and the tibial insert, the lower the contact pressure, producing fewer polyethylene wear particles. However, when the contact area is larger, the constraint between the femoral component and tibial insert becomes higher, and the mechanical stress, such as edge loading of the polyethylene insert, increases during knee rotation, making polyethylene prone to mechanical failure. Conversely, if the contact surface area between the femoral component and the tibial insert is smaller, there is reduced mechanical stress, such as edge loading during knee rotation; however, the contact pressure increases, leading to a higher generation of polyethylene wear particles. Therefore, the articular surface shape of TKA must have a "moderate" contact area and constraint. The total condylar type has a "moderate" contact area and restraint and remains a widely used TKA design. In 1978, mobile bearings were introduced as an alternative solution. It achieves a larger contact area and fewer constraints by increasing the contact area on the upper surface of the tibial insert and reducing the constraints for knee rotation on the lower surface of the tibial insert, using a mobile-bearing mechanism. Roles were divided between the upper and lower surfaces. A concern with this design is the potential increase in the number of joint surfaces that produce polyethylene wear particles to two surfaces: the upper and lower surfaces of the tibial insert. The medial pivot type was first introduced in 1994. The medial condyle was designed as a ball-socket joint, similar to THA to maximize the contact area and reduce polyethylene wear particle generation. Conversely, the lateral condyle was designed to be less constrained to allow medial pivot motion and reduce mechanical stress during knee rotation. Roles were divided between the medial and lateral condyles.

4 Femoral and Tibial Component Material and Polyethylene Wear

A rough or damaged surface of the femoral component increases polyethylene wear. To reduce polyethylene wear, the material of the femoral component requires low surface roughness and high scratch resistance. Therefore, cobalt-chromium alloys are widely used as femoral components. In the mobile-bearing type, where the undersurface of the tibial insert is also articular, a cobalt-chromium alloy is also used for the tibial component. Additionally, alumina ceramic, zirconia ceramic, zirconium oxide, and ceramic-coated components have been introduced as materials with lower surface roughness and scratch resistance than cobalt-chromium alloys (Fig. 31.4).

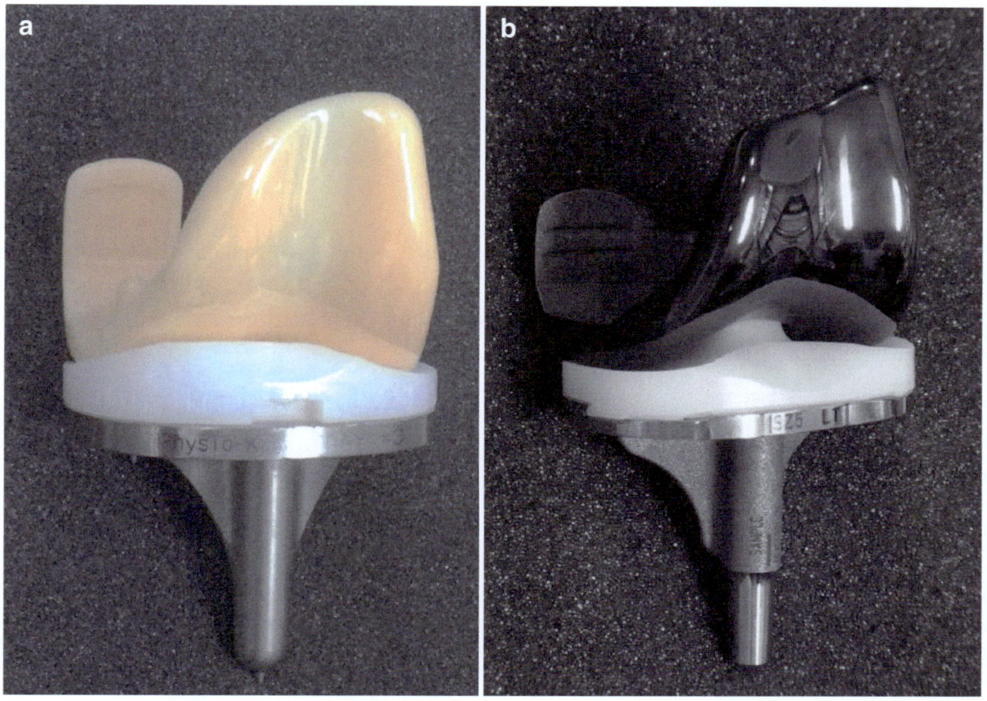

Fig. 31.4 Femoral components made of alumina ceramic and zirconium oxide

5 Clinical Evaluation of Polyethylene

5.1 Early Feedback of Polyethylene Wear

Various designs and materials for articulating surfaces have been tested in basic laboratory experiments and clinically applied. However, upon reviewing the history of TKA, including Poly II and heat press-like polyethylene, the newly introduced design and material does not necessarily reduce in vivo polyethylene wear particle generation, nor does it lead to excellent clinical results. However, it takes decades for feedback on polyethylene wear and clinical results of the newly introduced TKA prostheses to become available. Even if more polyethylene wear and poor clinical results are reported decades later, the issue becomes challenging to address because these prostheses would have already been used in many patients. For newly introduced TKA prostheses, early feedback on polyethylene wear is clinically important.

5.2 Polyethylene Wear Measurement by X-Ray

THA has a ball-in-socket joint, allowing for accurate evaluation of the amount of polyethylene wear from postoperative X-ray images. However, most TKA do not have a ball-in-socket joint, posing challenges to accurately measure polyethylene wear using X-ray images in TKA. Nonetheless, there have been reports of polyethylene wear volume measurements using 3D/2D matching [6, 7].

5.3 In Vivo Polyethylene Wear Particle Analysis [8]

When a polyethylene insert is worn at the interface with another material, polyethylene wear particles are produced (Fig. 31.1). These particles are mainly produced from the articulating surface between the femoral component and upper surface of the tibial insert, locking interface between the lower surface of the tibial insert and tibial tray in a fixed-bearing prosthesis, articulating surface between the lower surface of the tibial insert and tibial tray in a mobile-bearing prosthesis, and interface between the patella and femoral components. These polyethylene wear particles are released into the joint fluid and are captured by macrophages in the joint capsule. Macrophages release various chemical mediators, such as TNFα, IL-1β, IL-6, RANKL, and M-CSF, which induce osteoclasts, cause osteolysis around the prosthesis, and lead to loosening. Therefore, polyethylene wear particles are clinically important.

For polyethylene wear particles, there are three important factors: (1) the greater the number, [9], (2) the smaller the size [10], and (3) the more elongated the shape [11], the higher the biological activity and the higher the likelihood of osteolysis and loosening. Therefore, reducing the number of small, sharp polyethylene wear particles is clinically important.

In vivo polyethylene wear particle analysis was introduced for early feedback on polyethylene wear particle generation [8]. Joint fluid was collected from the knee joint with a well-functioning TKA prosthesis in the early clinical stage, with IRB approval and patient informed consent. After protein degradation by a strong hydroxide, polyethylene wear particles in the joint fluid were isolated by undergoing ultracentrifugation twice. The morphology of the particles was observed using an electron microscope and analyzed using an image analyzer. The number (counts in the joint), size (equivalent circle diameter), and shape (roundness and aspect ratio) were calculated.

5.4 Influence of the Polyethylene Manufacturing Method on In Vivo Polyethylene Wear Particle Generation

Remelted HXLPE (Prolong; Zimmer) generated fewer, smaller, and rounder particles than conventional polyethylene with the same prosthesis design (NexGen CR; Zimmer) [12].

Sequentially annealed, highly cross-linked polyethylene (X3; Stryker) generated fewer particles than conventional polyethylene (N2Vac; Stryker) with the same prosthesis design (Scorpio NRG PS; Stryker). However, no significant difference in the size or shape of the polyethylene wear particles was observed [13]. Vitamin E-doped HXLPE (E1; Zimmer Biomet) generated more, smaller, and rounder particles than conventional polyethylene (ArCom; Zimmer Biomet) with the same prosthesis design (Vanguard PS; Zimmer Biomet) [14].

5.5 Influence of the Articulating Design on In Vivo Polyethylene Wear Particle Generation

Medial pivot prothesis (Advance MP; Wright Medical) generated fewer particles than PS prosthesis (IB-II, Zimmer), but no significant difference in the size or shape of the polyethylene wear particles was observed [15]. There was no significant difference in the number, size, and shape of polyethylene wear particles between mobile-bearing CR prosthesis (MBK; Zimmer) and PS prosthesis (IB-II,

Zimmer) [16]. There was no significant difference in the number, size, and shape of polyethylene wear particles between mobile-bearing PS prosthesis (Vanguard RP; ZimmerBiomet) and fixed-bearing PS prosthesis (Vanguard PS; ZimmerBiomet) [17].

5.6 Influence of Femoral Component Material on In Vivo Polyethylene Wear Particle Generation

Alumina ceramic femoral component (MPK; Kyocera) generated fewer and rounder particles than Co-Cr alloy femoral component (Advance MP; Wright Medical) [18]. There was no significant difference in the number, size, and shape of polyethylene wear particles between oxidized zirconium (Genesis II; Smith & Nephew) and Co-Cr alloy femoral component (Genesis II; Smith & Nephew) [19].

In vivo polyethylene wear particle analysis showed that polyethylene manufacturing method, articulating surface design, and material of femoral component influence in vivo polyethylene wear particle generation.

- Medial pivot design generates less polyethylene wear particles than PS prosthesis.
- Mobile-bearing prosthesis generates similar polyethylene wear particles to fixed-bearing prosthesis.
- Alumina femoral component generates less polyethylene wear particles than Co-Cr alloy femoral component.
- HXLPE generates less polyethylene wear particles than conventional polyethylene.
- Vitamin E-doped HXLPE generates more polyethylene wear particles than conventional polyethylene.

Regarding Vitamin E-doped HXLPE, we need careful follow up for long term.

6 Clinical Performance

6.1 Polyethylene Manufacturing Method

An analysis of 950 knees using a CR TKA prosthesis with conventional polyethylene reported an increase in the revision rate with gamma irradiation in air and prolonged shelf life [20]. A comparison between remelted HXLPE (Prolong; Zimmer) and conventional polyethylene using the same design (NexGen CR; Zimmer) showed no osteolysis, aseptic loosening, or polyethylene failure 2 years postoperatively, as well as no significant difference in clinical outcomes [21]. Although numerous reports have shown no difference in revision rates between HXLPE and conventional polyethylene, the Australian National Joint Registry showed that HXLPE (4.9%) exhibited less loosening than normal polyethylene (7.8%), along with a lower revision rate at 15 years [22]. This difference was more pronounced in active patients aged <65 years, with revision rates of 7.0% for HXLPE and 11.7% for conventional polyethylene at 15 years.

When using the same model (Vanguard PS; Zimmer Biomet), there was no osteolysis, loosening, or polyethylene failure 2 years postoperatively and no significant difference in clinical outcomes between Vitamin E-doped HXLPE (E1; Zimmer Biomet)and conventional polyethylene (ArCom; Zimmer Biomet). There are few reports on the clinical results of HXLPE-containing antioxidants, and further research findings are awaited [23].

6.2 Articulating Surface Design

For the articulating surface shape, a mobile-bearing PS prosthesis (Vanguard RP; Zimmer Biomet) and a fixed-bearing PS prosthesis (Vanguard PS; Zimmer Biomet), both using identical polyethylene materials, were compared 2 years postoperatively. The comparison showed no osteolysis, loosening, or polyethylene failure, and no significant difference in the clinical results [17]. Australian national registry data showed that the revision rate for fixed-bearing prostheses remained lower up to 7 years postoperatively, but the revision rate was lower for mobile-bearing prostheses at 7 years postoperatively. Medial pivot prostheses have been reported to be free from failure due to polyethylene wear after 10 years [24, 25]. According to Australian national registry data, the revision rate of pivot prostheses, excluding old and non-currently used prostheses (Advance, Wright Medical), was lower than that of posterior-stabilized prostheses and equivalent to minimally stabilized prostheses [22].

7 Conclusion

Unlike THA, TKA involves a variety of articular surface shapes, high contact pressure on articular surfaces, and a high risk of delamination wear and fracture owing to oxidation because the articular surface is exposed to harsh conditions such as rolling, sliding, and rotational movement. Therefore, polyethylene material science requirements for TKA exceed those for THA.

In the future, the number of patients with TKA is expected to increase, particularly with an increasing proportion of young people with high activity. The development of polyethylene and articular surface designs that balance considerations such as wear powder, oxidation, fatigue resistance, and stability is awaited.

References

1. Kondo M, et al. Arthroscopy for evaluation of polyethylene wear after total knee arthroplasty. J Orthop Sci. 2008;13:433–7.
2. Akisue T, et al. Paratibial cyst associated with wear debris after total knee arthroplasty. J Arthroplast. 2001;16:389–93.
3. Niki Y, et al. Gigantic popliteal synovial cyst caused by wear particles after total knee arthroplasty. J Arthroplast. 2003;18:1071–5.
4. Kobayashi S, et al. Rapid tibial polyethylene failure in porous-coated anatomic total knees as a cause of clinical failure. Contemp Orthop. 1993;26:567–76.
5. Hirakawa K, et al. Relationship between wear debris particles and polyethylene surface damage in primary total knee arthroplasty. J Arthroplast. 1999;14:165–71.
6. Fukuoka Y, et al. A simple radiographic measurement method for polyethylene wear in total knee arthroplasty. IEEE Trans Rehabil Eng. 1999;7:228–33.
7. Hoshino A, et al. Accurate in vivo measurement of polyethylene wear in total knee arthroplasty. J Arthroplast. 2002;17:490–6.
8. Minoda Y, et al. In vivo analysis of polyethylene wear particles after total knee arthroplasty: the influence of improved materials and designs. J Bone Joint Surg Am. 2009;91(Suppl 6):67–73.
9. Kobayashi A, et al. Number of polyethylene particles and osteolysis in total joint replacements: a quantitative study using a tissue-digestion method. J Bone Joint Surg Br. 1997;79:844–8.
10. Matthews JB, et al. Evaluation of the response of primary human peripheral blood mononuclear phagocytes to challenge with in vitro generated clinically relevant UHMWPE particles of known size and dose. J Biomed Mater Res. 2000;52:296–307.
11. Yang SY, et al. Diverse cellular and apoptotic responses to variant shapes of UHMWPE particles in a murine model of inflammation. Biomaterials. 2002;23:3535–43.
12. Iwakiri K, et al. In vivo comparison of wear particles between highly crosslinked polyethylene and conventional polyethylene in the same design of total knee arthroplasties. J Biomed Mater Res B Appl Biomater. 2009;91:799–804.
13. Minoda Y, et al. Sequentially annealed highly cross-linked polyethylene reduced in vivo wear particle generation in total knee arthroplasty. J Orthop Surg (Hong Kong). 2017;25:2309499017718909.
14. Orita K, et al. Vitamin E-infused highly cross-linked polyethylene did not reduce the number of in vivo wear particles in total knee arthroplasty. Bone Joint J. 2020;102:1527–34.
15. Minoda Y, et al. Polyethylene wear particles in synovial fluid after total knee arthroplasty. Clin Orthop Relat Res. 2003;410:165–72.
16. Minoda Y, et al. Characteristics of polyethylene wear particles isolated from synovial fluid after mobile-bearing and posterior-stabilized total knee arthroplasties. J Biomed Mater Res B Appl Biomater. 2004;71:1–6.
17. Minoda Y, et al. Comparison of in vivo polyethylene wear particles between mobile- and fixed-bearing TKA in the same patients. Knee Surg Sports Traumatol Arthrosc. 2017;25:2887–93.
18. Minoda Y, et al. Polyethylene wear particle generation in vivo in an alumina medial pivot total knee prosthesis. Biomaterials. 2005;26:6034–40.
19. Minoda Y, et al. No difference in in vivo polyethylene wear particles between oxidized zirconium and cobalt-chromium femoral component in total knee arthroplasty. Knee Surg Sports Traumatol Arthrosc. 2014;22:680–6.
20. Urban JA, et al. Relationship between product demand, tibial polyethylene insert shelf age, and total knee arthroplasty survival: retrospective review of total knees of one design. J Arthroplast. 2006;21:330–7.
21. Minoda Y, et al. Comparison between highly cross-linked and conventional polyethylene in total knee arthroplasty. Knee. 2009;16:348–51.
22. Australian Orthopedic Association Nasional Joint Replacement Registry https://aoanjrr.sahmri.com/
23. Takemura S, et al. Comparison of a vitamin E-infused highly cross-linked polyethylene insert and a conventional polyethylene insert for primary total knee arthroplasty at two years postoperatively. Bone Joint J. 2019;101:559–64.
24. Nakamura S, et al. Clinical results of alumina medial pivot total knee arthroplasty at a minimum follow-up of 10years. Knee. 2017;24:434–8.
25. Ueyama H, et al. Long-term clinical outcomes of medial pivot total knee arthroplasty for Asian patients: a mean 10-year follow-up study. Knee. 2020;27:1778–86.

Cementless Fixation

32

Hiroshi Horiuchi and Shaw Akizuki

1 Cementless TKA

Cementless TKA is a surgical procedure that does not use bone cement for component fixation. The advantages of cementless TKA include bone preservation, the absence of polyethylene wear or bone resorption caused by cement debris, and biological fixation between the prosthesis and host bone. In recent years, this technique has attracted attention because of its ability to provide biological fixation for a long period, especially in young patients with good bone quality. However, the lack of bone cement has raised concerns about achieving initial fixation. Research has been conducted to improve the component design and shape of the bone contact surface and materials(porous or hydroxyapatite (HA)-coated). In addition, it has been found that achieving appropriate soft tissue balance intraoperatively is also important in achieving initial fixation [1].

2 History of Development

Cementless fixation was first introduced in hip arthroplasty. It has evolved as a method for managing highly active patients, especially younger patients, owing to concerns regarding the deterioration of cemented fixation over time. The Kodama-Yamamoto knee joint prosthesis developed by Kodama and Yamamoto at Okayama University was the pioneer of cementless TKA [2]. However, this series (Mark II, Mark III, Mizuho) used the press-fit method of fixation, where the components were fixed by direct contact with the bone without cement. In contrast, in the late 1970s, Low Contact Stress (LCS, Depuy Synthes) with a porous structure and Porous Coated Anatomic (PCA, Howmedica) were launched. LCS has been in continuous clinical use for more than 40 years. However, PCA was a breakthrough model in TKA because a technique for osteotomizing the proximal tibia to 3° varus in anatomical alignment and a surgical guide were developed. However, the porosity was not suitabe for cementless TKA, and there were problems with the polyethylene treatment method, resulting in many cases of early failure. Therefore, it was only briefly in clinical use.

Later, the Miller-Galante TKA (Zimmer Biomet), which has a titanium alloy (Ti-6 Al-4 V) fiber metal structure, became available for clinical use, expanding the options for cementless TKA. In addition, components coated with hydroxyapatite-tricalcium phosphate (HA-TCP), which reduces the micromotion of the tibial component and improves the fixation of the femoral and tibial components, are now available, and stable clinical results have been reported. In recent years, components made of tantalum, which mimics the structure of cancellous bone, and components with a surface structure made by three-dimensional (3D)-printing titanium have been introduced and are expected to achieve more stable initial fixation.

3 Basic Research to Stabilize Initial Fixation

It is clinically important to achieve a secure initial fixation in cementless TKA, especially in the tibial component, and ensure long-term durability. If initial fixation is not achieved, the tibial component may lift off anteriorly during flexion, and there is a risk of proximal tibial fracture [3]. Basic research has shown that the use of screws, sleeves, and keels improves the initial fixation of cementless tibial components.

Verifying the effect of the screw and sleeve in the initial fixation of the tibial component of cementless TKA [4]. Biomechanical analysis revealed that fixation with a sleeve

H. Horiuchi (✉)
Department of Rehabilitation Medicine, Shinshu University Hospital, Matsumoto, Japan
e-mail: horiuchih@shinshu-u.ac.jp

S. Akizuki
Hitsujigaoka Hospital, Sapporo, Japan
e-mail: akizuki@grn.janis.or.jp

S. Matsuda (ed.), *Advances in Total Knee Arthroplasty*, https://doi.org/10.1007/978-981-97-4920-1_32

and four cancellous bone screws suppressed micromotion of the tibial component, using screws prevented lift-off even when axial pressure was applied, and the combination of both is useful for obtaining initial fixation. In addition, Ortholoc II and Ortholoc Modular (Wright Medical) tibial component fixation in autopsy cases have been reported [5]. This study shows that screws are useful for achieving initial fixation; however, their role ends once biological fixation is established.

4 Tibial Bone Grafting in Cementless TKA

If there is a bone defect when cementless TKA is performed, autologous cancellous bone grafting will be used; however, because it is an unfavorable environment for achieving initial fixation, bone grafting (size and procedure) may be confusing. Although there are few studies in this area, there is one report on the micromotion of the tibial component after bone grafting for a posterior medial tibial defect using a nonlinear three-dimensional finite element model [6]. This study indicates that bone grafting to compensate for bone defects is not the preferred technique for achieving stable initial fixation with cementless tibial components and requires a strong bone graft.

5 Postoperative Course of Initial Gap

In cementless TKA, regardless of the accuracy of the osteotomy, there is an initial gap (IG) between the cutting surface and the component. Porous structures alone only fill IGs less than 0.5 mm with new bone; however, HA and porous tantalum are expected to fill IGs even better.

The effect of HA-TCP coating has been reported for the same tibial component [7]. Ti was applied to the surface of the porous structure in NexGen Cruciate Retaining (CR) TKA (Zimmer Biomet). The tibial component is fixed without screws. The HA-TCP group comprised 46 knees in which Ti tibial components with HA-TCP (70% HA, 30% TCP) applied to the surface of the porous structure in the NexGen CR TKA were fixed without screws (Fig. 32.1). Meanwhile, the control group consisted of 46 knees wherein components with only a porous structure and no HA-TCP application were fixed with screws. The IGs of both groups were evaluated by fluoroscopic analysis up to 1 year postoperatively. The anteroposterior and lateral views were examined under the same conditions. On the femoral side, the IG disappeared in all cases until 9 months after surgery in the HA-TCP group. On the tibial side, the frequency of IG remaining in the control group was 32.6% at 1 year after surgery, whereas in the HA-TCP group, it was 2% (1/46). This finding suggests that HA-TCP helps to achieve initial fixation in cementless TKA.

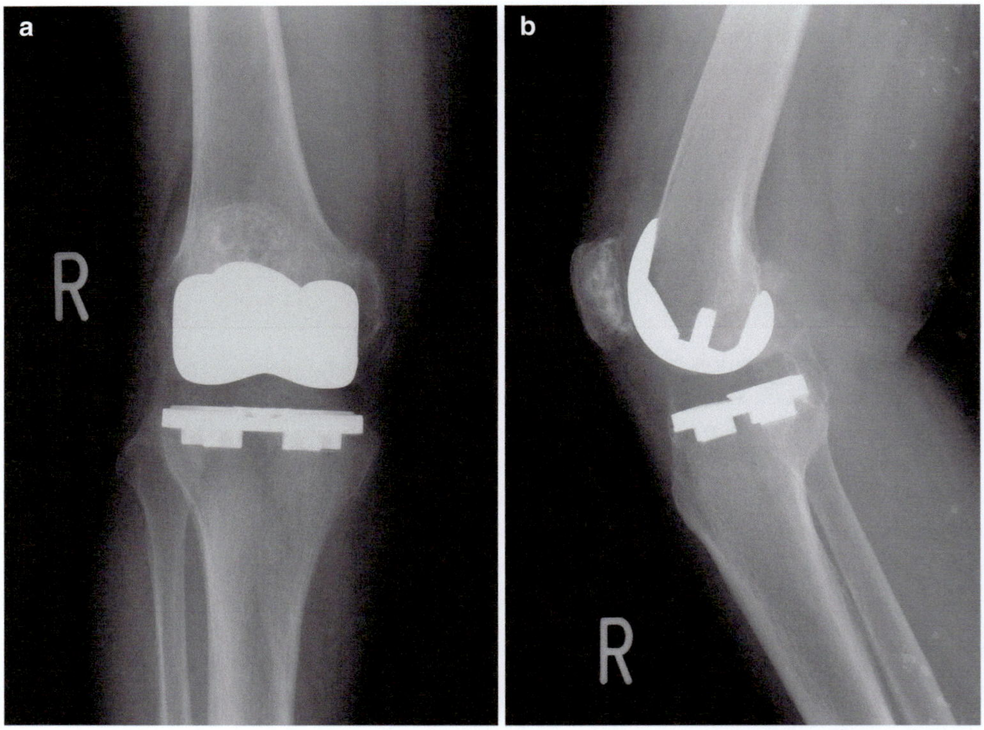

Fig. 32.1 Cementless CR-TKA. Age 63 years at surgery, 12 years postoperatively. (**a**): Anteroposterior radiograph. (**b**): Lateral radiograph. This model does not use screws, only the proximal tibial osteotomy surface including the four pegs to fix the component but maintains good fixation with good knee function. No radiolucent line is observed on radiographs

Radiographic changes in IGs in HA-applied femur parts in posterior-stabilized (PS)-TKA have been reported over time [8]. 82 PS-TKA (Scorpio NRG, Stryker) was used to evaluate the transition of the IGs on the femoral side. A cuneiform space was present in 23% of postoperative cases. However, IG disappeared in all cases in the anterior chamfer region; only one knee remained IG in the distal femur, indicating the effect of HA application in the femoral part of PS-TKA.

In addition, in 90 cases of CR-type cementless TKA and 125 knees using porous tantalum NexGen Trabecular Metal Monoblack tibial components (Zimmer Biomet) [9], a radiolucent line was observed at the tip of the inner peg in 19% of cases 3 years postoperatively. However, no clinical problems, including revision, were present, hence indicating the usefulness of porous tantalum tibial components.

6 Postoperative Trabeculae and Bone Density Changes in Cementless Tibial Parts

The fixation of cementless components, especially on the tibial side, depends on the bone quality and bone mineral density (BMD) of the host bone. However, stress shielding and other factors have been shown to reduce the BMD around the knee after TKA, and materials and implant designs have been improved to prevent this.

6.1 Trabeculae and BMD Changes in Porous Tantalum Tibial Component

The clinical results of porous tantalum mono-block tibial components (Zimmer Biomet) have been reported, with emphasis on radiographic evaluation [10]. A cementless TKA 32 knee with a mono-block tibial component fixed with one medial and one lateral peg was followed up for a mean of 7.8 years postoperatively. Osteogenesis was classified into three groups based on the anteroposterior radiographic view. Stress shielding was observed on the medial side of the tibia in 21.9%, but there was no relationship with the position of the peg or functional axis. Bone atrophy due to proximal tibial stress shielding leads to component loosening and should be further investigated.

A few reports have compared the bone density in the proximal tibia of cemented PS-TKA and cementless PS-TKA with porous tantalum tibial components [11, 12]. 28 porous tantalum tibial components (Trabecular Metal, Zimmer Biomet) and cemented CoCr tibial components of the 28 knees (PFC Sigma RP, DePuy) were evaluated. In addition, the BMD was compared in three regions (medial, middle, and lateral) of the tibia after surgery. BMD decreased over time (1, 5, and 11.4 years postoperatively at the time of the final study) in all measurement areas. In addition, there were no differences in clinical outcomes such as BMD change, KSS, and knee range of motion (ROM) between the two groups at the time of the final follow-up. Therefore, the clinical efficacy of using porous tantalum in the tibial component should be clarified, and further long-term postoperative observations are required.

A report has also compared CR and PS-TKA using a tantalum tibial component. CR in 22 knees and PS in 22 knees, clinical results, and 3D multidetector-row computed tomography were used to evaluate the bone density around the tibial peg over the postoperative period. The clinical outcomes were not different between the two groups for up to a mean of 5.5 years postoperatively. Furthermore, there was no difference in bone density around the peg between the CR and PS groups [13].

6.2 Effect of Parathyroid Hormone on Tibial BMD

One study reported that patients undergoing cementless TKA were treated with PTH once a week [14]. In cementless TKA using a trabecular tantalum modular tibial component (NexGen CR-Flex, Zimmer Biomet), a comparison of BMD trends up to 12 months postoperatively in the PTH (n = 20) and non-PTH (n = 20) groups showed that the PTH group had significantly higher BMD in the medial observation area.

7 Long-Term Clinical Results of Cementless TKA

Currently, there are fewer reports on the long-term clinical results of cementless TKA than those of cement TKA; however, there are some reports from Japan. Osteonics series 3000 cementless CR TKA(Stryker) was performed on 76 knees in 54 patients. Clinical results of the Osteonics 3000 series CR-TKA 10–13 years after surgery have been reported [15] Although five patients were not followed up (follow-up: 90.7%), only one patient was revised owing to tibial component loosening at 10.5 years. Additionally, the survival rate by the Kaplan–Meier method was 100% and 96.7% at 10 and 13 years after surgery, respectively. Although this series continues to be followed at Nagano Matsushiro General Hospital, there have been no cases of revision, and this series has a survival rate of 96.7% at 20 years postoperatively. In addition, the knee ROM and clinical score improved significantly compared to that before the surgery for >10 years postoperatively, and good knee function was maintained.

8 Histopathological Examination Around Cementless Tibial Components

Although biological fixation, the greatest advantage of cementless TKA, is difficult to assess histologically in postoperative clinical cases, a report from Japan has verified biological fixation by histological analysis of autopsy cases [15, 16].

An examination of autopsy cases at 5.5 years postoperatively on the Osteonics 3000 series, which had a CoCr porous structure with good knee function, showed that no new bone formation had formed on the proximal tibia. Moreover, only fibrous tissue was observed around the two screws used for initial fixation. This case was considered fibrous fixation [15]. In addition, a histological evaluation of an autopsy case of RA with a Miller-Galante II TKA (Zimmer Biomet) with an HA-TCP coated tibial component on a fiber-metal structure at 2 years postoperatively was also reported [16]. Osteogenesis was observed in 77.7% of the proximal tibial osteotomy surfaces, and fibrous tissue was identified in only 1%. In addition, Osteogenesis was observed in 81.1% of the prosthesis surfaces, including within the fiber metal, and only 4.5% was fibrous tissue. No osteolysis or fibrous tissue was observed around the screws. Analyzing the postoperative resorption process of HA and HA-TCP is challenging. However, histological examination of this autopsy case 2 years postoperatively showed residual HA-TCP on the surface of the fiber metal surface, indicating that it remained unabsorbed for at least 2 years postoperatively.

9 Anti-Bleeding Measures

Bleeding control is one of the most important issues in cementless TKA. Compared with cemented TKA, in which the exposed bone marrow is covered with bone cement, cementless TKA involves more postoperative bleeding. To reduce postoperative bleeding, a method of administering hemostatic agents (e.g., sodium carbazochrome sulfonate preparations and tranexamic acid) into the joint (drain-clamp method) has been reported [17] (Fig. 32.2).

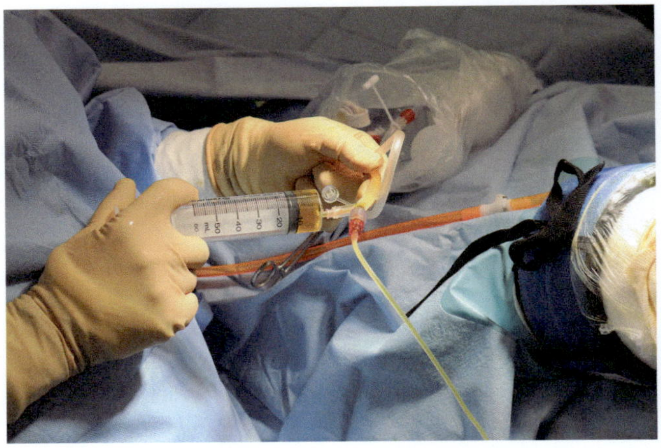

Fig. 32.2 Administration of intra-articular hemostatic agents. Because a drain is usually placed in cementless TKA, hemostatic agents (such as sodium carbazochrome sulfonate and/or tranexamic acid) is administered retrogradely and intraarticularly immediately before the release of the tourniquet. If no drain is used, the drug is administered directly to the joint. The usefulness of tranexamic acid in TKA has been reported in many clinical trials in terms of bleeding control and prevention of complications

10 Challenges of Cementless TKA

Physiological kinematics can be reproduced even with cementless CR-TKA if appropriate soft tissue balance can be obtained by surgery with precise procedures [1, 18]. In the future, it will be necessary to develop new biomaterials and implant designs that can improve patient satisfaction, including knee ROM, while reducing micromotion and achieving initial fixation.

References

1. Akizuki S, et al. Measured (anatomical reference) resection technique for cruciate-retaining total knee arthroplasty. In: Matsuda S, et al., editors. Soft tissue balancing in total knee arthroplasty. Berlin: Springer; 2017. p. 1196–202.
2. Yamamoto S, et al. A follow-up study of an uncemented knee replacement. J Bone Joint Surg Br. 1989;71:505–8.
3. Shimagaki H, et al. Stability of initial fixation of the tibial component in cementless total knee arthroplasty. J Orthop Res. 1990;8:64–71.

4. Miura H, et al. Effects of screws and a sleeve on initial fixation in uncemented total knee tibial components. Clin Orthop. 1990;259:160–8.
5. Matsuda S, et al. Evaluation of tibial component fixation in specimens retrieved at autopsy. Clin Orthop. 1999;363:249–57.
6. Totoribe K, et al. Finite element analysis of the tibial bone graft in cementless total knee arthroplasty. J Orthop Surg Res. 2018;13:113.
7. Gejo R, et al. Fixation of the NexGen HA-TCP-coated cementless, screwless total knee arthroplasty: comparison with conventional cementless total knee arthroplasty of the same type. J Arthroplast. 2002;17:449–56.
8. Kuriyama S, et al. Bone-femoral component interface gap after sagittal mechanical axis alignment is filled with new bone after cementless total knee arthroplasty. Knee Surg Sports Traumatol Arthrosc. 2018;26:1478–84.
9. Mitsuzxaki H, et al. Radiolucent lines are decreased at 3 years following total knee arthroplasty using trabecular metal tibial components. J Intern Med Res. 2018;46:1919–27.
10. Hayakawa K, et al. Mid-term results of total knee arthroplasty with a porous tantalum monoblock tibial component. Knee. 2014;21:199–203.
11. Minoda Y, et al. Comparison of bone mineral density between porous tantalum and cemented tibial total knee arthroplasty components. J Bone Joint Surg Am. 2010;92:700–6.
12. Minoda Y, et al. Periprosthetic loss of bone mineral density after cementless porous tantalum and cemented total knee arthroplasties: a mean of 11-year concise follow-up of a previous report. J Arthroplast. 2020;35:3156–60.
13. Kaneko T, et al. Midterm comparison of tibial fixation between posterior cruciate-retaining and substituting porous tantalum total knee arthroplasty: three-dimensional computed tomography analysis. J Knee Surg. 2021;34:47–56.
14. Kaneko T, et al. Weekly injection of teriparatide for bone ingrowth after cementless total knee arthroplasty. J Ortho Surg. 2016;24:16–21.
15. Watanabe H, et al. Survival analysis of a cementless, cruciate-retaining total knee arthroplasty: clinical and radiographic assessment 10 to 13 years after surgery. J Bone Joint Surg Br. 2004;86:824–9.
16. Akizuki S, et al. Fixation of a hydroxyapatite-tricalcium phosphate-coated cementless knee prosthesis: clinical and radiographic evaluation seven years after surgery. J Bone Joint Surg Br. 2003;85:1123–7.
17. Akizuki S, et al. A new method of hemostasis for cementless total knee arthroplasty. Bull Hosp Joint Dis. 1997;56:222–4.
18. Horiuchi H, et al. In vivo kinematic analysis of cruciate-retaining total knee arthroplasty during weight-bearing and nonweight-bearing deep knee bending. J Arthroplast. 2012;27:1196–202.

Cement Fixation

33

Hiroshi Takagi

The fixation method is an important factor that affects the clinical outcomes of TKA. Bone cementation is a technique used for fixation. Bone cement (PMMA) primarily used in orthopedics for implant fixation. In 1960, Charnley used cement fixation for femoral implants and reported good results [1]. Since then, cement fixation has become the typical fixation method used in joint replacement surgery. In particular, cement fixation has many advantages in TKA, including the ability to cover irregularities and imperfections in the osteotomy surface and to obtain firm initial fixation. However, aseptic loosening has been reported to occur between the implant-cement interface and cement-bone interface [2, 3]. Therefore, it is important to fully understand the types and characteristics of bone cement used in cement TKA and to learn the techniques of cement fixation. Understanding the effects of cement fixation on the human body is also essential.

1 Types and Characteristics of Bone Cement

Bone cement is comprises approximately 40 g of powdered polymer and 20 mL of liquid monomer. The polymer contains barium sulfate as an X-ray contrast agent and a small amount of polymerization accelerator. Their composition and quantity differ depending on the manufacturing company. The viscosity and working time of cement vary depending on the particle size of the polymer. Cemex RX (TOKIBO) is classified as having low viscosity, and Simplex P (Stryker) is classified as moderate. Endurance bone cement (DePuy Johnson & Johnson) and Biomet Bone Cement R (Zimmer) are classified as having high viscosities. Cemex RX is considered to have a low polymerization heat.

H. Takagi (✉)
Adashi Medical Center, Tokyo Women's Medical University, Tokyo, Japan
e-mail: takagi.hiroshi@twmu.ac.jp

The component composition (%w/w) of endurance bone cement, which I mainly use, is 67% polymethyl methacrylate, 21% methyl metacrate/styrene copolymer, 1.9% benzoyl peroxide, and 10% barium sulfate; 98% methyl methacrylate and 2% or less of N,N-dimethyl-p-toluidine for the liquid monomer, and the curing time is 11–13 min under a controlled environment (temperature 23 °C, humidity 50%). The curing time of Simplex P is approximately 12 min when the temperature of the bone cement is 20 °C. Temperature (room temperature and bone cement temperature) is the major factor influencing the curing time. When the temperature is high, the curing time is reduced. If there are more powdered polymers and fewer liquid monomers, the curing time will be shortened. When the stirring speed is increased, the curing time will be shortened. In general, the mixing state of the cement depends on the do time (mixing time), working time, and setting time. Do time is the time from mixing and stirring until the bone cement reaches a state where it is no longer touchable, and working time is the time when the cement is applied to the implants and bones and fixed. Therefore, it is important for the surgeon to understand the characteristics of the cement used, the do time, working time, and setting time, and perform cement fixation in the same standardized manner each time.

Bone cement containing antibiotics has been widely used in Western countries for a long time, and it has been reported that the infection rate in joint replacement surgery is lower with the use of bone cement containing antibiotics than with the use of cement without antibiotics [4, 5].

In Japan, bone cement containing antibiotics was first released in 2014. Currently, it is covered by health insurance in the second stage of a two-stage revision surgery associated with postoperative infection of joint replacement surgery, and Simplex P tobramycin (Stryker), Endurance Gentamicin Bone Cement (DePuy Johnson & Johnson), and Rifovacin Bone Cement R Gentamicin (Zimmer) can be used. The author has been adding a maximum of 2 g of antibiotic to 40 g of powdered polymer, which is considered a range that does not compromise strength when the surgeon includes an

S. Matsuda (ed.), *Advances in Total Knee Arthroplasty*, https://doi.org/10.1007/978-981-97-4920-1_33

antibiotic, such as vancomycin, in regular bone cement after MRSA infection.

2 Actual Usage of Bone Cement

First, the inside of the joint was cleaned to remove bone debris and prepare for the surface of osteotomy. Osteosclerosis often persists on the osteotomy surface on the medial side of the tibia. The appropriate depth of bone-cement penetration is 3–4 mm [6]. Because hardened bone has difficulty penetrating cement, which is disadvantageous for cement fixation, anchor holes were prepared at intervals of approximately 5–7 mm using a 3.5 mm diameter drill. Drilling increases bone cement penetration and reduces the radiolucent line (RLL) on X-rays [7].

Pulse cleaning of the osteotomy surface was performed because inclusions such as adipose tissue, blood, and bone scraps between the cement and bone reduce the strength of fixation. Compared with cleaning using a syringe, pulse cleaning is advantageous in terms of the depth of bone cement penetration and the strength of fixation between the bone and cement [8].

Similarly, on the femur side, if the sclerotic part of the medial condyle remained, drilling was performed, after which gauze and suction were used to remove as much water and inclusions as possible from the surface of the osteotomy and the stem hole.

If air is mixed during cement agitation, it can lead to the generation of air bubbles; vacuum mixing under negative pressure suppresses the formation of air bubbles and can maintain the strength of cement when compared with hand mixing [9, 10]. While it is ideal to use a vacuum mixing system to perform stirring, the author does not currently use it because of costs and other problems. Therefore, to avoid air bubbles as much as possible, the spatula was first mixed as if the powder and liquid were sliced and then slowly stirred. Bone cement (endurance bone cement) stored in a refrigerator was brought to room temperature (approximately 20 °C) prior to use, and after stirring for about 2 min, it was poured into a catheter tip with the tip cut short for using as a cement syringe [11]. The cement was first applied to the tibial component of the syringe. Two methods are used for applying cement to the tibial component: full cementation, in which cement is applied to the stem, and surface cementation, in which cement is not applied to the stem. At this point, there is no clear conclusion as to which method is superior, and the choice between the two influenced by the design, material, and concept of the implant [12]. The authors preferred cementation from a syringe and applying the cement evenly to all surfaces of the components using hand. However, care should be taken to avoid touching the adhesive surface of the component. Subsequently, the cement was applied using a syringe while applying pressure to the tibial osteotomy surface (Fig. 33.1). Compared with hand packing, cement fixation under pressure with a cement gun or cement syringe has been reported to be superior in terms of bone cement penetration depth and the presence or absence of RLL [11]. In our case, bone cement penetration was significantly improved by applying pressure to the osteotomy surface with a syringe, compared to the method of applying bone cement using a spatula to fix the tibial component. After inserting the tibial component into the hand, a hammer was used to compress and fix it. Thereafter, fixation was performed on the femoral side. Cement was applied to the femoral component using a syringe and it was evenly applied to all surfaces of the component manually. If excess cement is applied to the posterior condyle of the femoral component, it may be extruded backward during insertion, which will be difficult to remove. Therefore, cement was also applied to the distal osteotomy and upper chamfer surfaces. No cement was applied to the surface of the posterior condylar osteotomy. The femoral components were inserted manually after first checking the direction of extension-flexion and then driven with a hammer. After removing the excess cement, a tibial insert was placed, the knee was extended and compressed, and the excess cement was removed again. During extension, pressure is applied at 0°, ensuring an even medial and lateral force from the heel in the proximal direction. After applying cement to the patellar osteotomy surface using a syringe, pressure was applied using hand packing. The patellar component was inserted into the peg hole, and pressure was

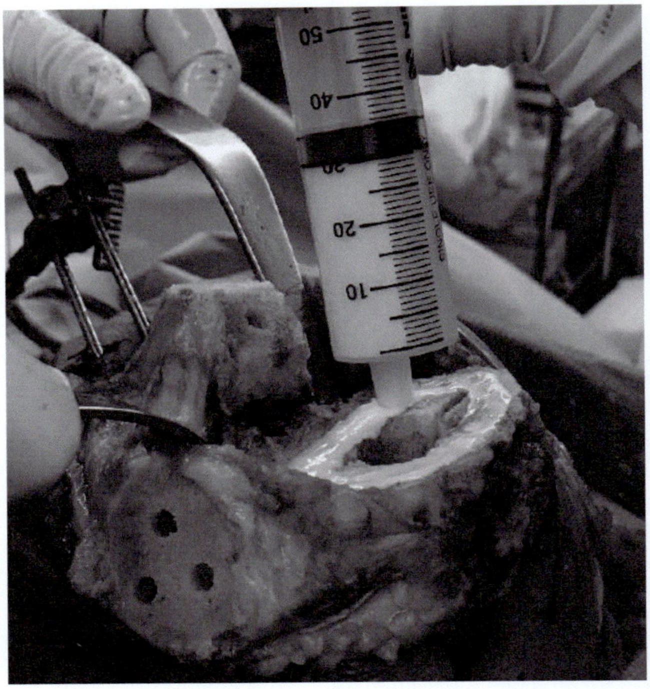

Fig. 33.1 Cement is applied to the tibial osteotomy surface while applying pressure with a syringe

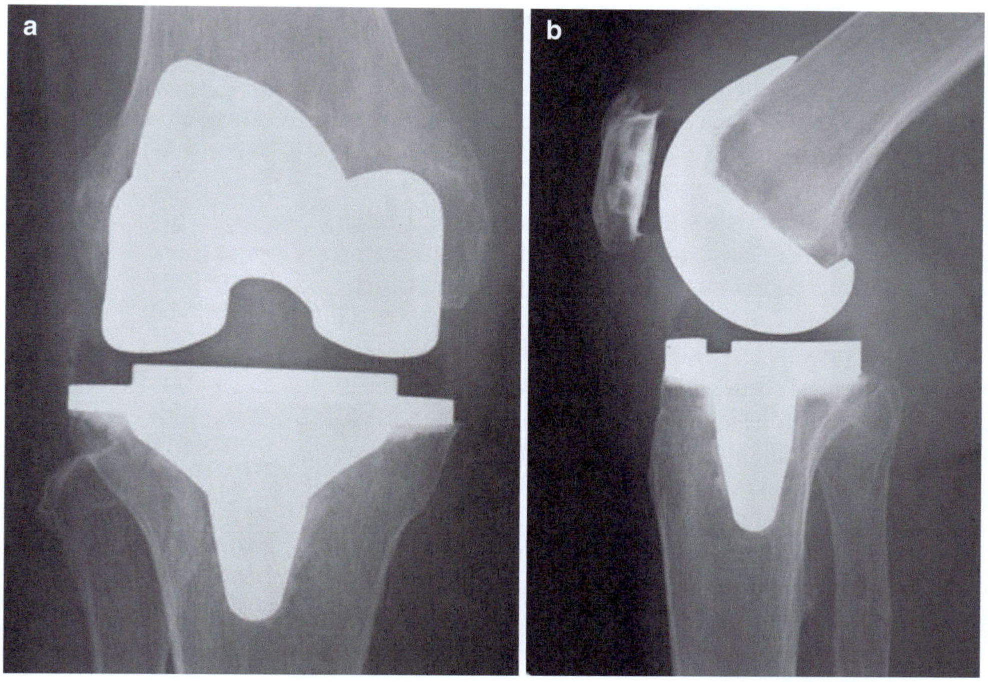

Fig. 33.2 Postoperative X-ray. Cement penetration was observed in the tibial osteotomy site and around the stem in the anteroposterior view (**a**), and cement penetration was observed in the anterior femur and patellar adhesive surface in the lateral view (**b**)

applied using a patellar clamp. Excess cement was removed, and more pressure was applied. While maintaining the pressure for approximately 13 minutes, the implant was washed using a solution consisting of 1.5 L saline and 45 mL povidone-iodine. Based on the report [13] that approximately 4 L of pulse washing is effective in removing bone debris and PMMA debris in the cement-fixed TKA, a total of 4.5 L of washing solution (3 L of saline and 1.5 L of saline with povidone iodine) was used for the entire surgery. It is important to make efforts to follow the same procedure for fixing the bone cement under the same conditions, such as room temperature and when to take bone cement out of the refrigerator. (Fig. 33.2).

3 Clinical Results of Cement-Fixed TKA

Regarding the long-term results of cement-fixed TKA, many studies have shown good results, including reports of a 90.77% survival rate at 21 years [14] and 87% at 18 years after surgery [15]. However, there have been some recent reports of loosening due to debonding after cement-fixed TKA. Debonding occurs when the cement does not stick to the implant. It has been reported that in 25 of 1337 knees (1.9%), debonding occurred on the tibial side during the relatively early period of 13–95 months postoperatively [16]. In addition, debonding was observed in 9 of the 175 knee revision TKAs: 1 knee on the femoral side, 3 knees on the tibial side, and 5 knees on both sides of the femoral tibia [17]. Debonding may be caused by multiple factors, such as cement type, cement technique, patient factors, and implant design (coating, locking mechanism of inserts, length of keel, and pegs).

4 Effect of Cement-Fixed TKA on the Human Body

4.1 Bone Density

On average, 2 years postoperatively, the bone density in the distal anterior femoral component was lower in cemented TKA than in cementless TKA [18]. Bone density of the lateral tibia was similarly lower in tibial component TKA than in cemented TKA [19]. However, there are additional reports that the bone density of the tibial component did not change for either cemented or cementless TKA over an average of 11 years in the same series; long-term changes in bone density are not yet clear [20]. The effect of cemented TKA on bone density differs depending on the type of TKA, and it has been found that mobile-bearing TKA does not reduce bone density in the distal femur more than fixed-bearing TKA at either 2 years or 11 years postoperatively [21, 22]. Therefore, from the viewpoint of changes in bone density around the implant, cemented TKA has disadvantages compared with cementless TKA and may affect long-term loosening and revision surgery.

4.2 Coincident Heat

The heat of cement polymerization is also affected by the storage temperature of the cement, thickness of the cement layer, and temperature of the bone surface. When used at room temperature, the heat of cement polymerization does not increase sufficiently to cause tissue necrosis, [23] but if the temperature of the cement itself is high, the heat of cement polymerization may be excessive [24]. In addition, simulation studies have reported that osteonecrosis is observed approximately 2 mm from the bone-cement surface; therefore, it is better to make the cement layer as thin as possible and lower the temperature of the osteotomy surface before cement fixation [25]. When using bone cement, we considered the effect of the heat of polymerization. It is better to cool the bone itself or thoroughly wash the bone surface before fixation to eliminate inclusions and simultaneously reduce the temperature of the bone surface.

4.3 Toxicity

PMMA monomers are toxic and cause a rapid decrease in blood pressure if they enter the bloodstream. This is known as bone cement implantation syndrome (BIS), and cardiac arrest and intraoperative deaths have been reported. This is attributed to pulmonary embolism (PE), neurovascular reflex, platelet aggregation in a vein due to peripheral vasodilating effects, and allergic reactions due to the absorption of monomers into blood vessels during polymerization and direct toxicity to the cardiopulmonary system. It was also reported to increase transiently when measured [26].

As described above, it is necessary to perform cement TKA while considering the effects of fixing the implant with cement on the human body.

References

1. Charnley J. Anchorage of the femoral head prostheses of the shaft of the femur. J Bone Joint Surg Br. 1960;42:28–30.
2. Kutzner I, et al. Early aseptic loosening of a mobile-bearing total knee replacement. Acta Orthop. 2018;89:77–83.
3. Dahabreh Z, et al. The effect of application time of two types of bone cement on the cement bone interface strength. Eur J Orthop Surg Traumatol. 2015;25:775–81.
4. Parvizi J, et al. Efficacy of antibiotic-impregnated cement in total hip replacement: a meta-analysis. Acta Orthop. 2008;79:335–41.
5. Espehaug B, et al. Antibiotic prophylaxis in total hip arthroplasty: review of 10905 primary THA reported to Norweigian register, 1897 to 1995. J Bone Joint Surg Br. 1997;79:590–5.
6. Walker PS, et al. Control of cement penetration in total knee arthroplasty. Clin Orthop Relat Res. 1984;185:155–64.
7. Miskovsky C, et al. The cemented unicondylar knee arthroplasty: an in vitro comparison of three cement techniques. Clin Orthop Relat Res. 1992;284:215–20.
8. Maisterelli GL, et al. Cement penetration with pulsed lavage versus syringe irrigation in total knee arthroplasty. Clin Orthop Relat Res. 1995;312:261–5.
9. Lewis G, et al. Effect of mixing method on select properties of acrylic bone cement. J Biomed Master Res. 1997;38:221–8.
10. Lidgren L, et al. Bone cement improved by vacuum mixing and chilling. Acta Orthop Scand. 1987;57:27–32.
11. Lutz MJ, et al. The effect of cement gun and cement syringe use on the tibial cement mantle in total knee arthroplasty. J Arthroplast. 2009;24:461–7.
12. Anders MR, et al. Cementing technique for primary knee arthroplasty: a scoping review. Acta Orthopedica. 2019;90:582–9.
13. Niki Y, et al. How much sterile saline should be used for efficient lavage during total knee arthroplasty? Effects of pulse lavage irrigation on removal of bone and cement debris. J Arthroplast. 2007;22:95–9.
14. Font-Rodriguez DE, et al. Survivorship of cemented total knee arthroplasty. Clin Orthop Relat Res. 1997;345:79–86.
15. Metsovitis SR, et al. Rotaglide total knee arthroplasty: a long-term follow-up study. J Bone Surg Am. 2011;93:878–84.
16. Arsoy D, et al. Aseptic tibial debonding as a cause of early failure in a modern total knee arthroplasty design. Clin Orthop Relat Res. 2013;471:94–101.
17. Sadauskas A, et al. Implant interface debonding after total knee arthroplasty: a new cause for concern? Arthroplast Today. 2020;6:972–5.
18. Seki T, et al. Is bone density in the distal femur affected by use of cement and by femoral component design in total knee arthroplasty? J Orthop Sci. 1994;4:180–6.
19. Minoda Y, et al. Comparison of bone mineral density between porous tantalum and cemented tibial total knee arthroplasty components. J Bone Joint Surg Am. 2010;92:700–6.
20. Minoda Y, et al. Periprosthetic loss of bone mineral density after cementless porous tantalum and cemented total knee arthroplasties: a mean of 11-year concise follow-up of a previous report. J Arthroplast. 2020;35:3156–60.
21. Minoda Y, et al. A cemented mobile-bearing total knee replacement prevents periprosthetic loss of bone mineral density around the femoral component: a matched cohort study. J Bone Joint Surg Br. 2010;92:794–8.
22. Minoda Y, et al. A cemented mobile-bearing total knee prosthesis prevents peri-prosthetic bone mineral density loss around the femoral component: a consecutive follow-up at a mean of 11 years. Knee Surg Sports Traumatol Arthrosc. 2021; https://doi.org/10.1007/s00167-021-06448-4.
23. Jeffriss CD, et al. Thermal aspects of self-curing polymethylmethacrylate. J Bone Joint Surg Br. 1975;57:511–8.
24. Meyer PR, et al. On the setting properties of acrylic bone cement. J Bone Joint Surg Am. 1973;55:149–56.
25. Fukushima H, et al. Conduction analysis of cement interface temperature in total knee arthroplasty. Kobe J Med Sci. 2002;78:63–72.
26. Kinoshita H, et al. The use of bone cement induces an increase in serum astroglial S-100B protein in patients undergoing total knee arthroplasty. Anesth Analg. 2003;97:1657–60.

Part V

Implant Design

Articular Surface Geometry

34

Tatsuhiko Kutsuna

1 Articular Surface Design in Total Knee Arthroplasty (TKA)

Design of the articular surface in TKA plays a crucial role in determining the stability of the knee joint and significantly impacts postoperative clinical outcomes. Historically, the focus of articular surface geometry in TKA has been on ensuring durability of the implant. However, in recent years, there has been an increasing demand for TKA implants that not only provide durability but also enhance patient satisfaction by allowing deep knee flexion and facilitating high levels of physical activity. In this section, we will discuss the fundamental concepts of femoral-tibial articular surface designs in TKA.

2 Basic Concept of Articular Surface Design

The stability of the knee joint relies on the geometry of the articular surface, static stabilizers (ligaments), and dynamic stabilizers (muscle-tendon units) [1]. In TKA, compensating for dysfunctional or excised soft tissue structures is essential, and articular surface geometry must provide an appropriate level of constraint to achieve a proper ligament balance. However, there is a lack of clear clinical evidence supporting the definition of an optimal level of constraint of femoral-tibial articular surface design in TKA.

The ideal articular surface design for TKA aims to reduce mechanical loosening and polyethylene wear while allowing for an extensive range of motion (ROM). A joint surface design that prioritizes stability, such as a high conformity design with a significant degree of curvature matching between the femoral and tibial radii, increases the contact area and minimizes stress concentration at the bearing surface. However, concerns regarding the concentration of loads transmitted during activity of daily living (ADLs) at the fixation surface (implant-bone interface); that may potentially lead to mechanical loosening persist.

Femoral rollback is essential for achieving deep knee flexion. If the contact surface between the femoral component and polyethylene insert becomes a point contact during deep knee flexion, the polyethylene is subjected to severe contact conditions. To ensure long-term durability while maintaining both ROM and knee joint stability, the bone surface must receive adequate loading to promote remodeling. Therefore, it is important to select an appropriate articular surface geometry in combination with the surgical technique, based on a comprehensive understanding of the design concept.

3 Articular Surface Shape: Coronal Plane

Articular surface geometry in the coronal plane of the femoral-tibial joint can be classified into three combinations: curve-on-flat, flat-on-flat, and curve-on-curve (Fig. 34.1). The curve-on-flat design is mainly utilized in unicondylar knee arthroplasty (UKA) as it relies on joint stability provided by the ligaments due to its low constraint and small contact area. The flat-on-flat design offers a large contact area with less constraint on the coronal plane but is no longer used because of concerns about abnormal polyethylene wear caused by edge loading when lift-off occurs. On the other hand, the curve-on-curve design provides the advantage of greater constraint and a larger contact area than the other two designs, making it a widely used approach in modern TKA implants.

The design concepts for the articular surface in the coronal plane of a curve-on-curve design include single-radius and double-radius designs. The single-radius design in the coronal plane, with a single curve-on-curve geometry across the medial and lateral articular surfaces, raises the possibility

T. Kutsuna (✉)
Department of Bone and Joint Surgery, Ehime University Graduate School of Medicine, Ehime, Japan
e-mail: kutsuna@m.ehime-u.ac.jp

S. Matsuda (ed.), *Advances in Total Knee Arthroplasty*, https://doi.org/10.1007/978-981-97-4920-1_34

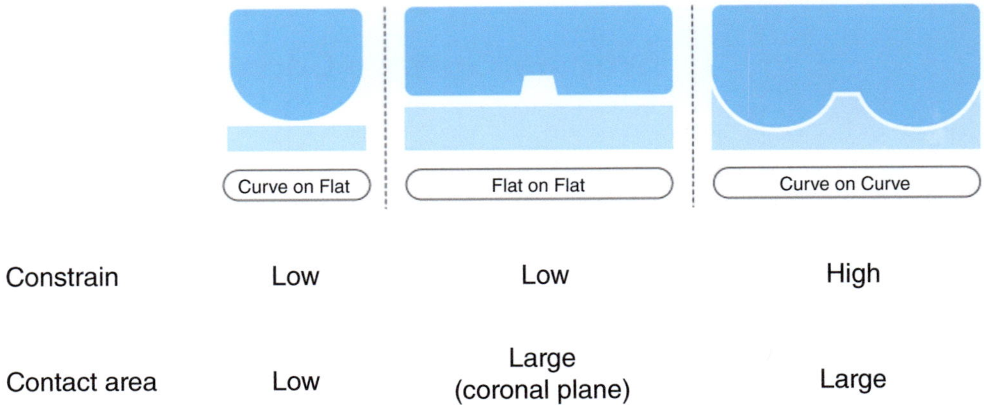

Fig. 34.1 Articular surface geometry in the coronal plane of the femoral-tibial joint

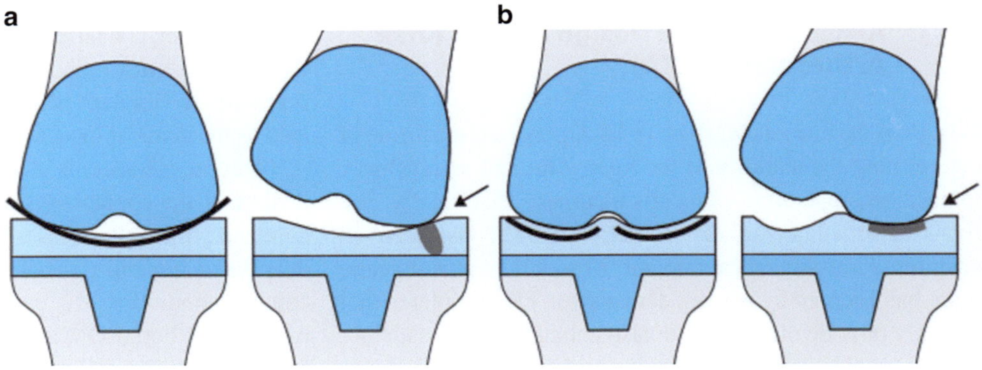

Fig. 34.2 Design concepts for the articular surface in the coronal plane. (**a**): Single-radius design. (**b**): Double-radius design

of edge loading when lift-off occurs (Fig. 34.2). In contrast, the double-radius design is currently the main coronal design because it is expected to reduce the risk of polyethylene wear by preventing edge loading and maintaining the contact area even when lift-off occurs (Fig. 34.2).

Clinical comparison has shown that the flat-on-flat design (MG1) exhibits an increased varus alignment and polyethylene wear, whereas the curve-on-curve design (Nexgen-LPS) shows no significant changes in alignment and polyethylene thickness over time [2]. A biomechanical analysis of contact pressure also revealed that the curve-on-curve design exhibited a large contact area and effectively reduced the increase in contact pressure at the femoral-tibial joint under both 0° and 5° varus lift-off conditions. Furthermore, the curve-on-curve design is expected to maintain the contact area even in cases of implant malposition, thereby enhancing the durability of the polyethylene insert and reducing the risk of revision surgery.

The selection of articular surface geometry in the coronal plane should also consider the distinction between cruciate retaining (CR) and posterior stabilized (PS) TKA. For a CR TKA, a double-radius design that allows some degree of freedom is preferred to avoid interfering with the function of the retain posterior cruciate ligament (PCL). Conversely, for a PS TKA, a double-radius design with a deep-dish surface is preferred to enhance varus/valgus stability following PCL resection and to maintain femoral-tibia joint surface contact during lift-off (Fig. 34.3). Additionally, other designs, such as medial pivot-type designs with asymmetric constraint on the medial and lateral joint surfaces (high medial conformity) and anatomical designs equipped with physiological joint surface inclination, are expected to effectively improve the clinical outcomes.

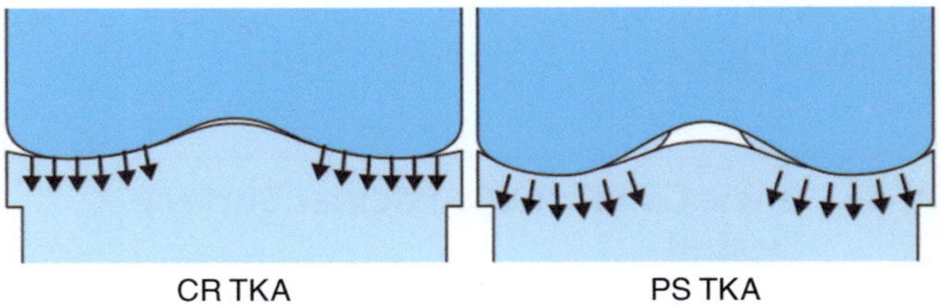

Fig. 34.3 Selection of articular surface geometry in coronal plane

4 Articular Surface Shape: Sagittal Plane

Kinematics of the knee joint in the sagittal plane are controlled by the femoral condyle geometry, anteroposterior articular surface shape of the polyethylene insert, and ligament balance. In a normal knee, during knee extension, the femoral-tibial joint forms a contact surface between the flat tibial and the femoral condylar surfaces with a large radius of

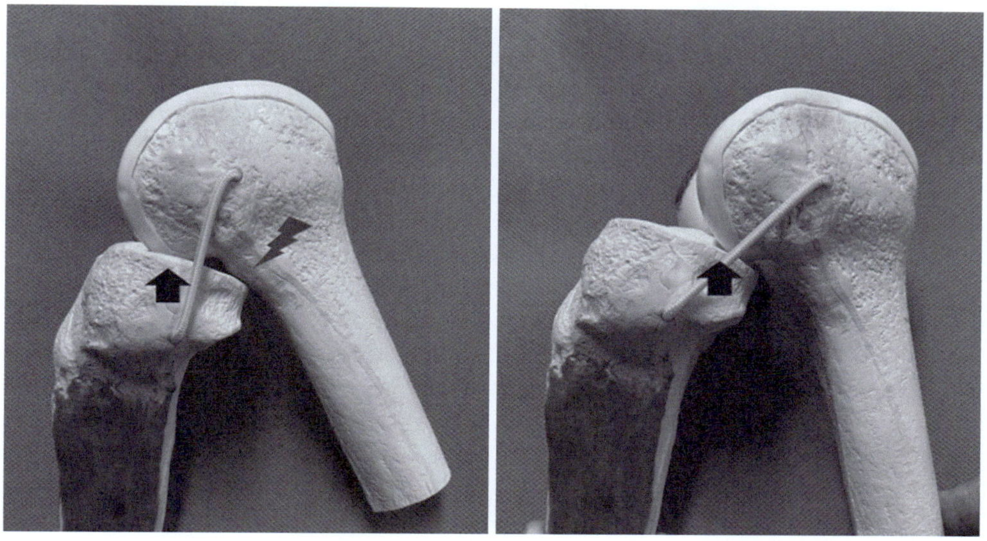

Fig. 34.4 Femoral rollback

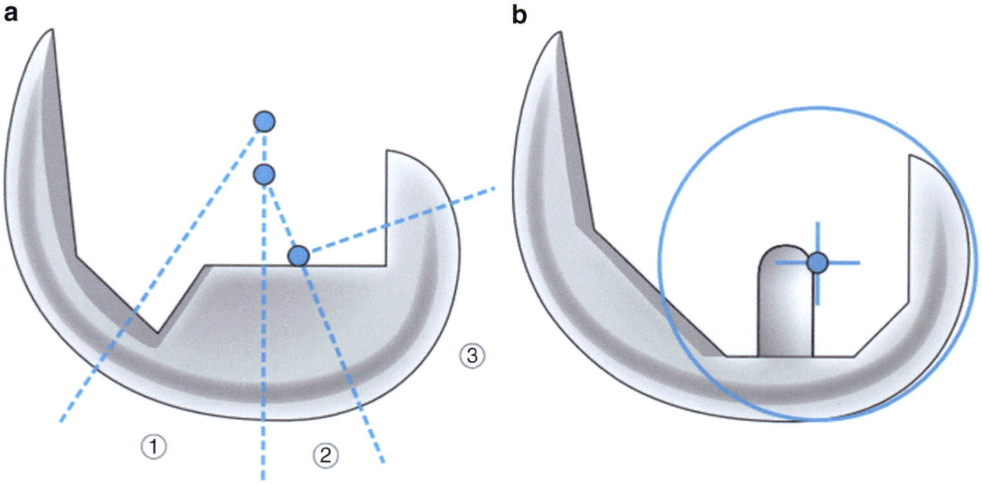

Fig. 34.5 Sagittal curvature of the femoral component. (**a**): Multi-radius design. (**b**): Single-radius design

curvature. In contrast, during knee flexion, the contact surface shifts to the posterior condyle with a small radius of curvature. During deep knee flexion, the small radius of curvature of the posterior femoral condyle, along with the function of the PCL, induces femoral rollback, allowing sufficient ROM in knee flexion (Fig. 34.4). Femoral rollback also increases the quadriceps moment arm, thereby improving the strength of knee extensor muscles [4]. In TKA, reproducing the rollback ability of a normal knee is crucial for achieving favorable postoperative outcomes.

4.1 Sagittal Curvature of the Femoral Component

The sagittal curvature of the femoral component can be classified into two designs: the multi-radius (MR) design and the single-radius (SR) design (Fig. 34.5).

The MR design is based on the concept that in a normal knee, during knee flexion from extension, the axis is initially posterior and gradually shifts distally as flexion progresses, resulting in femoral rollback. Curvature between the femoral component and polyethylene insert provides significant congruency at early stages of knee flexion, and the radius of curvature of the femoral component decreases as the insert extends towards the posterior femoral condyle, promoting femoral rollback. However, a drawback of the MR design is the occurrence of "anterior paradoxical motion" observed in TKAs lacking the anterior cruciate ligament (ACL) or both ACL and PCL [3–5].Anterior paradoxical motion occurs due to the poor fit of the femoral-tibial joint surface caused by the changing radius of curvature and can lead to anterior knee pain and instability in the mid-flexion range, raising concerns about a potential negative effect of the postoperative clinical outcomes.

On the other hand, the SR design was developed based on the theory that shapes of the medial and lateral femoral condyles may differ but can be approximated by two circles. If the center of the approximate circle is considered the femoral reference point during knee flexion and extension, the femoral axis remains fixed and unchanged throughout the knee joint ROM. The SR design offers biomechanical advantages: (1) reduced occurrence of anterior paradoxical motion because the sagittal curvature of the femur remains constant during knee flexion and extension. (2) proper alignment of the femoral component along the flexion-extension axis, leading to maintained tension on the medial and lateral collateral ligaments throughout the knee joint ROM; that is expected to improve the clinical outcomes [6–8].

4.2 Clinical Results: Multi-radius Versus Single-Radius Design

Numerous studies have compared the clinical outcomes of MR and SR designs, but the superiority of the SR design has not been definitively established. A prospective cohort study involving more than 200 knees in each group, treated with standardized surgical procedures and rehabilitation protocols, revealed no significant difference in postoperative clinical scores (Knee Society Score [KSS], SF-36, ROM, and patient satisfaction) between the MR and SR designs [9]. Similarly, a meta-analysis comparing MR and SR design also demonstrated no significant difference in the KSS, knee flexion angles, complication rates, or survival rates; however, the MR design showed significantly better knee joint ROM [10]. The failure of the SR design to translate into superior outcomes despite its biomechanical advantages may depends on several factors. These factors include a small difference in the radius of curvature between the two designs within the range of flexion required for basic ADLs and models with limited compatibility between the femoral component and polyethylene insert. This may result in insufficient anteroposterior stability in the intermediate flexion position [10]. However, a matched case-control study evaluating the clinical outcomes of these two designs with high medial congruence (MR design: Persona Medial Congruent and SR design: GMK Sphere) reported no significant differences in the clinical outcomes (Oxford knee score, KSS) between the two implants. Interestingly, the MR design exhibited a better knee flexion angle, consistent with the results of the meta-analysis [11]. The variation in the recommended posterior tilt angle of the tibial component between these two implants (Persona Medial Congruent: 5°, GMK Sphere: 0°) was discussed as a possible contributing factor to better knee flexion angles observed with the MR designs.

As various factors, such as insert shape and implant alignment, influence the clinical outcomes of TKA, it becomes challenging to solely attribute differences in these outcomes to the sagittal curvature of femoral components. Therefore, fully comprehending the design concept of the chosen implant, making appropriate selections based on the surgical technique, and executing a surgical procedure meticulously to effectively reproduce the design concept is crucial.

4.3 Gradually Reducing Femoral Radius Design

The gradually reducing femoral radius (GRR) design is specifically developed to address the issue of anterior paradoxical motion that arises in the mid-flexion range due to changes in the femoral radius, a concern observed in the MR design. The primary objective of the GRR design is to achieve enhanced knee joint stability throughout the ROM by implementing a gentle reducing effect in the femoral radius according to flexion. Notably, gradient radius designed implants, such as ATTUNE and MERA Quest Knee System, have been introduced for clinical use. Computer simulation analyses have demonstrated the effectiveness of the GRR design in suppressing anterior paradoxical motion. When compared to the MR design, the GRR design resulted in a 21% reduction in anterior paradoxical motion when using a less constrained insert, and a 68% reduction when combined with a more constrained insert [12]. Furthermore, intraoperative kinematics analyses by navigation [13] and fluoroscopy kinetic analysis [14] have provided supportive evidence for the benefits of the GRR design. The GRR design (ATTUNE CR) effectively suppressed paradoxical anterior motion on the medial side when compared to the MR design (PFC SIGMA CR), thereby exhibiting a knee kinematics pattern that closely resembles that of a normal knee joint.

The GRR design shows promise in improving anterior-posterior instability in the mid-flexion range that may lead to high-performance activities and reduced polyethylene wear. However, while the GRR design addresses the issue of anterior paradoxical motion associated with changes in the femoral radius, it may not necessarily exhibit superior kinematics compared to all other MR design implants, particularly in cases where the MR design involves multiple femoral radius points. For future research, a comparison of kinematics between the MR design (with multiple femoral radius points) and the GRR design would be valuable in providing further insights into their kinematic advantages and limitations. Such studies could contribute to a better understanding of how different design features impact knee joint biomechanics and influence the clinical outcomes, aiding in the ongoing improvement and refinement of TKA implant design.

5 Design for Deep Knee Flexion and Long-Term Durability

Knee joint ROM of approximately 120° is necessary to carry out daily activities without difficulty. However, to engage in Japanese-specific lifestyles, such as getting up from the floor and using a futon on a tatami, an extensive ROM is required [15]. Achieving deep knee flexion is essential, as it allows individuals to perform various tasks and activities comfortably. Nevertheless, deep flexion places significant stress on the knee joint. Therefore, when designing knee implants, optimizing the ability to achieve deep knee flexion and prioritizing long-term durability is essential. Striking a balance between providing sufficient flexion capacity while ensuring the durability of the implant is imperative for promoting both short-term functionality and long-term performance and patient satisfaction.

A high-flexion design was modified in three points to facilitate deep knee flexion: (1) to achieve deep knee flexion without edge contact and to minimize the risk of abnormal polyethylene wear (Fig. 34.6), the thickness of the posterior femoral condyle was augmented, and the sagittal curvature was enhanced in the deep flexion region. (2) The frontal surface of the polyethylene insert was specifically modified to prevent patella impingement during deep knee flexion. This adjustment ensures smooth knee motion and reduces the likelihood of discomfort or complications related to patellar impingement. (3) To prevent post-cam dislocation during deep knee flexion, the post-cam mechanism was meticulously designed to establish contact at a low position [16]. This design feature ensures a safe jumping height of the post-cam, promoting stability and minimizing the risk of instability during knee movements. These three modifications in the high flexion design aim to provide patients with the ability to achieve deep knee flexion safely and effectively, thereby enabling the performance of various activities that require a wide range of motion while reducing the risk of adverse effects associated with extreme knee flexion.

A meta-analysis showed that the high-flexion design had better postoperative ROM than the conventional design, regardless of PS and CR TKA. However, the clinical outcomes (KSS and Hospital for Special Surgery score) were not significantly different between these two designs [17]. A study on patients who underwent TKA for over 10 years reported that the revision rate and appearance rate of the radiolucent line (RLL) at the posterior femoral condyle were significantly better in the high-flexion design as compared to the conventional design [18]. Additionally, the high-flexion design showed a favorable survival rate of patients (99.1%) when revision surgery due to mechanical loosening was used as the endpoint, indicating a robust long-term durability [19]. A finite element model analysis study that simulated knee joint ROM suggested that in the range of 120–145° of knee flexion, the tensile force at the implant-bone junction surface increases, potentially leading to a higher frequency of RLL occurrence in the femoral component [20]. Therefore, special attention should be given to implant fixation methods, particularly in the cementing technique, when using the high flexion design to ensure long-term stability and reduce the risk of complications associated with RLLs.

In the context of the Japanese lifestyle, achieving deep flexion is considered crucial in improving patient satisfaction as it facilitates engagement in high flexion activities [21]. Although the high-flexion design is anticipated to improve the angle of knee flexion, it has not consistently translated into improved clinical outcomes, and there is an ongoing debate regarding its long-term durability. Carrying out deep knee flexion after TKA inevitably increases stress on the joint surfaces of the femoral component. Therefore, it is important for the surgeon to exercise caution in determining the appropriate extent of deep knee flexion to be allowed in each individual case. Balancing the desire for improved

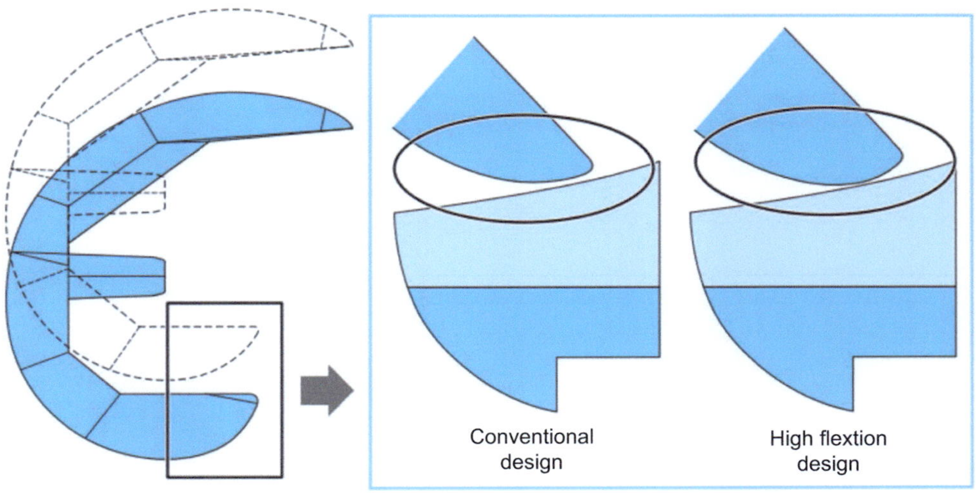

Fig. 34.6 High-flexion design

ROM and high flexion activities with the potential concerns related to long-term durability is a critical aspect of optimizing surgical outcomes and patients' satisfaction after TKA. Further research and clinical experience will be instrumental in guiding surgeons toward making informed decisions in this regard.

6 Future Challenges

TKA has undergone significant advancements, leading to improved clinical outcomes and favorable long-term results. However, to further improve patient satisfaction, maintaining a delicate balance between high activity levels (including sports activities) and ensuring long-term durability is imperative. Addressing these challenges requires a comprehensive approach that involves not only implant designs but also an exploration of complementary surgical techniques. The in vivo environment of artificial joint prostheses changes constantly, based on various physiological conditions and with the passage of time. TKA surgeons must bear this in mind, and engage in suitably frequent, long-term follow-up of the adopted prosthesis. It is essential not to be unduly influenced by short-term clinical results but to continually monitor and assess the progress of the patient, thus influencing the development of future implant designs and establishment of optimal surgical techniques.

References

1. Bartel DL, et al. The effect of conformity, thickness, and material on stresses in ultrahigh molecular weight components for total joint replacement. J Bone Joint Surg Am. 1986;68:1041–51.
2. Nishikawa K, et al. Improved design decreases wear in total knee arthroplasty with varus malalignment. Knee Surg Sports Traumatol Arthrosc. 2014;22:2635–40.
3. Dennis DA, et al. In vivo anteroposterior femorotibial translation of total knee arthroplasty: a multicenter analysis. Clin Orthop Relat Res. 1998;356:47–57.
4. Dennis DA, et al. Multicenter determination of in vivo kinematics after total knee arthroplasty. Clin Orthop Relat Res. 2003;416:37–57.
5. Banks SA, et al. Implant design affects knee arthroplasty kinematics during stair-stepping. Clin Orthop Relat Res. 2004;426:187–93.
6. Ostermeier S, et al. Quadriceps force after TKA with femoral single radius. Acta Orthop. 2011;82:339–43.
7. Kessler O, et al. Sagittal curvature of total knee replacements predicts in vivo kinematics. Clin Biomech (Bristol, Avon). 2007;22:52–8.
8. Stoddard JE, et al. The kinematics and stability of single-radius versus multiradius femoral components related to mid-range instability after TKA. J Orthop Res. 2013;31:53–8.
9. Hinarejos P, et al. No differences in functional results and quality of life after single-radius or multiradius TKA. Knee Surg Sports Traumatol Arthrosc. 2016;24:2634–40.
10. Liu S, et al. Meta-analysis of outcomes of a single-radius versus multiradius femoral design in total knee arthroplasty. J Arthroplast. 2016;31:646–54.
11. Indelli PF, et al. No clinical differences at the 2-year follow-up between single radius and J-curve medial pivot total knee arthroplasty in the treatment of neutral or varus knees. Knee Surg Sports Traumatol Arthrosc. 2020;28:3949–54.
12. Clary CW, et al. The influence of total knee arthroplasty geometry on mid-flexion stability: an experimental and finite element study. J Biomech. 2013;46:1351–7.
13. Takagi H, et al. Case series report of navigation-based in vivo knee kinematics in total knee arthroplasty with a gradually reducing femoral radius design. Ann Med Surg (Lond). 2017;17:33–7.
14. Pfitzner T, et al. Modifications of femoral component design in multiradius total knee arthroplasty lead to higher lateral posterior femoro-tibial translation. Knee Surg Sports Traumatol Arthrosc. 2018;26:1645–55.
15. Niki Y, et al. Is greater than 145° of deep knee flexion under weight-bearing conditions safe after total knee arthroplasty?: a fluoroscopic analysis of Japanese-style deep knee flexion. Bone Joint J. 2013;95:782–7.
16. Guild GN 3rd, et al. Clinical outcomes in high flexion total knee arthroplasty were not superior to standard posterior stabilized total knee arthroplasty: a multicenter, prospective, randomized study. J Arthroplast. 2014;29:530–4.
17. Wang Z, et al. Comparison of high-flexion and conventional implants in total knee arthroplasty: a meta-analysis. Med Sci Monit. 2015;21:1679–86.
18. Angerame MR, et al. Ten-year follow-up of high-flexion versus conventional total knee arthroplasty: a matched-control study. J Arthroplast. 2021;36:2795–800.
19. Baek JH, et al. Long-term survivorship of total knee arthroplasty with a single-radius, high-flexion posterior stabilized prosthesis. Knee. 2021;30:275–82.
20. Zelle J, et al. Does high-flexion total knee arthroplasty promote early loosening of the femoral component? J Orthop Res. 2011;29:976–83.
21. Han HS, et al. A high degree of knee flexion after TKA promotes the ability to perform high-flexion activities and patient satisfaction in Asian population. BMC Musculoskelet Disord. 2021;22:565.

Mobile Bearing

35

Masahiro Hasegawa

1 History of Development

In mobile-bearing TKA, the polyethylene insert can move, and the tibial component is not fixed. The Oxford unicompartmental knee (Zimmer Biomet) was developed as the first mobile-bearing approach to increase the conformity of the articular surface to reduce polyethylene wear and to make it mobile to reduce restraint. Moreover, total replacement low contact stress (LCS, Johnson & Johnson DePuy) was introduced. The conformity of the articular surfaces was extremely high at 40° flexion; however, the radius of curvature of the posterior condyle decreased; thus, the conformity decreased [1]. The 20-year survival rate was extremely good at 97%, according to the developer's report [2]. Later, LCS-like multiple-radius-shaped Press Fit Condylar (PFC) Sigma (Johnson & Johnson DePuy), Legacy (Zimmer Biomet), Genesis II (Smith & Nephew), NexGen LPS-Flex Mobile (Zimmer Biomet) (Fig. 35.1a), Attune (Johnson & Johnson DePuy), Vanguard RP Hi-Flex (Zimmer Biomet), and Vanguard ROCC (Zimmer Biomet) (Fig. 35.1b) have been developed. Moreover, single-radius Rotaglide (Corin), Mobile Bearing Knee (MBK; Zimmer Biomet), and Scorpio (Stryker) have been developed to provide full conformity from extension to flexion. There are meniscal bearing (MB) and anterior–posterior glide with the PCL preserved and rotating platform (RP) with the PCL sacrificed, and the latter is widely used.

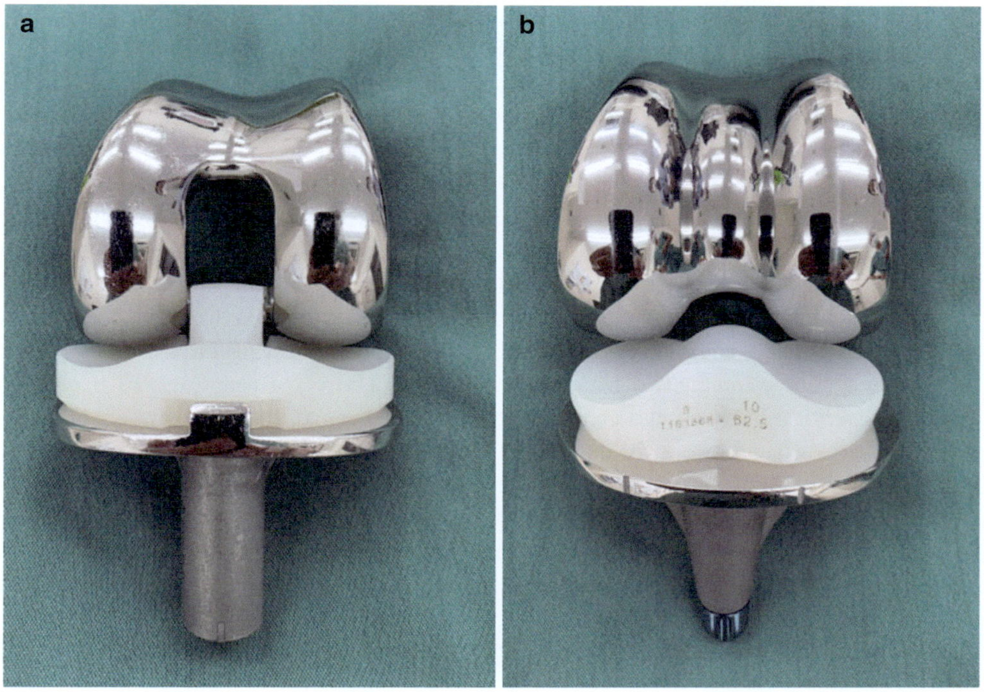

Fig. 35.1 Mobile-bearing TKA. (**a**): NexGen LPS Flex Mobile (Zimmer Biomet). (**b**): Vanguard ROCC (Zimmer Biomet)

M. Hasegawa (✉)
Department of Orthopaedic Surgery, Mie University Graduate School of Medicine, Tsu, Japan
e-mail: masahase@clin.medic.mie-u.ac.jp

S. Matsuda (ed.), *Advances in Total Knee Arthroplasty*, https://doi.org/10.1007/978-981-97-4920-1_35

2 The Strength of Theory

The articular surface of the polyethylene insert was designed to increase the contact area and reduce wear, and the tibial tray surface was designed to move and rotate backward to reduce the shear force and prevent loosening. Furthermore, an increase in the range of motion (ROM) was expected; this is useful for preventing mismatches between the femoral and tibial components, and tibial coverage may be prioritized. In a study using LCS RP, the mean coverage area of the tibial component was 82.7% [3].

3 Kinematics

Kinematic evaluation after mobile-bearing TKA has been reported in Japan. In a study using a cadaver implanted with Tricon II knee system (Smith & Nephew), the contact stress on the undersurface of the polyethylene insert was small when the rotation error of the tibial component was <30°; however, the stress increased as a result of the rotation error of the 45° internal rotation [4]. In a study using a knee simulator in a dynamic condition that considered the difference depending on walking speed, MBK and LCS RP had a small amount of insert rotation, with MBK having less insert rotation [5]. A finite element analysis showed that mobile bearings have less shear force even in deep flexion than that of fixed bearings [6]. In a comparison of the load axis before and after TKA, both LCS MB and RP improved in the mediolateral direction; however, controversial reports revealed that the anteroposterior direction worsened [7] and demonstrated no difference in the MB case [8]. In TKA using the PCL-preserving MB and PCL-sacrificed RP, the varus–valgus laxities of the MP were 2.9° under anesthesia and 3.0° postoperatively; the degrees of adduction of the MB were 4.4° under anesthesia and 3.6° postoperatively; the degrees of abduction of the RP were 3.8° under anesthesia and 3.5° postoperatively; and the degrees of adduction were 4.6° under anesthesia and 4.0° postoperatively. No significant differences between the two groups were noted [9]. The anterior/posterior movement was greater in the MB, and the femur moved further posteriorly during the stair-climbing motion. The RP showed no significant kinematic changes even when the activity was increased [10]. After TKA using LCS RP, the kinematics were compared between loaded and unloaded, and the movement was similar except for the initial rotation and central pivot pattern [11]. Postoperative kinematic evaluation using the NexGen LPS Flex Mobile showed a medial pivot pattern and rotation between the polyethylene insert and tibial component [12]. With flexion, the femur moved posteriorly with a mean of 4.7 mm on the medial side and 11.6 mm on the lateral side, with a mean rotation of 9° [13]. When squatting was performed 1 year after TKA with the same implant, the mean rotation of the femoral and tibial components was 6.3°; the mean rotation between the femoral component and insert was 5.5°; and the mean rotation between the insert and tibial component was 1.6°, with the insert and tibial component rotation being low [14]. In a comparison of NexGen LPS Flex mobile and fixed bearings in terms of postoperative kinematics, the external femoral rotation of the mobile bearings increased with flexion; however, that of the fixed bearings increased by 90°. No difference in the amount of anteroposterior movement was noted between the two groups; however, the mobile bearing moved significantly anteriorly on the lateral side [15]. In a similar comparison, the internal rotation of the tibia (external femur) was larger in the mobile group, and the anteroposterior movement was larger in the fixed group [16]. No difference in the amount of rotation was noted between the implant and stair-climbing motion. The mobile bearing showed rotation of the polyethylene insert, and the anteroposterior positions were located more posteriorly in the fixed bearing [17]. Postoperative kinematic evaluation using the Vanguard RP Hi-Flex showed a medial pivot pattern in the loading position [18]. After TKA using the same implant, paradoxical anterior movement was not observed, post/cam contact was noted from a mean flexion of 33.8°, and all patients were in contact up to 120°; however, there was no contact at 130° in two patients [19]. After TKA using PFC Sigma RP-F, the central pivot pattern was observed from 0° to 80°. From 80° to 120°, bicondylar rollback was observed, and the polyethylene insert rotated approximately 8° [20]. In a study on PFC Sigma RP using CR and RP, the femoral component moved anteriorly from 80° to 40° and posteriorly from 40° to 10° of flexion, and external rotation was observed in both components [21]. In CR Mobile's Dual Bearing Knee (DBK prosthesis, Finsbury), the ROM was improved by reducing the diameter of the posterior condyle and moving the lowest point of the polyethylene insert posterior [22].

In summary, many studies have shown that the kinematics in normal knees have not been obtained, although some reports have shown a medial pivot pattern after mobile-bearing TKA.

Regarding PFJ tracking, although RCT results show no difference between mobile and fixed bearings [23], it has been shown that mobile bearings are more convenient and have less contact stress [24, 25].

4 Clinical Performance

The bone mineral content of the distal femur 2 years after cemented TKA using PFC Sigma RP for mobile bearings and NexGen LPS Flex Fixed for fixed bearings was measured using DEXA, and the bone mineral density was higher in the mobile bearings [26]. In TKA using LCS RP, the ROM was better in the knee with a rectangular flexion gap than in that with a trapezoidal gap [27]. No difference in ROM was

Table 35.1 Comparisons of mobile and fixed bearings

Authors	Mobile	Fixed	ROM	Clinical results
Minoda et al. [33, 34]	Vanguard RP Hi-Flex	Vanguard PS	NS	NS
Watanabe et al. [35]	Rotaglide	NexGen CR	NS	NS
Hasegawa et al. [36]	PFC Sigma RP	PFC Sigma Fixed	NS	NS
Matsuda et al. [37]	NexGen LPS Flex Mobile	NexGen LPS Flex Fixed	NS	NS

ROM range of motion, *NS* not significant

noted 2 years after TKA using the MB and RP of the LCS; however, the ROM at 6 months after surgery for MB and 3 months after surgery for RP could predict ROM at 2 years [28]. There was no correlation between patellar tendon strain and ROM for both [29]. Furthermore, the patellar tendon length showed no significant effect on medium- to long-term performance [30]. When LCS MB or RP was immobilized on the surface cementation (cemented only on the underside of the tibial tray), subsidence was observed in 2% of the patients 5 years after surgery, none of revision [31]. The results of 38 TKA patients using LCS with PCL sparing on one side and PCL sacrifice on the other side 10 years after surgery showed no revisions, and the clinical outcomes were not different between the two components [32].

In the results of TKA using mobile and fixed bearings (Table 35.1), no difference in clinical evaluation was noted between the mobile-bearing Vanguard RP Hi-Flex and fixed-bearing Vanguard PS at >2 years postoperatively and at a mean postoperative period of 3.3 years (Table 35.1) [33, 34]. However, the anteroposterior instability at 30° and 75° flexion was lower in the knees with the Vanguard RP Hi-Flex at a mean of 3.3 years postoperatively [34]. Moreover, 8 years after TKA with mobile-bearing Rotaglide on one side and fixture-bearing NexGen CR on the opposite side, no difference was noted between both sides [35]. There was no difference in the results at 3 years postoperatively using PFC Sigma in a comparative study of mobile and fixed bearings using the same femoral component on the left and right sides [36]. In another comparative study of mobile and fixed bearings using the NexGen LPS Flex, there was no difference in the results at 5 years postoperatively [37]. A meta-analysis of medium- and long-term results showed no differences in clinical results between mobile and fixed bearings [38].

5 Complications

Bearing dislocation and spinout can occur when the flexion gap is too large; unilateral condylar deviation is defined as spinout, and bilateral condylar deviation is defined as dislocation [39]. Proper soft tissue tension is an important factor in mobile-bearing TKA, and a laxity in the coronal plane of 3–4° is recommended to prevent dislocation with LCS [40]. LCS, PFC Sigma RP, Vanguard RP Hi-Flex, and Vanguard ROCC (Fig. 35.1b) do not have a rotation stop mechanism; however, the Rotaglide NexGen LPS Flex Mobile (Fig. 35.1a) has a rotation stop mechanism; thus, this complication has been prevented. A case of dislocation of Scorpio due to the breakdown of the locking ring has been reported [41]. Moreover, breakage of the polyethylene insert has been reported. The cone fracture not post occurred 27 months postoperatively in the Vanguard RP Hi-Flex [42].

Patellar clunk syndrome is a complication of PS TKA [43] and is not specific to mobile bearings. It describes the painful catching or jumping of the patella when the knee moves from flexion to extension. It is caused by the growth of fibrous scars proximal to the superior pole of the patellar components, and the scars are caught in a box with deep flexion. There is a report on PFC Sigma RP for mobile bearings, with an increased risk of occurrence in models with a relatively long box intercondylar box ratio of ≥0.7 [44].

Regarding the anterior of the post-impingement, when 100 N of posterior stress was applied using the Vanguard RP Hi-Flex, PFC Sigma RP, and NexGen LPS Flex Mobile, PFC Sigma RP did not establish contact. The compressive yield stress of the polyethylene exceeded 10 MPa [45]. Moreover, NexGen LPS Flex Mobile post-fractures have been reported [46].

There are two sliding surfaces of the polyethylene—the joint surface and undersurface—and the shapes of the wear particles are different. A third body can enter the undersurface [47]. In a study using postoperative joint fluid 1 year after surgery, a comparison of MBK in mobile bearings and PS in fixed bearings showed no difference in the size, shape, and amount of polyethylene wear powder, and low wear of mobile bearings was not expected [48]. However, the theoretical usefulness of mobile bearings has not been demonstrated in basic research or clinical studies.

References

1. Stiehl JB. Mobile bearings in total knee arthroplasty. In: Scott WN, editor. Insall & Scott surgery of the knee. 4th ed. Philadelphia: Churchill Livingstone; 2006. p. 1558–73.
2. Buechel FF Sr, et al. Twenty-year evaluation of meniscal bearing and rotating platform knee replacements. Clin Orthop Relat Res. 2001;388:41–50.
3. Ishii Y, et al. Tibial component coverage and rotational alignment accuracy after mobile-bearing total knee arthroplasty. Eur J Orthop Surg Traumatol. 2018;28:1143–9.
4. Matsuda S, et al. Contact stress analysis in meniscal bearing total knee arthroplasty. J Arthroplast. 1998;13:699–706.
5. Hamai S, et al. Kinematic analysis of mobile-bearing total knee arthroplasty using a 6-DOF knee simulator. J Orthop Sci. 2008;13:543–9.

6. Anuar MA, et al. Dynamic finite element analysis of mobile bearing type knee prosthesis under deep flexional motion. Sci World J. 2014;2014:586921.
7. Ishii Y, et al. Changes in lower extremity 3-dimensional load-bearing axes before and after mobile-bearing total knee arthroplasty. J Arthroplast. 2012;27:1203–9.
8. Ishii Y, et al. Changes in the three-dimensional load-bearing axis after mobile-bearing total knee arthroplasty. Eur J Orthop Surg Traumatol. 2016;26:537–43.
9. Ishii Y, et al. Effect of soft tissue tension on measurements of coronal laxity in mobile-bearing total knee arthroplasty. J Orthop Sci. 2005;10:496–500.
10. Mikashima Y, et al. Does mobile-bearing knee arthroplasty motion change with activity? Knee. 2013;20:422–5.
11. Matsumoto K, et al. In vivo kinematics of a low contact stress rotating platform total knee arthroplasty system under weight bearing and nonweight bearing condition. J Orthop Sci. 2014;19:750–5.
12. Tanaka A, et al. Three-dimensional kinematics during deep-flexion kneeling in mobile-bearing total knee arthroplasty. Knee. 2011;18:412–6.
13. Nakamura E, et al. Three-dimensional tibiofemoral kinematics during deep flexion kneeling in a mobile-bearing total knee arthroplasty. J Arthroplast. 2009;24:1120–4.
14. Hoshi K, et al. Mobile-bearing insert used with total knee arthroplasty does not rotate on the tibial tray during a squatting activity: a cross-sectional study. J Orthop Surg Res. 2020;15:114.
15. Watanabe T, et al. Matched comparison of kinematics in knees with mild and severe varus deformity using fixed- and mobile-bearing total knee arthroplasty. Clin Biomech (Bristol, Avon). 2012;27:924–8.
16. Shi K, et al. Kinematic comparison between mobile-bearing and fixed-bearing inserts in NexGen legacy posterior stabilized flex total knee arthroplasty. J Arthroplast. 2008;23:164–9.
17. Okamoto N, et al. In vivo kinematic comparison between mobile-bearing and fixed-bearing total knee arthroplasty during step-up activity. J Arthroplast. 2014;29:2393–6.
18. Kage T, et al. In vivo kinematics of a newly updated posterior-stabilized mobile-bearing total knee arthroplasty in weight-bearing and nonweight-bearing high-flexion activities. Knee. 2021;29:183–9.
19. Kurita M, et al. In vivo kinematics of high-flex mobile-bearing total knee arthroplasty, with a new postcam design, in deep knee bending motion. Int Orthop. 2012;36:2465–71.
20. Futai K, et al. In vivo kinematics of mobile-bearing total knee arthroplasty during deep knee bending under weight-bearing conditions. Knee Surg Sports Traumatol Arthrosc. 2011;19:914–20.
21. Murakami K, et al. Kinematic analysis of stair climbing in rotating platform cruciate-retaining and posterior-stabilized mobile-bearing total knee arthroplasties. Arch Orthop Trauma Surg. 2017;137:701–11.
22. Tsuji S, et al. Effect of posterior design changes on postoperative flexion angle in cruciate retaining mobile-bearing total knee arthroplasty. Int Orthop. 2011;35:689–95.
23. Pagnano MW, et al. Rotating platform knees did not improve patellar tracking: a prospective, randomized study of 240 primary total knee arthroplasties. Clin Orthop Relat Res. 2004;428:221–7.
24. Sawaguchi N, et al. Mobile-bearing total knee arthroplasty improves patellar tracking and patellofemoral contact stress: in vivo measurements in the same patients. J Arthroplast. 2010;25:920–5.
25. Kuroda Y, et al. Mobile-bearing insert reduced patellar contact force at knee flexion during posterior stabilized total knee arthroplasty. Clin Biomech (Bristol, Avon). 2020;76:105022.
26. Minoda Y, et al. A cemented mobile-bearing total knee replacement prevents periprosthetic loss of bone mineral density around the femoral component: a matched cohort study. J Bone Joint Surg Br. 2010;92:794–8.
27. Tsukeoka T, et al. The rectangular flexion gap is associated with an increased knee flexion angle in a cruciate-sacrificing rotating platform mobile-bearing total knee arthroplasty. J Orthop Sci. 2017;22:313–7.
28. Ishii Y, et al. Prediction of range of motion 2 years after mobile-bearing total knee arthroplasty: PCL-retaining versus PCL-sacrificing. Knee Surg Sports Traumatol Arthrosc. 2011;19:2002–8.
29. Ishii Y, et al. Patellar tendon length during knee flexion of meniscal-bearing and rotating total knee arthroplasty implants. Knee Surg Sports Traumatol Arthrosc. 2017;25:3773–8.
30. Ishii Y, et al. Insall–Salvati ratio stabilizes one year after mobile-bearing total knee arthroplasty and does not correlate with mid-to-long-term clinical outcomes. Knee Surg Sports Traumatol Arthrosc. 2019;27:1604–10.
31. Ishii Y, et al. Mobile-bearing total knee arthroplasty implants combined with surface cementation produced satisfactory clinical and radiographic outcomes at the 5-year follow-up. Knee Surg Sports Traumatol Arthrosc. 2019;27:2181–8.
32. Ishii Y, et al. Comparison of long-term clinical outcomes after bilateral mobile-bearing total knee arthroplasties using PCL-retaining and PCL-substituting implants in the same patients. Knee Surg Sports Traumatol Arthrosc. 2017;25:3711–7.
33. Minoda Y, et al. Mobile-bearing prosthesis and intraoperative gap balancing are not predictors of superior knee flexion: a prospective randomized study. Knee Surg Sports Traumatol Arthrosc. 2015;23:1986–92.
34. Minoda Y, et al. Mobile-bearing TKA improved the anteroposterior joint stability in mid-flexion range comparing to fixed-bearing TKA. Arch Orthop Trauma Surg. 2016;136:1601–6.
35. Watanabe T, et al. Comparison between mobile-bearing and fixed-bearing knees in bilateral total knee replacements. Int Orthop. 2005;29:179–81.
36. Hasegawa M, et al. Staged bilateral mobile-bearing and fixed-bearing total knee arthroplasty in the same patients: a prospective comparison of a posterior-stabilized prosthesis. Knee Surg Sports Traumatol Arthrosc. 2009;17:237–43.
37. Matsuda S, et al. Mobile-bearing prosthesis did not improve mid-term clinical results of total knee arthroplasty. Knee Surg Sports Traumatol Arthrosc. 2010;18:1311–6.
38. Wen Y, et al. A meta-analysis of the fixed-bearing and mobile-bearing prostheses in total knee arthroplasty. Arch Orthop Trauma Surg. 2011;131:1341–50.
39. Beverland DE, et al. LCS rotating platform dislocation and spin-out – etiology, diagnosis and management. In: Hamelynck KJ, et al., editors. LCS mobile bearing knee arthroplasty: a 25years worldwide review. Berlin: Springer; 2002. p. 35–40.
40. Matsuda Y, et al. Effect of flexion angle on coronal laxity in patients with mobile-bearing total knee arthroplasty prostheses. J Orthop Sci. 2005;10:37–41.
41. Kobayashi H, et al. Spontaneous dislocation of a mobile-bearing polyethylene insert after posterior-stabilized rotating platform total knee arthroplasty: a case report. Knee. 2011;18:496–8.
42. Tanikake Y, et al. Nontraumatic tibial polyethylene insert cone fracture in mobile-bearing posterior-stabilized total knee arthroplasty. Arthroplast Today. 2016;2:157–63.
43. Hozack WJ, et al. The patellar clunk syndrome: a complication of posterior stabilized total knee arthroplasty. Clin Orthop Relat Res. 1989;241:203–8.
44. Fukunaga K, et al. The incidence of the patellar clunk syndrome in a recently designed mobile-bearing posteriorly stabilized total knee replacement. J Bone Joint Surg Br. 2009;91:463–8.
45. Kuwashima U, et al. Contact stress analysis of the anterior tibial post in bicruciate stabilized and mobile-bearing posterior stabilized total knee arthroplasty designs. J Mech Behav Biomed Mater. 2016;60:460–7.

46. Lee CS, et al. Early nontraumatic fracture of the polyethylene tibial post in a NexGen LPS-flex posterior stabilized knee prosthesis. J Arthroplast. 2009;24:1292.e5-9.
47. Hirakawa K, et al. Characterization of debris adjacent to failed knee implants of 3 different designs. Clin Orthop Relat Res. 1996;331:151–8.
48. Minoda Y, et al. Characteristics of polyethylene wear particles isolated from synovial fluid after mobile-bearing and posterior-stabilized total knee arthroplasties. J Biomed Mater Res B Appl Biomater. 2004;71:1–6.

PCL Substitution: PS, BCS, CS

36

Satoshi Hamai

1 Historical Background of the Development of Post and Cam Mechanism

The post-cam mechanism was first introduced by Insall and Burstein, and the first model using this mechanism was developed by Insall in 1978. Prior to this, a total condylar model had been designed with a high lip of the tibial insert to combat instability following anterior and posterior cruciate ligament (ACL and PCL) resection. However, this model was not stable, owing to forward sliding of the femur in the knee flexion during stair movements. In addition, other side effects, such as impingement of the posterior lip of the polyethylene in the deep flexion leading to ROM restriction, have been noted. The central protrusion of the insert, called POST, corresponds to a bearing called CAM, provided located posterior femoral position at the flexion, and compensates for PCL function. The PS model was designed to improve braking performance in the flexion by adding this post-cam mechanism, and to improve ROM by inducing posterior femoral rollback.

2 Stability and Rollback by the Post and Cam Mechanism

In studies on cadaveric knees, posterior rollback of the femur associated with flexion was not reported to occur in models with a high lip of the tibial insert. Furthermore, the rotational tolerance was reduced in the extension position. However, in models with a post-cam mechanism, rollback occurred, and the quadriceps muscle strength required for knee extension was small [1]. In addition, in a simulator study based on *in vivo* data, I/B 2 experienced temporary forward movement between 30° and 60° of flexion, but rollback was observed between 60° and 90° of flexion, where post-cam contact occurred [2].

In the dynamic analysis using X-ray images, the six degrees-of-freedom parameters that represent the relative relationships between metal components were evaluated *in vivo*. In addition, the contact states of the sliding surfaces of the insert and post-cam mechanism were evaluated using a CAD model of the polyethylene insert. According to the *in vivo* analysis conducted while performing kneeling motion, the post-cam mechanism of the PS type (NexGen LPS, Zimmer) remains in contact (Fig. 36.1), and induction of a stable femoral rollback is recognized [3]. The contact points on the sliding surfaces of the insert move backward with flexion and the lateral contact point is located on the posterior edge, whereas the contact point of the post-cam mechanism moves to the inner edge of the posterior post with external rotation of the femur. Similarly, analysis during upright sitting (Seiza position) showed that the contact point of the insert sliding surface was located on its posterior edge, indicating that subluxation in 8.3% of cases [4]. The contact point of the post-cam mechanism was located below the posterior inner edge of the post, and contact was not observed in 50% of cases. In addition, the flexion angle was 148° and the internal rotation angle of the tibia was 13°. Models with helical post-cam mechanism (EMK, Japan Medical Materials) has allowed for a wide contact area between the post and cam, in line with the implant design concept [5]. The internal rotation angles of the tibia during standing extension, forward knee thrusting, and kneeling were 1.4°, 7.5°, and 7.9°, respectively. In a fixed-bearing TKA (Attune, Depuy Synthes) with a s-curve design in the sagittal surface of the post, post-cam contact is reported to occur from 86° flexion during squat. Further, the variation in anteroposterior position before and after contact, as well as between patients, was relatively small, and stable anteroposterior movement was observed throughout the range of flexion [6]. In mobile-bearing PS (Vanguard RP Hi-Flex, Biomet) with a highly compatible saddle-shaped post-cam mechanism, contact occurs from 34° at the early flexion. Following this, the

S. Hamai (✉)
Department of Orthopedic Surgery, Graduate School of Medical Sciences, Kyushu University, Fukuoka, Japan
e-mail: hamai.satoshi.075@m.kyushu-u.ac.jp

S. Matsuda (ed.), *Advances in Total Knee Arthroplasty*, https://doi.org/10.1007/978-981-97-4920-1_36

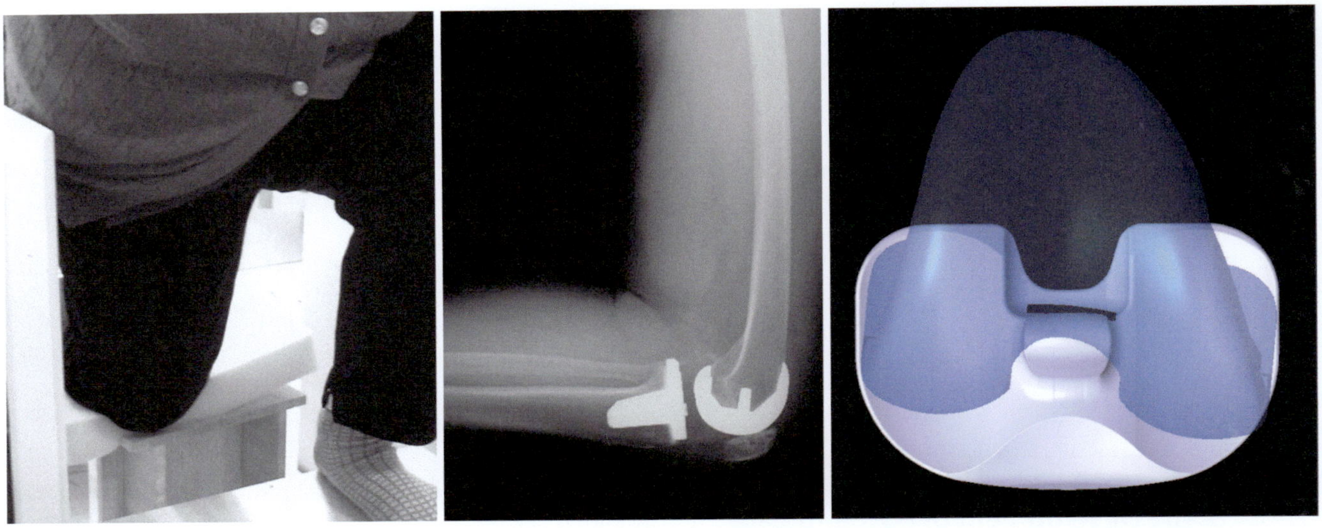

Fig. 36.1 3D dynamic analysis of PS TKA during kneeling: Contact evaluation of the post-cam mechanism

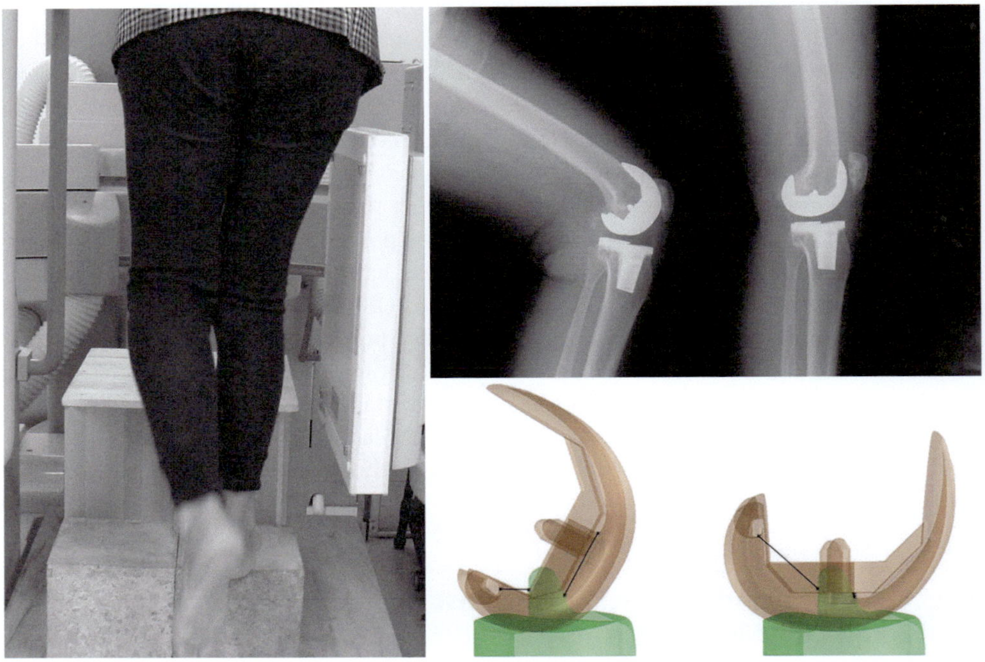

Fig. 36.2 Analysis of contact between post-cam mechanism and anterior post impingement during stair climbing of PS TKA

femoral component simultaneously undergoes external rotation and rollback of 5.1° and 2 mm, respectively, resulting in a medial pivot of up to 90° of flexion [7]. Contact with the post and cam was observed up to 120° of flexion; however, when flexion was 130° or more, no contact occurred.

The timing of post-cam contact varies depending on the model; however, owing to its history of development, this function was intended to be active during stair-climbing movements. However, the angle of flexion between implants was 10° smaller than the corresponding angle between bone axes, due to the influence of implant placement (femoral component flexion and tibial component posterior tilt) in the sagittal plane. It has been reported that the implant flexion angle during stair climbing ranged from 66° to −6° and that a PS model (NexGen LPS, Zimmer) with angle of 75° at the start of post-cam contact could not demonstrate post-cam contact (Fig. 36.2) [8]. Where the posterior tilt of the tibial component was particularly large compared to the baseline value (6.9° vs. 3.1°), instability was observed due to paradoxical translation of the femur in the mid-flexion. In a knee simulation study, a similar tendency was observed for the same model and motion [9]. In rotating platform-type

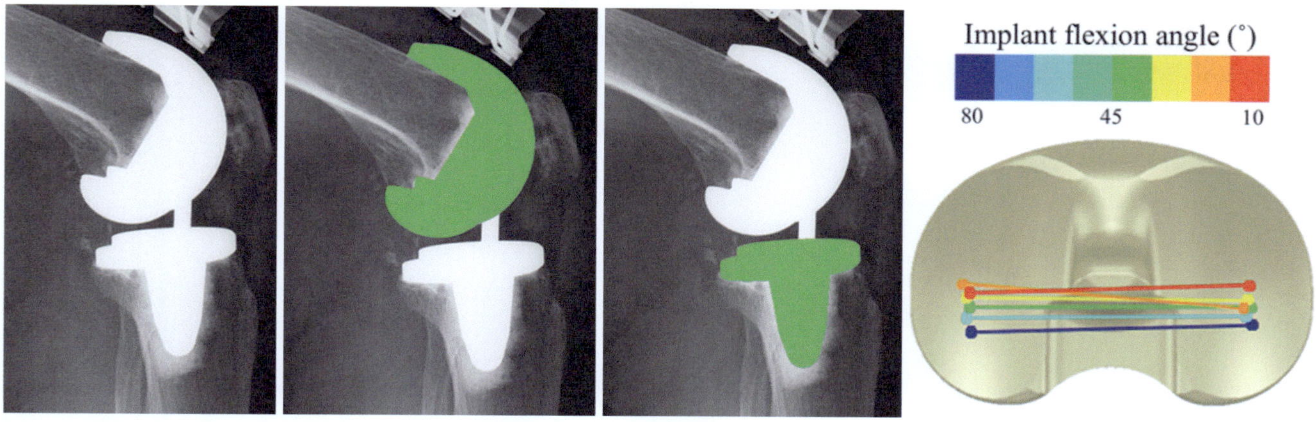

Fig. 36.3 Analysis of anteroposterior and rotational motions of PS TKA using image-matching method during stair climbing

mobile bearing (PFC Sigma, DePuy Synthes), which has relatively high restraint on the sliding surface, stable longitudinal movement was observed even without post-cam contact in the flexion [10]. Similarly, paradoxical translation in the intermediate flexion was not observed, even in the absence of post-cam contact during stair climbing (Fig. 36.3) [6]. The flexion angle at which post-cam contact occurs is different for each model, and it was reported that, depending on the model (ACTIYAS, Kyocera), contact was observed in 83% of cases during stair climbing [11].

3 Anterior Post Impingement of PS

In some types of PS, intercondylar notch of the femoral component collides with anterior post in the knee extension position. This is because, in TKA without an ACL, the tibia moves forward in extension (paradoxical translation). This leads to a reduction in the knee flexion angle during the stance phase (loss of double knee action), which reduces the external knee flexion moment (quadriceps avoidance gait). In addition, even with normal implant installation, the implant flexion angle is hyperextended by approximately 10° more than the bone axis flexion angle. The relative relationship between the femoral component intercondylar notch and the anterior post based on implant shapes also affects the decrease in clearance before the post.

A separation of 13 mm was reported between the femoral component intercondylar notch and the anterior post in the swing phase. In the mid-stance phase, contact occurred at the medial and lateral margins of the anterior post, frequently leading to anterior post impingement (NexGen LPS, Zimmer) (Fig. 36.4) [12]. A similar form of contact was demonstrated in a study that used a knee simulator for artificial knee joints [13]. The sagittal plane is placed at 5.6° flexion for the femoral component and 5.9° backward tilt for the tibial component, and the flexion angle during gait was 6° between bone shafts and −7° between implants at mid-stance. To prevent wear and breakage due to unintentional anterior post impingement, the shapes of the intercondylar notch and post-anterior surfaces should be considered. Compared to the similar model (NexGen LPS-flex, Zimmer), a single-radius sagittal shaped femur component with wide clearance ahead of the post (NRG, Stryker) did not experience anterior post impingement during the stance phase or paradoxical translation during the swing phase [14].

In rotating platform-type mobile-bearing (PFC Sigma, DePuy Synthes), anterior post impingement during knee extension was not observed during stair climbing [10]. In examinations of the contact state using the simulator, the same model was not shown to cause anterior post impingement, owing to the relative position and height of the post and femoral component intercondylar notch [15]. A model with a slightly flat insert shape (NexGen LPS, Zimmer) exhibited paradoxical translation during knee extension in stair climbing. Anterior post impingement has been reported in 33% of patients using this specific model (Fig. 36.2) [8]. This may function as a compensatory mechanism for the ACL, and contact design that takes this mechanism into account is recommended. Major cause of anterior impingement has been reported to be the posterior tilt of the tibial component (7.5° vs. 2.8°). In a report (ACTIYAS, Kyocera) examining the frequency and timing of anterior post impingement during stair climbing, impingement occurred from the latter half of the stance phase to foot detachment. Cases with anterior post impingement [11] showed a large implant extension angle during knee extension, a more posteriorly located femoral component, and a more posteriorly tilted tibial component. The impingement may have acted as an anterior stabilizer; therefore, the posterior tilt of the tibial component should be targeted at <5° to avoid the impingement.

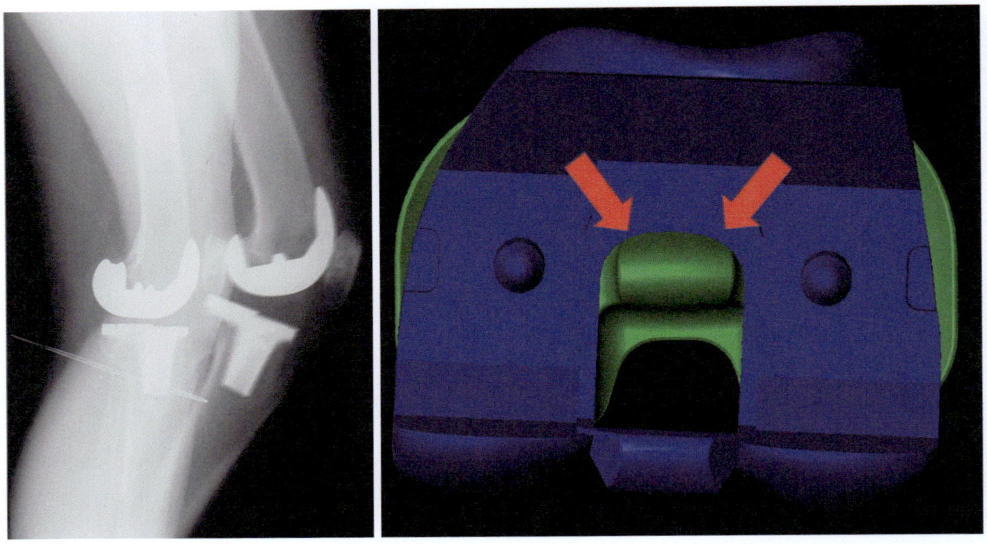

Fig. 36.4 Anterior post impingement (red arrow) during gait of PS TKA

4 Dynamic Analysis of BCS: Dual Post and Cam Mechanism

The bicruciate stabilized (BCS) knee system (Smith & Nephew) is an implant with features such as a dual post-cam mechanism that reproduces the function of the ACL and PCL, an asymmetrical component shape, and a polyethylene insert with built-in medial concave and lateral convex surfaces and a 3° inward inclination. It differs from the PS type in having a dual post-cam that compensates for ACL function. This mechanism involves contacting the anterior cam of the femoral intercondylar canopy in the extension position.

Although the dynamics of squat and kneeling in Journey 1 BCS differed from the physiological kinematics of normal knees, the post-TKA kinematics were reported to be closer to OA than to CR and PS types [16]. The successor model (Journey 2) introduced in Japan showed good rollback in both squat and stair-climbing activities, as well as no paradoxical translation in the extension and mid-flexion ranges [17]. The frequency of anterior post-contact for squat and stair-climbing movements was 14% and 27%, respectively, which was lower than that of PS type anterior impingement (63% [19] and 33% [8], respectively), and contact initiation was 4.3° and 2° extension, respectively. In Journey 2, 2.5° external rotation of the femur during a squatting motion was smaller than 10° in Journey 1 [16]. A computer simulation of the stair-climbing motion reported that the posterior tilt of the tibial component should be less than 10° to avoid paradoxical translation [18].

During gait, the femoral component is located almost anteriorly in the midstance phase and posteriorly in the swinging phase; therefore, many cases of physiological double-knee action have been reported [19]. In addition, anterior/posterior post-contact was observed in 74% and 30% of patients, respectively. The inclination of the tibial component relative to the ground was 3° greater during the stance phase of walking than during the standing position. This occurred owing to the outward movement of the center of gravity, resulting in a 5° lateral inclination in a PS type. Because the implant joint surfaces were medially inclined by 3°, the BCS type accommodated the lateral tilt of the joint line. Physiological knee function and load distribution may be advantageous for long-term performance [20].

A posterior femoral rollback ratio of 38% was observed at an active maximum flexion of 124°, indicating that rollback was reproduced owing to a compensatory mechanism of the ACL and PCL [21]. Although no difference in rollback was observed, the lateral area was located farther back in the load position [22]. In addition, the anterior post-contact in the loaded and unloaded positions was 55% and 36%, respectively, and contact at 10° of flexion was observed in both positions. Posterior contact was observed in all cases; however, the contact start angles varied (68° and 43° flexion).

In a study that conducted prospective kinetic analysis of squat movements before and after surgery and examined the effects of preoperative dynamics and surgical techniques on postoperative dynamics, the postoperative flexion angle and anteroposterior movement were reported to increase; however, the rotation angle did not significantly change. Moreover, the postoperative flexion and rotation angles were affected by the preoperative flexion and rotation angles (Fig. 36.5) [23]. In addition, in patients with small postoperative rotation angles, the rotation mismatch between the femur and tibial components exceeded 5°. This indicates that accurate rotational alignment is necessary.

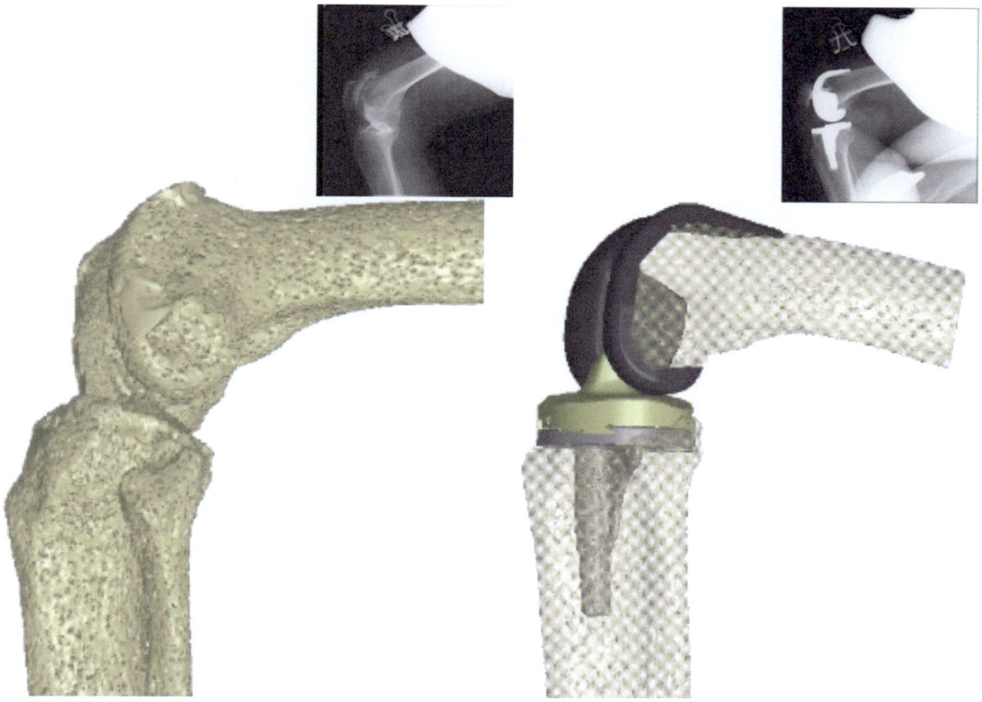

Fig. 36.5 Dynamic analysis before and after BCS TKA; contact evaluation of dual post-cam mechanism

5 Contact Conditions in Anterior and Posterior Post

Examination of contact conditions in the anterior and posterior post and sliding surfaces [24, 25] using a simulator for artificial knee joints and pressure-sensitive films has been reported. Furthermore, analyses of contact stress distribution using finite element models (FEM) and virtual knee simulators based on multibody dynamics have been reported as well.

5.1 Posterior Post

In a study on contact position, area, and contact pressure in the posterior post and cam parts at 90°, 120°, and 150° flexion and rotation of 0° and 10° respectively, 500 N was applied forward to the tibial component when the femoral component was fixed. Although models with a curved shape had a mild increase in contact pressure (Scorpio, Stryker) even when the tibia rotated 10° internally, a significant increase in contact pressure has been reported in models with a flat shape (NexGen LPS-flex, Zimmer) [26]. Contact area was measured by lifting the femoral component 3 mm proximally to mimic the so-called lift-off phenomenon in which the femoral component is lifted off the tibia. When this was done, the round type (TRAC, Alpina, Biomet) with a large radius of curvature of the cam showed significantly reduced contact area [27]. FEM analysis showed that models with square post-cam mechanisms (NexGen LPS-flex, Zimmer) and circular models (Actiyas, Kyocera Medical) have similarly reduced stress concentrations in the posterior post in flexion position and in the anterior post in hyperextension position [28]. Analysis of walking, stair climbing, chair sitting, and squat movements using LS-DYNA shows high abrasion resistance and low rotation tolerance in situations with high compatibility between the femur and tibial components and the post and cam [29, 30].

5.2 Anterior Post: PS, BCS

Because impingement in the anterior post was not anticipated during development, the contact state and shape and area of the contact surface were not adequately examined. Therefore, high contact stress may be applied, which has led to wear and breakage being reported.

A study measured the contact state in the anterior post in fixed-bearing PS; 100 N was added backward with an implant flexion angle of −15° to 5°, a rotation of 0°, and an external rotation of 5°. NexGen LPS-flex (Zimmer) had two contact zones on the inner and outer edges of the anterior post, and Genesis II (DePuy Synthes) and Scorpio NRG (Stryker) found a horizontal band-like contact zone in front of the post [13]. Although contact pressure varied with implant and flexion/rotation conditions, it exceeded the yield stress of polyethylene in some cases, resulting in insufficient contact morphology between the anterior post and the femoral component intercondylar notch. Therefore, design modifications are required to provide a larger contact area during anterior impingement. In the mobile-bearing PS type, PFC Sigma RP (DePuy Synthes) did not allow contact under any condition, and TRAC (Biomet), VANGUARD RP (Biomet), and LPS Mobile (Zimmer) did not change contact pressure during rotation. However, LPS Mobile demonstrated an increase in tactile pressure under several conditions [15]. The contact state of the BCS-type Journey II (Smith & Nephew) has been reported to have a wider contact area than the PS types; furthermore, no large increase in contact pressure during rotation was observed. In simulations, stress concentration was observed in the anterior post when the posterior tilt of the tibial component was ≥10° [18].

6 CS

The cruciate-substituting (CS) implant is characterized by an increase in the anterior lip of the insert that suppresses anterior translation of the femur due to resection or degeneration of the PCL and leads to posterior stability. Types of CS implants include the Bi-Surface Knee (Kyocera), which has a unique ball-and-socket joint structure at the posterior center of the intercondylar area, and the medial pivot type (Advance medial pivot knee, Wright Medical Technology), in which the ball-in-the-socket structure enhances the congruity of the medial compartment.

A study using a Bi-surface Knee examined the trajectory and area of the contact surface between the femoral and tibial components, and the trajectory showed no transient forward movement, but rollback during flexion [2]. The ball-and-socket joint has been reported to function normally from 60° flexion. Above 90°, a wide contact area was observed at the ball-and-socket joint, and the main load was transferred to the same part. In a dynamic analysis study using X-ray images, the medial and lateral condyles of the femur showed rollback of 8.7 mm and 11 mm, respectively, and rotation of 5.7° at a maximum flexion of 126°. A normal rotational pattern was observed in 85% of patients [31]. Posterior rollbacks of the medial and lateral condyles of the femur were induced after ball-and-socket contact, and no change was observed in rotational motion [32]. Contact between the ball and socket has been reported to occur in the socket and contribute to posterior stability of the tibia. Furthermore, equivalent anteroposterior and rotational motions between CS and PS implants have been reported [33]. Although paradoxical translation occurred at extension, the medial pivot was observed to from 30° to 60° of flexion, followed by bicondylar rollback up to the maximum flexion position [34]. Post-cam contact, which was not observed in stair-climbing movements [35], was reported to occur in 80% of cases at 57° flexion when rising and in 60% of cases at 60° flexion when sitting. Analysis of the CS type kneeling motion showed no anterior subluxation in most cases, and 7.7% of patients showed forward movement of ≥2 mm [36]. While the flat and dish types showed equivalent anteroposterior and rotational movements, posterior rollback and rotation at a deep flexion of 150° in the flat type were greater than that in the dish type. This may have occurred due to this type of implant having less constraint of sliding surfaces.

In a study on cadaveric knees with a medial pivot type, although bicondylar rollback was observed, medial pivot motion was not observed at >60° flexion with PCL preservation. Furthermore, kinematic conflict occurred at the posterior edge of the socket structure [37]. PCL dissection showed medial pivot motion; however, no rollback was induced in the absence of a post-cam compensatory mechanism. In a study of flexion motion using X-ray images, the tibia tended to rotate externally up to 20° and then rotate internally, indicating medial pivot motion without anteroposterior movement of the medial condyle of the femur [38].

References

1. Matsuda S, et al. Knee kinematics of posterior cruciate ligament sacrificed total knee arthroplasty. Clin Orthop Relat Res. 1997;341:257–66.
2. Akagi M, et al. Functional analysis of the effect of the posterior stabilizing cam in two total knee replacements. J Bone Joint Surg Br. 2002;84:561–5.
3. Hamai S, et al. Kinematic analysis of kneeling in cruciate-retaining and posterior-stabilized total knee arthroplasties. J Orthop Res. 2008;26:435–42.
4. Niki Y, et al. Is greater than 145{degrees} of deep knee flexion under weight-bearing conditions safe after total knee arthroplasty?: A fluoroscopic analysis of Japanese-style deep knee flexion. Bone Joint J. 2013;95:782–7.
5. Tsumori Y, et al. Analysis of weight-bearing kinematics of posterior-stabilized total knee arthroplasty with novel helical post-cam design. J Arthroplast. 2011;26:1556–61.
6. Ushio T, et al. Does gradual-radius femoral component with s-curve post-cam provide stable kinematics at mid-flexion after total knee arthroplasty? J Orthop Sci. 2022;27:665–71.
7. Kurita M, et al. In vivo kinematics of high-flex mobile-bearing total knee arthroplasty, with a new post-cam design, in deep knee bending motion. Int Orthop. 2012;36:2465–71.
8. Hamai S, et al. Continuous sagittal radiological evaluation of stair-climbing in cruciate-retaining and posterior-stabilized total knee arthroplasties using image-matching techniques. J Arthroplast. 2015;30:864–9.
9. Okamoto S, et al. Effect of tibial posterior slope on knee kinematics, quadriceps force, and patellofemoral contact force after posterior-stabilized total knee arthroplasty. J Arthroplast. 2015;30:1439–43.
10. Murakami K, et al. Kinematic analysis of stair climbing in rotating platform cruciate-retaining and posterior-stabilized mobile-bearing total knee arthroplasties. Arch Orthop Trauma Surg. 2017;137:701–11.
11. Watanabe T, et al. Anterior tibial post impingement during stair climbing: a kinematic analysis and clinical outcomes. J Arthroplast. 2019;34:379–84.
12. Hamai S, et al. Evaluation of impingement of the anterior tibial post during gait in a posteriorly stabilized total knee replacement. J Bone Joint Surg Br. 2008;90:1180–5.

13. Hamai S, et al. Contact stress at the anterior aspect of the tibial post in posterior-stabilized total knee replacement. J Bone Joint Surg Am. 2010;92:1765–73.
14. Tamaki M, et al. Factors in high-flex posterior stabilized fixed-bearing total knee arthroplasty affecting in vivo kinematics and anterior tibial post impingement during gait. J Arthroplast. 2013;28:1722–7.
15. Kuwashima U, et al. Contact stress analysis of the anterior tibial post in bicruciate stabilized and mobile-bearing posterior stabilized total knee arthroplasty designs. J Mech Behav Biomed Mater. 2016;60:460–7.
16. Kuroyanagi Y, et al. In vivo knee kinematics during stair and deep flexion activities in patients with bicruciate substituting total knee arthroplasty. J Arthroplast. 2012;27:122–1228.
17. Murakami K, et al. Knee kinematics in bicruciate stabilized total knee arthroplasty during squatting and stair-climbing activities. J Orthop. 2018;15:650–4.
18. Hada M, et al. Bicruciate stabilized total knee arthroplasty can reduce the risk of knee instability associated with posterior tibial slope. Knee Surg Sports Traumatol Arthrosc. 2018;26: 1709–16.
19. Murakami K, et al. In vivo kinematics of gait in posterior-stabilized and bicruciate-stabilized total knee arthroplasties using image-matching techniques. Int Orthop. 2018;42:2573–81.
20. Murakami K, et al. Preoperative tibial mechanical axis orientation and articular surface design influence on the coronal joint line orientation relative to the ground during gait after total knee arthroplasties. Knee Surg Sports Traumatol Arthrosc. 2018;26: 3368–76.
21. Iriuchishima T, et al. A comparison of rollback ratio between bicruciate substituting total knee arthroplasty and Oxford unicompartmental knee arthroplasty. J Knee Surg. 2018;31:568–72.
22. Kono K, et al. Bicruciate-stabilized total knee arthroplasty provides good functional stability during high-flexion weight-bearing activities. Knee Surg Sports Traumatol Arthrosc. 2019;27: 2096–103.
23. Kiyohara M, et al. Comparison of in vivo knee kinematics before and after bicruciate-stabilized total knee arthroplasty during squatting. BMC Musculoskelet Disord. 2021;12:772.
24. Shiramizu K, et al. Tibiofemoral contact areas and pressures in six high flexion knees. Int Orthop. 2009;33:403–6.
25. Murakami K, et al. Variable tibiofemoral articular contact stress in fixed-bearing total knee arthroplasties. Orthop Traumatol Surg Res. 2018;104:177–83.
26. Nakayama K, et al. Contact stress at the post-cam mechanism in posterior-stabilized total knee arthroplasty. J Bone Joint Surg Br. 2005;87:483–8.
27. Akasaki Y, et al. Contact stress analysis of the conforming post-cam mechanism in posterior-stabilized total knee arthroplasty. J Arthroplast. 2008;23:736–43.
28. Watanabe T, et al. Post-cam design and contact stress on tibial posts in posterior-stabilized total knee prostheses: comparison between a rounded and a squared design. J Arthroplast. 2017;32:3757–62.
29. Anuar MAM, et al. Mechanics of posterior stabilized total knee arthroplasty during daily activities in relation to prosthesis condylar and post-cam design: a finite element study. J Mech Eng SI. 2017;4:211–28.
30. Anuar MAM, et al. Computational muscle driven knee simulator for assessment of total knee replacement post-cam mechanics. Int J Eng Technol (UAE). 2018;7:165–8.
31. Nakamura S, et al. In vivo femoro-tibial kinematic analysis of a tri-condylar total knee prosthesis. Clin Biomech (Bristol, Avon). 2014;29:400–5.
32. Nakamura S, et al. 3D in vivo femoro-tibial kinematics of tri-condylar total knee arthroplasty during kneeling activities. Knee. 2014;21:162–7.
33. Nakamura S, et al. Can post-cam function be replaced by addition of a third condyle in PS TKA? J Arthroplast. 2014;29:1871–6.
34. Mine T, et al. Kinematic analysis of posterior-stabilized total knee arthroplasty during standing up from and sitting down on a chair. J Orthop Surg Res. 2016;11:142.
35. Mine T, et al. In vivo kinematics of a unique posterior-stabilized knee implant during a stepping exercise. J Orthopedic Surg Res. 2016;11:18.
36. Nakamura S, et al. In vivo kinematic effects of ball and socket third condyle as a post-cam mechanism in tri-condylar knee implants. Knee. 2015;22:237–42.
37. Omori G, et al. The effect of geometry of the tibial polyethylene insert on the tibiofemoral contact kinematics in advance medial pivot total knee arthroplasty. J Orthop Sci. 2009;14:754–60.
38. Miyazaki Y, et al. Analysis of the kinematics of total knee prostheses with a medial pivot design. J Arthroplast. 2011;26:1038–44.

Medial Pivot

37

Takashi Sato

1 History and Development of Each Medial Pivot Model (Fig. 37.1)

The medial pivot (MP) TKA traces its roots back to the Freeman-Samuelson (FS)-1000 system (Finsbury, UK) [1] developed by Freeman et al. in 1994. It was one of the models developed in the history of TKA which used the functional approach [2], which places emphasis on the wear resistance of polyethylene inserts. It was first introduced in Japan in 1998 by Wright Medical Inc.(USA) as the Advance® Medial Pivot Knee, and was subsequently introduced in 2001 by Kyocera Inc.(Japan) as the MPK®. The former was replaced with its successor model, Evolution®, which was introduced in 2010, and the latter was succeeded by the Physio Knee (JMM) in 2005. Furthermore, in 2011, the MGK Sphere (Medacta, Switzerland), which was developed based on the aforementioned research by Freeman et al. [3], was introduced in Japan. Currently (November 2021), there are four MP-type TKA models that being utilized in Japan: Advance, Evolution, Physio Knee, and MGK sphere.

In recent years, the number of models with increased restraint through the insert shape of the medial compartment has been on the rise. However, the MP TKAs defined in this chapter refer to models that have been developed based on the aforementioned background and concept, and are compatible with the ball-in-socket mechanism described below.

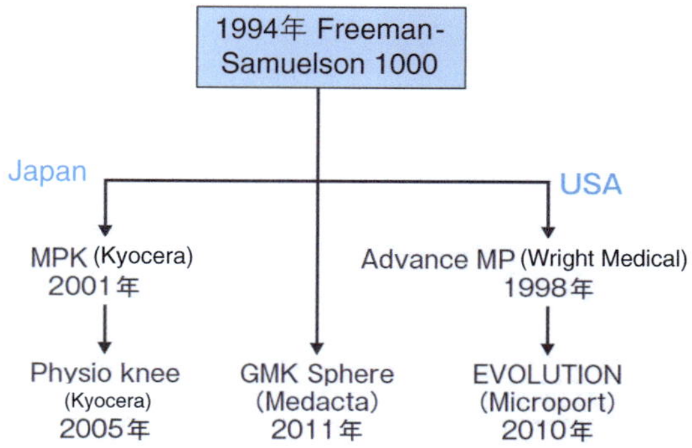

Fig. 37.1 Roots of the MP TKA and three models currently available in Japan. (Reference: Amin et al. [1])

2 Geometric Features

MP TKAs feature a spherical shape on the bearing surface of the femoral implant and a socket shape on the polyethylene insert, which corresponds to the spherical shape of the femoral side in the medial compartment. Its ball-in-socket mechanism has a high conformity (Fig. 37.2). This design was responsible for reducing polyethylene wear and achieving joint stability, while reducing conformity on the lateral side, allowing for rotational motion. This difference in conformity between the medial and lateral sides of the joint explains why the medial pivot motion tends to occur in this design.

T. Sato (✉)
Department of Orthopaedic Surgery, Niigata Medical Center, Nigata, Japan

S. Matsuda (ed.), *Advances in Total Knee Arthroplasty*, https://doi.org/10.1007/978-981-97-4920-1_37

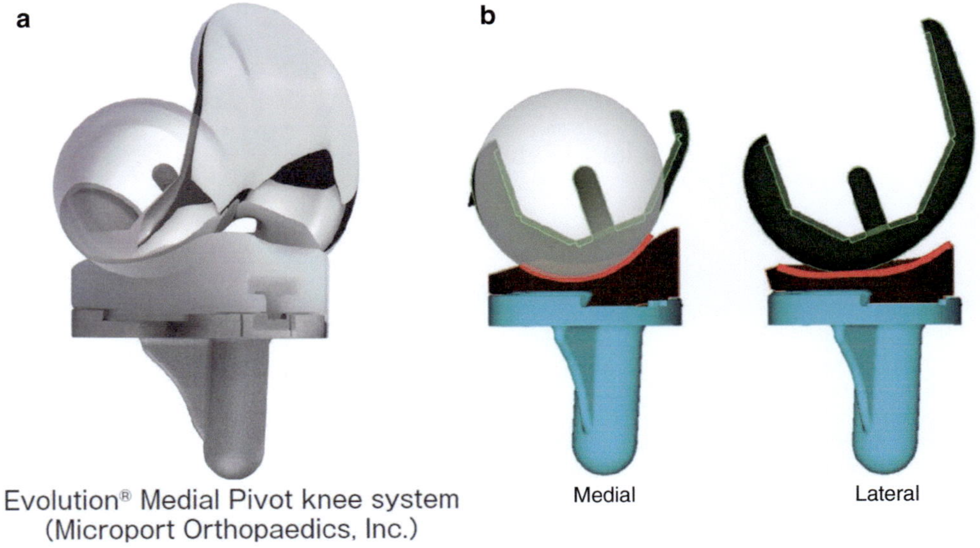

Fig. 37.2 Geometric characteristics of MP-type TKA. (Left image: courtesy of Microport Orthopedic (licensed), http://osimplantes.com.br/produtos/files/microport/joelho/Evolution%20Medial%20Pivot.pdf)

Table 37.1 Quantity, size, and shape of polyethylene wear particles in various total knee prostheses*

Prosthesis type	Total number of particles	Equivalent circle diameter (μm)	Aspect ratio	Roundness
Posterior stabilized	1160 ± 570 × 10^5	0.78 ± 0.08	2.30 ± 0.22	2.52 + 0.36
Mobile-bearing	1750 ± 1020 × 10^5	0,81 ± 0.12	1.94 ± 0.13	1.92 + 0.18
Metal medial-pivot	570 ± 282 × 10^5	0.66 ± 0.06	1.88 ± 0.11	1.75 + 0.12
Alumina-ceramic medial-pivot	71.0 ± 28.6 × 10^5	0.78 ± 0.04	1.52 ± 0.05	1.34 + 0.05
Cruciate-relating with a highly cross-linked polyethylene insert	2.8 ± 1.2 × 10^5	0.64 ± 0.07	1.33 ± 0.10	1.11 + 0.05

Reference: Minoda et al. [8]
*The values are given as the mean and standard error of the mean. The CR type with the MP type and highly crosslinked polyethylene inserts had less wear powder than the PS and mobile-bearing types

3 The Strength of Theory

3.1 Wear Resistance of Polyethylene Inserts

The main objective of MP design development is to increase the contact area and minimize polyethylene wear by employing the ball-in-socket mechanism on the medial side. Long-term results in actual clinical cases are very good [4–7]. Minoda et al. analyzed the joint fluid wear particle after TKA surgery and found that the CR type using the MP type and highly crosslinked polyethylene insert had fewer wear particles than the PS and mobile-bearing types. They reported that it was theoretically advantageous for osteolysis [8] (Table 37.1).

3.2 Joint Stability

In general, for low-conformity implants, direct stability in the anteroposterior direction depends only on the PCL or the posterior cam mechanism, as the anterior cruciate ligament is resected and the PCL is either preserved (CR type) or resected (PS type). Therefore, in the case of the CR type, achieving anterior stability of the tibia in all ROMs is theoretically difficult (Fig. 37.3a). On the other hand, for the PS type, obtaining anteroposterior stability from extension to mid-flexion is difficult due to the absence of a direct stabilization mechanism (Fig. 37.3b). Considering the variability of soft tissue properties among individuals, obtaining optimal anteroposterior stability in all cases by adjusting only the gap and soft tissue is theoretically challenging.

However, in the case of MP TKA, the shape of the medial ball socket leads to high conformity, resulting in anticipated anteroposterior stability across all ROMs under load (Fig. 37.3c). As a result, the dependence on soft tissues is lower compared to low-conformity implants. Therefore, the implant does not require fine-tuning of the soft-tissue balance and remains unaffected by individual variations in soft-tissue properties. This aspect is extremely important for achieving a certain level of clinical success, particularly in situations where surgeons with varying skill levels perform TKA on a large number of patients.

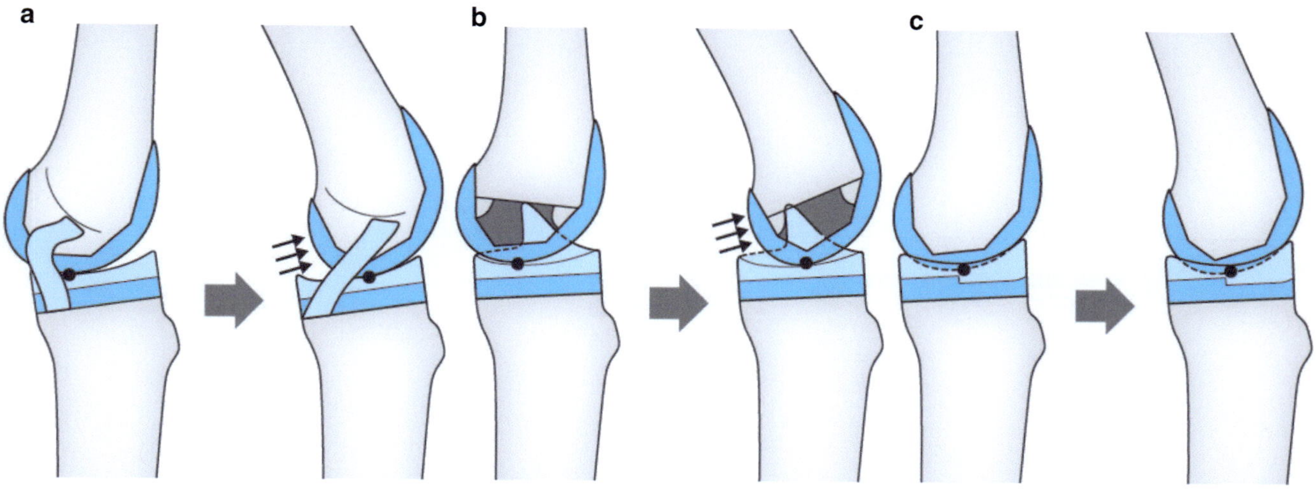

Fig. 37.3 Longitudinal stability from extension position to intermediate flexion position in low conformity type and MP type. (**a**): In the CR extension position, the femur is located posteriorly owing to ACL failure and posterior inclination of the tibia, and the PCL is relaxed; from this state to the intermediate flexion position, the femur moves forward, and the lowest point of the femoral implant (contact point: ●) moves forward. (**b**): In the PS-type post-cam mechanism, flaccidity occurs anteriorly from the forward contact in the extension position to the posterior contact after the intermediate flexion position, and forward movement occurs at the lowest point of the femoral implant. (**c**): In the MP-type, because the ball-in-socket mechanism has longitudinal braking in all ROMs; the lowest point of the femoral implant does not move

3.3 Joint Dynamics

Normal knees show medial pivot movements under load, which has been reported in numerous studies [9–11]. Miyazaki et al. [12] reported that after TKA surgery, the MP type (Advance®, Wright Medical Inc.) showed medial pivot movement during *in vivo* 3D motion analysis (Fig. 37.4). The authors performed TKA using the MP model (Evolution®, Microport) in 21 patients with a favorable postoperative course and analyzed the three-dimensional dynamics of 21 knees using the 2D-3D image matching technique. On average, from extension to 70° of flexion, both the medial and lateral femoral condyles moved slightly in the anteroposterior direction. However, from 70° to 120° of flexion, a medial pivot movement was observed, and in deep flexion, a bi-condylar rollback movement was observed, with both the medial and lateral condyles moving backward (Fig. 37.5). Focusing on individual cases, all knees showed high anteroposterior stability on the medial side from the extension position to 120° of flexion; however, there tended to be greater variation on the lateral side. Furthermore, both the study by Miyazaki et al. and our study showed a decreased amount of rotation throughout knee motion compared with normal knee motion [11]. This may be because the anteroposterior motion on the lateral side is not necessarily uniform, and some knees do not exhibit large anteroposterior movements. The high conformity of the MP TKA on the medial side and low conformity on the lateral side, enable medial pivot movement owing to the difference in conformity. However, the design does not actively induce the same movement observed in the normal knee. The amount of rotation and lateral movement varies depending on the soft tissue conditions in each case. It should be interpreted that rotational motion can be allowed if the situation requires it in various knee movements in daily life. Therefore, if anteroposterior stability on the medial side is acquired, the original design concept can be considered appropriate, even if there is variation in movement on the lateral side.

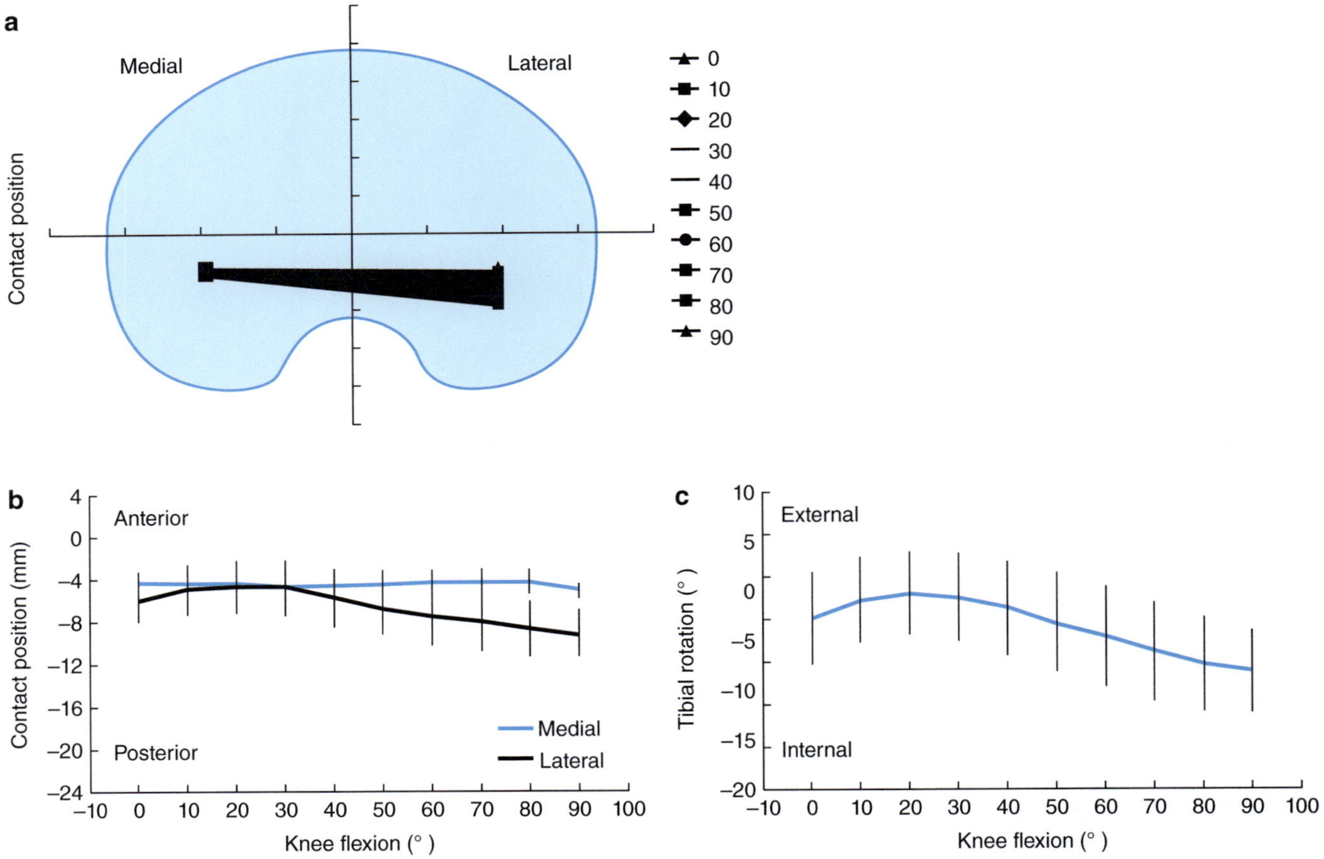

Fig. 37.4 Joint dynamics in Advance TKA. (**a**): Medial pivot motion centered on the inner rotation. (**b**): The interior of the solid line remains almost constant in the longitudinal direction and the outer line moves backward from approximately 40° of flexion. (**c**): External rotation volume associated with knee flexion is less than 10°. (Reference: Miyazaki et al. [12])

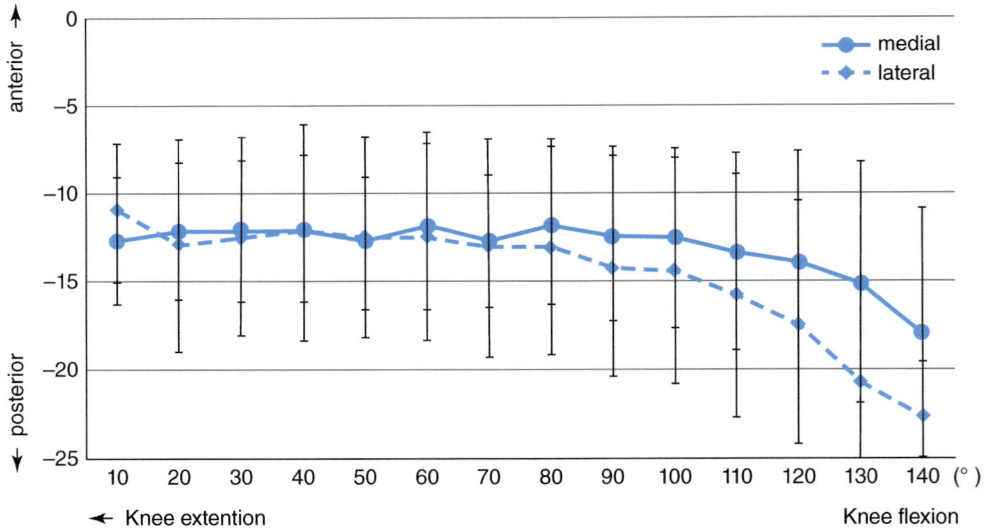

Fig. 37.5 Joint dynamics in MP-type TKA (self-study example). The solid line shows the medial epicondyle of the femur and the dotted line shows the change in the anteroposterior position associated with knee flexion, as seen in the tibia of the lateral epicondyle of the femur. The medial epicondyle remained almost constant from the extension position to approximately 120° of flexion and then moved backward in the deep flexion region, whereas the lateral epicondyle moved backward from approximately 70° of flexion

References

1. Amin A, et al. The early radiological follow-up of a medial rotational design of total knee arthroplasty. Knee. 2008;15:222–6.
2. Robinson RP. The early innovators of today's resurfacing condylar knees. J Arthroplast. 2005;20(Suppl 1):2–26.
3. Freeman MAR, et al. The movement of the normal tibio-femoral joint. J Biomech. 2005;38:197–208.
4. Fitch DA, et al. Mid- to long-term outcomes of a medial-pivot system for primary total knee replacement: a systematic review and meta-analysis. Bone Joint Res. 2014;3:297–304.
5. Karachalios T, et al. A mid-term clinical outcome study of the Advance medial pivot knee arthroplasty. Knee. 2009;16:484–8.
6. Macheras GA, et al. A long term clinical outcome of the medial pivot knee arthroplasty system. Knee. 2017;24:447–53.
7. Chinzei N, et al. Satisfactory results at 8 years mean follow-up after ADVANCE medial-pivot total knee arthroplasty. Knee. 2014;21:387–90.
8. Minoda Y, et al. In vivo analysis of polyethylene wear particles after total knee arthroplasty: the influence of improved materials and designs. J Bone Joint Surg Am. 2009;91:67–73.
9. Iwaki H, et al. Tibiofemoral movement 1: the shapes and relative movements of the femur and tibia in the unloaded cadaver knee. J Bone Joint Surg Br. 2000;82:1189–95.
10. Pinskerova V, et al. The knee in full flexion: an anatomical study. J Bone Joint Surg Br. 2009;91:830–4.
11. Tanifuji O, et al. Three-dimensional in vivo motion analysis of normal knees using single-plane fluoroscopy. J Orthop Sci. 2011;16:710–8.
12. Miyazaki Y, et al. Analysis of the kinematics of total knee prostheses with a medial pivot design. J Arthroplast. 2011;26:1038–44.

38 Patellofemoral Joint: Patellar Component

Takeshi Mochizuki

The design of patellar components varies depending on the model. However, artificial joint models are often selected based on the concept of femoral component design and inserts. PFJs must have appropriate internal and external heights and an asymmetrical pulley structure that considers the anatomical structure [1].

PFJ problems involving patellar components often cause pain and require revision surgery.

1 Patella Component Shape-1

1.1 Onlay Type

The Onlay type is currently the mainstream patellar component; and osteotomy of the entire articular surface is performed considering the thickness of the component to be installed. There are several designs, including the following (Fig. 38.1 and Table 38.1):

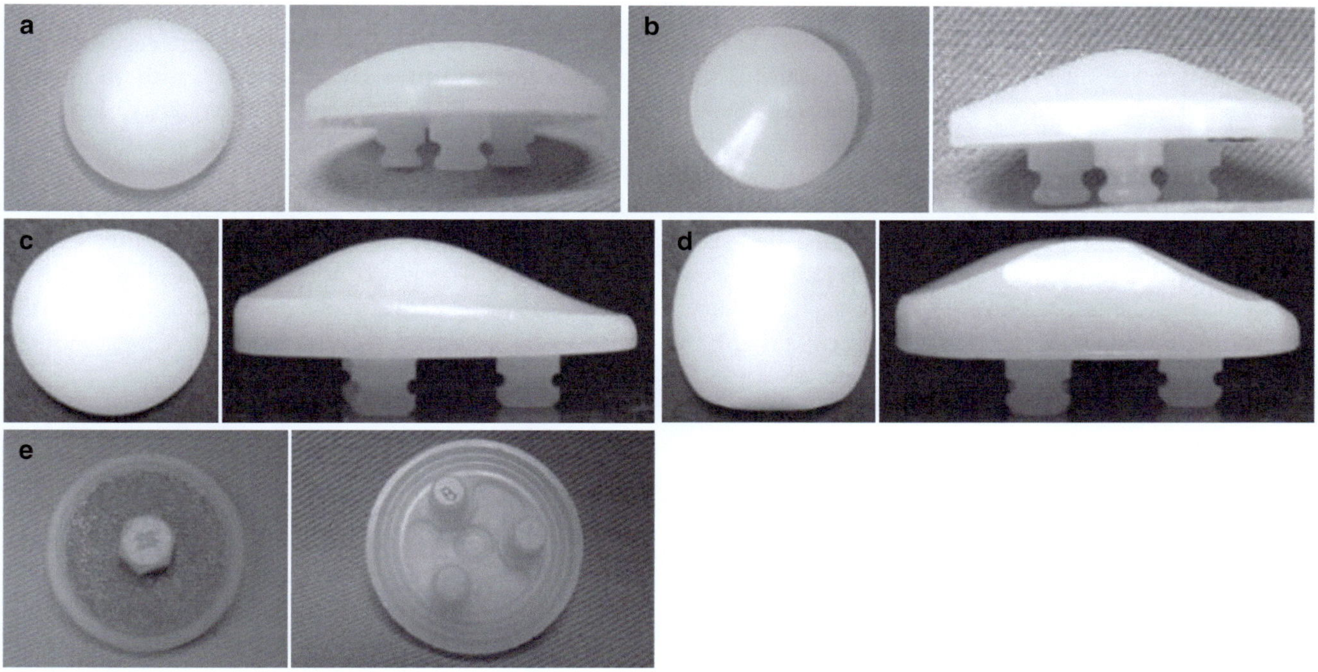

Fig. 38.1 Main shape. (**a**): Dome type (front, side). (**b**): Cone type (front, side). (**c**): Medialized dome type (front, side). (**d**): Anatomic type (front, side). (**e**): Peg (3 pegs, 1 peg)

T. Mochizuki (✉)
Kamagaya General Hospital, Chiba, Japan

S. Matsuda (ed.), *Advances in Total Knee Arthroplasty*, https://doi.org/10.1007/978-981-97-4920-1_38

Table 38.1 List of patella components of major models (in alphabetical order)

Company	Model	Shape	Size, mm	Thickness, mm	Number of pegs	Length of pegs, mm
Depuy	ATTUNE	Medialized Dome	29–41	8.5–10.5	3	5.1
		Medialized Anatomic	29–41	8.5–10.5	3	5.1
	PFC Sigma	Dome	32–41	7.8–11.4	Oval: 3 Round: 1	5.1 5.7
MDM	Balanced knee(TriMax)	Dome	29–35	7.5–9	3	5
NAKASHIMA	FINE	Dome	24–26	8	1	7
	FINE	Dome	28–34	7–9	3	5
	FUTURE KNEE	Dome	24–36	7–9	2–3	4.5–5
Smith & Nephew	JOURNEYII, GENESISII	Modified Dome	23–26	8.9–9.9	1	3.1–4.1
		Modified Dome	26–35	9 7.5	3	4.4
Stryker	Triathlon	Asymmetric	29–38	9–11	3	5
		Symmetric	27–33	8–9	3	5
Zimmer-Biomet	Vanguard	Dome	28–34	8–8.5	3	6
	Vanguard (THIN Patella)	Dome	25–34	6.2–7.8	3	6
	NexGen Persona	Modified Dome	26–41	7.5–10	3	5.6
	NexGen, (TM Patella) Persona	Modified Dome	32–41	10	1	7

1. Dome type

   When knee flexion exceeds 90°, medial and outer contact pressures increase and problems such as patellar fracture, medial and external wear, and sagging have been reported [2].
2. Modified dome type (Cone type, Gaussian type).

   Developed to alleviate the clinical problems of the dome type described above and was designed to have a concave edge to reduce the contact pressure and shear force during deep flexion [3].
3. Medialized (Offset) dome type

   Placing the patellar component inward reduces the Q-angle and improves tracking [4]. Considering the contact pressure and kinematics in cadavers, it is recommended to install the patellar component inward by approximately 2.5 mm [5]. Moving the apex of the Modified dome mold inward increases the coverage area and improves tracking upon installation.
4. Anatomic type

   This shape was expected to reduce the sagittal slope and improve the affinity of the extension mechanism [6].

1.2 Inset Type

In the Inset type, only the installation part of the patellar component was drilled, the surrounding articular cartilage was preserved, and most were cemented using a single peg.

2 Shape of the Patellar Component 2

The number of pegs is three pegs and one peg, respectively. The risk of patellar fractures for one peg, and three pegs have been shown to reduce the frequency of fractures [7]. Three pegs were thicker near the articular surface to prevent damage. In the clinical results, with an average observation period of 5.4 years, the frequency of patellar fractures was 4.7% for one peg and 2.1% for three pegs. The Hospital for Special Surgery Knee Score was 85 points for one peg and 89 points for three pegs, respectively [8]. Currently, the use of three pegs is increasing; recently, clinical results of total knee arthroplasty (TKA) have been considered to indicate the performance of the three pegs.

3 Shape of the Patellar Component 3

In addition to the previously mentioned classifications of the articular surface, there are circular and oval shapes. Due to the natural variation in patellar shape among individuals, it is necessary to determine whether the patella is round or oval using the preoperative CT. By using oval components for the oval patella, the coverage area can be increased, and the medial ridge can be better reproduced [9]. However, unlike a circular patella, care must be taken in the direction of installation.

4 Kinematics and Clinical Results Due to Design Differences

4.1 Comparison of Dome Type and Anatomic Type

In an in vivo comparison between the medialized dome and the medialized anatomy (Depuy, ATTUNE), it was found that the dome type resulted in higher contact pressure with the femoral components. In addition, quadriceps muscle strength did not differ between the two in knee extension motion, but the anatomic type was superior at a knee flexion of 75° in the lunge motion. The patellar flexion angle associated with knee flexion showed movement closer to that of the normal knee in the anatomic type [10]. We compared the dome and anatomical types during surgery using the same biological knee and a similar model. The dome type had a rotation angle of 120° knee flexion compared with the anatomic type. Flexion angles were significantly greater at knee flexion angles of 90° and 120°, and the lateral tilt was significantly greater at a knee flexion angle of 30° (Fig. 38.2). Differences in the design of the patellar components suggest different patellar kinematics [11]. In the anatomic type, the mobile-bearing types were closer to normal kinematics than the fixed-type. The fixed- and mobile-bearing kinematics of the dome types were not significantly different [12]. Differences in patellar component design and differences between Fixed and Mobile-bearing types were found to affect kinematics.

Based on the aforementioned report, we compared the clinical outcomes of the dome and anatomic types. 100 mm improved by 39.7 ± 15.2 mm and 40.7 ± 25.8 mm, respectively. According to the Japanese version of the Japan Knee Osteoarthritis Measure (JKOM), the JKOM score improved by 20.1 ± 21.8 points and 16.3 ± 22.8 points, respectively [13]. In contrast, the anatomic type was significantly higher in the Knee Injury and Osteoarthritis Outcome Score (KOOS) Function than in the dome type, and the frequency of walking pain and difficulty walking was significantly lower (Fig. 38.3) [14].

4.2 Comparison of the Onlay Type and Inset Type

In a comparison between the Onlay and Inset types in cadaver knees, the lateral shift and tilt tended to be larger in the Inset type, particularly when the lateral shift and tilt had a flexion of 70° [15]. However, there was no difference in clinical outcomes based on patient-based evaluation [16].

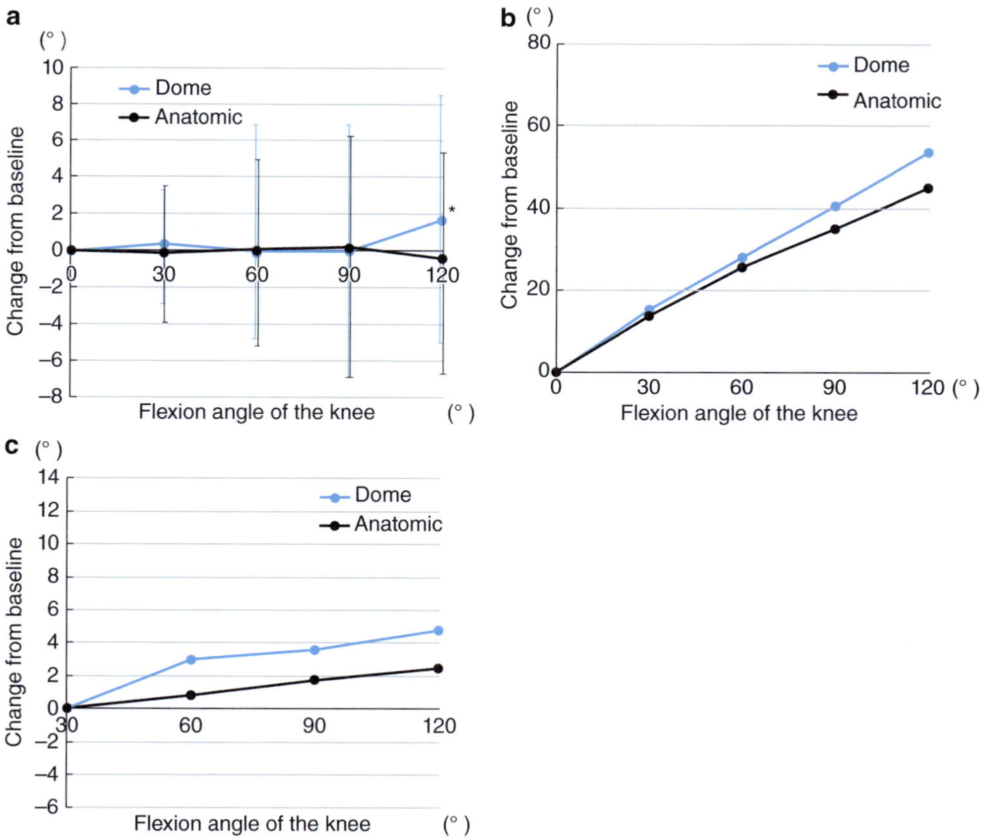

Fig. 38.2 Dome and anatomic kinematics. (**a**): rotation angle. (**b**): flexion angle. (**c**): lateral tilt

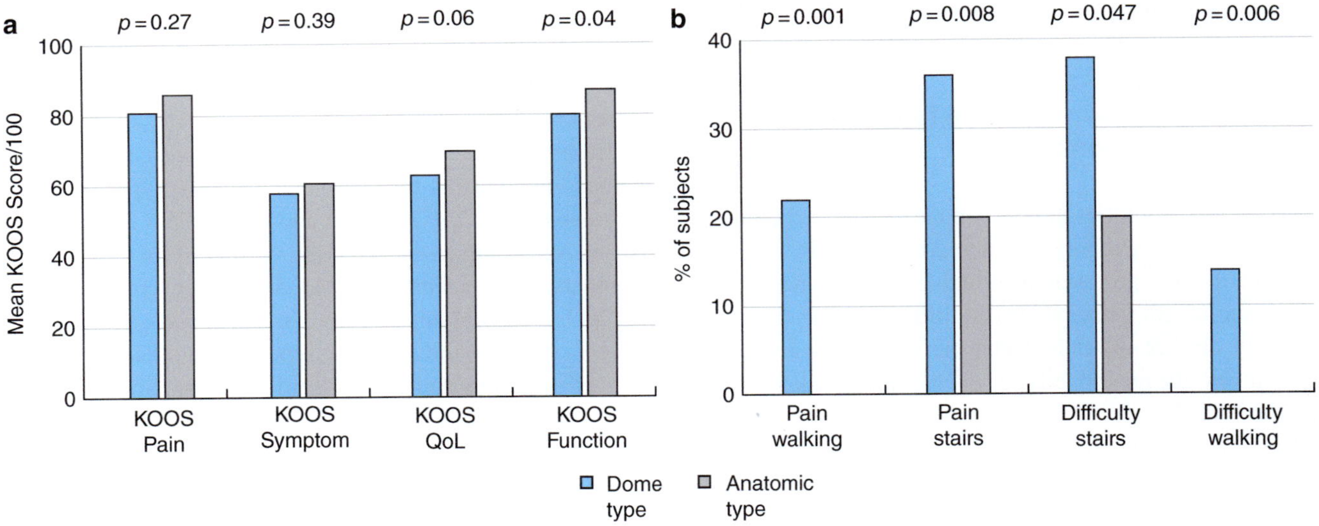

Fig. 38.3 Comparison of clinical outcomes between the dome and anatomic type. (**a**): KOOS score. (**b**): Percentage of people with pain and difficulty walking and stair movements

We believe that the clinical results of different patellar component designs need to be re-examined for a variety of movements, including other implants.

5 Added Value of Materials

Highly crosslinked ultrahigh-molecular-weight polyethylene is increasingly used as a material for these components. This technology also reduces wear and osteolysis. However, free radicals are formed during this cross-linking process, creating the risk of damage and wear owing to oxidation. The addition of vitamin E provides oxidative protection and reduces wear [17, 18]; however, other reports state that there is no difference in wear [19, 20]. Future studies are required to determine how this affects the long-term clinical outcomes.

6 Future Prospects and Challenges

Although non-replacement is an option for the patella in TKA, long-term results from the perspective of the revision surgery rate have reported that non-patellar replacement is higher than patellar replacement and that non-conformity of patellar components has a higher revision surgery rate in patellar replacement [21]. More suitable patellar components, including those compatible with the femoral components, must be selected and developed.

To improve the surgical outcomes of TKA, it is important to consider the role of patellar components and investigate impact of the PFJ on surgical outcomes. This includes evaluating the influence of PFJ shape, and durability through clinical studies.

References

1. Werth L, et al. The need for secondary resurfacing is affected by trochlear height in total knee arthroplasty. Knee Surg Sports Traumatol Arthrosc. 2017;25:3818–23.
2. Schindler OS. The controversy of patellar resurfacing in total knee arthroplasty: Ibisne in medio tutissimus? Knee Surg Sports Traumatol Arthrosc. 2012;20:1227–44.
3. Singerman R, et al. Effect of femoral component rotation and patellar design on patellar forces. Clin Orthop Relat Res. 1997;334:345–53.
4. Yoshii I, et al. The effect of patellar button placement and femoral component design on patellar tracking in total knee arthroplasty. Clin Orthop Relat Res. 1992;275:211–9.
5. Anglin C, et al. Biomechanical consequences of patellar component medialization in total knee arthroplasty. J Arthroplast. 2010;25:793–802.
6. Abo-Alhol TR, et al. Patellar mechanics during simulated kneeling in the natural and implanted knee. J Biomech. 2014;47:1045–51.
7. Huang CH, et al. Failure of the all-polyethylene patellar component after total knee arthroplasty. J Arthroplast. 1999;14:940–4.
8. Larson CM, et al. One-peg versus three-peg patella component fixation in total knee arthroplasty. Clin Orthop Relat Res. 2001;392:94–100.
9. Yang CC, et al. Patellar component design influences size selection and coverage. Knee. 2017;24:460–7.
10. Ali AA, et al. In vivo comparison of medialized dome and anatomic patellofemoral geometries using subject-specific computational modeling. J Orthop Res. 2018;36:1910–8.
11. Mochizuki T, et al. Effect on patellar kinematics of the different patellar component designs in total knee arthroplasty: intraoperative measurement of dome type versus anatomic type. Eur J Orthop Surg Traumatol. 2020;30:419–24.
12. Smith LA, et al. Anatomic vs dome patella: is there a difference between fixed- vs mobile-bearing posterior-stabilized total knee arthroplasties? J Arthroplast. 2021;36:3773–80.
13. Mochizuki T, et al. Difference in patient-reported outcomes of various patellar component designs in total knee arthroplasty: a randomized clinical study. J Orthop Surg (Hong Kong). 2021;29:2309499021996068.

14. Ahearn N, et al. Conforming patellar button design improves outcome of total knee arthroplasty. Knee. 2021;29:399–404.
15. Ezzet KA, et al. Patellar tracking in total knee arthroplasty: inset versus onset design. J Arthroplast. 2001;16:838–43.
16. Gharaibeh MA, et al. Characteristics of three different patellar implant designs in total knee arthroplasty. ANZ J Surg. 2020;90:1303–9.
17. Minoda Y, et al. Sequentially annealed highly cross-linked polyethylene reduced in vivo wear particle generation in total knee arthroplasty. J Orthop Surg (Hong Kong). 2017;25:2309499017718909.
18. Lambert B, et al. Effects of vitamin E incorporation in polyethylene on oxidative degradation, wear rates, immune response, and infections in total joint arthroplasty: a review of the current literature. Int Orthop. 2019;43:1549–57.
19. Takemura S, et al. Comparison of a vitamin E-infused highly cross-linked polyethylene insert and a conventional polyethylene insert for primary total knee arthroplasty at two years postoperatively. Bone Joint J. 2019;101:559–64.
20. Orita K, et al. Vitamin E-infused highly cross-linked polyethylene did not reduce the number of in vivo wear particles in total knee arthroplasty. Bone Joint J. 2020;102:1527–34.
21. Lewis PL, et al. Outcome of prosthesis matched and unmatched patella components in primary and revision total knee replacement. Knee. 2017;24:1227–32.

Patellofemoral Joint: Femoral Component

39

Naoki Nakano

Approximately one in five patients who undergo total knee arthroplasty (TKA) will experience postoperative pain, which serves as a contributor to decreased satisfaction after TKA. Although the exact etiology and underlying mechanism remain unknown, postoperative pain may be attributable to patellofemoral joint (PFJ) disorder, and severe PFJ injury may warrant revision surgery [1]. In the early period after its introduction, only compatibility between the femur and tibia was considered in TKA, and thorough evaluation of the PFJ was not performed. Therefore, patients frequently presented with PFJ disorders, such as postoperative patellar clunk syndrome (PCS) and anterior knee pain (AKP). In recent years, this design has been improved to avoid post-TKA PFJ injury. The goal is to realize the natural patellar glide across all ranges of motion (ROMs) of the knee joint by bringing the implant shape closer to the original PFJ shape.

In this section, we review PCS, a major complication associated with post-TKA PFJ failure and PFJ pressure, which is associated with postoperative AKP, together with a literature review focusing on the shape and design of the femoral component.

1 Patellar Clunk Syndrome

1.1 Patellar Clunk Syndrome

PCS was first described in 1989 as a PFJ-derived symptom observed after TKA [2]. PCS is characterized by a snagging and creaking sensation perceived anterior to the knee during knee extension. This symptom is attributable to excessive growth of a fibrous nodule that extends between the quadriceps tendon and the proximal patella after TKA; this complication often occurs between 6 months and 4 years after TKA [3]. Although some Japanese studies have reported PCS after cruciate-retaining TKA [4], most cases occur after posterior-stabilized (PS)-TKA. The incidence of this complication varies across studies and is reportedly 0.0–20.0% [5–7]. PCS is characterized by the formation of a fibrous nodule that gets trapped in a flexed position between the box and post in the intercondylar region, and the knee is locked in a flexed state. However, knee extension from approximately 30–45° releases the nodule from the intercondylar box, which causes a jerky snagging sensation accompanied by a "clunk" and pain (Fig. 39.1). This condition is diagnosed based on physical evaluation and imaging findings. Gradual extension of the knee joint (which contains the fibrous nodules) from a flexed position is accompanied by patellar catching, followed by release of the nodules. Magnetic resonance imaging and ultrasonography are useful to identify fibrous nodules [2, 8].

With regard to treatment, conservative therapies such as quadriceps muscle stretching are of limited benefit, and in many cases, arthroscopic resection of fibrous nodules becomes necessary [9, 10]. Recent Japanese studies have reported removal of all fibrous nodules in the internal and external tibiofemoral joints, PFJ, femoral intercondylar area, and subpatellar fat body using the following technique [11]: Creation of anterolateral, anteromedial, and proximal superomedial portals under general anesthesia, followed by insertion of a camera through the anterolateral portal and introduction of a shaver or a similar device through the anteromedial and proximal superomedial portals.

Overall, the postoperative results were good, although several reports have described the following findings:

- Arthroscopic surgery led to significant improvement in pain and in the Knee Society Score (KSS) and KSS function score in 15 cases of PCS [10].
- Arthroscopic surgery led to significant improvement in the KSS and KSS function scores but without any improvement in stair climbing and standing up on the sub-scales in 12 cases of PCS [12].

N. Nakano (✉)
Department of Orthopaedic Surgery, Kobe University Graduate School of Medicine, Kobe, Japan

S. Matsuda (ed.), *Advances in Total Knee Arthroplasty*, https://doi.org/10.1007/978-981-97-4920-1_39

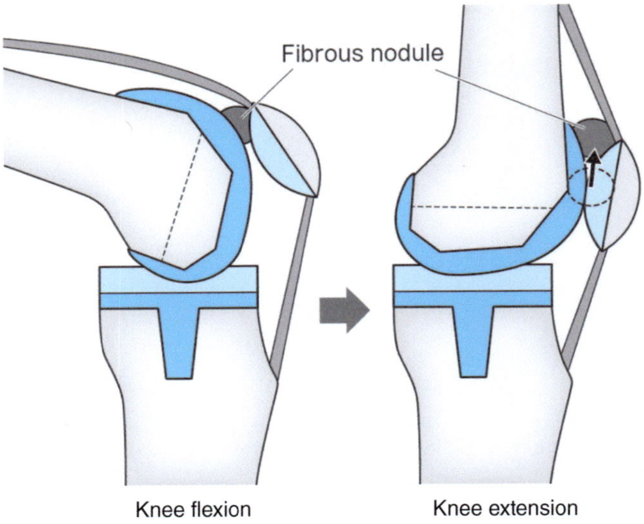

Fig. 39.1 Mechanism of Patellar clunk syndrome

- Arthroscopic surgery led to significant improvement in the KSS and The Western Ontario and McMaster Universities Arthritis Index, although multiple arthroscopic surgeries were required in 17.0% of 46 cases of PCS [13].

1.2 Etiological Factors

Currently, PCS is considered a multifactorial condition [6, 7, 14].

1. Femoral component design

   Recent research has conclusively established that the design of the intercondylar box of the femoral component is associated with development of PCS. The quadriceps tendon tends to interfere with the box component during flexion, which triggers formation of fibrous nodules. Therefore, the box size and design were modified in subsequent models. We reviewed previous studies, including those from Japan, which have discussed the femoral component design and PCS development.

   - In TKA using the AGC and Y/S II prostheses without patellar resurfacing, PCS-like patellar snagging was observed intraoperatively in 11/192 cases in the AGC group and in 0/257 cases in the Y/S II group, which is attributable to the fact that the intercondylar box ratio of the Y/S II device was smaller than that of the AGC model [15].
   - Of the 113 patients who underwent PS-TKA using the Press Fit condylar (PFC) Sigma RP prosthesis, 15 had postoperative PCS and 6 had accompanying pain. PCS occurred more frequently in cases of a large postoperative patellar tilt angle [16].
   - In PS-TKA using the Vanguard Complete Total Knee System, the incidence of postoperative PCS was significantly higher in the nonpatellar replacement group than in the patellar replacement group (33.3% vs. 4.8%). We conclude that patellar replacement is necessary when this model is used [17].
   - Comparison between the Insall-Burstein II and NexGen Legacy PS prostheses showed that implantation of the former was associated with a 4.0% risk of PCS in contrast to a 0.0% risk associated with implantation of the latter. The patellar groove shows a more diagonal design and smoother patellar glide in the latter model [18].
   - Comparison between the PFC Sigma RP and NexGen Legacy PS groups showed that the incidence of PCS was higher in the former group, which was attributable to a high intercondylar box ratio observed in the PFC Sigma RP device [14].
   - Comparison between the PFC Sigma RP and Attune PS groups in cases of PS-TKA showed that the incidence of PCS was higher in the former group, which was attributable to the small anterior flange and the small intercondylar box ratio observed in the Attune device [1].
   - The incidence of PCS in patients who underwent TKA using the AMK-Congruency, AMK-PS, and PFC Sigma RP devices was 13.5%, 3.8%, and 0.0%, respectively, which suggests that PCS is more likely to occur in patients in whom a large and wide intercondylar box is used [19].

   Although studies have reported the overall trend of the association between the femoral component design and PCS onset, the parameters, which are most strongly associated with PCS remain unknown; therefore, further research is warranted to gain deeper insight into this topic.

2. Other factors

   Postoperative patellar shortening, use of small femoral implants, placement of femoral implants in a flexed state, use of thick tibial inserts, and anterior placement of tibial inserts are implicated as risk factors for PCS [6, 16, 20]. Specific attention is necessary to monitor postoperative patellar shortening because it directly causes the patella to easily fit into the box. Postoperative PCS is a concern in patients with preoperative patella baja [21]. PCS is more likely to occur in patients in whom post-TKA deep flexion can be performed, although this is preferred in itself [22]. The risk of PCS increases with increased patellar tilt, use of small-diameter patellar implants, a thin patella, and insertion of thin patellar implants, all of which are considered to be secondary to the short physical distance between the trochlear groove and greater scope for contact because of the short quadriceps tendon

[16, 20]. With regard to PCS prevention, some surgeons recommend complete removal of synovial tissue proximal to the patella intraoperatively and patellar replacement in all cases [17]. In contrast, some surgeons do not recommend patellar replacement to prevent PCS based on their experience with patellar complications, including PCS [23]. However, definitive evidence is currently unavailable.

2 Patellofemoral Joint Pressure After Total Knee Arthroplasty

Postoperative AKP occurs in approximately 45.0% of cases, including mild cases, and high post-TKA PFJ pressure is considered a likely contributor [24]. The shape of the femoral and patellar implants and patellar thickness affect the post-TKA PFJ pressure. We measured the internal and external PFJ contact pressures across each knee flexion angle between 0° and 135° using a pressure transducer placed intraoperatively between the patellar implant and the patellar osteotomy surface in 55 patients who underwent mobile-bearing PS-TKA using the PFC Sigma device previously introduced by DePuy and its successor, the currently available Attune model [25]. We observed that the PFC Sigma group showed significantly higher medial PFJ pressure than that observed in the Attune group at 120° of knee flexion and at 120° and 135° knee flexion for lateral PFJ pressure. Based on the fact that the PFC Sigma device was associated with frequent PFJ injuries in the past [26], the trochlear shape of the groove was approximated to the anatomical shape, and the design was improved to make the anterior flange of the femoral implant thinner and narrower [1], which enabled more stable sliding along the patellar groove across the full flexion ROM of the knee joint, which reduced the PFJ pressure.

The following findings have been reported (notably all studies were performed in Japan) regarding post-TKA PFJ pressure:

- PFJ pressure was compared in 10 cadaver knees using a digital electronic sensor with various implants. A porous coated anatomic prosthesis with a smooth patellar groove does not increase PFJ pressure in the mid-to-deep flexion range. The patella does not get trapped in the intercondylar area in the model with posterior extension of the patellar groove (Profix), which prevents concentration of stress in the PFJ during deep flexion [27].
- Among 157 patients who underwent TKA without patellar resurfacing for end-stage osteoarthritis, postoperative patellar sclerosis was compared using radiography, and PFJ pressure was compared using the finite element method among several implants. Postoperative patellar sclerosis was more severe in the Genesis II than in the Low-Contact-Stress and NexGen groups. The Low-Contact-Stress and NexGen devices have a deeper trochlear design than that of Genesis II, which may have contributed to the results in our study, which concur with those reported previously [28].

Further research is warranted to address unresolved issues with regard to the PFJ, including PCS and AKP. We are hopeful that high-quality research for improvements in femoral implant design and development of novel implants will continue in Japan.

References

1. Martin JR, et al. Femoral implant design modification decreases the incidence of patellar crepitus in total knee arthroplasty. J Arthroplast. 2017;32:1310–3.
2. Hozack WJ, et al. The patellar clunk syndrome: a complication of posterior stabilized total knee arthroplasty. Clin Orthop Relat Res. 1989;241:203–8.
3. Beight JL, et al. The patellar "clunk" syndrome after posterior stabilized total knee arthroplasty. Clin Orthop. 1994;299:139–42.
4. Niikura T, et al. Patellar clunk syndrome after TKA with cruciate retaining design: a report of two cases. Orthopedics. 2008;31:90.
5. Frye BM, et al. Effect of femoral component design on patellofemoral crepitance and patellar clunk syndrome after posterior-stabilized total knee arthroplasty. J Arthroplast. 2012;27:1166–70.
6. Dennis DA, et al. The John Insall award: control matched evaluation of painful patellar crepitus after total knee arthroplasty. Clin Orthop Relat Res. 2011;469:10–7.
7. Hoops HE, et al. Control-matched computational evaluation of tendo-femoral contact in patients with posterior-stabilized total knee arthroplasty. J Orthop Res. 2012;30:1355–61.
8. Geannette C, et al. Sonographic evaluation of patellar clunk syndrome following total knee arthroplasty. J Clin Ultrasound. 2017;45:105–7.
9. Costanzo JA, et al. Patellar clunk syndrome after total knee arthroplasty: risk factors and functional outcomes of arthroscopic treatment. J Arthroplast. 2014;29:201–4.
10. Dajani KA, et al. Arthroscopic treatment of patellar clunk and synovial hyperplasia after total knee arthroplasty. J Arthroplast. 2010;25:97–103.
11. Sekiya H, et al. Painful knee after total knee arthroplasty is not a frequent complication and could be treated by arthroscopic debridement. Orthop Proc. 2016;98:152.
12. Koh YG, et al. Arthroscopic treatment of patellofemoral soft tissue impingement after posterior stabilized total knee arthroplasty. Knee. 2008;15:36–9.
13. Gholson JJ, et al. Management of painful patellar clunk and crepitance: results at a mean follow-up of five years. Iowa Orthop J. 2017;37:171–5.
14. Choi WC, et al. Painful patellar clunk or crepitation of contemporary knee prostheses. Clin Orthop Relat Res. 2013;471:1512–22.
15. Shoji H, et al. Patellar clunk syndrome in total knee arthroplasty without patellar resurfacing. J Arthroplast. 1996;11:198–201.
16. Fukunaga K, et al. The incidence of the patellar clunk syndrome in a recently designed mobile-bearing posteriorly stabilized total knee replacement. J Bone Joint Surg Br. 2009;91:463–8.

17. Ogawa H, et al. Effect of patellar resurfacing on patellofemoral crepitus in posterior-stabilized total knee arthroplasty. J Arthroplast. 2016;31:1792–6.
18. Clarke HD, et al. The influence of femoral component design in the elimination of patellar clunk in posterior-stabilized total knee arthroplasty. J Arthroplast. 2006;21:167–71.
19. Pollock DC, et al. Synovial entrapment: a complication of posterior stabilized total knee arthroplasty. J Bone Joint Surg Am. 2002;84:2174–8.
20. Yau WP, et al. Patellar clunk syndrome after posterior stabilized total knee arthroplasty. J Arthroplast. 2003;18:1023–8.
21. Figgie HE 3rd, et al. The influence of tibial-patellofemoral location on function of the knee in patients with the posterior stabilized condylar knee prosthesis. J Bone Joint Surg Am. 1986;68:1035–40.
22. Schroer WC, et al. Association of increased knee flexion and patellar clunk syndrome after mini-subvastus total knee arthroplasty. J Arthroplast. 2009;24:281–7.
23. Matsumoto T, et al. Patella resurfacing-never. In: Scott WN, editor. Insall & Scott Surgery of the knee. Elsevier; 2017. p. 1630–6.
24. Campbell DG, et al. Patellar resurfacing in total knee replacement: a ten-year randomized prospective trial. J Bone Joint Surg Br. 2006;88:733–4.
25. Nakano N, et al. Modern femoral component design in total knee arthroplasty shows a lower patellar contact force during knee flexion compared with its predecessor. Knee. 2021;30:1–8.
26. Ranawat AS, et al. Patellar crepitation in the P.F.C. Sigma total knee system. Orthopedics. 2006;29:S68–70.
27. Matsuda S, et al. Contact stresses with an unresurfaced patella in total knee arthroplasty: the effect of femoral component design. Orthopedics. 2000;23:213–8.
28. Takahashi A, et al. Patellar morphology and femoral component geometry influence patellofemoral contact stress in total knee arthroplasty without patellar resurfacing. Knee Surg Sports Traumatol Arthrosc. 2012;20:1787–95.

Part VI

Surgical Goals in Alignment and Soft Tissue Balance

40 Coronal Alignment: Mechanical Alignment

Shuichi Matsuda

1 Anatomical Features of the Lower Limbs and Coronal Alignment in Total Knee Arthroplasty

1.1 Neutral Alignment

The basic goal of total knee arthroplasty (TKA) coronal alignment is to achieve a neutral alignment of the lower limb functional axis connecting the hip and ankle centers through the center of the knee joint (Fig. 40.1). The angle between the functional axes of the femur and tibia (i.e., the hip–knee–ankle [HKA] angle) is 0°. This goal can be achieved by aligning implants perpendicular to the functional axes of the femur and tibia, which is called mechanical alignment. Clinical performance deteriorates when the neutral alignment exceeds ±3°; therefore, the goal has been to maintain the results within ±3°. However, recent reports suggest that the results do not deteriorate even when the alignment exceeds ±3° [1], and discussions on the appropriate alignment are ongoing.

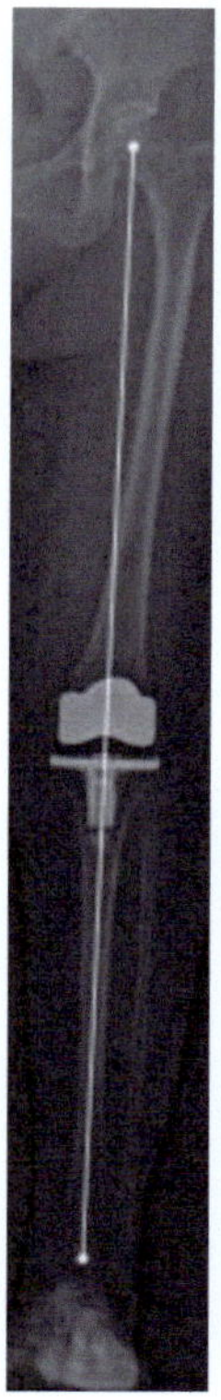

Fig. 40.1 Neutral alignment. The mechanical axis passes through the middle of the knee joint

1.2 Constitutional Varus

Bellemans et al. [2] proposed the concept of constitutional varus. In their study, the average HKA of healthy adults in Belgium was 1.3° varus, and 32% of men and 17% of women had varus alignment ≥3, raising the question of whether aiming for neutral alignment in all patients is appropriate when performing TKA. The alignment of healthy people was also evaluated in Japan, wherein the average HKA was 2.3° varus, and 35.2% of men and 36.4% of women had a varus alignment of 3°. The tendency toward varus was stronger than that in the United States, Hong Kong, and the aforementioned Belgian study [3]. On the basis of the detailed examinations of healthy human alignment described above, various arguments have been made, including that mild varus should be improved after TKA, and that the goal should be personalized for each patient. One method of achieving native alignment (e.g., constitutional varus) is kinematic alignment.

1.3 Management of Diaphyseal Deformity

In healthy Japanese participants and patients with osteoarthritis (OA), the lateral bowing of the femur strengthens with age (Fig. 40.2). This tendency is particularly prominent in patients with OA [4]. Therefore, managing such deformities

S. Matsuda (✉)
Department of Orthopaedic Surgery, Kyoto University, Kyoto, Japan
e-mail: smat522@kuhp.kyoto-u.ac.jp

S. Matsuda (ed.), *Advances in Total Knee Arthroplasty*, https://doi.org/10.1007/978-981-97-4920-1_40

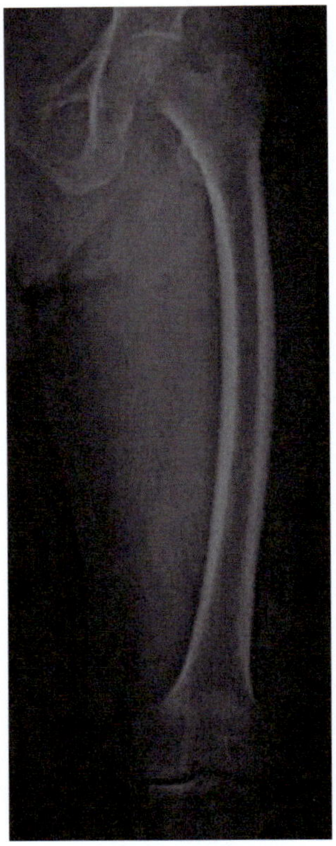

Fig. 40.2 Varus deformity of the femur. The diaphysis of the femur has lateral bowing showing varus deformity

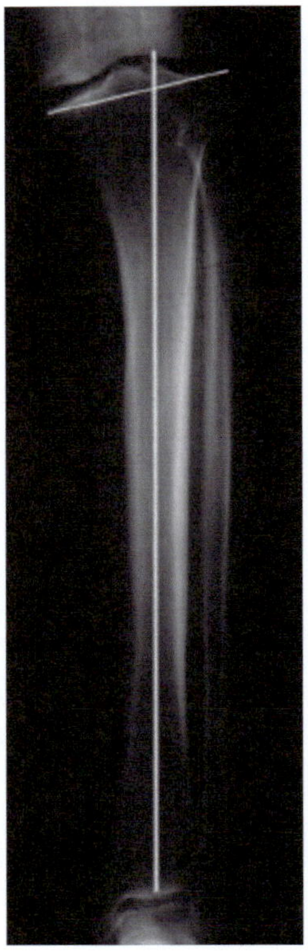

Fig. 40.3 Varus deformity of the tibia. Varus deformity of the tibia occurs proximally, and the intercondylar eminence of the tibia deviates inward from the diaphysis axis

during TKA is often difficult. In cases of strong femoral lateral bowing, the femoral component is often placed in varus with conventional surgical guides [5], and navigation is reportedly useful for achieving targeted alignment, even in patients with femoral deformity [6]. However, deformity at the diaphysis of the tibia is not significant, and varus deformity often occurs around the articular surface proximal to the tibial articular surface [7]. The center of the articular surface of the tibia may deviate inward, and careful attention should be paid to preoperative planning (Fig. 40.3) [8, 9].

2 Coronal Alignment and Biomechanics

Malalignment in the coronal plane has the following mechanical effects.

2.1 Increase in Articular Surface Stress Due to Varus Malalignment (Effect on Polyethylene Wear)

Computer simulations have shown that postoperative HKA values of 0°, 2°, 6°, and 10° increase the force exerted on the medial compartment during walking (Fig. 40.4), and that the maximum contact pressure applied to polyethylene also gradually increases, especially in cases of varus ≥6° [10].

When the relationship between alignment and polyethylene wear was examined in TKA using the Miller–Galante I prosthesis, varus alignment progressed to medial polyethylene wear; further, the varus alignment progressed with the postoperative course [11]. However, in the Miller–Galante I study, the coronal shape of the implant was flat. In varus alignment, the polyethylene is overloaded by edge loading [12]. Therefore, we examined a prosthesis (NexGen) with a curved coronal surface. First, the contact pressure was measured at the intermediate and 5° varus positions. With the 5° varus position, the increase in contact pressure was suppressed in the curved-on-curved design compared with that in the flat-on-flat design [13]. In the clinical results, compared with the flat design, the curved design significantly reduced polyethylene wear in patients with similar varus alignment [13]. The reduction in wear due to such improvements in shape may contribute to better clinical results, even in cases of coronal malalignment.

2.2 Increased Load on the Bone Due to Varus Malalignment (Risk of Tibial Component Collapse)

In the aforementioned study using computer simulations, we examined the stress applied to the tibia just below the tibial component. The postoperative alignment was approximately

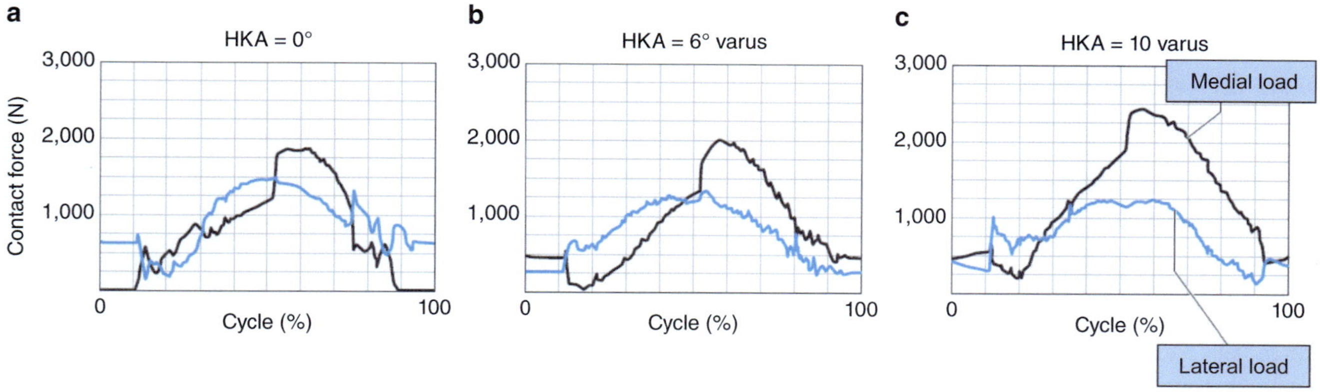

Fig. 40.4 Changes in loading due to varus alignment. As varus alignment becomes stronger, the load on the medial compartment increases. (Nakamura S et al. The effects of kinematically aligned total knee arthroplasty on stress at the medial tibia. Bone Joint Res. 2017;6(1):43–51. © 2017 Nakamura S et al. (https://creativecommons.org/licenses/by-nc/4.0/) [10])

26 MPa for 0° or 2° varus; however, stress >40 MPa occurred when the postoperative varus alignment was >6° [10]. A study on the relationship between alignment and clinical outcomes in 6070 patients reported that the medial tibial component had a higher risk of collapse in the varus alignment group [14]. Although the stress on the articular surface can be reduced by improving the design, alleviating the load on the bone is difficult owing to the varus alignment. Therefore, patients with osteoporosis or excessive obesity should not have a varus alignment of >5–6°.

2.3 Alignment and Kinematics

Dynamic analysis of deep flexion motion using fluoroscopy has shown that external rotation of the femur tends to increase in the varus position of the tibial component [15]. A study evaluated the effects of varus and valgus of the femoral component and varus of the tibial component using computer simulations [16]. When the femoral component was in varus, the femur tended to internally rotate during knee flexion, and when the tibial component was in varus, the sliding motion of the femur in the medial direction increased [16]. Changes in alignment modify not only the load on implants and bones but also the kinematics due to changes in the direction of the force applied to the joint.

3 Coronal Alignment and Postoperative Symptoms and Functions

To date, most studies have examined the effects of coronal alignment on long-term durability; however, recently, an increasing number of studies has examined the relationship between coronal alignment and postoperative symptoms, functions, and satisfaction. In addition to comparing neutral alignment within ±3° with other goals, recent studies have further elucidated varus and valgus alignments. A study reported that varus alignment up to 6° was comparable to neutral alignment within ±3°, but the function score decreased when the varus was >6° [17]. This finding suggests that excessive varus alignment should be avoided from a functional perspective. We examined 500 knees using the New Knee Society Scoring System and found that the degree of fulfillment of expectations was reduced in patients with varus alignment [18]. The expectation of some patients of lower-extremity correction was considered a possible explanation. In contrast, valgus alignment has been reported to reduce the function score 17 [19], and computer simulations have shown abnormalities in kinematics, such as anterior femoral translation in valgus alignment [20]. Therefore, valgus alignment should be avoided, even if it is mild. On the basis of postoperative symptoms and functions, a valgus alignment of approximately 3° to a varus of 5° appears to be within the acceptable range.

4 Alignment and Long-Term Durability

Historically, many studies have reported that postoperative results are poor with malalignment of ≥3°; however, in 2010, the Mayo Clinic reported that the 15-year revision rate did not worsen in modern TKA, even with malalignment of >3°, and that the revision rate at 20 years also did not worsen [1]. However, even in relatively recent studies, varus alignment >3° was more frequently associated with loosening [21] and migration of the tibial component [22]. In a study comparing mild varus ≤3° with a varus of 6°, the revision rate was higher in the varus ≥6° group [23]. These results suggest that even in modern TKA, an alignment >3° does not affect long-term durability. Furthermore, the aforementioned computer simulations have shown that a medial alignment of >6°places a greater load on the bone supporting the tibial component; further, the polyethylene is also exposed to the contact pressure. Considering that osteoporosis progresses over a long period after surgery, avoiding a varus of ≥6° is recommended to prevent sinking of the tibial component.

5 Coronal Alignment and Ligament Balance

5.1 Medial Tightness

Alignment and ligament balance are often discussed separately; however, they are closely related and should always be considered together. The challenge with the mechanical alignment method is that it causes imbalance in the ligaments [24]. For example, cutting the femur perpendicular to the mechanical axis in a varus-deformed knee increases the amount of bone resection at the lateral condyle and may increase ligamentous imbalance in the direction of the varus (lateral opening) [25]. Medial release has often been used to manage ligament imbalance, but excessive medial ligament release results in medial instability in flexion, even though balance is achieved during extension [26]. Medial instability reduces patient activity and satisfaction [27, 28]. Therefore, excessive medial release should be avoided. One factor that may deteriorate function after mechanical alignment is medial instability during flexion due to excessive medial dissection. Notably, there is almost no contracture of the medial soft tissue, even if the varus deformity is severe [29]. If a medial gap cannot be acquired, osteophyte or additional bone resection should be performed instead of medial release.

5.2 Lateral Looseness

When the varus deformity becomes stronger, there is no medial contracture, but the lateral structure is slacked, and the lateral opening ligament imbalance becomes stronger [29]. An adverse effect of lateral opening ligament imbalance is lateral lift-off (Fig. 40.5). Reportedly, polyethylene wear increases when lift-off occurs. Using computer simulations to examine the effects of varus alignment and ligament balance on lateral liftoff, we found that even with normal ligament balance, stronger varus alignment caused lateral liftoff; in contrast, alignment within 2° from neutral did not cause lift-off even when the lateral ligaments were loose (Fig. 40.6) [30]. Fluoroscopic analysis of patients with the mechanical axis passing through the middle one-third of the joint in 90% of cases [31] demonstrated no association between ligament balance and lift-off. Therefore, if the alignment is close to neutral, lift-off is unlikely to occur, even if the laxity of the lateral ligaments is caused by a varus deformity. In our study, among the 100 TKA cases operated with the mechanical alignment method without any medial release, 54 cases were balanced within 3° of extension, 21 cases were imbalanced between 3° and 6°, and 25 cases were imbalanced over 6° [32]. However, according to the 2011 Knee Society Scoring System, 2 years after surgery, no difference among the three groups was observed, suggesting that lateral laxity may be acceptable to some extent if neutral alignment is achieved.

In addition, many reports have shown that intraoperative varus alignment worsens further after surgery 11 [33, 34], and whether varus alignment further worsens lateral laxity remains unclear. In a study that examined joint laxity using stress X-ray over a period of ≥5 years after surgery, joint laxity did not change in the neutral alignment group and varus group of ≥3° (mean 5.8°) [35].

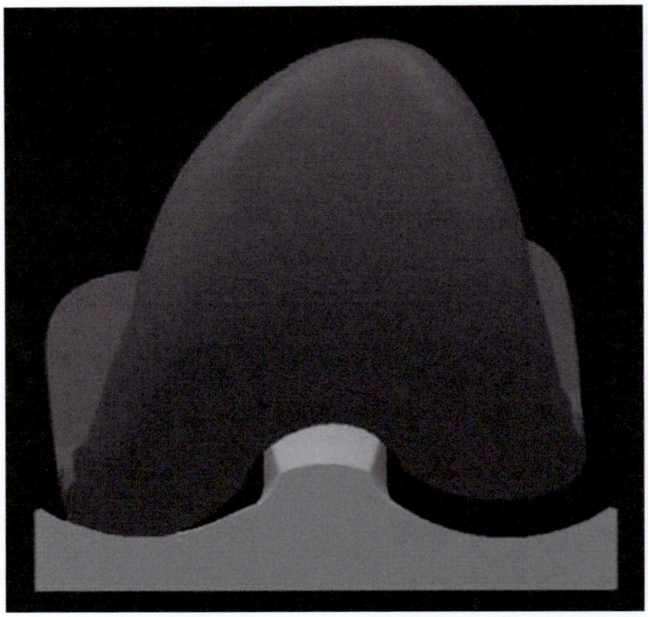

Fig. 40.5 Lateral lift-off

LCL Laxity (mm)	Femoral varus alignment (°)					
	0	1	2	3	4	5
0	–	–	–	1.3	2.8	4.4
1	–	–	1.1	2.8	4.3	6.6
2	–	–	1.7	3.7	5.9	9.9
3	–	–	1.8	4.7	7.2	10.5
4	–	–	1.8	5.5	8.6	12.3
5	–	–	1.8	6.8	10.1	15.3

Fig. 40.6 Effect of varus alignment and lateral laxity on lift-off motion. (Created by Kuriyama et al. [30])

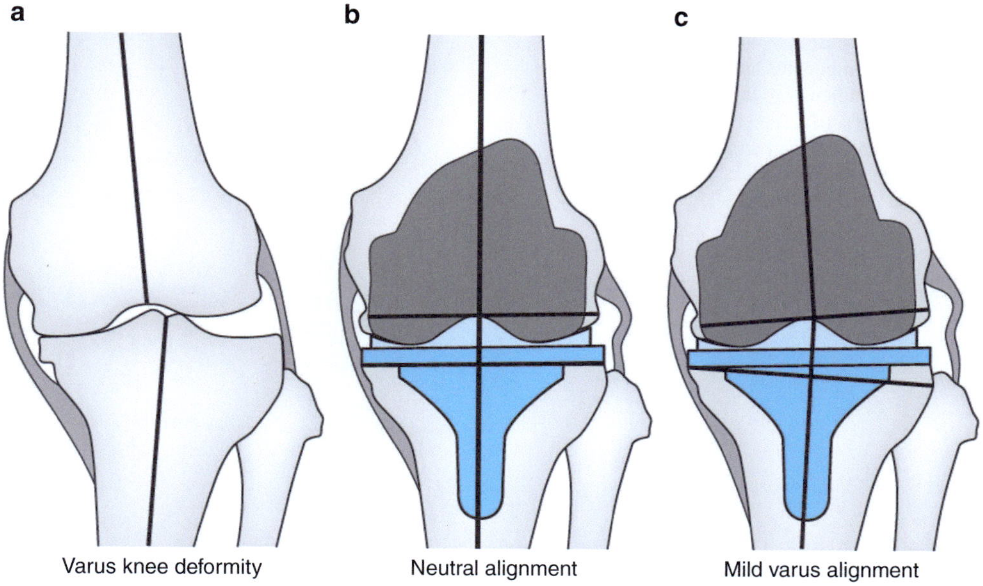

Fig. 40.7 Alignment and ligament balance. When total knee arthroplasty (TKA) was performed on a varus knee (**a**) with neutral alignment, lateral laxity remained (**b**). Mild varus alignment decreased lateral laxity; however, it was more lax than normal because the lateral structure was stretched out by osteoarthritis (OA) (**c**)

LCL Laxity (mm)	Femoral varus alignment (°)					
	0	1	2	3	4	5
0	–	–	–	1.3	2.8	4.4
1	–	–	1.1	2.8	4.3	6.6
2	–	–	1.7	3.7	5.9	9.9
3	–	–	1.8	4.7	7.2	10.5
4	–	–	1.8	5.5	8.6	12.3
5	–	–	1.8	6.8	10.1	15.3

Fig. 40.8 Alignment adjustment and ligament balance. When total knee arthroplasty (TKA) was performed with neutral alignment, +3 mm of laxity was assumed to occur in addition to the normal lateral laxity (leftmost ○ mark). When aligned in 1° varus, the lateral laxity improved by approximately 1° (second ○ mark from the left). After making these adjustments, the patient achieved nearly normal lateral laxity with 3° varus alignment (rightmost circle). In either case, no lift-off motion occurred, and the adjustments were safe for preventing lift-off motion. (Created by Kuriyama et al. [30])

5.3 Ligament Balancing by Change of Alignment

If TKA is performed on a mildly deformed knee, it is more likely to achieve near-normal ligament balance by targeting the natural alignment of the knee. However, as the deformity becomes more severe, estimating the "native" alignment in each patient becomes more difficult. Even if "native" alignment can be obtained, the lateral ligament is lax when the varus deformity is severe [29], and the "native" soft tissue condition cannot be acquired (Fig. 40.7). In such cases, the lateral laxity can be decreased by placing them slightly more varus than their native alignment (Fig. 40.8); however, excessive varus alignment should be avoided for the durability of the implants. The optimal laxity of the lateral soft tissue appears to vary depending on the knee implant, and discovering an optimal target for lateral laxity will be a major research theme in the future.

6 Tilt of the Articular Surface

The kinematic alignment method reproduces the native alignment and articular surface inclination; however, we considered the effect of articular surface inclination alone in neutral alignment (Fig. 40.9). In the mechanical alignment method, the articular surface was inclined at 0° with respect to the mechanical axis to align the implant perpendicular to the mechanical axis.

The effects of the joint surface inclination on the kinematics and other parameters were also analyzed. In the walking on a flat floor analysis after TKA using the mechanical alignment method, the tibial component was tilted by approximately 5° to the ground. Therefore, the joint surface is closer to being parallel to the ground when the joint surface incorporates an inward inclination [36]. The average joint surface is tilted 3–4° inward in participants with normal conditions [4]. By placing the femoral component in a 3° valgus position and the tibial component in a 3° varus position or by using a design that incorporates a 3° joint surface tilt in the implant shape itself, a combination of neutral alignment and a 3° joint surface tilt can be achieved.

In the dynamic analysis of a mechanically aligned TKA when climbing stairs, the joint surface tilted outward by

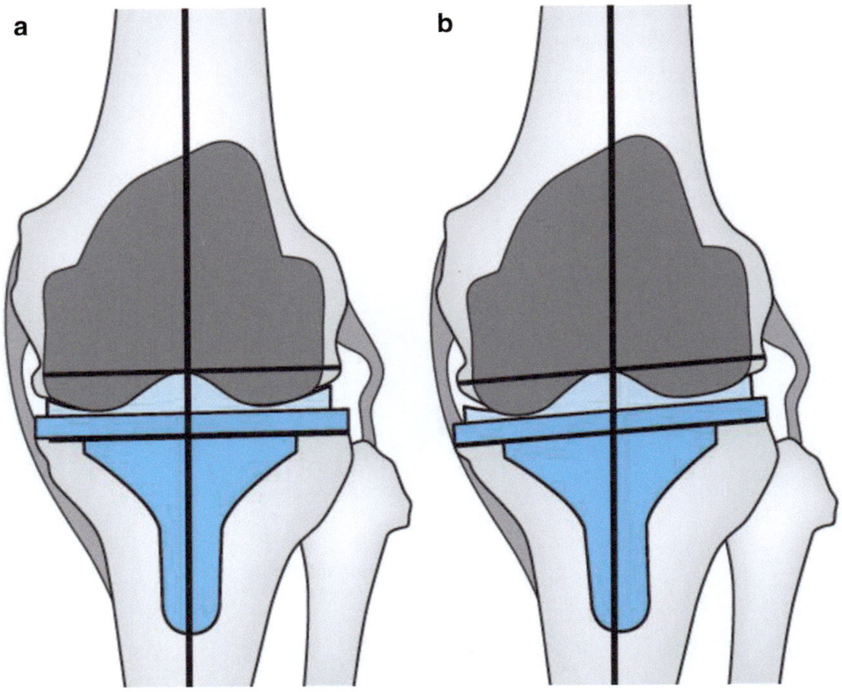

Fig. 40.9 Articular surface inclination in neutral alignment. Components are aligned perpendicular to the mechanical axis (**a**). The components are aligned with a 3° articular surface tilt toward the mechanical axis (**b**)

approximately 5° under a load [37]. Computer simulations have also shown that the external rotational motion of the femur in flexion increases when the components are aligned with a 3° inclination [38]. Near-normal kinematics may be achieved by replacing the joint surfaces more closely with the preoperative joint surface geometry. Clinically, whether changes in articular surface inclination alone can improve postoperative function or durability in the same alignment of the lower leg remains to be concluded. This is an issue for consideration in the future.

7 Prospects

TKA alignment is one of the most important factors influencing postoperative outcomes. Until now, it has been treated as a single factor; however, the importance of discussing it in conjunction with ligament balance is increasingly evident. To further improve TKA performance, setting a single surgical goal for all patients is likely insufficient. First, research determining the safe range is required. Then, individual goals must be set to obtain the best possible functionality within a range that does not exceed durability.

References

1. Abdel MP, et al. Effect of postoperative mechanical axis alignment on survival and functional outcomes of modern total knee arthroplasties with cement: a concise follow-up at 20 years. J Bone Joint Surg Am. 2018;100:472–8.
2. Bellemans J, et al. The Chitranjan Ranawat award:is neutral mechanical alignment normal for all patients? The concept of constitutional varus. Clin Orthop Relat Res. 2012;470:45–53.
3. Wanezaki Y, et al. Lower limb alignment in healthy Japanese adults. J Orthop Sci. 2021;28:200–3. https://doi.org/10.1016/j.jos.2021.10.016.
4. Matsumoto T, et al. A radiographic analysis of alignment of the lower extremities – initiation and progression of varus-type knee osteoarthritis. Osteoarthr Cartil. 2015;23:217–23.
5. Umatani N, et al. Femoral bowing affects varus femoral alignment but not patient satisfaction in mechanically aligned total knee arthroplasty. Eur J Orthop Surg Traumatol. 2021; https://doi.org/10.1007/s00590-021-03164-0.
6. Kobayashi H, et al. Influence of coronal bowing on the lower alignment and the positioning of component in navigation and conventional total knee arthroplasty. Orthop Traumatol Surg Res. 2017;103:251–6.
7. Fukagawa S, et al. Anterior border of the tibia as a landmark for extramedullary alignment guide in total knee arthroplasty for varus knees. J Orthop Res. 2011;29:919–24.
8. Matsuda S, et al. Tibial shaft axis does not always serve as a correct coronal landmark in total knee arthroplasty for varus knees. J Arthroplast. 2003;18:56–62.
9. Takasaki M, et al. Accuracy of image-free navigation for severely deformed knees. Knee Surg Sports Traumatol Arthrosc. 2010;18:763–8.
10. Nakamura S, et al. The effects of kinematically aligned total knee arthroplasty on stress at the medial tibia: a case study for varus knee. Bone Joint Res. 2017;6:43–51.
11. Matsuda S, et al. Changes in knee alignment after total knee arthroplasty. J Arthroplast. 1999;14:566–70.
12. Matsuda S, et al. The effect of varus tilt on contact stresses in total knee arthroplasty: a biomechanical study. Orthopedics. 1999;22:303–7.
13. Nishikawa K, et al. Improved design decreases wear in total knee arthroplasty with varus malalignment. Knee Surg Sports Traumatol Arthrosc. 2014;22:2635–40.

14. Fang DM, et al. Coronal alignment in total knee arthroplasty: just how important is it? J Arthroplast. 2009;24:39–43.
15. Sekiguchi K, et al. Varus alignment after total knee arthroplasty results in greater axial rotation during deep knee bend activity. Clin Biomech (Bristol, Avon). 2020;77:105051.
16. Watanabe M, et al. Varus femoral and tibial coronal alignments result in different kinematics and kinetics after total knee arthroplasty. Knee Surg Sports Traumatol Arthrosc. 2017;25:3459–66.
17. Nishida K, et al. Remaining mild varus limb alignment leads to better clinical outcome in total knee arthroplasty for varus osteoarthritis. Knee Surg Sports Traumatol Arthrosc. 2017;25:3488–94.
18. Matsuda S, et al. Postoperative alignment and ROM affect patient satisfaction after TKA. Clin Orthop Relat Res. 2013;471:127–33.
19. Kuroda Y, et al. Influence of limb alignment and prosthetic orientation on patient-reported clinical outcomes in total knee arthroplasty. J Orthop Sci. 2019;24:668–73.
20. Nishitani K, et al. Valgus position of the femoral component causes abnormal kinematics in the presence of medial looseness in total knee arthroplasty: a computer simulation model of TKA for valgus knee osteoarthritis. Knee Surg Sports Traumatol Arthrosc. 2019;27:2051–60.
21. Lee BS, et al. Femoral component varus malposition is associated with tibial aseptic loosening after TKA. Clin Orthop Relat Res. 2018;476:400–7.
22. van Hamersveld KT, et al. The effect of coronal alignment on tibial component migration following total knee arthroplasty: a cohort study with long-term radiostereometric analysis results. J Bone Joint Surg Am. 2019;101:1203–12.
23. Oh SM, et al. Impact of preoperative varus deformity on postoperative mechanical alignment and long-term results of "mechanical" aligned total knee arthroplasty. Orthop Traumatol Surg Res. 2019;105:1061–6.
24. Niki Y, et al. Mechanically aligned total knee arthroplasty carries a risk of bony gap changes and flexion-extension axis displacement. Knee Surg Sports Traumatol Arthrosc. 2017;25:3452–8.
25. Nagamine R, et al. Distal femoral cut perpendicular to the mechanical axis may induce varus instability in flexion in medial osteoarthritic knees with varus deformity in total knee arthroplasty: a pitfall of the navigation system. J Orthop Sci. 2004;9:555–9.
26. Mullaji A, et al. Quantification of effect of sequential posteromedial release on flexion and extension gaps: a computer-assisted study in cadaveric knees. J Arthroplast. 2009;24:795–805.
27. Tsukiyama H, et al. Medial rather than lateral knee instability correlates with inferior patient satisfaction and knee function after total knee arthroplasty. Knee. 2017;24:1478–84.
28. Azukizawa M, et al. Intraoperative medial joint laxity in flexion decreases patient satisfaction after total knee arthroplasty. Arch Orthop Trauma Surg. 2018;138:1143–50.
29. Okamoto S, et al. Lateral soft tissue laxity increases but medial laxity does not contract with varus deformity in total knee arthroplasty. Clin Orthop Relat Res. 2013;471:1334–42.
30. Kuriyama S, et al. No condylar lift-off occurs because of excessive lateral soft tissue laxity in neutrally aligned total knee arthroplasty: a computer simulation study. Knee Surg Sports Traumatol Arthrosc. 2016;24:2517–24.
31. Hamai S, et al. No influence of coronal laxity and alignment on lift-off after well-balanced and aligned total knee arthroplasty. Knee Surg Sports Traumatol Arthrosc. 2014;22:1799–804.
32. Tanaka Y, et al. Intraoperative physiological lateral laxity in extension and flexion for varus knees did not affect short-term clinical outcomes and patient satisfaction. Knee Surg Sports Traumatol Arthrosc. 2020;28:3888–98.
33. Kuroda Y, et al. Varus deformity in the proximal tibia and immediate postoperative varus alignment result in varus progression in limb alignment in the long term after total knee arthroplasty. Knee Surg Sports Traumatol Arthrosc. 2020;28:3287–93.
34. Okamoto R, et al. Yearly changes of limb alignment toward varus as measured by the femorotibial angle after total knee arthroplasty. Bull Hosp Jt Dis Orthop Inst. 1991;51:230–5.
35. Hatayama K, et al. Does residual varus alignment cause increasing varus laxity at a minimum of five years after total knee arthroplasty? J Arthroplast. 2017;32:1808–13.
36. Murakami K, et al. Preoperative tibial mechanical axis orientation and articular surface design influence on the coronal joint line orientation relative to the ground during gait after total knee arthroplasties. Knee Surg Sports Traumatol Arthrosc. 2018;26:3368–76.
37. Yabu T, et al. Static mediolateral tilt of the joint line after total knee arthroplasty does not reflect dynamic tilt during a stair ascent activity. J Knee Surg. 2021; https://doi.org/10.1055/s-0041-1740999.
38. Ishikawa M, et al. Kinematic alignment produces near-normal knee motion but increases contact stress after total knee arthroplasty: a case study on a single implant design. Knee. 2015;22:206–12.

41 Coronal Alignment-Kinematic Alignment

Yasuo Niki

1 History of the KA-TKA Method

In the 1970s, two alignment methods gained popularity in total knee arthroplasty (TKA): the mechanical alignment (MA) method proposed by Insall et al. and the anatomical alignment (AA) method introduced by Hungerford et al., which aimed to follow the original anatomy of the lower extremity. The AA method sought to reproduce the natural joint line, with an average of 87° of medial proximal tibial angle (MPTA) relative to the tibial mechanical axis in the coronal plane. However, the hip-knee-ankle (HKA) angle averaged 0°. Unfortunately, 7% of the AA prosthesis, known as porous-coated anatomic (PCA) prosthesis, exhibited polyethylene wear within 4.5 years after TKA, leading to the need for revision surgery. This premature failure was attributed to the poor wear resistance of heat-pressed polyethylene at that time, resulting in a high rate of delamination-type wear. Consequently, despite the promising concept of AA, it was eventually abandoned [1].

In 2006, there was a resurgence of interest in the AA concept due to improvements in the wear resistance of polyethylene. Howell et al. further evolved the AA method by incorporating patient-specific instruments (PSI) and introduced a new technique called kinematically aligned (KA)-TKA. This approach preserved the flexion-extension axis of the femur by osteotomizing only the thickness of the implant from the articular surface [2]. As a result, approximately 20,000 cases of KA-TKA were performed in the United States by 2009. However, fully KA-TKA by PSI, which attempted to perfectly reproduce the original joint line without adjusting HKA, carried a risk of poor clinical outcomes and potential premature prosthetic loosening [3]. Consequently, in 2009, the FDA banned KA-TKA using PSI. At that time, the concept of mechanically aligned TKA (MA-TKA) was considered the gold standard, and it was recommended to keep the HKA within ±3° of neutral alignment. Any outliers exceeding ±3° were unambiguously considered technical failures. Subsequently, Howell et al. developed a surgical technique using calipers to remove only the thickness of the implant from the joint surface while correcting the amount of cartilage wear during surgery. This modified approach was reported to yield excellent or good clinical outcomes, and more than 2800 surgeries were performed between 2006 and 2014 [2].

2 KA-TKA Concept

As mentioned earlier, the full KA-TKA approach aims to minimize soft tissue release, especially for the medial collateral ligament (MCL). It achieves this by selectively cutting only the necessary thickness of the implant from the articular surface after resecting osteophytes, thereby preserving the flexion-extension axis of the femur. Consequently, the functional axis of the lower limb after TKA may deviate significantly from neutral alignment in some cases.

In cases where the knee has excessive varus alignment or presents a trapezoidal gap with significant lateral laxity, the knee adduction moment (KAM) is increased. This elevated KAM may lead to early implant loosening and joint instability. Therefore, during clinical practice, particular attention should be given to the component gap and the final limb alignment, specifically the HKA. To address these concerns, several alignment concepts derived from KA-TKA have been proposed (Table 41.1). One such concept gaining attention is restricted KA-TKA (rKA-TKA), which allows for up to 3° of HKA and 5° of medial tibial inclination. Recently, with the introduction of robotic-assisted technology, it has become possible to place implants with precise joint space gap, joint line inclination, and final limb alignment while minimizing the extent of MCL release (potentially down to the deep fibers). This new approach is referred to as functional alignment.

Y. Niki (✉)
Center for Musculoskeletal Surgery and Research, Fujita Medical Innovation Center, Tokyo, Japan
e-mail: yasuo.niki@fujita-hu.ac.jp

S. Matsuda (ed.), *Advances in Total Knee Arthroplasty*, https://doi.org/10.1007/978-981-97-4920-1_41

Table 41.1 Various alignment techniques derived from the KA concept

Technique	Concept	Method
Full KA	Cutting parallel to the articular surface while considering the cartilage thickness	Caliper
Reverse gap balancing KA	Full KA bone cut for femur. Gap balancing technique for tibia	Gap tensor
Restricted KA	Tibia limited to 5°inclination, HKA up to 3°varus	Optical navigation PSI Robot
Restricted KA (tibia consistently 3°)	Tibia is set as 3°inclination, HKA up to 3°varus	Optical navigation PSI Robot
Functional alignment	Monitoring gap, joint line inclination, and HKA at preop. Planning	Robot

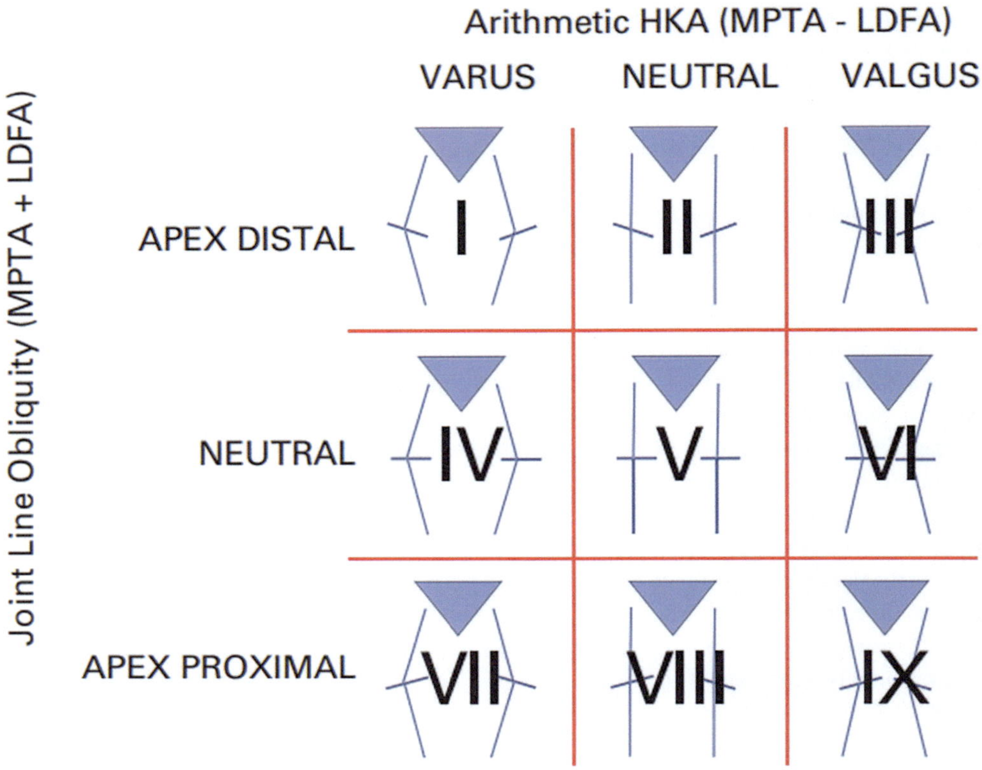

Fig. 41.1 CPAK Classification: The aHKA values and joint line obliquity were categorized into nine types. The distribution within the OA cohort is presented below. (Reproduced with permission from MacDessi SJ et al. Coronal Plane Alignment of the Knee (CPAK) classification. Bone Joint J. 2021;103-B(2):329–337. © 2021 MacDessi SJ et al. (http://creativecommons.org/licenses/by-nc-nd/4.0/)) [4]

3 Relationship Between Joint Line Inclination and Limb Alignment

With the increasing acceptance of KA-TKA, the constitutional varus and inclination of the joint line have become subjects of interest. MacDessi et al. conducted a comprehensive analysis of the constitutional alignment in 250 healthy subjects from Belgium and 500 OA patients from Australia, categorizing them into 9 different alignment patterns [4]. They referred to this classification as the coronal plane alignment of the knee (CPAK) classification (Fig. 41.1). The classification uses a two-dimensional graph with joint line direction (distal/proximal) on the vertical axis and coronal limb alignment (varus/valgus) on the horizontal axis. However, the joint line convergence angle (JLCA) was not considered in this coronal alignment, nor were changes in alignment due to cartilage wear or tibiofemoral joint subluxation because of radiographic assessment. The coronal limb alignment was expressed as an arithmetic calculation of the HKA, simply calculated by subtracting the lateral distal femoral angle (LDFA) from the medial proximal tibial angle (MPTA). Consequently, there was no significant difference in the nine types of distribution between OA patients and healthy subjects. Constitutional varus was found in 32% of normal subjects and 30% of OA patients, while the joint line was inclined medially in 76% of normal subjects and 67% of OA patients.

In cases where OA patients underwent either mechanically aligned (MA) or KA-TKA, and the intraoperative soft tissue balance of each patient was measured using the VERASENSE Knee System (OrthoSensor Inc., Dania Beach, FL), mediolateral gap balance was significantly better with KA-TKA for the category of varus alignment and MPTA≦90° (types I, II, and IV). Therefore, knees falling under types I, II, and IV are considered good indications for KA-TKA. It is worth noting that knees belonging to type II, which is the most common type in Caucasians, closely resemble the alignment and joint line of the anatomical alignment (AA) concept proposed by Hungerford. This observation suggests that Hungerford's concept was correct from the perspective of the CPAK classification. Given that type I knees were overwhelmingly the most common among Japanese OA patients, some degree of varus alignment of the lower limb remains even when full KA-TKA is applied to Japanese patients. It remains a matter of debate to what extent varus alignment can be allowed while ensuring promising prosthetic longevity.

4 Clinical Results of KA-TKA

According to several previous meta-analyses comparing KA-TKA and MA-TKA, KA-TKA outperforms MA-TKA in the Knee Society Function Score, WOMAC score, and Oxford Knee Score [6, 7]. Gao et al. reviewed 11 randomized controlled trials (RCTs) and reported that soft tissue balance was clearly superior in the KA-TKA group, with less MCL release, improved flexion angles, and better clinical outcomes [8]. These findings align with the results of MacDessi et al., as mentioned earlier. Long-term results over 10 years by Howell et al. showed a survival rate of 97.5% in their 220 KA-TKA series using a caliper technique, where postoperative coronal alignment did not significantly affect clinical outcomes [5]. Theoretically, full KA-TKA may lead to an increase in the number of outliers in coronal alignment, but it maintains the same medial and lateral joint contact force [9]. This is attributed to the osteotomizing of only the thickness of the prosthesis in parallel to the original joint line while preserving the patient's original limb alignment. Although good soft-tissue balancing in KA-TKA may satisfy surgeons, there are limited reports regarding patient satisfaction after the procedure. In a study using the Knee Society Score 2011 at 30 months after surgery, the functional score was significantly better in the KA-TKA group than in the MA-TKA group, but there was no significant difference in patient satisfaction [10]. Sub-analysis showed a notable difference in the advanced activity score, particularly for climbing stairs and stepladders. Multicenter research from different institutions is eagerly awaited to provide further insights into the postoperative results of KA-TKA, building on the extensive reports from Howell's group.

5 KA-TKA Biomechanics

Gait analysis, which measures kinematic parameters like walking speed, joint flexion angle change, and ground reaction force (GRF), is valuable for predicting durability, stability, mechanical load, and energy efficiency after TKA. Numerous studies have compared gait kinematic parameters between KA-TKA and MA-TKA. Blakeney et al. reported the results of gait analysis for 45 restricted KA-TKAs (rKA-TKAs) and 45 MA-TKAs using an optical computer navigation system and caliper measurements of resected bone [11]. The definition of rKA-TKA involved limiting bone cuts in the coronal plane to 5° of the neutral mechanical axis, with the final HKA set within 3° of neutral alignment. Most gait parameters in KA-TKA more closely resembled those in normal knees than those in MA-TKA. MA-TKA showed a decrease in range of motion (ROM) and an increase in knee adduction angle compared to healthy knees. Yeo et al. performed robotic-assisted rKA-TKA, setting component alignment to 2° of MPTA, 2° of LDFA, and 2° internal rotation to transepicondylar axis [12]. At an 8-year follow-up, gait analysis revealed that knee varus angle and mediolateral GRF (horizontal GRF) were significantly reduced in KA-TKA compared to MA-TKA. These results indicate that the medially inclined joint line after KA-TKA orients parallel to the ground, positioning the center of pressure (COP) more inward, leading to a more vertical center of mass (COM)-COP line and a smaller length of the lever arm to the knee center and horizontal GRF compared to MA-TKA. McNair et al. compared gait profiles of 14 KA-TKAs performed by PSI with those of 15 MA-TKAs performed by optical computer navigation [13]. In the sagittal knee kinetics, KA-TKA exhibited a larger knee flexion moment than MA-TKA, while in the coronal plane kinetics, KAM was comparable between the two TKAs. Principal component analysis showed differences in the transverse plane kinetics, and MA-TKA tended to have a larger internal rotation moment, although the clinical significance of this difference was small. Chen et al. used computer simulation techniques to calculate the mechanical load and contact pressure applied to the ligaments and polyethylene insert during gait [14]. The load on the MCL increased by about 8% in KA-TKA compared to MA-TKA, while the load on the LCL and PCL decreased by 50% and 70%, respectively. Patellofemoral joint contact pressure increased by up to 40%. Similarly, Ishikawa et al. reported in computer simulations that while the contact pressure of the patellofemoral joint increased in KA-TKA, the external rotation of the femur associated with flexion was significantly larger than that of MA-TKA [15]. Two-dimensional (2D) to three-dimensional (3D) registration analysis of the medial pivot prosthesis at 1 year after surgery also indicated that KA-TKA exhibited larger femoral external rotation relative to the tibia

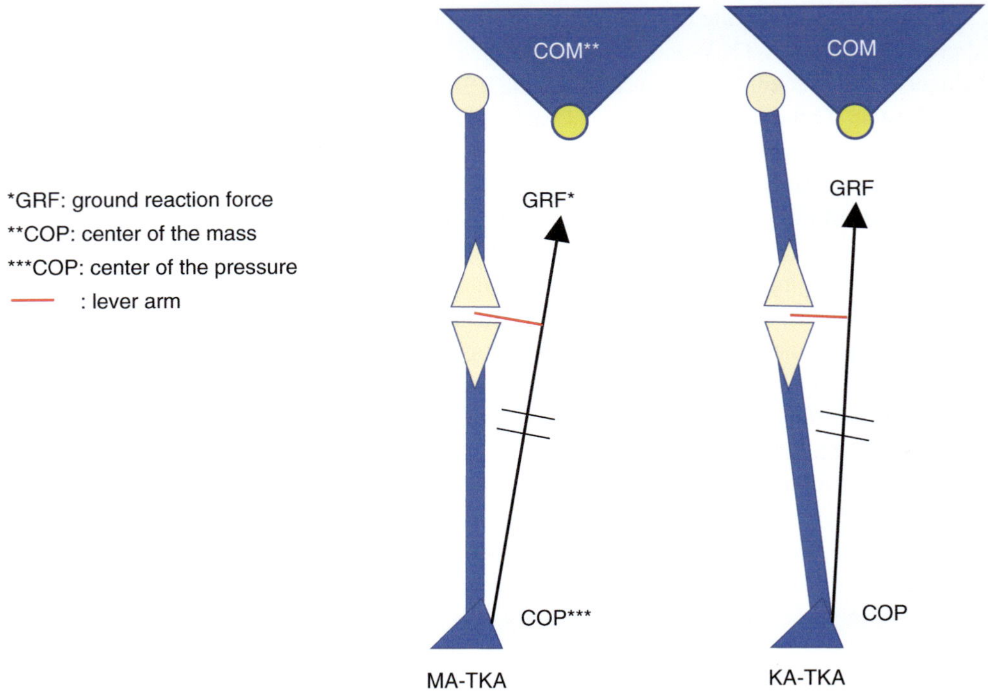

Fig. 41.2 Relationship between joint line inclination and lever arm. (Reproduced with permission from Reference Niki et al. [17]

than MA-TKA [16]. In a comparative study for gait analyses between 21 KA- and 21 MA-TKAs, convincing evidence has been reported that the degree of KAM was significantly reduced after KA-TKA compared to MA-TKA [17]. Significant factors affecting KAM include HKA, length of the lever arm, horizontal GRF, knee adduction angle, center of gravity position, and toe-out angle. In KA-TKA, it can be considered that when the articular surface is inclined medially (i.e., MPTA<90), the lever arm of the KAM is shortened because the joint surface is parallel to the ground during landing (Fig. 41.2).

6 Challenges and Future Direction for KA-TKA

The patellofemoral joint poses certain disadvantages in KA-TKA. Patellofemoral tracking is theoretically exacerbated by the internal rotation of the femoral component. Simulations have indicated an increase in patellofemoral pressure and potential instability with extension [15]. Although there have been no symptoms observed in our experience, cases with lateral patellar deviation on postoperative merchant view have been encountered [18]. Depending on the shape of the femoral component groove, the patellofemoral joint can be problematic and must be carefully considered in cases of poor preoperative patellar alignment.

The faithful reproduction of the original joint line and limb alignment (i.e., HKA) in KA-TKA remains a matter of debate. With the widespread use of robotic technology, surgeons are increasingly focused on achieving patient-specific physiological joint line and alignment. Robotic-assisted TKA allows for precise bone cuts and reduces the number of outliers, enabling surgeons to perform rKA-TKA, limited to 3° of HKA and 5° of medial tibial inclination. Moreover, implants are now being placed in a soft tissue-friendly strike zone, with a focus on preserving the original soft tissue balance as much as possible. By adhering to the original soft tissue balance and alignment, our goal should aim to improve patient satisfaction after TKA, a feat that conventional MA-TKA may not have achieved.

References

1. Tsao A, et al. Failure of the porous-coated anatomic prosthesis in total knee arthroplasty due to severe polyethylene wear. J Bone Joint Surg Am. 1993;75:19–26.
2. Howell SM, et al. Kinematic alignment in total knee arthroplasty. Definition, history, principle, surgical technique, and results of an alignment option for TKA. Arthropaedia. 2014:44–53.
3. Klatt BA, et al. Custom-fit total knee arthroplasty (Otis knee) results in malalignment. J Arthroplast. 2008;23:26–9.
4. MacDessi SJ, et al. Coronal plane alignment of the knee (CPAK) classification. Bone Joint J. 2021;103:329–37.
5. Howell SM, et al. Does varus alignment adversely affect implant survival and function six years after kinematically aligned total knee arthroplasty? Int Orthop. 2015;39:2117–24.
6. Courtney PM, Lee GC. Early outcomes of kinematic alignment in primary total knee arthroplasty: a meta-analysis of the literature. J Arthroplast. 2017;32:2028–32.
7. Lee YS, et al. Kinematic alignment is a possible alternative to mechanical alignment in total knee arthroplasty. Knee Surg Sports Traumatol Arthrosc. 2017;25:3467–79.

8. Gao ZX, et al. Comparison of kinematic alignment and mechanical alignment in total knee arthroplasty: a meta-analysis of randomized controlled clinical trials. Orthop Surg. 2020;12:1567–78.
9. Shelton TJ, et al. Do varus or valgus outliers have higher forces in the medial or lateral compartments than those which are in-range after a kinematically aligned total knee arthroplasty? Limb and joint line alignment after kinematically aligned total knee arthroplasty. Bone Joint J. 2017;99:1319–28.
10. Niki Y, et al. Joint line modification in kinematically aligned total knee arthroplasty improves functional activity but not patient satisfaction. J Arthroplast. 2018;33:2125–30.
11. Blakeney W, Clement J, Desmeules F, Hagemeister N, Riviere C, Vendittoli PA. Kinematic alignment in total knee arthroplasty better reproduces normal gait than mechanical alignment. Knee Surg Sports Traumatol Arthrosc. 2019b;27(5):1410–7.
12. Yeo JH, et al. No difference in outcomes and gait analysis between mechanical and kinematic knee alignment methods using robotic total knee arthroplasty. Knee Surg Sports Traumatol Arthrosc. 2019;27:1142–7.
13. McNair PJ, et al. A comparison of walking gait following mechanical and kinematic alignment in total knee joint replacement. J Arthroplast. 2018;33:560–4.
14. Chen Z, et al. Biomechanics and wear comparison between mechanical and kinematic alignments in total knee arthroplasty. Proc Inst Mech Eng H. 2018; https://doi.org/10.1177/0954411918811855.
15. Ishikawa M, et al. Kinematic alignment produces near-normal knee motion but increases contact stress after total knee arthroplasty: a case study on a single implant design. Knee. 2015;22:206–12.
16. Kaneda K, et al. Kinematically aligned total knee arthroplasty using medial pivot knee prosthesis enhances medial pivot motion: a comparative kinematic study with mechanically aligned total knee arthroplasty. Arthroplast Today. 2021;13:24–8.
17. Niki Y, et al. Kinematically aligned total knee arthroplasty reduces knee adduction moment more than mechanically aligned total knee arthroplasty. Knee Surg Sports Traumatol Arthrosc. 2018;26:1629–35.
18. Nagai K, et al. Radiographic evaluation of patellofemoral alignment in kinematically aligned total knee arthroplasty: a comparative study with mechanically aligned total knee arthroplasty. J Orthop Sci. 2021;26:1043–50.

42 Sagittal Alignment

Kohei Nishitani

1 Sagittal Plane Alignment of Femoral Components

1.1 Setting the Axis Used in the Sagittal Plane

To measure the alignment of femoral component, several axes of the femur are defined as references in the sagittal plane. One of the most commonly used axes is the distal femoral shaft axis obtained from a lateral knee radiograph. This axis is the line connecting the middle of the anterior and posterior diameters at two points in the distal femur [1–4] (Fig. 42.1). The anterior cortical axis of the femur is also used as an index for implant position and alignment [5] (Fig. 42.1). The mechanical axis of the femur in the sagittal plane is obtained using a full-length lateral radiograph of the femur and a projection image on CT. The proximal point is uniformly used as the center point of the femoral head, whereas several distal points are used, including the antero-posterior central point of the femoral condyle, posterior point of the femoral patellar sulcus (the most anterior point of the Blumensaat's line), most distal point of the femoral condyle, midpoint of the SEA, midpoint of the TEA, etc. [6–9]. In addition, the distal femoral axis connecting the medullary center of the middle of the femur and the insertion point of the intramedullary rod is also used for femoral component implantation; this axis represents the intramedullary rod alignment in the sagittal plane [7]. Thus, multiple axes settings are present in the sagittal plane of the femur, and surgeons should know which one is used as a reference for measuring the implant alignment.

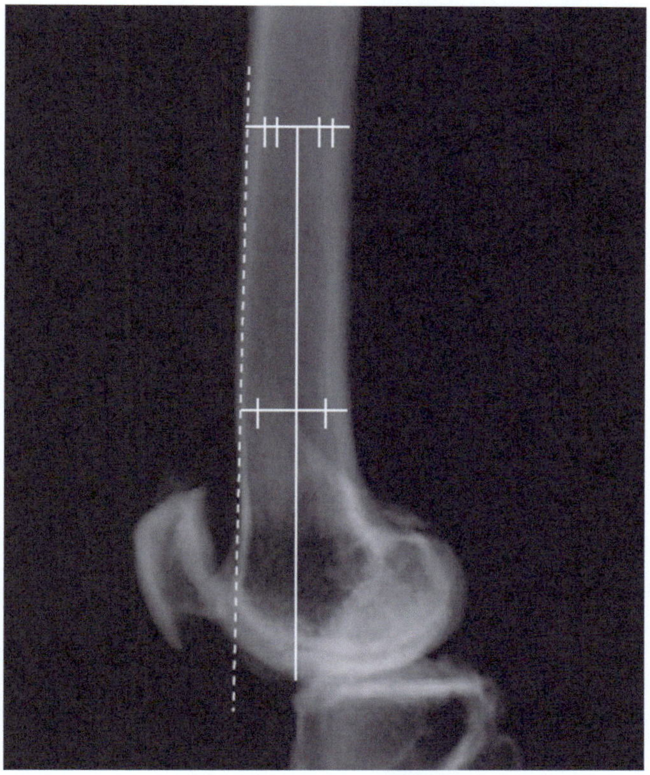

Fig. 42.1 Distal femoral shaft axis (white line) and distal anterior cortical axis of the femur (white dotted line)

1.2 Femoral Component Angle and Notch Formation

In the sagittal positioning and alignment of femoral components, surgeons should avoid forming an anterior notch. Studies on cadaveric knees have shown that fractures may occur in cases of notch formation due to an excessive load on the notch when large forces act on the femur in the flexion direction [10]. Therefore, implant positions that avoid anterior notching have been recommended. However, the method of dealing with notch formation differs between the conventional intramedullary rod method and methods using CAS, such as navigation or

K. Nishitani (✉)
Department of Orthopaedic Surgery, Kyoto University, Kyoto, Japan
e-mail: nkohei@kuhp.kyoto-u.ac.jp

S. Matsuda (ed.), *Advances in Total Knee Arthroplasty*, https://doi.org/10.1007/978-981-97-4920-1_42

robots. In conventional osteotomy using an intramedullary rod, distal osteotomy is performed perpendicular to the intramedullary rod. As the implant flexion angle is determined by the intramedullary rod insertion position, the insertion point significantly affects the implant flexion angle. Previously, an intramedullary rod inserted 7–10 mm anterior to the BlumenSaat's line was thought to be aligned close to the distal femoral shaft and parallel to the anterior femoral cortex [7]; however, data from patients in Thailand suggest that a 12-mm anterior insertion is more appropriate [11]. Care should be taken to avoid insertion of the intramedullary rod too anteriorly which could result in the femoral component being positioned in extension and an increased risk of notch formation. In addition, the anterior bowing and distal curve of the femur vary greatly among patients. Therefore, the insertion position of the intramedullary rod should be confirmed preoperatively in each patient. Minoda et al. investigated multiple settings of the femoral mechanical axis by changing the index of the distal femoral point to the insertion of the intramedullary rod (navigation femoral axis; NFA1), the anteroposterior center of the femoral condyle (NFA2), and the most distal point of the femoral condyle (NFA3) to avoid notch formation during computer navigation surgery (Fig. 42.2). In healthy young males, NFA1 showed 0.2° of extension from the distal femoral shaft axis, while NFA2 and NFA3 showed 2.7° and 3.0° of flexion, respectively [6]. In elderly females, NFA1 showed 1.2° of extension from the distal femoral shaft axis, while NFA2 and NFA3 showed 0.6° and 1.2° of flexion, respectively [9]. In contrast to the anterior cortex, NFA1 showed 1.7° of extension, NFA2 showed 0.1° of extension, and NFA3 showed 0.7° of flexion [5]. Therefore, if distal femoral osteotomy is performed perpendicular to the functional axis indicated by navigation, anterior notching may occur; thus, care must be taken to avoid this phenomenon. The risk of notch formation can be reduced by flexing the osteotomy by 3° from the functional axis indicated by the navigation. Another 3D analysis study showed that using the most anterior part of the BlumenSaat's line as the distal index point of the functional axis and implanting the femoral component 3° internally rotated from the TEA resulted in a notch formation rate of approximately 14%. However, setting the distal index point more anterior than the aforementioned point or setting the femoral component rotation to parallel or externally rotated from the TEA increased the risk of notch formation [12]. Regardless of the method used, to avoid notch formation, osteotomy should be performed after sufficient preparation and analysis.

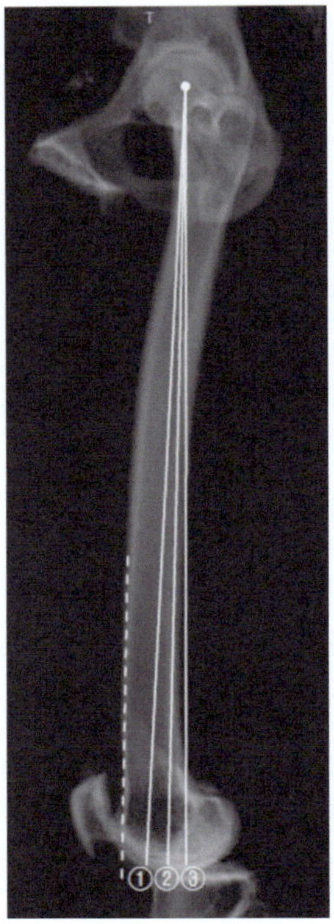

Fig. 42.2 The distal anterior cortical axis of the femur (white dotted line) and the three navigation femoral axes (NFA) used in [6]. Proximal point of NFA is always the center of the femoral head, and the distal points are (1) the intramedullary rod insertion position (NFA1), (2) the anteroposterior center of the femoral condyle (NFA2), and (3) the most distal point of the femoral condyle (NFA3)

1.3 Femoral Component Angle and Implant Size

In addition to the aforementioned cutting guide position, femoral component malalignment can also be caused by errors during osteotomy, such as movement of the cutting guide and blurring of the bone saw in the cutting slot [13]. According to an evaluation using intraoperative navigation, osteotomy errors in the sagittal plane average 1.6° (−2.5° to 5.5°) in extension [14]. Osteotomy errors can cause size mismatches due to over- or undersizing of the implant relative to the size of the original anatomy of the patient. A 5° osteotomy error in extension increases the anteroposterior length by an average of 3.1 mm, whereas a 5° osteotomy error in flexion decreases it by an average of 2.6 mm [4]. Considering that most implants have a size variation of 2 mm, a difference of more than two sizes from the expected size may indicate a malalignment of ≥5°. This factor requires surgeons to proceed with caution. In addition, the femoral component is flexed from the osteotomy surface during the final implantation in 23% of the cases of malalignment [15]. Care must also be taken to prevent implant flexion throughout the procedure.

1.4 Femoral Component Position and Clinical Results

The relationship between the extension and flexion malposition of femoral components and clinical outcomes has been reported in multiple studies. The flexion position of the femoral component has been found as an independent factor for flexion contracture after TKA. According to one study, a group with flexion contractures had a flexion angle of the femoral component of 7.2° ± 1.2° relative to the anterior femoral cortical axis. In contrast, the group without flexion contractures had an average flexion position of 4.2° ± 1.4°. The group with flexion contractures showed inferior clinical scores such as the 2011KSS function score [2]. In a study of 156 patients who underwent TKA with the Bi-Surface Knee System, the group in which the femoral component was implanted in excessive flexion of ≥8.5° had inferior 2011KSS symptom, satisfaction, expectation, and functional activities scores, compared with those of the group whose femoral component was implanted between 2.5° and 5.5° of flexion. According to the quadratic curve approximation, the best satisfaction, expectation, and function were obtained with a slightly flexed position of approximately 3–4° from the distal femoral shaft axis. Some reports have cautioned against extension implantation. Approximately 25% of the patients who underwent TKA with extension implantation presented with anterior knee pain 10 years postoperatively. Patients with residual anterior pain had a femoral component with an extension of −0.6° (95% CI −1.5° to 0.3°) compared with a flexion of 1.4° (0.9–2.0°) in patients without anterior knee pain [16]. Regarding long-term results, Kim et al. found a revision rate of 3.3% in the group with femoral components flexed ≥3°. Conversely, the revision rates were 0% and 0.9% in the 0–3° flexion and extension groups, respectively [17]. These results suggest that clinical outcomes may be inferior when the femoral component is placed in excessive flexion or extension, and that no major problems occur when the femoral component is placed in slight flexion.

1.5 Flexion Alignment of the Femoral Component and Biomechanics

A relationship has been observed between the flexion position of the femoral component and an increase in the posterior offset [3]. A 2° increase in flexion of the femoral component causes a decrease in the flexion gap of 1–2 mm and may contribute to posterior tightness [8, 18]. In addition, a study using computer simulations showed that the contact force of the PF joint decreases and the kinematics improve with implantation of the femoral component in the flexion position. However, hyperflexion was not examined in the study [19]. Conversely, in computer simulations of the PS TKA, 3° of flexion of the femoral component did not increase the contact pressure of the polyethylene insert. However, the 7° flexion position caused anterior impingement during knee extension, and the contact pressure on the polyethylene insert significantly increased [20].

1.6 Anteroposterior Position of Femoral Components

When the femoral component is implanted anteriorly, notch formation is less likely to occur; however, anterior tightness occurs. When the component is implanted posteriorly, the posterior offset increases and posterior tightness occurs. Patients with end-stage OA who undergo TKA also develop large osteophytes in the anterior condylar region of the femur. The anterior condylar height (ACH) from the anterior cortical surface of the femur has been significantly correlated with the flexion restriction of the knee joint, with approximately 1.2° of flexion restriction for each increase in medial ACH of 1 mm [21]. In addition, the change in ACH before and after TKA correlates well with the change in flexion resulting from TKA, and a decrease in ACH of 1 mm is associated with a 1.7° improvement in flexion [22] (Fig. 42.3). The group with an increase in ACH of ≥5 mm had an average reduction of 21° of flexion, and the group with a decrease in ACH of ≥5 mm had an average improvement of 15° of flexion. Computer simulations have shown that the PF joint contact force increases and decreases due to an increase or decrease in postoperative ACH, respectively, and that changes in contact force and improvement in flexion are strongly correlated ($r = 0.96$) [22] (Fig. 42.3). During femoral implant sizing in the measured resection, an anterior reference may affect the flexion gap, whereas a posterior reference may result in anterior femoral notch formation or an increased anterior offset. When the implant size is increased with the anterior reference using a PS implant, the posterior offset of the femur is increased, and the contact force of the PF joint subsequently increased when the knee is flexed ≥120° [23]. In contrast, a prospective study examining the effects of the anterior and posterior references when the femoral implant sizing guide showed the middle of two implant sizes revealed that the anterior reference resulted in significantly smaller anterior and posterior offsets than the posterior reference; however, there was no difference in clinical outcomes between the two groups [24]. Although there is a guide can assist surgeons in choosing anterior or posterior references during surgery, they must know the characteristics of the cutting guide during preoperative planning. This will facilitate appropriate decisions such as whether the implant size should be determined by the anterior or posterior reference and whether the center of rotation of the sizing guide should be the center or medial pivot.

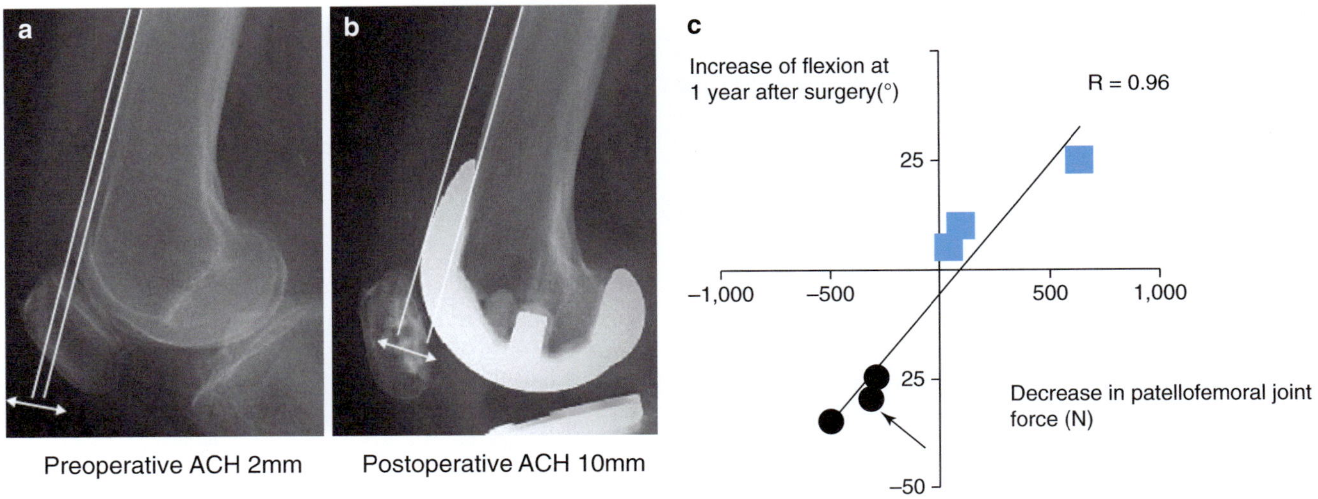

Fig. 42.3 A 65-year-old woman. (**a**, **b**): The anterior condylar height (ACH) from preoperatively (**a**) to postoperatively (**b**) increased by 8 mm, and the flexion angle decreased by 30°. (**c**): Computer simulation shows a strong correlation between the decrease in patellofemoral (PF) joint contact force and improvement in flexion range of motion (ROM). (Created from Nishitani et al. [22]. The arrow in (**c**) indicates the case of (**a**, **b**))

2 Sagittal Plane Alignment of Tibial Components

2.1 Tibial Axes Used in the Sagittal Plane

Several reference axes have been established in the sagittal plane of the lower leg. The functional axis of the tibia is defined by a line connecting the anteroposterior center of the proximal tibial plateau and the center of the tibial plafond [6]. As the axes of the lower leg that can be measured using plain radiography of the knee, anatomical axes at the proximal part of the tibia [3], anterior cortical axis of the tibia [9], and the axis of the fibula are used as indices of the sagittal plane of the lower leg [25] (Fig. 42.4). The axis of the central part of the tibia is also used and can be measured using a full-length radiograph of the lower limb [26]. Recently, 3D analysis has led to the definition of the bony axis of the tibia in 3D by connecting two points at the center of the proximal and distal medullary canals using CT or other methods [27].

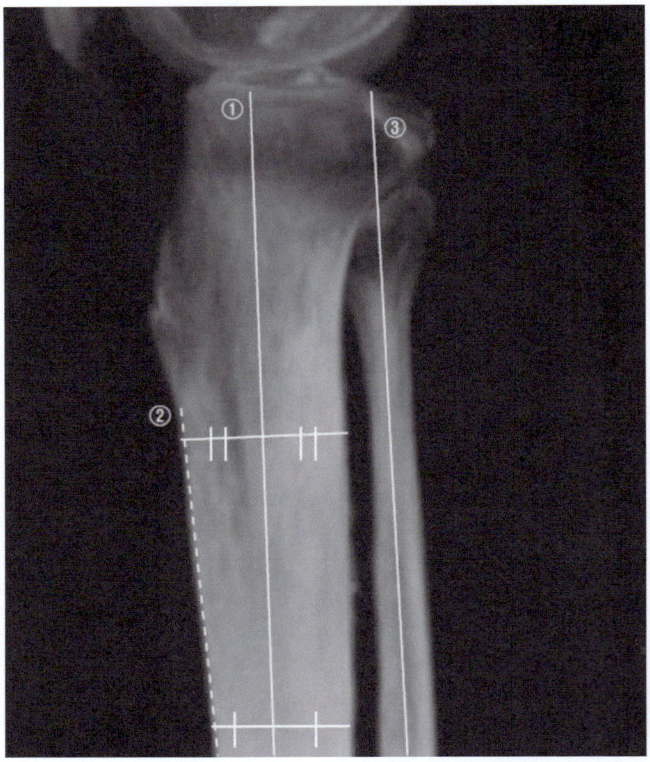

Fig. 42.4 Commonly used axes of the proximal lower leg. (1) Proximal tibial axis (white line); (2) anterior cortical tibial axis of the tibia (white dotted line); and (3) proximal fibular axis (white line)

2.2 Anatomical Study of the Native Posterior Tibial Slope

The preoperative posterior slope of the proximal tibia is commonly referenced during TKA. The angle between the tibial plateau and the axis of the lower leg, or that between the tibial plateau and the perpendicular of the axis of the lower leg, is often measured using a plain lateral radiograph [28]. As mentioned above, the posterior tibial slope angle varies depending on the axis adopted as the lower leg axis. A study examining the differences in the posterior tibial slope among races in more than 10,000 cases showed that Asians had a posterior tibial slope of 9.5° ± 4.5°, which was greater than that of Caucasians in Europe (6.2° ± 3.8°) and North America (7.3° ± 3.6°) [29]. The posterior tibial slope is often expressed as a single value as a projection of the medial and lateral condyles. In normal knees or those with mild degeneration, the medial and lateral posterior tibial slopes can be measured separately; however, in knees with advanced degeneration, distinguishing between the slopes using plain radiography may be difficult. Matsuda et al. measured the posterior slope of the medial and lateral condyles separately using MRI and found that the posterior slope of the medial condyle was 10.7° (range: 5–15.5°) in the normal knee and 9.9° in the varus knee (range: 1.5–19°), and the posterior slope of the lateral condyle was 7.2° in the normal knee (range: 0–14.5°) and 6° in the varus knee (range: 1–13°) [30]. Since the publication of that study, MRI and CT have been used to investigate the posterior slope of the medial and lateral condyles separately. According to a study that examined the posterior slope of the medial condyle by sex for Japanese individuals with normal knees, the posterior slope of the medial condyle was 9.2° ± 2.4° in males and 10.2° ± 1.8° in females, and the posterior slope of the lateral condyle was 8.8° ± 1.7° in males and 10.3° ± 2.2° in females; in general, females exhibited a greater posterior slope than males [31]. When referencing either the medial or lateral posterior tibial slope, as in UKA, it is also useful to measure the respective slopes using CT or MRI.

2.3 Posterior Inclination of the Tibial Component During CR/BCR TKA

In TKA, which preserves the cruciate ligament, the sagittal alignment of the tibial component is very important. Specifically, in CR TKA, the sagittal alignment of the tibial component affects the flexion gap and alters the tightness of the PCL. Excessive tension of the PCL in CR TKA causes an increase in the contact pressure between the tibia and femur, which may lead to long-term failure of the polyethylene. In CR TKA, an increase in the posterior inclination of the tibial component has been thought to reduce PCL overstrain and lead to favorable postoperative outcomes. A study using cadaveric knees showed that increasing the posterior inclination of the tibial component led to a decrease in PCL tension, suppressing excessive rollback and providing a clinical advantage [32]. Conversely, results from other studies raised concern that a 10° posterior tilt of the tibial component may result in an excessive decrease in PCL tension, preventing rollback of the femur and thus not demonstrating the true advantage of CR [33]. In postoperative patients who underwent high medial constraint CR TKA, a group with a tibial posterior tilt of the implant of 5.6° ± 1.5° was compared with a group with a tibial posterior tilt of 9.8° ± 1.8°. The latter group showed improved flexion ROM without particularly abnormal kinematics [34]. Another study found that in patients who could preoperatively flex their knee more than 120°, a posterior tibial component tilt of ≤6° was an independent factor for decreased knee flexion postoperatively, indicating a favorable posterior tilt of ≥7° [35]. In a study in which the posterior inclination angle of the tibial component was changed by ±2° using computer simulations, the force applied to the PCL decreased with an increase in the posterior inclination. In addition, an increase in the slight posterior inclination of approximately 2° did not cause abnormal kinematics [36]. Therefore, increasing the posterior inclination of the tibial component may be clinically advantageous. However, tibial osteotomy with an increased posterior tilt may injure the PCL attachment site. In a study using a cadaveric knee, researchers found that if the bone island at the PCL attachment site could not be preserved, a PS implant should be considered for osteotomies >10 mm from the lateral articular surface and a posterior tibial slope of ≥5° to preserve at least 50% of the PCL attachment site [37]. In a 2D–3D registration analysis using fluoroscopy, patients who underwent TKA were divided into two groups according to the posterior tibial component inclination; the group with an average of 10° of posterior tibial component inclination showed a posterior shift of the medial femoral condyle during loading compared with that of the group with an average of 4.6° of posterior inclination [38]. In the recently proposed kinematically aligned TKA, the posterior tilt of the tibial component is targeted at the preoperative posterior tibial slope, aiming to reconstruct the preoperative articular surface [39]. In BCR TKA, the posterior inclination of the implant is aimed at the preoperative posterior slope to preserve the function of both the ACL and PCL [40]. For the preservation of the cruciate ligament, it is beneficial to reproduce the patient's inherent posterior tibial slope to preserve the PCL attachment during TKA and thus maintain its function. However, in Asian patients, the preoperative posterior slope may be larger than 10°. The extent of posterior inclination of tibial implants which is acceptable in TKA remains unclear. Furthermore, in CR, the femur may shift posteriorly due to ACL removal, and a postoperative inclination of >10° should be avoided in conjunction with concerns about PCL dysfunction. Okazaki et al. placed a plate with a 5° tilt on the tibial osteotomy surface and examined the flexion gap with a 5° decrease and 5° increase in the tibial posterior tilt angle. They found that a 5° change in tibial posterior tilt caused changes of 2 mm in the flexion gap in the CR and 1 mm in the PS implant [41]. Reducing the posterior tilt by 5° (to 10°) in a patient with a preoperative posterior slope of 15° will result in tightness of the flexion gap of approximately 2 mm. In such cases, the PCL should be loosened or the femoral component should be adjusted to reduce the posterior offset by 2 mm.

2.4 Posterior Inclination of Tibial Components During PS/BCS-TKA

There is no ideal standard posterior inclination angle for the tibial component in all TKA procedures. For many years, a posterior tilt of approximately 0–7° in the articular plane has been used in PS TKA, including any inherent tilt of the polyethylene, depending on the implant mechanism [42]. When this posterior inclination is used, some patients may need a decrease in the posterior tibial slope compared to that before surgery. In the case of CR, the PCL is preserved, and tightness during flexion may occur. However, in the case of PS, the flexion gap is opened by PCL resection by approximately 3 mm, and even when the posterior inclination angle is slightly reduced, tightness during flexion does not occur [43]. The effect of the posterior inclination of the tibia in PS implants is less obvious than that in CR implants. PS TKA has shown a positive correlation between the gap at 90° and 135° knee flexion and the posterior inclination angle of the tibial component [44]. A computer simulation analysis showed that a posterior tibial component tilt of >10° caused paradoxical anterior slide and impingement of the post and femoral components during extension. Fluoroscopic imaging studies have shown that an increase in the posterior tilt angle of the tibial component in PS TKA causes paradoxical anterior translation in the midflexion range as the femur is positioned posteriorly in extension and flexion, as well as post-and-cam impingement in the extended position [45]. Using an insert trial that increases the posterior tilt angle of the tibial component by 10° intraoperatively does not have a positive effect on the intraoperative knee joint ROM [46]. The bicruciate stabilized (BCS) implant is a new type of implant that compensates for both the PCL and ACL. Computer simulations showed that even when using a BCS implant, a posterior tilt of the tibial component of >10° caused paradoxical anterior sliding of the femur in the stair-climbing motion, and a posterior tilt of >6° caused contact between the anterior post and the femur [47, 48]. Thus, we consider that the advantage of increasing the posterior inclination of the tibial component is minimal when PS implants are used. Referring to the aforementioned data [41], even when the posterior tilt is reduced by 10° (to 5°) in a patient with a preoperative posterior slope of 15°, a decrease in the flexion gap of approximately 2 mm occurs. However, this effect is canceled and no particular flexion tightness occurs because the resection of the PCL increases the gap by approximately 3 mm.

2.5 Method for Determining the Posterior Inclination of the Tibial Component

Navigation and robotic surgery are effective tools for increasing the accuracy of implant placement. The cortical axis of the anterior tibia can be touched subcutaneously and can be directly referenced intraoperatively. The advantages of the intraoperative reference of the anterior tibial cortical axis have been confirmed. As a method of reproducing the posterior slope without CAS, preoperative measurement of the length from the anterior surface of the tibia to the extramedullary rod at two points in the middle and distal lower leg and intraoperative reproduction of this measurement resulted in a mean deviation from the installed posterior inclination of 1.6° [49]. In addition, as the posterior inclination of the medial condyle is larger than that of the lateral one [50], the amount of lateral resection would be too large if cut according to the medial inclination; therefore, planning the tibial osteotomy according to the native posterior slope of the lateral condyle is also considered [51].

2.6 Effect of Sagittal Alignment of Both Components

The flexion position of the femoral component and the posterior inclination of the tibial component both affect the flexion gap. In a combined study of both in CR TKA, flexion of the femoral component and posterior inclination of the tibial component, along with the preoperative flexion angle, were independently associated with postoperative ROM [25]. In 1091 recent TKA studies using machine learning, the sagittal alignment of the implant at which the patients were satisfied or felt that the knee was functioning well, was a flexion angle of 0–7° for the femoral component. Furthermore, the change from the posterior slope of the tibial component was within 2° compared with that of the preoperative native slope; therefore, the posterior inclination of the tibial component should not be changed from the native slope [52]. However, patients sometimes have a preoperative posterior tibial slope of more than 10°, and reducing the posterior tilt of the component is necessary. In such cases, changing the posterior tibial slope based on an understanding of the characteristics of CR and PS TKA is necessary.

References

1. Ewald FC. The knee society total knee arthroplasty roentgenographic evaluation and scoring system. Clin Orthop Relat Res. 1989;248:9–12.
2. Okamoto Y, et al. Sagittal alignment of the femoral component and patient height are associated with persisting flexion contracture after primary total knee arthroplasty. J Arthroplast. 2019;34:1476–82.
3. Nishitani K, et al. Excessive flexed position of the femoral component was associated with poor new knee society score after total knee arthroplasty with the bi-surface knee prosthesis. Bone Joint J. 2020;102-B:36–42.
4. Nakahara H, et al. Sagittal cutting error changes femoral anteroposterior sizing in total knee arthroplasty. Clin Orthop Relat Res. 2012;470:3560–5.
5. Minoda Y, et al. The risk of notching the anterior femoral cortex with the use of navigation systems in total knee arthroplasty. Knee Surg Sports Traumatol Arthrosc. 2010;18:718–22.

6. Minoda Y, et al. TKA sagittal alignment with navigation systems and conventional techniques vary only a few degrees. Clin Orthop Relat Res. 2009;467:1000–6.
7. Asada S, et al. Influence of the sagittal reference axis on the femoral component size. J Arthroplast. 2013;28:943–9.
8. Tsukeoka T, et al. Sagittal flexion of the femoral component affects flexion gap and sizing in total knee arthroplasty. J Arthroplast. 2012;27:1094–9.
9. Sugama R, et al. Conventional or navigated total knee arthroplasty affects sagittal component alignment. Knee Surg Sports Traumatol Arthrosc. 2012;20:2454–9.
10. Lesh ML, et al. The consequences of anterior femoral notching in total knee arthroplasty: a biomechanical study. J Bone Joint Surg Am. 2000;82:1096–101.
11. Wangroongsub Y, et al. Proper entry point for femoral intramedullary guide in total knee arthroplasty. J Med Assoc Thail. 2009;92(Suppl 6):S1–5.
12. Minoda Y, et al. Theoretical risk of anterior femoral cortex notching in total knee arthroplasty using a navigation system. J Arthroplast. 2013;28:1533–7.
13. Otani T, et al. Cutting errors in preparation of femoral components in total knee arthroplasty. J Arthroplast. 1993;8:503–10.
14. Nakahara H, et al. Cutting error of the distal femur in total knee arthroplasty by use of a navigation system. J Arthroplast. 2012;27:1119–22.
15. Catani F, et al. Alignment deviation between bone resection and final implant positioning in computer-navigated total knee arthroplasty. J Bone Joint Surg Am. 2008;90:765–71.
16. Scott CEH, et al. Association between femoral component sagittal positioning and anterior knee pain in total knee arthroplasty: a 10-year case–control follow-up study of a cruciate-retaining single-radius design. J Bone Joint Surg Am. 2019;101:1575–85.
17. Kim YH, et al. The relationship between the survival of total knee arthroplasty and postoperative coronal, sagittal and rotational alignment of knee prosthesis. Int Orthop. 2014;38:379–85.
18. Tsukeoka T, et al. The effect of a sagittal cutting error of the distal femur on the flexion-extension gap difference in total knee arthroplasty. J Arthroplast. 2013;28:1099–102.
19. Kang KT. Flexed femoral component improves kinematics and biomechanical effect in posterior stabilized total knee arthroplasty. Knee Surg Sports Traumatol Arthrosc. 2019;27:1174–81.
20. Watanabe M, et al. Impact of intraoperative adjustment method for increased flexion gap on knee kinematics after posterior cruciate ligament-sacrificing total knee arthroplasty. Clin Biomech. 2019;63:85–94.
21. Nishitani K, et al. High and varied anterior condyle of the distal femur is associated with limited flexion in varus knee osteoarthritis. Cartilage. 2021;13(1 Suppl):1487S–93S.
22. Nishitani K, et al. A greater reduction in the distal femoral anterior condyle improves flexion after total knee arthroplasty in patients with osteoarthritis. Knee. 2019;26:1364–71.
23. Kawahara S, et al. Upsizing the femoral component increases patellofemoral contact force in total knee replacement. J Bone Joint Surg Br. 2012;94:56–61.
24. Nojiri S, et al. Which is better? Anterior or posterior referencing for femoral component position in total knee arthroplasty. J Orthop Surg (Hong Kong). 2021;29:23094990211002324.
25. Fujimoto E, et al. Significant effect of the posterior tibial slope and medial/lateral ligament balance on knee flexion in total knee arthroplasty. Knee Surg Sports Traumatol Arthrosc. 2013;21:2704–12.
26. Yoo JH, et al. Anatomical references to assess the posterior tibial slope in total knee arthroplasty: a comparison of 5 anatomical axes. J Arthroplast. 2008;23:586–92.
27. Miura M, et al. Interobserver and intraobserver reliability of computed tomography-based three-dimensional preoperative planning for primary total knee arthroplasty. J Arthroplast. 2018;33:1572–8.
28. Faschingbauer M, et al. Can the tibial slope be measured on lateral knee radiographs? Knee Surg Sports Traumatol Arthrosc. 2014;22:3163–7.
29. Meric G, et al. Tibial slope is highly variable in patients undergoing primary total knee arthroplasty: analysis of 13,546 computed tomography scans. J Arthroplast. 2015;30:1228–32.
30. Matsuda S, et al. Posterior tibial slope in the normal and varus knee. Am J Knee Surg. 1999;12:165–8.
31. Endo Y, et al. Difference in sex ánd the effect of a dominant lower extremity in the posterior tibial slope angle in healthy Japanese subjects. Asia Pac J Sports Med Arthrosc Rehabil Technol. 2021;23:8–12.
32. Jojima H, et al. Effect of tibial slope or posterior cruciate ligament release on knee kinematics. Clin Orthop Relat Res. 2004;426:194–8.
33. Takatsu T, et al. The function of posterior tilt of the tibial component following posterior cruciate ligament-retaining total knee arthroplasty. Bull Hosp Jt Dis. 1998;57:195–201.
34. Fujito T, et al. Influence of posterior tibial slope on kinematics after cruciate-retaining total knee arthroplasty. J Arthroplast. 2018;33:3778–3782.e1.
35. Sugitani K, et al. Factors affecting range of motion after total knee arthroplasty in patients with more than 120 degrees of preoperative flexion angle. Int Orthop. 2015;39:1535–40.
36. Kuriyama S, et al. Posterior tibial slope and femoral sizing affect posterior cruciate ligament tension in posterior cruciate-retaining total knee arthroplasty. Clin Biomech. 2015;30:676–81.
37. Onishi Y, et al. The influence of tibial resection on the PCL in PCL-retaining total knee arthroplasty: a clinical and cadaveric study. J Orthop Sci. 2016;21:798–803.
38. Fujimoto E, et al. Significant effect of the posterior tibial slope on the weight-bearing, midflexion in vivo kinematics after cruciate-retaining total knee arthroplasty. J Arthroplast. 2014;29:2324–30.
39. Howard JL, et al. Reproducing the native posterior tibial slope in cruciate-retaining total knee arthroplasty: technique and clinical implications. Orthopedics. 2019;43:e21–6.
40. Takasago T, et al. Insufficient lateral joint laxity after bicruciate-retaining total knee arthroplasty potentially influences kinematics during flexion: a biomechanical cadaveric study. Knee. 2021;28:311–8.
41. Okazaki K, et al. Influence of the posterior tibial slope on the flexion gap in total knee arthroplasty. Knee. 2014;21:806–9.
42. Gromov K, et al. What is the optimal alignment of the tibial and femoral components in knee arthroplasty? Acta Orthop. 2014;85:480–7.
43. Kadoya Y, et al. Effects of posterior cruciate ligament resection on the tibiofemoral joint gap. Clin Orthop Relat Res. 2001;391:210–7.
44. Oka S, et al. The influence of the tibial slope on intra-operative soft tissue balance in cruciate-retaining and posterior-stabilized total knee arthroplasty. Knee Surg Sports Traumatol Arthrosc. 2014;22:1812–8.
45. Hamai S, et al. Continuous sagittal radiological evaluation of stair-climbing in cruciate-retaining and posterior-stabilized total knee arthroplasties using image-matching techniques. J Arthroplast. 2015;30:864–9.
46. Takemura S, et al. The posterior tibial slope does not influence the flexion angle in posterior-stabilized mobile-bearing total knee arthroplasty. Arthroplasty. 2021;3:28.
47. Hada M, et al. Bi-cruciate stabilized total knee arthroplasty can reduce the risk of knee instability associated with posterior tibial slope. Knee Surg Sports Traumatol Arthrosc. 2018;26:1709–16.

48. Hada M, et al. Posterior tibial slope and anterior post-cam contact can change knee kinematics in extension in bi-cruciate stabilized total knee arthroplasty. Bone Joint Res. 2020;9:761–7.
49. Tomite T, et al. The usefulness of planning using a preoperative lateral leg image to determine accurate posterior tibial slope in total knee arthroplasty. J Orthop. 2019;16:25–30.
50. Nishitani K, et al. Plain radiography may underestimate the steep posterior tibial slope of the medial condyle in elderly Japanese with varus knee osteoarthritis undergoing arthroplasties. J Joint Surg Res. 2023:139–44.
51. Kuwano T, et al. Importance of the lateral anatomic tibial slope as a guide to the tibial cut in total knee arthroplasty in Japanese patients. J Orthop Sci. 2005;10:42–7.
52. Farooq H, et al. Machine learning algorithms identify optimal sagittal component position in total knee arthroplasty. J Arthroplast. 2021;36:S242–9.

Rotational Alignment

43

Shinya Kawahara

1 Rotational Alignment of the Femoral Components

1.1 Anatomical Indicators

Conventionally, femoral rotational alignment differs depending on whether it is performed using the measured or gap technique. The measured technique determines rotational alignment based on anatomical parameters, and the gap technique adjusts the soft tissue balance in the flexed position with femoral rotational alignment. Several anatomical landmarks have been identified (Fig. 43.1).

1.1.1 Surgical Epicondylar Axis (SEA) [1]

This axis connects the tip of the lateral epicondyle and the sulcus of the medial epicondyle. It is anatomically located at the femoral attachments of the medial collateral ligament (MCL) and the lateral collateral ligament (LCL), and is reported to approximate the central axis of extension and flexion of the normal knee [2, 3]. This is the most frequently used landmark.

1.1.2 Clinical Epicondylar Axis (CEA)

This axis connects the tip of the lateral epicondyle to the tip of the medial epicondyle. The CEA is externally rotated by approximately 3° relative to the SEA [4]. The term "transepicondylar axis (TEA)" is sometimes used to mean either SEA or CEA; therefore, it is necessary to confirm the definitions and interpret them in each study.

1.1.3 Whiteside Line [5]

The axis connects the deepest part of the trochlear groove to the middle of the intercondylar notch.

1.1.4 Posterior Condylar Axis (PCA)

This is the axis connecting the bilateral posterior condylar margins and is the easiest to refer to during surgery. However, it is more internally rotated than the other three axes, which usually affects patellar tracking. The PCA is used as a landmark in the kinematic alignment method; however, no alignment abnormality occurs during flexion because the coronal alignment of the proximal tibia is also varus with respect to the functional axis. Details of the kinematic alignment method are omitted.

The SEA, CEA, and Whiteside lines are generally recommended because most of the prostheses currently used are designed based on these reference axes. Surgeons should understand which landmarks to refer to for prostheses with an asymmetrical shape of the femoral condyle.

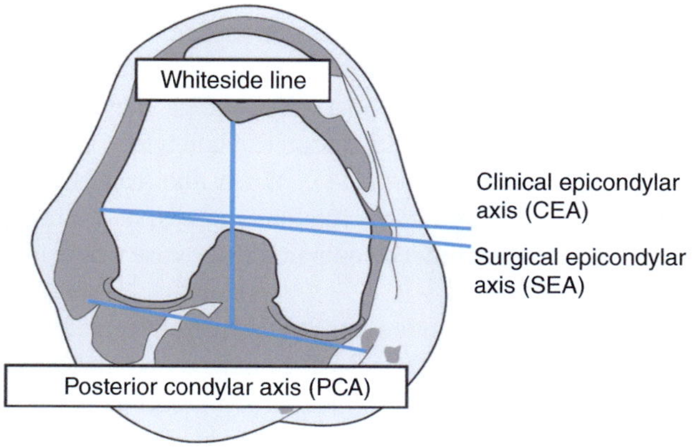

Fig. 43.1 Anatomical landmark of femoral rotational alignment

S. Kawahara (✉)
Orthopedic Surgery, Kyushu University Hospital, Fukuoka, Japan
e-mail: kawahara.shinya.310@m.kyushu-u.ac.jp

S. Matsuda (ed.), *Advances in Total Knee Arthroplasty*, https://doi.org/10.1007/978-981-97-4920-1_43

1.2 Extension and Flexion Alignment and Soft Balance

When the medial and lateral soft tissue balance in the flexion position is adjusted by the rotational alignment of the femur, the femoral component is often excessively externally rotated because of lateral looseness in the case of a varus knee, resulting in varus alignment in the flexion position. Hanada et al. reported a higher stress in the medial compartment in a cadaver study [6]. In addition, the medial extension and flexion gaps are difficult to maintain because the medial flexion gap tends to increase. In a normal knee, the lateral soft tissue balance is physiologically loose [7, 8] and maintaining the balance of the medial extension and flexion gaps has been reported to lead to good results [9, 10]. Consequently, the so-called "medial stabilizing TKA" has been recommended, which does not release the medial soft tissue as much as possible, stabilizes the medial extension and flexion gap balance, and allows lateral gap looseness in the flexion position [11].

1.3 Preoperative Planning

It is often difficult to accurately identify the SEA or CEA during surgery [5], and the external rotation angle from the PCA is measured in advance using CT or magnetic resonance imaging (MRI) before surgery. Although identification of the tip of the lateral epicondyle is usually easy, it has been reported that the medial epicondyle sulcus in SEA is difficult to identify in more than 25% of cases [4, 12, 13]. In this case, the external rotation angle should be measured by subtracting approximately 3° from the CEA.

When measuring using CT, care must be taken when handling the measured values because the residual cartilage is not evaluated. Although there are many reports from Japan, the cartilage of the lateral posterior condyle may remain comparable to that of the medial posterior condyle in patients with medial osteoarthritis (OA). When the cartilage is included in the evaluation, the rotation angle is, on average, 1.1° to 1.7°internally rotated compared to the case where it is not included [14–18].

In addition, detailed attention should be paid to measurement methods. Because the SEA is projected onto the distal resected plane perpendicular to the femoral mechanical axis during surgery, the measurement should be performed accordingly. The original three-dimensional (3D) coordinate system of computed tomography (CT) or MRI is determined by the position of the imaging limb and the slice levels of the medial and lateral epicondyles are often different. Therefore, it is necessary to measure 3D or multiplanar reformation (MPR) images in which the 3D coordinate system is adjusted to the femoral functional axis. Okamoto et al. reported that the rotation angle could be evaluated more internally if measurements were performed directly from the original images [19].

In lateral OA, the lateral posterior condyle is generally hypoplastic and the external rotation angle is larger than that medial OA [20].

1.4 Surgical Techniques

When measuring the external rotation angle using CT, the degree of residual cartilage in the lateral posterior femoral condyle should be evaluated, as mentioned above, and the value should be slightly smaller than the measured value. Alternatively, the cartilage of the lateral posterior condyle may be shaved with a file or bone saw before placement of the femoral sizer. Depending on the incision size, palpation of the medial and lateral epicondyles and an approximate SEA mark on the distal resection surface are useful. The Whiteside line is marked in the same manner. It is important to place the femoral sizer at an external rotation angle from the PCA and to comprehensively evaluate the rotational alignment with the above landmarks to prevent large rotational errors (Fig. 43.2).

When placing the femoral sizer, it is important to take sufficient care not to tuck soft tissues, such as the bilateral meniscus with the posterior condyle paddle, and to contact both the distal rection surface and posterior condyle. Surgeons should confirm the characteristics of the femoral sizer in detail, whether the external rotation angle is built-in, whether the angle can be adjusted, and whether the rotation center is located medially, laterally, or in the center, although most of the sizers currently available on the market have a

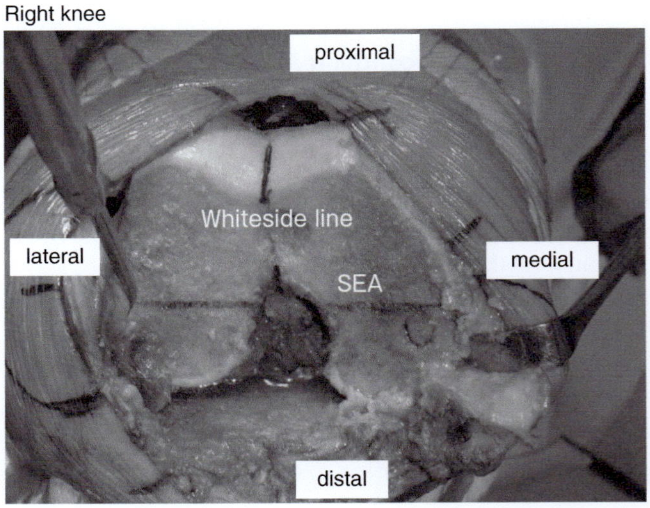

Fig. 43.2 Marking the Whiteside line and SEA during surgery and comprehensively determining rotation by combining the external rotation angle from the PCA

built-in external rotation angle [21]. The flexion gap depends on whether cruciate retaining (CR) or posterior stabilized (PS) total knee arthroplasty (TKA) is performed. The author mainly uses PS, and a medial rotational-type sizer is chosen to maintain the medial extension and flexion gaps. Next, the rotational alignment and anteroposterior position for the placement of a four- or 5-plane cutting guide are fixed with two pin insertions, which also requires careful attention. We always drill while pressing the two pins together because even the slightest misalignment when drilling the two pins causes rotational errors (Fig. 43.3).

Because the cutting guide is generally placed using the anterior referencing technique at our institution, the anterior flange length of the planned femoral component is marked on the anterior femur to avoid an anterior notch (Fig. 43.4). In contrast, Nishitani et al. reported that the postoperative flexion angle decreases when the anterior part becomes thicker after surgery [22, 23], and careful anterior resection is required. When parallel to the SEA, it has been reported that the medial width relative to the lateral one in the anterior resection surface is small, the so-called "grand-piano sign," which is one of the methods to check the rotational alignment [24]. Ohmori et al. reported medial/lateral (M/L) ratio is 0.62 to 0.67 in neutral rotation and 0.8 and 0.5 in 5° internal and external rotation of [25]. During posterior condyle and posterior chamfer resections, it is important to ensure the retraction of the MCL in the medial and posterior vascular nerves in the lateral direction using elevators or other means.

Right knee

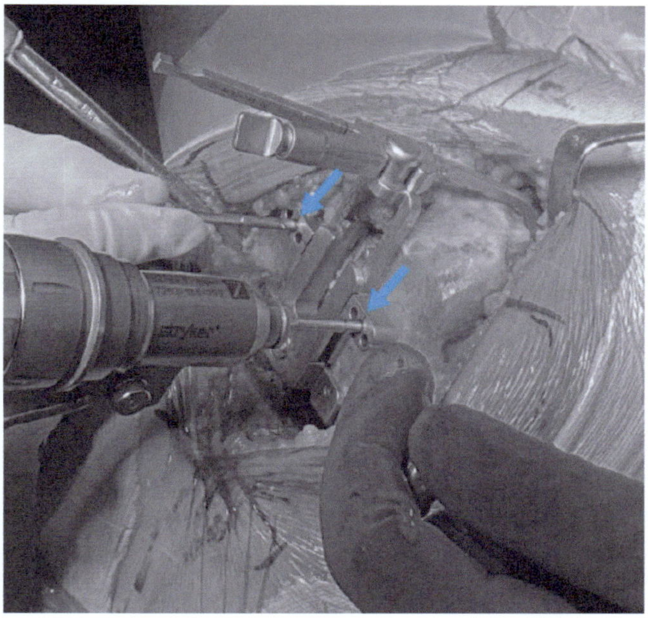

Fig. 43.3 Always drill while pressing two pins (arrows)

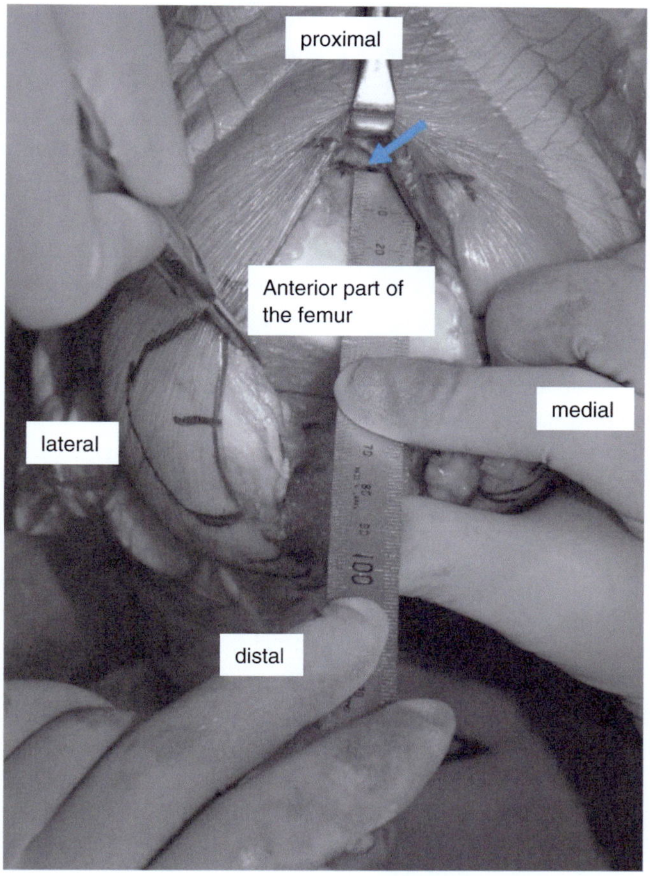

Fig. 43.4 The planned size of the anterior flange width was marked in advance (arrow), and care was taken not to cause an anterior notch

1.5 Clinical Results

Femoral rotational alignment directly affects soft-tissue balance during flexion and patellar tracking. Regarding soft-tissue balance in flexion, Hatayama et al. measured the femoral rotational alignment relative to the CEA with a radiograph in flexion and found a correlation with the flexion balance (lift-off angle); they reported that rotational placement based on the CEA is preferable to reduce the lift-off angle [26]. Release of the lateral retinaculum is sometimes added in cases of patellar maltracking. However, there are concerns regarding postoperative pain, delayed rehabilitation, and decreased blood flow to the patella. Therefore, it is important to reconstruct an anatomical rotational alignment that does not require lateral release. In 1999, Akagi et al. reported that patellar tracking was better with a femoral rotational alignment of 3–5° external rotation than with the PCA, and the rate of lateral release was lower (34% vs. 6%) [12], suggesting the current anatomical placement of the SEA. Terashima et al. measured the contact pressure on the patellofemoral (PF) joint during TKA and reported that the contact pressure tended to decrease as the alignment shifted from internal to external rotation [27]. We examined the effects of femoral rotational alignment on each item of the 2011 Knee Society Score. The Functional Activity Score was significantly lower in the group with more internal rotation than in the neutral rotation group, with no significant difference in walking and standing, which can be performed with relative knee extension, and significantly lower in Standard and Advanced Activities that require knee flexion [28]. When the femoral component is internally rotated, the patella is pulled laterally with a valgus alignment in the flexion position, resulting in patellar maltracking and decreased quadriceps muscle strength affected the score.

From the above, it is recommended that the SEA be used as a rotational landmark in principle, avoid internal rotation, and limit external rotation to a range that does not exceed the CEA.

2 Rotational Alignment of Tibial Components

Insall et al. stated that the medial 1/3 of the tibial tubercle is a landmark of tibial rotation [29], and is still widely used worldwide; however, they do not specify the posterior landmark. Most surgeons would use the medial 1/3 of the tibial tubercle as an anterior landmark when the component is placed at the center of the resection surface. If interpreted in this manner, it is a relatively reasonable landmark. However, definition of an "axis" needs at least two points connecting each other. "Aligning the proximal anteroposterior axis of the tibia with ○○" is a vague definition, and the definition "the axis connecting ○○ and △△" is appropriate.

Methods for determining the rotational alignment of the tibia include: (1) referring to anatomical landmarks, (2) the range of motion (ROM) method, (3) prioritizing component coverage over anatomical landmarks using mobile-bearing inserts, and (4) placing an asymmetric component so that coverage is good. Tibial rotational alignment directly influences the MCL and patellar tendon tension, patellar tracking, and the degree of component coverage, while being influenced by femoral rotational alignment. Because of the posterior tilt during proximal tibial resection, the tibial coronal alignment will be affected if the correct proximal tibial anteroposterior axis is not referenced [29]. A consistent rotational axis should be considered during surgery. Additionally, because the proximal tibia is internally twisted relative to the distal one [30], surgeons must adjust the distal part of the tibial extramedullary rod in the direction of the proximal tibial rotational axis [31]. The author does not refer to any anatomical landmarks in the distal tibia or foot because they may be greatly affected by foot deformity and rotation [32].

2.1 Anatomical Landmarks (Fig. 43.5)

Although several anatomical landmarks have been proposed, the anteroposterior (AP) axis proposed by Akagi et al. (Akagi

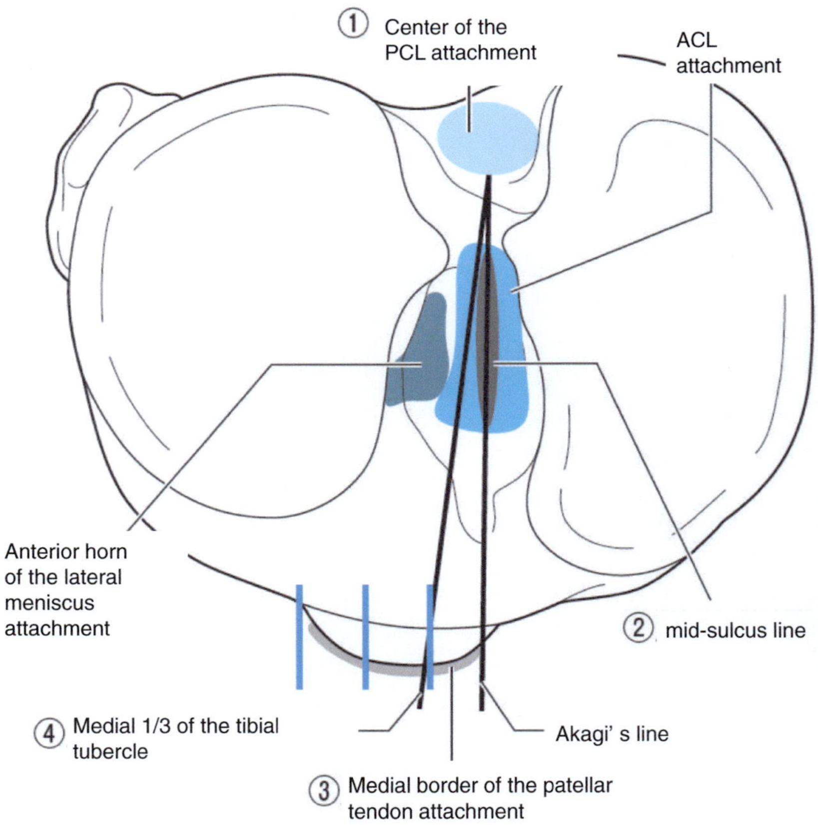

Fig. 43.5 Anatomical landmark of tibial rotational alignment

line) is frequently used, and its usefulness has been reported in the clinical results described later.

1. The AP axis connects the medial edge of the patellar tendon attachment to the center of the PCL attachment (Akagi line).

   Akagi et al. examined the axes perpendicular to the SEA in the knee extension position in normal knees and reported that the axis connecting the above two points was a reliable landmark [33, 34]. The authors performed similar measurements on medial and lateral knees with OA and reported that the relationship between SEA and Akagi's line was preserved and that Akagi's line was useful even in knees with OA [35].
2. Mid-sulcus line [36].

   This indicates the groove of the tibial intercondylar eminence; however, it is often ambiguous due to osteophytes and is used as an adjunct to the Akagi line only when there is little deformation.
3. Medial 1/3 of the tibial tubercle

   As mentioned above, this varies greatly depending on where the posterior reference landmark is set, and it has been reported that the AP axis is externally rotated when a PCL attachment is used as the posterior landmark [34].

2.2 ROM Method

After setting the femoral trial component, the tibial tray with the trial insert was placed without fixation, and patellar tracking was performed with knee flexion and extension. It is a commonly used self-alignment technique in which the knee determines its rotational position depending on the femoral component's rotational position of the femoral component, knee extensor mechanism, and mediolateral soft tissue balance. On the other hand, it should be recognized that it is also a method that depends on the accuracy described above, and it is recommended to use it in conjunction with the confirmation of anatomical landmarks [37, 38].

2.3 Mobile-Bearing Insert

The use of the mobile-bearing insert compensates for the femorotibial rotational mismatch by polyethylene insert rotation on the tibial component, allowing alignment of the tibial component without concern for slight rotational placement errors [39]. However, as mentioned above, surgeons should recognize that the great difference in the rotational axis during surgery may affect the tibial coronal alignment.

2.4 ML Asymmetrical Components

The tibial resection surface generally has a large AP width medially relative to laterally, and in recent years, an asymmetrical anatomical design has been adopted in anatomical design [40]. The asymmetrical components generally fit better than the symmetrical components. Ma et al. reported that when the asymmetrical component is placed with the best fit in the center of the resection plane, the anterior landmark is located at 3.3° internal rotation from the medial 1/3 of the patellar tendon attachment [41], which is almost consistent with the medial 1/6 discussed below [42]. Therefore, the method of fitting asymmetric components to the shape of the resection surface can be considered almost identical to that of the Akagi line.

2.5 Preoperative Planning

Three-dimensional preoperative planning helps in understanding component coverage, and large errors can be prevented during surgery. To establish a 3D coordinate system, the rotational axis must first be fixed. All of the above anatomical parameters could be identified intraoperatively.

Caution should be exercised in cases of knee OA after corrective osteotomy, such as high tibial osteotomy (HTO). The Akagi line is the tibial axis perpendicular to the SEA to avoid a rotational mismatch between the femoral and tibial components in the knee extension position. However, anatomical deformity occurs in the proximal part of the tibia after osteotomy. In closed-wedge and open-wedge HTOs, the tibial tubercle is located on the distal side, whereas in open-wedge distal tibial osteotomy, the tibial tubercle is located on the proximal side. Closing and opening procedures during osteotomy may result in rotational changes between the fragments; therefore, it is important to evaluate the accuracy of each anatomical landmark preoperatively. If a rotational mismatch with the SEA occurs, one option is to use a mobile-bearing insert that prioritizes coverage, as described above.

2.6 Surgical Techniques

As mentioned previously, it is necessary to have a consistent concept of the rotational axis during surgery. The Akagi line was used in this study. During PS TKA, the PCL attachment position was confirmed by firmly inserting a PCL retractor along the tibial posterior groove. When placing the extramedullary tibial rod, the rotational fixation pin was first inserted parallel to the Akagi line. The coronal/sagittal alignment and amount of osteotomy can be stably adjusted by first inserting the pin of the rotational fixation. The tibial anterior border was used as a landmark for adjusting the coronal and sagittal plane alignments (Fig. 43.6) [43, 44]. The alignment rod was set to confirm the approximate coronal alignment after tibial resection, and a consistent rotational direction should be considered even in this procedure. (Fig. 43.7).

Careful attention should be paid to the size and rotational placement of the tibial component. The center of the PCL attachment is generally difficult to identify after proximal tibial resection, and it is difficult to reproduce the Akagi line. One method is to leave the pin identifying the Akagi line that was inserted during proximal tibial resection; however, this often occurs during the process. The authors examined where the axis parallel to the Akagi line crossed anteriorly through the center of the resection surface for both medial and lateral OA and reported that the anterior position was approximately 4 mm lateral from the medial edge of the patellar tendon attachment, and the ratio was medial 1/6 of the patellar tendon attachment [42] (Fig. 43.8). In other words, placing the anterior center of the component toward the medial 1/6 of the patellar tendon attachment was sufficient when aligning the tibial component. As mentioned above, good coverage of the asymmetrical components is ensured in this manner. The projection of the Akagi line onto the resection plane was slightly medial to the center; therefore, the anterior landmark was slightly lateral to the medial border of the patellar tendon attachment. The author often sees surgeons confusing the Akagi line with the anterior landmark when placing tibial components, however, the rotational axis for component placement is "the axis connecting the center of the resection surface and an anterior landmark such as the medial 1/6 of the patellar tendon attachment," which should be understood to be parallel to the Akagi line. If the anterior center is between the medial edge and medial 1/3 of the patellar tendon attachment in the alignment of the tibial component, it can be judged that there is at least no major rotational error.

Some implants recommend the ROM method, and are occasionally used in combination. However, it rarely deviates significantly from the aforementioned anatomical landmarks, and if it does, it is necessary to check for abnormal rotation of the femoral component.

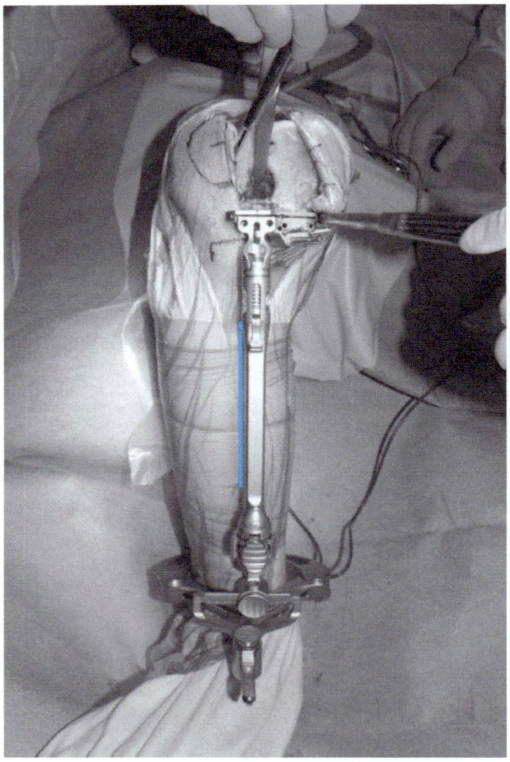

Fig. 43.6 Using the tibial anterior border (proximal 1/3 to distal 1/3, solid line) as a landmark, the coronal and sagittal alignments are adjusted

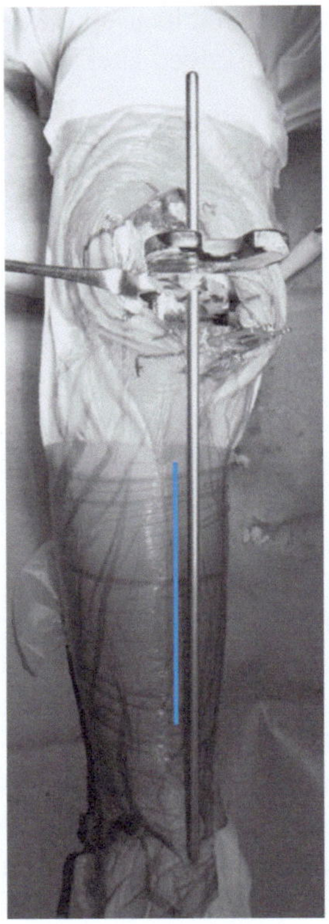

Fig. 43.7 Alignment check with rod. It is important to be aware of the rotational direction

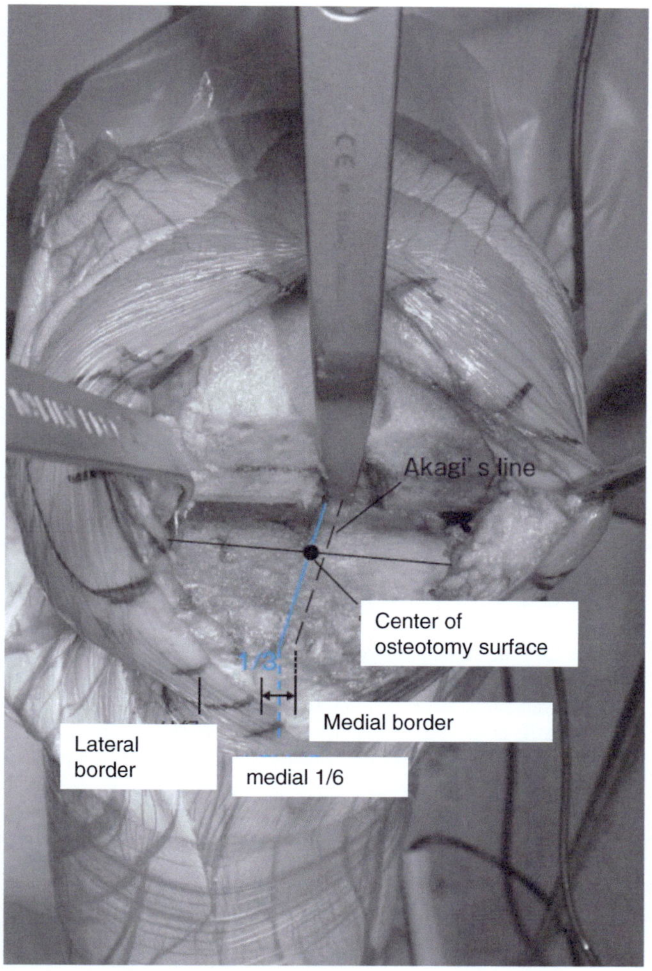

Fig. 43.8 Rotational axis at the tibial resection surface. The Akagi line is difficult to identify after tibial resection because the PCL attachment cannot be identified. If an axis (blue line) is drawn parallel to the Akagi line through the center of the resection surface, the anterior passes through the medial 1/6 of the patellar tendon attachment

2.7 Clinical Results

The rotational error tends to be greater on the tibial side than on the femoral side, and in some cases, an error of 20° or more may occur [45, 46]. This tendency was also observed when using navigation systems [47]. Excessive internal rotation installation leads to residual pain, knee contracture, and patellar maltracking [45]. Kuriyama et al. examined the effects of the internal/external rotational alignment of the tibial component up to 15° in a computer simulation that reproduced a musculoskeletal model. They reported that internal rotation increased MCL tension and contact stress at the tibiofemoral joint, which are considered factors in the pain and contracture associated with internal rotation placement [48]. Osano et al. used the finite element method to evaluate the effect of internal and external rotations of tibial components on the maximum stress on the insert and reported that both internal and external rotations increased the stress [49].

There are various reports on the rotational mismatch caused by tibial rotational alignment errors. It has been reported that tibial rotational alignment error was the main cause of rotational mismatch between both components [46]. However, it has also been reported that rotational mismatch mainly occurs between the two bones, and that a bone-to-bone mismatch of 10° or more is a risk factor for poor functional outcome in the KSS 2011 [50]. Another report stated that a mismatch occurred between the components and bones [51]. Because this depends on the conformity of the insert, the tendency may differ depending on the implant design. However, rotational malalignment that deviates greatly from anatomical landmarks, particularly internal rotation alignment, should be avoided.

Ikeuchi et al. evaluated rotational alignment using both the ROM method and anatomical landmarks during surgery and reported that it is not appropriate to use the ROM method alone because the ROM method tends to cause internal rotational alignment with great variation [37]. Kawaguchi et al. reported an average of 4.6° external rotation and 11.7% outliers over 10°malalignment for the ROM method alone and an average of 3° external rotation and 3% outliers over 10° for the ROM method combined with anatomical landmarks [38]. Therefore, it is recommended to refer to anatomical landmarks when using the ROM method. For mobile-bearing TKA, Ishii et al. reported good HSS scores with a rotational mismatch within 10° while obtaining good tibial coverage [39] and which seems to be a valid option. It has been reported that symmetrical components can have large rotational errors when aligned with coverage priority, so care must be taken [52]. On the other hand, it has been reported that asymmetrical anatomical components fit well and can be aligned in the correct rotational position [41].

References

1. Berger RA, et al. Determining the rotational alignment of the femoral component in total knee arthroplasty using the epicondylar axis. Clin Orthop Relat Res. 1993;286:40–7.
2. Yoshioka Y, et al. The anatomy and functional axes of the femur. J Bone Joint Surg Am. 1987;69:873–80.
3. Asano T, et al. The functional flexion-extension axis of the knee corresponds to the surgical epicondylar axis: in vivo analysis using a biplanar image-matching technique. J Arthroplast. 2005;20:1060–7.
4. Yoshino N, et al. Computed tomography measurement of the surgical and clinical transepicondylar axis of the distal femur in osteoarthritic knees. J Arthroplast. 2001;16:493–7.
5. Arima J, et al. Femoral rotational alignment, based on the anteroposterior axis, in total knee arthroplasty in a valgus knee: a technical note. J Bone Joint Surg Am. 1995;77:1331–4.

6. Hanada H, et al. Bone landmarks are more reliable than tensioned gaps in TKA component alignment. Clin Orthop Relat Res. 2007;462:137–42.
7. Okazaki K, et al. Asymmetry of mediolateral laxity of the normal knee. J Orthop Sci. 2006;11:264–6.
8. Tokuhara Y, et al. The flexion gap in normal knees. An MRI study. J Bone Joint Surg Br. 2004;86:1133–6.
9. Tsukiyama H, et al. Medial rather than lateral knee instability correlates with inferior patient satisfaction and knee function after total knee arthroplasty. Knee. 2017;24:1478–84.
10. Azukizawa M, et al. Intraoperative medial joint laxity in flexion decreases patient satisfaction after total knee arthroplasty. Arch Orthop Trauma Surg. 2018;138:1143–50.
11. Matsuda S, Ito H. Ligament balancing in total knee arthroplasty-medial stabilizing technique. Asia Pac J Sports Med Arthrosc Rehabil Technol. 2015;2:108–13.
12. Akagi M, et al. Effect of rotational alignment on patellar tracking in total knee arthroplasty. Clin Orthop Relat Res. 1999;366:155–63.
13. Akagi M, et al. Relationship between frontal knee alignment and reference axes in the distal femur. Clin Orthop Relat Res. 2001;388:147–56.
14. Miyasaka T, et al. Impact of posterior femoral condylar cartilage and posterior intercondylar distance on rotation of femoral component in total knee arthroplasty. BMC Musculoskelet Disord. 2020;21:498.
15. Asada S, et al. Effects of cartilage remnants of the posterior femoral condyles on femoral component rotation in varus knee osteoarthritis. Knee. 2012;19:185–9.
16. Fujii T, et al. Posterior condylar cartilage may distort rotational alignment of the femoral component based on posterior condylar axis in total knee arthroplasty. Surg Radiol Anat. 2012;34:633–8.
17. Hamada D, et al. The posterior condylar cartilage affects rotational alignment of the femoral component in varus knee osteoarthritis. J Med Investig. 2017;64:24–9.
18. Tashiro Y, et al. Articular cartilage of the posterior condyle can affect rotational alignment in total knee arthroplasty. Knee Surg Sports Traumatol Arthrosc. 2012;20:1463–9.
19. Okamoto S, et al. Two-dimensional planning can result in internal rotation of the femoral component in total knee arthroplasty. Knee Surg Sports Traumatol Arthrosc. 2016;24:229–35.
20. Matsuda S, et al. Anatomical analysis of the femoral condyle in normal and osteoarthritic knees. J Orthop Res. 2004;22:104–9.
21. Kawahara S, et al. Femoral sizer design can increase anterior notching during total knee arthroplasty. Knee. 2016;23:890–4.
22. Nishitani K, et al. A greater reduction in the distal femoral anterior condyle improves flexion after total knee arthroplasty in patients with osteoarthritis. Knee. 2019;26:1364–71.
23. Nishitani K, et al. High and varied anterior condyle of the distal femur is associated with limited flexion in varus knee osteoarthritis. Cartilage. 2021;13(1 Suppl):1487S–93S.
24. Kokubu Y, et al. "Grand-piano sign" as a femoral rotational indicator in both varus and valgus knees: a simulation study of anterior resection surface in total knee arthroplasty. Knee Surg Sports Traumatol Arthrosc. 2023;31(8):3259–67.
25. Ohmori T, et al. Usefulness of the "grand-piano sign" for determining femoral rotational alignment in total knee arthroplasty. Knee. 2018;25:15–24.
26. Hatayama K, et al. Relationship between femoral component rotation and total knee flexion gap balance on modified axial radiographs. J Arthroplast. 2011;26:649–53.
27. Terashima T, et al. External rotation of the femoral component decreases patellofemoral contact stress in total knee arthroplasty. Knee Surg Sports Traumatol Arthrosc. 2015;23:3266–72.
28. Kawahara S, et al. Internal rotation of femoral component affects functional activities after TKA-survey with the 2011 knee society score. J Arthroplast. 2014;29:2319–23.
29. Insall JN, Easley ME. Surgical techniques and instrumentation in total knee arthroplasty. In: Insall JN, Scott WN, editors. Surgery of the knee. 3rd ed. New York: Churchill Livingstone; 2001. p. 1553–620.
30. Nagamine R, et al. Medial torsion of the tibia in Japanese patients with osteoarthritis of the knee. Clin Orthop Relat Res. 2003;408:218–24.
31. Mizu-uchi H, et al. The effect of ankle rotation on cutting of the tibia in total knee arthroplasty. J Bone Joint Surg Am. 2006;88:2632–6.
32. Yoshioka Y, et al. Tibial anatomy and functional axes. J Orthop Res. 1989;7:132–7.
33. Akagi M, et al. Variability of extraarticular tibial rotation references for total knee arthroplasty. Clin Orthop Relat Res. 2005;436:172–6.
34. Akagi M, et al. An anteroposterior axis of the tibia for total knee arthroplasty. Clin Orthop Relat Res. 2004;420:213–9.
35. Kawahara S, et al. Relationship between the tibial anteroposterior axis and the surgical epicondylar axis in varus and valgus knees. Knee Surg Sports Traumatol Arthrosc. 2012;20:2073–7.
36. Dalury DF. Observations of the proximal tibia in total knee arthroplasty. Clin Orthop Relat Res. 2001;389:150–5.
37. Ikeuchi M, et al. Determining the rotational alignment of the tibial component at total knee replacement. J Bone Joint Surg Br. 2007;89:45–9.
38. Kawaguchi K, et al. A new technique for determining the rotational alignment of the tibial component during total knee arthroplasty. Knee. 2021;29:323–31.
39. Ishii Y, et al. Tibial component coverage and rotational alignment accuracy after mobile-bearing total knee arthroplasty. Eur J Orthop Surg Traumatol. 2018;28:1143–9.
40. Uehara K, et al. Anthropometry of the proximal tibia to design a total knee prosthesis for the Japanese population. J Arthroplast. 2002;17:1028–32.
41. Ma Y, et al. Effects of tibial baseplate shape on rotational alignment in total knee arthroplasty: three-dimensional surgical simulation using osteoarthritis knees. Arch Orthop Trauma Surg. 2018;138:105–14.
42. Kawahara S, et al. Medial sixth of the patellar tendon at the tibial attachment is useful for the anterior reference in rotational alignment of the tibial component. Knee Surg Sports Traumatol Arthrosc. 2014;22:1070–5.
43. Fukagawa S, et al. Anterior border of the tibia as a landmark for extramedullary alignment guide in total knee arthroplasty for varus knees. J Orthop Res. 2011;29:919–24.
44. Kawahara S, et al. Anterior tibial border as a landmark for extramedullary alignment guides for total knee arthroplasty in valgus knees. J Orthop Res. 2015;33:1897–9.
45. Matsuda S, et al. Effect of femoral and tibial component position on patellar tracking following total knee arthroplasty: 10-year follow-up of Miller-Galante I knees. Am J Knee Surg. 2001;14:152–6.
46. Watanabe S, et al. Change in tibiofemoral rotational alignment during total knee arthroplasty. J Orthop Sci. 2014;19:571–8.
47. Mizu-uchi H, et al. The evaluation of postoperative alignment in total knee replacement using a CT-based navigation system. J Bone Joint Surg Br. 2008;90:1025–31.
48. Kuriyama S, et al. Malrotated tibial component increases medial collateral ligament tension in total knee arthroplasty. J Orthop Res. 2014;32:1658–66.
49. Osano K, et al. The effect of malrotation of tibial component of total knee arthroplasty on tibial insert during high flexion using a finite element analysis. Sci World J. 2014;2014:695028.
50. Ueyama H, et al. Malrotation of the fixed-bearing posterior stabilized total knee prosthesis causes a postoperative rotational mismatch between the femur and tibia. Knee Surg Sports Traumatol Arthrosc. 2020;28:3810–20.
51. Nakahara H, et al. Rotational alignment of the tibial component affects the kinematic rotation of a weight-bearing knee after total knee arthroplasty. Knee. 2015;22:201–5.
52. Hirakawa M, et al. Evaluation of the rotational alignment of the tibial component in total knee arthroplasty: position prioritizing maximum coverage. Eur J Orthop Surg Traumatol. 2017;27:119–24.

Soft Tissue Balance/Joint Stability: Medial-Lateral, Flexion-Extension Balance

44

Tomoyuki Matsumoto

1 Soft Tissue Balance

Although total knee arthroplasty (TKA) is generally recognized as a good procedure in terms of pain relief and long-term outcomes, problems remain in terms of achieving good long-term outcomes and patient satisfaction. In the surgical technique, it is especially important to perform accurate osteotomies and component placement and to obtain an appropriate soft tissue balance. The accuracy of osteotomy and component placement has been improved by 3D preoperative planning, improvements in surgical instruments, and the spread of navigation/robotic systems. However, the evaluation and acquisition of soft tissue balance still often depend on the surgeon's experience and skill. However, in recent years, reproducible quantitative evaluations and analyses have become widespread and are being performed, contributing to the examination and improvement of surgical procedures.

To evaluate the intraoperative soft tissue balance in TKAs, it is important to quantitatively measure soft tissue balance while reproducing the state of the postoperative knee, that is, with femoral components inserted and patellofemoral (PF) joints reduced and provisionally repaired. Offset-type tensors have been developed to enable this type of measurement. Various types of tensors have been developed for use in Japan. Intraoperative measurements using these tensors revealed the effects of intraoperative and preoperative factors, including various surgical techniques, on intraoperative soft tissue balance; differences in the effects between CR-TKA and PS-TKA; and the effects of intraoperative soft tissue balance on postoperative range of motion (ROM), clinical scores, and kinematics. Furthermore, these reports have led to increased emphasis on postoperative knee stability and its effect on postoperative outcomes over time.

T. Matsumoto (✉)
Department of Orthopaedic Surgery, Kobe University Graduate School of Medicine, Kobe, Japan
e-mail: matsun@m4.dion.ne.jp

This section outlines the effects of intraoperative factors, type of prosthesis, implant design, surgical technique, and preoperative factors on soft tissue balance; the effects of soft tissue balance on postoperative ROM and clinical score; and the effects of knee joint stability. Detailed evaluation methods for soft-tissue balance, including the types of tensors, measurement methods, and measurement conditions, are described in a separate chapter (Chap. 8, Sect. 4).

2 Effects of Intraoperative Factors, Type of Prosthesis, Implant Design, and Surgical Techniques on Soft Tissue Balance

2.1 Intraoperative Factors

Firstly, intraoperative measurements reflecting the soft tissue balance of the postoperative knee, that is, measurements of PF joint reduction and femoral components in position, are essential, and these effects have been reported. In an intraoperative analysis of posterior stabilized TKA (PS-TKA) using a measured resection technique for 30 medial osteoarthritis (OA) knees, it has been reported that the component gap (CG) increases with flexion when the PF joint is evertd and decreases with deep flexion when the PF joint is reduced and provisionally repaired [1]. In an examination of 20 PS-TKA knees, this phenomenon was similarly reproduced, and the tension of the extension mechanism, including the patellar tendon, was affected [2]. However, because the patella cannot be everted during minimally invasive surgery (MIS), the CG decreases compared to the reduced or provisionally repaired state in the flexion position; therefore, care must be taken [3].

Recently, it has become possible to measure soft tissue balance using navigation systems. The correlation between the values obtained with navigation and those obtained with tensors showed a significant correlation between the two measurements in extension and flexion. However, the

S. Matsuda (ed.), *Advances in Total Knee Arthroplasty*, https://doi.org/10.1007/978-981-97-4920-1_44

correlation coefficients in the condition with the PF joint repaired or temporarily repaired were higher than with the PF joint everted [4]. Regarding the effect of femoral component placement, in the examination of 80 knees with PS-TKAs based on the measured resection method, the CG was significantly smaller than with the osteotomy gap (OG) corrected in the extension position, and the valgus balance (valgus angle) was also significantly smaller. This is thought to be due to the tension of the posterior capsule caused by the overhang of the posterior condyle of the femoral component and the reduction of the lateral gap caused by the externally rotated femoral component [5]. When a uniquely fabricated femoral trial component 4 mm larger was installed, a reduction in the medial/lateral extension gap of 2.1/2.8 mm was observed [6], In 259 knees with PS-TKAs, a component-induced decrease in the extension gap was also observed, and it was reported that a 2 mm larger OG in extension was necessary to create an equalized rectangular gap [7]. Similarly, a study analyzing the relationship between the postoperative posterior condylar offset (PCO) and intraoperative soft tissue balance in PS-TKA showed a negative correlation between CG (0°), CG difference (10–0°), and PCO. It has also been reported that protrusion of the posterior condyle of the femoral component is also affected [8]. Thus, it has been shown that it is important to position femoral components and reduce and temporarily repair PF joints when measuring intraoperative soft tissue balance.

The joint distraction force during intraoperative soft tissue balance measurement is extremely important. In PS-TKA for medial compartment OA knees, both the CG and varus/valgus balance increased with increasing joint distraction forces of 20, 40, and 60 lb. However, CG and the varus/valgus balance difference (90–0°) were not affected by the joint distraction force, and it has been reported that the varus/valgus balance difference (90–0°) serves as an indicator for determining the external rotation angle during femoral bone cutting as a medial preserving gap technique that emphasizes medial stability [9].

2.2 Type of Prosthesis

There is also an effect of the prosthesis type on soft tissue balance. In a comparative study of intraoperative soft tissue balance between PS-TKA and cruciate-retaining TKA (CR-TKA) using the measured resection method for varus OA knees, the CG generally increased up to 10° flexion and decreased after increasing to 90° in PS-TKA, while CR-TKA tended to stabilize thereafter [10]. The lateral laxity tended to decrease in CR-TKA, while lateral laxity gradually increased in PS-TKA [11]. The CG and valgus balance (varus angle, lateral laxity) in PS-TKA were larger than those in CR-TKA in the flexion region, owing to the influence of the posterior cruciate ligament (PCL). Intraoperative soft tissue balance measurements of 70 CR-TKA and 70 PS-TKA patients using the medial preserving gap technique described above showed no significant differences in the medial gap from extension to deep flexion between CR-TKA and PS-TKA. PS-TKA and CR-TKA, using the medial-preserving gap technique, provide intraoperative medial stability [12].

2.3 Implant Design

Recent implant designs have also been reported to be related to the anterior condylar axis. When determining femoral rotation with reference to the posterior condylar axis, not only the conventional central reference but also a model that can refer to the medial compartment can be used. A study examining navigated PS-TKA in 57 knees with OA reported that a smaller CG and internal rotation of the tibia could be maintained in the flexion position in the medial reference group [13]. In addition, when selecting a femoral component, it is now possible to use a model that allows the selection of a narrow component with the same anterior/posterior diameter when the medial/lateral diameter overhangs compared with the anterior/posterior diameter. In a study examining PS-TKA in 30 varus OA knees to determine the effect of this difference on intraoperative soft tissue balance, CG and valgus balance (varus angle, lateral laxity) tended to be large [14].

2.4 Surgical Techniques

In addition, TKA surgical technique is also one of the factors that affect soft tissue balance. There are two types of osteotomy methods in TKAs: the measured resection method and the gap technique. In the tibia-first gap technique, the comparison between the soft tissue balance of CR-TKA after tibial osteotomy and the final value of the component placed after femoral osteotomy showed no significant difference between the reference value and the final value of the flexion position, indicating the advantage of predicting the final soft tissue balance before femoral osteotomy [15]. Furthermore, in the comparison of intraoperative soft tissue balance, PS-TKA was performed using the measured resection method and the gap technique, and the CG and valgus balance (varus angle) in PS-TKA were similarly larger than those in CR-TKA in the flexion range [16]. In analyses including CR-TKA, it has been reported that the acquisition rate of an equalized rectangular gap in extension and flexion is superior to measured resection with the gap technique and superior to PS-TKA with CR-TKA [17]. Furthermore, in a comparison between 127 knees treated using the medial preserving gap technique and 148 knees treated using the measured resection technique that underwent PS-TKA for medial OA, the medial preserving gap technique showed significantly less change in the mediolateral joint gap from extension to deep flexion. With the use of the medial preserving gap technique in PS-TKA, a more stable joint gap could be achieved [18]. When assessing postoperative knee joint stability using stress radiographs, the medial-preserving gap technique was significantly better than the measured resection technique regarding medial stability during extension and flexion from 6 months to 3 years postoperatively [19]. The impact of MIS must also be considered. In a comparison of intraoperative soft tissue balancing in TKA using the quad-sparing (MIS) and medial parapatellar approaches, an increase in CG and an increase in varus balance in the flexion region was observed in the MIS group throughout the ROMs. It has been reported that this may be affected by procedural difficulties due to an insufficient visual field [20].

3 Effects of Preoperative Factors on Soft Tissue Balance

Preoperative factors, particularly the degree of knee deformity before surgery, patellar height, and tibial posterior tilt, have also been reported to affect intraoperative soft tissue balance.

Regarding the degree of knee deformity before surgery, it was reported that there was no significant difference in the full ROM between the three CG groups, which were categorized based on the degree of varus deformity (< 10°, 11° to <20°, and ≥ 21°). In contrast, the varus balance (medial tilt) of the medial-lateral balance was significantly greater at each flexion angle in the group with a varus deformity of ≥21°. In TKA, where excessive medial release is avoided to maintain the joint line, patients with severe varus deformities tended to have a residual varus balance in the lax lateral supporting tissues [21]. Similarly, in the measurement of the medial-lateral compartment gap in 90 knees with varus OA, it was reported that there was no significant difference in the medial gap between the three groups categorized by the degree of varus deformity (< 10°, 11° to <20°, and ≥ 21°), whereas the lateral gap was significantly larger in the group with a varus deformity of ≥21° [22].

Regarding patellar height, CR-TKA showed no correlation between the Insall–Salvati index (ISI) and CG at each flexion angle, and PS-TKA showed a positive correlation between ISI and CG at 90° flexion and 135° flexion. In PS-TKA, it has been reported that preoperative evaluation of the position of the patella is an indicator of appropriate soft tissue balance [23, 24].

Regarding tibial posterior tilt, CR-TKA showed no correlation between the tibial posterior tilt and CG at each flexion angle, whereas PS-TKA showed a positive correlation between the CG at the tibial posterior tilt and flexion of 90° and the CG difference (90–0°, 135–0°) at 90° flexion and 135° flexion. It has been suggested that the flexion-extension gap may increase due to an increase in the tibial posterior tilt in PS-TKA [25].

4 Effect of Soft Tissue Balance on Postoperative ROM

Factors influencing postoperative ROM include patient factors, artificial joint design, surgical techniques, postoperative rehabilitation, and soft tissue. It is necessary to understand CR-TKA- and PS-TKA regarding soft tissue balance and kinematics, which affect the postoperative extension and flexion angles.

4.1 Extension Angle

Analysis of PS-TKA for 75 knees with varus OA showed that the CG in extension correlated with flexion contractures at 1 year postoperatively. The CG in extension with a 40 lbs. joint distraction force required a laxity of 1 mm or more with the insert in mind [26]. Similarly, PS-TKA in 73 knees showed a positive correlation with the postoperative active extension angle and with both the CG and medial compartment gap in extension [27].

4.2 Flexion Angle

In an one-year postoperative stress radiography study of 40 CR-TKAs and 40 PS-TKAs, it was reported that the flexion ROM was particularly improved in the balanced group within 2° of the varus balance difference [28] and that attention should be given to PCL tension to obtain a good ROM in CR-TKAs [29]. In contrast, in a study of 85 PS-TKAs, no association was observed in the balanced group with a varus/valgus balance difference of ≤2° in extension, but a good ROM was obtained in the rectangular gap group in flexion achieved [30], In 80% of all cases, there was no correlation between varus/valgus balance and flexion angle in balanced TKA with a varus/valgus balance difference of ≤3° in extension [31]. In addition, in the examination of optimal tension between intraoperative osteotomy surfaces in 504 PS-TKAs and varus/valgus balances in 298 PS-TKAs, there was no correlation between optimal tension for extension and flexion, the varus/valgus balance in flexion, and the postoperative flexion angle [32, 33]. Therefore, varus/valgus and extension/flexion balances are important.

In contrast, in a study of OG measurement comparing CR-TKA in 31 mobile-bearing knees and 45 fixed-bearing knees, a positive correlation was observed between the CG difference (90–0°) and postoperative flexion angle only in fixed-bearing knees [34]. Intraoperative soft tissue balance analysis of CR-TKA for 30 varus OA knees showed a positive correlation between the preoperative flexion angle, CG (90°, 120°), CG difference (90–0°, 120–0°), and postoperative flexion angle, and CR-TKA was required to avoid tight flexion gaps to obtain the flexion angle [35]. Analysis of the 25 PS-TKAs showed a negative correlation between CG (90°), CG difference (90–0°), and postoperative flexion angle, and it has been reported that it is important to obtain a moderate component gap difference so that the flexion gap does not become too loose in PS-TKA [36]. In contrast, in a study of intraoperative measurements of 246 PS-TKAs, the CG difference (90–0°) and CG difference (120–0°) were factors that affected the postoperative flexion angle, and it was important to secure the gap in the deep flexion position rather than in the mid-range flexion position [37]. This difference is thought to be due to the patient population and surgical technique. It is important to obtain the postoperative flexion angle by not making the flexion gap too loose and the deep flexion gap not becoming too tight in PS-TKAs.

In recent years, it has become possible to measure the intraoperative tibial anteroposterior and internal rotation movements using a navigation system, and its effects on soft tissue balance have been reported. In an analysis of intraoperative soft tissue balance and tibial anterior translation/tibial internal rotation, positive correlations were found between varus balance at 60° and 90° knee flexion and between CG at 90° knee flexion and tibial internal rotation [38]. In the same series of analyses of 58 navigated CR-TKAs, a positive correlation was reported between the postoperative flexion angle and tibial internal rotation [39]. In addition, a positive correlation was found between varus balance (varus angle, lateral laxity) and the lateral compartment gap at 90° knee flexion and the postoperative flexion angle in 55 navigated CR-TKAs, [40], suggesting that a certain degree of lateral laxity in the flexion range induces internal rotation of the tibia, which is advantageous for the acquisition of the postoperative flexion angle.

This phenomenon was also reported in an analysis of flexion balance and postoperative flexion angles using postoperative stress imaging in 41 CR-TKA cases, which showed a similar positive correlation [41]. A recent intraoperative analysis of guided-motion bicruciate-stabilized TKA in 95 knees similarly showed a relationship between tibial internal rotation in the flexion position and the medial gap [42]. In a study comparing the minimal medial dissection group in which only osteophytes were resected with the medial dissection group up to the semimembranosus tendon in navigated CR-TKA for 50 varus OA knees, it was reported that the minimal medial dissection group had an advantage in achieving a high postoperative flexion angle by maintaining the tibial internal rotation [43].

In an analysis of navigated PS-TKAs of 54 varus OA knees, tibial rotation patterns were classified into four types, of which the group showing internal rotation movement during deep flexion was superior to the other groups [44]. In a study of 100 PS-TKA cases, preoperative internal rotation deformities and flexion contractures were suggested to be involved as factors affecting postoperative flexion ROM [45]. In a study of 25 PS-TKAs, the tibial internal rotation decreased after PCL dissection and further decreased after component placement; the smaller the amount of change, the more favorable it was for obtaining a postoperative flexion angle [46]. In an analysis of 55 other navigated PS-TKAs, a negative correlation was found between the medial compartment gap and tibial internal rotation at 60° and 90° flexion, and a positive correlation between tibial internal rotation and postoperative flexion angle, suggesting that medial stability (lateral laxity) in the flexed position and tibial internal rotation motion are important for obtaining the postoperative flexion angle, as in CR-TKA [47]. Table 44.1 shows the relationship between these soft tissue balance evaluations and postoperative flexion angles.

Although it is difficult to summarize all these reports owing to differences in surgical techniques, measurement methods, measurement conditions, and prosthesis types, the following can be said to have led to the acquisition of a good flexion ROM.

1. Both CR-TKA and PS-TKA should aim for equal balance in extension/flexion and varus/valgus rotation.
2. Reproduction of physiological knee kinematics.
3. Maintaining medial stability while allowing moderate lateral laxity.

Table 44.1 Studies reporting relationship between these soft tissue balance evaluations and postoperative flexion angles

Authors	Number of cases CR/PS	Evaluation methods	Results
Matsuda [28]	40 CR/40 PS	Stress X-ray Varus/valgus balance (extension)	Better flexion angle in balanced knee (extension) (Varus/valgus difference≦2°, 6°≦total laxity≦10°)
Tsukeoka [30]	85 PS	Stress X-ray Varus/valgus balance (extension, flexion)	Better flexion angle in rectangular gap (flexion) No difference even in balanced knee (extension)(varus/valgus difference≦2°)
Ishii [31]	82 CR/122 PS	Stress X-ray Varus/valgus balance (extension)	No correlation between Varus/valgus/varus+valgus balance and flexion angle(balanced knee 80%)
Tsukada [32, 33]	504 PS 298 PS	Intraoperative tensor Appropriate tension (extension, flexion) Intraoperative tensor Varus/valgus balance (extension, flexion)	No difference between tension (90°) and (90–0°) and flexion angle Np difference between lateral laxity (90°) and flexion angle
Higuchi [34]	31 MB CR/45 FB CR	Intraoperative tensor Osteotomy gap, Varus/valgus balance	Positive correlation between OG(90–0°) and flexion angle only in FB group
Takayama [35]	30 CR	Intraoperative tensor Component gap, Varus/valgus balance	Positive correlation with flexion angle: CG90°, CG120°, CG(90–0°), CG(120–0°)
Matsumoto [36]	25 PS	Intraoperative tensor Component gap, Varus/valgus balance	Negative correlation with flexion angle: CG90°, CG(90–0°)
Minoda [37]	246 PS	Intraoperative tensor Component gap, Varus/valgus balance	Positive correlation with flexion angle: CG(90–0°), CG(120–0°)
Matsuzaki [39]	58 CR	Navigation Component gap, Varus/valgus balance	Positive correlation between flexion angle and IR (120–60°)
Nakano [40]	55 CR	Intraoperative tensor Component gap, Varus/valgus balance	Positive correlation with flexion angle: VA90°, LCG90°
Kobayashi [41]	41 CR	Stress X-ray Varus/valgus balance (extension, flexion)	Positive correlation between flexion angle and VA90°
Ishida [44]	54 PS	Navigation Component gap, Varus/valgus balance	Better flexion angle in IR pattern during deep flexion
Kamenaga [47]	55 PS	Intraoperative tensor, navigation Component gap, Varus/valgus balance	Negative correlation with IR(120–60°): MCG60°, MCG90° Positive correlation between flexion angle and IR(120–60°)

5 Effect of Intraoperative Soft Tissue Balance on Postoperative Clinical Score

There are few reports on the relationship between intraoperative soft tissue balance and postoperative clinical scores. In the 2-year postoperative evaluation of CR-TKA and PS-TKA in the four groups performed using the measured resection method and the gap technique, respectively, despite the differences in intraoperative soft tissue balances between the four groups, there was no significant difference among the four groups in the Knee Society Score (KSS), Knee Score and Function Score, [17].

However, in recent years, various evaluation methods focusing on patient-based evaluation, including patient satisfaction, have attracted attention, and there have been several reports on their relationships with clinical scores. In a study of CR-TKA in 25 OA knees, each score of the 2011 KSS assessment 1 year after surgery, including patient satisfaction, showed positive correlations with CG difference (90–0°) and CG (120–0°), suggesting that securing the component gap in flexion with respect to extension is important for improving patient satisfaction [48]. Similarly, a CR-TKA study reported that a moderate CG difference (90–0°) was advantageous for improving the Western Ontario and McMaster Universities Osteoarthritis Index (WOMAC) functional score [49]. Furthermore, a study of 55 CR-TKAs

Table 44.2 Studies reporting relationship between these soft tissue balance evaluations and postoperative clinical scores

Authors	Number of cases CR/PS	Evaluation methods	Results
Matsumoto [48]	35 CR	Intraoperative tensor Component gap, Varus/valgus balance	Positive correlation with objective indicator, symptom, satisfaction and expectation in 2011 KSS: CG(90–0°), CG(120–0°)
Fujimoto [49]	30 CR	Intraoperative tensor Component gap	Better WOMAC in CG(90–0°)≧3 mm
Kamenaga [50]	55 CR	Intraoperative tensor Component gap, Varus/valgus balance	Negative correlation between MCG(0°) and satisfaction/FJS Negative correlation with MCG(90°): satisfaction, functional activity in 2011KSS and FJS
Nakahara [51]	94 PS	Stress X-ray Varus/valgus balance (extension)	No correlation between varus/valgus/varus-valgus balance and 2011KSS
Tsukiyama [52]	50 PS	Stress X-ray Varus/valgus balance (extension, flexion)	Negative correlation with medial laxity(90°): Objective indicator, symptom, satisfaction, and functional activity in 2011KSS
Azukizawa [53]	40 PS	Intraoperative tensor Component gap, Varus/valgus balance	Worse 2011KSS in medial laxity (0°) ≧4 mm
Watanabe [54]	40 PS	Stress X-ray Varus/valgus balance (extension, flexion)	Negative correlation with medial laxity (80°): satisfaction, expectation in 2011KSS
Tanaka [55]	100 PS	Intraoperative tensor Component gap, Varus/valgus balance	No correlation between 2011KSS and parameters including CG and varus/valgus balance

also reported a negative correlation between patient satisfaction in the 2011 KSS and the Forgotten Joint score and medial compartment gap in extension and flexion, which considered inserts [50].

A study of 94 PS-TKAs found no correlation between varus balance on stress radiographs in extension and postoperative 2011 KSS endpoints [51]. A study of 50 PS-TKAs assessing stress radiographs in flexion found a negative association between medial rather than lateral laxity in flexion and each sub-score of the 2011 KSS endpoints, including postoperative satisfaction [52]. A study of intraoperative soft tissue balance measurements of 46 PS-TKAs showed a negative correlation between the 2011 KSS, including patient satisfaction and medial laxity in flexion [53], and a similarly good functional 40 PS-TKA study showed a negative correlation between laxity and clinical scores [54]. Furthermore, in a study of 100 PS-TKAs without substantial medial dissection, intraoperative soft tissue balance did not affect the 2011 KSS [55], which can be interpreted as indicating that a certain degree of lateral laxity does not affect postoperative results, as this tendency is masked in cases in which medial stability is achieved.

A recent study of 43 guided-motion bicruciate-retaining TKAs showed that varus/valgus laxity in flexion was positively correlated with the 2011 KSS, including patient satisfaction [56], which may reflect the tight nature of the implant.

Table 44.2 reports on the relationship between these soft tissue balance assessments and postoperative clinical scores.

These points lead to good postoperative clinical results.

1. On the premise that both CR-TKA and PS-TKA aim for varus/valgus and extension/flexion balance,
2. Try to maintain medial stability with moderate lateral laxity.

6 Postoperative Changes in Knee Joint Stability

The surgeon performs intraoperative measurements based on the assumption that the intraoperative soft tissue balance is maintained to some extent after surgery. In the evaluation of varus/valgus balance in 113 navigated TKAs, a correlation was reported between balance under joint separation forces and balance after installation and closure of all final components, and the correlation was observed with stronger joint separation forces and lower BMIs [57], suggesting the importance of intraoperative evaluation. This has been important for a long time, and a study of 53 CR-TKAs and 24 PS-TKAs 5 years after surgery showed no difference between the two groups, and lateral laxity remained in both groups [58]. In addition, the intraoperative valgus balance was maintained in both CR-TKA and PS-TKA during the intraoperative evaluation and at 6 months postoperatively [59]. Based on this recognition, PS-TKA, which allowed lateral laxity in 71 knees with varus OA, showed that the medial gap in extension did not change over time up to 1 year after surgery, whereas the lateral gap converged to some extent 3 months after surgery [60]. This report confirms that medial laxity increases after surgery in the intraoperative balance evaluation performed under the same conditions; that is, the relative lateral laxity decreases [61].

In contrast, in a study that evaluated varus/valgus balances postoperative stress radiography for 30 CR-TKAs and 30 PS-TKAs, CR-TKA and PS-TKA did not differ over time at 6 months, 1 year, and 2 years after surgery [62], varus/valgus balance in 121 CR TKAs did not change between 1 and 5 years after surgery [63], and 49 knees showed that lateral laxity in extension and flexion remained even after 10 years [64]. In a study investigating the relationship between intraoperative soft tissue balance and knee stability rating by stress radiography 5 years after surgery in CR-TKA and PS-TKA, CG and valgus balance (varus angle) in CR-TKA correlated with intraoperative measurements in both extension and flexion and only the extension position correlated in PS-TKA [65].

Interestingly, it has been reported that intraoperative evaluation of varus/valgus balance in flexion showed a stronger correlation with postoperative evaluation in the kneeling condition than in the non-weight-bearing condition and that intraoperative evaluation under joint distraction force reflects the functional knee state rather than the non-weight bearing condition [66]. Although these reports show some differences owing to different measurement conditions, it is well known that intraoperative evaluation is extremely important for predicting and acquiring stability.

7 Future Prospects

With the aim of improving the postoperative outcomes of TKAs, we will continue to examine the relationship between soft tissue balance and preoperative and intraoperative factors and their influence on clinical outcomes from various perspectives. Therefore, it is essential to accumulate evidence using a unified intraoperative soft tissue balance evaluation method. Examining the influence of soft tissue balance, as described in this chapter, is expected to contribute to the improvement of surgeons' skills and the development of surgical equipment, which in turn will lead to improvements in postoperative outcomes, including further patient satisfaction.

References

1. Matsumoto T, et al. Joint gap kinematics in posterior-stabilized total knee arthroplasty measured by a new tensor with the navigation system. J Biomech Eng. 2006;128:867–71.
2. Gejo R, et al. Joint gap changes with patellar tendon strain and patellar position during TKA. Clin Orthop Relat Res. 2008;466:946–51.
3. Oka S, et al. The influence of patellar orientation on soft tissue balance in minimally incision total knee arthroplasty. Knee Surg Sports Traumatol Arthrosc. 2012;20:1064–8.
4. Matsumoto T, et al. Soft tissue balance measurement in posterior-stabilized total knee arthroplasty with a navigation system. J Arthroplast. 2009;24:358–64.
5. Muratsu H, et al. Femoral component placement changes soft tissue balance in posterior-stabilized total knee arthroplasty. Clin Biomech (Bristol, Avon). 2010;25:926–30.
6. Mitsuyasu H, et al. Enlarged postoperative posterior condyle tightens extension gap in total knee arthroplasty. J Bone Joint Surg Br. 2011;93:1210–6.
7. Minoda Y, et al. Decreased extension gap and valgus alignment after implantation of total knee prosthesis in primary varus knees. Knee Surg Sports Traumatol Arthrosc. 2016;24:3642–7.
8. Tsubosaka M, et al. Posterior condylar offset influences the intraoperative soft tissue balance during posterior-stabilized total knee arthroplasty. J Orthop Sci. 2017;22:1071–6.
9. Nagai K, et al. The influence of joint distraction force on the soft-tissue balance using modified gap-balancing technique in posterior-stabilized total knee arthroplasty. J Arthroplast. 2017;32:2995–9.
10. Matsumoto T, et al. The intraoperative joint gap in cruciate-retaining compared with posterior-stabilised total knee replacement. J Bone Joint Surg Br. 2009;91:475–80.
11. Matsumoto T, et al. Soft-tissue tension in cruciate-retaining and posterior-stabilized total knee arthroplasty. J Arthroplast. 2011;26:788–95.
12. Tsubosaka M, et al. Comparison of intraoperative soft tissue balance between cruciate-retaining and posterior-stabilized total knee arthroplasty performed by a newly developed medial preserving gap technique. J Arthroplast. 2018;33:729–34.
13. Ishida K, et al. Posterior reference position affects intraoperative kinematic and soft tissue balance in navigated posterior-stabilized total knee arthroplasty. J Arthroplast. 2018;33:2851–7.
14. Ishida K, et al. Influence of narrow femoral implants on intraoperative soft tissue balance in posterior-stabilized total knee arthroplasty. J Arthroplast. 2020;35:388–93.
15. Matsumoto T, et al. Soft tissue balance using tibia first gap technique with navigation system in cruciate-retaining total knee arthroplasty. Int Orthop. 2012;36:975–80.
16. Matsumoto T, et al. Different pattern in gap balancing between the cruciate-retaining and posterior-stabilized total knee arthroplasty. Knee Surg Sports Traumatol Arthrosc. 2013;21:2338–45.
17. Matsumoto T, et al. Soft tissue balancing in total knee arthroplasty: cruciate-retaining versus posterior-stabilized, and measured resection versus gap technique. Int Orthop. 2014;38:531–7.
18. Nagai K, et al. Intraoperative soft tissue balance using novel medial preserving gap technique in posterior-stabilized total knee arthroplasty: comparison to measured resection technique. Knee Surg Sports Traumatol Arthrosc. 2018;26:3474–81.
19. Tsubosaka M, et al. Knee stability following posterior-stabilized total knee arthroplasty: comparison of medial preserving gap technique and measured resection technique. J Knee Surg. 2021;36:95–104. https://doi.org/10.1055/s-0041-1729968. Online ahead of print
20. Matsumoto T, et al. Soft tissue balance measurement in minimal incision surgery compared to conventional total knee arthroplasty. Knee Surg Sports Traumatol Arthrosc. 2011;19:880–6.
21. Matsumoto T, et al. The influence of preoperative deformity on intraoperative soft tissue balance in posterior-stabilized total knee arthroplasty. J Arthroplast. 2011;26:1291–8.
22. Okamoto S, et al. Lateral soft tissue laxity increases but medial laxity does not contract with varus deformity in total knee arthroplasty. Clin Orthop Relat Res. 2013;471:1334–42.
23. Sasaki H, et al. The influence of patella height on intraoperative soft tissue balance in posterior-stabilized total knee arthroplasty. Knee Surg Sports Traumatol Arthrosc. 2012;20:2191–6.
24. Nishizawa Y, et al. The influence of the patella height on soft tissue balance in cruciate-retaining and posterior-stabilized total knee arthroplasty. Int Orthop. 2013;37:421–5.

25. Oka S, et al. The influence of the tibial slope on the intraoperative soft tissue balance in cruciate-retaining and posterior-stabilized total knee arthroplasty. Knee Surg Sports Traumatol Arthrosc. 2014;22:1812–8.
26. Okamoto S, et al. Extension gap needs more than 1-mm laxity after implantation to avoid postoperative flexion contracture in total knee arthroplasty. Knee Surg Sports Traumatol Arthrosc. 2014;22:3174–80.
27. Nagai K, et al. Influence of intraoperative soft tissue balance on postoperative active knee extension in posterior-stabilized total knee arthroplasty. J Arthroplast. 2015;30:1155–9.
28. Matsuda Y, et al. Varus-valgus balance and range of movement after total knee arthroplasty. J Bone Joint Surg Br. 2005;87:804–8.
29. Matsuda Y, et al. Effect of flexion angle on coronal laxity in patients with mobile-bearing total knee arthroplasty prosthesis. J Orthop Sci. 2005;10:37–41.
30. Tsukeoka T, et al. The rectangular flexion gap is associated with increased knee flexion angle in a cruciate-sacrificing rotating platform mobile-bearing total knee arthroplasty. J Orthop Sci. 2017;22:313–7.
31. Ishii Y, et al. Mediolateral coronal laxity does not correlate with knee range of motion after total knee arthroplasty. Arch Orthop Trauma Surg. 2019;139:851–8.
32. Tsukada S, et al. Intraoperative soft tissue tension and postoperative range of motion in posterior stabilized total knee arthroplasty. Arch Orthop Trauma Surg. 2013;133:243–51.
33. Tsukada S, et al. Impact of soft tissue imbalance on knee flexion angle after posterior stabilized total knee arthroplasty. J Arthroplast. 2017;32:2399–403.
34. Higuchi H, et al. Relationship between joint gap difference and range of motion in total knee arthroplasty: a prospective randomized study between different platforms. Int Orthop. 2009;33:997–1000.
35. Takayama K, et al. Influence of intra-operative joint gap on postoperative flexion angle in cruciate-retaining total knee arthroplasty. Knee Surg Sports Traumatol Arthrosc. 2012;20:532–7.
36. Matsumoto T, et al. Influence of intraoperative joint gap on post-operative flexion angle in osteoarthritis patients undergoing posterior-stabilized total knee arthroplasty. Knee Surg Sports Traumatol Arthrosc. 2007;15:1013–8.
37. Minoda Y, et al. Joint gap in mid-flexion is not a predictor of postoperative flexion angle after total knee arthroplasty. J Arthroplast. 2018;33:735–9.
38. Matsuzaki T, et al. Tibial international rotation is affected by lateral laxity in cruciate-retaining total knee arthroplasty: an intraoperative kinematic study using a navigation system and offset-type tensor. Knee Surg Sports Traumatol Arthrosc. 2014;22:615–20.
39. Matsuzaki T, et al. Kinematic factors affecting postoperative knee flexion after cruciate-retaining total knee arthroplasty. Int Orthop. 2013;37:803–8.
40. Nakano N, et al. Postoperative knee flexion angle is affected by lateral laxity in cruciate-retaining total knee arthroplasty. J Arthroplast. 2016;31:401–5.
41. Kobayashi T, et al. Lateral laxity in flexion increases the postoperative flexion angle in cruciate-retaining total knee arthroplasty. J Arthroplast. 2016;27:260–5.
42. Inui H, et al. The relationship between soft-tissue balance and intraoperative kinematics of guided motion total knee arthroplasty. J Knee Surg. 2019;32:91–6.
43. Matsumoto T, et al. Semimembranosus release reduces Tibial internal rotation and flexion angle in cruciate-retaining total knee arthroplasty. J Arthroplast. 2015;30:1537–41.
44. Ishida K, et al. Navigation-based femorotibial rotation pattern correlated with flexion angle after total knee arthroplasty. Knee Surg Sports Traumatol Arthrosc. 2016;24:89–95.
45. Ishida K, et al. Factors affecting intraoperative kinematic patterns and flexion angles in navigated total knee arthroplasty. Knee Surg Sports Traumatol Arthrosc. 2015;23:1741–7.
46. Ishida K, et al. Tibiofemoral rotational alignment affects flexion angles in navigated posterior-stabilized total knee arthroplasty. Knee Surg Sports Traumatol Arthrosc. 2018;26:1532–9.
47. Kamenaga T, et al. Medial knee stability at flexion increases the tibial internal rotation and knee flexion angle after posterior-stabilized total knee arthroplasty. Clin Biomech (Bristol, Avon). 2019;68:16–22.
48. Matsumoto T, et al. Relative loose flexion gap improves patient-reported clinical scores in cruciate-retaining total knee arthroplasty. J Knee Surg. 2018;31:573–9.
49. Fujimoto E, et al. Intraoperative gaps affect outcome and postoperative kinematics in vivo following cruciate-retaining total knee arthroplasty. Int Orthop. 2016;40:41–9.
50. Nakahara H, et al. Does knee stability in the coronal plane in extension affect function and outcome after total knee arthroplasty? Knee Surg Sports Traumatol Arthrosc. 2015;23:1693–8.
51. Tsukiyama H, et al. Medial rather than lateral knee instability correlates with inferior patient satisfaction and knee function after total knee arthroplasty. Knee. 2017;24:1478–84.
52. Azukizawa M, et al. Intraoperative medial joint laxity in flexion decreases patient satisfaction after total knee arthroplasty. Arch Orthop Trauma Surg. 2018;138:1143–50.
53. Kamenaga T, et al. The influence of postoperative knee stability on patient satisfaction in cruciate-retaining total knee arthroplasty. J Arthroplast. 2018;33:2475–9.
54. Watanabe T, et al. Coronal and sagittal laxity affects clinical outcomes in posterior-stabilized total knee arthroplasty: assessment of well-functioning knees. Knee Surg Sports Traumatol Arthrosc. 2020;28:1400–9.
55. Tanaka Y, et al. Intraoperative physiological lateral laxity in extension and flexion for varus knees did not affect short-term clinical outcomes and patient satisfaction. Knee Surg Sports Traumatol Arthrosc. 2020;28:3888–98.
56. Kaneko T, et al. Greater postoperative relatively medial loose gap at 90°of flexion for varus knees improves patient-reported outcome measurements in anatomical bi-cruciate retaining total knee arthroplasty. Knee. 2020;27:1534–41.
57. Yamagami R, et al. Navigation-based analysis of associations between intraoperative joint gap and mediolateral laxity in total knee arthroplasty. Knee. 2021;30:314–21.
58. Ishii Y, et al. Coronal laxity in extension in vivo after total knee arthroplasty. J Orthop Sci. 2003;8:538–42.
59. Ishii Y, et al. Effect of soft tissue tension on measurements of coronal laxity in mobile-bearing total knee arthroplasty. J Orthop Sci. 2005;10:496–500.
60. Sekiya H, et al. Postoperative lateral ligamentous laxity diminishes with time after TKA in the varus knee. Clin Orthop Relat Res. 2009;467:1582–6.
61. Tsukeoka T, et al. Residual medial tightness in extension is corrected spontaneously after total knee arthroplasty in varus knees. Knee Surg Sports Traumatol Arthrosc. 2019;27:692–7.
62. Takeda M, et al. Changes in varus-valgus laxity after total knee arthroplasty over time. Knee Surg Sports Traumatol Arthrosc. 2012;20:1988–93.
63. Hatayama K, et al. Does residual Varus alignment cause increasing Varus laxity at a minimum of five years after total knee arthroplasty? J Arthroplast. 2017;32:1808–13.
64. Yoshihara Y, et al. Assessing coronal laxity in extension and flexion at a minimum of 10 years after primary total knee arthroplasty. Knee Surg Sports Traumatol Arthrosc. 2016;24:2512–6.
65. Matsumoto T, et al. Intraoperative soft tissue balance reflects minimum 5-year midterm outcomes in cruciate-retaining and posterior-stabilized total knee arthroplasty. J Arthroplast. 2012;27:1723–30.
66. Moro-oka TA, et al. Modified gap-balancing technique in total knee arthroplasty: evaluation of the post-operative coronal laxity. Knee Surg Sports Traumatol Arthrosc. 2010;18:375–80.

Soft Tissue Balance/Joint Stability: Anteroposterior Stability

45

Yoshinori Ishii

1 Introduction: Anteroposterior Stability

Adequate anteroposterior stability plays a crucial role in total knee arthroplasty (TKA), as it enables rollback or backward sliding of the femoral component associated with flexion, which theoretically improves range of motion (ROM). The cruciate retaining (CR) design aims to prevent femur impingement with the posterior part of the tibia in early flexion by using the posterior cruciate ligament (PCL) tension with knee flexion to induce rollback. Conversely, the cruciate substituting (CS) design aims to prevent early impingement by preventing excessive roll forward or forward sliding through the post-cam mechanism. Recent designs, such as bicruciate-retaining and bicruciate-stabilizing designs, also rely on the movement of the remaining ligaments or their compensatory shapes to prevent impingement between the femur and tibia during early flexion. Therefore, good clinical results, especially large knee flexion angles, are expected owing to moderate anteroposterior stability after TKA, regardless of the design. An adequate flexion angle not only benefits the activities of the daily life of the patient but also contributes to increased satisfaction [1]. Therefore, the surgeon must understand the optimal stability requirements associated with the chosen designs.

Several studies have highlighted the importance of achieving optimal postoperative anteroposterior stability after TKA [2–24]. Several factors have been identified as influencing stability, including design characteristics such as the constraint between the femoral and tibial components based on their geometry and the use of mobile or fixed tibial inserts. Other factors include the soft tissue composition, the degree of soft tissue tension with knee position changes, and the load on the knee during measurement. Various instruments have been used to evaluate the system stability under consistent loading conditions [2, 8, 22], including Telos arthrometers [3, 4, 15, 20, 21], besides KT-1000 and KT-2000 devices [5–10, 12–17, 19, 23, 24].

In this chapter, we extensively discuss anteroposterior stability following TKA, incorporating the findings from biomechanical analyses and highlighting our good clinical cases. Furthermore, we describe intraoperative procedures aimed at achieving optimal anteroposterior stability after TKA while considering the insights gained from previous studies.

Y. Ishii (✉)
Ishii Orthopaedic & Rehabilitation Clinic, Saitama, Japan
e-mail: ishii@sakitama.or.jp

2 Biomechanical Analysis

2.1 Static Stability

A KT2000 was used to assess stability, with anterior and posterior drawer forces of 133 N and 89 N, respectively. Since the initial contact position between the femur and tibia differed for each knee joint, stability was evaluated based on the total movement rather than separate forward and posterior movements.

2.1.1 Genesis I (Fixed Tibial Insert Design)

Among the 74 patients (82 knees) who achieved favorable clinical outcomes at a minimum five-year follow-up [25], those who consented to undergo measurement (CR vs. CS, 53 vs. 24 knees) were examined. The mean ± standard deviation (SD) measurements for CR at flexion angles of 30°/75° were 5.8 ± 2.9/4.8 ± 2.3 mm, while those for CS were 5.3 ± 3.2/3.4 ± 1.5 mm [6]. This design is semi-constrained and characterized by relatively low conformity between the components (Fig. 45.1a, b). The decreasing movement during flexion observed in the CR group was attributed to increased PCL tension. In a study by Matsuda et al. [14], a knee with less movement at 75° flexion compared with 30° flexion after TKA was defined as a functional PCL knee, whereas a knee with substantial movement at 75° flexion was classified as a non-functional PCL knee. An observational

S. Matsuda (ed.), *Advances in Total Knee Arthroplasty*, https://doi.org/10.1007/978-981-97-4920-1_45

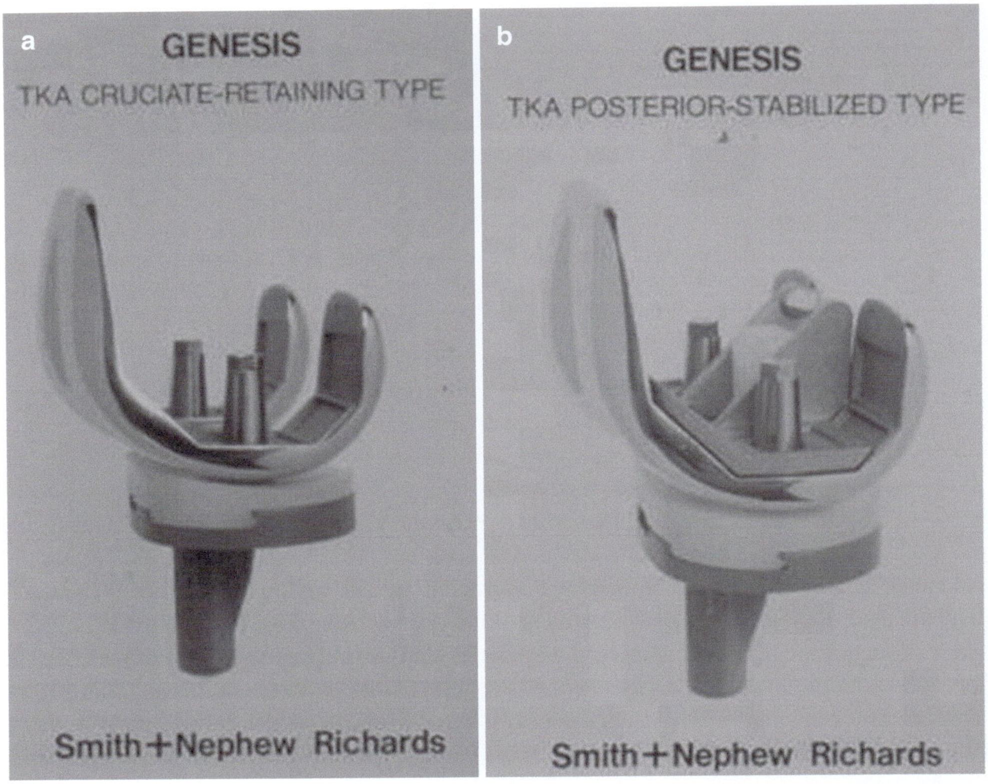

Fig. 45.1 Design characteristics of Genesis TKA System (Smith & Nephew, Memphis, TN, USA). This design is semi-constrained and characterized by relatively low conformity between the components. Anteroposterior movement was controlled by posterior cruciate ligament (**a**) CR type or post-cam mechanism (**b**) CS type

study using a magnetic resonance imaging reported that the PCL relaxes in extension and becomes tense at 90° flexion [26]. However, it was inferred that the decreasing movement during CS flexion was controlled by changes in the tension in the patellar tendon [27] and the medial and lateral ligaments [28] during flexion, as well as an increase in muscular defense (specifically the biceps femoris muscle) [29], in addition to the post-cam mechanism of PCL substitution. A TKA knee that achieved favorable clinical outcomes reflected the design concept accurately, with the optimal sagittal plane stability for CR and CS in this design falling within a range of 3–6 mm, indicative of good joint function.

2.1.2 LCS (Mobile Tibia Insert Design)

The study included 74 knee joints (one-side meniscal-bearing CR vs. the other-side rotating CS) from 37 cases of bilateral knee arthroplasty, demonstrating favorable results with an average follow-up period of approximately 10 years [7]. Based on median values, the results indicated that CR had a stability measurement of 11.8/11.7 mm, while CS had 7.1/7.9 mm at flexion angles of 30°/75°. In the low contact stress (LCS) design, CR had no constraint in the anteroposterior or the rotational direction (Fig. 45.2a), whereas CS had no constraint only in the rotational direction (Fig. 45.2b). Total contact and high constraint were observed between the components up to 30° of flexion, after which the curvature of the femur decreased with flexion, resulting in decreased constraint. By comparing the angles of each design, it can be inferred that CR showed similar stability due to a trade-off between geometric constraint and PCL relaxation at 30° (Fig. 45.2c) and low constraint with increased PCL tension at 75° (Fig. 45.2d). Although CS showed a higher stability value at 75° due to reduced geometric constraints, the changes in soft tissue tension, including tendons, ligaments, and muscles [27–29], likely played an essential role in the absence of a significant difference. However, when comparing the designs at each angle, the presence or absence of anteroposterior constraints was evident, with significantly larger values observed in the unconstrained anteroposterior direction of CR. Additionally, it can be inferred that the difference in stability between angles was greater at 30° when the PCL was not taut and significant at 75° when the PCL was tensed. The comparison within the same patients, minimizing anatomical biases between patients, demonstrated that anteroposterior stability after TKA was determined by the specific TKA design, as the design with higher constraints showed reduced movement and the lower constraints showed more movement. Therefore, surgeons need to understand the optimal anteroposterior stability inherent in TKA designs, as it plays a vital role in achieving favorable postoperative outcomes. In this design, the optimal sagittal plane stability range associated with good joint function was 7–12 mm.

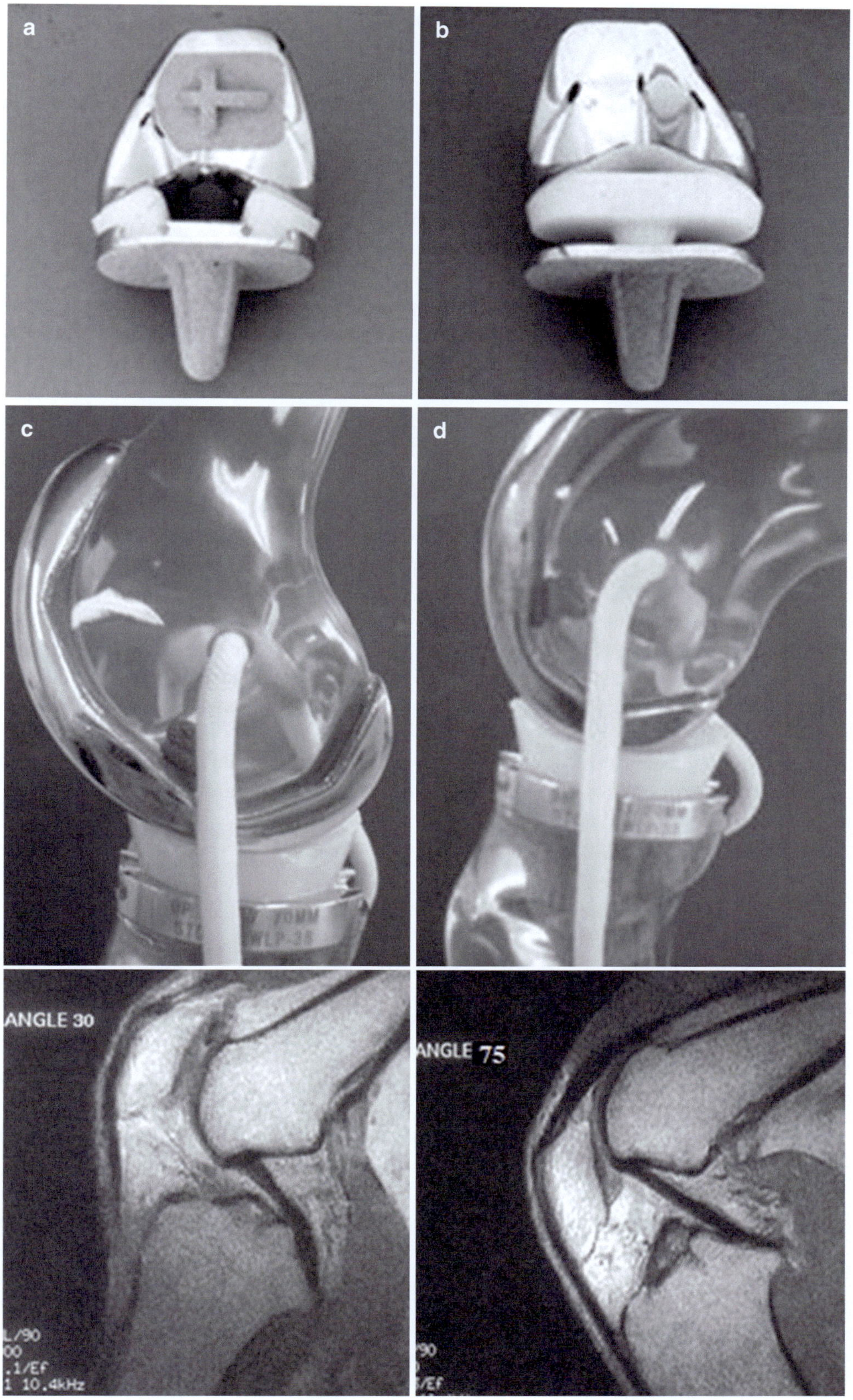

Fig. 45.2 Design characteristics of the LCS TKA System (DePuy, Warsaw, IN, USA). CR had no constraint in the anteroposterior or the rotational direction (**a**), whereas CS had no constraint only in the rotational direction (**b**). At 30° of flexion, the geometric constraint was high, but the PCL tension was low (**c**). At 75° of flexion, the geometric constraint was low, but the PCL tension was high (**d**)

2.2 Anteroposterior Movement Stability

Knee movement, including anteroposterior stability, is influenced by three key factors: joint shape, soft tissue composition, and soft tissue tension [30]. Therefore, it is crucial to consider analysis results that examine not only the variations in knee movement caused by different geometries [31, 32] but also the differences in soft tissue balance [33] and muscle tension [34] around the knee joint.

2.2.1 Genesis I (Smith & Nephew, Memphis, TN, USA) (Fixed Tibial Insert Type)

In patients who had undergone TKA and had no clinical complaints with an average of 2 years postoperatively, the rollback motion in the CR and CS designs was analyzed as anteroposterior stability during active knee motion in the sitting position. This analysis was conducted using a 6-degree-of-freedom electrical goniometer [31], and the results revealed that CR and CS had movements of 8.8 ± 10.9 mm and 10.5 ± 8.5 mm, respectively. In comparison, the normal control group exhibited a movement of 18.3 ± 8.8 mm, which was significantly different from the TKA group [31]. The study concluded that the disparity in movement, with TKA showing approximately half the amount compared to the normal knee, could be attributed to the differences in geometry and friction coefficient between the femur and tibia in TKA [31]. In gait analysis [35], the maximum anteroposterior movement values during the swing phase were evaluated to be 10.8 ± 8.6, 9.0 ± 5.2, and 1 ± 12.2 for the CR, CS, and normal control groups, respectively. The presence or absence of a load (axial force) on the knee joint strongly influenced anteroposterior stability [36]. Based on the observed distal movement between the components during the swinging phase, it was determined that there was no axial force applied between the components. Therefore, the difference between TKA and normal knees in terms of anteroposterior movement could be attributed to the presence or absence of ACLs, which control such movement, rather than the properties of the articular surface. Although in these two knee joint motion analyses, the stability observed in the normal control group was not fully reproduced in the TKA group, similar results were obtained regarding dynamic stability between the designs, aligning with favorable clinical results.

2.3 LCS (DePuy, Warsaw, IN, USA) (Tibial Insert Mobile Type)

Using fluoroscopic imaging analysis, we evaluated the amount of anteroposterior movement of the medial and lateral condyles during the rollback phenomenon occurring between maximum extension and flexion in the standing position in a patient with CR and no clinical complaints. The mean movement was observed to be 6.4 ± 1.7 and 6.2 ± 3.4 mm for the medial and lateral femoral condyles, respectively. Additionally, our study found that the anteroposterior and abduction or adduction static soft tissue balances, evaluated before motion analysis, had minimal impact on dynamic stability (knee movement) [33]. Three types of knee joint motion analyses were conducted to investigate the effect of differences in muscle tension on stability: active motion in the sitting position, standing motion, and stair climbing motion. These analyses involved patients with bilateral knee arthroplasty, with one side having CR and the other having CS designs [34]. In the CR design, the movements of the medial and lateral condyles were measured as 6/6, 2/1, and 5/3 mm, respectively, across the three load conditions. No significant differences were observed among the three load conditions, with measurements ranging from 3–4/2–4 mm in the CS design. Meanwhile, in the CR design, the contact point between the femur and tibia moved backward with an increasing muscle tension, resulting in an increased arm movement that favored the knee extension mechanism. Conversely, compared to the CR design, the CS design was found to be relatively less affected by muscle tension, likely due to its higher constraints. These results highlight the importance of considering not only the degree of geometric constraint but also the presence or absence of load and differences in soft tissue tension, particularly the degree of muscle tension when evaluating dynamic anteroposterior stability.

3 Discussion

3.1 Previous Studies of Anteroposterior Stability and Clinical Outcomes (ROMs)

Anteroposterior stability in various TKA designs has been evaluated, often correlating it with the maximum knee flexion angle achieved ROM postoperatively [2, 7, 11, 13, 20, 22, 24] (Table 45.1). Finding the right balance (not too stiff and not too loose) is the key to achieving optimal anteroposterior stability after TKA. Warren et al. [22] concluded that an anteroposterior stability of 5 mm or more was desirable, whereas Seon et al. [20] recommended a stability of 10 mm or less. Previous studies have shown that measurements and conditions, including the devices and their loading protocols, besides design variations, have significant variance. However, based on the available literature, an optimal value for static measurement for anteroposterior stability falls within the range of approximately 3–13 mm [2–14].

However, there is no consensus from previous clinical studies [2, 7, 11, 13, 20, 22, 24] on the involvement of anteroposterior stability in achieving a good knee flexion angle after TKA. Some studies [11, 13, 20, 22] showed a signifi-

Table 45.1 Preferred values for postoperative anteroposterior movement

Study	Implant design	PCL	AP translation
Chouteau et al. [2]	Innex® PCL-retaining design	Retaining	12–13 mm
Ishii et al. [7]	LCS®, PCL-substituting, rotating-platform design	Retaining Substituting	11.8 mm, 30° flexion 11.7 mm, 75° flexion 7.1 mm, 30° flexion 7.9 mm, 75° flexion
Itokazu et al. [11]	Miller-Galante® PCL-retaining design	Retaining	5.05 mm
Jones et al. [13]	PCA® or Duracon® prostheses	Retaining	5 to 10 mm
Seon et al. [20]	e.motion® PCL-retaining design	Retaining	7.1 mm
Warren et al. [22]	Insall-Burstein™ posterior stabilized knee Kinemax® condylar knee Oxford meniscal knee	Substituting Retaining Retaining	Greater than 5 mm
Yamakado et al. [24]	Yoshino or Shoji-4 PCL-retaining design and anatomically graduated components shoji PCL-retaining design	Retaining	9.71 mm

PCL posterior cruciate ligament, *AP* anteroposterior
Innex®, Zimmer, Winterthur, Switzerland; Miller-Galante®, Zimmer, Warsaw, IN, USA; PCA®, Howmedica, Rutherford, NJ, USA; Duracon®, Howmedica, Rutherford, NJ, USA; e.motion®, Aesculap, Tuttlingen, Germany; Insall-Burstein™, Johnson & Johnson, New Milton, Hampshire, UK; Kinemax®, Howmedica International, Staines, Middlesex, UK; Oxford, Biomet Ltd., Swindon, Wiltshire, UK; Yoshino/Shoji-4, Biomet, Warsaw, IN, USA; LCS®, DePuy, Warsaw, IN, USA

cant correlation, while others [2, 7, 24] found no such correlation. These conflicting results can be partially attributed to the fact that all of these studies conducted measurements while the patients were awake. The level of soft tissue tension among the patients may have influenced the results. In other words, the "degree of relaxation" of each patient at the time of measurement may have affected the voluntary and involuntary contractions of the soft tissues. Quantifying this "degree" is challenging, if not impossible, and it may have contributed to the discrepancies observed among the studies. Our study also found no correlation [7]. However, despite this lack of correlation, the clinical scores and ROM were good, and most cases showed consistent stability of 7–8 mm in CS and 11–12 mm in CR. Therefore, our results should not be interpreted to indicate that anteroposterior movement or stability is generally unimportant. Contrastingly, achieving proper stability and balance in TKAs is imperative for optimizing ROM and functional outcomes.

3.2 Effect of Soft Tissue Tension on the Anteroposterior Stability of the Patient

To examine the impact of soft tissue tension on the anteroposterior stability mentioned earlier, we compared the stability between patients under consciousness and general anesthesia, which can be assumed to be in a condition without spontaneous soft tissue tension. This study evaluated patients with favorable clinical outcomes [8, 9]. The measurements under general anesthesia were performed on the first knee during the second knee surgery in a scheduled two-stage bilateral TKA procedure.

For the CS design, the movement at 30° flexion was 5.1 ± 2.1 vs. 6.7 ± 2.7 mm (conscious vs. anesthetized), while at 75° flexion, it was 7.0 ± 2.0 vs. 7.7 ± 3.4 mm [8]. Regarding the CR design, the movement at 30° flexion was 10.5 ± 4.6 vs. 13.8 ± 5.4 mm, while at 75° flexion, it was 10.4 ± 4.5 vs. 12.7 ± 5.2 mm [9]. In both designs, significantly larger movements were observed under anesthesia. These studies confirmed that soft tissue tension, particularly the degree of relaxation under the awareness of the patient, plays a role in anteroposterior stability after TKA.

4 For Intraoperative Acquisition of the Target Anteroposterior Stability

Studies [8, 9] demonstrated that anteroposterior stability is greater under anesthesia than that under consciousness. To further investigate how to achieve optimal intraoperative stability as envisioned by the surgeon, we compared the anteroposterior stability in the 75° flexion position during CS in LCS TKA under conscious anesthesia [10]. As previously mentioned, measurements under general anesthesia were performed on the first knee during the second knee surgery in a scheduled two-stage bilateral TKA procedure. The stability measurements revealed a significant correlation between the two conditions, with the median value approximately 2 mm larger (7.3 vs. 9.2 mm) under anesthesia. Assuming that the measurements under anesthesia in this study accurately reflected the soft tissue condition of the patient during the TKA procedure, it suggests that surgical manipulation should be performed to achieve more anteroposterior motion (approximately 2 mm in this particular design) than that of the postoperative motion target.

5 Key Points of TKA Anteroposterior Stability

1. Achieving optimal anteroposterior stability is vital for obtaining a good ROM after TKA, as it helps prevent early impingement between the femoral and tibial components.
2. TKA knees with favorable clinical outcomes often exhibit anteroposterior stability, reflecting the design concept.
3. The optimal values of anteroposterior stability in each TKA design differ based on the static and dynamic kinematic analyses, including previous reports, and should be determined based on each design concept understanding.
4. To achieve the optimal postoperative anteroposterior stability envisioned by the surgeon, intraoperative maneuvers should be performed such that the movement is slightly larger than the postoperative target value.

References

1. Matsuda S, et al. Postoperative alignment and ROM affect patient satisfaction after TKA. Clin Orthop Relat Res. 2013;471:127–33.
2. Chouteau J, et al. Sagittal laxity after posterior cruciate ligament-retaining mobile-bearing total knee arthroplast. J Arthroplast. 2009;24:710–5.
3. Han HS, et al. Anterior-stabilized TKA is inferior to posterior-stabilized TKA in terms of postoperative posterior stability and knee flexion in osteoarthritic knees: a prospective randomized controlled trial with bilateral TKA. Knee Surg Sports Traumatol Arthrosc. 2020;28:3217–25.
4. Hinarejos P, et al. Increased posterior translation but similar clinical outcomes using ultracongruent instead of posterior stabilized total knee arthroplasties in a prospective randomized trial. Knee Surg Sports Traumatol Arthrosc. 2022;30:3041–8.
5. Inui H, et al. The relationship between anteroposterior stability and medial-lateral stability of the bicruciate stabilized total knee arthroplasty. Knee. 2018;25:1247–53.
6. Ishii Y, et al. Sagittal laxity in vivo after total knee arthroplasty. Arch Orthop Trauma Surg. 2005;125:249–53.
7. Ishii Y, et al. Anteroposterior translation and range of motion after total knee arthroplasty using posterior cruciate ligament-retaining versus posterior cruciate ligament -substituting prostheses. Knee Surg Sports Traumatol Arthrosc. 2017;25:3536–42.
8. Ishii Y, et al. Effect of voluntary soft tissue tension and articular conformity after total knee arthroplasty on in vivo anteroposterior displacement. Knee. 2011;18:11–4.
9. Ishii Y, et al. In vivo anteroposterior translation after meniscal-bearing total knee arthroplasty: effects of soft tissue conditions and flexion angle. Eur J Orthop Surg Traumatol. 2014;24:967–71.
10. Ishii Y, et al. Anteroposterior translation does not correlate with knee flexion after total knee arthroplasty. Clin Orthop Relat Res. 2014;472:704–9.
11. Itokazu M, et al. Influence of anteroposterior and mediolateral instability on range of motion after total knee arthroplasty: an ultrasonographic study. Orthopedics. 2000;23:49–52.
12. Jin QH, et al. No difference in the anteroposterior stability between the GRADIUS and multi-radius designs in total knee arthroplasty. Knee. 2020;27:1197–204.
13. Jones DP, et al. The effect of sagittal laxity on function after posterior cruciate-retaining total knee replacement. J Arthroplast. 2006;21:719–23.
14. Matsuda S, et al. Knee stability in posterior cruciate ligament retaining total knee arthroplasty. Clin Orthop Relat Res. 1999;366:169–73.
15. Matsuda Y, et al. In vivo laxity of low contact stress mobile-bearing prostheses. Clin Orthop Relat Res. 2004;419:138–43.
16. Minoda Y, et al. Mobile-bearing TKA improved the anteroposterior joint stability in mid-flexion range comparing to fixed-bearing TKA. Arch Orthop Trauma Surg. 2016;136:1601–6.
17. Mizu-Uchi H, et al. Anteroposterior stability in posterior cruciate ligament-retaining total knee arthroplasty. J Arthroplast. 2006;21:592–8.
18. Schuster AJ, et al. Anteroposterior stability after posterior cruciate-retaining total knee arthroplasty. Knee Surg Sports Traumatol Arthrosc. 2011;19:1113–20.
19. Seah RB, et al. Evaluation of the relationship between anteroposterior translation of a posterior cruciate ligament-retaining total knee replacement and functional outcome. J Bone Joint Surg Br. 2012;94:1362–5.
20. Seon JK, et al. The effect of anteroposterior laxity on the range of movement and knee function following a cruciate-retaining total knee replacement. J Bone Joint Surg Br. 2010;92:1090–5.
21. Seon JK, et al. In vivo stability of total knee arthroplasty using a navigation system. Int Orthop. 2007;31:45–8.
22. Warren PJ, et al. Laxity and function in knee replacements: a comparative study of three prosthetic designs. Clin Orthop Relat Res. 1994;305:200–8.
23. White SH, et al. Sagittal plane laxity following knee arthroplasty. J Bone Joint Surg Br. 1991;73:268–70.
24. Yamakado K, et al. Influence of stability on range of motion after cruciate-retaining TKA. Arch Orthop Trauma Surg. 2003;123:1–4.
25. Ishii Y, et al. Primary total knee arthroplasty using the genesis I total knee prosthesis: a 5- to 10-year follow-up study. Knee. 2005;12:341–5.
26. Pinskerrova V, et al. The shapes and relative movement of the femur and tibia in the unloaded cadaveric knee: a study using MRI as an anatomical tool. In: Insall JN, et al., editors. Surgery of the knee. 3rd ed. New York: Churchill Livingstone; 2001. p. 255–83.
27. Gejo R, et al. Joint gap changes with patellar tendon strain and patellar position during TKA. Clin Orthop Relat Res. 2008;466:946–51.
28. Hosseini Nasab SH, et al. Elongation patterns of the collateral ligaments after total knee arthroplasty are dominated by the knee flexion angle. Front Bioeng Biotechnol. 2019;7:323.
29. Iversen BF, et al. Implications of muscular defense in testing for the anterior drawer sign in the knee: a stress radiographic investigation. Am J Sports Med. 1989;17:409–13.
30. Lafortune MA, et al. Three-dimensional kinematics of the human knee during walking. J Biomech. 1992;25:347–57.
31. Ishii Y, et al. Comparison of knee joint functional laxity after total knee replacement with posterior cruciate retaining and substituting prosthesis. Knee. 1995;2:195–9.
32. Ishii Y, et al. Influence of total knee replacement design on screw home movement: comparison of five designs for total knee replacement prostheses. J Orthop Sci. 1996;1:313–7.
33. Ishii Y, et al. Effect of knee laxity on in vivo kinematics of meniscal-bearing knee prostheses. Knee. 2007;14:269–74.
34. Mikashima Y, et al. Do mobile-bearing knee arthroplasty motions change with activity? Knee. 2013;20:422–5.
35. Ishii Y, et al. Gait analysis after total knee arthroplasty: comparison of posterior cruciate retention and substitution. J Orthop Sci. 1998;3:31031–7.
36. Markolf KL, et al. The role of joint load in knee stability. J Bone Joint Surg Am. 1981;63:570–85.

46 Soft Tissue Balance/Joint Stability: Mid-Flexion Stability

Kazunori Hino

The acquisition of an appropriate soft tissue balance leads to good postoperative outcomes and high patient satisfaction in total knee arthroplasty (TKA). Moreover, an inappropriate soft tissue balance can cause polyethylene wear, postoperative pain, and implant migration. However, the boundary between appropriate and inappropriate soft tissue balance is vague, as it is difficult to determine the proper target balance in TKA. Furthermore, standardized variations exist among surgeons [1], some of whom often rely on subjectivity by using hand sensations; others utilize objective measurements such as tension meters, navigation systems, and pressure sensors. Therefore, accurate evaluation is difficult unless the reproducibility of the amount of stress and flexion angle of the knee joint at the time of measurement is ensured. In addition, research on mid-flexion stability has been the focus of many researchers because it is required for daily life movements.

To avoid instability caused by inappropriate soft tissue balance, a widely recognized goal is to adjust the soft tissue envelope to feel stable in 0° extension and 90° flexion gap. However, it is unclear how stability at 0° extension and 90° flexion affects stability acquisition in the mid-flexion range. Moreover, the influence of the method based on the gap (axial direction) on the mid-flexion stability of the 6-degrees of freedom (6-DOF) remains unclear. The concept of mid-flexion instability was first reported by Martin and Whiteside in 1990 [2]. They reported the risk of mid-flexion instability due to the elevation of the joint line in a study of 10 cadaver knees [2]. Since then, many researchers have discussed mid-flexion stability. Their research methods can be roughly divided into three types: computational, cadaver, and clinical studies. Fourteen risk factors have been identified as causes of mid-flexion instability, which can be broadly classified into three factors: patient factors, technique factors, and implant factors [3] (Table 46.1).

K. Hino (✉)
Department of Joint Reconstruction, Ehime University School of Medicine, Ehime, Japan

Table 46.1 Risk factors for mid-flexion instability

Patient factor
Motion/activity
Preoperative stability
Technique factor
Joint line elevation
Method of ligament balancing (reverse gap vs. classical gap balance)
Position of the femoral component
Concept for alignment (mechanical vs. anatomical)
Ligament balance between 0° and 90°
Implant selection(CR vs. PS)
Implant factor
Size of the femoral component
Bearing design
Bicruciate-retaining design
Radius of the femoral component
Conformity of the implant

1 Stability After TKA Surgery

1.1 Varus and Valgus Stability

In a study of varus-valgus stability in patients with mobile-bearing TKA, Matsuda et al. reported that CR knees had significantly less stability at 75° flexion than in extension, although no significant differences were noted for PS knees [4]. Using knees from cadavers, Van Damme showed that the joint stability of CR knees decreased at 30 and 90 degrees of flexion, despite no changes in stability in full extension [5]. Meanwhile, Siston et al. employed a surgical navigation system to intraoperatively measure joint stability undergoing primary PS-TKA, and found no significant differences in the range of motion before and after surgery [6]. Varus-valgus stability at 0 degrees was restored after PS-TKA, but not at 30 degrees of flexion [7]. However, owing to differences in evaluation and research methods, there is no obvious consensus on the difference between the CR and PS-TKA. A total of 34 knees under-

S. Matsuda (ed.), *Advances in Total Knee Arthroplasty*, https://doi.org/10.1007/978-981-97-4920-1_46

went TKA with computer navigation, during which the investigator applied a maximum varus-valgus stress to the knee while steadily moving the leg from full extension to flexion both before and after prosthetic implantation. The femorotibial angle was measured simultaneously by the navigation system at every 10° throughout the range of movement. It was found that PS-TKA had more varus-valgus laxity than CR-TKA at all angles examined, and the differences were statistically significant at 10°, 20°and 30° of flexion [8]. Saeki et al. showed in cadaver knees that TKA with medial collateral ligament release resulted in significantly higher laxity with excision of the PCL [9]. Similar results have been reported in a recent cadaver study by Athwal et al. [10]. This may be due to the influence of the second stabilization mechanism on the valgus stress due to the presence or absence of the posterior cruciate ligament. This result is significant as a theoretical basis for the fact that mid-flexion laxity increases with the amount of MCL release in PS-TKA and is considered to be a crucial factor for obtaining mid-flexion stability. However, unlike previous results, a cadaver study using single-radius TKA reported no difference between the PS and CR types. Thus, it is important to consider that the results of cadaver studies differ depending on the study methods, subjects, and TKA designs employed, and that the subjects of cadaver studies differ from those involving patients with end-stage knee OA. Two distinct types of TKA design characteristics exist: the single-radius type with a single sagittal center of rotation and a multi-radius type with a stepwise change in the center of rotation. The mid-flexion laxity of these two designs was investigated based on the hypothesis that laxity generated at the change point of the multi-radius type leads to mid-flexion instability. However, contrary to this hypothesis, no significant difference was observed between the two radial types [11]. Conversely, in a study that evaluated mid-flexion laxity in a model that continuously changed the center of the radius by applying an insert matched to the radial shape of the femur, improvements in stability were confirmed [12]. Although these concepts are still in the research phase and have primarily been investigated using cadavers, it has been suggested that mid-flexion laxity may not solely depend on the choice between CR and PS types. Instead, it appears to be influenced by component geometry and size matching between the femur and tibia. Additionally, the intraoperative varus–valgus laxity of bi cruciate substituting (BCS)-type TKA with a guided motion mechanism showed proper stability even during mid-flexion flexion, suggesting that component design, rather than the simple classification of only PS and CR types, contributes to mid-flexion stability [13].

1.2 Flexion Stability (Gap Balance)

PCL is the major stabilizer, and its presence or absence significantly affects gap balance stability. Tsuneizumi et al. highlighted the tendency for a further flexion gap in the PS type compared to the CR type when using weights [14]. Cadaver studies have also shown that the flexion gap expands following PCL dissection [15]. In recent years, Oshima et al. has reported that the flexion gap did not increase as much as previously reported with minimum medial soft tissues release in PS-TKA [16]. (Fig. 46.1). Matsumoto et al. evaluated the gap changes in TKA using an offset-type tensor and clarified that the PS type had larger gaps than the CR type in the flexion range [17]. Moreover, Minoda et al. also conducted a gap balance evaluation, including mid-flexion, using a tensor device of the same shape as the insert. They reported that the 30° mid-flexion gap was larger than the extension and flexion gap, and that the 60° mid-flexion gap increased following an increase in the extension–flexion gap imbalance [18, 19].

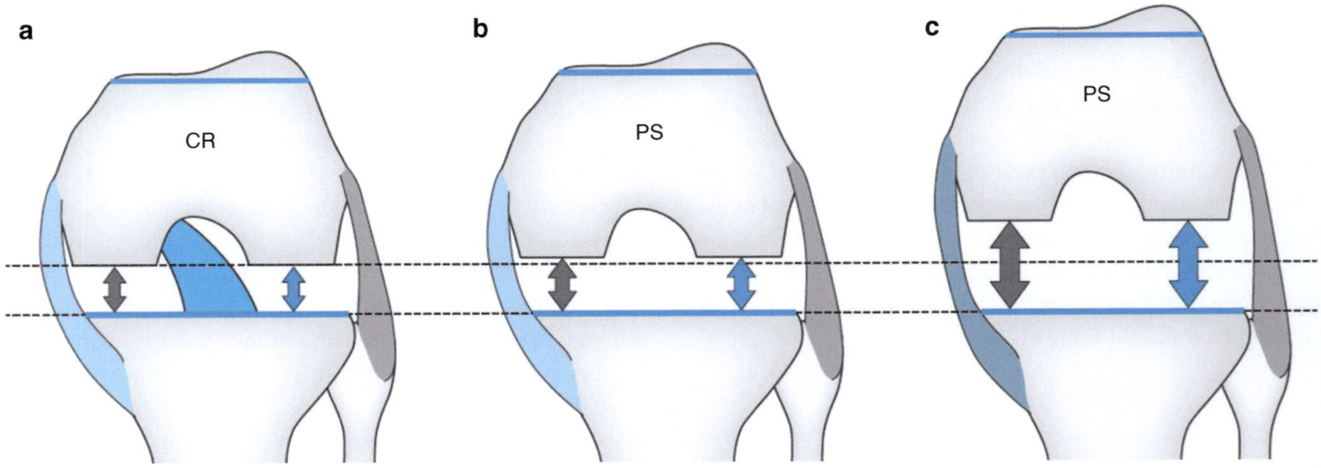

Fig. 46.1 Change in flexion gap due to PCL dissection and MCL release. (**a**): CR-type TKA. (**b**): PS-type TKA with minimal MCL release. (**c**): PS-type TKA with extensive MCL release

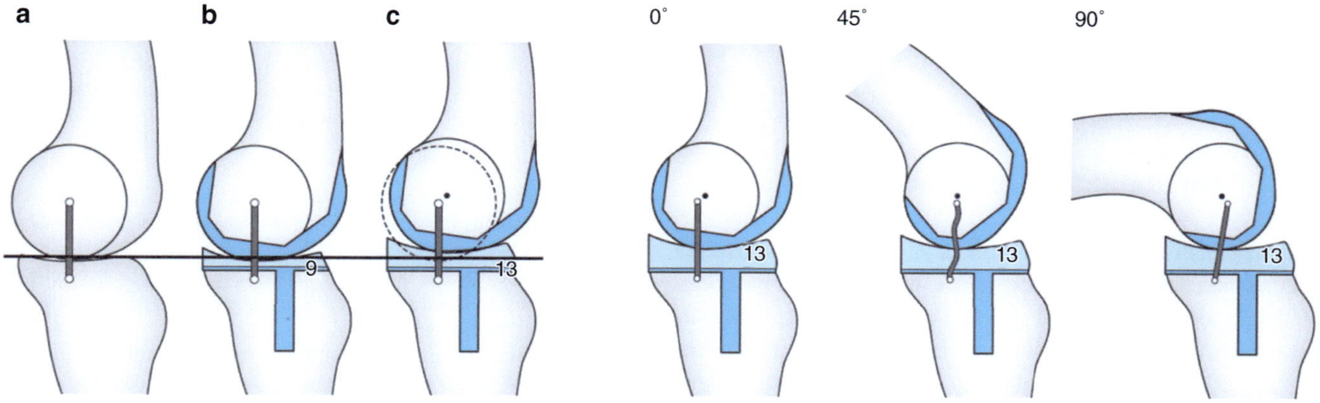

Fig. 46.2 Rise of the joint line and MCL deflection. (**a**): Normal knee. (**b**): Joint line unchanged. (**c**): Joint line with change. If the joint line changes, there is a risk that the MCL will deflect and cause mid-flexion instability. (From Luyckx et al. [20])

MCL release depends on changes in the joint line, resulting in mid-flexion instability. However, there exist conflicting reports on this matter, and these discrepancies may arise due to variations in measurement methods, surgical methods, and models used in different studies [20, 21] (Fig. 46.2). Changes in the posterior condylar offset (PCO), amount of medial soft tissue release, postoperative alignment, and gap balancing methods have also been reported as factors contributing mid-flexion stability [22–24]. In addition, Kaneko et al. reported that a preferable gap balance between the medial and lateral compartments was obtained during mid-flexion in guided-motion TKA using a sensor of the same shape as the insert [25]. Accordingly, component design may also be a factor for mid-flexion stability.

1.3 Anteroposterior (AP) Stability

Nabeyama et al. used the KT-2000 arthrometer to evaluate changes in anteroposterior stability at 30°and 75°flexion before and 1 month and 3 years after surgery. The results revealed that the CR type exhibited no significant changes throughout the observation period and demonstrated stability comparable to that of a normal knee. Conversely, the PS type showed significantly larger laxity 3 years after surgery [26]. Another study reported that the CR type maintained anteroposterior stability even 5 years after surgery [27]. In addition, Mobile-bearing TKA improved the anteroposterior stability in mid-flexion range, and the use of medial- high constrained-type inserts improved AP stability in CR-TKA [28–30].

1.4 Rotational Stability

Almost all TKA surgical methods require cruciate ligament removal. However, there are still uncertainties regarding the change in mid-flexion rotational laxity before and after surgery. The post-operative internal rotational laxity was statistically significantly increased compared to the preoperative level at 30°, 45°, 60°, and 90° of flexion. The post-operative external rotational laxity was statistically significantly decreased compared to the preoperative level at 45° and 60° of flexion. The post-operative internal–external rotational laxity was statistically significantly increased compared to the preoperative level only at 30° of flexion in a computer navigation study. Overall, postoperative internal–external laxity significantly increased only at 30° flexion after PS-TKA. Furthermore, a significantly high correlation was observed between preoperative and postoperative rotational laxity. Therefore, surgeons should consider a large preoperative rotational laxity increased the risk of a large post-operative laxity, especially at the initial flexion range in PS-TKA [31].

2 Clinical Results and Mid-Flexion Stability

In terms of patient-reported outcomes, it has been found that 60° AP stability is correlated with Knee Injury and Osteoarthritis Outcome Score (KOOS) pain score [32]. Additionally, AP stability at 30° flexion has been found to be associated with postoperative subjective stability [33], suggesting that appropriate AP stability at the mid-flexion angle is directly related to patient satisfaction. The lift-off of the femoral condyle and paradoxical anterior movement of the femur relative to the tibia during the mid-flexion range in CR-type TKA have been identified as critical issues in clinical studies using the fluoroscopic image-matching method [34, 35]. These issues are recognized as problems that need to be addressed, and imbalance at mid-flexion range appears to be a key contributing factor. Paradoxical anterior motion has also been observed in PS-TKA, but not in BCS-TKA or highly constrained mobile-bearing TKA, even during daily activities, including the mid-flexion angle. These results clarify that the component design also affects postoperative AP kinematics [36–40].

3 For Stability Throughout All ROMs

Conventionally, the term "rectangular gap," which equalizes the medial and lateral or flexion and extension gaps, has been widely recognized as a gold standard concept for achieving balanced soft tissue tension. However, it remains unclear whether the acquisition of a rectangular gap is the best way to achieve all-range stability. A computer navigation study that evaluated the correlation between varus–valgus laxity at 0° extension, 90° flexion, and the mid-flexion range in PS-TKA indicated that laxity at 90° flexion was highly correlated with laxity at a wide range of mid-flexion, whereas laxity at 0° extension was correlated with only a narrow range [41]. In particular, no strong correlation was observed between laxity at 90° flexion and 0° extension and the initial flexion range (20–40°). The approach to achieving stability across all ranges, including the initial flexion range, requires further investigation [41].

Recently, an association between medial stability and good clinical outcomes was reported especially in Japan [42, 43]. Some researchers are particularly focused on the lack of evidence supporting the conventional medial release balancing method that has been historically performed for a long time [44]. Instead, they have been studying the achievement of medial stability throughout the entire range, like the soft tissue balance observed in a normal knee [45], rather than solely pursuing a rectangular gap.

References

1. Lombardi AV, et al. Why knee replacements fail in 2013: patient, surgeon, or implant? Bone Joint J. 2014;96:101–4.
2. Martin JW, Whiteside LA. The influence of joint line position on knee stability after condylar knee arthroplasty. Clin Orthop Relat Res. 1990:146–56.
3. Vajapey SP, et al. Risk factors for mid-flexion instability after total knee arthroplasty: a systematic review. J Arthroplast. 2020;35:3046–54.
4. Matsuda Y, et al. Effect of flexion angle on coronal laxity in patients with mobile-bearing total knee arthroplasty prostheses. J Orthop Sci. 2005;10:37–41.
5. Van Damme G, et al. What should the surgeon aim for when performing computer-assisted total knee arthroplasty? J Bone Joint Surg Am. 2005;87(Suppl 2):52–8.
6. Siston RA, et al. Coronal plane stability before and after total knee arthroplasty. Clin Orthop Relat Res. 2007;463:43–9.
7. Casino D, et al. Intraoperative evaluation of total knee replacement:kinematic assessment with a navigation system. Knee Surg Sports Traumatol Arthrosc. 2009;17:369–73.
8. Hino K, et al. Mid-flexion laxity is greater after posterior-stabilised total knee replacement than with cruciate-retaining procedures: a computer navigation study. Bone Joint J. 2013;95:493–7.
9. Saeki K, et al. Stability after medial collateral ligament release in total knee arthroplasty. Clin Orthop Relat Res. 2001:184–9.
10. Athwal KK, et al. The superficial medial collateral ligament is the primary medial restraint to knee laxity after cruciate-retaining or posterior-stabilised total knee arthroplasty: effects of implant type and partial release. Knee Surg Sports Traumatol Arthrosc. 2016;24:2646–55.
11. Stoddard JE, et al. The kinematics and stability of single-radius versus multi-radius femoral components related to mid-range instability after TKA. J Orthop Res. 2013;31:53–8.
12. Clary CW, et al. The influence of total knee arthroplasty geometry on mid-flexion stability: an experimental and finite element study. J Biomech. 2013;46:1351–7.
13. Hino K, et al. Bi-cruciate substituting total knee arthroplasty provides varus-valgus stability throughout the midflexion range. Knee. 2018;25:897–902.
14. Tsuneizumi Y, et al. Evaluation of joint laxity against distal traction force upon flexion in cruciate-retaining and posterior-stabilized total knee arthroplasty. J Orthop Sci. 2008;13:504–9.
15. Mihalko WM, Krackow KA. Posterior cruciate ligament effects on the flexion space in total knee arthroplasty. Clin Orthop Relat Res. 1999:243–50.
16. Oshima Y, et al. The influence of posterior cruciate ligament resection on tibiofemoral joint gap in varus osteoarthritic knees. J Knee Surg. 2022;35:323–30.
17. Matsumoto T, et al. Soft tissue tension in cruciate-retaining and posterior-stabilized total knee arthroplasty. J Arthroplast. 2011;26:788–95.
18. Minoda Y, et al. Intraoperative assessment of midflexion laxity in total knee prosthesis. Knee. 2014;21:810–4.
19. Minoda Y, et al. Midflexion laxity after implantation was influenced by the joint gap balance before implantation in TKA. J Arthroplast. 2015;30:762–5.
20. Luyckx T, et al. Raising the joint line in TKA is associated with mid-flexion laxity: a study in cadaver knees. Clin Orthop Relat Res. 2018;476:601–11.
21. Minoda Y, et al. Joint line elevation is not associated with mid-flexion laxity in patients with varus osteoarthritis after total knee arthroplasty. Knee Surg Sports Traumatol Arthrosc. 2020;28:3226–31.
22. Matziolis G, et al. Changes of posterior condylar offset results in midflexion instability in single-radius total knee arthroplasty. Arch Orthop Trauma Surg. 2017;137:713–7.
23. Incavo SJ, et al. Total knee arthroplasty using anatomic alignment can produce mid-flexion laxity. Clin Biomech (Bristol, Avon). 2013;28:429–35.
24. MacDessi SJ, et al. Restoring the constitutional alignment with a restrictive kinematic protocol improves quantitative soft-tissue balance in total knee arthroplasty: a randomized controlled trial. Bone Joint J. 2020;102-B:117–24.
25. Kaneko T, et al. Bi-cruciate substituting total knee arthroplasty improved medio-lateral instability in mid-flexion range. J Orthop. 2017;14:201–6.
26. Nabeyama R, et al. Changes in anteroposterior stability following total knee arthroplasty. J Orthop Sci. 2003;8:526–31.
27. Mizu-Uchi H, et al. Anteroposterior stability in posterior cruciate ligament-retaining total knee arthroplasty. J Arthroplast. 2006;21:592–8.
28. Minoda Y, et al. Mobile-bearing TKA improved the anteroposterior joint stability in mid-flexion range comparing to fixed-bearing TKA. Arch Orthop Trauma Surg. 2016;136:1601–6.
29. Wang XH, et al. Motion type and knee articular conformity influenced mid-flexion stability of a single radius knee prosthesis. Knee Surg Sports Traumatol Arthrosc. 2019;27:1595–603.
30. Tsubosaka M, et al. Mid-flexion stability in the anteroposterior plane is achieved with a medial congruent insert in cruciate-

retaining total knee arthroplasty for varus osteoarthritis. Knee Surg Sports Traumatol Arthrosc. 2021;29:467–73.
31. Hino K, et al. Assessment of the midflexion rotational laxity in posterior-stabilized total knee arthroplasty. Knee Surg Sports Traumatol Arthrosc. 2017;25:3495–500.
32. Matsumoto K, et al. Postoperative anteroposterior laxity influences subjective outcome after total knee arthroplasty. J Arthroplast. 2017;32:1845–9.
33. Mochizuki T, et al. Association between anteroposterior laxity in mid-range flexion and subjective healing of instability after total knee arthroplasty. Knee Surg Sports Traumatol Arthrosc. 2017;25:3543–8.
34. Dennis DA, et al. Femoral condylar lift-off in vivo in total knee arthroplasty. J Bone Joint Surg Br. 2001;83:33–9.
35. Dennis DA, et al. In vivo anteroposterior femorotibial translation of total knee arthroplasty: a multicenter analysis. Clin Orthop Relat Res. 1998:47–57.
36. Shimizu N, et al. The effect of weight-bearing condition on kinematics of a high-flexion, posterior-stabilized knee prosthesis. J Arthroplast. 2011;26:1031–7.
37. Kono K, et al. Bicruciate-stabilised total knee arthroplasty provides good functional stability during high-flexion weight-bearing activities. Knee Surg Sports Traumatol Arthrosc. 2019;27:2096–103.
38. Murakami K, et al. Knee kinematics in bi-cruciate stabilized total knee arthroplasty during squatting and stair-climbing activities. J Orthop. 2018;15:650–4.
39. Murakami K, et al. In vivo kinematics of gait in posterior-stabilized and bicruciate-stabilized total knee arthroplasties using image-matching techniques. Int Orthop. 2018;42:2573–81.
40. Kage T, et al. In vivo kinematics of a newly updated posterior-stabilised mobile-bearing total knee arthroplasty in weight-bearing and non-weight-bearing high-flexion activities. Knee. 2021;29:183–9.
41. Hino K, et al. Varus-valgus stability at 90 degrees flexion correlates with the stability at midflexion range more widely than that at 0 degrees extension in posterior-stabilized total knee arthroplasty. Arch Orthop Trauma Surg. 2017;137:1429–34.
42. Nagai K, et al. Intraoperative soft tissue balance using novel medial preserving gap technique in posterior-stabilized total knee arthroplasty: comparison to measured resection technique. Knee Surg Sports Traumatol Arthrosc. 2018;26:3474–81.
43. Azukizawa M, et al. Intraoperative medial joint laxity in flexion decreases patient satisfaction after total knee arthroplasty. Arch Orthop Trauma Surg. 2018;138:1143–50.
44. Hunt NC, et al. Lack of evidence to support present medial release methods in total knee arthroplasty. Knee Surg Sports Traumatol Arthrosc 2014;22:3100–12.
45. Okazaki K, et al. Asymmetry of mediolateral laxity of the normal knee. J Orthop Sci 2006;11:264–6.

Part VII

Preoperative Planning

47 2D/3D Planning

Hideki Mizu-uchi

Accurate preoperative planning is important in achieving optimal implant placement for target alignment in total knee arthroplasty (TKA). Generally, radiographs and computed tomography (CT) slices are used for two-dimensional (2D) preoperative planning to ensure reproducible intraoperative procedures. However, caution should be exercised because measurements using radiographs and CT slices are affected by limb position and imaging direction due to the internal and external rotation of the lower limbs and flexion contractures. Detailed preoperative measurements using 2D planning is difficult because of one-slice evaluation. In contrast, three-dimensional (3D) preoperative planning can enable a more detailed preoperative planning with the use of 3D analysis software, such as 3D templates and bone models reconstructed from CT scans. 3D planning enables the accurate setting of 3D spatial coordinates according to the TKA surgical technique and detailed planning to reproduce intraoperative osteotomy and implant placement. By implementing an accurate preoperative planning, difficulties experienced during the actual operation, such as "Determination of implant size and cutting bone volume is different from preoperative planning: Do you trust preoperative planning or surgical technique?"

In this chapter, specific methods and precautions for 2D planning and evaluation of surgical techniques using 3D planning to achieve mechanical neutral alignment are described.

1 2D Planning

1.1 Coronal Alignment

2D planning is typically performed using either an anteroposterior radiograph of the knee joint or a full-length anteroposterior radiograph. An anteroposterior radiograph of the knee joint is a short film. When the short film is used, the plan is based on the femorotibial angle (FTA), which consists of the distal femoral and proximal tibial anatomical axes. However, the FTA varies from person to person, and there is a risk that femoral and tibial bowing might be missed. A plan should be made using full-length anteroposterior radiographs if the target alignment of the surgeon is based on the mechanical axes. Surgeons should consider that the mechanical and anatomical axes can change because measurements using radiographs are affected by the limb position and imaging direction due to the internal and external rotation of the lower limbs and flexion contractures. If a full-length anteroposterior radiograph is difficult to obtain, an anteroposterior radiograph of the entire femoral and tibial bones is needed to check and confirm the mechanical axis.

1.1.1 Femoral Side

The direction of insertion of the intramedullary guide is parallel to the femoral anatomical axis. Surgeons should measure the femoral valgus angle, which is the angle between the femoral mechanical and anatomical axes, as well as the thickness of the cut bone of the medial and lateral distal femurs. The cutting level is based on the thickness of the implant, and the remaining articular cartilage is approximately 2 mm thick. If the anatomical axis of the femur is defined as the straight line between the center of the femoral shaft located by bisecting the proximal and distal parts of the femur, the insertion position of the intramedullary guide (extension of the anatomical axis) is slightly medial to the center of the femoral intercondylar notch; therefore, the distance from the center of the femoral intercondylar notch should be measured in preparation for the intraoperative insertion position of the intramedullary guide. When the insertion position of the intramedullary guide is at the center of the femoral intercondylar notch, the insertion direction changes depending on the location of the proximal reference point; thus, preoperative planning should be done to achieve intraoperative reproducibility. When a long intramedullary guide is used, inserting it parallel to the femoral anatomical

H. Mizu-uchi (✉)
Department of Orthopaedic Surgery, Saiseikai Fukuoka General Hospital, Fukuoka, Japan
e-mail: mizuuchi.hideki.243@m.kyushu-u.ac.jp

S. Matsuda (ed.), *Advances in Total Knee Arthroplasty*, https://doi.org/10.1007/978-981-97-4920-1_47

axis can be difficult because of lateral bowing of the femoral bone; therefore, inserting it slightly medial to the center of the femoral intercondylar notch is recommended (Fig. 47.1).

1.1.2 Tibial Side

When an intramedullary guide is used, planning should be performed such that the position and direction of insertion can be reproduced intraoperatively as accurately as in the femur. The difference in the angle between the tibial anatomical and mechanical axes (the line connecting the centers of the ankle joint and tibial joint surface) is not as large as that of the femur, but even a small difference should be measured before surgery. When an extramedullary guide is used, the placement position is parallel to the mechanical axis. Therefore, it should not be performed using the anatomical and shaft axes used in preoperative planning. The postoperative tibial mechanical axis is a line connecting the center of the tibial cutting surface and the center of the ankle joint, which is different from the line connecting the center of the tibial joint surface and the center of the ankle joint in preoperative planning. To ensure that the implant position is perpendicular to the postoperative tibial mechanical axis, preoperative planning should be performed such that the perpendicular bisector of the planned osteotomy line (cutting plane) passes through the center of the ankle joint [1] (Fig. 47.2). The cutting level is based on the thickness of the implant, and the remaining articular cartilage is approximately 2 mm thick.

1.2 Sagittal Alignment

The use of a full-length radiograph is not as common as the use of coronal alignment, and planning is often difficult because the images are easily affected by rotation depending on limb position and imaging direction. Although confirming the mechanical axis using a lateral radiograph of the knee joint is not possible, planning the implant size as well as the insertion position and direction of the intramedullary guide by determining the anatomical axis around the joint is possible. Sagittal alignment is easily affected by cutting errors, which means that the intraoperative component size can differ frequently from the size defined in preoperative planning.

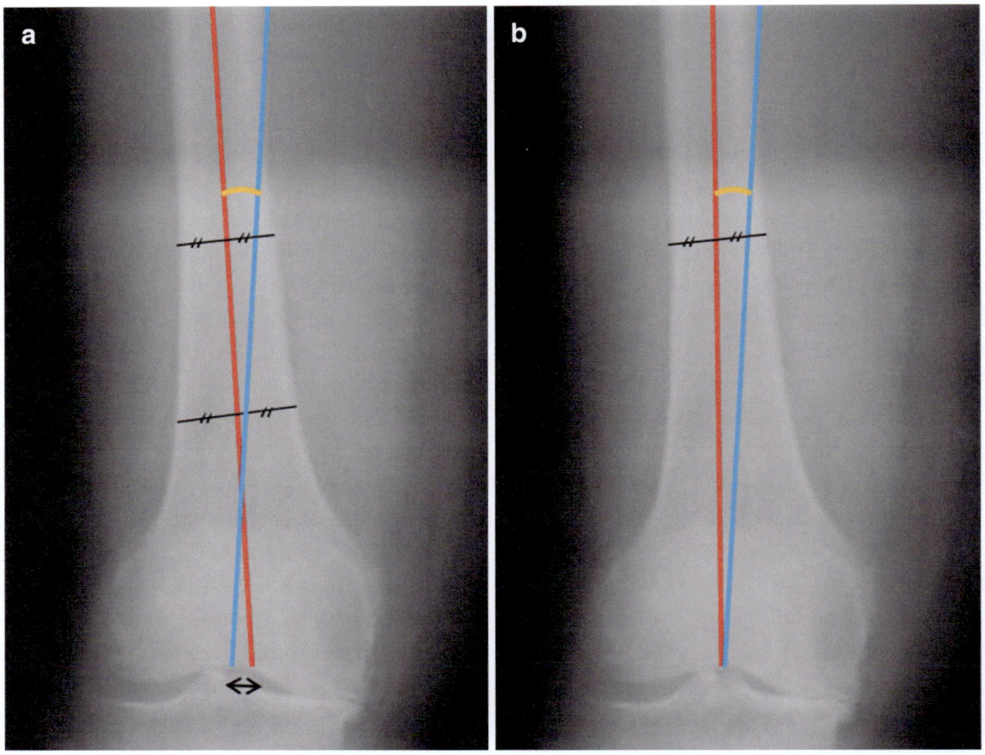

Fig. 47.1 Femoral coronal alignment: 2D planning. (**a**): Insertion direction of the intramedullary guide (the anatomical axis of the femur is defined as the straight line between the center of the femoral shaft located by bisecting the proximal and distal parts of the femur). (**b**): Insertion position of the intramedullary guide (center of the femoral intercondylar notch). The valgus angle should not be measured as if the guide is inserted from the center of the femoral intercondylar notch in preoperative planning; the guide is inserted from the slightly medial side from the center of the femoral intercondylar notch during surgery. Measuring the distance between the insertion position and the center of the femoral intercondylar notch is important when the anatomical axis of the femur is defined as the straight line between the center of the femoral shaft located by bisecting the proximal and distal parts of the femur

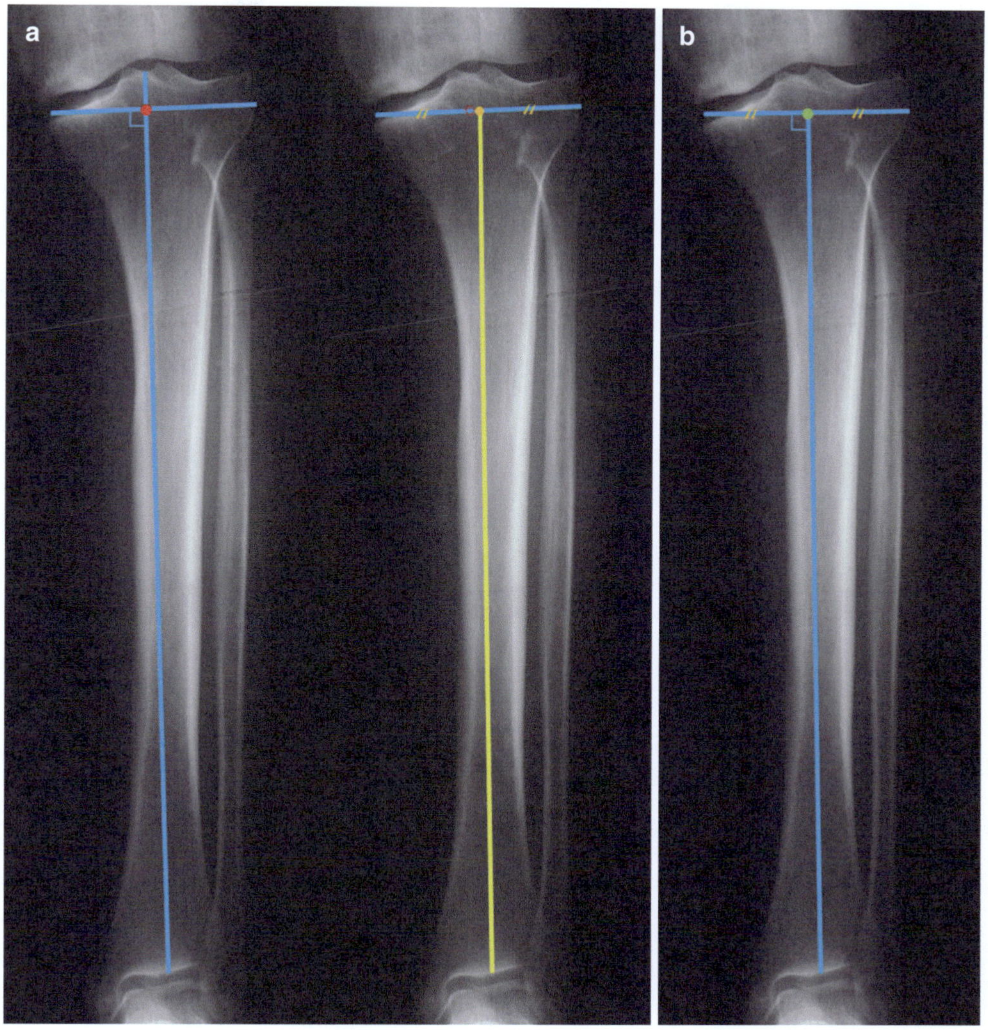

Fig. 47.2 Tibial coronal alignment: 2D planning. (**a**): Left: before cutting the proximal tibia; right: after cutting the proximal tibia. In this preoperative plan, the tibial mechanical axis is defined as the line connecting the center of the tibial joint surface and the center of the ankle joint. The intersection point of the mechanical axis and the cutting line (red circle) is not the center of the cutting plane, which is the center of the tibial component (yellow circle). Therefore, the postoperative tibial mechanical alignment (the line connecting the center of the tibial component and the center of the ankle joint) is not perpendicular to the cutting plane. (**b**): Ideal preoperative planning: the preoperative plan should be performed such that the perpendicular bisector of the planned cutting line (cutting plane) passes through the ankle joint

1.2.1 Femoral Side

Surgeons should check the insertion direction and position of the intramedullary guide. If the insertion direction of the intramedullary guide is parallel to the anatomical axis in the coronal plane, the insertion direction in the sagittal plane is also parallel to the anatomical axis. Measuring the distance between the distal position of the anatomical axis and the anterior part of the intercondylar notch (insertion position) and the angular difference between the anatomical axis and the anterior cortex (insertion direction, e.g., looking for a parallel line that serves as a reference) are useful for intraoperative reproducibility (Fig. 47.3). Because the cutting plane is easily changed by the cutting direction and errors in the sagittal plane, the size of the component should also be well planned by templating.

1.2.2 Tibial Side

The posterior tibial slope is measured using the proximal tibial anatomical axis on the lateral radiographs of the knee joint. The posterior tibial slope varies greatly from case to case, especially before osteoarthritis (OA) surgery, and the posterior tibial slope often exceeds 10° due to the influence of degeneration. The target angle of the posterior tibial slope should be determined based on the implant design used for the surgery. The same angle as that before surgery is desirable for the posterior cruciate ligament (PCL)-retaining design but not in excess of 10°, even if the preoperative posterior tibial slope is more than 10°. An excessive posterior tibial slope should be avoided in the posterior-stabilized design, and a smaller angle than that before surgery is desirable. The lateral radiograph of the full-length tibia is not affected by the imaging conditions as much as that of the femoral side. It is useful because it is possible to define the tibial mechanical axis and compare it with anatomical landmarks such as the anterior border of the tibia [2].

1.3 Rotational Alignment

CT slices are primarily used for preoperative planning; however, radiographs can be substituted on the femoral side.

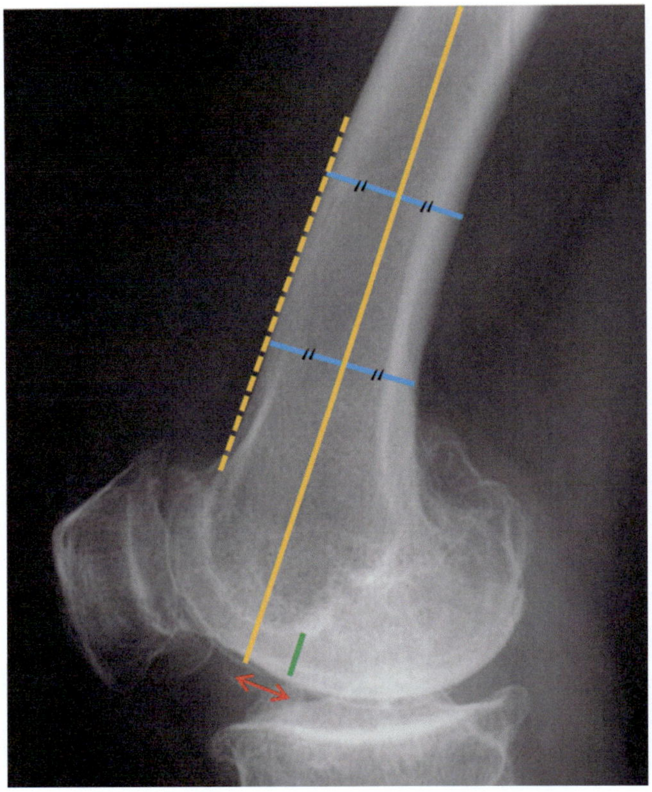

Fig. 47.3 Femoral sagittal alignment: 2D planning. Insertion position of the intramedullary guide: For intraoperative reproducibility, the distance between the distal position of the anatomical axis and the anterior part of the intercondylar notch should be measured. Insertion direction of the intramedullary guide: checking the relationship between the anatomical axis and the anterior cortex preoperatively is recommended

Magnetic resonance imagine (MRI) can consider cartilage conditions that cannot be evaluated by CT, allowing planning that more closely resembles the actual surgery [3]. From MRI studies [3] and intraoperative measurements [4, 5], the influence of the cartilage of the lateral posterior condyle of the femur on rotational alignment averages 1.5–1.7°, with large individual differences. Use of CT slices can evaluate not only the rotational alignment but also the bone anatomy (e.g., anteroposterior/mediolateral diameter ratio), the position of cysts and osteophytes, and the state of the PFJ in the knee extension position, which are difficult to identify by radiographs; hence, surgeons should use CT slices as much as possible. However, similar to radiographs, evaluation using CT slices can be easily affected by the degree of deformity, limb position, and imaging direction; therefore, surgeons should not fully trust the information from CT images as they are taken. If possible, it is desirable to perform accurate planning to define a plane perpendicular to the mechanical axis using 3D planning or to use images superimposed on several CT slices [6].

1.3.1 Femoral Side

When anatomical landmarks are used to define rotational alignment, CT and/or MRI are usually used to measure the angle between the posterior condylar line (connecting the medial and lateral posterior condyles) and the transepicondylar axis. The transepicondylar axis has two types: the surgical epicondylar axis (SEA) and the clinical epicondylar axis. The SEA is defined as the line connecting the lateral epicondylar prominence and the sulcus of the medial epicondyle. The clinical epicondylar axis is defined as the line connecting the lateral epicondylar prominence to the most prominent point of the medial epicondyle. Because the angle between the posterior condylar line and transepicondylar axis changes depending on the degree of degeneration and patient anatomy, surgeons should avoid using the same angle in all cases, which increases the risk of implant malposition. In lateral OA, the Whiteside line should also be used because the angle can be large owing to hypoplasia of the lateral condyle. In some cases, identifying the sulcus of the medial epicondyle is difficult when checking for an SEA. In such cases, measuring the angle between the clinical epicondylar axis and the posterior condylar line and subtracting approximately 3°, which is the difference between the clinical epicondylar axis and SEA, is reasonable.

Computed tomography (CT) scans are frequently used for the evaluation of rotational alignment; however, it has problems such as radiation exposure and cost; furthermore, the facilities that can be used are limited. It is better if radiographs can be substituted, and the kneeling view [7] (a method of kneeling at approximately 80° on the X-ray cassette and a posteroanterior projection parallel to the articular surface) and epicondylar view [8] (a method in which the knee joint is flexed 90° while sitting, the X-ray cassette is placed forward, and the X-ray beam is directed at a 10° upward angle) are useful for drawing the transepicondylar axis and posterior condyle lines.

1.3.2 Tibial Side

Tibial rotational alignment is typically expressed as the anteroposterior axis. The Akagi line is a useful anteroposterior axis for avoiding rotational mismatch when the femoral component is placed parallel to the SEA. The Akagi line connects the middle of the PCL to the medial border of the patellar tendon attachment site [9]. Preoperative planning of the Akagi line using CT slice images is not required because this axis is an anatomical landmark that can be defined intraoperatively. Similar to the femoral side, evaluation using CT slices can be easily affected by limb position and imaging direction, and preoperative planning from CT images should be avoided because of discrepancy with intraoperative findings.

2 3D Planning

2.1 Coronal Alignment

2.1.1 Tibial Side

Even if accurate preoperative planning can be performed using 2D planning, determining the exact center of the ankle joint is often difficult when an extramedullary guide is used during actual surgery. In addition, the tibia has torsion, which means a rotational mismatch between the proximal part of the tibia and the distal end of the tibia (the ankle joint) is present; attention should be paid to this torsion to align the guide. On average, the ankle joint is more externally rotated than the proximal part of the tibia. When the center of the ankle joint is determined in the direction of the ankle joint as seen from the anteroposterior axis of the ankle joint (the distal end of the guide is an extension of the anteroposterior axis of the ankle joint), the proximal tibia is unintentionally cut in the varus position with respect to the mechanical axis. To avoid tibial malalignment, it is important to first place the guide at right angles to the proximal tibial anteroposterior axis and then place the distal end of the guide at the center of the ankle joint. The center of the ankle joint should be aligned with the proximal tibial anteroposterior axis rather than with the anteroposterior axis of the ankle joint [10] (Fig. 47.4). This technique of determining the center of the ankle joint in terms of the proximal anteroposterior axis of the tibia is safe because, even if the distal end of the guide cannot be placed as targeted, a deviation of approximately 1 cm from the ideal position is within the range of 1.4° valgus to 1.7° varus [11]. In addition, a line connecting the proximal and distal one-thirds of the anterior border of the tibia is almost parallel to the mechanical axis and is useful as a reliable landmark for intraoperative use of the guide [12, 13]. This line can be used for both preoperative deformities: varus and valgus OA [14]; even with this method, it is important that the guide be positioned parallel to the line connecting the proximal and distal one-thirds of the anterior border of the tibia, while the rotational direction of the guide is adjusted to the anteroposterior axis of the proximal tibia [12–14].

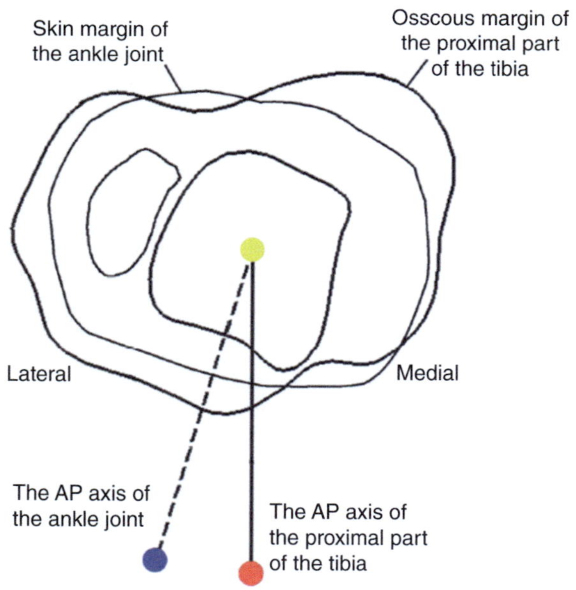

Fig. 47.4 The center of the ankle joint is judged in the direction of the ankle joint as seen from the anteroposterior (AP) axis of the ankle joint (the distal end of the guide is an extension of the AP axis of the ankle joint). The distal end of the guide shifted laterally in the direction of the knee joint, as observed from the AP axis of the knee joint (proximal tibia), which may have resulted in varus cutting of the proximal part of the tibia with respect to the mechanical axis

2.2 Sagittal Alignment

2.2.1 Femoral Side

Sagittal alignment has a significant impact on component size because the cutting plane is easily changed by the cutting direction and errors. From a 3D computer simulation using templating system, when the planned osteotomy is designed to align perpendicular to the anatomic axis, the anteroposterior diameter from the cutting plane is 1.7–3.0 mm larger if the osteotomy is extended 3° to 5° from the plan and 1.4–2.5 mm smaller if the osteotomy is flexed 3–5° from the plan [15]. The cutting direction in the sagittal plane may be in extension if the intraoperative judgment of the component size defined from the cutting plane is larger than that from the preoperative planning.

2.2.2 Tibial Side

Computer simulations using 3D bone models have shown that the line connecting the proximal and distal one-thirds of the anterior border of the tibia is tilted anteriorly by an average of 4.2° with respect to the mechanical axis, making it a useful and palpable anatomical landmark, such as the coronal plane [12]. Measuring the angle between the anterior border of the tibia and the articular surface using 3D planning is useful when the tibial component is placed according to the posterior slope of the articular surface.

2.3 Rotational Alignment

2.3.1 Femoral Side

Compared with the method accurately planned using the 3D bone model, 2D planning using CT slices results in a difference of more than 3° in the measurement of the angle between the posterior condylar line and the transepicondylar axis at a rate of approximately 10%, mainly because the

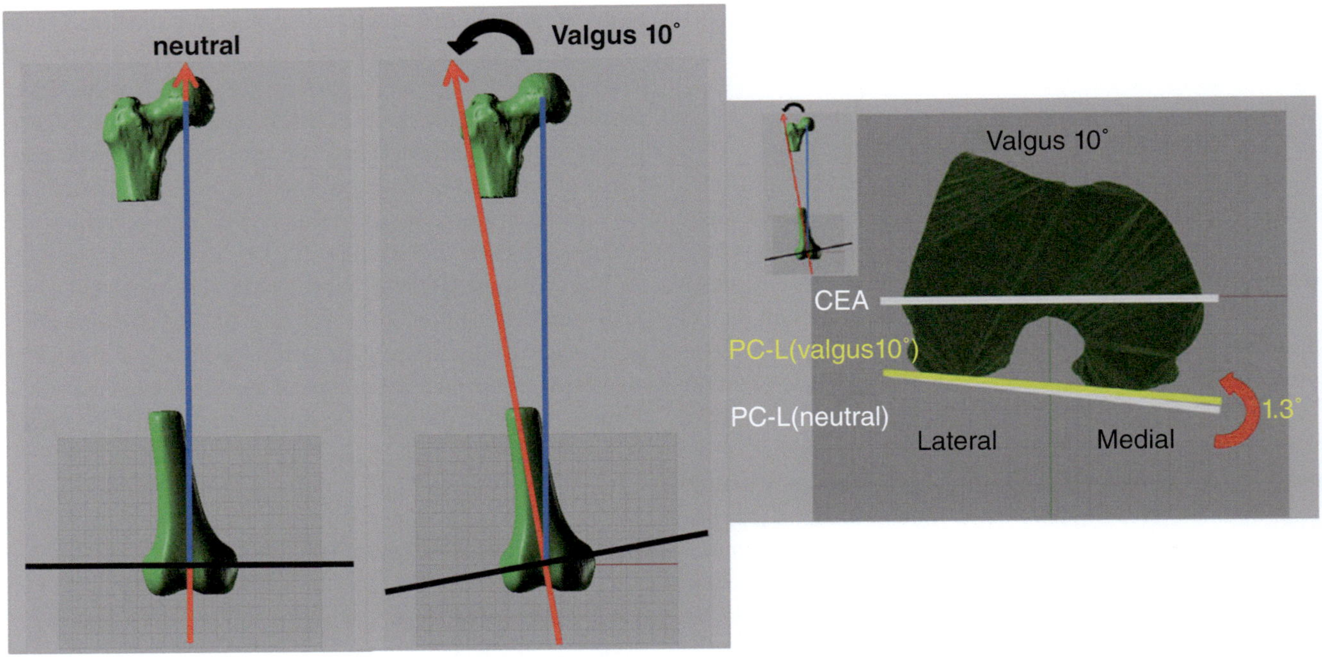

Fig. 47.5 Influence of the CT scanning direction on the angle between the posterior condylar line and the transepicondylar axis. Neutral: The CT scanning direction is parallel to the femoral mechanical axis (blue line). Valgus 10: The CT scanning direction is inclined in the valgus direction by 10° to the femoral mechanical axis (red line). The CT scanning direction could easily change the posterior condylar line in the CT slice

posterior condylar line is drawn differently depending on the scanning direction (Fig. 47.5) [16]. The CT slice planes tended to tilt, depending on the direction of CT scanning. In most cases, the scanning direction is perpendicular to the femoral anatomical axis rather than the femoral mechanical axis (TKA standard axis). Accurate planning in order to define a plane perpendicular to the mechanical axis and accurately measure the angles using 3D planning is recommended.

2.3.2 Tibial Side

The Akagi line was originally proposed based on a 2D plan for healthy knees [9], and it has been reported to be the most reproducible anteroposterior axis with the least rotational mismatch, even in varus OA knees with flexion contracture and varus deformity [17]. 3D planning enables detailed evaluation of the tibial cutting surface according to the actual surgical technique. (1) The line connecting the center of the cutting surface (center of maximum mediolateral length) with the medial one-sixth of the patellar tendon attachment site (4–5 mm lateral to the medial border) can be used as a substitute axis if the Akagi line is difficult to use, such as when the center of the PCL attachment part cannot be recognized [17, 18], (2) unexpected internal malrotation of the tibial component can occur if the anteroposterior axis is determined by the shape of the tibial cutting surface without using anatomical landmarks [19], and (3) the shape of the tibial component can affect the maximum cutting surface coverage easily; thus, surgeons should fully understand the features of the baseplate design concept and rotational alignment for an ideal tibial rotational position [20, 21].

References

1. Matsuda S, et al. Tibial shaft axis does not always serve as a correct coronal landmark in total knee arthroplasty for varus knees. J Arthroplast. 2003;18:56–62.
2. Tomite T, et al. The usefulness of planning using a preoperative lateral leg image to determine accurate posterior tibial slope in total knee arthroplasty. J Orthop. 2018;16:25–30.
3. Tashiro Y, et al. Articular cartilage of the posterior condyle can affect rotational alignment in total knee arthroplasty. Knee Surg Sports Traumatol Arthrosc. 2012;20:1463–9.
4. Fujii T, et al. Posterior condylar cartilage may distort rotational alignment of the femoral component based on posterior condylar axis in total knee arthroplasty. Surg Radiol Anat. 2012;34:633–8.
5. Miyasaka T, et al. Impact of posterior femoral condylar cartilage and posterior intercondylar distance on rotation of femoral component in total knee arthroplasty. BMC Musculoskelet Disord. 2020;21:498.
6. Oshima Y, et al. A computed-tomography-scan-based template to place the femoral component in accurate rotation with respect to the surgical epicondylar axis in total knee arthroplasty. Knee. 2018;25:195–202.
7. Takai S, et al. Kneeling view: a new roentgenographic technique to assess rotational deformity and alignment of the distal femur. J Arthroplast. 2003;18:478–83.

8. Kanekasu K, et al. Axial radiography of the distal femur to assess rotational alignment in total knee arthroplasty. Clin Orthop Relat Res. 2005;434:193–7.
9. Akagi M, et al. An anteroposterior axis of the tibia for total knee arthroplasty. Clin Orthop Relat Res. 2004;420:213–9.
10. Mizu-uchi H, et al. The effect of ankle rotation on cutting of the tibia in total knee arthroplasty. J Bone Joint Surg Am. 2006;88:2632–6.
11. Mizu-uchi H, et al. The adjustment of the rotational alignment of the distal end of the extramedullary guide to the anteroposterior axis of the proximal tibia in total knee arthroplasty. J Knee Surg. 2021; https://doi.org/10.1055/s-0040-1722660.
12. Fukagawa S, et al. Anterior border of the tibia as a landmark for extramedullary alignment guide in total knee arthroplasty for varus knees. J Orthop Res. 2011;29:919–24.
13. Nishikawa K, et al. Accuracy of proximal tibial bone cut using anterior border of tibia as bony landmark in total knee arthroplasty. J Arthroplast. 2015;30:2121–4.
14. Kawahara S, et al. Anterior tibial border as a landmark for extramedullary alignment guides for total knee arthroplasty in valgus knees. J Orthop Res. 2015;33:1897–9.
15. Nakahara H, et al. Sagittal cutting error changes femoral anteroposterior sizing in total knee arthroplasty. Clin Orthop Relat Res. 2012;470:3560–5.
16. Okamoto S, et al. Two-dimensional planning can result in internal rotation of the femoral component in total knee arthroplasty. Knee Surg Sports Traumatol Arthrosc. 2016;24:229–35.
17. Ma Y, et al. Bony landmarks with tibial cutting surface are useful to avoid rotational mismatch in total knee arthroplasty. Knee Surg Sports Traumatol Arthrosc. 2019;27:1570–9.
18. Kawahara S, et al. Medial sixth of the patellar tendon at the tibial attachment is useful for the anterior reference in rotational alignment of the tibial component. Knee Surg Sports Traumatol Arthrosc. 2014;22:1070–5.
19. Ushio T, et al. The anteroposterior axis of the proximal tibia can change after tibial resection in total knee arthroplasty: computer simulation using Asian osteoarthritis knees. J Arthroplast. 2017;32:1006–12.
20. Ma Y, et al. Effects of tibial baseplate shape on rotational alignment in total knee arthroplasty: three-dimensional surgical simulation using osteoarthritis knees. Arch Orthop Trauma Surg. 2018;138:105–14.
21. Hirakawa M, et al. Evaluation of the rotational alignment of the tibial component in total knee arthroplasty: position prioritizing maximum coverage. Eur J Orthop Surg Traumatol. 2017;27:119–24.

Part VIII

Basic Surgical Techniques

48 Important Aspects of Surgical Techniques

Toru Suguro

To obtain excellent postoperative results, it is necessary to estimate the degree of preoperative deformities, rotational abnormalities, and contractures. The goal of the preoperative evaluation is to predict whether surgery will result in superior function. In TKA, it is important to obtain near-normal joint function (kinematics) by acquiring correct alignment, fixation, and good stability with respect to the functional axis of the lower limb. Therefore, it is important to establish reliable surgical techniques as well as component design and materials.

1 Preoperative Planning Is Directly Linked to Surgical Techniques

Various elements are included in knee joint deformities. The basic forms of joint deformities are varus, valgus, and flexion deformities. Preoperative information should include plain radiographs (in two directions), plain radiographs under weight bearing, and CT images (MDCT Reconstruction images). A meticulous physical examination, which considers the relationship between the deformity and soft tissues from physical examinations, such as femoral deformity, osteophytes, position of the patella, tibia bowing, malposition, and possibility of correcting the deformity, ensures optimal surgical techniques. Additionally, it is important to confirm the degree of manual correction before surgery. Understanding the preoperative condition of the patient is important to perform the surgery smoothly and obtain excellent postoperative outcomes.

T. Suguro (✉)
Japan Research Institute of Artificial Joint, Chiba, Japan
e-mail: smat522@kuhp.kyoto-u.ac.jp

2 Important Aspects of the Surgical Technique

2.1 The Choice of Implant Determines the Basic Surgical Technique

As the surgical technique differs for the CR and PS types, the implant model should be selected preoperatively. CR implants utilize the measured resection technique and anatomical measured resection method, while PS-type implants are based on the gap balancing technique (modified gap balancing technique).

2.2 Important Aspects of Surgical Technique

2.2.1 Incision

Skin incisions play a crucial role in joint exposure. To ensure a good surgical field of view, a midline or medial incision is recommended; however, a midline incision is preferred for valgus deformities (Fig. 48.1). Treatment of the subcutaneous tissue is also important, and the superficial thin facial layer should be protected to ensure sutureability. This approach protects the blood volume of the skin, minimizing postoperative surgical layer issues and allowing for postoperative skin mobility.

2.2.2 Joint Exposure

The methods can be classified into medial parapatellar, midvastus, subvastus, and lateral parapatellar approaches.

The medial parapatellar approach provides the widest view. It is important to have a maximal field of view for anatomical understanding. The key is to separate the quadriceps tendon and make an incision so that the stitch can be firmly resutured. Patellar eversion can be easily performed by excising the lateral plica, moving the patella inward, and flipping it directly above the femur. Visualizing the entire joint without excessive soft tissue loading is the key. (Fig. 48.2).

S. Matsuda (ed.), *Advances in Total Knee Arthroplasty*, https://doi.org/10.1007/978-981-97-4920-1_48

2.2.3 Checking the Bone Shape

In particular, it is necessary to confirm the structure of the condyle from the distal region, estimate the state of the cartilage and bone, and confirm the pathological lesion (Fig. 48.3). This information allows to predict the patient's post-implantation condition.

2.2.4 Importance of Osteotomy (Table 48.1)

Both independent and dependent cuts are used, each serving different purposes. Independent cuts aims to reproduce bone morphology, while dependent cuts involve reconstruction based on the structure of the implant. The osteotomy angle is set from the load axis observed from the front; however, rotational alignment should be carefully monitored because it affects the range of motion. Excessive external or internal rotation during replacement can change the center of rotation and limit flexion. This is because, when the center of rotation moves, the length of the moment arm (i.e., the distance from the articular surface) is affected, thereby limiting the ROM, contributing to poor performance.

- **Important aspects for preparation before osteotomy:** Performing the osteotomy while checking anatomical landmarks ensures a nearly accurate execution based on the preoperative plan. However, preoperative ligament balance is often not sufficiently restored. In many cases, the MCL is shortened and tightly medial because of varus deformities. The causes are varus deformity of the proximal tibia, medial displacement of the tibia, and osteophytes, which often cause overstraining of the MCL. Reduction osteotomy, which involves removing excess bone proximal to the tibia, helps to relieve this tension. Osteophyte resection and MCL release are performed to ensure simultaneous posterior clearance and flexion space.
- **Important aspects of femoral osteotomy:** It is also important to check the rotational position of the femur. The most important aspect of femoral osteotomy is determining the center of rotation. Therefore, it is necessary to identify the anatomical landmarks.
- The femoral landmarks include the (1) anterior femoral cortex, (2) clinical epicondyle axis (CEA), (3) surgical epicondyle axis (SEA), (4) posterior condyle axis (PCA), (5) intercondylar notch, and (6) patellar groove.
- It is also important to accurately determine the SEA, which is the line connecting the center of the basin-shaped sulcus of the MCL attachment to the epicondyle of the LCL attachment. The center of the MCL is easily identified by firmly pressing the sulcus with the palm. During osteotomy, care should be taken to avoid injuring the popliteus tendon during resection of the lateral condyle, as it regulates the external rotation of the tibia. Damage to the popliteus tendon results in excessive external rotation and poor axis balance. The posterior condylar axis should also be checked simultaneously. In most cases, the PCA is internally rotated with respect to the SEA by approximately 3°. Because the anterior surface of the femur is also a landmark, the osteotomy guide is adjusted to confirm whether the grand piano sign was formed after the osteotomy. A distal osteotomy can be performed as planned by checking the amount of bone resection in the medial and lateral condyles. In femoral osteotomies, checking the distal, anterior-posterior, and medial-lateral sides of the femur is crucial to make comprehensive decisions.
- Osteotomy is performed sequentially in the distal, medial anterior, lateral anterior, and posterior condyles and Schanfer. As a rule, cutting starts from the medial condyle

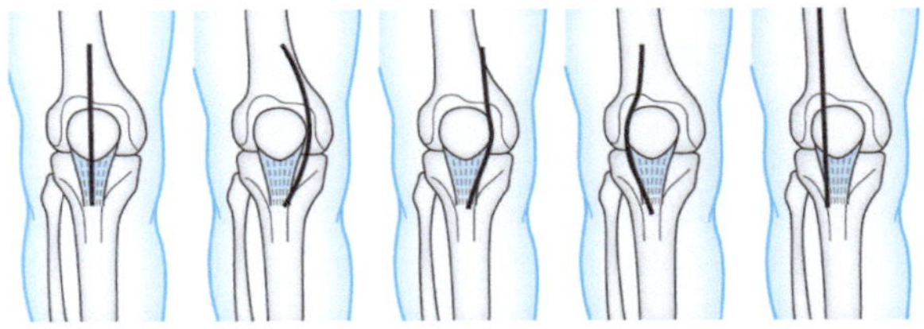

- Varus Knee: A gentle arcuate skin incision on the inside is commonly used.This is suitable for Japanese people who often kneel.
- Valgus Knee: Midline incision is used. Easy to approach to the lateral compartment.

Fig. 48.1 Basic incision

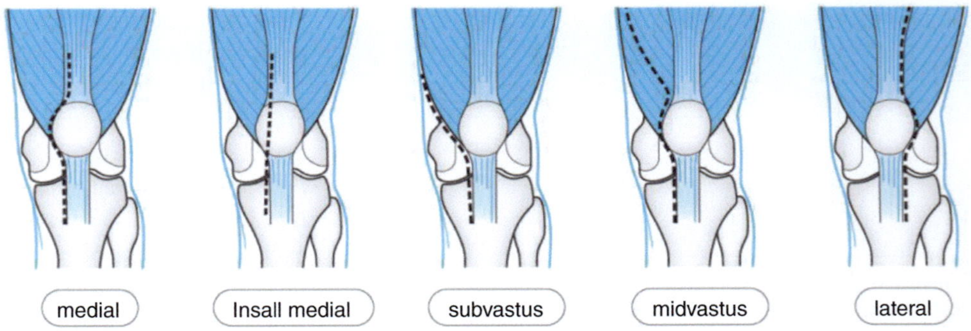

Fig. 48.2 Basic joint exposure method

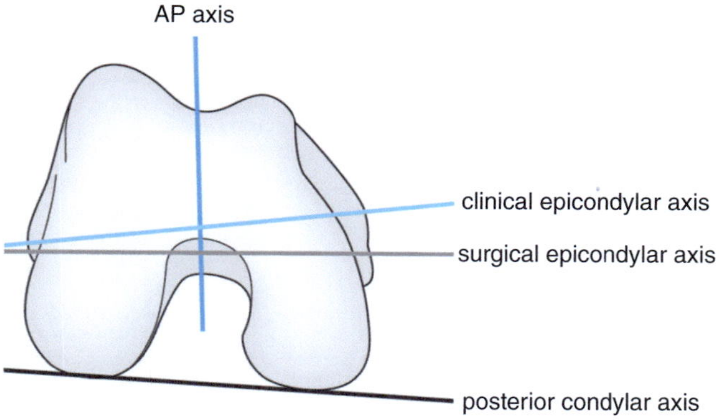

Fig. 48.3 Anatomical structure

Table 48.1 Check points in femoral bone resection

Important points in femoral bone resection
1. Find surgical epicondylar axis
2. Confirm PC line, SEA, and grand piano sign
3. Confirm the attachments of MCL, LCL, PCL, ACL, and popliteus tendon
4. Osteotomy in the distal femur
5. Osteotomy in the anterior part of the femur
6. Osteotomy in the posterior condyle of thefemur
It is important to proceed with basic procedures

(priority should be given to the function of the MCL). Osteotomy of the posterior condyle (the amount of bone resection and degree of external rotation should be checked). Care should be taken not to cut the popliteal tendon when cutting the lateral schanfer because the popliteus tendon plays an important role in restraining external rotation motion. This is noteworthy because the tibia becomes externally rotated when the popliteus tendon is cut. If rotational alignment is correct, bone resection parallel to the SEA results in a ground piano shape in the anterior cortex of the femur (Fig. 48.4). Therefore, it is important to verify its shape. The posterior offset must also be secured and an optimal implant should be selected during preoperative planning. Bellemans et al. [1] defined the distance from the posterior cortex of the femoral shaft to the posterior condylar offset and reported that when this distance decreased, the posterior condylar angle of flexion reduced and the femoral condyle quickly contacted the tibial insert.

- **Important aspects of tibial osteotomy** (Fig. 48.5)**:** The tibia is similarly treated with medial osteophyte resection, followed by confirmation of the tibial axis and placement of the osteotomy guide. Many instruments are tilted posteriorly; therefore, it is important to align the guide to the exact frontal plane, that is, the A-P axis (Fig. 48.5). Rotational malpositioning of the guide results in an oblique osteotomy with a posterior tilt. However, this important aspect requires further investigation. Subsequently, the osteophytes around the PCL are removed, crucial for CR-type models. Excellent range of motion is achieved by ensuring normal PCL length and tension.
- **Important aspects of patellar resurfacing** (Table 48.2)**:** There are two surgical techniques for patellar resurfacing: onset and inset. For inset, confirming the central ridge is vital, while onset requires accurate osteotomy. The landmarks for osteotomy are the patellar tongue and quadriceps tendon attachment sites, which should be confirmed using both surgical techniques.

Additionally, patellar tracking should be checked to avoid postoperative anterior pain. By verifying these points, it is possible to obtain good range of motion and clinical results.

2.2.5 Important Aspects of Soft Tissue Balancing

After optimal bone resection, ligament balance is checked with the inserted implants. Maintaining the MCL function is essential for TKA. Correction of soft tissue tension is essential for any deformity (varus deformity, valgus deformity, flexion deformity, etc.).

In particular, for the frequently observed varus deformities, maintaining soft-tissue balance is critical during correction. Soft-tissue release should be performed sequentially to achieve ligament balance on the medial side. The tip first releases the deep MCL and meniscal capsular ligament with the minimum amount necessary and then advances while checking the overall balance. The condition of the osteophytes should be checked, considering that the formation of a large medial osteophyte and deformed proximal tibia resulting from remodeling can stress the MCL. In most cases, MCL release and reduction osteotomy achieve an optimal balance (Fig. 48.6). This is an important consideration.

Therefore, planning the amount of MCL release from the beginning and considering the MCL relaxation effect of reduction osteotomy help minimize MCL release, limiting it up to approximately 5 cm from the articular surface. If this cannot be corrected, the periosteum is released in a sleeve-like manner. It is important to recognize that there are flexion contractures. It should also be noted that there was relative shortening of the PCL. In osteoarthritis, posterior osteophytes may impair extension, requiring complete resection.

1. In the CR type, the interaction between the MCL and PCL is involved in joint stability and physiological joint movement; therefore, a PCL release procedure is also required. If the balance between the MCL and PCL is optimal, natural medial pivot motion can be reproduced. Because the LCL has some slack physiologically, that is, the LCL changes its distance slightly by internal and external rotation of the tibia, it is necessary to keep in

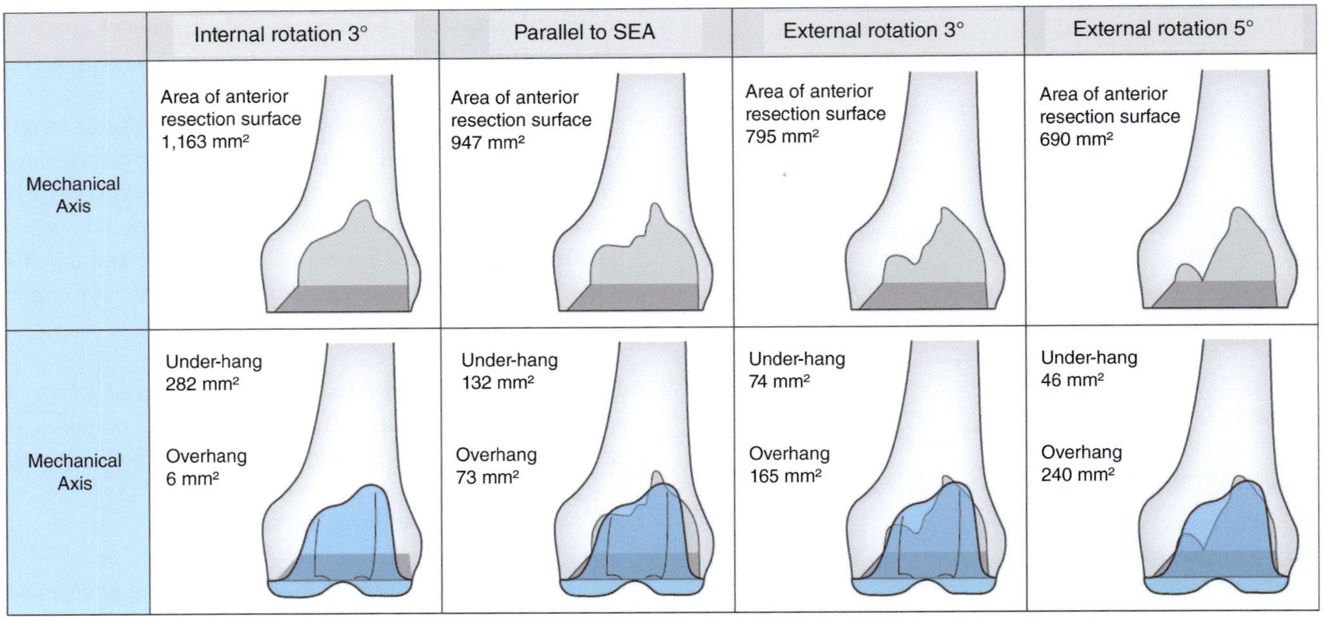

Fig. 48.4 SEA and PC line in frontal and distal images

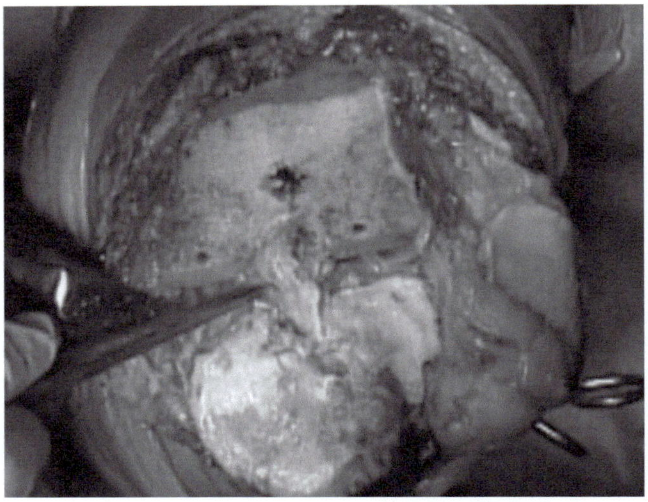

Fig. 48.5 Full exposure of the tibia enables safe osteotomy

Table 48.2 Check points in tibial bone resection

Important points in femoral bone resection
1. Confirm rotational alignment; A-P axis, Akagi's line
2. Confirm the tibial axis
3. Confirm the PCL attachment
4. Confirm the medial and lateral eminence of the tibia
5. Confirm the lateral edge of the tibia and Gerdy tuberocity
6. Resection of osteophytes at the medial and lateral side
7. Assessment of bone defect and the shape of the proximal tibia
8. MCL is released when it is tight (not over-released!).
9. Clear view of the tibia is obtained by drawing the tibia forward in flexion and external rotation.
10. Confirm the bone axis in the sagittal view(parallel to the fibula)

mind that the slack state of the LCL is physiologically in an intermediate position.

2. For the PS type, the gaps between the extension and flexion positions must be equal. When the PCL is resected, the gap widens, resulting in an imbalance between extension and flexion. To avoid this, starting the osteotomy from the tibia at a right angle to the tibial axis and then performing the femoral osteotomy is essential. Performing a slightly larger osteotomy of the distal femur is also a good idea. Proper observation of POST/CAM engagement is crucial for proper medial pivot motion. Posterior sagging of the tibia can cause problems in the patellofemoral joint, warranting attention. When rotational abnormalities are observed, it is necessary to investigate and address their causes. A relatively common problem is the ITT overstrain. The key point is to optimize the tension by paying attention to the difference in function between the anterior and posterior ITT.

2.2.6 Important Aspects in Closing the Wound

Once proper soft tissue tension has been achieved through the improvement of bony alignment, the tension of the medial and lateral retinaculum at the time of wound closure should be monitored. The tension of the retinaculum in flexion and extension positions should be checked during suturing to ensure optimal tension. Closing the joint in the flexion position is useful for reducing pain, because the tension of the retinaculum in the flexed position is consistent with the tension of the soft tissues in the flexed position after surgery.

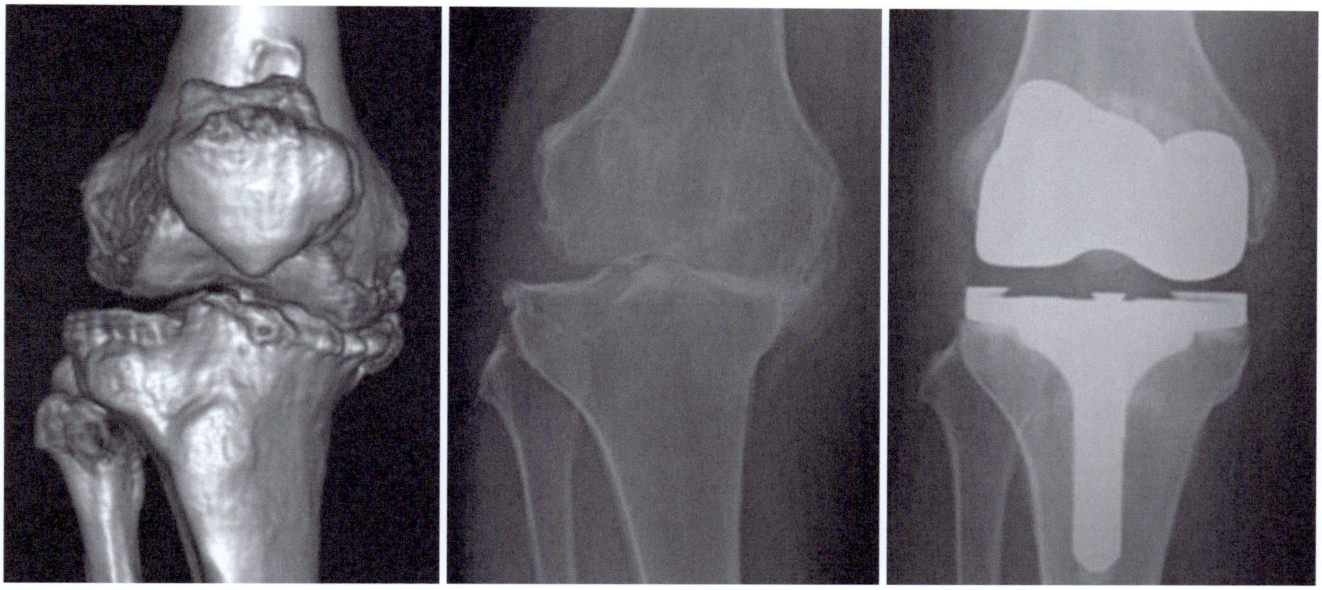

Fig. 48.6 Reduction osteotomy of the tibia and osteophyte resection (showing in MDCT) improves the balance of the MCL

Suturing of the subcutaneous tissue is important. The superficial fascial layer promotes skin with mobility, blood flow, tissue flexibility and prevents scarring, thereby reducing postoperative discomfort. Therefore, proper anatomical reconstruction is crucial for successful outcomes.

Reference

1. Bellemans J, et al. Fluoroscopic analysis of the kinematics of deep flexion in total knee arthroplasty. Influence of posterior condylar offset. J Bone and Joint Surg Br. 2002;84:50–3.

49 Surgical Approach

Kengo Harato

In TKA, the position of the incision and the approach method are extremely important, and a skilled surgeon can quickly perform skin incision to exposure [1]. Orthopedic surgeons need to be familiar with the differences in the position of the incision and various approach methods.

1 Incision

1.1 Concept

The most important point when considering the position of the incision is the running of the arterial network around the knee and the infrapatellar branch of saphenous nerve (Figs. 49.1 and 49.2). The main skin incisions in TKA are (1) midline, (2) anteromedial, and (3) anterolateral. Incision-induced damage to the infrapatellar branch of the saphenous nerve can cause difficulty in kneeling after TKA surgery, which is known to contribute to decreased postoperative satisfaction. In an RCT conducted in 2018, it was reported that the area of cutaneous hypesthesia was significantly smaller in the anterolateral skin incision group than the anteromedial skin incision group, and anterolateral skin incision may provide less cutaneous hypesthesia and better kneeling ability after TKA compared with anteromedial skin incision [2]. Although transverse incision has been shown to be more advantageous for wound difficult as it is impossible to extend the incision toward the longitudinal direction [3]. TKA was attempted to be performed with a medial 8–12 cm skin incision as minimally invasive surgery (MIS) [4]. However, skin problems may occur due to poor exposure and forceful intraoperative skin traction in MIS. Therefore, it is better to perform a less invasive TKA as much as possible without forceful skin traction, based on the competence of the surgeon.

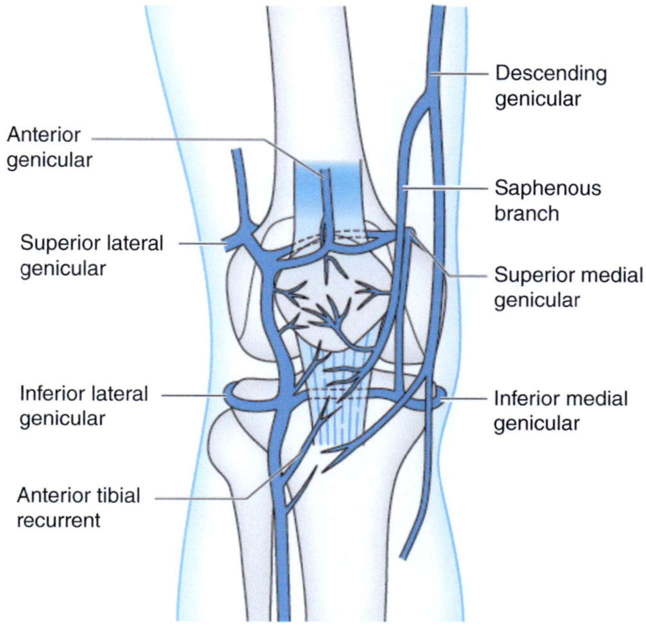

Fig. 49.1 Arterial network around the knee

1.2 Procedure

The most common skin incision in TKA for medial OA is anteromedial or midline, and is often performed from 2 to 3 cm proximal to the patella to 1 cm distal to the tibial tubercle. A skin incision from the proximal patellar border to the tibial tuberosity is sufficient for UKA. When performing an anterolateral skin incision for medial OA, a length of 4–5 cm proximal to the proximal patellar border and 2 cm distal to the tibial tuberosity is required (Fig. 49.3). In the case of obese patients or stiff knee with restricted range of motion,

K. Harato (✉)
Department of Orthopaedic Surgery, Keio University School of Medicine, Tokyo, Japan
e-mail: harato@keio.jp

S. Matsuda (ed.), *Advances in Total Knee Arthroplasty*, https://doi.org/10.1007/978-981-97-4920-1_49

longer incision is necessary. Forceful skin traction of the edge of the incision, especially the distal end, should be avoided intraoperatively, and it is better to perform a sufficiently long skin incision for good exposure.

Because the oxygen tension of the skin is lower in the lateral and distal areas of the skin incision, even if the incision is small [5], subcutaneous dissection should be performed sharply with a scalpel to minimize the damage for the subcutaneous tissue. It is also believed that the tourniquet should be used at the lower pressure, since a high-pressure tourniquet may cause skin necrosis due to the lower oxygen tension on the lateral side of the incision (Fig. 49.4). The characteristics of each skin incision is found in Fig. 49.5.

1.3 Problems

Problems around the surgical wound include hypertrophic scarring, numbness on the lateral side, and anterior knee pain during kneeling. Although there are many factors affecting the wound condition, such as patient factors, postoperative rehabilitation and anticoagulant therapy, the position and length of the skin incision also affect it. Thus, the surgeon should be familiar with the advantages and disadvantages of various skin incisions in TKA.

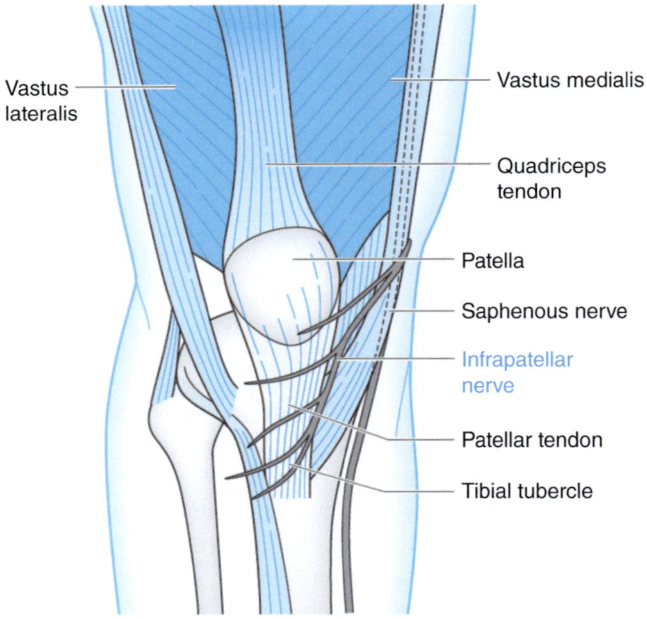

Fig. 49.2 Infrapatellar branch of saphenous nerve

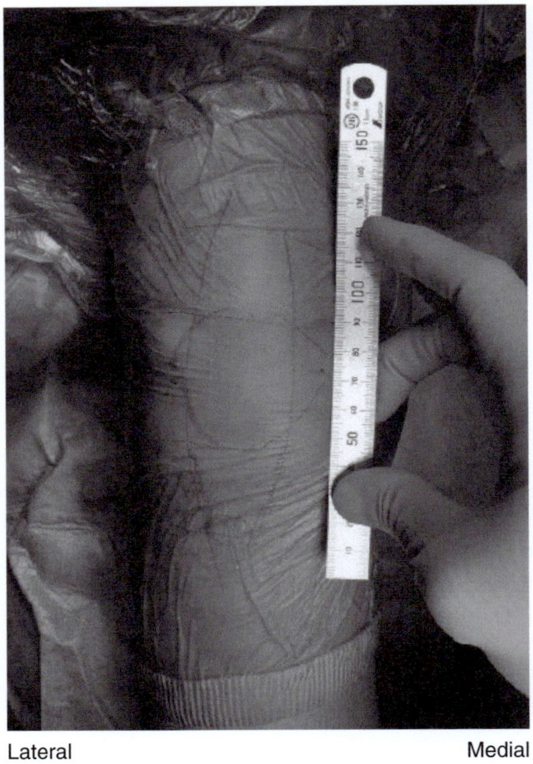

Fig. 49.3 Medial and lateral skin incision in the right knee

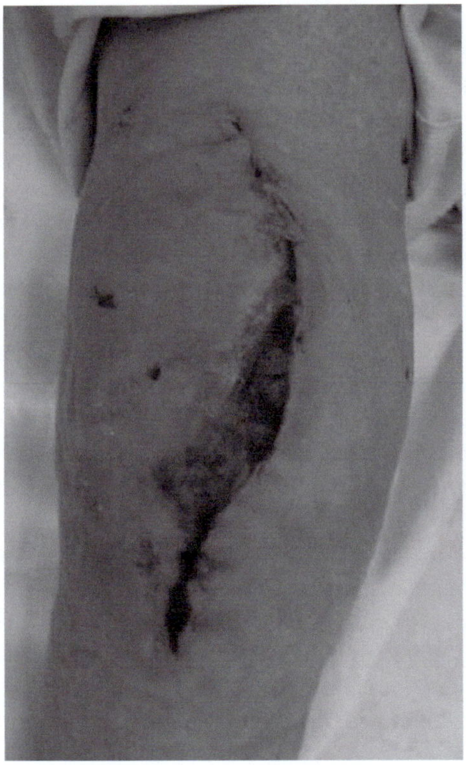

Fig. 49.4 Example of skin necrosis. Necrosis occurs on the lateral side of the skin

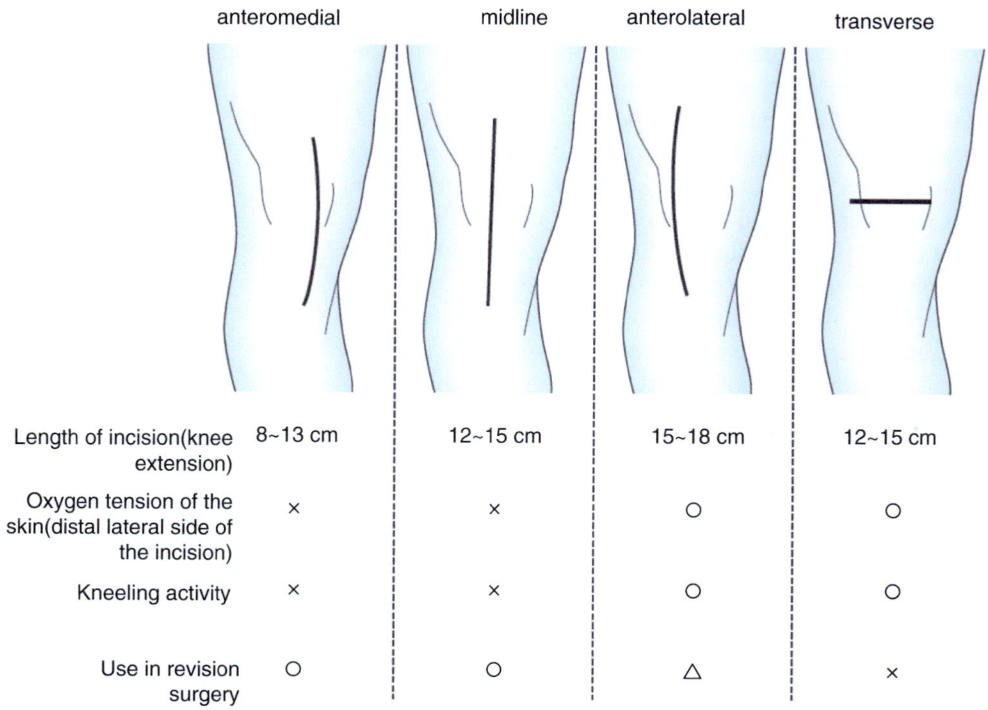

Fig. 49.5 Characteristics of each skin incision

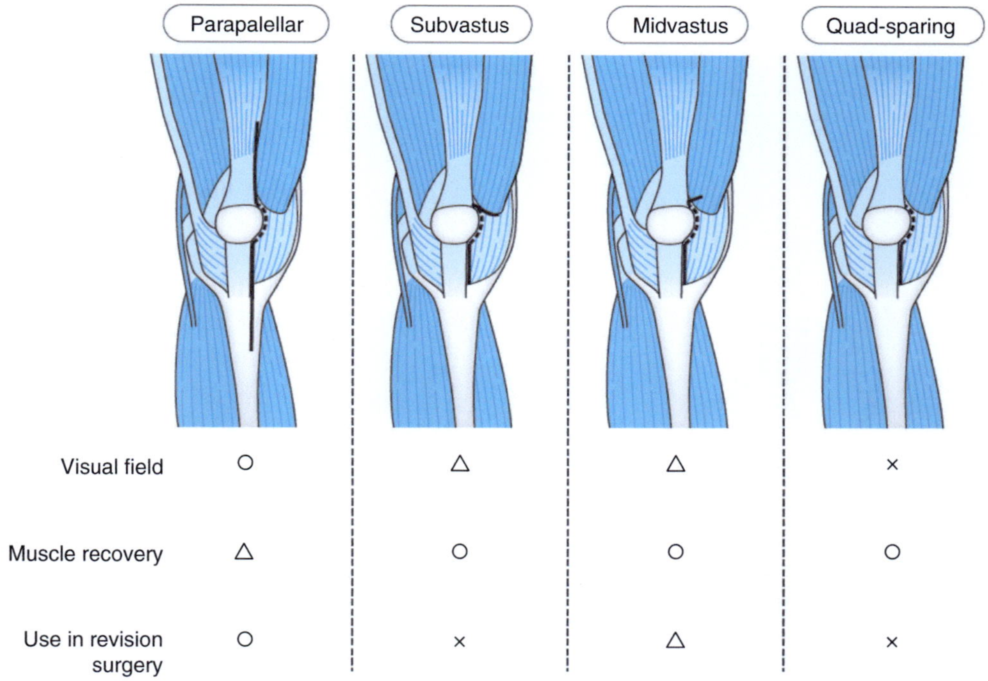

Fig. 49.6 Characteristics of each approach

2 Approach

2.1 Concept

There are four approaches that are usually used in TKA: parapatellar, midvastus, subvastus, and quadriceps sparing (Fig. 49.6) [6, 7]. Parapatellar has good exposure and an excellent surgical field. Quadriceps sparing is advantageous for postoperative recovery, but the visual field is poor, the procedure is complicated, and the surgical time is long, compared to others. Further, revision TKA using quadriceps sparing is impossible. Lateral parapatellar approach through a lateral skin incision may be used for lateral OA [8]. In case of severe contracture, rectus snip may be necessary, and turndown or tibial tuberosity osteotomies is sometimes used in revision cases (Fig. 49.7). Rectus snip does not require delayed postoperative rehabilitation and is a reasonable procedure for complex primary TKA, in which exposure is difficult. Accurate component placement and proper ligament balance are the primary goals in TKA. Therefore, similar to the skin incision, it is necessary to avoid longer operation time and malalignment due to poor exposure [9], and surgeons should consider less invasive surgery as much as possible according to their surgical skill. Although there is no clear definition of MIS, less invasive surgery is more advantageous for functional recovery than conventional methods [10, 11].

2.2 Procedure

Basically, it is desirable to perform TKA using a surgical approach selected by surgeon based on the situation with the enough knowledge of each approach. Considering the final wound closure, capsule is cut about 5 mm apart from the medial edge of the patellar tendon and the patella (Fig. 49.8). Infrapatellar fat pad may cause impingement during flexion [12] and interfere with the visual field, and thus, it should be resected to a certain extent. The patellar eversion during surgery depends on the surgeon's preference, but it may reduce oxygen tension of the surgical wound and cause patellofemoral joint problems, which leads to worse wound healing [13, 14]. As primary TKA can be performed without patellar eversion, surgeons should avoid patellar eversion if possible. Although TKA can be done without tourniquet, it is also important to note that the patellar tracking may differ depending on whether the tourniquet is turned on or not [15].

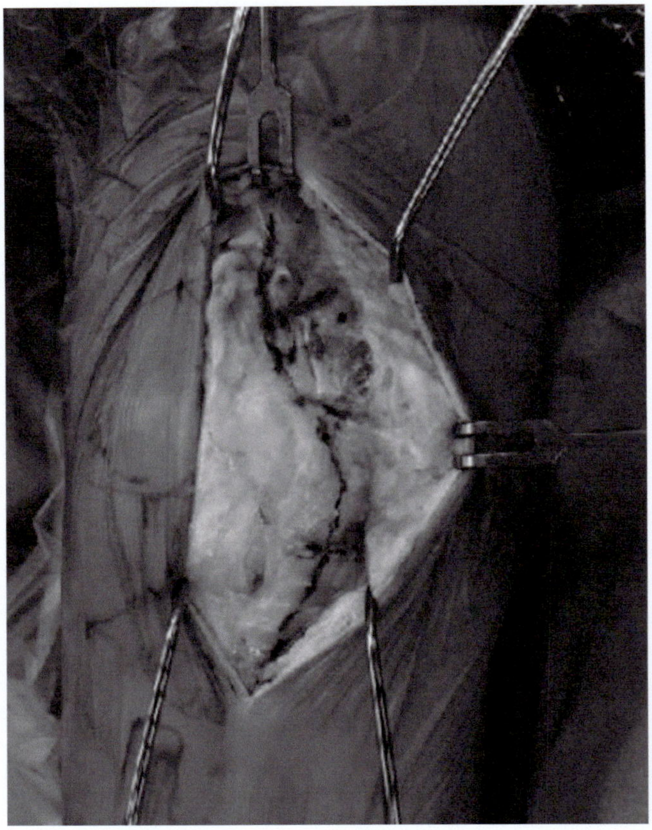

Fig. 49.8 Medial approach in the right knee. The assumed lines of parapatellar, midvastus, and subvastus are shown

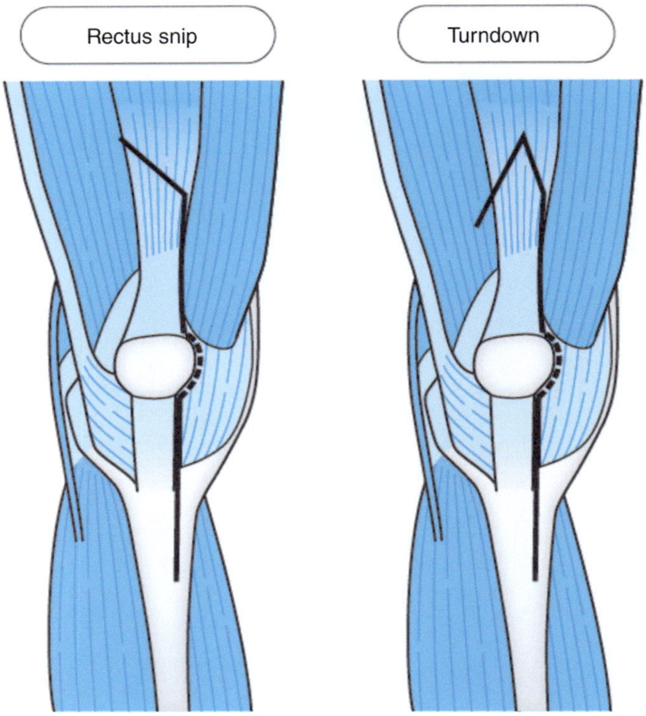

Fig. 49.7 Approach method used in complex cases

2.3 Problems

Quadriceps sparing may be advantageous for early postoperative recovery, but it may cause poor exposure. If surgeons are not familiar with this technique, traction of the wound edge will lead to skin damage (Figs. 49.9 and 49.10).

Fig. 49.9 Quadriceps sparing approach in the right knee. It is difficult to see the lateral femoral condyle

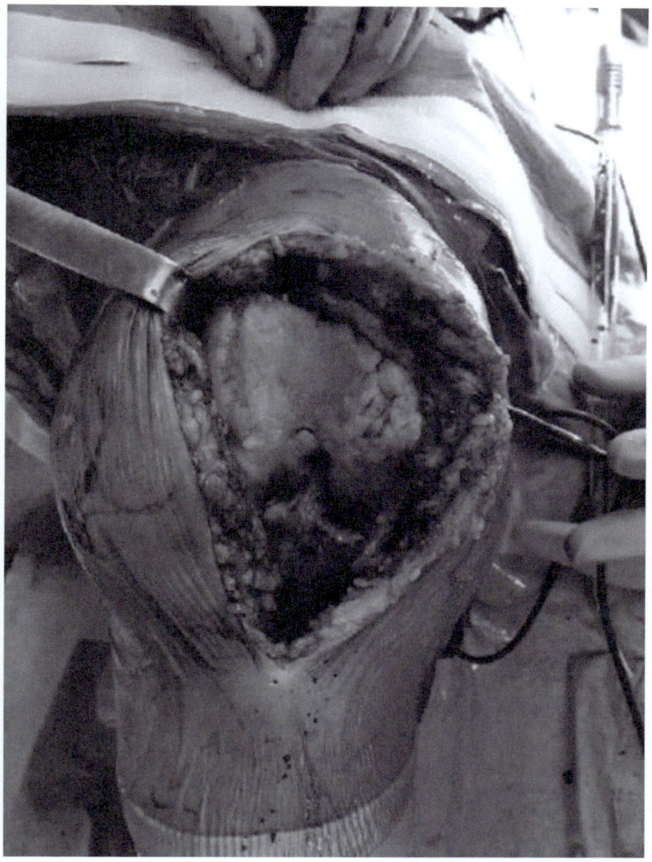

Fig. 49.10 Parapatellar approach in the right knee (same case as Fig. 49.9). A sufficient field of view can be secured up to the lateral femoral condyle

References

1. Harato K, et al. What are the important maneuveres for beginners to minimize surgical time in primary total knee arthroplasty? Knee Surg Sports Traumatol Arthrosc. 2016;24:2704–9.
2. Tsukada S, et al. Cutaneous hypesthesia and kneeling ability after total knee arthroplasty: a randomized controlled trial comparing anterolateral and anteromedial skin incision. J Arthroplast. 2018;33:3174–80.
3. Ojima T, et al. Transverse incision advantages for total knee arthroplasty. J Orthop Sci. 2011;16:524–30.
4. Watanabe T, et al. Is a minimally invasive approach superior to a conventional approach for total knee arthroplasty? Early outcome and 2- to 4-year follow-up. J Orthop Sci. 2009;14:589–95.
5. Aso K, et al. Transcutaneous oxygen tension in the anterior skin of the knee after minimal incision total knee arthroplasty. Knee. 2012;19:576–9.
6. Niki Y, et al. Is minimally invasive surgery in total knee arthroplasty really minimally invasive surgery? J Arthropl. 2009;24:499–504.
7. Niki Y, et al. Effects of four different surgical approaches on intraoperative joint gap in posterior-stabilized total knee arthroplasty. Knee Surg Sports Traumatol Arthrosc. 2012;20:2026–31.
8. Sekiya H, et al. Lateral approach is advantageous in total knee arthroplasty for valgus deformed knee. Eur J Orthop Surg Traumatol. 2014;24:111–5.
9. Niki Y, et al. Accuracy of implant positioning for minimally invasive total knee arthroplasty in patients with severe varus deformity. J Arthroplast. 2010;25:381–6.
10. Tashiro Y, et al. Minimally invasive versus standard approach in total knee arthroplasty. Clin Orthop Relat Res. 2007;463:144–50.
11. Tsuji S, et al. Is minimally invasive surgery-total knee arthroplasty truly less invasive than standard total knee arthroplasty? A quantitative evaluation. J Arthroplast. 2010;25:970–6.
12. Maeno S, et al. Patellar impingement against the tibial component after total knee arthroplasty. Clin Orthop Relat Res. 2006;452:265–9.
13. Harato K, et al. What are the important surgical factors affecting the wound healing after primary total knee arthroplasty? J Orthop Surg Res. 2016;11:7.
14. Hasegawa M, et al. Changes to patellar blood flow after minimally invasive total knee arthroplasty. Knee Surg Sports Traumatol Arthrosc. 2009;17:1195–8.
15. Komatsu T, et al. The effect of surgical approaches and tourniquet application on patellofemoral tracking in total knee arthroplasty. J Arthroplast. 2003;18:308–12.

Surgical Instruments

50

Tadashi Tsukeoka

1 Intramedullary Femoral Cutting Guide

1.1 Intramedullary Rod Insertion

The alignment of the femoral component can be affected by the direction of rod insertion [1]. Coronal alignment influences the postoperative outcomes. Additionally, sagittal plane alignment is crucial as it affects the size of the femoral component [2, 3] and gap balance [4], making it an important factor that cannot be disregarded.

1.1.1 Distal Femoral Entry Point

The insertion point is a crucial factor in determining the direction of an intramedullary rod. It is commonly reported to be slightly anterior and medial to the intercondylar notch. However, owing to individual variations, it is desirable to determine the insertion point during preoperative planning (Fig. 50.1). By utilizing three-dimensional (3D) planning software, the insertion point and depth can be easily planned, resulting in component placement accuracy comparable to that achieved with accelerometer-based portable navigation [5]. A jig-engaged 3D preoperative planning system is also commercially available.

1.1.2 Measures for Complications Associated with Intramedullary Rod Insertion

The risk of embolism can be reduced by irrigation and suction [6]. Additionally, a careful drilling technique involving expansion of the entry hole, followed by slow insertion of the rod targeting the anterior superior iliac spine, has the potential to minimize invasion of the medullary canal to a level similar to that of extramedullary guides [7]. Blood loss can be minimized by using bone plugs at the entry site [8].

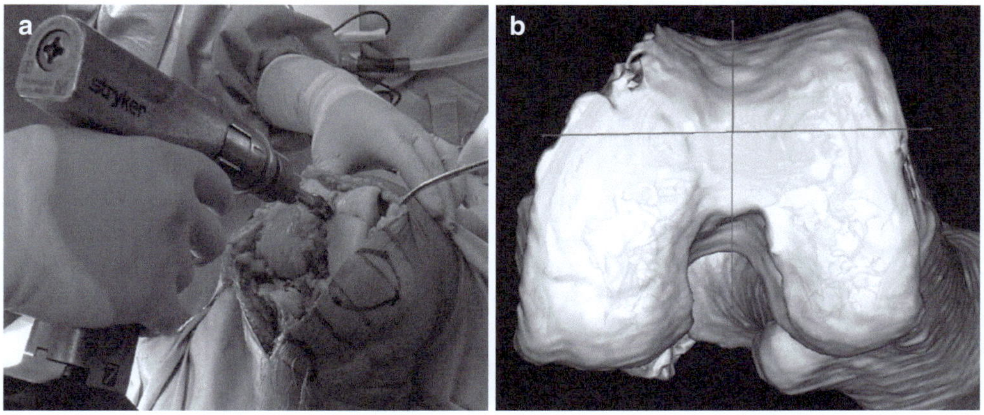

Fig. 50.1 Intramedullary Guide Insertion Point. The insertion point planned in three dimensions: Drawing a reference line using electrocautery

T. Tsukeoka (✉)
Department of Orthopaedic Surgery, Chiba Rehabilitation Center, Chiba, Japan
e-mail: mdddd940@ybb.ne.jp

S. Matsuda (ed.), *Advances in Total Knee Arthroplasty*, https://doi.org/10.1007/978-981-97-4920-1_50

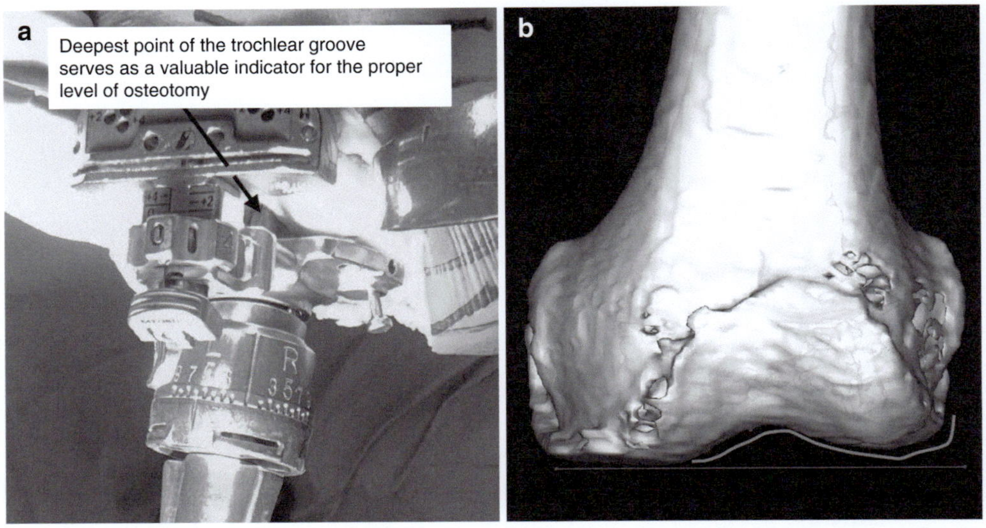

Fig. 50.2 Anticipated Gap between Intramedullary Guide and Confirmation during Surgery Considering the presence of cartilage, the anticipated gap is estimated

1.2 Distal Osteotomy

1.2.1 Determination of Osteotomy Level

Many systems involve cutting to a thickness equal to that of the femoral component from the most distal point of the medial or lateral condyle. If the distal cutting guide touches an osteophyte, it can result in inadequate bone resection and decrease the extension gap. Therefore, it is necessary to remove the osteophytes prior to placing the distal resection guide. Additionally, cartilage wear and bone deformities can elevate the joint line. To accurately reproduce the joint line, a 'sulcus cut' technique has been proposed as a convenient intraoperative reference [9]. This technique involves performing an osteotomy based on the deepest point of the trochlear groove, and serves as a valuable indicator for confirming the proper level of osteotomy (Fig. 50.2).

1.2.2 Considerations for Setting the Resection Jig

When inserting the intramedullary guide, the alignment guide usually contacts either the medial or lateral condyle, leaving a gap on the opposite side. Preoperative planning should consider the cartilage thickness and anticipate this gap, which can serve as a confirmation of proper placement during the installation (Fig. 50.2). It is also important to ensure that the rotation of the distal bone-cutting guide is parallel to the desired rotation of the femur, as deviations can affect alignment in the coronal and sagittal planes. Therefore, caution should be exercised in this regard [10].

1.2.3 Resection Error Checking

Measure the thickness of the medial and lateral resected bone fragments and ensure that there are no significant deviations from the plan. Additionally, it is common to experience extension positioning after initial bone cut [11]. Extension resection errors can result in overestimation of femoral size [2, 3]. Pay attention to the deflection of the bone saw and repeat the resection process for better accuracy.

1.3 Sizing Guide (Fig. 50.3)

1.3.1 Types and Considerations

Anterior reference and posterior reference guides are available (Fig. 50.3). When measuring for an intermediate size, the former results in a change in posterior condyle bone resection, whereas the latter maintains the same amount of posterior condyle bone resection; however it causes the anterior component position to shift anteriorly or posteriorly. The anterior boom should be positioned near the most proximal part of the component at the lateral cortex bone prominence. Additionally, the posterior reference guide can be divided into three types based on the location of the rotational center to determine the external rotation of the component. In the medial-centered type, as external rotation increases, the amount of bone resection remains constant on the medial side: however, it decreases on the lateral side, resulting in an overall increase in size. In the lateral-centered type, as external rotation increases, the amount of bone resection remains constant on the lateral side: however, it increases on the medial side, resulting in an overall decrease in size. In the central-centered type, as external rotation increases, the amount of bone resection increases on the medial side and decreases on the lateral side, resulting in a minimal change in the overall size. When using the posterior reference guide, it is important to determine the type and consider the effects of external rotation before proceeding with the surgery [12, 13].

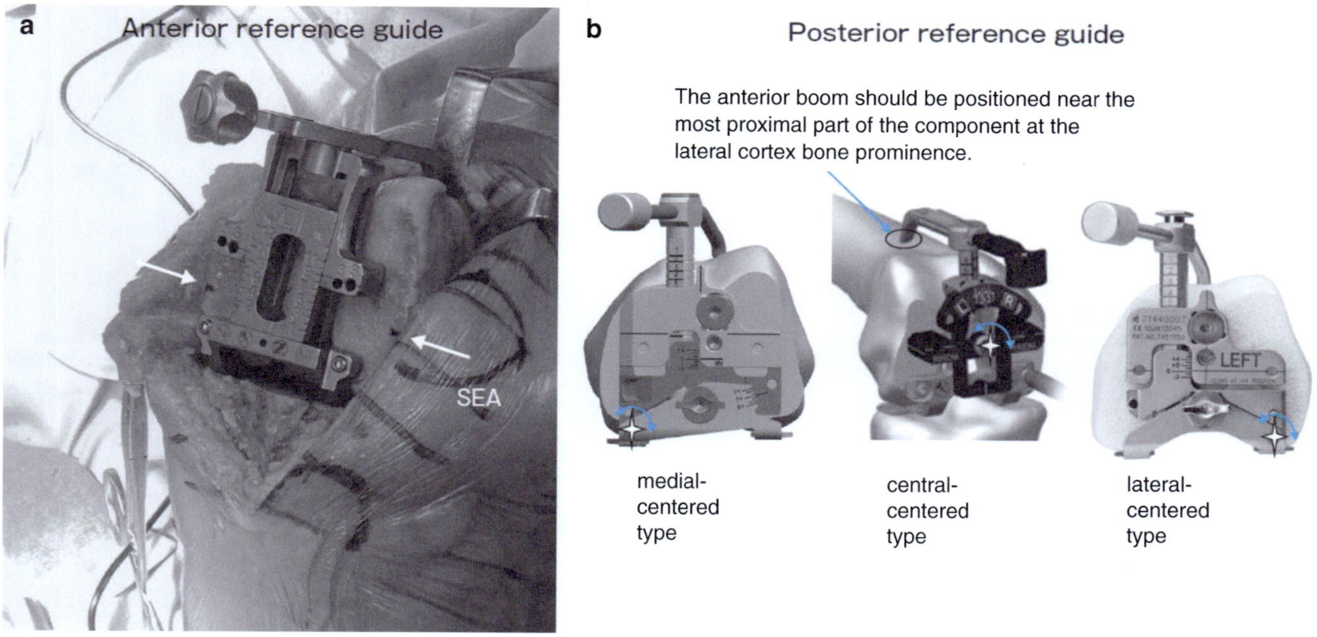

Fig. 50.3 Femoral sizing guide

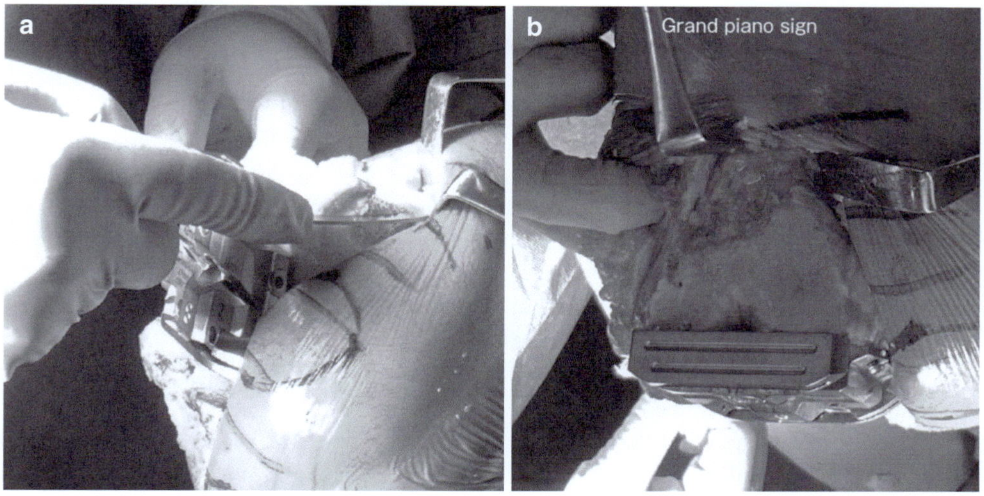

Fig. 50.4 Anterior Femur Cut. (**a**): Notch avoidance with an "angel wing." (**b**): With proper rotation during bone cutting, the anterior femur exhibits a bimodal peak, with the height on the lateral side approximately twice that on the medial side (grand-piano sign)

1.3.2 Determination of External Rotation Angle

While the posterior condyle is often used as a reference for setting the external rotation angle, it is important to also consider anatomical landmarks such as the Whiteside line and SEA to avoid significant errors in cases where the contact of the posterior condylar paddle is poor [14](Fig. 50.3). Additionally, in another technique called "gap balancing," the rotation of the femoral component is set based on the soft tissue tension at 90°of flexion, using the tibial osteotomy surface as a reference. Care should also be taken to minimize deviations from the anatomical landmarks in this technique.

1.4 4-in-1 Cutting Guide

The anterior femur, posterior femur, anterior chamfer, and posterior chamfer cuts are all made through the "4-in-1 cutting guide."

1.4.1 Guide Placement

Ensure that no gaps are present on the osteotomy surface.

Carefully check whether the resection guide causes a notch in the anterior cortical bone, as this may lead to postoperative supracondylar fracture (Fig. 50.4a). Many systems allow the anterior movement of the guide in cases where notch formation is likely to occur. In posterior osteotomies,

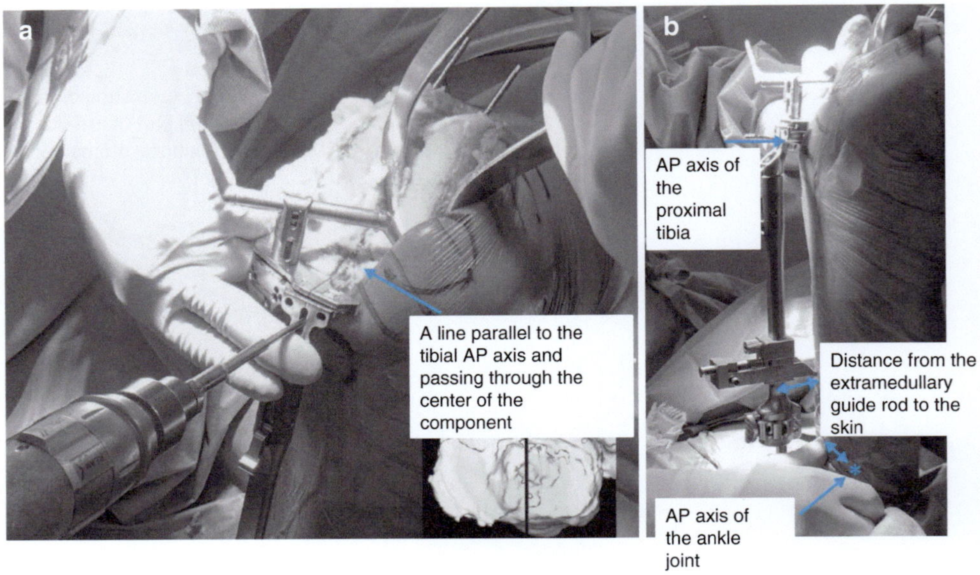

Fig. 50.5 Tibial Resection Guide Placement. (**a**): Alignment with the tibia's anterior-posterior axis is crucial. (**b**): The anteroposterior axis of the tibia differs from that of the ankle. The midpoint of the ankle in relation to its axis (*) is positioned laterally with reference to the tibial axis. (The photograph shows the anterior crest identified by the thumb and used as a distal reference point)

ensure to cut more medially than laterally to prevent internal rotation.

1.4.2 Considerations During Osteotomy

In the anterior cut, attention must be paid to notch formation, whereas, in the posterior cut, caution is required to avoid damage to the PCL and MCL.

On the anterior cutting surface, the presence of the "grand-piano sign" can be observed in case the rotation position is appropriate [15] (Fig. 50.4b).

2 Femoral Extramedullary Guide

Although the femoral extramedullary guide was a common method in the early days of TKA, accurately determining the position of the femoral head center using radiography was challenging, resulting in low placement precision. Consequently, intramedullary guides have gradually replaced this approach. Recently, a new extramedullary guide system was introduced. This system achieves higher accuracy by confirming the position of the femoral head preoperatively using radiographs, applying markers to the skin under tension, and paying close attention to the rotation and angle of the extramedullary guide in the sagittal plane [16]. In addition, ultrasonography has been proposed as a method for confirming the position of the femoral head before and during surgery [17].

3 Tibial Extramedullary Guide (Fig. 50.5)

3.1 Setting of Anteroposterior Axis

A discrepancy between the anteroposterior axis of the guide and that of the tibia can result in varus or valgus bone resection errors, even if the distal alignment is precise [18]. This effect becomes more pronounced when a posterior slope is introduced [19].

3.2 Adjustment of Varus-Valgus Alignment

3.2.1 Proximal Positioning

The postoperative mechanical axis does not always pass through the center of the knee. Therefore, in the preoperative planning, a line parallel to the tibial anteroposterior axis is determined to pass through the center of the component. The extramedullary rod is then positioned to align with the extension of this line [20] (Fig. 50.5a). Although it may receive less emphasis than distal positioning, poor exposure can result in a tendency toward medial deviation at the proximal end, which can contribute to varus alignment errors. It is recommended to consider the relationship with the tibial tuberosity during the determination process.

3.2.2 Distal Positioning

Traditionally, there are landmarks such as the second metatarsal, a few millimeters medial to the center of the ankle joint. However, relying solely on the second metatarsal as a single reference is not recommended because of its susceptibility to influence from the foot position [21]. Furthermore, the anteroposterior axis of the ankle joint is externally rotated compared with the anteroposterior axis of the proximal tibia. This means that, when considering the center of the ankle joint based on its anteroposterior axis, it is positioned laterally relative to the anteroposterior axis of the proximal tibia (Fig. 50.5b). This discrepancy in the anteroposterior axes is considered to be the cause of varus alignment errors [22]. By aligning the extramedullary guide with the anteroposterior axis of the tibia, the mismatch between the anteroposterior axes of the ankle joint and tibia can be eliminated, resulting in an error within 2 °of varus or valgus alignment, even with a deviation of up to 10 mm in the medial-lateral position at the distal end [23]. New landmarks include the extensor hallucis longus tendon [24, 25] and dorsalis pedis artery [25]. These landmarks have been shown to closely correlate with the center of the ankle joint in magnetic resonance imaging (MRI) and ultrasound studies.

3.2.3 Tibial Crest as an Alignment Landmark

The tibial crest is a useful landmark for varus-valgus alignment [26–29]. When used in conjunction with other landmarks, such as the dorsalis pedis artery, it can help prevent significant alignment errors [30].

3.3 Posterior Slope

3.3.1 Lateral Anatomic Tibial Slope as a Reference for the Posterior Slope

The lateral anatomical tibial slope has been proposed as a reference for determining the posterior slope [31]. However, significant individual variations are observed, and when preoperative planning reveals a substantial posterior slope, it is necessary to reduce it to avoid posterior subluxation. Confirming that the thickness of the resected bone fragment is equal in the anterior and posterior aspects indicates the achievement of the desired posterior slope. Nevertheless, its use as a guide placement reference can be challenging.

3.3.2 Fibular Axis

The fibular axis is commonly used to assess posterior slope. It typically exhibits an average posterior slope of 2–3°compared to the mechanical axis [27, 32]. However, there is substantial individual variation, and its reliability is not high [32].

3.3.3 Distance from the Extramedullary Cutting Guide Rod to the Skin Surface (Fig. 50.5b)

It is a measurement method that replicates the planned rod-skin distance at a specific distance from the osteotomy level, as determined during preoperative planning [33]. 3D preoperative planning has been recognized as superior to accelerometer-based portable navigation [34]. A more effective approach, utilizing plain X-ray imaging to measure and replicate the planned rod-skin distance using the same technique as in 3D preoperative planning, has been reported as an alternative to relying solely on the fibular axis [35].

3.3.4 Cutting Errors

An insufficient posterior slope after the initial bone cut is common. Similar to the femur, it is important to be mindful of the deflection of the bone saw and to perform repeated resection processes for better accuracy.

4 Tibial Intramedullary Guide

The axis of the medullary canal does not necessarily align parallel to the mechanical axis [20]; therefore, thorough preoperative planning is essential, and in some cases, adjustment of the alignment guide angle may be required. Additionally, similar to the femoral intramedullary guide, determination of the insertion point significantly affects the alignment of the guide. The ideal insertion position of the tibial intramedullary guide varies substantially among individuals, emphasizing the importance of preoperative planning [36].

References

1. Haruta Y, et al. Deviation of femoral intramedullary alignment rod influences coronal and sagittal alignment during total knee arthroplasty. Knee. 2018;25:644–9.
2. Tsukeoka T, et al. Sagittal flexion of the femoral component affects flexion gap and sizing in total knee arthroplasty. J Arthroplast. 2012;27:1094–9.
3. Nakahara H, et al. Sagittal cutting error changes femoral anteroposterior sizing in total knee arthroplasty. Clin Orthop Relat Res. 2012;470:3560–5.
4. Tsukeoka T, et al. The effect of a sagittal cutting error of the distal femur on the flexion-extension gap difference in total knee arthroplasty. J Arthroplast. 2013;28:1099–102.
5. Tsukeoka T, et al. An accelerometer-based navigation did not improve the femoral component positioning compared to a modified conventional technique of preoperatively planned placement of intramedullary rod in total knee arthroplasty. Arch Orthop Trauma Surg. 2019;139:561–7.
6. Cornel C, et al. The reamer/irrigator/aspirator reduces femoral canal pressure in simulated TKA. Clin Orthop Relat Res. 2009;467:805–9.

7. Kaneko S, et al. NaF PET assessment of bone metabolic changes around the femoral canal by intramedullary femoral alignment technique in total knee arthroplasty. Clin Case Rep. 2019;7:1211–4.
8. Kumar N, et al. Plugging the intramedullary canal of the femur in total knee arthroplasty: reduction in postoperative blood loss. J Arthroplast. 2000;15:947–9.
9. Kuriyama S, et al. Is a "sulcus cut" technique effective for determining the level of distal femoral resection in total knee arthroplasty? Knee Surg Sports Traumatol Arthrosc. 2014;22:3060–6.
10. Gangadharan R, et al. Distal femoral resection at knee replacement: the effect of varying entry point and rotation on prosthesis position. Knee. 2010;17:345–9.
11. Nakahara H, et al. Cutting error of the distal femur in total knee arthroplasty by use of a navigation system. J Arthroplast. 2012;27:1119–22.
12. Minoda Y, et al. Posterior reference guides do not always maintain the size of posterior femoral condyles in TKA. Knee Surg Sports Traumatol Arthrosc. 2016;24:2489–95.
13. Ishida K, et al. Posterior reference position affects intraoperative kinematic and soft tissue balance in navigated posterior-stabilized total knee arthroplasty. J Arthroplast. 2018;33:2851–7.
14. Tsukeoka T, et al. Postoperative femoral component rotation using posterior condylar referencing is difficult to predict preoperatively in total knee arthroplasty. Knee. 2023;41:380–8.
15. Moyad TF, et al. "Grand piano sign," a marker for proper femoral component rotation during total knee arthroplasty. Am J Orthop (Belle Mead NJ). 2011;40:348–52.
16. Matsumoto K, et al. Accuracy of a novel extramedullary femoral alignment guide system in primary total knee arthroplasty. Arch Orthop Trauma Surg. 2015;135:1743–8.
17. Matsuda Y, et al. Identifying the center of the femoral head using ultrasonography to assess the higher accuracy of femoral extramedullary guides in TKA. J Orthop Sci. 2004;9:6–9.
18. Tsukeoka T, et al. The effect of rotational fixation error of the tibial cutting guide and the distance between the guide and the bone on the tibial osteotomy in total knee arthroplasty. J Arthroplast. 2013;28:1094–8.
19. Tsukeoka T, et al. The effect of the posterior slope of the tibial plateau osteotomy with a rotational error on tibial component malalignment in total knee replacement. Bone Joint J. 2013;95:1201–3.
20. Matsuda S, et al. Tibial shaft axis does not always serve as a correct coronal landmark in total knee arthroplasty for varus knees. J Arthroplast. 2003;18:56–62.
21. Tsukeoka T, et al. Accuracy of the second metatarsal as a landmark for the extramedullary tibial cutting guide in total knee arthroplasty. Knee Surg Sports Traumatol Arthrosc. 2014;22:2969–74.
22. Mizu-uchi H, et al. The effect of ankle rotation on cutting of the tibia in total knee arthroplasty. J Bone Joint Surg Am. 2006;88:2632–6.
23. Mizu-uchi H, et al. The adjustment of the rotational alignment of the distal end of the extramedullary guide to the anteroposterior axis of the proximal tibia in total knee arthroplasty. J Knee Surg. 2021;35:1273–9.
24. Hino M, et al. Extensor hallucis longus tendon is a new distal landmark for coronal tibial component alignment in total knee arthroplasty: a study of magnetic resonance imaging. J Orthop Surg (Hong Kong). 2020;28:2309499020912340.
25. Sugimura N, et al. The dorsal pedis artery as a new distal landmark for extramedullary tibial alignment in total knee arthroplasty. Knee Surg Sports Traumatol Arthrosc. 2014;22:2618–22.
26. Fukagawa S, et al. Anterior border of the tibia as a landmark for extramedullary alignment guide in total knee arthroplasty for varus knees. J Orthop Res. 2011;29:919–24.
27. Tsukeoka T, et al. The tibial crest as a practical useful landmark in total knee arthroplasty. Knee. 2014;21:283–9.
28. Nishikawa K, et al. Accuracy of proximal tibial bone cut using anterior border of tibia as bony landmark in total knee arthroplasty. J Arthroplast. 2015;30:2121–4.
29. Kawahara S, et al. Anterior tibial border as a landmark for extramedullary alignment guides for total knee arthroplasty in valgus knees. J Orthop Res. 2015;33:1897–9.
30. Ikeuchi M, et al. Improving tibial component coronal alignment using clearly defined anatomical structures in total knee arthroplasty. Acta Orthop Belg. 2013;79:694–8.
31. Kuwano T, et al. Importance of the lateral anatomic tibial slope as a guide to the tibial cut in total knee arthroplasty in Japanese patients. J Orthop Sci. 2005;10:42–7.
32. Kuroda Y, et al. Fibular axes are not a reliable landmark for tibial mechanical axes of osteoarthritic knees that underwent total knee arthroplasty. Knee Surg Sports Traumatol Arthrosc. 2015;23:3362–7.
33. Tsukeoka T, et al. The distance from the extramedullary cutting guide rod to the skin surface as a reference guide for the tibial slope in total knee arthroplasty. Knee. 2016;23:314–7.
34. Tsukeoka T, et al. Preoperative planned distance between the skin surface and the guide rod provides accurate posterior tibial slope in total knee arthroplasty. Arch Orthop Trauma Surg. 2019;139:1133–9.
35. Kim HJ, et al. A sagittal reference line using the preoperative radiograph in total knee arthroplasty. J Orthop Surg (Hong Kong). 2020;28:2309499020924143.
36. Oh SM, et al. The entry point of intramedullary tibia cutting guide should vary according to the individual tibia morphology in TKA. Arch Orthop Trauma Surg. 2020;140:391–400.

Evaluation of Soft Tissue Balance

51

Hirotsugu Muratsu and Tomoyuki Matsumoto

1 Importance of Soft Tissue Balance Measurement

Since there has been no standard quantitative measurement method, the acquisition of appropriate soft tissue balance in TKA largely depends on the experience and skill of the surgeon, and it is still largely unsolved.

2 Biomechanical Issues of Intraoperative Soft Tissue Balance Measurement

In general, soft tissue balance is evaluated by the distance between reference points (gap) and the state of varus and valgus equilibrium (balance) when a certain joint distraction force is applied to the knee joint, and the osteotomy angle and the amount of osteotomy have a great influence on soft tissue balance. Moreover, the initial length and elongation of all soft tissues around the knee joint determine the gap and balance. However, since soft tissues have viscoelastic mechanical properties and because stress relaxation and creep elongation occur, a certain compromise must be clinically considered because the measured values vary depending not only on the magnitude of the force applied but also on the loading time duration and number of tensions. In addition to the MCL/LCL and ACL/PCL, the joint capsule, retinaculum, knee extensor mechanisms, and hamstrings affect the soft tissue balance. After the implant is placed, the shape and volume of the implant affects the running route and tension of soft tissue, especially with extension and flexion of the knee. Therefore, soft tissue balance is based on the alignment of bone tissue and implants and comprehensively evaluated as a soft tissue envelope.

H. Muratsu (✉)
Hyogo Prefectural Harima-Himeji General Medical Center, Himeji, Japan

T. Matsumoto
Kobe University Graduate School of Medicine, Kobe, Japan

3 Soft Tissue Balance Measurement Using Offset Type Tensor

3.1 Development of the Offset Type Tensor

Conventional soft tissue balance assessment was performed only in extension and flexion between osteotomy surfaces (osteotomy gap: Fig. 51.1a). The knee was in a nonphysiological condition with the patella everted without the femoral component, which was different from the postoperative TKA knee. Therefore, a new offset type tensor was developed [1, 2] to enable quantitative measurement of soft tissue balance throughout the ROM with reproduction of the postoperative TKA knee. (Fig. 51.1b) The femoral trial component was placed, and the patellofemoral (PF) joint was temporally reduced while controlling for femorotibial alignment (component gap: Fig. 51.1c). Since 2009, Zimmer has made it possible to use OFR[R] tensor n Japan (Fig. 51.2), and component gap measurement has been added as a new soft tissue balance index. A number of studies were subsequently conducted regarding the relationship between the component gap, preoperative factors, and postoperative clinical outcomes in Japan (for details, refer to Part 6, Chapter [3]. In addition, the FuZion[R] tensor has been provided globally by Zimmer since 2014 and has been used in research [4].

3.2 Features of the Offset Type Tensor and Points to Note in the Measurement

As an overview of the offset type tensor (OFR[R] tensor), the intra- and extra-articular parts are connected through the arthrotomy with an offset connection arm (Fig. 51.2a), enabling measurement after PF joint reduction and temporary repair (Fig. 51.1c). The extra-articular main body is

S. Matsuda (ed.), *Advances in Total Knee Arthroplasty*, https://doi.org/10.1007/978-981-97-4920-1_51

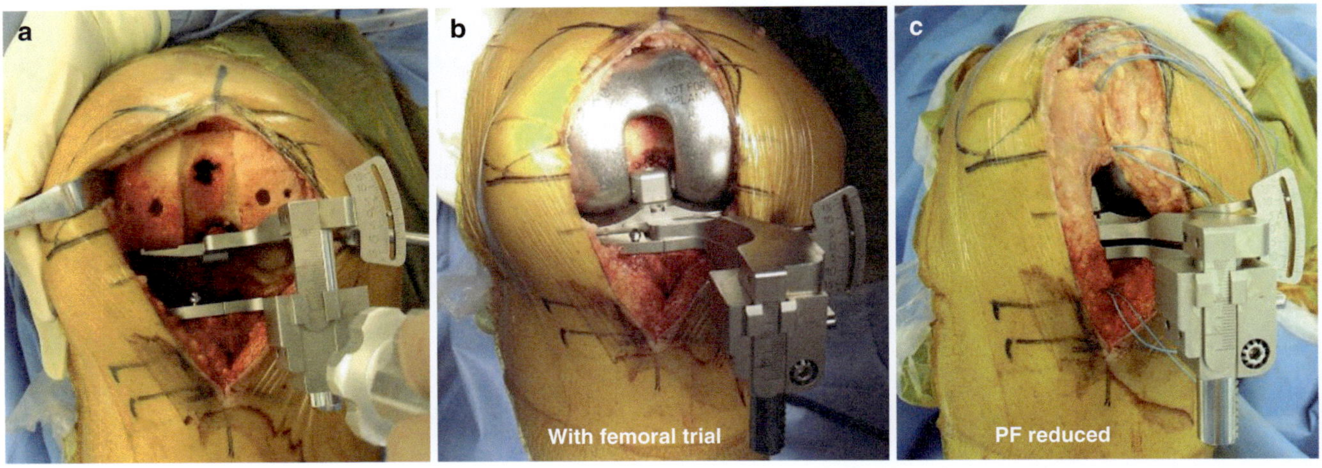

Fig. 51.1 Osteotomy gap and component gap measurements. (**a**): Osteotomy gap measurement. (**b**): Post-cam engagement. (**c**): Component gap measurement

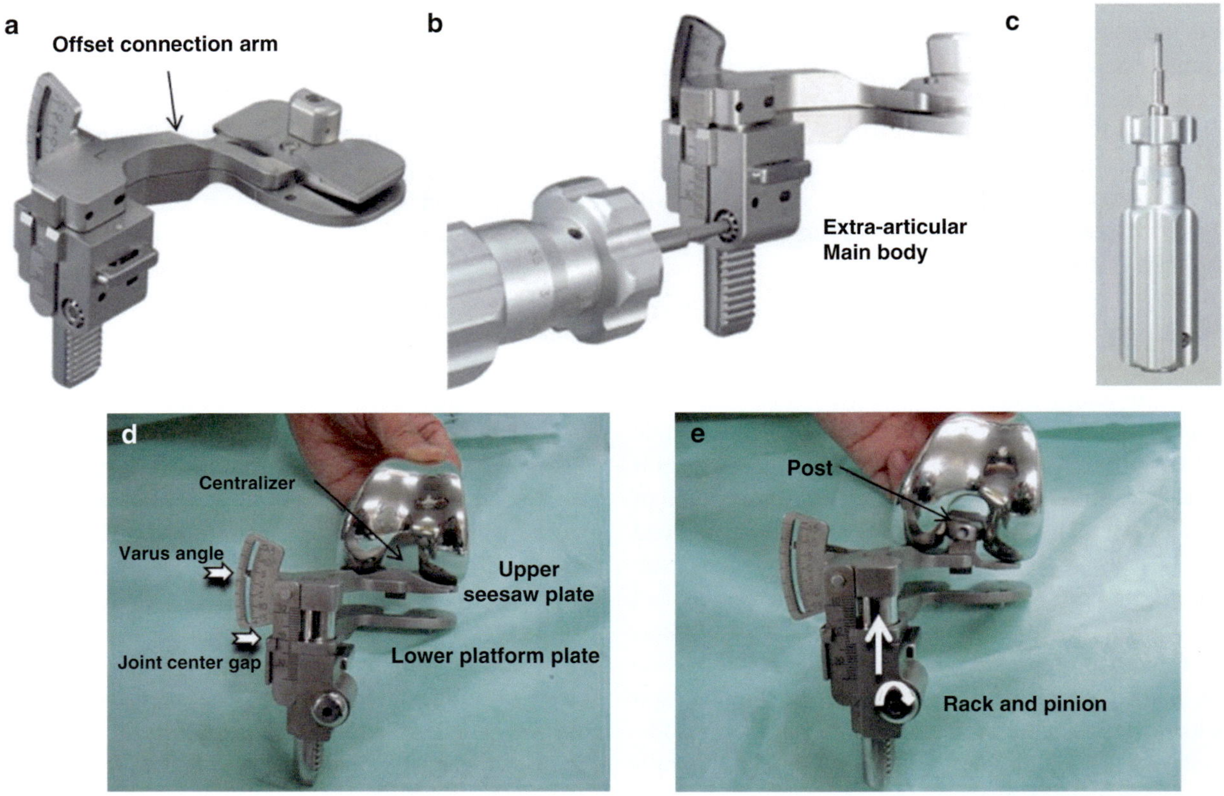

Fig. 51.2 Offset Repo(OFR^R)Tensor. (**a**): Offset connection arm passes through the arthrotomy. (**b**): Rack and pinion mechanism in the extra-articular main body. (**c**): Sterilizable variable torque driver. (**d**): Configuration for CR-TKA. (**e**): Configuration for PS-TKA

designed to convert torque to the intra-articular distraction force while containing the rack and pinion mechanism (Fig. 51.2b). A new sterilizable variable torque driver was developed to adjust torque value to generate the indicated joint distraction force. Depending on the set value, any joint distraction force from 20 pounds (9.1 kg) to 60 pounds (27.3 kg) can be applied between the lower platform plate on the tibial side and the upper seesaw plate on the femoral side with an accuracy of ±5% or less (Fig. 51.2c). The joint center gap (mm) in the middle of the joint and the varus-valgus angle in the coronal plane between the upper seesaw plate and lower platform plate are displayed on the extra-articular scale (Fig. 51.2d). The lower platform plate was fixed to the proximal osteotomy surface of the tibia with spikes and additional screws. It is important to always place the tensor at the middle of the tibia while paying attention to the rota-

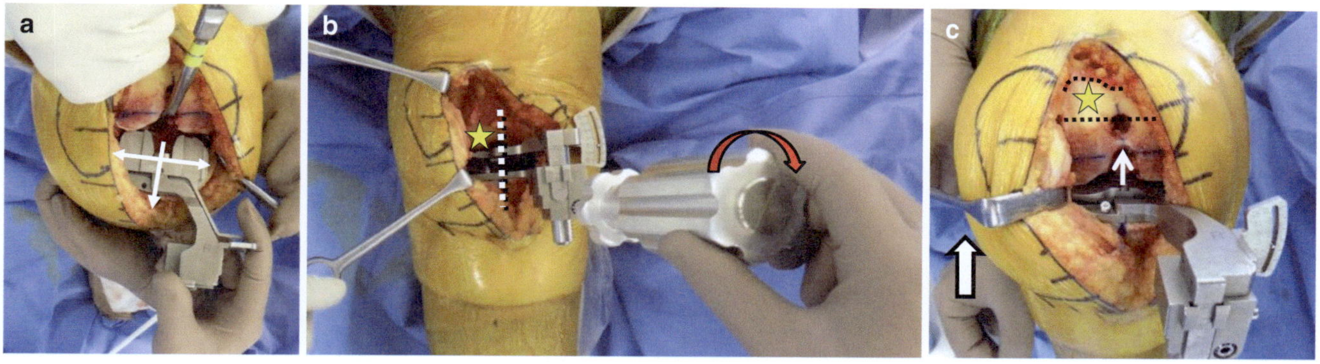

Fig. 51.3 Points to note in offset repo (OFR^R) tensor installation. (**a**): Install the lower platform plate at the middle of the tibial osteotomy surface with attention to the rotational alignment. (**b**): Measurement of the extension gap (osteotomy gap). Pay attention to the positional relationship between the tibia, tensor, and femur. Torque is applied at least three times to decrease creep elongation of soft tissues. (**c**): Measurement of flexion gap before femoral posterior condylar osteotomy. Align the middle of the femur with the middle of the tensor. Lift the thigh so that thigh weight is not applied on the knee. ☆: Preliminary resection of anterior chamfer of the femur. Avoid interference of the anterior edge of the distal femoral osteotomy and connection arm of the tensor

tional orientation to reproduce the tibial component placement (Fig. 51.3a). A centralizer for CR-TKA (Fig. 51.2d) and a post for PS-TKA (Fig. 51.2e) were installed on the upper seesaw plate, and after the femur trial was installed, the joint alignment was controlled by mating the intercondylar notch and cam of the femoral component, respectively, to reproduce the tibiofemoral alignment after TKA (Fig. 51.2d, e). When measuring the extension osteotomy gap, it is important that the center of the tibia, tensor, and femur are aligned in a row (Fig. 51.3b). In the flexion osteotomy gap measurement, a centralizer (+post) is attached to the seesaw plate and the femoral intercondylar area is positioned in the center of the upper seesaw plate (Fig. 51.3c). In the osteotomy gap measurement at extension before femoral posterior condylar osteotomy in the modified gap technique, it is necessary to precut the anterior femoral chamfer part to avoid interference between the anterior edge of femoral osteotomy and the offset connection arm (Fig. 51.3b, c).

It is also important for the assistant to lift the thigh so that the weight of the thigh does not apply on the knee joint and to hold the thigh's adjusted knee alignment in the sagittal plane of the knee joint so that the external varus-valgus moment does not apply to the joint (Fig. 51.3c).

3.3 Effect of Femoral Component Placement

Regarding the effect of femoral component placement, in a study of 80 knees with PS-TKA using the measured resection technique, the component gap at extension was significantly smaller than the predicted value from the osteotomy gap (i.e., extension gap—distal thickness of femoral component). The varus balance was also significantly smaller with the femoral trial component than without it. It is thought that the lateral gap is reduced with increased tension of the posterior soft tissue due to the lateral posterior femoral condyle overhang, which would be more than that at the medial side because of the externally rotated placement of the femoral component [3]. In addition, when a uniquely fabricated femoral trial component with 4 mm thick posterior condyles was placed in 80 PS-TKA, a 2.1/2.8 mm reduction in the extension gap was observed on the medial/lateral side [5]. In 259 PS-TKA, a similar reduction in the extension gap was observed, and it was reported that a 2 mm larger osteotomy gap in extension is necessary to create a rectangular extension and flexion equal component gap [6]. In the modified gap technique, the posterior soft tissue release is insufficient prior to femoral posterior condylar osteotomy, which affects the extension component gap. For this reason, a technique in which a precut of the femoral posterior condyles is performed to facilitate posterior release, and a precut trial is inserted in order to reproduce the original femoral posterior condyles for soft tissue balance measurement [7].

3.4 Importance of Patellar Reduction

Intraoperative soft tissue balance measurements that reflect the postoperative knee after TKA must be performed with the PF joint reduced and the trial femoral component in place. In the intraoperative soft tissue balance measurements in PS-TKA using a measured resection technique for 30 varus type OA knees, the component gap increases with knee flexion when the patella is laterally everted. However, with the PF joint reduced and temporally repaired, it decreased with deep flexion [2] (Fig. 51.4). Furthermore, in a study of 20 PS-TKA, an increased tension in the extensor mechanism,

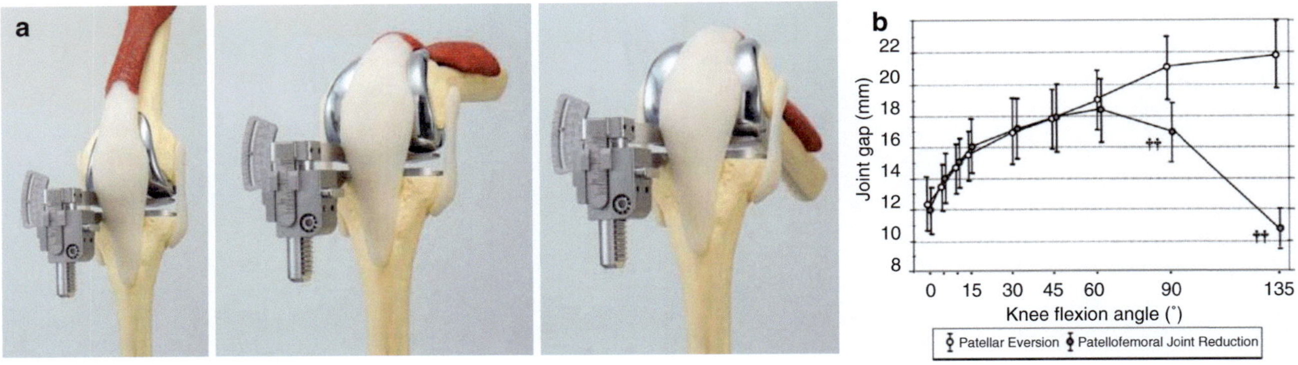

Fig. 51.4 Component gap kinematic curve. (**a**): Component gap measurements with different flexion angle. (**b**): Component gap kinematic curve [2]

including the patellar tendon, was observed with knee flexion after PF joint reduction [8]. In addition, in PS-TKA and CR-TKA, the effect of patellar reduction on the tibio-femoral joint stress was observed only in PS-TKA [9]. On the contrary, the patella is difficult to evert with knee flexion in the minimally invasive technique, so the component gap is decreased after PF joint reduction compared to patella eversion [10]. In recent years, it has become possible to measure soft tissue balance using the navigation system. A significant correlation between the values obtained by navigation and tensor at both extension and flexion has been reported; however, the correlation coefficient with the PF joint reduced is higher than that with the patella everted, and the reduced PF joint is important for the reproducibility of soft tissue balance measurement [11].

3.5 Effects of Joint Distraction Force

The joint distraction force during intraoperative soft tissue balance measurement is also extremely important. In PS-TKA using the modified gap technique for 65 varus type OA knees, varus balance increased at both extension and flexion before femoral posterior condylar osteotomy in response to increasing joint distraction forces of 20, 40, and 60 pounds. This is biomechanically caused by the stiffness difference between the medial and lateral soft tissues of the knee joint. Since the lateral side of the knee is less stiff than the medial side in the varus type OA knee, the joint gap increases more at lateral than medial with the higher joint distraction force, which results in the increase of varus balance with a larger difference between the medial and lateral gaps [12].

The modified gap technique (parallel cut technique) assumes perfect ligament balance at extension before femoral posterior condylar osteotomy, but with the increase of joint distraction force, the varus angle increases and additional medial release is needed. Moreover, the rotational osteotomy angle of the femur is based on the varus angle at flexion before the posterior condylar osteotomy, but the varus angle increases with the increase of distraction force at the time of evaluation. Therefore, it is difficult to quantify this surgical technique without clear information on the distraction force. Although skilled surgeons may empirically adjust the joint distraction force during evaluation, there is no clear definition, which causes confusion for many surgeons and is a barrier to learning the gap technique. The difference of varus angle between flexion and extension before femoral posterior condylar osteotomy in the modified gap technique was examined. The difference was not significantly affected by joint distraction force and correlated with condylar twist angle on CT. This finding led to the development of the "medial preserving gap technique" that emphasizes medial stability, using the varus angle difference between flexion and extension as an indicator of the external rotation angle of femoral posterior condylar osteotomy [13].

3.6 Component Gap Kinematics(Fig. 51.4)

The relationship between intraoperative soft tissue balance and postoperative clinical outcomes has rarely been reported in the analysis of the osteotomy gap. However, in recent years, clinical outcomes have centered around patient-reported measurement, including patient satisfaction. Many correlations between clinical scores and component gap (CG) changes with knee motion (component gap kinematics) have been reported.

In a study of CR-TKA for 35 varus type OA knees, a positive correlation was found between CG difference (90–0°) and CG (120–0°) to the patient satisfaction score in the 2011 KSS 1 year after surgery. In CR-TKA, it was reported that an adequate flexion gap in relation to an extension gap was important for improving patient satisfaction [14], and similarly, in a study of CR-TKA, a moderate CG difference (90–0°) was advantageous for improving WOMAC function scores [15].

4 Future Prospects

The practical application of robotic technology to TKA is progressing, and robotic tensor is being developed for soft tissue balance measurement [16]. Furthermore, the time is approaching when optimal surgery will be proposed using artificial intelligence by analyzing data that is aggregated in the cloud by implantable sensors and IOT technology. As surgeons, we must remember that we can only make further progress if we have the ability to scientifically confront each of our operations and ruminate on our experiences.

References

1. Muratsu H, et al. Patellar eversion affects soft tissue balance in total knee arthroplasty. Trans Orthop Res Soc. 2003;28:242.
2. Matsumoto T, et al. Joint gap kinematics in posterior-stabilized total knee arthroplasty measured by a new tensor with the navigation system. J Biomech Eng. 2006;128:867–71.
3. Muratsu H, et al. Femoral component placement changes soft tissue balance in posterior-stabilized total knee arthroplasty. Clin Biomech. 2010;25:926–30.
4. Cho BW, et al. The effect of hamstrings tightness on intraoperative extension gap in posterior stabilized total knee arthroplasty. Sci Rep. 2021;11:3655.
5. Mitsuyasu H, et al. Enlarged post-operative posterior condyle tightens extension gap in total knee arthroplasty. J Bone Joint Surg Br. 2011;93:1210–6.
6. Minoda Y, et al. Decreased extension gap and valgus alignment after implantation of total knee prosthesis in primary varus knees. Knee Surg Sports Traumatol Arthrosc. 2016;24:3642–7.
7. Kanayama R, et al. Improvement of the gap adjustment in total knee arthroplasty using the posterior condylar pre-cut technique. J Knee Surg. 2019;32:1001–7.
8. Gejo R, et al. Joint gap changes with patellar tendon strain and patellar position during TKA. Clin Orthop. 2008;466:946–51.
9. Yoshino N, et al. The influence of patellar dislocation on the femoro-tibial loading during total knee arthroplasty. Knee Surg Sports Traumatol Arthrosc. 2011;19:1817–22.
10. Oka S, et al. The influence of patellar orientation on soft tissue balance in minimally incision total knee arthroplasty. Knee Surg Sports Traumatol Arthrosc. 2012;20:1064–8.
11. Matsumoto T, et al. Soft tissue balance measurement in posterior-stabilized total knee arthroplasty with a navigation system. J Arthroplast. 2009;24:358–64.
12. Nagai K, et al. The influence of joint distraction force on the soft-tissue balance using modified gap-balancing technique in posterior-stabilized total knee arthroplasty: The novel index for the femoral component rotation. J Arthroplast. 2017;32:2995–9.
13. Nagai K, et al. Intraoperative soft tissue balance using novel medial preserving gap technique in posterior-stabilized total knee arthroplasty: comparison to measured resection technique. Knee Surg Sports Traumatol Arthrosc. 2018;26:3474–81.
14. Matsumoto T, et al. Relative loose flexion gap improves patient-reported clinical scores in cruciate-retaining total knee arthroplasty. J Knee Surg. 2018;31:573–9.
15. Fujimoto E, et al. Intra-operative gaps affect outcome and postoperative kinematics in vivo following cruciate-retaining total knee arthroplasty. Int Orthop. 2016;40:41–9.
16. Okamoto J, et al. Robotic tensor device for optimal soft tissue balance in TKA. Annu Int Conf IEEE Eng Med Biol Soc. 2013;2013:6695–8.

Part IX

Surgical Techniques and Clinical Outcomes

CR-TKA (Measured Resection)

52

Shuichi Matsuda

1 CR-TKA

CR-TKA preserves the PCL, in addition to the MCL and LCL, to achieve knee joint stability. CR-TKA seeks stability in the patient's ligaments and is performed using a measured resection technique that seeks stability in the patient's ligaments and removes bone by the thickness of the implant [1].

2 Indications for CR-TKA

Although the indications for CR-TKA do not differ majorly from those for TKA, a few points should be considered to achieve the full function of the PCL.

2.1 Posterior Stability

Although degenerative rupture of the PCL in OA is very rare, it should be noted in cases, such as posttraumatic OA. A preoperative manual examination should confirm that posterior stability is maintained, and MRI should be performed if possible.

2.2 Joint Deformity

If the tibial osteotomy level is set more distal than usual because of bone loss, CR should be avoided because of the significant loss of PCL attachment of the tibia.

Patients with a strong posterior tilt of the tibia are unsuitable for CR, which maintains the preoperative posterior tilt. Patients with a preoperative posterior tilt of 12–13° or greater were considered unsuitable for CR if the postoperative tilt was aimed at less than 10°.

2.3 Flexion Contracture

When there is a flexion contracture of >30°, a larger-than-usual osteotomy of the distal femur may be used to obtain an adequate extension gap. In such cases, it is difficult to maintain the joint line, and proper tension of the PCL may not be maintained; therefore, PS implants should always be prepared.

2.4 Valgus Deformity

In cases of severe valgus deformity with large laxity in the medial soft tissue, it is difficult to achieve medial stability with CR; therefore, it is necessary to prepare a constrained condylar knee (CCK) or other highly constrained implants.

2.5 RA

A few reports have indicated that if posterior stability is achieved at the time of surgery, instability does not occur afterward, and RA is not necessarily an exclusion factor. However, patients with highly active disease may have a loss of the PCL; therefore, it is advisable to confirm this by MRI and manual examination.

2.6 After HTO

A few reports of CR after HTO demonstrated the occurrence of postoperative instability [2]. However, few reports suggest no difference in CR results after HTO when compared to those of conventional TKA. PS is safer when excessive valgus correction causes medial laxity or when the posterior tilt angle changes to an extent that PCL function is unlikely to be maintained.

The original version of the chapter has been revised. A correction to this chapter can be found at https://doi.org/10.1007/978-981-97-4920-1_89

S. Matsuda (✉)
Department of Orthopaedic Surgery, Kyoto University, Kyoto, Japan
e-mail: smat522@kuhp.kyoto-u.ac.jp

S. Matsuda (ed.), *Advances in Total Knee Arthroplasty*, https://doi.org/10.1007/978-981-97-4920-1_52

3 Basic Concept of CR-TKA: The Purpose of Preserving PCL

3.1 Rollback of the Femur

The rollback of the femur enables a deep flexion angle to be obtained without impingement of the femur on the posterior tibia. In addition, the lever arm during knee joint extension becomes greater, and less force is required for extension [3] (Fig. 52.1). There are advantages and disadvantages to whether the rollback mechanism and posterior stability of the femur are achieved by a post-cam mechanism, as in PS, or by preserving the PCL. The post-cam mechanism is more reproducible for obtaining femoral rollback; however, there is a risk of post-wear and tear. In the case of CR, the rollback could not be achieved without appropriate PCL tension. Therefore, the rollback of the femur should be examined intraoperatively.

3.2 Bone Preservation

CR-TKA does not require creating box space for the post-cam mechanism, and this allows for relatively good bone preservation. Therefore, CR is suitable for patients with severe fragility and is also suitable for relatively young patients who are more likely to undergo revision.

3.3 Others

The preservation of the PCL may lead to good proprioception; however, its clinical benefits remain unclear. For the patellofemoral joint, CR has the advantage of extending the patellar groove of the femur more posteriorly; however, its clinical advantages are not well understood.

4 Notes in CR-TKA

The significance of CR-TKA is based solely on the assumption that the PCL tension level is maintained appropriately. Historically, high tension in the PCL has caused flexion loss. It was believed that over-tension of the PCL should be avoided first, and to avoid over-tension, the PCL should be released. Paradoxical anterior motion in which the femur moves forward during flexion has been reported to occur after CR in a few cases (Fig. 52.2).

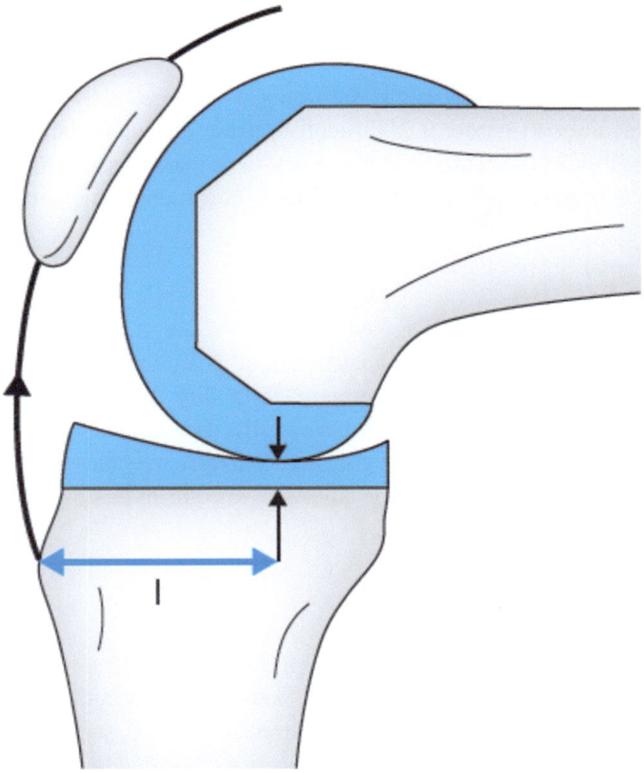

Fig. 52.1 Rollback of the femur. The femur moves posteriorly during flexion to avoid posterior impingement and the lever arm (l) of the extension mechanism increases

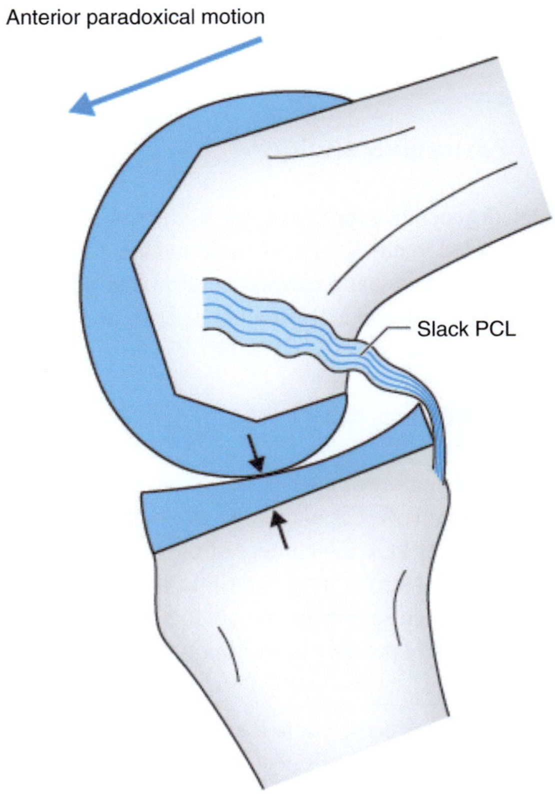

Fig. 52.2 Anterior paradoxical motion. When the PCL is lax, anterior paradoxical motion occurs, which moves the femur forward during flexion

We conducted a study, 9 years postoperatively, of CR-TKAs performed in the late 1980s using the KT-2000 at 75° of flexion. In this study, PCL insufficiency was found in 7 of the 15 cases, and the KSS was lower in cases with PCL insufficiency [4].

Therefore, to obtain good postoperative results, it is necessary to not only avoid over-tension of the PCL but also perform surgery to preserve PCL function.

5 Concept of Surgery

5.1 Joint Line

A consensus has been reached that medial stability is important for excellent clinical outcomes after TKA, and the surgical technique to achieve medial stability in CR involves obtaining adequate tension in the MCL and PCL by adjusting the thickness of bone resection in the femur and tibia. Since most knees with varus deformity have distal cartilage loss, the cartilage is usually preserved posteriorly. Therefore, the thickness of bone resection needs to be adjusted so that the resection is thicker in the posterior part than in the distal part (Fig. 52.3). On the tibial side, if the medial joint line is aligned with the native joint line, a conventional tibial component with equal medial and lateral thicknesses would result in overly thick tibial bone resection, making it difficult to preserve the PCL attachment. Therefore, when CR is performed with a normally symmetrical medial/lateral implant, it is practical to slightly raise the medial joint line to the same level as the lateral joint line.

5.2 Ligament Balance

As it is impossible to accurately predict the effect of MCL release on mediolateral soft tissue balance and there is a risk of occasional medial instability, extensive release of the MCL is recommended. Various methods have been proposed for adjusting the PCL tension.

6 Surgical Techniques

The surgical techniques of CR-TKA using the mechanical alignment method for medial OA are as follows:

6.1 Deep Skin Incision

A medial parapatellar approach was adopted through a median longitudinal incision. The medial capsular attachment of the tibia was slightly released so that the osteophytes could be resected.

Even in severe knee injuries, medial contracture usually does not exist, and it is possible to secure sufficient medial space without extensive release [5]. Dissection of the semimembranosus muscle in CR results in less internal rotation of the tibia during flexion and decreased ROM [6].

6.2 Femoral Bone Resection

The osteotomy was performed such that the distal femur was perpendicular to the mechanical axis of the coronal plane. The thickness of the distal medial bone resection was indexed

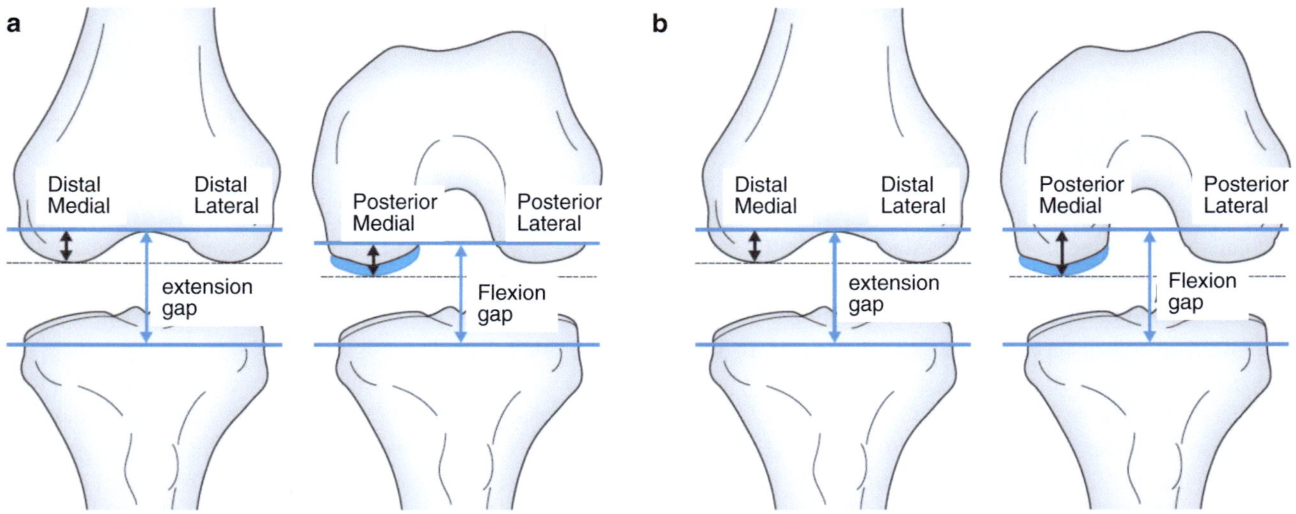

Fig. 52.3 Concept of femoral osteotomy. (**a**): If the articular cartilage of the distal medial femur is lost and the posterior medial articular cartilage is preserved, the flexion gap is tight if the distal and posterior bone resections are the same. (**b**): By increasing the posterior bone resection by increasing the thickness of the cartilage (approximately 2 mm), the size of the extension and flexion gaps become closer

to that of the implant. In the sagittal plane, the distal femur is resected perpendicular to the distal femoral anatomical axis, and unlike PS, post-cam impingement during extension is not a concern. After distal osteotomy, the thickness of the resected bone was measured to confirm that the osteotomy was performed as planned. If the osteotomy was too small, an additional osteotomy was considered and the next step was performed.

The rotational alignment is determined by bony landmarks, such as the surgical epicondyle axis (SEA). If the articular cartilage of the medial posterior condyle remains, resection of the medial posterior condyle should be approximately 2 mm thicker than the thickness of the implant. For this purpose, a posteromedial reference guide was used (Fig. 52.4).

Fig. 52.4 Posteromedial reference guide. Cutting guide using the medial posterior condyle as reference. The gauge inserted into the osteotomy was 2 mm thicker than that of the implant

The disadvantage of a thicker posterior condyle resection is that the posterior condylar offset (PCO) decreases because the medial joint line is raised by 2 mm, but the lateral PCO is the same as before surgery. However, there are reports demonstrating that the ROM decreases with a smaller PCO; therefore, an excessively smaller PCO than the preoperative PCO should be avoided.

In CR, unlike PS, the amount of bone resection and rotational alignment of the femoral component are not adjusted to manage the increased flexion gap and inclination of the gap due to the resection of the PCL.

6.3 Tibial Resection

In the coronal plane, the proximal tibia was cut perpendicular to the mechanical axis, with an osteotomy height of approximately 10 mm from the lateral articular surface.

The tibial posterior tilt angle should be the same as that before surgery. PCL tension increases when the posterior tilt angle is smaller than the preoperative level and decreases when the posterior tilt angle is larger [7]. Similarly, a larger or smaller posterior tilt angle increases or decreases the flexion gap and affects knee kinematics. Postoperative kinematic analysis showed that at a posterior tilt angle of 8°, the medial contact position under load was shifted posteriorly if the articular surface of the implant was relatively flat [8], and no effect of the posterior tilt angle was observed if the joint surface had high medial conformity [9]. Although the basic principle is to match the preoperative tibial slope, it is recommended not to use an excessive posterior tilt >10° in CRs in which the ACL is resected.

It is also important to preserve PCL attachment (Fig. 52.5). When the lateral osteotomy was 10 mm, it was difficult to preserve more than 50% of the tibial attachment of the PCL if the posterior tilt angle was >7°. Although a few reports suggest that the loss of bony attachment does not increase posterior instability if soft tissue continuity is maintained [10], it is safer to preserve the bone island to avoid ligamentous tissue damage.

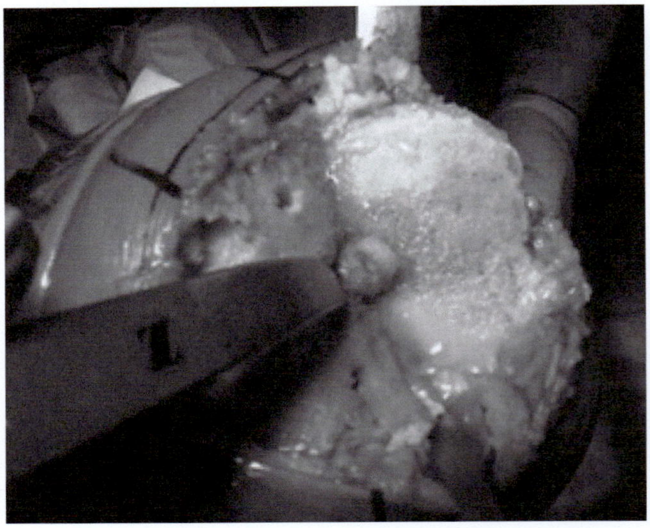

Fig. 52.5 Tibial PCL attachment site. The bone near the PCL attachment site is preserved as a bone island

When the tibia is cut with a posterior tilt, the rotational direction of the cutting guide should match the rotational alignment of the tibial component. The cutting guide is aligned with bony landmarks, such as the Akagi line, and markings on the bone or pins are left to serve as indicators for component placement.

6.4 Soft Tissue Balance

6.4.1 Medial/Lateral Balance

In the evaluation of the ACL-resected knee, the medial gap was almost constant throughout knee flexion; however, the lateral gap increased with flexion. Therefore, lateral laxity in the flexed position is considered physiological while the PCL remains. In the CR, the basic concept of ligament balance is to achieve medial stability through extension and flexion, thereby allowing lateral laxity. However, because lateral laxity is greater in severe varus deformities than in the normal knee [5], it is necessary to carefully consider whether lateral laxity will worsen postoperative outcomes. Studies have reported that lateral laxity at 90° correlated with internal rotation movement of the tibia in intraoperative evaluation [11] and with good postoperative ROM [12]. Postoperative stress radiography has reported that greater lateral laxity at 80° is associated with better postoperative ROM [13]. Although lateral laxity during flexion may improve results, determining the most appropriate degree of laxity is a subject for future research.

6.4.2 PCL Evaluation

The evaluation of PCL should be carefully performed in patients undergoing CR-TKA. When the flexion gap is measured by applying a distraction force in the CR, the tibia may come forward, resulting in a larger measurement of the flexion gap (Fig. 52.6). A tight flexion gap has been reported to decrease flexion angles and clinical score [14]. When measuring the flexion gap, a flexion gap tighter than the extension gap should be avoided, especially in the CR.

Rather than measuring the gap, dynamic evaluation of trial components is more important. The POLO test is a well-known method in which a loose PCL allows the insert to be pulled out (pullout), whereas a tight PCL causes the front of the insert to lift off (liftoff) (Fig. 52.7). Therefore, surgeons should ensure that these conditions do not occur. Even if the POLO test is positive, it was insufficient to confirm that the tension of the PCL was adequate. The intraoperative rollback position is also an indicator because the rollback is larger if the PCL is tight and smaller if the PCL is not tight [15] (Fig. 52.8).

The optimal rollback level for the CR is not well understood. In a few reports, the intraoperative rollback was greater in the group with decreased ROM than in the group

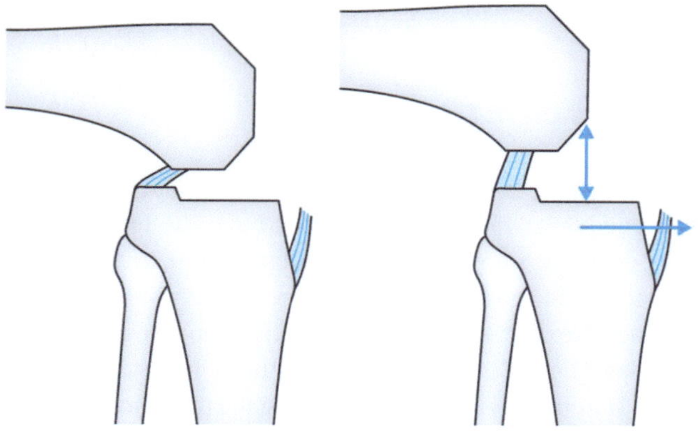

Fig. 52.6 Gap measurement in CR. Compared with the postoperative position (left), when the joint is opened with a tensioner, the tibia may move forward and be measured as a larger gap (right) (Reprinted with permission. © 2007 The British Editorial Society of Bone & Joint Surgery. All rights reserved. Christen B et al: J Bone Joint Surg[Br] 2007; 89-B(8): 1046–1050) [36]

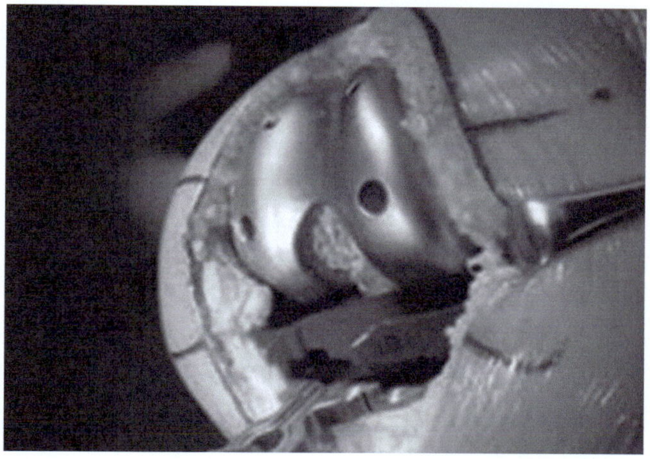

Fig. 52.7 Lift-off of tibial insert. A tight PCL raised the anterior aspect of the tibial insert in the flexed position

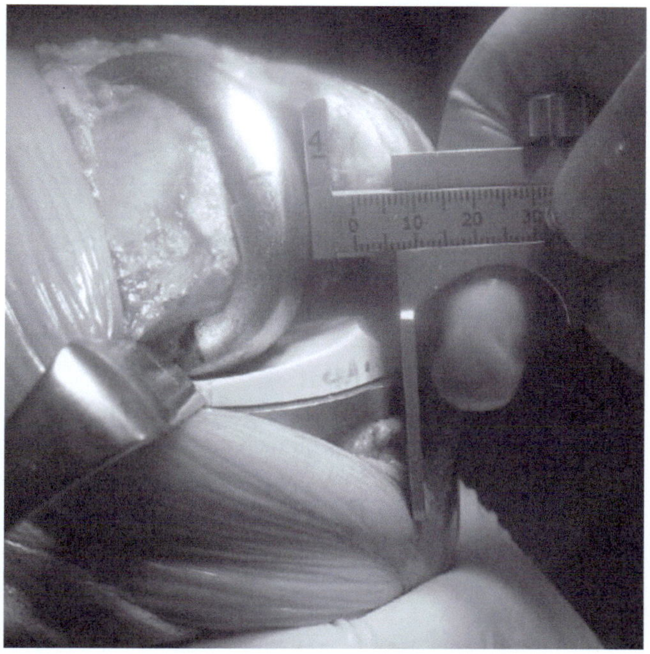

Fig. 52.8 Intraoperative rollback measurement. The distance between the most distal femoral component and the anterior margin of the tibial component was measured

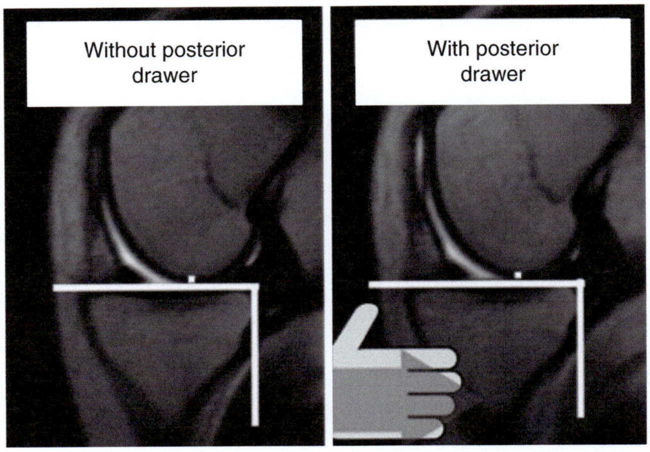

Fig. 52.9 Contact position of the medial compartment. Medial contact position at 90° flexion (MRI) The femur is in contact with the medial tibial articular surface at the posterior 1/3 position without (left) and with (right) a posterior force

with improved or maintained ROM, and excessive rollback should be avoided. During surgery, the normal knee is in 90° flexions, and the medial condyle of the femur is in contact with the posterior 1/3 of the tibial articular surface [16]; this may serve as a reference (Fig. 52.9).

6.4.3 PCL Tension Adjustment

Experiments using cadaveric knees have shown that anterior-posterior laxity is increased by releasing the femoral attachment of the PCL and the tibial attachment to the bone [17]. Akizuki et al. proposed a method to adjust the tension of the PCL by osteotomizing the tibial attachment in a V-shape [18] and reported that no paradoxical motion of the femur was observed in a dynamic analysis using fluoroscopy [19].

Although PCL release is a relatively simple procedure, its effect is sometimes difficult to predict, and caution must be exercised to avoid excessive dissection. Causes of high PCL tension in the surgical technique include an insufficient posterior tilt angle of the tibia and insufficient osteotomy at the medial femoral condyle [7]. Therefore, recutting the tibia with an additional posterior tilt angle or downsizing the femur is a useful method for adjusting the PCL tension.

6.4.4 Patella Replacement

When the patella is to be replaced, an osteotomy is performed such that the thickness of the patella does not change from the preoperative thickness, and the implant is placed.

6.4.5 Implant Fixation: Closed Wounds

Once the soft tissues were adjusted and the size of the component was selected, the knee joint was thoroughly cleaned, and the components were fixed with or without cement The capsule was sutured and the subcutaneous layer and skin were closed to complete the procedure.

7 Clinical Results

7.1 Long-Term Durability

A high survival rate of 96.7% at 10 years and 96.7% at 13 years was reported in 76 cementless CR knees using Osteonics 3000 [20]. In an analysis using the LFA-1 alumina ceramic femoral component, the revision at 5.5 years after surgery was 1.8% [21]. The results of this registry and other studies indicate that the long-term durability of CR is comparable to that of other models.

7.2 Clinical Symptoms and ROM

In a study comparing CR and PS, no differences were found in the symptoms, satisfaction, and activity of the KSS2011 at 8 years postoperatively [22].

In a meta-analysis, ROM was improved in the extension angle, but not in the flexion angle, compared with preoperatively [23]. In a study which CR was performed on one knee and PS on the opposite knee in the same patients [24], and in the other international multicenter study [25], the clinical scores of PS and CR were comparable; however, the improvement in ROM was better with PS. In the RCTs of NexGen CR-flex and Legacy PS, there was no difference in postoperative knee scores; however, postoperative ROM and satisfaction were higher with PS [26]. These results suggest that the postoperative ROM was better in PS than in CR, on average, partly because the degree of tension in the PCL affected the flexion angle.

A few studies have examined the influence of surgical techniques on postoperative outcomes in CR. CR-TKA using a medial stabilizing technique with almost no release of the medial soft tissue has been reported to have a better postoperative ROM than CR-TKA with medial release to balance the medial and lateral gaps [27].

7.3 Implant Design

A high-flexion model with modification of the posterior femoral condyle shape was developed so that the femoral condyle does not impinge, even in deep flexion. However, no difference in postoperative flexion angles was observed between NexGen CR-flex (high-flexion model) and normal NexGen CR [28]. However, it has been reported that the ROM is improved by 6.5° by lowering the posterior lip of the tibia and simultaneously reducing the size of the posterior femoral condyle [29].

Constraining the shape of the joint surface is also important. In terms of medial constraint, the FINE Knee design with increased medial constraint and increased lateral freedom was reported to have superior ROM and clinical scores compared to the historical control Hi-Tech Knee II [30]. In addition, intraoperative evaluation of kinematics showed that a design with increased medial constraints suppressed anterior femoral translation in the mid-flexion position [31]. To achieve stability at shallow flexion angles where the PCL is not functioning, it is recommended that the implant be constrained to some degree.

7.3.1 Joint Stability Evaluation and Dynamic Analysis

It was reported that stability in the anteroposterior direction by KT-2000 at 30° flexions and in the varus/valgus direction by stress radiography in extension were not related to ROM or clinical scores [32]. In contrast, medial stability in extension and flexion contributed to patient satisfaction with postoperative stress X-ray evaluation [33].

Several fluoroscopic analyses of postoperative motion have been performed in Japan. Ten patients with an average flexion angle of 130° after CR showed that lateral rollback was the same as that preoperatively, whereas medial rollback showed an anterior shift in only one of 10 patients [34]. Although paradoxical motion after CR-TKA is a problem with CR, it was reported that there was no difference in anterior-posterior motion between PS and CR-TKA performed relatively recently (2007–2012) [35].

8 Future Prospectives

Although the surgical technique of CR has focused on avoiding overstraining of the PCL, it has become clear that the constraints of the articular surface and PCL tension level are necessary to maintain medial stability. In recent years, a few studies have demonstrated good stability of CR. The next question is what amount of PCL tension is adequate for achieving excellent clinical results. To answer this question, a combination of detailed intraoperative measurements, postoperative dynamic analyses, and patient-reported outcome measures were used.

References

1. Matsuda S, et al. Ligament balancing in total knee arthroplasty: medial stabilizing technique. Asia Pac J Sports Med Arthrosc Rehabil Technol. 2015;2:108–13.
2. Mizu-Uchi H, et al. Anteroposterior stability in posterior cruciate ligament-retaining total knee arthroplasty. J Arthroplast. 2006;21:592–8.
3. Matsuda S, et al. Knee kinematics of posterior cruciate ligament sacrificed total knee arthroplasty. Clin Orthop Relat Res. 1997:257–66.
4. Matsuda S, et al. Knee stability in posterior cruciate ligament retaining total knee arthroplasty. Clin Orthop Relat Res. 1999:169–73.
5. Okamoto S, et al. Lateral soft tissue laxity increases but medial laxity does not contract with varus deformity in total knee arthroplasty. Relat Res. 2013;471:1334–42.
6. Matsumoto T, et al. Semimembranosus release reduces Tibial internal rotation and flexion angle in cruciate-retaining total knee arthroplasty. J Arthroplast. 2015;30:1537–41.
7. Kuriyama S, et al. Posterior tibial slope and femoral sizing affect posterior cruciate ligament tension in posterior cruciate-retaining total knee. Clin Biomech (Bristol, Avon). 2015;30:676–81.
8. Fujimoto E, et al. Significant effect of the posterior tibial slope on weight-bearing, midflexion in vivo kinematics after cruciate-retaining. J Arthroplast. 2014;29:2324–30.
9. Fujito T, et al. Influence of posterior Tibial slope on kinematics after cruciate-retaining total knee arthroplasty. J Arthroplast. 2018;33:3778–82.
10. Aoki M, et al. Functional evaluation in cruciate-retaining-type TKA: anatomical relationship between tibial osteotomy level and PCL attachment. Eur J Orthop Surg Traumatol. 2014;24:1531–7.
11. Matsuzaki T, et al. Tibial internal rotation is affected by lateral laxity in cruciate-retaining total knee arthroplasty: an intraoperative kinematic. Knee Surg Sports Traumatol Arthrosc. 2014;22:615–20.
12. Nakano N, et al. Postoperative knee flexion angle is affected by lateral laxity in cruciate-retaining total knee arthroplasty. J Arthroplast. 2016;31:401–5.
13. Kobayashi T, et al. Lateral laxity in flexion increases the postoperative flexion angle in cruciate-retaining total knee arthroplasty. J Arthroplast. 2012;27:260–5.
14. Matsumoto T, et al. Relatively loose flexion gap improves patient-reported clinical scores in cruciate-retaining total knee arthroplasty. J Knee Surg. 2018;31:573–9.
15. Shoifi Abubakar M, et al. Influence of posterior cruciate ligament tension on knee kinematics and kinetics. J Knee Surg. 2016;29:684–9.
16. Fukagawa S, et al. Posterior displacement of the tibia increases in deep flexion of the knee. Clin Orthop Relat Res. 2010;468:1107–14.
17. Arima J, et al. Effect of partial release of the posterior cruciate ligament in total knee arthroplasty. Clin Orthop Relat Res. 1998:194–202.
18. Akizuki S, et al. Measured (anatomical reference) resection technique for cruciate-retaining total knee arthroplasty. In: Matsuda S, editor. Soft tissue balancing in total knee arthroplasty. Berlin: Springer; 2017. p. 1196–202.
19. Horiuchi H, et al. In vivo kinematic analysis of cruciate-retaining total knee arthroplasty during weight-bearing and nonweight-bearing deep. J Arthroplast. 2012;27:1196–202.
20. Watanabe H, et al. Survival analysis of a cementless, cruciate-retaining total knee arthroplasty: clinical and radiographic assessment 10 to 13. J Bone Joint Surg Br. 2004;86:824–9.
21. Majima T, et al. Clinical results of posterior cruciate ligament retaining TKA with alumina ceramic condylar prosthesis: comparison to Co-Cr alloy. Knee Surg Sports Traumatol Arthrosc. 2008;16:152–6.
22. Kawakami Y, et al. Intermediate-term comparison of posterior cruciate-retaining versus posterior-stabilized total knee arthroplasty using the. Orthopedics. 2015;38:e1127–32.
23. Sumino T, et al. Does cruciate-retaining total knee arthroplasty enhance knee flexion in Western and east Asian patient populations? A meta-analysis. Knee. 2013;20:376–83.
24. Maruyama S, et al. Functional comparison of posterior cruciate-retaining versus posterior stabilized total knee arthroplasty. J Arthroplast. 2004;19:349–53.
25. Harato K, et al. Midterm comparison of posterior cruciate-retaining versus -substituting total knee arthroplasty using the genesis II prosthesis: a. Knee. 2008;15:217–21.
26. Yagishita K, et al. High-flex posterior cruciate-retaining versus posterior cruciate-substituting designs in simultaneous bilateral total knee. J Arthroplast. 2012;27:368–74.

27. Ishibashi K, et al. Medial stabilizing technique preserves anatomical joint line and increases range of motion compared with the gap-balancing. Knee. 2020;27:558–64.
28. Minoda Y, et al. Range of motion of standard and high-flexion cruciate retaining total knee prostheses. J Arthroplast. 2009;24:674–80.
29. Tsuji S, et al. Effect of posterior design changes on postoperative flexion angle in cruciate retaining mobile-bearing total knee arthroplasty. Int Orthop. 2011;35:689–95.
30. Nakamura J, et al. A comparative study of flat surface design and medial pivot design in posterior cruciate-retaining total knee arthroplasty: a matched pair cohort study of two years. BMC Musculoskelet Disord. 2018;19:234.
31. Tsubosaka M, et al. Mid-flexion stability in the anteroposterior plane is achieved with a medial congruent insert in cruciate-retaining total knee. Knee Surg Sports Traumatol Arthrosc. 2021;29:467–73.
32. Yamakado K, et al. Influence of stability on range of motion after cruciate-retaining TKA. Arch Orthop Trauma Surg. 2003;123:1–4.
33. Kamenaga T, et al. The influence of postoperative knee stability on patient satisfaction in cruciate-retaining total knee arthroplasty. J Arthroplast. 2018;33:2475–9.
34. Kitagawa A, et al. In vivo comparison of knee kinematics before and after high-flexion posterior cruciate-retaining total knee arthroplasty. J Arthroplast. 2010;25:964–9.
35. Murakami K, et al. Kinematic analysis of stair climbing in rotating platform cruciate-retaining and posterior-stabilized mobile-bearing total knee arthroplasties. Arch Orthop Trauma Surg. 2017;137:701–11.
36. Christen B, Heesterbeek P, Wymenga A, Wehrli U. Posterior cruciate ligament balancing in total knee replacement: the quantitative relationship between tightness of the flexion gap and tibial translation. J Bone Joint Surg Br. 2007;89(8):1046–50.

Posterior Stabilized Total Knee Arthroplasty (PS-TKA) (Gap-Balancing Technique)

53

Yukihide Minoda

1 PS-TKA

PS-TKA uses a post–cam mechanism that compensates for posterior cruciate ligament (PCL) function to improve the range of motion (ROM) and anteroposterior instability of total condylar TKA, an anterior cruciate ligament (ACL)/PCL resection-type TKA. PS-TKA has good long-term durability even when osteotomy is performed to the thickness of the femoral implant (measured resection technique). However, the measured resection technique often results in a large flexion gap because the PCL is resected [1–4] A large flexion gap induces midflexion laxity [5, 6] and instability. A surgical technique that adjusts the osteotomy of the posterior condyle (gap balancing technique) before that of the posterior condyle [7] can reduce laxity in the midflexion position.

2 Indications for Surgery

Since the indications for surgery do not depend on the condition of the ACL/PCL, there is no need to consider the indication for surgery as with models that preserve the ACL or PCL.

1. Bone loss

   If there is a bone defect of ≥10 mm in the femur or tibia, prepare a stem extension or metal block as a backup.
2. Medial cruciate ligament(MCL), lateral cruciate ligament (LCL)

   MCL and LCL functions are not provided by prosthesis, so MCL and LCL insufficiency are inapplicable.
3. Preoperative ROM

   There is no specific knee joint ROM limitation. However, genu recurvatum due to neurogenic diseases (polio and hemiplegia among others) cannot be controlled without a rotating hinge. Thus, it is inapplicable.
4. Others

   In cases where it is difficult to adjust the ligament balance of MCL and LCL, such as fixed valgus deformity, severe deformity, and extraarticular deformity, a constrained type should be prepared in advance as a backup.

3 Basic Concept of PS-TKA

The most significant advantage of PS-TKA is that it has fewer important decisions that affect the outcome than other models.

1. No need to evaluate ACL and PCL: Since the indication for surgery does not depend on the condition of ACL and PCL, there is no need to consider the indication for surgery as in models that preserve the ACL and PCL.
2. No need to adjust the ligament balance of PCL: The post–cam mechanism ensures that an appropriate amount of femoral rollback is obtained regardless of the surgical technique, and a larger knee flexion angle can be easily obtained.
3. No need to adjust the posterior tilt angle: The posterior tilt angle of the tibia can be adjusted according to the implant design; thus, there is no need to adjust it for each case, as is the case with the cruciate retaining (CR) and bicruciate retaining (BCR) prostheses. If the posterior tilt angle is greater than the implant design; anterior tibial post impingement may occur.
4. No ACL or PCL function in the midflexion position until the post-cam mechanism engages: The post–cam mechanism compensates for the PCL function by initiating engagement at 40–70° of flexion, depending on the model. In other words, the ACL and PCL functions are not present in the midflexion position until the post–cam mechanism starts to engage. Stability in the midflexion position is reportedly lower than that of CR-TKA [8]. Therefore, a surgical technique to achieve stability in the midflexion position is desirable.

Y. Minoda (✉)
Department of Orthopaedic Surgery, Osaka Metropolitan University, Osaka, Japan
e-mail: yminoda@omu.ac.jp

S. Matsuda (ed.), *Advances in Total Knee Arthroplasty*, https://doi.org/10.1007/978-981-97-4920-1_53

4 Problems with PS-TKA

4.1 Patellar Clunk Syndrome (Fig. 53.1)

The soft tissue formed at the superior border of the patella is stuck in the box of the femoral component and is dislodged when the knee is extended from the flexed position, resulting in a clunking sound and pain. Patellar clunk syndrome is common in models with a large longitudinal length of the femoral component box (models with a large intercondylar box ratio) [9]. Recently designed models have a smaller intercondylar box ratio, so this syndrome is rarely encountered with them. Resection of the synovial membrane above the patella is a preventive measure.

4.2 Patellar Post Impingement (Fig. 53.2)

This occurs when the patellar and posterior tibial implants collide in flexion. It may reduce flexion ROM, increase patellofemoral (PF) pressure, and cause anterior knee pain. Patellar post impingement may result from the short distance between the osteotomy surface and the inferior margin of the patellar component (tibial-patellar clearance) [10, 11]. The distal femur and proximal tibia osteotomies affect tibia–patella clearance. The amount of distal femoral and proximal tibial osteotomies should be adjusted to ensure a tibia–patella clearance of ≥30 mm.

4.3 Anterior Post Impingement (Fig. 53.3)

If the femoral/tibial component is placed in an excessively flexed position beyond the design of the implant, the anterior surface of the post may collide with the anterior edge of the femoral box in knee extension, resulting in wear and fracture of the post. The contact area and contact pressure of the anterior surface of the post depend on the design of the post and box [12, 13]. Designing a post/box that does not increase the contact pressure is desirable.

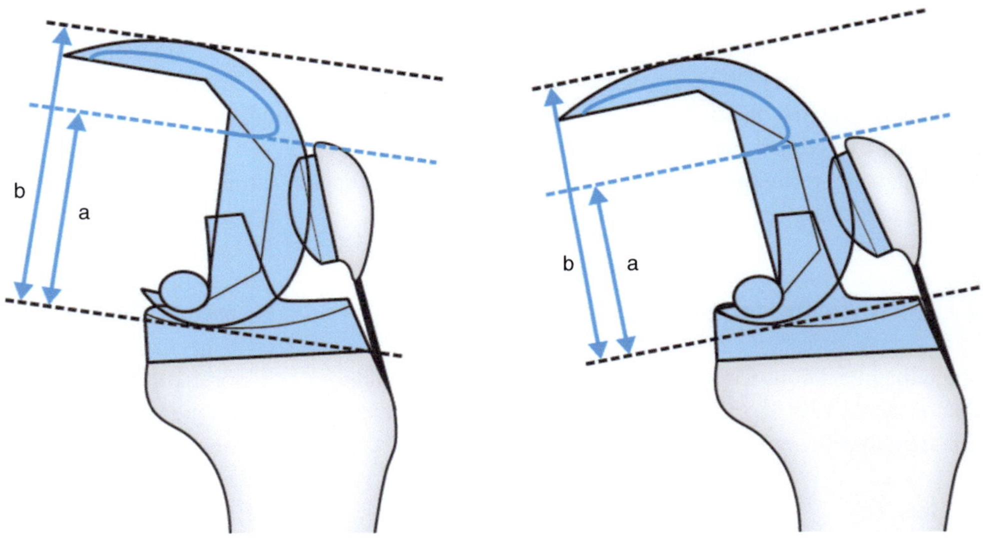

Fig. 53.1 Relationship between intercondylar box ratio (**a/b**) and patellar clunk syndrome. When the intercondylar box ratio is large, the knee joint angle where the suprapatellar pole meets the box angle is small, meaning that the soft tissue of the suprapatellar pole is stuck in the box at a small flexion angle. If the intercondylar box ratio is small, the soft tissue at the suprapatellar pole is less likely to fit into the box

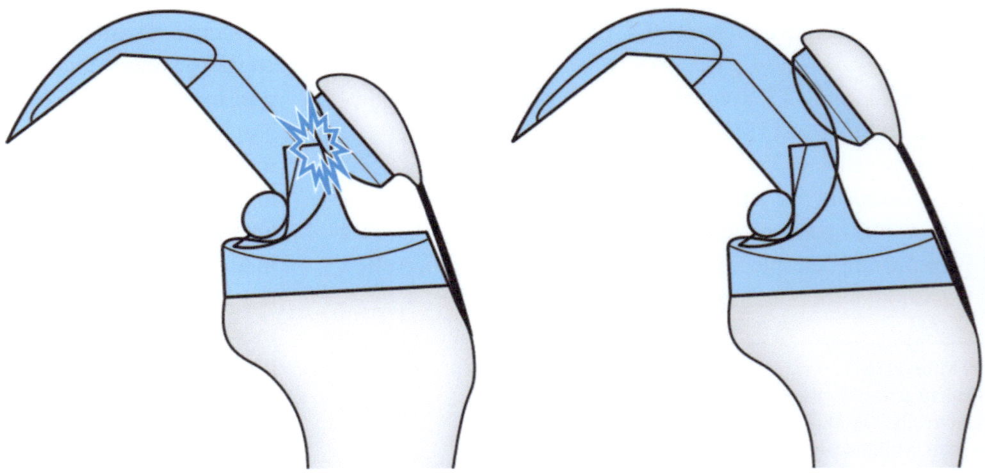

Fig. 53.2 Relationship between patellar height and patellar post impingement. If the distance between the osteotomy surface and the inferior border of the patellar component (tibia–patella clearance) is short, patellar post impingement is likely to occur

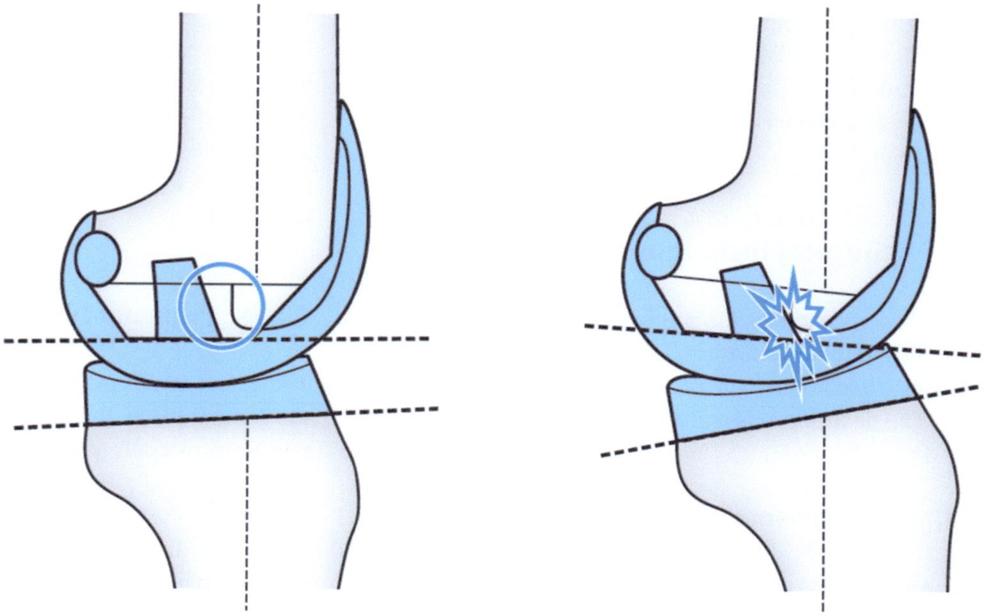

Fig. 53.3 If the femoral and tibial components are placed in excessive flexion, there is a risk that the anterior post and the anterior edge of the femoral box will collide in knee joint extension (anterior post impingement), resulting in wear and fracture of the post

5 Concept of Surgery

5.1 Coronal Plane Alignment

The goal in the coronal plane is to place the implant perpendicular to the functional axis of both the femur and tibia. The reason is that PS-type implants themselves are designed that way. Errors of ≥3° have been considered outliers. Recent studies have shown that "a little bit" of outliers does not affect long-term durability due to advances in materials and design [14]. However, computer simulation studies have reported that the varus alignment of the tibial component of PS-TKA causes edge-loading of the femoral component and tibial insert [15]. Even with the latest navigation systems, completely eliminating outliers is impossible [16]. Considering surgical errors, the goal is to achieve a placement perpendicular to the functional axis.

5.2 Ligament Balance

Lateral laxity in extension can be tolerated to some extent. It has a high tolerance for alignment and ligament balance. It has a better intraoperative view of the posterior femur and tibia than other types of prosthesis because of PCL resection. If medial laxity persists or lateral laxity exceeds the acceptable range, selecting an insert (PS Plus) with a high varus/valgus constraint is possible.

In terms of reducing laxity in the midflexion position, it is preferable to use a surgical technique that adjusts the amount of osteotomy of the posterior condyle (gap-balancing technique) before osteotomy of the posterior condyle.

The gap-balancing technique consists of two methods:

1. A flexion–extension gap-balancing technique in which a flexion gap is created by osteotomy of the proximal tibia and posterior femur first, and the amount of distal femoral osteotomy is determined according to the size of the flexion gap.
2. The modified gap-balancing technique, in which an extension gap is created by osteotomy of the proximal tibia and distal femur first, and the amount of posterior femoral condyle osteotomy is determined according to the size of the extension gap.

The latter is preferable in terms of avoiding joint-line elevation.

Additional distal femoral osteotomy is often performed during flexion contracture or when the extension gap is small. Elevation of the joint line by additional distal femoral osteotomy reportedly causes instability in the midflexion position in previous experiments using cadavers [17, 18]. However, a recent study in patients undergoing TKA showed that an additional 2 mm of distal femoral osteotomy did not cause laxity in the midflexion position [19]. However, an additional 4 mm of distal femoral osteotomy opened the extension gap by 2 mm after implant placement [20]. Additional distal femoral osteotomy should thus be limited to 2 mm.

5.3 Patellar Resurfacing

Patella resurfacing/nonresurfacing is still controversial. Caritilage wearing out of the vertical ridge in nonresurfacing patellar at the corner of the anterior margin of the femoral component box might cause anterior knee pain when preserving the patella. The Australian National Joint Registry showed that patellar nonresurfacing had a significantly higher revision rate than patellar resurface in PS-TKA, unlike other models, and patellar resurfacing is recommended in PS-TKA.

6 Surgical Techniques

This chapter describes the surgical technique of PS-TKA for medial-type osteoarthritis (OA) using the "modified gap-balancing technique," in which the extension gap is made first. The flexion gap is made based on the extension gap.

6.1 Distal Femoral Osteotomy

The osteotomy is made perpendicular to the femoral mechanical axis in the coronal plane. The amount of osteotomy at the distal femur should consider the worn-out medial condyle in the medial OA knee. The amount of osteotomy at the center of the intercondylar area, which is not affected by the stage of the disease, such as the so-called "sulcus cut " [21] or "butterfly sign," is also helpful in determining the amount of osteotomy at the distal femur. There is no clear evidence for the sagittal plane because there is no clear reference axis [22]. Excessive extension of femoral component causes notching of the anterior femoral cortex [23]. Conversely, excessive flexion of femoral component causes anterior post impingement, in which the anterior margin of the femoral box collides with the anterior post during knee extension (Fig. 53.3).

6.2 Proximal Tibia Osteotomy

The osteotomy is made perpendicular to the femoral axis in the coronal plane. In the gap-balancing technique, the proximal tibial osteotomy is the reference plane for ligament balance. The tibial alignment in the coronal plane significantly influences the extension and flexion gaps (Fig. 53.4). Therefore, the angle of the proximal tibial osteotomy in the coronal plane must be accurate when performing the gap -balancing technique. Conventional methods such as extramedullary guides have significant alignment errors in the proximal tibial osteotomy angle and large interoperator accuracy differences. However, the navigation system has fewer alignment errors and fewer interoperator accuracy differences [16]. Regarding the posterior tilt, some models have a built-in posterior tilt angle in the component itself, while

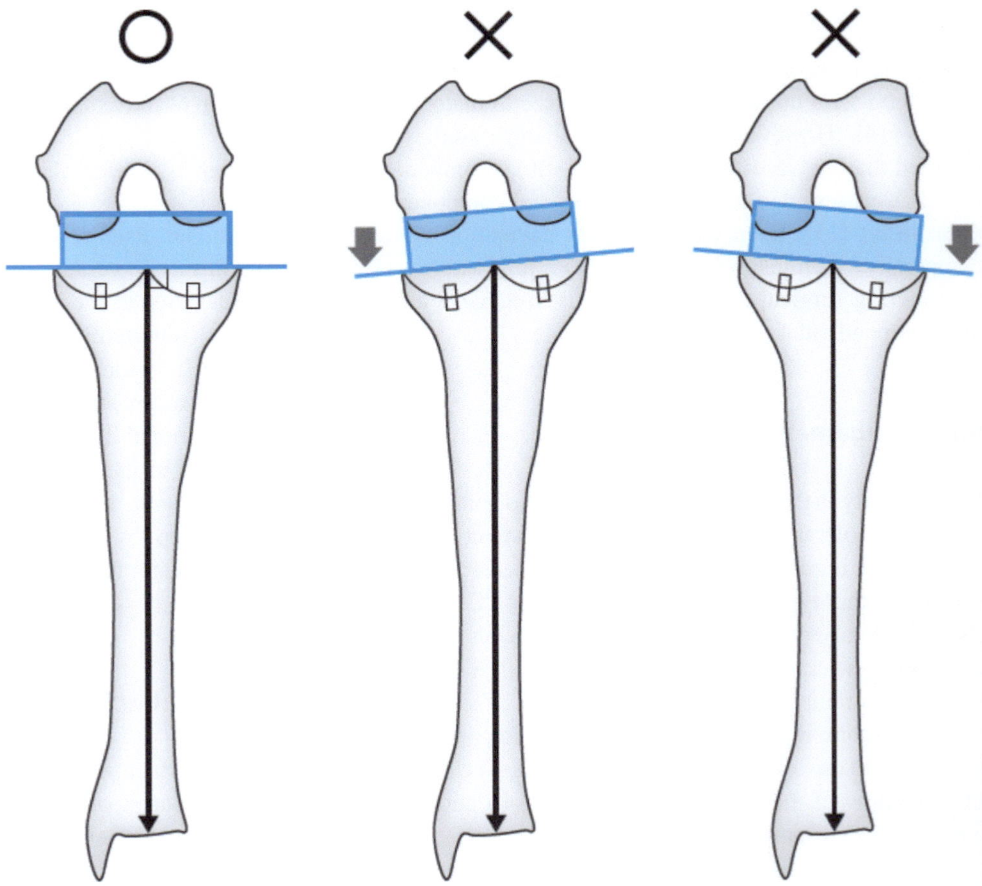

Fig. 53.4 Tibial alignment in the coronal plane has a significant effect on the extension gap and the flexion gap

others do not. Therefore, the recommended posterior tilt angle for the design of tibial components varies widely from model to model (0–7°). If the posterior tilt angle is greater than designed, anterior post impingement may occur in the extended position. The influence of the tibial posterior tilt angle on the flexion gap is small in PS-TKA, unlike in CR-TKA [24]. The tibial component posterior tilt angle should be adjusted according to the design of the model used.

6.3 Preparation of Extension Gap

At the femur condyle, osteophytes are resected as far as possible. Regarding tibia, tibial component size was decided and then medial osteophytes and bone not covered by the implant are resected from the tibia (Fig. 53.5). The extension gap is measured; tensor devices, spacer blocks, and laminar spreaders are some of the measurement methods. The spacer block is simple but subjective and does not display angles or numerical values. Tensor devices measure pull-off tension accurately and reproducibly, but there is no fixed viewpoint of pull-off tension. Each method has pros and cons and should be used cautiously.

If the medial soft tissue is tight in the varus knee, medial soft tissue release is performed. Excessive medial soft tissue dissection causes postoperative instability and should be minimal.

Tibial reduction osteotomy (placing a one-size-smaller tibial component outside the tibial osteotomy surface and osteotomy of the protruding medial tibia) is also effective in severe varus deformity knees [25].

The goal of the extension gap at this stage should not be rectangular. This is because the subsequent steps of femoral posterior condylar osteotomy and removal of posterior femoral osteophyte will make the extension gap 1° valgus [26], and the placement of the femoral component will make the extension gap 1° valgus [27]. Therefore, when creating the extension gap, a slight (approximately 2–5°) varus should be left before proceeding to the next step (e.g., posterior condylar osteotomy and creation of posterior clearance).

6.4 Preparation of Flexion Gap

The extension and flexion gaps before posterior condylar osteotomy are measured. The relative balance of the extension and flexion gaps is adjusted by adjusting the amount of osteotomy at the distal and posterior femoral condyles and the proximal tibia.

6.4.1 Size of Flexion Gap

Since ACL and PCL are resected in PS-TKA, the flexion gap tends to be larger [28]. Therefore, the key is addressing the loose flexion gap by adjusting the amount of osteotomy.

The relationship between osteotomy volume and extension/flexion gap can be summarized into three categories (Fig. 53.6).

1. The amount of osteotomy in the distal femur affects the extension gap.
2. The amount of osteotomy at the posterior femoral condyle affects the flexion gap.
3. The amount of osteotomy in the proximal tibia affects both the extension and flexion gaps.

Therefore, three specific ways to make the flexion gap relatively small exist.

1. Additional osteotomy of the distal femur: An additional osteotomy of the distal femur is performed to increase the extension gap.
2. Increase femoral component size: Reduce the amount of posterior femoral condyle osteotomy, select a femoral component one size larger, and reduce the flexion gap.
3. Slightly flexion placement of the femoral component: The flexion gap can be reduced by approximately 2 mm by placing the femoral component in a 2° flexion position [29]. An osteotomy is performed by inserting a stylus in front of the femoral 4-plane cut block and then fixing cut block. A slight gap between the implant and bone is observed at the distal femur, but this is not a clinical problem if the implant is fixed with cement.

Combining these three methods can accommodate a loose flexural gap of up to 6 mm.

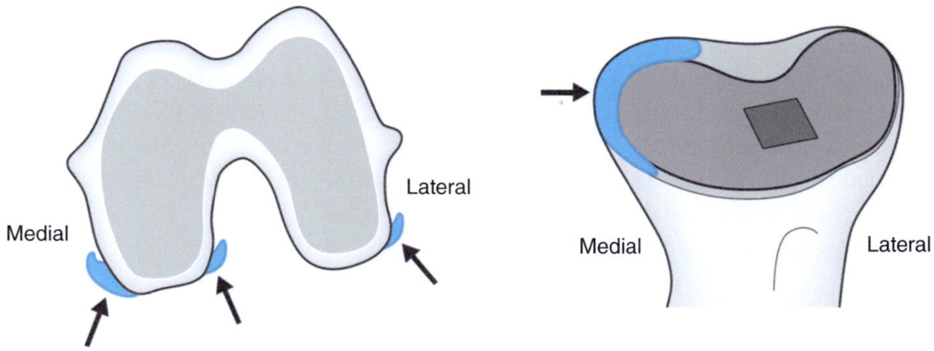

Fig. 53.5 Before measuring the extension gap, the femur is resected for osteophytes at the condyle, and the tibia is resected for medial osteophytes and bone not covered by the implant by determining component size

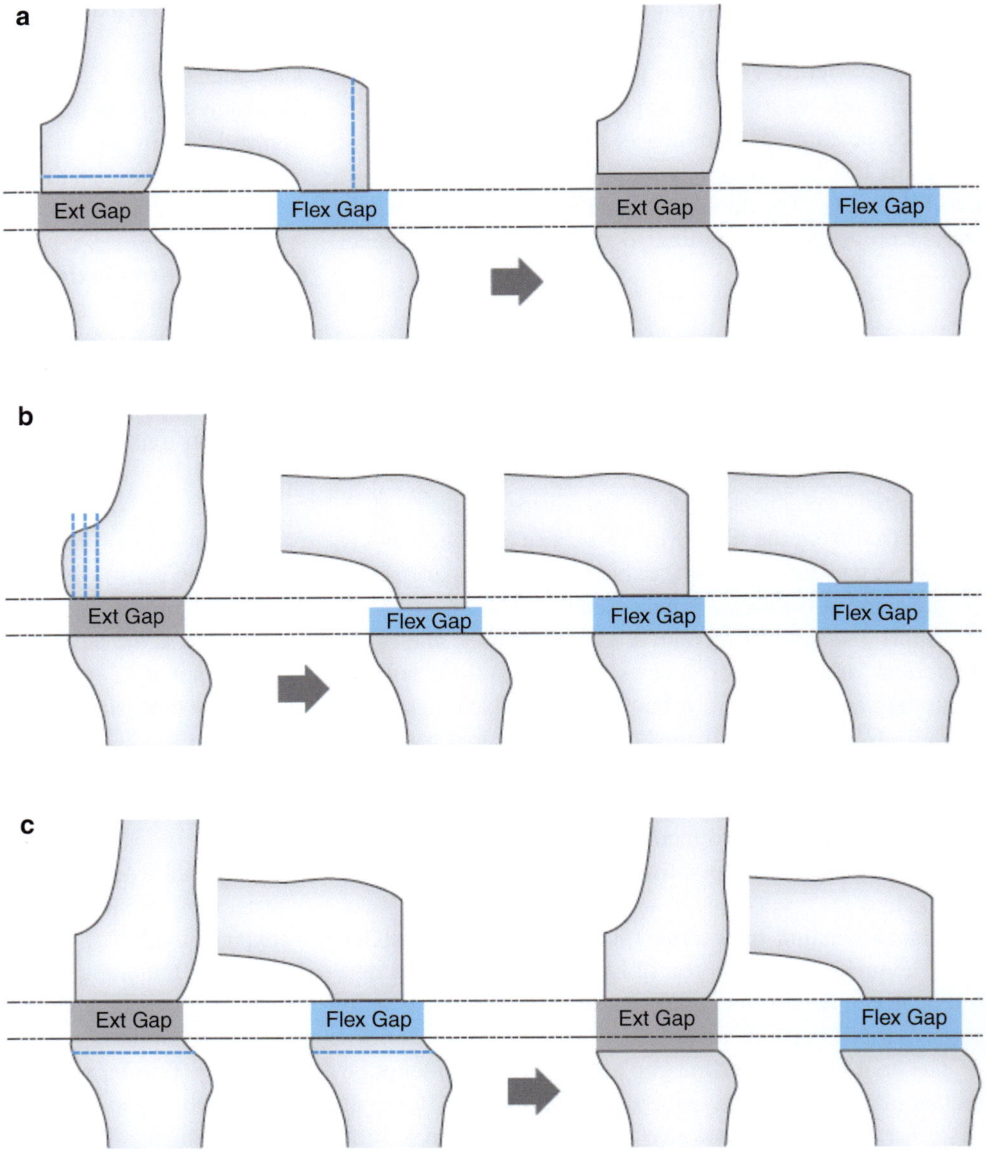

Fig. 53.6 The relationship between osteotomy volume and extension/flexion gap. (**a**): The amount of osteotomy in the distal femur affects the extension gap. (**b**): The amount of osteotomy at the posterior femoral condyle affects the flexion gap. (**c**): The amount of osteotomy in the proximal tibia affects the extension and flexion gaps

6.4.2 Shape of Flexion Gap (Fig. 53.7)

The rotational position of the femoral component determines the shape of the flexion gap. A rectangular flexion gap can be created by osteotomy of the posterior femoral condyle parallel to the tibial osteotomy plane. However, if the rotational position of the femoral component is placed far from the anatomical landmarks, a rotation mismatch between the femur and tibia is caused in extension [30]. Excessive internal rotation of the femoral component causes patellar maltracking, and excessive external rotation causes anterior femoral cortical notching [23]. The rotation of the femoral component should be adjusted between the surgical epicondylar axis (SEA) and the clinical epicondylar axis (CEA).

6.5 Creation of Posterior Condylar Osteotomy and Posterior Clearance

The creation of posterior clearance after posterior condylar osteotomy (resection of the posterior osteophyte protruding from the implant) is important to prevent impingement of the polyethylene insert and posterior condyle during deep flexion (Fig. 53.8). This procedure opens the extension and flexion gaps by approximately 1 mm in PS-TKA. it does not affect the relative balance of the extension and flexion gaps, since there is no significant difference in the amount of increase in the gaps [26].

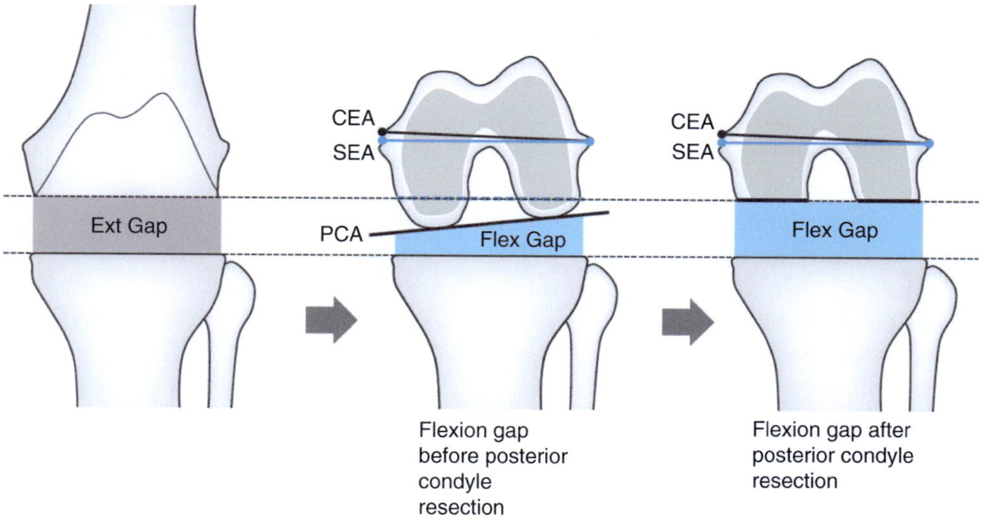

Fig. 53.7 Determining the amount of osteotomy at the posterior condyle considering the surgical epicondylar axis, clinical epicondylar axis, PCA, extension gap, and flexion gap

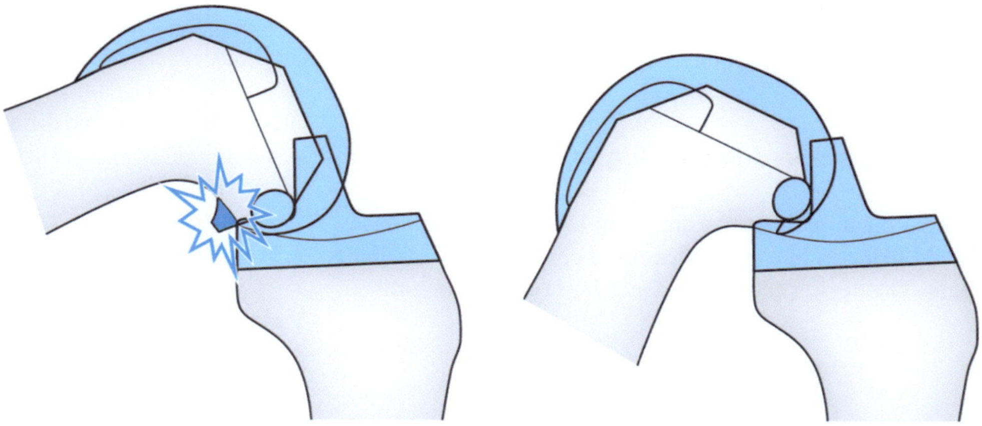

Fig. 53.8 Posterior condylar osteophytes protruding from the implant impinge on the polyethylene insert during deep flexion and prevent deep flexion

6.6 Patellar Replacement

The thickness of the patella is measured, and an osteotomy of the same thickness as the implant is performed. However, at least 13 mm should be left to avoid fractures. The synovium above the patella is excised and coagulated with an electrocautery scalpel to prevent patellar clunk syndrome.

6.7 Tibial Component Rotation Installation

In PS-TKA, patients with a rotation mismatch between the femur and tibia due to abnormal tibial component rotational placement by >10° have poor postoperative outcomes (New KSS functional activities) [30]. Tibial component rotational placement is difficult even with CT-based navigation guidance [31]. Patient-specific guides [32], asymmetric tibial components [33], and mobile-bearing mechanisms [34] effectively reduce abnormal rotational placement of the tibial component.

6.8 Trial

Patellar tracking, patellar post impingement, and total ROM confirm the balance of the internal and external ligaments. The joint gap at the osteotomy surface is not always the same as that after trial placement [2]. In Vanguard RP, the extension gap is 1° and 1 mm less than the osteotomy surface [27]. Therefore, flexion contracture may be observed after trial placement. In such cases, dissection of the intercondylar joint capsule opens the extension gap by 2 mm [35, 36], which improves flexion contracture. A PS+ insert with a slightly stronger constraint should be selected if internal or external instability remains in the extended position.

6.9 Periarticular Injection

Cocktail injections into the joint capsule should be performed. When epinephrine is included, injections into subcutaneous tissues should be avoided to prevent skin necrosis due to vasoconstriction. After injection, wash the joint to remove any leaked drug solution from the joint.

6.10 Cement Fixation

Anchor holes are drilled in the medial part of the tibial osteotomy surface where the cancellous bone does not protrude to increase fixation. After cleaning, the osteotomy should be thoroughly dried, and cement should be pressed into the cancellous bone. Both the implant and the bone should be applied with cement [37].

6.11 Wound Closure

Suture the joint watertight. After suturing, inject tranexamic acid into the joint and release the air tourniquet. Suture the retinaculum, subcutaneous tissue, and dermis.

6.12 Postoperative Radiography (Fig. 53.9)

Frontal plane radiographs are taken in extension and in the lateral plane in maximum flexion to confirm implant alignment. The flexion angle under anesthesia can be determined.

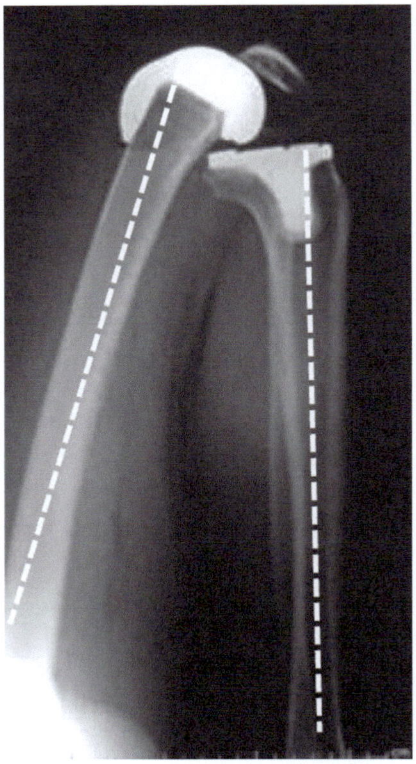

Fig. 53.9 Immediate postoperative lateral radiographic image. Share the postoperative flexion angle with the rehabilitation staff

The postoperative flexion angle under anesthesia correlates with the flexion angle 2 years postoperatively [38]. If the implant is flexed more than this during the rehabilitation, there is a risk of patellar fracture or patellar tendon rupture, so information should be shared with the rehabilitation staff.

7 Clinical Results

7.1 Long-Term Durability

Good results have already been reported in the 1990s, with survival rates of 92–98% at 10–16 years [39].

7.2 Clinical Symptoms and ROM

Although there are reports of no difference between PS- and CR-TKA regarding ROM [40], a meta-analysis [41] and a systematic review [42] suggest that PS-TKA is better than CR-TKA. However, no difference between PS- and CR-TKA was observed in clinical scores.

7.3 Implant Design and Joint Stability

Various models have been introduced to solve PS-TKA midflexion instability. Vanguard RP (Zimmer Biomet), which has a stronger midflexion anteroposterior constraint, exhibits less anteroposterior movement 2 years postoperatively than the Vanguard PS (Zimmer Biomet), which has a normal insert design [43].

7.4 Implant Design and Joint Dynamics

PS-TKA did not show much physiological rotational motion [44]. Joint surface geometry has been used to imitate normal knee kinematics. The mobile-bearing Vanguard PSRP (Zimmer Biomet) with high midflexion anteroposterior constraint [45] and the Journey II BCS (Smith & Nephew) with improved post–cam mechanism and joint surface geometry [46] showed rotational motion closer to normal.

8 Future Tasks

PS-TKA is a well-established model with long-term durability, good ROM, and suitable surgical techniques. Improved design, surgical technique, and surgical devices have been used to improve clinical outcomes, such as patient-oriented evaluation. The advantages of PS-TKA should be maintained during these attempts.

References

1. Matsumoto T, et al. Soft-tissue balancing in total knee arthroplasty: cruciate-retaining versus posterior-stabilised, and measured-resection. Int Orthop. 2014;38:531–7.
2. Matsumoto T, et al. The intra-operative joint gap in cruciate-retaining compared with posterior-stabilised total knee replacement. J Bone Joint Surg Br. 2009;91:475–80.
3. Hananouchi T. Sagittal gap balancing with the concept of a single radius femoral component in posterior cruciate sacrificing total knee arthroplasty. Int Orthop. 2015;39:659–65.
4. Itokazu M, et al. Anatomical landmarks of the distal femoral condyles are not always parallel to the tibial bone cut surface in flexion during total knee arthroplasty. Knee. 2016;23:725–9.
5. Minoda Y, et al. Midflexion laxity after implantation was influenced by the joint gap balance before implantation in TKA. J Arthroplast. 2015;30:762–5.
6. Minoda Y, et al. Intraoperative assessment of midflexion laxity in total knee prosthesis. Knee. 2014;21:810–4.
7. Nagamine R, et al. Distance of the femorotibial joint gap can be controlled in the modified ligament dependent cut procedure during total knee arthroplasty. Fukuoka Igaku Zasshi. 2003;94:296–303.
8. Hino K, et al. Mid-flexion laxity is greater after posterior-stabilised total knee replacement than with cruciate-retaining procedures: a computer navigation study. Bone Joint J. 2013;95-B:493–7.
9. Fukunaga K, et al. The incidence of patellar clunk syndrome in a recently designed mobile-bearing posteriorly stabilised total knee replacement. J Bone Joint Surg Br. 2009;91:463–8.
10. Minoda Y, et al. Impingement of the patellar component against the tibial post depends on the design of the post-cam mechanism: comparison between 12 posterior stabilized total knee prostheses. Knee. 2018;25:676–81.
11. Minoda Y, et al. Impingement of the patellar component against the tibial post after posterior stabilized and constrained TKA: a pilot study. Arch Orthop Trauma Surg. 2017;137:861–5.
12. Hamai S, et al. Contact stress at the anterior aspect of the tibial post in posterior-stabilized total knee replacement. J Bone Joint Surg Am. 2010;92:1765–73.
13. Nakayama K, et al. Contact stress at the post-cam mechanism in posterior-stabilised total knee arthroplasty. J Bone Joint Surg Br. 2005;87:483–8.
14. Abdel MP, et al. Effect of postoperative mechanical axis alignment on survival and functional outcomes of modern total knee arthroplasties with cement: a concise follow-up at 20 years. J Bone Joint Surg Am. 2018;100:472–8.
15. Watanabe M, et al. Varus femoral and tibial coronal alignments result in different kinematics and kinetics after total knee arthroplasty. Knee Surg Sports Traumatol Arthrosc. 2017;25:3459–66.
16. Minoda Y, et al. Usefulness of an accelerometer-based portable navigation system for total knee arthroplasty: a multicenter prospective randomized controlled trial. J Bone Joint Surg Am. 2020;102:1993–2000.
17. Martin JW, et al. The influence of joint line position on knee stability after condylar knee arthroplasty. Clin Orthop Relat Res. 1990;259:146–56.
18. Luyckx T, et al. Raising the joint line in TKA is associated with mid-flexion laxity: a study in cadaver knees. Clin Orthop Relat Res. 2018;476:601–11.
19. Minoda Y, et al. Joint line elevation is not associated with mid-flexion laxity in patients with varus osteoarthritis after total knee arthroplasty. Knee Surg Sports Traumatol Arthrosc. 2020;28:3226–31.
20. Minoda Y, et al. Four-millimeter additional bone resection in the distal femur does not result in an equivalent increase in the extension joint gap in total knee arthroplasty. J Arthroplast. 2021;36:958–62.
21. Kuriyama S, et al. Is a "sulcus cut" technique effective for determining the level of distal femoral resection in total knee arthroplasty? Knee Surg Sports Traumatol Arthrosc. 2014;22:3060–6.
22. Minoda Y, et al. TKA sagittal alignment with navigation systems and conventional techniques vary only a few degrees. Clin Orthop Relat Res. 2009;467:1000–6.
23. Minoda Y, et al. Theoretical risk of anterior femoral cortex notching in total knee arthroplasty using a navigation system. J Arthroplast. 2013;28:1533–7.
24. Okazaki K, et al. Influence of the posterior tibial slope on the flexion gap in total knee arthroplasty. Knee. 2014;21:806–9.
25. Niki Y, et al. Effects of reduction osteotomy on gap balancing during total knee arthroplasty for severe varus deformity. J Arthroplast. 2015;30:2116–20.
26. Minoda Y, et al. The flexion gap preparation does not disturb the modified gap technique in posterior stabilized total knee arthroplasty. Knee. 2012;19:832–5.
27. Minoda Y, et al. Decreased extension gap and valgus alignment after implantation of a total knee prosthesis in primary varus knees. Knee Surg Sports Traumatol Arthrosc. 2016;24:3642–7.
28. Kadoya Y, et al. Effects of posterior cruciate ligament resection on the tibiofemoral joint gap. Clin Orthop Relat Res. 2001;391:210–7.
29. Tsukeoka T, et al. The effect of a sagittal cutting error of the distal femur on the flexion-extension gap difference in total knee arthroplasty. J Arthroplast. 2013;28:1099–102.
30. Ueyama H, et al. Malrotation of the fixed-bearing posterior stabilized total knee prosthesis causes a postoperative rotational mismatch between the. Knee Surg Sports Traumatol Arthrosc. 2020;28:3810–20.
31. Kuriyama S, et al. Tibial rotational alignment was significantly improved by the use of a CT-navigated control device in total knee arthroplasty. J Arthroplast. 2014;29:2352–6.
32. Yamamura K, et al. Design improvement in patient-specific instrumentation for total knee arthroplasty improved the accuracy of the tibial. Knee Surg Sports Traumatol Arthrosc. 2020;28:1560–7.
33. Minoda Y, et al. Asymmetric tibial component improved the coverage and rotation of the tibial component in a medial pivot total knee prosthesis. J Knee Surg. 2018;31:416–21.
34. Ueyama H, Minoda Y, Sugama R, Ohta Y, Takemura S, Nakamura H. Mobile-bearing prosthesis suppresses the postoperative rotational mismatch and improves patient-reported outcome measurements better than fixed-bearing prosthesis: rotational analysis by 3D measurement in total knee arthroplasty. Arch Orthop Trauma Surg. 2023;143:6781–90.
35. Okamoto Y, et al. Capsular release around the intercondylar notch increases the extension gap in posterior-stabilized rotating-platform total knee arthroplasty. Knee. 2016;2:730–5.
36. Kinoshita T, et al. Efficacy of posterior capsular release for flexion contracture in posterior-stabilized total knee arthroplasty. J Exp Orthop J Exp Orthop. 2021;8:102.
37. Refsum AM, et al. Cementing technique for primary knee arthroplasty: a scoping review. Acta Orthop. 2019;90:582–9.
38. Minoda Y, et al. Mobile-bearing prosthesis and intraoperative gap balancing are not predictors of superior knee flexion: a prospective randomized study. Knee Surg Sports Traumatol Arthrosc. 2015;23:1986–92.
39. Diduch DR, et al. Total knee replacement in young, active patients. Long-term follow-up and functional outcome. J Bone Joint Surg Am. 1997;79:575–82.
40. Kawakami Y, et al. Intermediate-term comparison of posterior cruciate-retaining versus posterior-stabilized total knee arthroplasty using the new knee scoring system. Orthopedics. 2015;38:e1127–32.
41. Verra WC, et al. Similar outcome after retention or sacrifice of the posterior cruciate ligament in total knee arthroplasty. Acta Orthop. 2015;86:195–201.

42. Longo UG, et al. Outcomes of posterior-stabilized compared with cruciate-retaining total knee arthroplasty. J Knee Surg. 2018;31:321–40.
43. Minoda Y, et al. Mobile-bearing TKA improved anteroposterior joint stability in the mid-flexion range compared to fixed-bearing TKA. Arch Orthop Trauma Surg. 2016;136:1601–6.
44. Victor J, et al. Kinematics of posterior cruciate ligament-retaining and -substituting total knee arthroplasty: a prospective randomised outcome. J Bone Joint Surg Br. 2005;87:646–55.
45. Kage T, et al. In vivo kinematics of a newly updated posterior-stabilised mobile-bearing total knee arthroplasty in weight-bearing and non-weight-bearing high-flexion activities. Knee. 2021;29:183–9.
46. Ishibashi T, et al. Kinematics of bicruciate and posterior stabilized total knee arthroplasty during deep knee flexion and stair climbing. J Orthop Res. 2021;39:1262–70.

BCS-TKA

54

Ken Okazaki

1 Features of Journey II BCS

The BCS-type TKA is the Journey II BCS by Smith & Nephew. The characteristics of this implant are as follows:

1.1 A PS-Type Post-cam Mechanism Is Present at the Front (Fig. 54.1a)

The post-cam mechanism of the PS type compensates the function of the PCL for posterior stability (the PS-type post-cam mechanism controls posterior translation of the tibia and promotes posterior translation of the femur in the flexed position) after resection of the ACL and PCL. The BCS also provides a contact surface on the anterior aspect of the post for the anterior wall of the femoral intercondylar notch, compensating for the anterior stability, which is a function of the ACL (braking the anterior movement of the tibia in extension and positioning the femur anteriorly). Hence, it is called Bi-Cruciate Stabilized. This mechanism allows the femoral component to be positioned more anteriorly on the tibial insert in extension and rollback posteriorly in flexion, resulting in greater anterior-posterior translation [1, 2].

1.2 Asymmetry in the Shape of the Inserted Joint Surface (Fig. 54.1a)

The articular surface of the polyethylene insert is concave on the medial side with a high posterior lip, with the lowest point approximately at the center of the anterior-posterior width. In contrast, the lateral side is convex, with two sulcal areas located anterior and posterior to the articular surface of the polyethylene insert. Combined with the stabilizing function of the BCS mechanism described above, the femur is positioned anteriorly in extension and on the anterior sulcus of the lateral tibial surface. In flexion, the posterior post-cam mechanism promotes the posterior roll-back of the femur, in particular, the lateral side moves posterior toward the posterior sulcus on the lateral tibial surface, increasing the flexion range of motion (ROM) [3]. In addition, the anterior positioning of the femur during extension decreases the tension of the posterior capsule and improves flexion contracture, contributing to its ability to fully extend.

1.3 Incorporation of a Physiological Medial Tilt of Approximately 3° into the Articular Plane (Fig. 54.1b)

The medial and lateral thicknesses of the implants were different; the tibia and the femur are 2.5 mm thicker and thinner on the lateral side, respectively. The medial thickness of the implant is 9.5 mm, and the lateral thickness is 12 mm when the thinnest insert is used for the tibia, while the medial condyle is 9.5 mm for both the distal and posterior condyles and the lateral condyle is 7 mm for both the distal and posterior condyles for the femur. Osteotomies for implant placement are performed perpendicular to the mechanical axis as usual, creating a total of 19 mm interosteotomy gap. Consequently, a physiological medial tilt of approximately 3° is incorporated into the articular surface. In flexion, the femoral posterior articular line incorporates approximately 3° of medial tilt, even when osteotomies for the femoral posterior condyles are performed parallel to the SEA or Whiteside line as usual. This allows the joint line to be aligned at normal height, both medially and laterally. A physiological medial tilt of the joint surface of approximately 3° has the effect of providing the joint surface horizontal to the ground during standing and walking [4].

K. Okazaki (✉)
Department of Orthopaedic Surgery, Tokyo Women's Medical University, Tokyo, Japan
e-mail: okazaki.ken@twmu.ac.jp

S. Matsuda (ed.), *Advances in Total Knee Arthroplasty*, https://doi.org/10.1007/978-981-97-4920-1_54

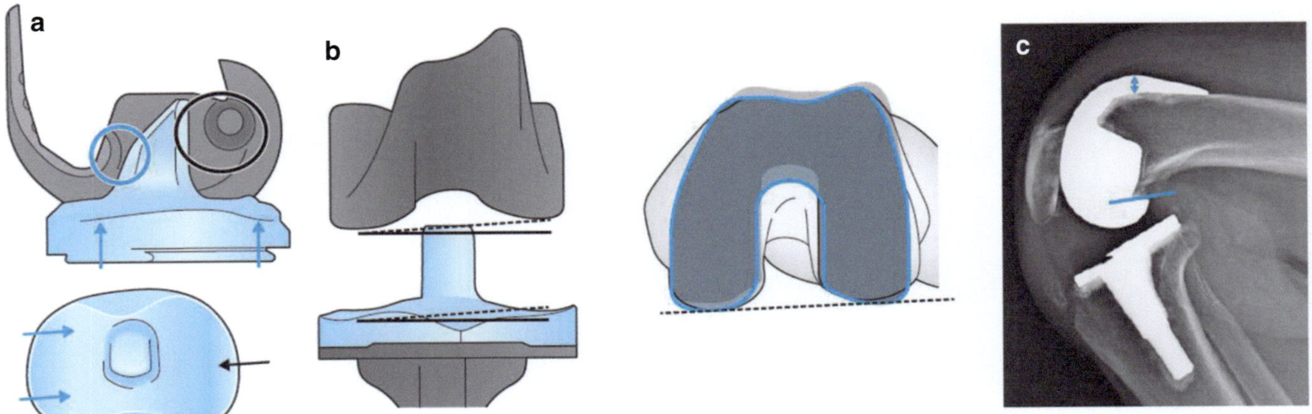

Fig. 54.1 Features of Journey II BCS. (**a**): A post-cam mechanism contacts both the anterior (blue circle) and posterior (black circle) sides to provide both anterior and posterior stability. The medial articular surface is concave, with the lowest point (black arrow) near the center, and the lateral articular surface is a convex shape in the center with the two lowest points located anteriorly and posteriorly (blue arrows). In extension, the femur is positioned anteriorly and the lateral side tends to move posteriorly when flexed. (**b**): Articular surface has a medial tilt of approximately 3° to form a physiological joint line. The posterior condyle also has a physiological joint plane tilt of approximately 3°, and even if the postoperative posterior condyle lines are aligned with the preoperative posterior condylar axis, they are not set internally in relation to the SEA or Whiteside line. (**c**): Deep-flexion design: The post-cam mechanism of BCS provides stable rollback of the femur. The anterior flange (blue arrow) is relatively thin, presumably reducing the stress on the quadriceps muscle. The osteotomy line at the posterior femoral condyle (blue line) is designed to be raised by 15°, which contributed to a reduction in the amount of osteotomy while maintaining the contact surface with the polyethylene insert in deep flexion. It also resists pulling-out stresses applied from the tibia to the femur under deep flexion loading

1.4 Thin Anterior Flange and Early Contact Timing of Post-cam During the Flexion Phase

The anterior femoral flange is thin (Fig. 54.1c), which helps decrease the quadriceps muscle tension in the flexed position [5]. The posterior post-cam begins to make contact at approximately 60° flexions, inducing rollback of the femur from an early stage [1]. This contributed to obtaining a good flexion ROM.

1.5 The Osteotomy Surface of the Posterior Femoral Condyle Has a 15° Cutout

In many other implants, the osteotomy surface of the posterior femoral condyle is perpendicular to the distal femoral surface. However, in Journey II, the osteotomy of the posterior condyle is 15° anterior to accommodate deep flexion (Fig. 54.1c). This feature offers two advantages. First, the implant covers the posterior femoral condyle to the point of contact with the tibial articular surface in deep flexion of 130° or greater while preventing an increase in the amount of bone resection of the posterior condyle. Second, in deep flexion loading positions, such as crouching, the stress transmitted from the tibia to the femoral posterior condyle is in the direction of pulling the femoral implant distal-anteriorly. Previously, it was reported that other deep-flexion implants had a high frequency of loosening at the posterior femoral condyle [6]. A 15° osteotomy surface can resist the distal anterior pull-out stress of the femoral implant.

2 Indications for Surgery

2.1 Advanced Osteoarthritis (OA) and Rheumatoid Joint Disease Are Included

BCS is indicated in patients with OA and rheumatoid arthritis (RA) for whom TKA is the treatment of choice, with no restriction on ROM. In cases of severe deformity where ligament balancing may be difficult, the option of a constrained insert as well as an extended stem or supplemental block in the tibia is available. However, the use of other systems that facilitate a transition to an implant with a more constrained insert is preferable in the following cases:

2.2 Severe Bone Defects and Collateral Ligament Insufficiency Are Excluded

In cases where there is a bone defect on the femoral side that needs to be replaced or where there is significant collateral ligament insufficiency, the use of a constrained model with a femoral stem should be considered.

2.3 Severe Valgus Knees That cannot Be Manually Corrected Are Excluded

BCS is also indicated for valgus knees that can be corrected manually. However, a severe valgus knee in which the medial and lateral ligaments cannot be balanced without extensive lateral soft tissue release should be excluded. Medial laxity with lateral loading may remain. Because the BCS has less constraint on the lateral articular surface, it should be avoided in cases where the load is predominantly lateral. In addition, knees with a valgus deformity of the proximal tibial joint surface also make little sense when using BCS with a medially sloping joint surface.

3 Surgical Planning

3.1 Implant Placement Perpendicular to the Mechanical Axis (Fig. 54.2)

The standard surgical plan aims for placement of the osteotomy perpendicular to the mechanical axis of the lower extremity. The implant incorporates an articular plane tilt of approximately 3°, resulting in an approximately 3° medial tilt of the articular plane relative to the mechanical axis of the lower extremity. However, in cases of strong femoral lateral bowing, we do not necessarily aim for a large valgus angle to the distal femoral shaft perpendicular to the mechanical axis because we consider this to be a constitutional varus knee.

3.2 Osteotomy Volume (Joint Line)

The optimal thickness of distal femoral and proximal tibial osteotomies should be confirmed. A physiological medial tilt of 3° is incorporated into the joint line in Journey II, allowing the postoperative joint line to be close to normal. In a varus knee, the osteotomy line for the tibia is 11–12 mm away from the lateral cartilage surface, which is equivalent to the thickness of the lateral side of the tibial implant. The distal femur is planned to be 7–8 mm from the lateral distal cartilage surface, and the actual joint line is fine-tuned according to the intraoperative extension and flexion gap measurements.

3.3 Femoral Rotation

The femoral rotational alignment may be determined in the same way as in other TKA models, either by alignment to the SEA or Whiteside line or by adjusting the rotational angle to create a rectangular gap using the gap balancing method. However, we use another concept of "reproducing the native joint line". "Reproduction of the native joint line" is a concept of the kinematic alignment method; however, when using a standard implant, the femoral rotation must be set parallel to the posterior condyle line and placed with a posterior reference, resulting in 3° more internal rotation than SEA. Problems with the patello-femoral joint were identified. However, the Journey incorporates an articular plane tilt of approximately 3° in the posterior condyle; therefore, the joint line of the posterior condyle can be reproduced by set-

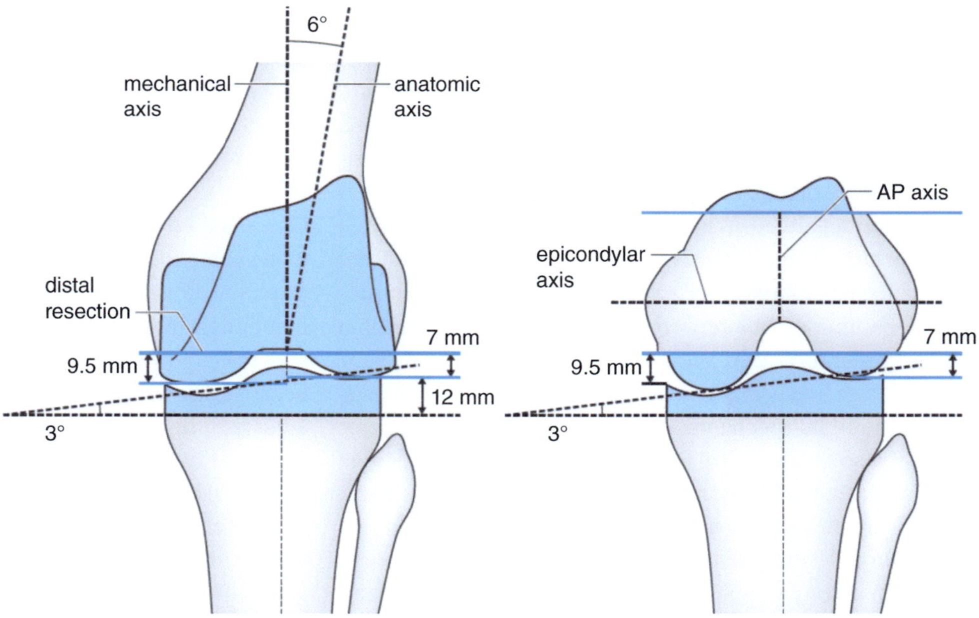

Fig. 54.2 Standard Installation Plan. The femoral implant has a medial thickness of 9.5 mm and a lateral thickness of 7 mm; the tibial implant has a medial thickness of 9.5 mm and a lateral thickness of 12 mm. In the case of a varus knee, an osteotomy perpendicular to the mechanical axis is performed, with a 7 mm osteotomy on the lateral distal femur and a 12 mm osteotomy on the lateral proximal tibia from the healthy cartilage surface, and a 3° external rotation of the femur from the posterior condyle line with a posterior reference to reproduce a physiological joint line in both extension and flexion

ting the posterior reference at 3° of external rotation from the posterior condyle line (Fig. 54.2). If the deformity of the posterior condyle is not strong, 3° external rotation is the basic position in most cases. However, if the lateral laxity exceeds 5 mm, or if there is medial laxity, the angle of external rotation should be adjusted.

3.4 Application to the Kinematic Alignment Method

In the kinematic alignment technique, the osteotomy plane is often inclined toward the mechanical axis when implants with uniform medial and lateral thicknesses are used to reproduce the native articular surface. In some cases, significant tilting may occur, and there is concern about excessive loading on the bone-implant interface. Journey II incorporates an approximately 3° slope on the articular surface, which reduces the osteotomy slope by 3°.

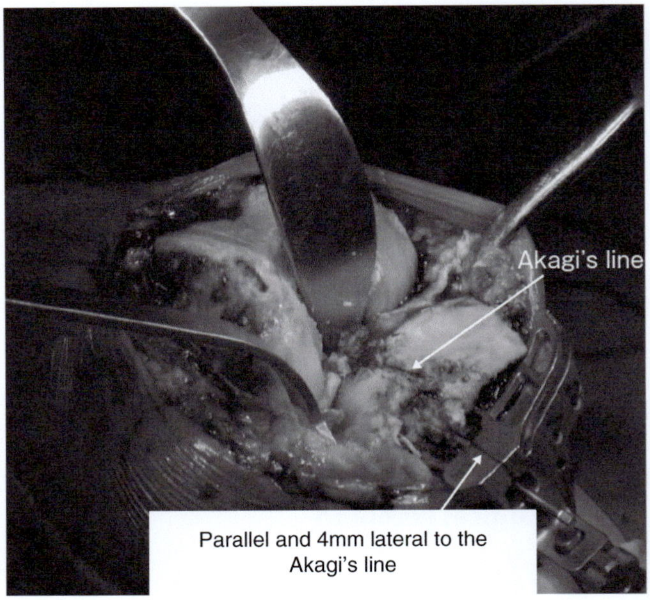

Fig. 54.3 Proximal landmark for tibial cut

4 Surgical Techniques

We used a standard procedure with a subvastus approach, tibia-first cut, osteotomy adjustment of the distal femur and posterior condyle considering the medial gap, femoral rotation of 3° external rotations from the posterior condyle line, and patellar replacement.

4.1 Tibia First Cut Technique

The osteophytes are removed at the joint margins and intercondylar area, posteriorly and fully. The collateral ligaments are not released in most cases. Both cruciate ligaments are resected from the attachments and the retractors were placed posterior to the PCL attachment to dislocate the tibia anteriorly. The posterior root attachments of the medial and lateral menisci are separated. When the posterior root of the meniscus is detached, the meniscus deviates posteriorly, and the tibia dislocates more easily anteriorly. An osteophyte posterior to the tibia can be removed with a chisel, and the tibia can be further displaced anteriorly.

The Akagi line is marked, and the exit of the articular surface of the tibial mechanical axis, as determined preoperatively, s also marked. To place the tibial osteotomy guide using an extramedullary rod, the anteroposterior axis mark of the osteotomy guide was first aligned with the Akagi line and rotation of the guide. The center pin of the osteotomy guide is inserted parallel to the Akagi line, just below the exit of the articular surface of the tibial mechanical axis. Usually, the center pin is placed 4 mm lateral to the Akagi line and below the lateral intercondylar ridge (Fig. 54.3) [7]. The distal portion of the extramedullary guide is aligned with the center of the medial and external ankle joints. We ascertain that the extramedullary guide is parallel to the line segment connecting the proximal 1/3 and distal 1/3 points on the anterior margin of the tibia, which represents the mechanical axis of the tibia [8]. For posterior inclination, a 3° osteotomy guide was used, and the relationship between the extramedullary guide and tibial axis is determined from the lateral view. Finally, the height of the osteotomy guide is determined. It is set at 11–12 mm from the external healthy cartilage. The cutting depth from the medial articular surface is measured to confirm no difference compared to the preoperative plan.

4.2 Measurement of Gaps (Fig. 54.4)

The standard surgical technique for Journey II BCS is the extension gap-first technique in which the distal femur and proximal tibia are first resected (Fig. 54.4a). However, as mentioned above, we use the tibia-first-cut technique to measure the gap using only the resected tibia (Fig. 54.4b). This is because the amount of distal femoral osteotomy can be adjusted after confirming the initial extension gap and the joint line can be normalized on the femoral side depending on the case.

The extension and flexion gaps were measured using a spacer block reflecting the thickness of the tibial implant. Based on these measurements, the extent of osteotomy of the distal femur and posterior condyle was examined. Unlike implants in which the medial and lateral sides are equal, the medial joint line is normalized when the spacer block is

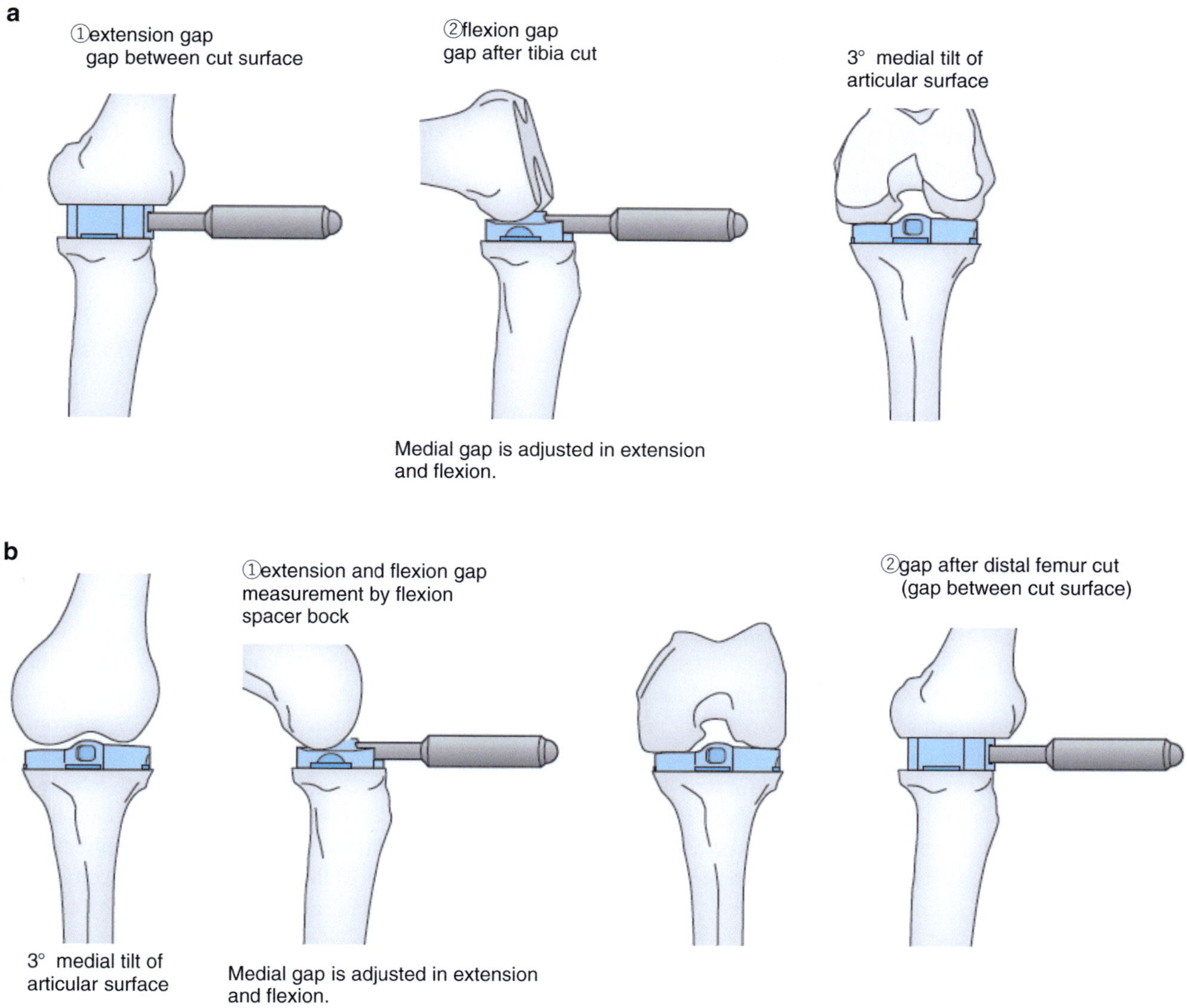

Fig. 54.4 (**a**): Standard surgical technique. (1) An extension gap is created and the interosteotomy gap is evaluated. (2) Evaluation of the flexion gap before posterior condylar osteotomy. The femoral size, rotation, and anteroposterior position were determined, considering the difference between the two gaps and the difference between the medial and lateral gaps. (**b**): Surgical technique for the tibia-first method. The proximal tibia was resected at the implant thickness, and the joint line was determined using a flexion spacer block. The extension and flexion gaps were measured, and the extent of osteotomy for the distal femur and posterior condyle was determined. Reconfirm the extension gap (between osteotomies) after distal femoral osteotomies. The difference between the extension and flexion gaps is used to determine the femoral size, rotation, and anteroposterior position. The difference from (**a**) is that the amount of distal femoral osteotomy (distal femoral joint line) can be adjusted with this method

placed. Therefore, in the case of a worn medial distal femur, there is more room for the medial extension gap. In such cases, the resection thickness of the distal femur can be reduced by the amount of wear required to normalize the distal femoral joint line. Conversely, if there is flexion contracture and the extension gap is insufficient, a decision can be made to adjust the resection thickness for the distal femur. If there is significant lateral laxity in the extension gap, the difference between the medial and lateral extension gaps can be reduced by placing the femur in a mild varus alignment. Even when placing a 3° varus angle on the femur, the femoral articular surface is still perpendicular to the femoral mechanical axis.

4.3 Distal Femoral Osteotomies

The distal femur is osteotomized at the preoperative measured valgus angle; Journey's implant is 9.5 mm thick medially and 7 mm thick laterally. As mentioned above, the previously measured extension and flexion gaps were considered to determine the thickness of bone resection on the distal femur; whether the femoral joint line should be moved to the normal position, aligned with the worn medial surface, or resected further. The normal position of the joint line is indicated by measurements on the lateral healthy cartilage surface. After the osteotomy, a spacer block with the total thickness of the femoral and tibial sides is used to check for the created extension gap (Fig. 54.4b②).

4.4 Femur Sizing and Rotation Determination (Fig. 54.5)

The angle of rotation is determined preoperatively and rechecked to ensure no significant deviation from the Whiteside's line or SEA. The decision of the angle of rotation is based on the surgeon's preference, but by using a 3° external rotation with a posterior reference, the postoperative posterior condylar line will match the preoperative posterior condylar line. This is a feature of the model, and a patient-specific joint line can be maintained without difficulty. However, if medial laxity is present in the flexion gap or if significant lateral laxity is present, the external rotation angle should be adjusted.

If there is a difference between the extension and flexion gaps, it can be reduced by adjusting the anteroposterior position. The goal is to achieve a medial flexion gap equal to or greater than the medial extension gap of up to 2 mm. If the flexion gap is small, it is shifted forward; if the flexion gap is larger than 3 mm and an intermediate size is shown, the larger size is selected and shifted posteriorly to reduce the flexion gap.

4.5 Five-Side Femoral Osteotomy

The Journey BCS has two anterior chamfers, with the posterior condyle cut at 15°. Osteotomies is started from slot #2 (the anterior chamfer). Slot 1 is an osteotomy of the anterior flange; however, when slot 2 is osteotomized first, it is easy to confirm that the bone saw does not make a mistake in slot 1 by cutting into the anterior cortex and creating an anterior notch.

4.6 Preparation for Installation of Tibial Baseplate

The center pin inserted during tibial osteotomy is used as an indicator of baseplate placement. The center pin is parallel to the Akagi line and inserted approximately 4 mm lateral to the Akagi line, with the center pin positioned toward the center of the osteotomy surface. If the medial osteophyte is completely removed and the transverse diameter perpendicular to the center pin is measured, the center pin is found to be almost at the center of the osteotomy surface. The center pin is removed by marking the osteotomy surface along the center pin, which is also an indicator of rotation. The appropriate size is selected without overhang, taking into consideration the anteroposterior coverage, and is placed with reference to the markings. As rotational alignment is important, the orientation of the tibial tubercle is also used as an indicator for double-checking. The ROM method is also recommended for determining the tibial rotational alignment by checking

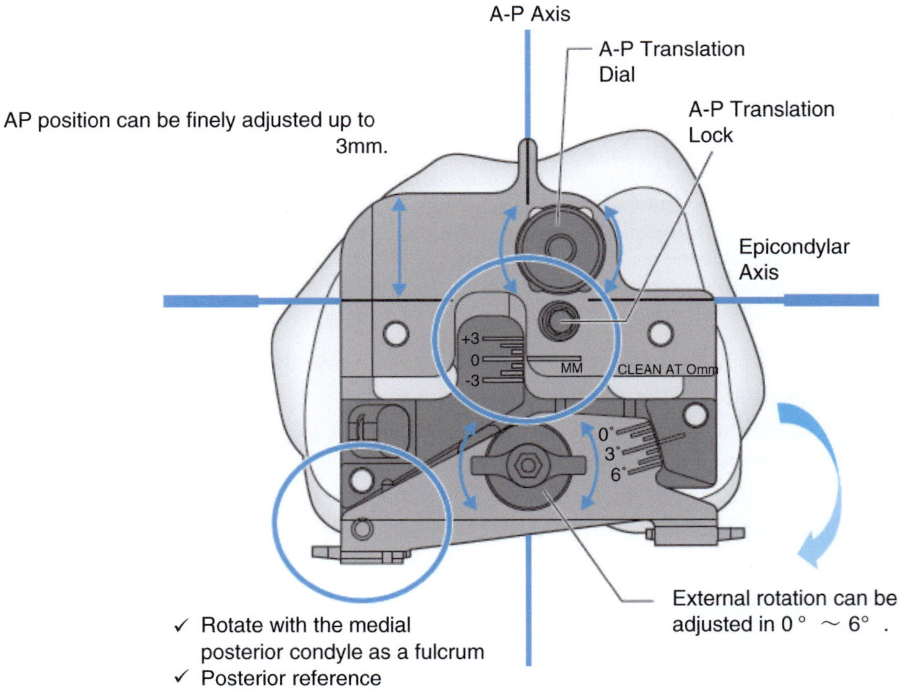

Fig. 54.5 Features of the femoral positioning and sizing guide. To adjust the external rotation angle, the entire guide is rotated with the medial posterior condyle as the pivot point. The guide can be easily checked against the reference axis of rotation, and the thickness of bone resection for the medial posterior condyle does not increase or decrease according to the angle of rotation. In addition, the guide can be moved anterior and posterior in 3 mm increments for easy adjustment of the medial flexion gap

the femoral trial, insert, and baseplate alignment in an extended position.

4.7 Femur Trial and Box Fabrication

Because the femoral posterior condyle is osteotomized with a 15° inclination, the femoral trial is also inserted with mild hyperflexion. After confirming the insertion of the posterior condyle, it is extended so that the anterior flange is aligned with the osteotomy surface. Therefore, there are two anterior chamfers. The medial and lateral positions are confirmed by trial and fixed using an anterior pin. An instrument for fabricating the intercondylar box is attached, and the box is fabricated using a reamer and box chisel.

4.8 Patellar Resurfacing

Subsequently, the patella is replaced. Onlays and inlays are also available. The thickness of the patella is measured If it is an onlay, a thickness of at least 12 mm remained, and the thickness of the implant is resected. The implant thickness is 9.5 mm or 7.5 mm.

5 Clinical Results

Many favorable clinical results for Journey II BCS have been reported. Inui et al. evaluated 113 patients using the Knee Society Score (KSS) 2011 and reported a mean satisfaction score of 28.2 (out of 40) [9]. Taniguchi et al. reported a mean total score of 140.3 (of 180) on the KSS 2011 and a mean ROM of 129° [10]. Iriuchishima et al. examined whether patients could climb stairs step-by-step or needed a handrail before and after surgery and compared the results with BCS, CR, and PS. Although there was no difference in preoperative stair-climbing ability, 89% of BCS patients were able to climb stairs within one step after surgery, which was better than that of CR (72%) and PS (58%) [11].

References

1. Murakami K, et al. Knee kinematics in bi-cruciate stabilized total knee arthroplasty during squatting and stair-climbing activities. J Orthop. 2018;15:650–4.
2. Iriuchishima T, Ryu K. A comparison of the rollback ratio between bicruciate substituting total knee arthroplasty and Oxford unicompartmental knee. J Knee Surg. 2018;31:588–72.
3. Kono K, et al. Bicruciate-stabilised total knee arthroplasty provides good functional stability during high-flexion weight-bearing activities. Knee Surg Sports Traumatol Arthrosc. 2019;27:2096–103.
4. Murakami K, et al. Preoperative tibial mechanical axis orientation and articular surface design influence on coronal joint line orientation relative to the ground during gait after total knee arthroplasties. Knee Surg Sports Traumatol Arthrosc. 2018;26:3368–76.
5. Itou J, et al. Anterior prominence of the femoral condyle varies among prosthesis designs and surgical techniques in total knee arthroplasty. BMC Musculoskelet Disord. 2021;22:784.
6. Han HS, et al. High incidence of loosening of the femoral component in legacy posterior stabilised-flex total knee replacement. J Bone Joint Surg Br. 2007;89:1457–61.
7. Kawahara S, et al. Medial sixth of the patellar tendon at the tibial attachment is useful for anterior reference in rotational alignment of the. Knee Surg Sports Traumatol Arthrosc. 2014;22:1070–5.
8. Fukagawa S, et al. Anterior border of the tibia as a landmark for extramedullary alignment guide in total knee arthroplasty for varus knees. J Orthop Res. 2011;29:919–24.
9. Inui H, et al. Comparison of intraoperative kinematics and their influence on the clinical outcomes between posterior stabilized total knee. Knee. 2020;27:1263–70.
10. Taniguchi H, et al. Noise after total knee arthroplasty has limited effect on joint awareness and patient-reported clinical outcomes. BMC Musculoskelet Disord. 2020;21:115.
11. Iriuchishima T, Ryu K. Bicruciate substituting total knee arthroplasty improves stair climbing ability when compared with cruciate-retain or Indian. J Orthop. 2019;53:641–5.

CS-TKA

55

Yoshinori Ishii

1 Background of the Low-Contact Stress System

The low-contact stress (LCS) CS design has unique characteristics that control the motion associated with knee joint flexion and extension with a relatively highly constrained geometry of the femur and tibia. The rotational shear forces between the tibial component and tibia that occur in the fixed type component are absorbed by the unrestrained movable tibial insert and tibial component, providing an advantage for both component and the tibia [1, 2]. This system, including the CR type, which has no constraint in the anteroposterior or rotational directions, has produced good long-term clinical outcomes for 40 years [3–9]. Conversely, poor outcomes in short-term follow-up studies have also been reported [10, 11]. We have used this design for more than 30 years, and in addition to observing favorable clinical outcomes [12–15], we have conducted a research to investigate the reasons behind the conflicting outcomes [16–22]. The research identified that this design has a wider "strike zone," than other designs, enabling compensation for suboptimal surgical techniques. If the femoral component and soft tissue balance are not ideally placed, the coronal geometry (Fig. 55.1a, b) is able to compensate for the suboptimal placement. Similarly, its mobile design compensates for the rotational placement of the tibial component [23, 24] if the malrotation is within 10°. Various anatomical landmarks have been proposed for placing the tibial component, including the Akagi line [25], and many reports have been published on the reliability of these landmarks. However, in some cases, the landmarks shown in preoperative images are difficult to clearly identify during the actual surgery and high interindividual variability can exist. Since surgery is performed manually by humans, achieving a perfect score of 100 points in every surgery is not always possible. We believe that the CS design avoids the occurrence of unacceptable outliers.

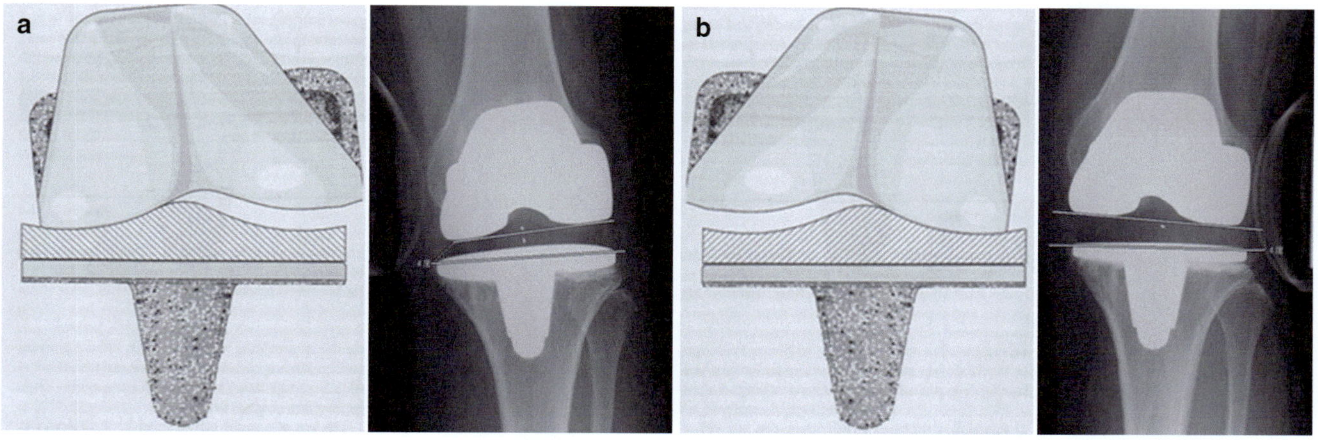

Fig. 55.1 The low-contact stress is highly conformed in the coronal plane even when tilted in the varus (**a**) and valgus (**b**) direction

Y. Ishii (✉)
Ishii Orthopaedic & Rehabilitation Clinic, Saitama, Japan
e-mail: ishii@sakitama.or.jp

S. Matsuda (ed.), *Advances in Total Knee Arthroplasty*, https://doi.org/10.1007/978-981-97-4920-1_55

With mobile total knee arthroplasty (TKA), the soft tissues are expected to control the movement of the tibial insert [26]; thus, greater skill is needed to achieve a proper soft tissue balance with a mobile TKA than that with a fixed TKA [27]. Particular attention is required to obtain good soft tissue balance when placing the femoral component using the gap technique utilized for this design. Various anatomical landmarks have been proposed for optimal placement of the tibial component. However, considering the anatomical variance among individuals, a definitive landmark is not feasible. Therefore, some reports recommend TKA with a mobile tibial insert, which allows for some malrotation of the tibial component [28, 29]. In this chapter, the author describes the concept of LCS TKA, key points of intraoperative manipulation, and the clinical outcomes of the procedure.

2 Concept of the Surgical Operation

For the basic surgical procedures, please refer to the manual [26]. When performing the gap technique using an LCS system, we should recognize that intraoperative soft tissue evaluation may not always align with the postoperative outcomes. Under anesthesia, without voluntary soft tissue tension, even if the necessary release is performed on the contracted side of the soft tissue (the medial side in the case of medial OA) to obtain a balance with the less contracted opposite side (many systems seem to confirm this using a tensioner with some load), this balance is not always reflected in the conscious state. In other words, if we compare the soft tissues to a rubber in the manipulation of a model bone of the knee joint, even if a balanced gap is created with great care using the rubber that has degenerated and hardened (contracted side) and rubber that remains elastic (noncontracted side) (i.e. under anesthesia), the response of each rubber will differ when subjected to a tensile or compressive load (i.e. in the conscious state); therefore, reproducing the balance during manipulation becomes challenging. It is easy to imagine that the response of each side of soft tissue would be different, and the balance achieved during the surgery cannot be reproduced. Therefore, we believe that confirming the balance only using a spacer block is sufficient, unless extreme contracture or instability necessitates an additional 2.5-mm spacer, as in the basic technique for LCS TKA. Notably, the LCS design maintains sufficient stability even when rotated in the varus–valgus direction in the coronal plane (Fig. 55.1a, b) and, to some extent, its stability is guaranteed by its geometry.

3 Key Points of the Surgery

3.1 Femoral Side [30]

The procedure begins with the tibia being cut first, with the osteotomy aimed at 90° to the functional axis of the tibia. Proper alignment is confirmed by palpating and ensuring that the transverse bar of the femoral cutting jig aligns with the medial and lateral femoral epicondyles at the time the spacer block is inserted (Fig. 55.2a). The stability of the femur in the flexion gap (varus, neutral, and valgus) is assessed with a spacer block in place. A certain amount of learning curve is necessary, but a moderate degree of laxity of 3° to 4° [31] is acceptable without requiring an additional 2.5-mm spacer (Fig. 55.2b).

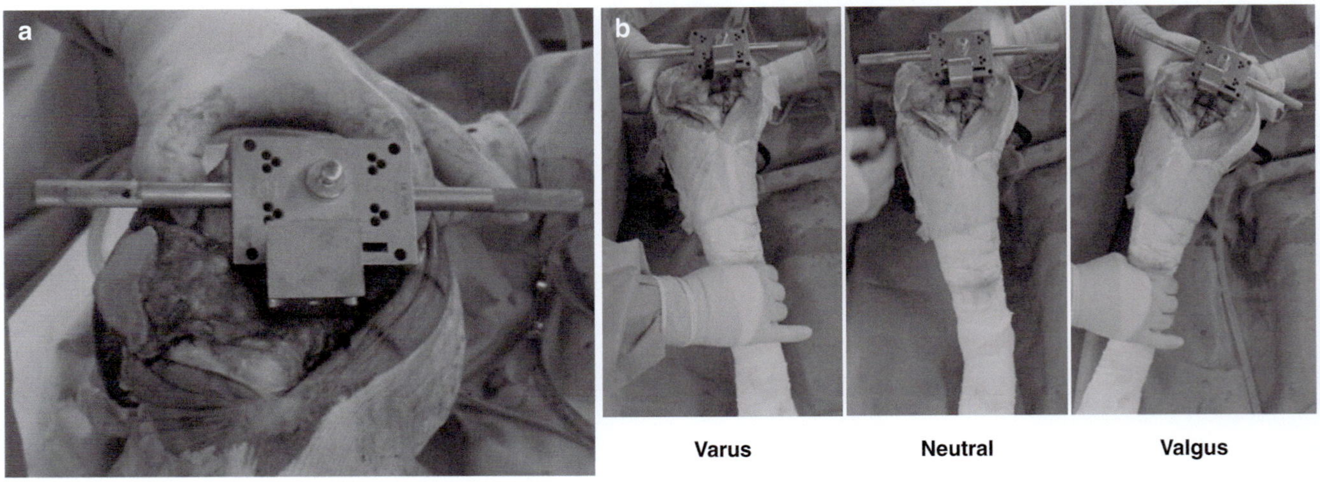

Fig. 55.2 Confirm proper alignment by palpating the transverse bar of the femoral cutting jig, ensuring it is almost in line with the medial and lateral femoral epicondyles upon insertion of the spacer block (**a**). Assess the stability of the flexion gap (varus, neutral, and valgus) with the spacer block inserted (**b**)

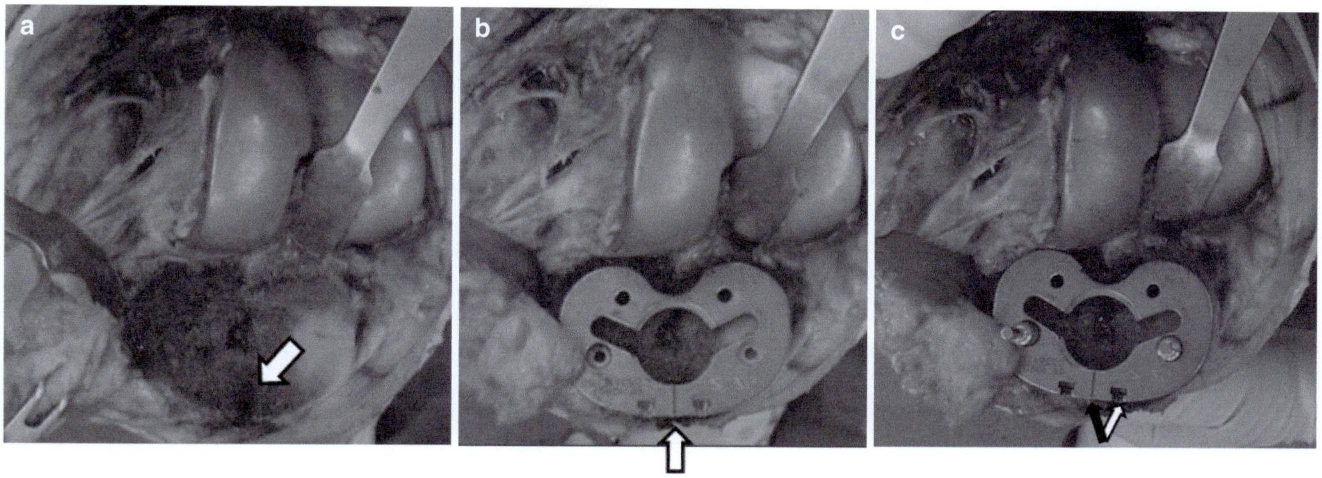

Fig. 55.3 Initially, check the landmark for the medial edge of the surface of the tibial tubercle (marking line: white arrow) (**a**). Subsequently, move the template in the anteroposterior direction and set it to obtain maximum coverage of the tibial osteotomized surface (the center of the tibial component is determined) (**b**). Finally, rotate the template internally and externally to further improve the coverage (rotation from white arrows to black arrows: external rotation direction in this case) (**c**)

3.2 Tibial Side [32]

Various anatomical landmarks for rotational tibial placement have been discussed. However, in some cases, identifying anatomical landmarks intraoperatively is difficult. In accordance with Insall et al. [33], we focused on the medial aspect of the tibial tubercle—which is relatively easy to identify—and used it as a landmark. The procedure involved the follows steps: (1) The medial margin of the tibial tuberosity was landmarked (Fig. 55.3a). The template was moved in the anteroposterior direction to align with this line and positioned to obtain the maximum extent of the tibial osteotomy surface (Fig. 55.3b). This would determine the center of the tibial component. (2) While maintaining the template's center, it was rotated internally and externally to maximize coverage (Fig. 55.3c). (3) Flexion and extension of the knee joint were repeated with this placement to confirm movement between the components, including patellar tracking. The basic idea is to ensure proper placement of the tibial component (coverage of the tibial osteotomy surface and patellar tracking), recognizing that the anatomic landmarks are only a guide and will vary between individual cases. Nevertheless, it is important to understand that achieving optimal rotation and maximum coverage often involves trade-offs; therefore, perfect placement may not be possible in all cases. The tibial insert of the mobile-bearing TKA design allows for up to 10° of internal/external rotation [23, 24] and, to a certain extent, its design compensates for inferior technique (malrotation).

3.3 Patella

As described above, good alignment of the patellofemoral joint is ensured through the placement of the femoral and tibial components, as well as tibial rotational tolerance. In principle, patellar replacement is not performed because this design offers good congruity of the femoral patellar groove, which reportedly reduces contact stress [34]. Lateral retinacular release had been performed in a few cases of severe lateral osteoarthritis; however, it is rarely necessary [35].

4 Verifying the Accuracy of Femoral and Tibial Component Placement and Tibial Osteotomized Surface Coverage of the Tibial Components

The accuracy of the femoral and tibial component setting and the coverage of the tibial osteotomized surface of the tibial component performed by the above procedure were investigated for 102 TKAs [30, 32]. The analysis was performed using a three-dimensional lower extremity alignment assessment system (Knee CAS, LEXI, Inc., Tokyo, Japan), which can be used for preoperative surgical planning and postoperative evaluation [36].

On the femoral side, the femoral rotation angle (Fig. 55.4), which is the angle between the posterior condylar line (posterior condylar surfaces) and transepicondylar axis (TEA), was measured as described in previous reports. On the tibial side, the tibial rotational angle, which was calculated from the transverse axis of the tibial component and superimposed TEA of the femur, was also evaluated (Fig. 55.5a, b). All values were expressed as negative (−) internal rotation relative to the transepicondylar axis and positive (+) external rotation relative to the TEA. In addition, the percentage of tibial component coverage, between the area of the tibial component and the osteotomized aspect of the tibia (tibial component/osteotomized aspect of the tibia × 100) was calculated (Fig. 55.5b).

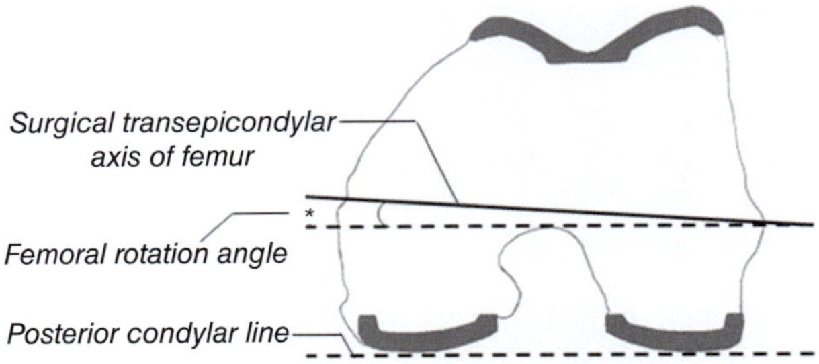

Fig. 55.4 Measurement of the postoperative femoral rotation angle (FRA), that is, the angle between the posterior condylar line (*dotted line*) and the transepicondylar axis (*solid line*)

For favorable clinical outcomes, the optimal laxity of LCS TKA in the flexed position is 3° to 4° in the varus–valgus plane [31]. Additionally, the tibial insert should rotate by 10° in the internal/external direction [23, 24]. If this range is considered acceptable, most cases were within these ranges, as shown in Table 55.1. Even on the tibial side, where anatomic variance is particularly high, the TRA was within 10° in all cases. Therefore, assuming that the TEA is the ideal axis of rotation, the gap technique was a reliable procedure

a

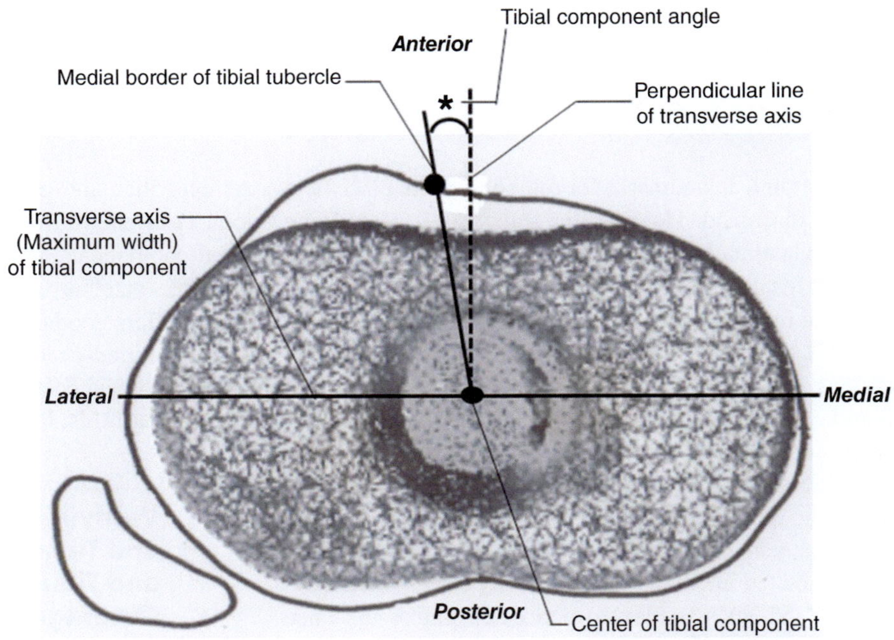

b

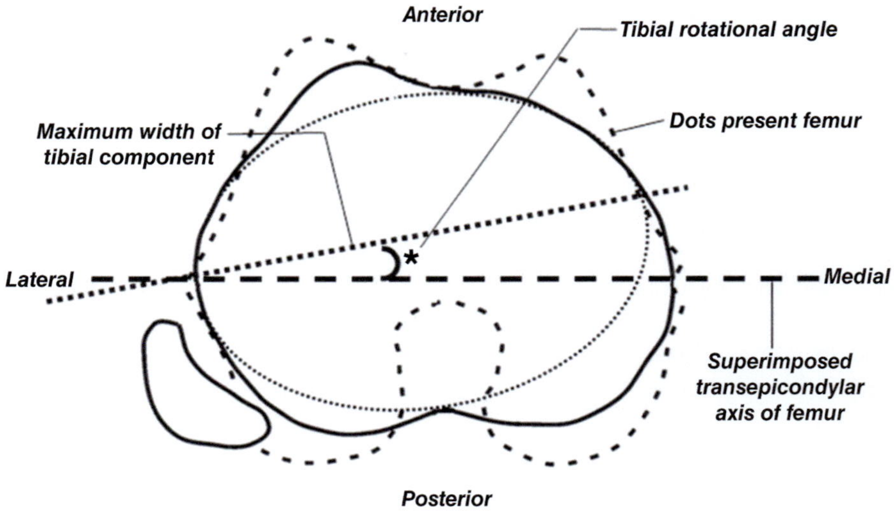

Fig. 55.5 The tibial component angle is determined by the angle between perpendicular line of the transverse axis of the tibial component and medial border of the tibial tubercle (**a**). The tibial rotational angle (TRA) is determined by the alignment of the transverse axis of the tibial component and superimposed transepicondylar axis of the femur. The tibial component coverage is calculated as the ratio of the tibial component coverage [dotted area] divided by tibial osteotomy surface area [solid area] × 100 (**b**)

Table 55.1 Femoral and tibial component placement index for the CS design

Variables	CS* (n = 102)
Preoperative FRA	−0.4° (−1.0°, 0.3°) [−3.6°– 3.2°]
Postoperative FRA	−0.1° (−2.5°, 2.8°) [−4.9°– 5.0°]
Difference between pre- and post-operative FRA	0.3° (−2.2°, 3.0°) [−5.7°– 7.0°]
TCA**	−1.6° (−4.3°, 3.1°) [−9.5°—9.9°]
TRA***	−2.1° (−4.8°, 2.7°) [−10.0 °– 9.7°]
Tibial component coverage	82.7% (80.6, 84.7%) [73.2–88.5%]

Minus (−) and plus (+) indicate internal and external rotation, respectively, relative to the perpendicular to the transverse axis of the tibial component in ** and to the transepicondylar axis of the femur in ***. CS; posterior cruciate ligament substituting design; * All values are expressed as the median (25th, 75th) [range]; *FRA* femoral rotation angle, *TCA* tibial component angle (TCA), *TRA* tibial rotation angle

for accurate femoral component placement in LCS TKA, and the medial border of the tibial tubercle could be a reliable anatomic indicator for tibial component placement.

Tibial coverage plays a significant role in knee joint stability: the greater the coverage, the greater is the advantage [37, 38]. This design is symmetrical in the medial and lateral condyles, which is disadvantageous for tibial coverage compared to asymmetrical or anatomical designs [39, 40]. However, if optimal rotation is prioritized when placing these designs, medial overhang could possibly occur, resulting in the need for size reduction. In contrast, with the current design, a slight overhang could be corrected within the range of allowable rotation, and downsizing can be avoided in some cases. In other words, there was a trade-off between optimal rotational placement and improved coverage, and this technique employed by the authors for TKA was deemed satisfactory in comparison with results of the previous studies [39–42].

5 Clinical Outcomes of the LCS CS Design

5.1 Long-Term Clinical Outcomes

We examined the long-term clinical outcomes over approximately 10 years of LCS CS design performed using the surgical techniques described above [14]. The sample included 38 knees that underwent a scheduled two-stage TKA. The preoperative HSS score improved from 43 to 95 following surgery, and the postoperative KSKSS was 132. Three knees had radiolucent lines around the tibial prostheses, which were less than 1 mm long and non-progressive. No significant differences in clinical or radiographic findings were observed between the contralateral CR and CS designs in the [38] knees [14].

5.2 Clinical Outcomes After a Modification in the Surgical Technique [13]

To date, the reports indicate that the cause of early sinking of the tibial component in tibial cementless cases is attributed not only to a malalignment of the lower limb and/or component placement but also to a deterioration in bone mineral density on the tibial side [43, 44], and this is also true for the current design [45]. Therefore, hybrid TKA using cement only on the tibial side is currently being developed for early stabilization of the osteotomized tibial surface [46]. In this design, only the osteotomized surface is cemented, which is advantageous to the fixation surface, based on the basic research findings [47]. No cement is used in the hole for the tibial stem; instead, cancellous bone is packed in the stem for biological fixation by osteogenesis. The outcomes using this technique have demonstrated good clinical scores and radiographic evaluations of 122 TKAs followed up for 5 years postoperatively [13].

5.3 Evaluation of Physical Function and Life Expectancy after LCS

We previously reported that the bone mineral density of the proximal femur was maintained and improved 2 years postoperatively owing to the improvement in physical function associated with the pain relief effect following TKA, although this study had a different design [48]. With regards to LCS, we also reported that calcaneal bone quality improved in 5-year follow-up cases [49] and significantly correlated with the bone mineral density of the femur and tibia [50]. Simultaneously, a prospective study using a gravicorder, which measures postural sway as the ability to balance the body, demonstrated an improvement in postural sway following TKA compared to that preoperatively [51]. Considering that more than 90% of femoral neck fractures are a consequence of osteoporosis-based falls [52], TKA may help prevent fall-related fractures by improving bone mineral density and body balance. However, quadriceps muscle strength was still inferior to that of controls of the same age group at 5 years postoperatively [53], leaving enough room for future rehabilitation interventions to improve muscle strength. Finally, a review of 326 patients for which 100% follow-up data were available demonstrated that the survival rate is 82% at 10 years and 67% at 15 years, indicating that patients who undergo TKA can have a life expectancy comparable to that of the general population in each age group [54]. It will become increasingly important for surgeons who perform TKA not only to study the improvement of patients' knee joint function, but also to examine the spillover effects of such improvement on patients' physical function, including the extension of healthy life expectancy.

References

1. Malinzak RA, et al. The effect of rotating platform TKA on strain distribution and torque transmission on the proximal tibia. J Arthroplast. 2014;29:541–7.
2. Small SR, et al. Micromotion at the tibial plateau in primary and revision total knee arthroplasty: fixed versus rotating platform designs. Bone Joint Res. 2016;5:122–9.
3. Buechel FF Sr, et al. Twenty-year evaluation of meniscal bearing and rotating platform knee replacements. Clin Orthop Relat Res. 2001;388:41–50.
4. Efstathopoulos N, et al. 10-year evaluation of a cementless low-contact stress rotating-platform total knee arthroplasty. J Long-Term Eff Med Implants. 2009;19:255–63.
5. Hooper G, et al. The low contact stress mobile-bearing total knee replacement: a prospective study with a minimum follow-up of ten years. J Bone Joint Surg Br. 2009;91:58–63.
6. Huang CH, et al. Long-term results of low contact stress mobile-bearing total knee replacements. Clin Orthop Relat Res. 2003;416:265–70.
7. Milligan DJ, et al. Twenty-year survivorship of a cemented mobile bearing total knee arthroplasty. Knee. 2019;26:933–40.
8. O'Brien S, et al. LCS mobile bearing total knee arthroplasty without patellar resurfacing. Does the unresurfaced patella affect survivorship at a minimum 10-year follow-up. Knee. 2012;19:335–8.
9. Van Ooij B, et al. Lower revision rates for cemented fixation in a long-term survival analysis of three different LCS designs. Knee Surg Sports Traumatol Arthrosc. 2022;30:2707–13.
10. Bert JM. Dislocation/subluxation of meniscal bearing elements after New Jersey low-contact stress total knee arthroplasty. Clin Orthop Relat Res. 1990;254:211–5.
11. Weaver JK, et al. Difficulties with bearing dislocation and breakage using a movable bearing total knee replacement system. Clin Orthop Relat Res. 1993;290:244–52.
12. Ishii Y, et al. Range of motion during the perioperative period in total knee arthroplasty. Arch Orthop Trauma Surg. 2008;128:795–9.
13. Ishii Y, et al. Clinical outcomes after mobile-bearing total knee arthroplasty with surface tibial cementation at 5-year follow-up. Knee Surg Sports Traumatol Arthrosc. 2019;27:2181–8.
14. Ishii Y, et al. Comparison of long-term clinical outcomes after bilateral mobile-bearing total knee arthroplasties using PCL-retaining and PCL-substituting implants in the same patients. Knee Surg Sports Traumatol Arthrosc. 2017;25:3711–7.
15. Ishii Y, et al. Prediction of range of motion two years after mobile-bearing total knee arthroplasty: PCL-retaining versus PCL-sacrificing. Knee Surg Sports Traumatol Arthrosc. 2011;19:2002–8.
16. Ishii Y, et al. Coronal laxity in extension in vivo after total knee arthroplasty. J Orthop Sci. 2003;8:538–42.
17. Matsuda Y, et al. In vivo laxity of low contact stress mobile-bearing prostheses. Clin Orthop Relat Res. 2004;419:138–43.
18. Ishii Y, et al. Sagittal laxity in vivo after total knee arthroplasty. Arch Orthop Trauma Surg. 2005;125:249–53.
19. Matsuda Y, et al. Varus-valgus balance and range of movement after total knee arthroplasty. J Bone Joint Surg Br. 2005;87:804–8.
20. Ishii Y, et al. Effect of voluntary soft tissue tension and constraint after total knee arthroplasty on in vivo anteroposterior displacement. Knee. 2011;18:11–4.
21. Takeda M, et al. Changes in varus-valgus laxity after total knee arthroplasty with a minimum five-year follow-up. Knee Surg Sports Traumatol Arthrosc. 2012;20:1984–9.
22. Mikashima Y, et al. Do mobile-bearing knee arthroplasty motions change with activity? Knee. 2013;20:422–5.
23. Banks SA, et al. Implant design affects knee arthroplasty kinematics during stair-stepping. Clin Orthop Relat Res. 2004;426:187–93.
24. Dennis DA, et al. Mobile-bearing total knee arthroplasty: do the polyethylene bearings rotate? Clin Orthop Relat Res. 2005;440:88–95.
25. Akagi M, et al. Variability of extraarticular tibial rotation references for total knee arthroplasty. Clin Orthop Relat Res. 2005;436:172–6.
26. Buechel FF. In: Hamelynck KJ, et al., editors. Surgical technique of the LCS. LCS mobile-bearing knee arthroplasty. 25 years of worldwide experience. Berlin: Springer; 2002. p. 121–35.
27. Vertullo CJ, et al. Mobile bearings in primary knee arthroplasty. J Am Acad Orthop Surg. 2001;9:355–64.
28. Matsui Y, et al. Rotational deformity in varus osteoarthritis of the knee: analysis with computed tomography. Clin Orthop Relat Res. 2005;433:147–51.
29. Matziolis G, et al. A prospective, randomized study of computer-assisted and conventional total knee arthroplasty: three-dimensional evaluation of implant and rotation. J Bone Joint Surg Am. 2007;89:236–43.
30. Ishii Y, et al. Retention of the posterior cruciate ligament does not affect femoral rotational alignment in TKA using a gap-balance technique. Knee Surg Sports Traumatol Arthrosc. 2014;22:3121–6.
31. Matsuda Y, et al. Effect of flexion angle on coronal laxity in patients with mobile-bearing total knee arthroplasty prostheses. J Orthop Sci. 2005;10:37–41.
32. Ishii Y, et al. Rotational alignment of tibial components in mobile-bearing TKA: posterior substituted vs PCL retaining. Arch Orthop Trauma Surg. 2015;135:1299–305.
33. Insall JN, et al. Surgical techniques and instrumentation in total knee arthroplasty. In: Insall JN, Scott WN, editors. Surgery of the knee. 3rd ed. New York: Churchill-Livingstone; 2001. p. 1553–620.
34. Takahashi A, et al. Patellar morphology and femoral component geometry influence patellofemoral contact stress in total knee arthroplasty without patellar resurfacing. Knee Surg Sports Traumatol Arthrosc. 2012;20:1787–95.
35. Ishii Y, et al. Patellar tendon length during knee flexion of meniscal bearing and rotating total knee arthroplasty. Knee Surg Sports Traumatol Arthrosc. 2017;25:3773–8.
36. Sato T, et al. Quantitative 3-dimensional analysis of preoperative and postoperative joint lines in total knee arthroplasty: a new concept for evaluation of component alignment. J Arthroplast. 2007;22:560–8.
37. Bourne RB, et al. The influence of tibial component intramedullary stems and implant-cortex contact on the strain distribution of the proximal tibia. Clin Orthop Relat Res. 1986;208:95–9.
38. Hsu HP, et al. Effect of knee component alignment on tibial load distribution with clinical correlation. Clin Orthop Relat Res. 1989;248:135–44.
39. Dai Y, et al. Anatomic tibial component design can increase tibial coverage and rotational alignment accuracy: a comparison of six contemporary designs. Knee Surg Sports Traumatol Arthrosc. 2014;22:2911–23.
40. Stulberg SD, et al. Which tibial tray design achieves maximum coverage and ideal rotation: anatomic, symmetric, or asymmetric? An MRI-based study. J Arthroplast. 2015;30:1839–41.

41. Clary C, et al. Tibial base design and patient morphology affecting tibial coverage and rotational alignment after total knee arthroplasty. Knee Surg Sports Traumatol Arthrosc. 2014;22:3012–8.
42. Martin S, et al. Maximizing tibial coverage is detrimental to proper rotational alignment. Clin Orthop Relat Res. 2014;472:121–5.
43. Barrack RT, et al. Winner of the 2003 James A. Rand young investigator's award. Early failure of cementless mobile-bearing total knee. J Arthroplast. 2004;19(7 Suppl 2):101–6.
44. James SJ, et al. Baseline bone mineral density and bone turnover in pre-operative hip and knee arthroplasty patients. Bone Joint Res. 2014;3:14–9.
45. Andersen MR, et al. Low preoperative BMD is related to high migration of tibia components in uncemented TKA-92 patients in a combined DEXA and RSA study with 2-year follow-up. J Arthroplast. 2017;32:2141–6.
46. Li MG, et al. The effect of preoperative bone quality on the fixation of the tibial component in total knee arthroplasty. J Arthroplast. 2000;15:744–53.
47. Luring C, et al. Micromotion in cemented rotating platform total knee arthroplasty: cemented tibial stem versus hybrid fixation. Arch Orthop Trauma Surg. 2006;126:45–8.
48. Ishii Y, et al. Changes in bone mineral density of the proximal femur after total knee replacement. J Arthroplast. 2000;15:519–22.
49. Ishii Y, et al. Bone quality of the calcaneus 5 years after total knee arthroplasty. J Arthroplast. 2015;30:196–8.
50. Ishii Y, et al. PCL retention or substitution during total knee arthroplasty does not affect long-term bone mineral density or quality. J Arthroplast. 2017;32:1148–52.
51. Ishii Y, et al. Changes of body balance before and after total knee arthroplasty in patients who suffered from bilateral knee osteoarthritis. Science. 2013;18:727–32.
52. Grisso JA, et al. Risk factor for falls as a cause of hip fracture in women. N Engl J Med. 1991;324:1326–31.
53. Ishii Y, et al. Quadriceps strength impairment in the mid- to long-term follow-up period after TKA. Knee Surg Sports Traumatol Arthrosc. 2017;25:3372–7.
54. Ishii Y, et al. Life expectancy of osteoarthritic patients after primary total knee arthroplasty. J Clin Orthop Trauma. 2017;8S:S57–61.

56 BCR-TKA

Hiroshi Inui

The ACL is always resected in conventional TKA. BCR-TKA was introduced to preserve all knee ligaments, including the ACL, in anticipation of preserving physiological kinematics, stability, and proprioception, to thereby achieve a favorable surgical outcome. However, for all ligaments to function effectively without kinematic conflicts, surgical techniques must be performed with consideration of constitutional alignment [1]. In addition, computer-assisted (navigation, robotic) surgery is desirable because the appropriate soft tissue balance of the medial and lateral compartments in all ranges impacts the surgical outcomes.

1 Indications for Surgery

All of the ligaments in the patient's knee joint, including the ACL, were preserved; however, the lesion was not confined to a single condyle, but was bicondylar or tri-condylar. When UKA is indicated, it should be preferentially performed instead of BCR-TKA.

1.1 Disease: OA, Osteonecrosis (ON) with Bicondylar or Tri-condylar Involvement Instead of Monocondylar Involvement

We encountered cases in which RA was diagnosed postoperatively in patients shown to possess a monocondylar lesion on imaging, but BCR-TKA was chosen instead of UKA because the patient reported involvement of the entire knee joint. Despite advancements in the treatment of RA and other inflammatory diseases, it is difficult to determine whether such disorders are ideal indications for BCR-TKA.

1.2 Preservation of All Knee-Joint Ligament Functions, Including that of the ACL

Although anteroposterior stability can be assessed by manual examination, MRI is essential to confirm that the ligaments are well-preserved and to exclude degenerative changes, such as mucoid degeneration of both cruciate ligaments. In addition, adequate consideration of the PCL condition is important. Sometimes, surgeons pay little attention to the PCL due to overarching concern regarding ACL preservation.

1.3 Good Joint ROM

If the function of all knee joint ligaments is well preserved, the joint ROM is naturally good. In particular, knees with flexion contracture often possess a contracted and dysfunctional ACL, even if the ACL appears visually intact. It is advisable to consider flexion contracture ≤10° as an indication for surgery.

1.4 Others

If all knee ligaments were well-preserved, BCR-TKA was indicated. However, to obtain stable and favorable surgical results with BCR-TKA, many knee surgeons consider the following indications for surgery: deformity of up to 180° ± 10° in the hip-knee-ankle angle, BMI ≤ 35 kg/m^2, and age at surgery ≤80 years [2].

2 Surgical Concept of BCR-TKA

The follow-up period for BCR-TKA is short, especially considering its relatively recent introduction in Japan. To date, there is no established surgical technique for BCR-TKA. The

H. Inui (✉)
Department of Orthopaedic Surgery, Saitama Medical University, Saitama Medical Center, Kawagoe, Saitama, Japan

S. Matsuda (ed.), *Advances in Total Knee Arthroplasty*, https://doi.org/10.1007/978-981-97-4920-1_56

following discourse is based on recent reports and our experience, with the understanding that a consensus may not necessarily be achieved.

2.1 Implant Shape

Simulation results have revealed that implants with anatomical surface shapes are desirable for good function of all preserved knee ligaments, including the ACL, and for physiological knee joint motion [3]. It has been reported that BCR-TKA using a non-anatomically shaped implant does not provide physiological knee joint motion [4, 5], and the clinical results are not favorable [6, 7]. Currently, anatomically shaped implants are preferred for BCR-TKA.

2.2 Target Alignment

Although there is a consensus that intraoperative medial stability is important in conventional TKA, intraoperative lateral stability (or laxity) is not considered important. However, in BCR-TKA, moderate lateral stability is necessary to achieve good postoperative outcomes [8]. Although lateral laxity has been reported to improve over time with conventional TKA [9], we encountered cases in which no improvement occurred with BCR-TKA (Fig. 56.1). The author suspects that the preserved ACL acts like a "stick" and prevents improvement in lateral laxity. Therefore, intraoperative lateral stability should be ensured during BCR-TKA. However, it is difficult to ensure both lateral and intraoperative stability when using the "mechanical alignment (MA) method," which is the conventional surgical technique for TKA. The "kinematic alignment (KA) method," which targets the native joint surface, is considered to be the appropriate surgical technique for BCR-TKA. The same concept is applied for the intraoperative planning of robot-assisted BCR-TKA.

We performed BCR-TKA using a surgical technique similar to the KA method. However, because it is difficult to plan the osteotomy on the tibial side preoperatively, the author uses the "functional alignment (FA) method" [10] in which the native joint surface is targeted, intraoperative soft tissue balance is considered, and fine adjustments are made using navigation.

3 Surgical Techniques

The following describes the surgical technique for BCR-TKA using the FA method with an anatomically shaped implant for the varus knee [10].

3.1 Approach

Medial dissection was maintained at a minimum (up to the deep MCL). The joint was exposed, and the condition of the ACL was evaluated. Osteophytes in the intercondylar region were resected as much as possible, but retractors were carefully used to prevent damage to the ligamentous attachments.

3.2 Determination of the Rotation Direction (Fig. 56.2)

The anteroposterior axis was difficult to identify on the tibial side because the tibial attachment of the PCL was not visible. An 18G catheter needle was inserted toward the tibial attachment of the PCL and from the medial border of the patellar tendon to the rectactors at the PCL attachment site to confirm the Akagi line [11] and to determine the anteroposterior axis

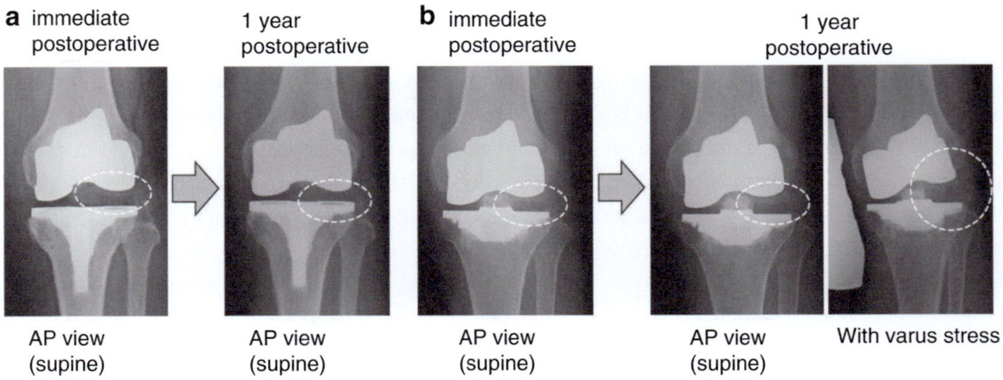

Fig. 56.1 Change in external flaccidity over time. (**a**) Conventional TKA: immediate postoperative lateral laxity is improved. (**b**) BCR-TKA: immediate postoperative lateral laxity persisted 1 year postoperatively

using Shakespeare's line [12] and the direction of the center of the head of the bone as indices. Whiteside's line was drawn from the femoral groove.

3.3 Distal Femoral Osteotomies

In FA, osteotomy is performed using the native articular surface. The anatomical surface shape of the implant available in Japan is 9.5 mm medial and 7 mm lateral to the distal femoral side. In most cases of BCR, lateral cartilage wear is not advanced, and the basic resection is 7 mm. Considering cartilage wear, 7–8 mm of the medial side was resected (Fig. 56.3). The varus/valgus angle of the resected surface did not exceed 3°.

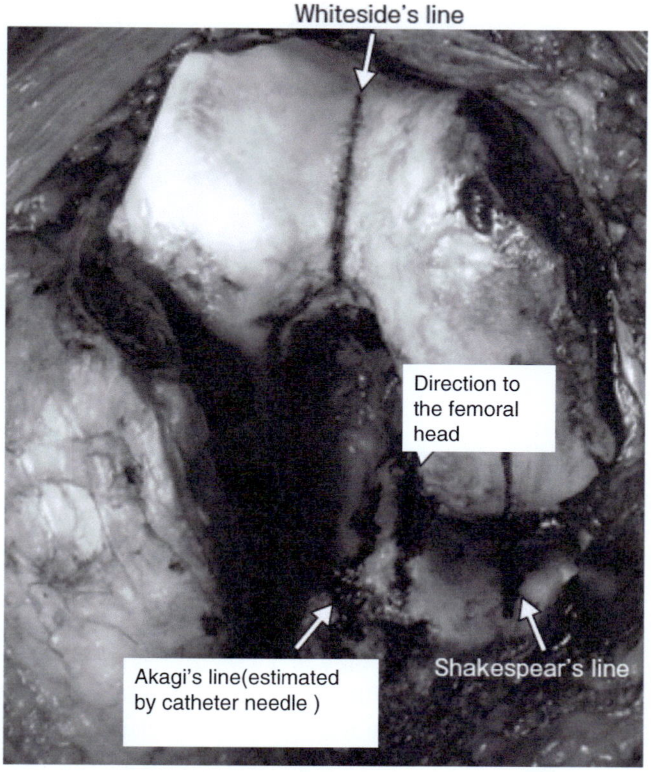

Fig. 56.2 Determination of the direction of rotation

3.4 Soft Tissue Balance Evaluation

A spacer block with a distal femur shape was inserted and full extension was confirmed. In addition, varus and valgus stresses were applied with the spacer block in place by using navigation to evaluate the balance of the medial and lateral compartments in the extended position before tibial osteotomy (Fig. 56.4).

3.5 Tibial Osteotomy

1. Coronal plane alignment

Since the medial side of the tibia is deficient in bone and cartilage in the medial tibia in the medial knee, the basic resection volume to reproduce the joint line was "lateral 11 mm" (tibial tray + thinnest insert thickness in Japanese anatomical shape implants), and the amount of medial osteotomy was determined by taking balance evaluation into consideration, as shown in Fig. 56.5.

2. Sagittal plane alignment

This study aimed to reproduce the posterior slope of the medial tibial plateau. However, an excessive posterior slope may cause postoperative overload of the ACL [13], invasion of the posterior medial tissues [14], and poor postoperative outcomes [15]. Therefore, we believe that the posterior slope should not exceed 10° (rotational alignment; Fig. 56.2).

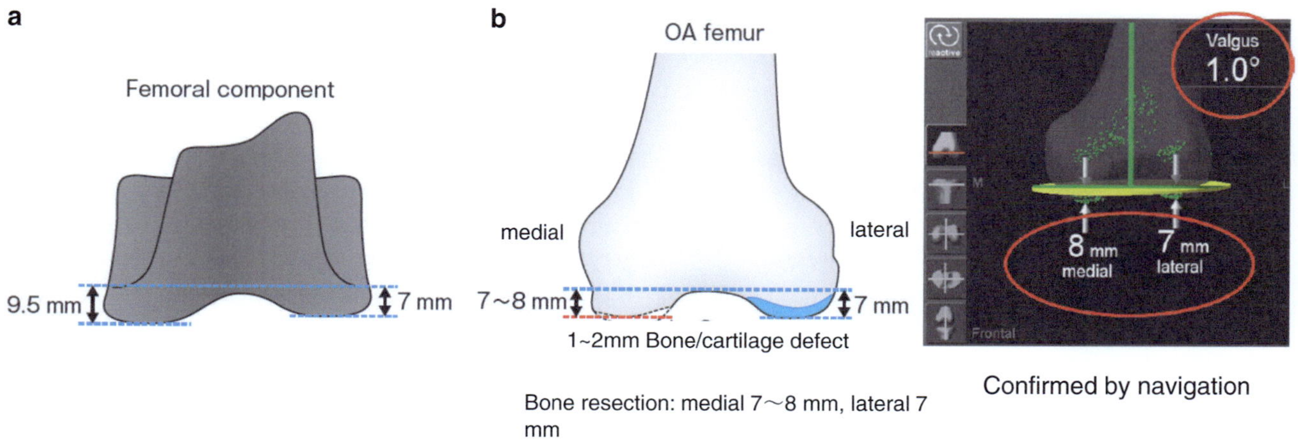

Fig. 56.3 Determination of distal femoral osteotomy volume

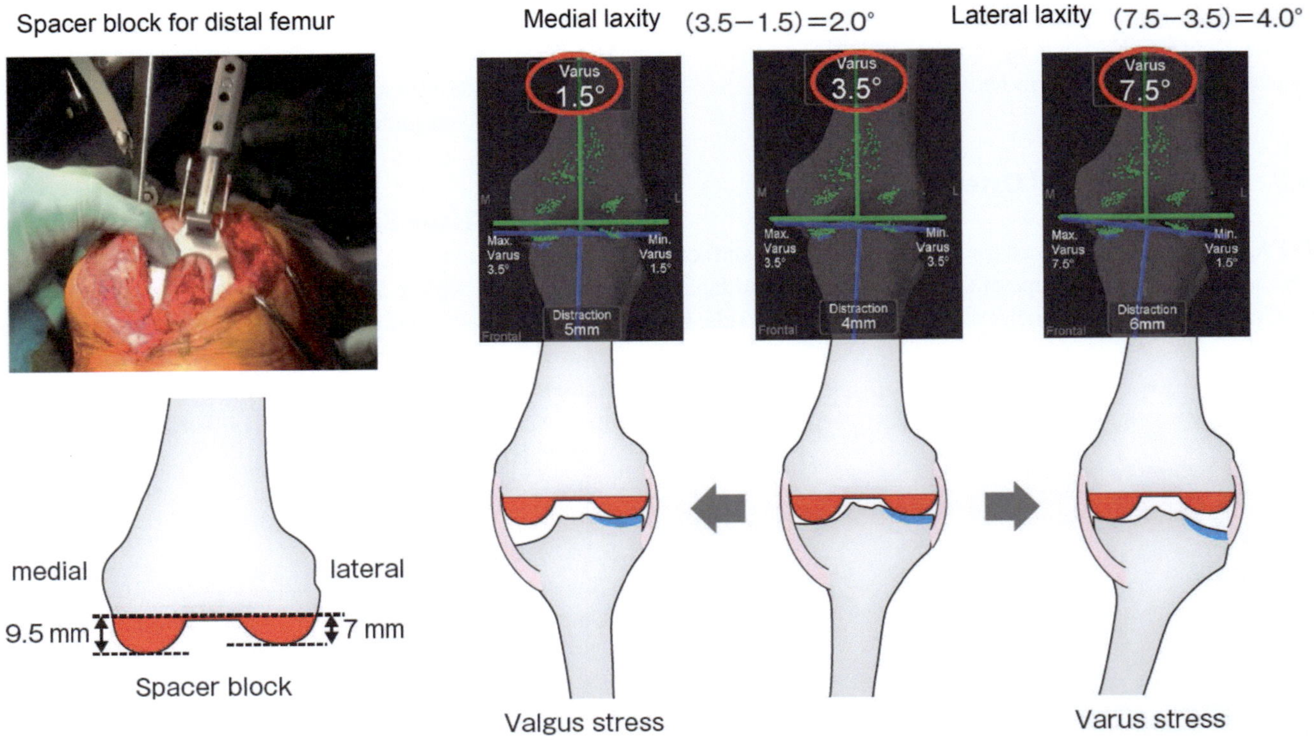

Fig. 56.4 Confirmation of soft-tissue balance before tibial osteotomy

3.6 Soft Tissue Balance Measurement and Posterior Condylar Osteotomy

The degree of posterior femoral condyle osteotomy was determined by measuring the extension/flexion gap at this point using a knee balancer to match the flexion/extension gap. Considering physiological lateral looseness, the lateral gap should be 1–2 mm larger than the medial gap in both extension and flexion. The angle of external rotation is based on the epicondylar axis. However, the amount of osteotomy that provides adequate soft tissue is more important. When osteotomizing the posterior condyle, a retractor should be inserted into the intercondylar area to protect the ACL, and care should be taken to prevent the bone saw from entering the intercondylar area.

3.7 Final Soft Tissue Balance Evaluation and Adjustment (Fig. 56.6)

1. Evaluation of medial and lateral compartments

In addition to bone gap measurements using a knee balancer, component gap measurements were performed. In computer-assisted surgery, it is possible to evaluate laxity under stress.

The following reports were published regarding the relationship between intraoperative soft-tissue balance and postoperative results of BCR-TKA using componentgap measurement [8]. The difference between the lateral and medial gap in extension and the difference between the lateral and medial gap in flexion at 140° were positively correlated with the 2011 KSS Symptoms; the difference between the lateral and medial gap at 140° flexion correlated negatively with the amount of improvement in the Knee Society Function Score and the ability to get up from a chair in the Patellar score [16], The difference between the medial gap at 140° flexion and extension was positively correlated with the Patellar score. In addition, it has been reported that the MA method can produce a medial/interior gap difference of 3–4 mm at each flexion angle from 10° to 90°, whereas the FA method can control the medial/interior gap difference to 1–2 mm 8, [10].

A report comparing the laxity evaluation by component gap measurement and computer-assisted surgery revealed that the amount of lateral joint laxity due to varus stress at 90° flexion was relatively larger than the amount of lateral joint laxity calculated by component gap measurement. Conversely, in computer-assisted (especially robotic) surgery, if surgical planning is performed using the value of varus/valgus stress as an index, the soft-tissue balance of the lateral compartment in the flexion position may be planned slightly tighter [10].

2. Evaluation of the front/rear direction (ACL/PCL)

In BCR-TKA, unlike conventional TKA, it is also important to ensure that adequate tension is maintained in the ACL/

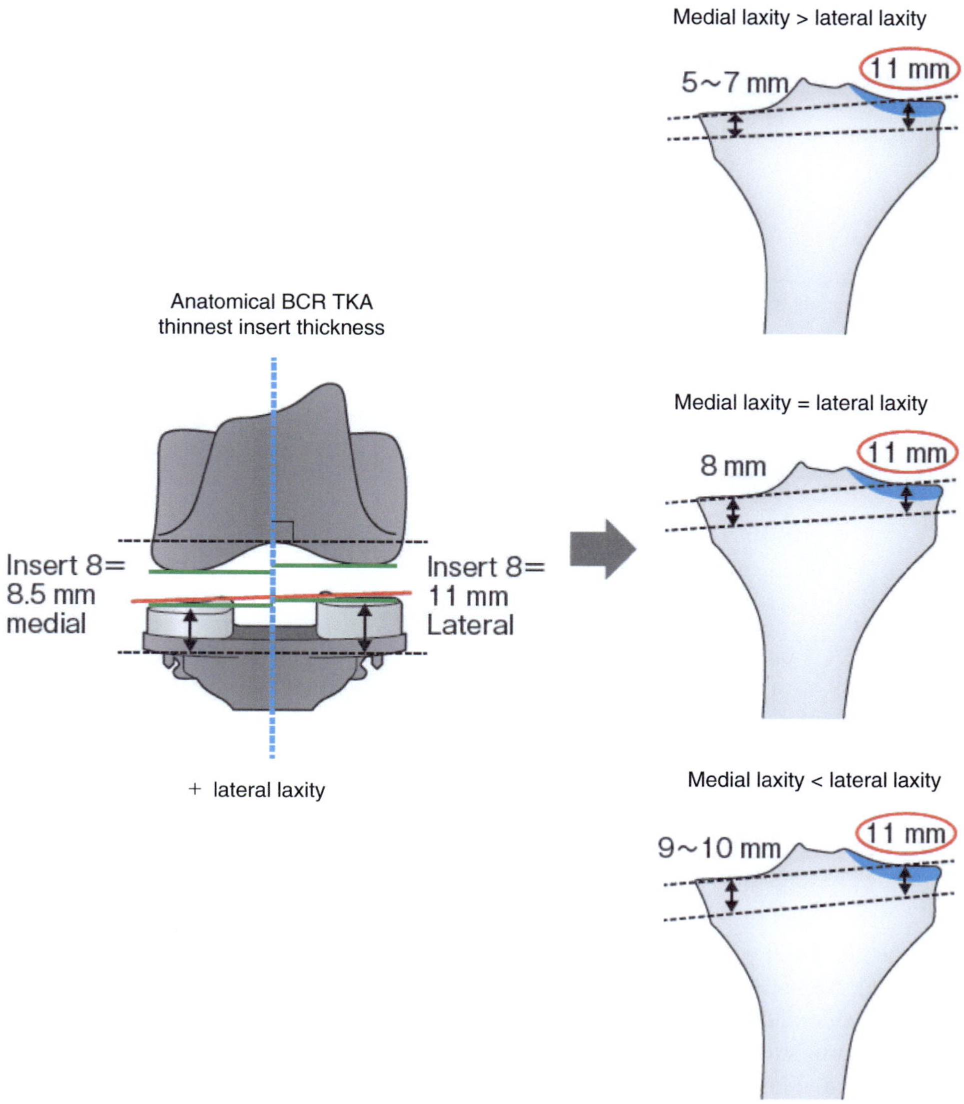

Fig. 56.5 Determination of tibial osteotomy volume

PCL. Although it is possible to evaluate the ACL/PCL intraoperatively by measuring the amount of anterior–posterior movement using manual stress, it is difficult to evaluate the tension of the ACL/PCL alone because anterior–posterior stability is complex and correlates with the medial support mechanism [17]. Currently, there is no established ACL/PCL evaluation method for BCR-TKA, and a sensory evaluation must be performed using fingers and probes. However, in most cases, if the joint line is reproduced using the FA method, the internal/external balance is moderately adjusted, the tibial native slope is reproduced, and ACL/PCL tension is naturally good.

3.8 Patellar Replacement

It is recommended that the patella be replaced in all cases. However, there are reports that PF pressure is lower in BCR-TKA than in CR-TKA and that PF pressure in BCR-TKA is not associated with postoperative anterior knee pain [18]; therefore, the option of non-replacement may be considered.

3.9 Implementation

Cementation techniques also play an important role. On the tibial side, the cement was thoroughly applied around the anterior keel. Because the posterior portion tends to float, it should be thoroughly checked to ensure that excess cement is pushed posteriorly and accurately positioned. A probe or similar tool can be used to remove excess cement from the posterior portion. A trial insert was placed, the femoropatellar component was cemented, and the patient was placed in an extension to allow the cement to set. However, at this time, the patient should be carefully positioned because of the risk of avulsion fracture at the ACL attachment site if the patella is pushed vigorously and forced into full extension. The insert thickness was finalized after cementation.

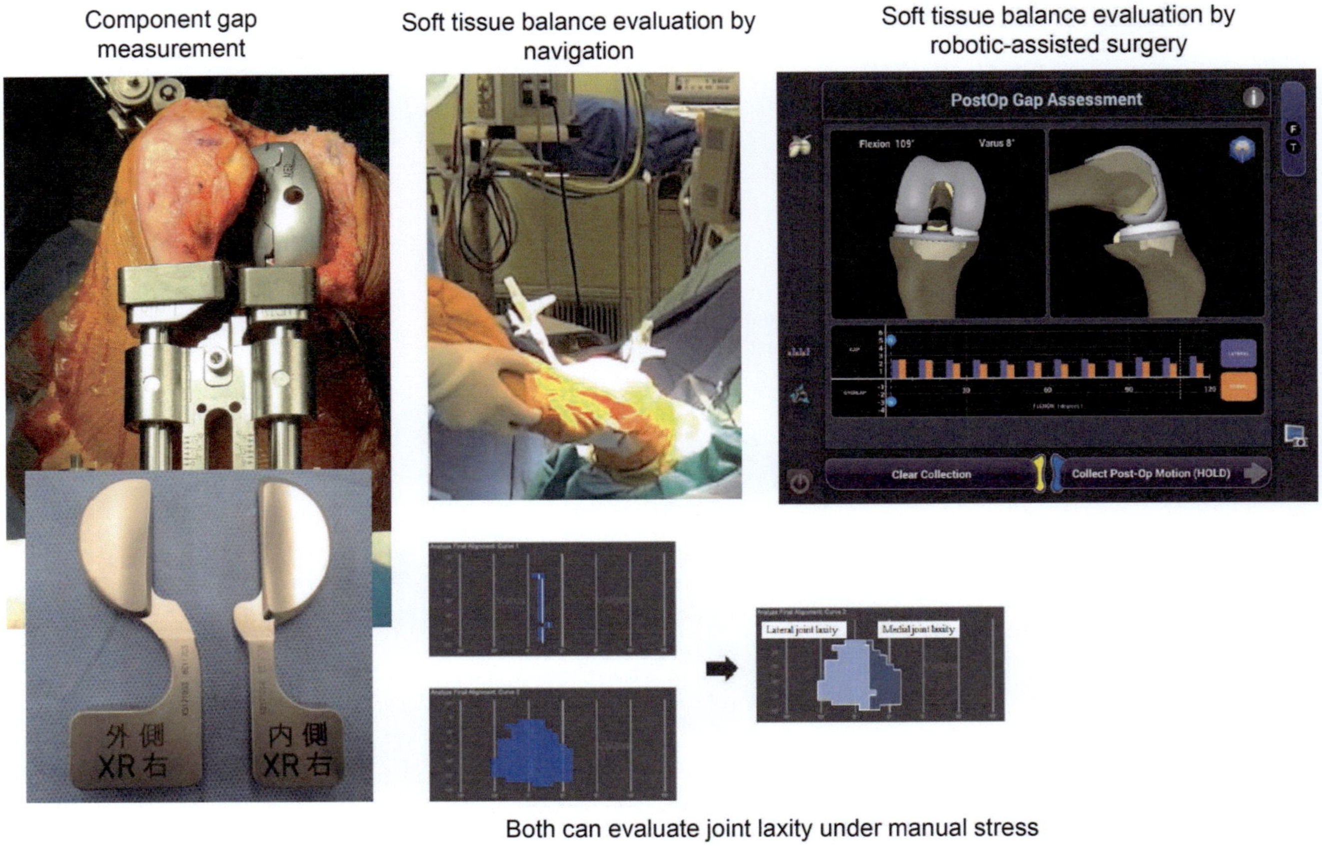

Fig. 56.6 Various soft tissue balance evaluations

Although the ideal joint laxity for BCR-TKA is not yet established, we believe that a laxity of approximately 1° on the medial side under valgus stress and 2–3° on the lateral side under varus stress is appropriate for postoperative stability of the medial knee. In some cases, inserts with different thicknesses may be installed on the medial and lateral sides, and a difference in thickness of up to 2 mm is considered acceptable in terms of implant design.

4 Clinical Results

4.1 Long-Term Durability

BCR-TKAs have been used in Europe and the United States since the 1970s. However, poor results have been reported, including a high revision rate [19]. CR-TKA and PS-TKA, which yield stable long-term results, have been widely used since the 1990s. Recently, however, reports of good survival rates of 89% at 23 years and 82% at 22 years, depending on the shape of the implant, have led to a re-evaluation of BCR-TKA [20, 21]. BCR-TKA was re-evaluated and introduced in Japan in 2015 as a new and improved modern BCR-TKA. However, the non-anatomical prosthesis of BCR-TKA, which was introduced earlier, has been reported to have high revision rates, with a 3-year survival rate of 88% [6, 7]. The presence of a mini-keel on the tibial side in anatomically shaped BCR-TKA, which was introduced in 2017, has increased fixation and reduced concerns regarding loosening. Therefore, medium- and long-term studies are required.

4.2 Clinical Results

Despite its development in the 1970s, the performance of BCR-TKA was avoided until recently because of the complexity of the surgical procedure, and early polyethylene wear [19, 21]. Moreover, the clinical results were not as good as those obtained with CR-TKA [22, 23], and were below expectations [22]. However, some reports suggest a high level of satisfaction, with 89.1% of the patients who underwent contralateral PS-TKA preferring the knee on the BCR-TKA side [24].

Results have been reported in Japan, mainly for anatomically shaped BCR-TKA and walking ability. Walking speed and stride length, of BCR-TKA patients at 6 weeks postoperatively was superior to that of BCS-TKA patients [25]. In a comparison of the clinical outcomes of approximately 20 BCR-TKA and UKA patients who underwent postoperative dynamic analysis, BCR-TKA patients exhibited inferior flexion ROM compared to UKA patients; however, other clinical outcomes were reported to be as good as those of UKA patients, with patient satisfaction scores of 30.8 (28.7 for UKA) and 10.3 (8.9 for UKA) on the 2011 NKSS [26]. In the present study, BCR-TKA was performed using the MA method. The FA method is also currently used at the same institution, and the satisfaction score at one year after surgery was 32.8 (28.0 for the MA group during the same time period). Further reports of good surgical results are expected in the future.

4.3 Differences in Surgical Outcomes According to Implant Design

In Japan, both anatomical and non-anatomical forms of BCR-TKA are available, as described above. However, as previously mentioned, the clinical results of non-anatomical BCR-TKA have not been favorable, and studies using cadaveric knees have reported that BCR-TKA using non-anatomical implants lost physiological rotation dynamics when the tibial component was placed [4], ACL tension during flexion was 2–6 times higher than that of normal knees [27], and the anatomical shape of the BCR-TKA implant was not effective for the treatment of knee fractures. However, it has also been reported that even with non-anatomically shaped implants, the use of a highly restrictive insert on the medial side can achieve rotation dynamics closer to physiological dynamics [28], and computer simulation analysis revealed that even with non-anatomically shaped BCR-TKA, kinematic conflict can be avoided if the implants are placed with KA [1]. We look forward to future reports on the clinical results.

4.4 Joint Stability Evaluation and Dynamic Analysis

Regarding the relationship between postoperative stability evaluation and postoperative outcomes of BCR-TKA, it was reported that lateral stability evaluated by the varus stress test at 30° and 90° flexion was positively correlated with patient expectations in the 2011 KSS, and medial laxity evaluated by the valgus stress test at 90° flexion was positively correlated with patient expectations and satisfaction [8]. In a report on the postoperative evaluation of the posterior–posterior stability of joints, the tension applied to ligaments was estimated from changes in the ligament length obtained by dynamic analysis of postoperative deep flexion movements (anatomically shaped implants). In addition, the ligament tension of the PCL increased with flexion, and there was no difference between the preoperative and postoperative periods [29]. The same report also demonstrated that the amounts of anteroposterior and posterior–posterior translations were similar in the ACL. In the same report, the relationship between the amount of anteroposterior translation and ligament tension was also evaluated. A strong positive correlation was observed between the ligament tension of the PCL and the amount of anteroposterior translation of the lateral femoral condyle in squatting motion [29].

Postoperative kinetic analysis of BCR-TKA, which has not yet been introduced in Japan, showed greater posterior migration than CR-TKA, with a pattern similar to that of the normal knee [30]; however, reports varied. Postoperative squatting motion of anatomically shaped implants has been analyzed in Japan, and a "medial pivot pattern" has occurred but, it has been reported that the "medial pivot pattern" does not reproduce the dynamics of the normal knee as well as the UKA in terms of the medial positional relationship between the femur and tibia in the extended position and the amount of lateral backward movement with flexion [29, 31]. However, these reports were the results of analyses of cases performed using the MA method, and the results of analyses of cases performed using the FA method are anticipated.

References

1. Watanabe M, et al. Abnormal knee kinematics caused by mechanical alignment in symmetric bicruciate-retaining total knee arthroplasty are alleviated by kinematic alginemnt. Knee. 2020;27:1385–95.
2. De Faoite D, et al. Indications for bi-cruciate retaining total knee replacement: an international survey of 346 knee surgeons. PLoS One. 2020;15:e 0234616.
3. Zumbrunn T, et al. Retaining native knee kinematics following joint arthroplasty: a novel biomimetric design with ACL and PCL preservation. J Arthroplast. 2015;30:2043–8.
4. Hamada D, et al. Native rotational kinematics are lost in bicruciate-retaining total knee arthroplasty when the tibial component is replaced. Knee Surg Sprots Traumatol Arthrosc. 2018;26:3438–43.
5. Tsai TY, et al. Bi-cruciate retaining total knee arthroplasty does not restore native tibiofemoral articular contact kinematics during gait. J Orthop Res. 2019;37:1929–37.
6. Christopher E, et al. Mean three-year survivorship of a new bicruciate-retaining total knee arthroplasty: are revisions still higher than expected? J Arthroplast. 2019;34:1957–62.
7. Christensen JC, et al. Higher frequency of reoperation with a new bicruciate-retaining total knee arthroplasty. Clin Orthop Relat Res. 2017;475:62–9.
8. Kaneko T, et al. Erratum to "greater postoperative relatively medial loose gap at 90° of flexion for varus knees improves patient-reported outcome measurements in anatomical bi-cruciate retaining total knee arthroplasty". Knee. 2020;27:1646–59.
9. Sekiya H, et al. Postoperative lateral ligament laxity diminishes with time after TKA in varus knee. Clin Orthop Relat Res. 2009;467:1582–6.
10. Inui H, et al. Comparison of the joint laxity of total knee arthroplasty evaluated by the distraction force and the varus-valgus force. Knee. 2022;34:98–107.

11. Akagi M, et al. An anteroposterior axis of the tibia for total knee arthroplasty. Clin Orthop Relat Res. 2004;420:213–9.
12. Shakespeare D, et al. The influence of the tibial sagittal cut on component position in the Oxford knee. Knee. 2005;12:169–76.
13. Klemt C, et al. Posterior tibial slope increases anterior cruciate ligament stress in bi-cruciate retaining total knee arthroplasty: in vivo kinematic. J Knee Surg. 2022;35:788–97.
14. Yamagami R, et al. Proximal tibial morphology is associated with risk of trauma to the posteromedial structures during tibial bone resection Yamagami R, et al. reproducing the anatomical posterior tibial slope in bicruciate-retaining total knee arthroplasty. Knee. 2022;36:1–8.
15. Peng Y, et al. Does component alignment affect patient reported outcomes following bicruciate retaining total knee arthroplasty? An in vivo three-dimensional analysis. J Knee Surg. 2020;33:798–803.
16. Fellar JA, et al. Patellar resurfacing versus retention in total knee arthroplasty. J Bone Joint Surg Br. 1996;78:226–8.
17. Inui H, et al. The relationship between anteroposterior stability and medial-lateral stability of a bi-cruciate stabilized TKA. Knee. 2018;25:1247–53.
18. Toyoda S, et al. Higher patellofemoral compressive force does not affect anterior knee pain in anoatomical bi-cruciate retaining total knee. Clin Biomech (Bristol, Avon). 2021;88:105444.
19. Rand JA, Coventry MB. Ten-year evaluation of geometric total knee arthroplasty. Clin Orthop Relat Res. 1988;232:168–73.
20. Pritchett JW. Bicruciate-retaining total knee replacement provides satisfactory function and implant survivorship at 23 years. Clin Orthop Relat Res. 2015;47:2327–33.
21. Sabouret P, et al. Total knee replacement with retention of both cruciate ligaments: a 22 year follow up study. Bone Joint J. 2013;95-B:917–22.
22. Migaud H, et al. Influence of the tibial slope on tibial translation and mobility of the non-constrained total knee prosthesis. Rev Chir Orthop Reparatrice Appart Mot. 1996;82:7–13.
23. Jenny JY, et al. Preservation of the anterior cruciate ligament in total knee arthoplasty. Arch Orthop Trauma Surg. 1998;118:145–8.
24. Pritchett JW. Patients prefer a bicruciate-retaining or a medial pivot total knee prosthesis. J Arthroplast. 2011;26:224–8.
25. Amemiya T, et al. Anatomical bi-cruciate retaining TKA improves gait ability earlier than bi-cruciate stabilized TKA based on triaxial accelerometery data: a prospective cohort study. Asia Pac J Sports Med Arthrosc Rehabil Technol. 2021;25:35–41.
26. Kono K, et al. Bicruciate-retaining total knee arthroplasty reproduces in vivo kinematics of normal knees to a lower extent than unicompartmental. Knee Surg Sports Traumatol Arthrosc. 2020;28:3007–15.
27. Okada Y, et al. ACL function in bicruciate-retaining total knee arthroplasty. J Bone Joint Surg. 2018;100:e114.
28. Wada K, et al. The medial constrained insert restores native knee rotational kinematics after bicruciate-retaining total knee arthroplasty. Knee Surg Sports Traumatol Arthrosc. 2019;27:1621–7.
29. Kono K, et al. In vivo kinematics and cruciate ligament forces in bicruciate-retaining total knee arthroplasty. Sci Rep. 2021;11:5645.
30. Moro-oka T, et al. Comparing in vivo kinematics of anterior cruciate-retaining and posterior cruciate-retaining total knee arthroplasty. Knee Surg Sports Traumatol Arthrosc. 2007;15:93–9.
31. Kono K, et al. In vivo kinematics of bicrciate-retaining total knee arthroplasty with anatomical articular surface under high-flexion conditions. J Knee Surg. 2021;34:452–9.

57 Medial Pivot TKA

Takashi Sato

MP TKA is used to obtain knee-joint stability by utilizing a ball-in-socket mechanism in the medial compartment. Through resection of the PCL [1–7] and the prevention of conflicts between the shape constraint of the implant and ligament tension, both stability and ROM can be achieved.

1 Indications and Contraindications

Although there are no reports that describe the indications for MP TKA in detail, theoretically, it can be used in almost all primary TKAs, except in special cases such as extremely severe valgus knees, morbid recurvatum knees, and collateral ligament dysfunction. As a result, it is suitable for a wide range of indications. In a report [8] comparing the postoperative clinical results of MP TKA for 15 external valgus knees with 156 varus knees, no significant differences were observed in ROM or various clinical scores, except in cases of deformity of 20° or greater. This indicates that this model is appropriate for moderate valgus knees. The authors used MP TKA for all primary TKA, except for valgus knees with flexion contracture of >15°, collateral ligament dysfunction, and morbid recurvatum knees.

2 Basic Concept of MP-TKA

1. In MP TKA, the medial side is the main load-bearing joint, and the surgical objective is to obtain stability on the medial side. Therefore, a certain degree of laxity on the lateral side should be considered acceptable.
2. The medial stability is achieved by the interaction between the ball-in-socket mechanism of the implant and the soft tissue tension. However, excessive soft tissue tension causes conflict with shape constraints and loss of proper joint laxity, resulting in decreased mobility. Therefore, it is important to obtain the appropriate medial soft tissue tension to achieve both stability and mobility.

3 Problems with MP TKA

The clinical outcomes of MP TKA [1–7, 9–11] have consistently been reported as excellent, without identification of any obvious problems. However, theoretically, if the medial soft tissue tension is too high, as described earlier, the high shape constraint may inhibit the rollback motion of the medial femoral condyle [12], which is necessary for deep flexion. Consequently, this may prevent the patient from obtaining deep flexion after surgery.

4 Preoperative Planning/Target Alignment

Numerous articles have reported favorable long-term durability and clinical outcomes [1, 2, 4, 6, 9, 13] and high patient satisfaction [1, 7, 11] with MP TKA. However, these results are derived from surgeries in which conventional mechanical alignment is the goal in the coronal plane. Additionally, if laxity in the lateral soft tissues is accepted based on the characteristics of MP TKA, varus alignment of the lower extremities should be avoided [14, 15], and neutral alignment is desirable from the viewpoint of joint stability in the coronal plane. As this chapter describes techniques that accept lateral soft tissue laxity in principle, it is assumed that a neutral alignment has been correctly obtained. Although there are no reports on sagittal or rotational alignment that have shown a relationship with clinical outcomes, and there is no consensus at this time, there are many reports that suggest femoral rotation angles. These include a femoral rotation angle parallel to the SEA [1], 3° [5], or 5° external rotation from the femoral posterior condylar axis [1], and 3° [1, 3] for posterior tibial implant tilt. In MP TKA with a ball-in-socket

T. Sato (✉)
Department of Orthopaedic Surgery, Niigata Medical Center, Nigata-shi, Niigata, Japan

S. Matsuda (ed.), *Advances in Total Knee Arthroplasty*, https://doi.org/10.1007/978-981-97-4920-1_57

mechanism, it is recognized that the greater the posterior tilt of the tibial implant, the more the posterior the femur becomes relative to the tibia in the upright extension position [16]. Thus, excessive posterior tilt should be avoided to prevent loss of joint stability caused by this mechanism.

5 Concept of Surgery

5.1 Target Soft Tissue Balance

Although no reports exist that examine the relationship between soft tissue balance and clinical outcomes in MP TKA, it is reasonable to aim for medial-dominant stability owing to the characteristics of the implants, as described earlier. Therefore, it is not necessary to aim for a rectangular gap with the goal of equal stability on the medial and lateral sides. As long as the surgery is performed with these considerations in mind, either the "measured resection technique" or the "gap balance technique" may be used.

5.2 Theory of Implantation Based on Geometric Features

MP TKA is referred to as a "single radius" type of implant in which the femoral implant has a bearing surface with a single radius of curvature and its rotation axis (implant flexion-extension axis). The SEA can be approximated as the optimal flexion-extension axis in normal knee joint motion from extension to approximately 80° of flexion [17, 18]. Within the same ROM, the radius of curvature of the medial femoral bearing surface is almost constant [19] when the SEA is used as the flexion-extension axis. Because the collateral ligaments are attached to both ends of the SEA, aligning the flexion-extension axis of the single-radius implant with the SEA theoretically minimizes the change in length and tension of the MCL during the flexion-extension movement, and physiological medial joint stability in the ROM is easily achieved (Fig. 57.1).

In contrast, the medial femoral condyle exhibits rollback motion together with the lateral condyle in deep flexion above 110° in the normal knee joint [12, 20, 21]. However, in MP TKA, high soft tissue tension in the medial part during flexion impairs medial rollback motion owing to a conflict with shape constraints. Therefore, it is necessary to provide a certain degree of laxity in the flexed position through PCL resection and adjust the size and anteroposterior position of the femoral implant to facilitate the acquisition of deep flexion.

5.3 Placement Concept on the Femoral Side

The design of femoral implants and bearing surface geometry for single-radius TKA, including MP TKA, was based on the relationship between the flexion and extension axis and

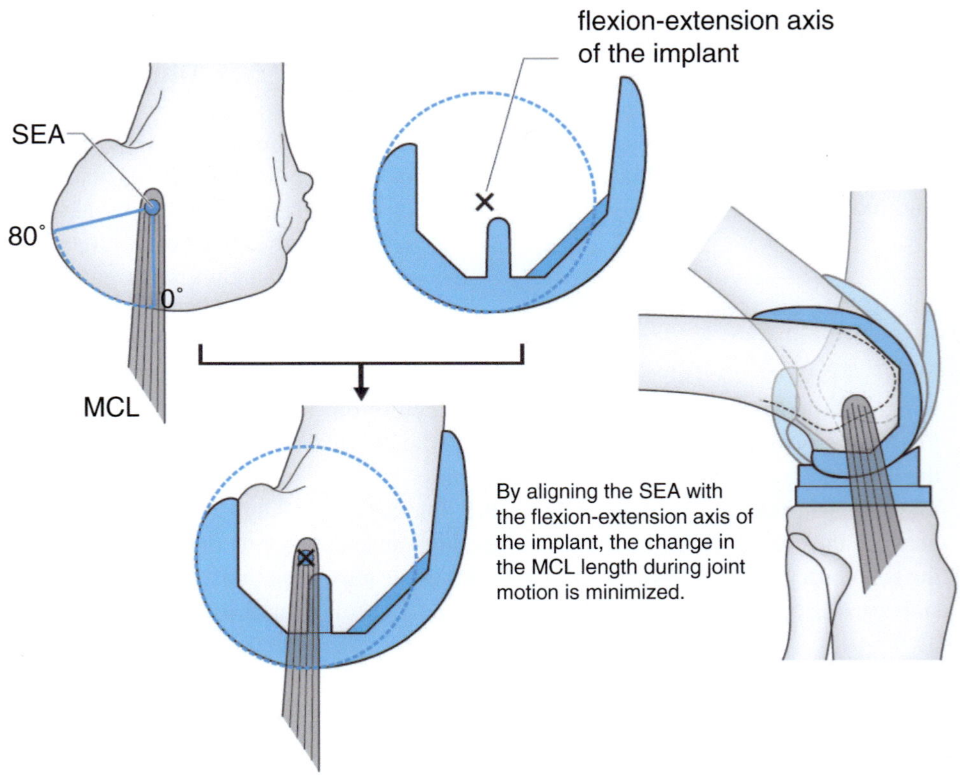

Fig. 57.1 Femoral implant placement concept. By aligning the SEA with the flexion-extension axis of the implant, the change in the MCL length due to joint motion from the extended position to approximately 80° is theoretically minimized, and joint stability is achieved

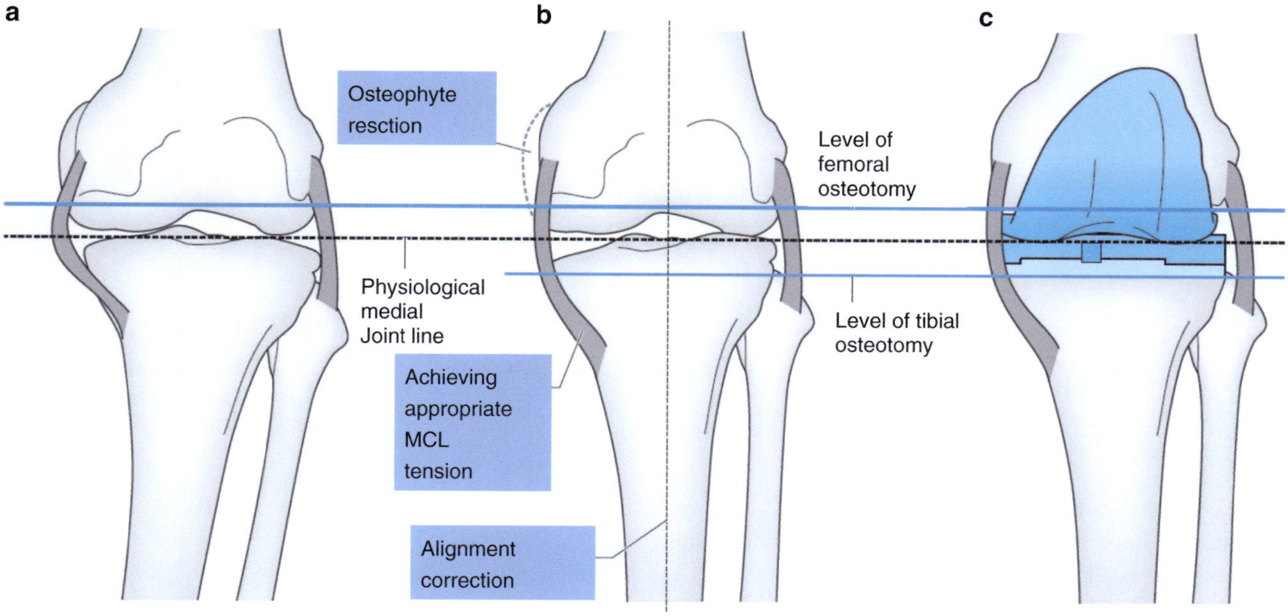

Fig. 57.2 Stretch-gap fabrication method. (**a**) Determination of the osteotomy level that preserves the physiological joint line (dashed line) of the medial femoral condyle. (**b**): Determination of the tibial osteotomy level with appropriate medial soft tissue tension achieved by osteophyte resection and alignment correction. (**c**) Implant placement maintaining the condition of (**b**)

bearing surface of the normal knee joint [22] (Fig. 57.1). Therefore, when the implant-bearing surface is positioned to match the pre-morbid physiological bearing surface, the implant flexion-extension axis closely aligns with the flexion-extension axis of the knee itself. Therefore, the amount of osteotomy at the distal and posterior condyles should be determined so that the preoperative joint line (bearing surface) is preserved as much as possible. The angle of rotation should be parallel to the SEA so that the SEA and flexion-extension axis of the implant are spatially close, and the installation is in accordance with the concept. Based on the above theory, the treatment of soft tissues, such as the collateral ligaments, should be minimized.

5.4 Placement Concept on the Tibial Side

In MP TKA, it is desirable to surgically restore the femoral joint line and knee flexion-extension axis to maximize medial ball-in-socket function. Therefore, when placing a tibial implant, osteotomy is performed at a level that allows the insertion of the tibial implant while preserving the physiological joint line of the distal femoral medial side in extension, ensuring correct coronal alignment, and achieving appropriate medial soft tissue tension through osteophyte resection (Fig. 57.2). In other words, the joint line on the tibial side is determined based on the femoral joint line in the extended position. With this method, the flexion gap tends to be larger than the extension gap, especially following PCL resection, resulting in laxity in the flexed position. However, if not extreme, the gap pattern in this model is desirable from the viewpoint of obtaining deep flexion.

5.5 Patellar Replacement

Several clinical studies using MP TKA have reported patella replacement [1–4, 7]. National registry data [13] reported significantly higher cumulative implant revision rates ("revision rates") for MP TKA in non-replaced cases compared to patella-replaced cases. However, no significant differences in revision rates were observed in patella-replaced cases compared to general low-constraint implants, though the revision rates for MP TKA were significantly higher in non-replaced cases. It should be noted that in the case of registry data, other factors may have intervened as biases in the decision to perform patellar replacement or non-replacement. Therefore, it is unclear whether patellar non-replacement was the direct cause of revision. However, the data suggested that patellar replacement may be advantageous in terms of long-term outcomes.

6 Key Points of Surgical Procedures

6.1 PCL Treatment

In MP TKA, in principle, the PCL should be resected to avoid conflict with high-shape constraints. To date, in many clinical studies the PCL has been resected [1–7]. However,

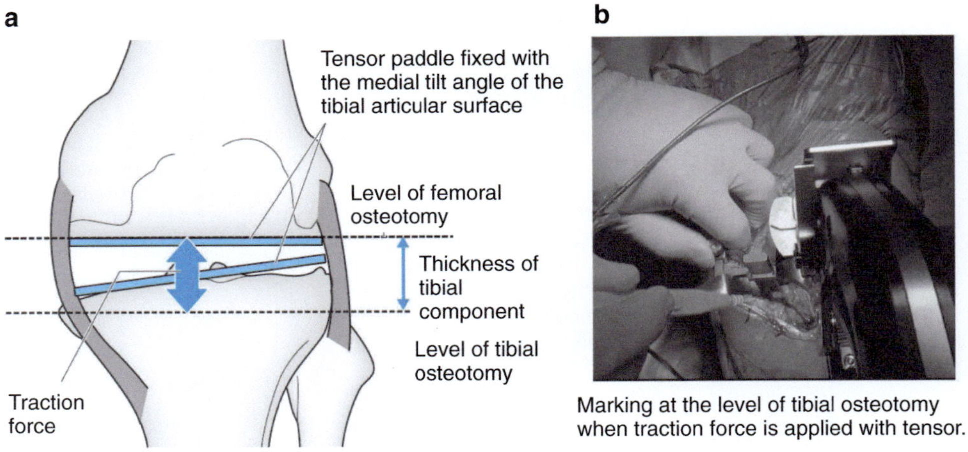

Fig. 57.3 Prediction of tibial osteotomy level using tensors, as performed by the authors

there have been some reports of PCL sparing [23, 24], and some studies have indicated no difference in clinical outcomes between PCL sparing and PCL preservation [24]. In this report, when tight flexion was observed after PCL preservation, partial resection of the PCL was performed and adjustments were made by increasing the posterior tibial tilt. In addition, a study [25] that examined the difference in dynamics between the presence and absence of the PCL and different insert shapes using MP TKA for the same cadaveric knee concluded that the PCL should be transected when a regular MP insert is used, and that an insert type with a reduced medial posterior lip height should be selected when preserving the PCL.

6.2 Soft Tissue Release

Only the deep layer of the MCL is released, and no additional soft tissue release is necessary to balance the medial and lateral soft tissues because the goal is not to achieve a rectangular gap equalizing the medial and lateral gaps.

6.3 Gap Fabrication Method Based on the Design Concept

In previous clinical studies, few reports have provided detailed methods for creating an extension and flexion gap, including the amount of osteotomy. Most of the reports [1, 3] have assumed that the amount of osteotomy is equal to the thickness of the implant, similar to general TKA. Therefore, the general method is considered acceptable. However, the following is a theoretically recommended method for creating a gap to take advantage of the MP TKA concept.

1. Creation of extension GAP

Osteophyte excision: All osteophytes that can be excised should be removed.

Determination of the osteotomy level on the distal femur: The amount of osteotomy should be determined considering the amount of cartilage and bone wear, so that the distal femur maintains the medial joint line as much as possible. If only cartilage wear is present, the amount of osteotomy should be reduced by 1 to 2 mm relative to the implant thickness. If obvious bone wear was observed, a sulcus-level cut should be performed (Fig. 57.2).

Determination of tibial osteotomy level: After osteotomy of the distal femur, the distance from the distal femur osteotomy level should be measured using a spacer block or tensor device with a traction force applied in the distal direction with the knee in extension; the osteotomy level at which the thinnest tibial implant (10 mm in this case) can be inserted should be used as the standard osteotomy amount (Fig. 57.3). In this case, a tensor equipped with a special paddle was inserted into the extension gap after distal femoral osteotomy and the osteotomy level at which a 10 mm tibial implant could be inserted while applying a 150 N traction force was marked (Fig. 57.3).

In cases of severe preoperative flexion contracture, marking in full extension may result in excessive osteotomy, which can be reduced by leaving approximately 5° of flexion.

If the extension gap is insufficient, the soft tissue tension in the flexed position should be determined using the "method of confirming the expected flexion gap" described below. If only the extension gap is insufficient, posteromedial soft tissue release or additional osteotomy of the distal femur should be performed. In cases where both the extension and flexion

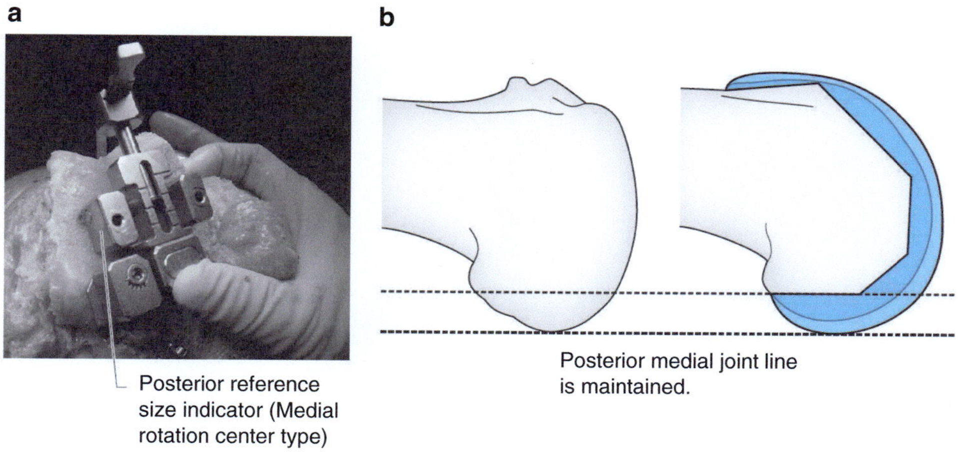

Fig. 57.4 Femur sizing and anterior-posterior positioning with a posterior reference sizer. The medial posterior condyle joint line should be preserved

gaps are insufficient, additional osteotomy of the tibia should be performed.

Confirmation of the extension gap: After the resection of the entire meniscus, the extension gap should be confirmed using a spacer block. Sufficient medial stability should be confirmed manually or with a tensor and lateral laxity should be accepted based on the assumption that the alignment is correct.

2. Femoral rotation angle determination and sizing

The rotational orientation of the femoral implant should be determined to be parallel to the SEA by placing the femoral sizing guide on the distal cutting surface. An anterior or posterior reference method may be used; however, the medial posterior condyle joint line should be preserved after implant placement, and the posterior cut level should be adjusted so that it does not become more posterior than preoperatively (i.e., the posterior condyle offset does not increase) (Fig. 57.4). If the size indicator shows a value between two sizes when an anterior reference guide is used, the smaller size should be selected.

3. Creation of flexion gap

The femoral size and anteroposterior position determined by the aforementioned method are based on the size and shape of the femoral condyle and can be considered as a concept aligned with the "measured resection technique." In normal cases, this method combined with PCL resection can provide an appropriate flexion gap. However, when the size and anteroposterior position are determined based on femoral geometry alone, there are cases in which the flexion gap is sometimes too large or too small, depending on the soft tissue condition of the individual case. Therefore, it is advisable to confirm whether the posterior osteotomy level determined by the sizer is appropriate for MP TKA from the viewpoint of soft-tissue tension in the flexed position. The following is a simple method used by the authors. The purpose was to avoid deep flexion limitations due to an undergap and prevent instability due to an overgap.

[Method for confirming expected flexion gap] This method is used to check soft tissue tension in flexion when the femur size and anteroposterior position are determined such that the osteotomy of the medial posterior femoral condyle is equal to the thickness of the implant. The knee was placed at 90° flexion, and a spacer block equivalent to the thickness of the tibial implant (e.g., 10 mm) was inserted into the flexion gap before femoral posterior condyle osteotomy. Medial laxity was confirmed by the amount of lift-off of the medial femoral posterior condyle from the spacer when valgus stress was manually applied at 90° flexion (Fig. 57.5). In principle, the lateral laxity in the flexed position should be considered acceptable.

1. If there is no liftoff or medial laxity even under stress, MP TKA may prevent deep flexion, and the posterior condylar osteotomy level should be anteriorized by approximately 2 mm. However, if a smaller size is selected using the anterior reference method for determining the femur size, the posterior condylar joint line should be determined, considering that it will be anteriorized by a few millimeters after implant placement.
2. A medial laxity of 2–3 mm was considered appropriate for obtaining deep flexion. However, if a smaller size was selected using the same anterior reference method, an excessive flexion gap occurred. Therefore, the osteotomy level should be shifted several millimeters posteriorly and, if necessary, the size should be increased by checking the anterior osteotomy level.
3. If the medial side of the osteotomy is >4 mm or if lift-off is observed even without external stress, the osteotomy is clearly excessive. The posterior osteotomy level should be posteriorized by at least 3 mm, and the size should be increased.

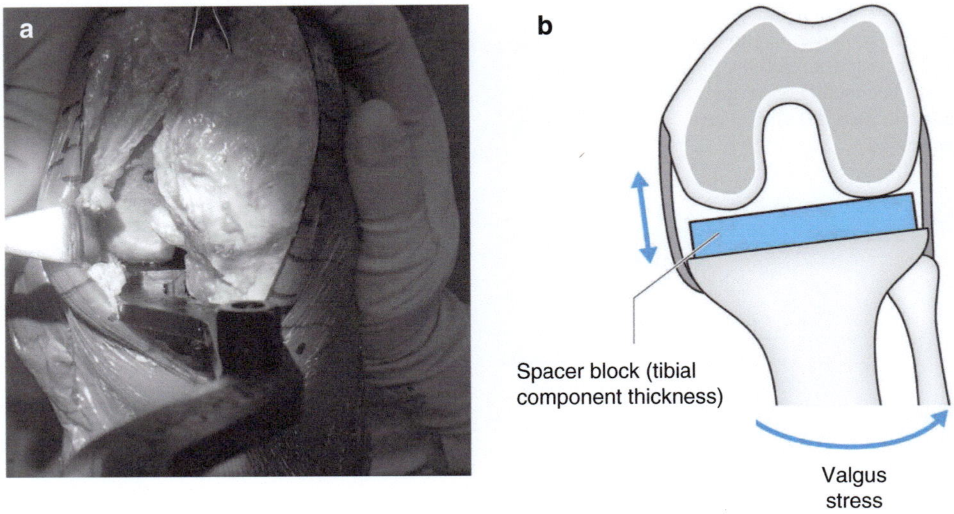

Fig. 57.5 Confirmation of medial soft tissue tension in flexion position using a spacer block

The authors examined the clinical results of 104 consecutive knees that underwent MP TKA (EVOLUTION medial pivot knee system, Microport, Inc.) using the gap creation method described above at least 1 year before the examination. The final joint laxity was calculated based on intraoperative measurements of the gap values in extension and 90° flexion and the insert thickness used. The final joint laxity was 1.2 mm on the medial side and 3.8 mm on the lateral side in extension and 2.3 mm on the medial side and 5.6 mm on the lateral side in flexion, indicating the target soft tissue balance pattern. Good joint stability, ROM, and patient satisfaction were achieved (see below).

6.4 Precautions During Femoral Osteotomy

MP TKA has a single radius up to approximately 100° of flexion, and, because of its geometrical characteristics, the cut line of the posterior chamfer is closer to the femoral attachments of the MCL/LCL and popliteus tendon than other curve implants. Therefore, when osteotomizing the posterior condyle and posterior chamfer, the ligaments should be adequately protected using a retractor or other means.

6.5 Determination of Tibial Implant Size and Placement Angle

The position of the tibial trial should be determined by either the anteroposterior tibial axis or the ROM method and the coverage condition should be checked. The rim of the implant should be positioned to ride on the medial cortex as much as possible because MP TKA is theoretically expected to medially apply a large force to the bone-implant boundary.

7 Clinical Results

7.1 Long-Term Durability

In TKA using the ADVANCE medial pivot knee system (Microport Inc.), the implant survival rates at 10 years postoperatively were 96.3% and 98.4%, respectively [2], with revision surgery and revision at 17 years postoperatively as the endpoints. In a systematic review [9] that summarized the mid- to long-term results of the same model, pooled component survivorship at 8 years was 99.2%, indicating excellent long-term durability. In the MPK (Kyocera Corporation) alumina femoral implant, the implant survival rates were 98.6% at 5 years [3] and 99.1% at 10 years [4], both showing extremely favorable outcomes. National registry data also reported a revision rate of 6% at 15 years in the Medial Pivot design, which was equivalent or better than that of minimally stabilized (6.1%) and posterior stabilized (7.5%) [13] implants. The long-term durability of MP TKA is comparable to or better than that of other conventional models.

7.2 Clinical Symptoms and ROM

In a study in which different TKA models were inserted into the right and left knees of the same patient and the patients were asked which knee was in better condition, the ACL + PCL-retaining and MP types were reported to be the best among the ACL + PCL-retaining, PCL-retaining, PS, MP, and mobile bearing types [7].

In a study examining the long-term results after TKA with ADVANCE, the Knee Society Knee Score (KSS) and the Knee Society Function Score (KSFS) were reported to be 87.8 and 90.3 points, respectively, [2] at an average of 10 years and 92.1 and 73.4 points, respectively, [1] at an average of 8 years. In addition, a systematic review [9] summarizing the mid- to long-term results of the same model reported that the cumulative points for the KSS and KSFS were 87.9 and 80.0 points, respectively, with both indicating favorable outcomes. In MPK, the KSS and KSFS were reported to be 90 and 76 points, respectively, at 5 years postoperatively [3] and 89 and 68 points, respectively, at 10 years postoperatively [4]. The postoperative ROM has been reported to be −1.1° to 0° in extension and 110.3° to 120° in flexion [1–4, 6].

We evaluated 104 consecutive knees that underwent TKA using EVOLUTION CS-type inserts for at least one year and found the 2011 Knee Society Score [3] was 75.3 (of which 91.3% were satisfactory), the ROM was an extension of −0.2°and flexion of 123.9°, and the incidence of subjective instability [26] was 10.6%.

8 Future Tasks

The best soft tissue balance in terms of knee joint function, such as ROM, and clinical symptoms, such as painlessness and comfort, is unknown for this unique model with a high medial constraint. Further studies are required to address this. In addition, single-radius implants may cause shape mismatch in the femoral condyle area, and further improvement of the implant design is needed to reduce overhang at each site and facilitate deep flexion.

References

1. Chinzei N, et al. Satisfactory results at 8 years mean follow-up after ADVANCE medial-pivot total knee arthroplasty. Knee. 2014;21:387–90.
2. Ueyama H, et al. Long-term clinical outcomes of medial pivot total knee arthroplasty for Asian patients: a mean 10-year follow-up study. Knee. 2020;27:1778–86.
3. Iida T, et al. Mid-term clinical results of alumina medial pivot total knee arthroplasty. Knee Surg Sport Traumatol Arthrosc. 2012;20:1514–9.
4. Nakamura S, et al. Clinical results of alumina medial pivot knee arthroplasty at a minimum follow-up of 10 years. Knee. 2017;24:434–8.
5. Ishida K, et al. No difference between double-high inert and medial-pivot inert in TKA. Knee Surg Sport Traumatol Arthrosc. 2014;22:576–80.
6. Macheras GA, et al. A long term clinical outcome of the medial pivot knee arthroplasty system. Knee. 2017;24:447–53.
7. Pritchett JW. Patients prefer a bicruciate-retaining or a medial pivot total knee prosthesis. J Arthroplast. 2011;26:224–8.
8. Iwakiri K, et al. Is medial pivot total knee arthroplasty suitable for patients with valgus knee osteoarthritis? Eur J Orthop Surg Traumatol. 2022;32:551–7.
9. Fitch DA, et al. Mid- to long-term outcomes of a medial pivot system for primary total knee replacement: a systematic review and meta-analysis. Bone Joint Res. 2014;3:297–304.
10. Karachalios T, et al. A mid-term clinical outcome study of the advance medial pivot knee arthroplasty. Knee. 2009;16:484–8.
11. Samy DA, et al. A retrospective comparison of a medial pivot and posterior-stabilized total knee arthroplasty with respect to patient-reported and radiographic outcomes. J Arthroplast. 2018;33:1379–83.
12. Tanifuji O, et al. Three-dimensional in vivo motion analysis of normal knees using single-plane fluoroscopy. J Orthop Sci. 2011;16:710–8.
13. Australian Orthopaedic Association National Joint Replacement Registry. Hip and Knee Arthroplasty Annual Report 2021. Australian Australian Orthopaedic Association National Joint Replacement Registry, Adelaide.
14. Sekiya H, et al. Postoperative lateral ligament laxity diminishes with time after TKA in the varus knee. Clin Orthop Relat Res. 2009;467:1582–6.
15. Kuriyama S, et al. No condylar lift-off occurs because of excessive lateral soft tissue laxity in neutrally aligned total knee arthroplasty: a computer simulation study. Knee Surg Sports Traumatol Arthrosc. 2016;24:2517–24.
16. Takahashi Y, et al. Postoperative femoral anteroposterior position while standing correlates with the posterior tibial slope and posterior femoral condylar offset in medial pivot total knee arthroplasty. Biomed Mater Eng. 2022;33:51–64.
17. Asano T, et al. The functional flexion-extension axis of the knee corresponds to the surgical epicondylar axis: in vivo analysis using a biplanar image-matching technique. J Arthroplast. 2005;20:1060–7.
18. Feng Y, et al. In-vivo analysis of flexion axes of the knee: femoral condylar motion during dynamic knee flexion. Clin Biomech. 2016;32:102–7.
19. Sato T, et al. Three-dimensional morphology of the distal femur based on the surgical epicondylar axis in a normal elderly population. Knee. 2021;30:125–33.
20. Iwaki H, et al. Tibiofemoral movement 1: the shapes and relative movements of the femur and tibia in the unloaded cadaver knee. J Bone Joint Surg Br. 2000;82:1189–95.
21. Pinskerova V, et al. The knee in full flexion: an anatomical study. J Bone Joint Surg Br. 2009;91:830–4.
22. Hollister AM, et al. The axes of rotation of the knee. Clin Orthop Relat Res. 1993;290:259–68.
23. Miyazaki Y, et al. Analysis of the kinematics of total knee prostheses with a medial pivot design. J Arthroplast. 2011;26:1038–44.
24. Bae DK, et al. Clinical outcome of total knee arthroplasty with medial pivot prosthesis a comparative study between the cruciate retaining and sacrificing. J Arthroplast. 2011;26:693–8.
25. Omori G, et al. The effect of geometry of the tibial polyethylene insert on the tibiofemoral contact kinematics in Advance medial pivot total knee arthroplasty. J Orthop Sci. 2009;14:754–60.
26. Knoop J, et al. Association of lower muscle strength with self-reported knee instability in osteoarthritis of the knee: results from the Amsterdam osteoarthritis cohort. Arthritis Care Res. 2012;64:39–44.

58 Kinematically Aligned TKA

Tomoyuki Matsumoto

1 Alignment of TKA

TKA is a well-established surgical technique that provides excellent pain relief and good mid- to long-term outcomes. However, patient satisfaction is not yet sufficient, and improvements in surgical equipment and techniques are expected. The key elements of the surgical technique are accurate osteotomy and component placement to reproduce lower-limb alignment and achieve an appropriate soft tissue balance. Although mechanically aligned TKA (MA-TKA), the current gold standard, has generally achieved good mid- to long-term clinical outcomes, there is still room for further discussion in the areas of alignment reproduction and appropriate soft tissue balance concerning improving patient satisfaction and reproducing postoperative knee function. This chapter reviews the concept, surgical technique, and clinical outcomes of kinematically aligned (KA)-TKA, which has attracted much attention in recent years.

2 KA-TKA Concept

Historically, the anatomical alignment method has been proposed to reproduce the joint line to reproduce the anatomy and function of a normal knee [1]. In the coronal plane, the tibia was osteotomized at approximately 3° varus, and the femur was osteotomized at approximately 3° valgus so that the joint lines were parallel when the patient was standing on one leg. The technical difficulty in osteotomizing the tibia to 3° varus and the durability of the porous-coated TKA system caused polyethylene wear in the early postoperative period, leading to its elimination. However, recent developments in navigation systems and other devices that enable accurate osteotomy, along with improvements in implant durability, have reaffirmed the superiority of the anatomical alignment method.

Furthermore, in a study of 250 lower-limb radiographs of normal volunteers, the average hip-knee-ankle angle was 1.9° varus in males and 0.9°varus in females, and 32% of males and 17% of females had knees with >3° varus [2]. This raises the question of whether the acquisition of neutral alignment in such cases of constitutional varus knees is ideal for postoperative alignment. A study of the 15-year survival rates for modern cemented MA-TKA reported no difference in the results between neutral alignment (varus/valgus less than 3°) and varus/valgus >3° [3]. Furthermore, in the 7.2-year results of MA-TKA for 132 varus knees, the resulting mild varus alignment was better than the neutral alignment [4], and there are now scattered reports that threaten the myth of neutral alignment in the coronal plane. These reports form the basis for the current promotion of KA-TKA by surgeons.

Howell et al., who popularized KA-TKA, developed an anatomical alignment technique using a patient-specific instrument (PSI) and established it as KA-TKA [5]. The key concept of this technique is the flexion-extension axis of the knee, which is defined as a cylindrical axis [6] that approximates the posterior femoral condyle as a cylinder and is centered at its center. They named it calipered KA-TKA because the cylindrical axis can be easily reproduced by an osteotomy of the same thickness as the implant medially and laterally from the articular surface, taking cartilage wear into account [7]. They observed 222 KA-TKA cases over 10 years and reported extremely good postoperative results with a survival rate of 97.5% [8]. There are two types of KA-TKA: the conventional technique using PSI and the generic instrumentation [9] combined with a dedicated guide and navigation. The latter is currently the mainstream technique owing to its simplicity. The concept of KA-TKA has undergone a transition. In conventional KA-TKA, depending on the degree of preoperative deformity, significant varus may remain after surgery owing to the unpredictability of the conventional anatomy, and there are concerns about catastrophic

T. Matsumoto (✉)
Department of Orthopaedic Surgery, Kobe University Graduate School of Medicine, Chuo-ku, Kobe, Japan
e-mail: matsun@m4.dion.ne.jp

S. Matsuda (ed.), *Advances in Total Knee Arthroplasty*, https://doi.org/10.1007/978-981-97-4920-1_58

failure. In such recognition, restricted KA-TKA with a limited osteotomy angle was reported to have good clinical results [10].

In Japan, in an alignment evaluation of 797 normal knees and 1251 OA knees, the tibial medial inclination remained at 3° to 4° in normal knees, regardless of age. In comparison, the more severe the OA knee deformity, the more the medial tilt progressed, reaching approximately 10° [11]. Howell et al. is to reproduce the natural tibial joint line; however, the contralateral knee is often similarly deformed, making it difficult to determine the tibial osteotomy angle [11]. In the normal knees of younger patients (15–39 years of age), males had an average of 2.7° varus and females 1.6° varus, indicating that Japanese lower limb alignment has more constitutional varus than previously reported [11]. In addition, a 2-year postoperative outcome study of 220 patients who underwent MA-TKA confirmed that the neutral alignment group (±3°) and the mild varus alignment group (3°-6° varus) showed better clinical outcomes than the valgus alignment group (>3°) and the severe varus alignment group (>6°) [12]. This suggests that, although mild varus deformities can be tolerated, residual severe varus deformities are dangerous [12]. In addition, a 10-year postoperative radiographic study of MA-TKA for varus OA knees confirmed that the mean 0.1° varus (10% outlier over 3°) immediately after surgery progressed to a mean 1.2° varus (24% outlier over 3°), with a tendency to return to a constitutional varus knee after MA-TKA [13]. Based on these results, a modified KA-TKA in which the femur is placed in a calipered cylindrical axis and the tibia in a 3° varus has been proposed, and good clinical results have been reported [14].

3 Surgical Techniques

There are two types of KA-TKA, as described above: conventional KA-TKA using PSI and KA-TKA using generic instrumentation combined with a dedicated guide and navigation system; the latter is becoming mainstream because of its simplicity. The latter is currently classified as (1) original full KA-TKA, (2) restricted KA-TKA, (3) modified KA-TKA, and (4) functional alignment TKA. Original fully KA-TKA is a technique in which there is literally no restriction on femur/tibia osteotomy angle and lower limb alignment, restricted KA-TKA is a technique in which the femur, tibia, and lower limb alignment are restricted to ±5° (or ± 3°) in coronal plane alignment, and modified KA-TKA is a technique in which the femur is osteotomized with a calipered cylindrical axis and the tibia is osteotomized with a 3°varus, as described above. Functional alignment is vaguely defined as using a robotic-assisted system to place the alignment and soft tissue balance according to the individual knee [15] and includes KA-TKA as a part of it.

Among these four techniques, the authors performed modified KA-TKA because of its safety and simplicity. In this technique, the most important concept of placing the femoral component in the cylindrical axis is the same as in the original full KA-TKA (not applied when the valgus angle is greater than 4°). The tibial component was placed at an average of 3° varus of the Japanese knee, as described above (as noted above, the proximal tibial plane is deformed to a medial tilt of approximately 10° in severe OA knees, making it difficult to determine the osteotomy angle individually), resulting in a lower limb alignment of 0°–3° varus, which allowed us to reproduce the lower limb alignment of a young patient as described above. Although there are no specific indications for the original full KA-TKA or restricted KA-TKA, the authors believe that selecting patients suitable for KA-TKA is important for obtaining stable and good clinical results. The surgical technique for modified KA-TKA is as follows: CR-TKA of varus OA knees, excluding knees with severe varus deformity greater than 20°, significant patellofemoral arthropathy or flexion contracture greater than 15°.

1. Approach

An approximately 10–12 cm median skin incision is made for entry. The joint was exposed through a medial parapatellar approach, and the patella is shifted laterally. Medial release is minimized, and on the tibial side, an elevatorium is inserted posterior to the superficial MCL, parallel to the articular surface of the tibia; only the deep MCL avulsion and osteophytes are resected. An elevatorium is inserted into the MCL sleeve on the femoral side, and the anterior, distal, and posterior osteophytes are resected. After ACL resection, measuring the amount of PCL rollback in flexion is recommended with a rollback gauge.

2. Distal femoral osteotomies

The distal femur is osteotomized 9 mm laterally and 7–8 mm medially, considering the thickness of the cartilage wear, according to the method described by Howell et al. An osteotomy guide is placed in combination with a navigation system to ensure an accurate osteotomy. The osteotomy guide placement is adjusted using a conventional navigation system (Fig. 58.1a), or when portable navigation is used, a 1 to 2 mm shim is placed inside the osteotomy guide, and the alignment is checked (Fig. 58.1b). Clinical outcomes have been reported to be poor when the femoral component is in the valgus position [12]. If over 4° valgus angle is shown in the femoral cut, this is not included for modifed KA-TKA. Therefore, confirming the consistency between the intraoperative osteotomy and preoperative planning is important.

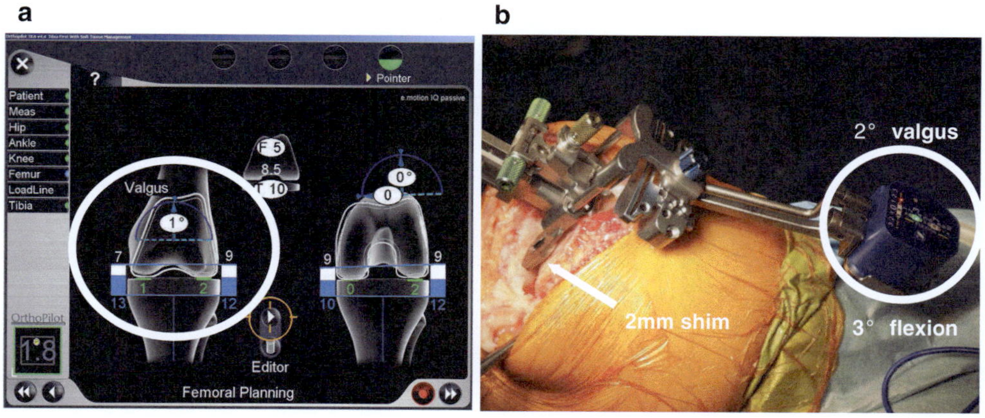

Fig. 58.1 Distal femoral osteotomy. (**a**) When using the navigation system (Orthopilot 4.2. B. Braun Aesculap), confirm that the femur is set at 7 mm medial and 9 mm lateral on the screen to obtain 0° to 3° valgus (1° valgus in this case), place the osteotomy guide, and perform osteotomy. (**b**) When using the portable navigation system (iASSIST Zimmer-Biomet), the osteotomy plane is set at 9 mm osteotomy and 3° flexion with a 2 mm shim medially attached to the guide (7 mm medial and 9 mm lateral). An osteotomy guide is placed, and an osteotomy is performed. (In this case, the femur was 2° valgus)

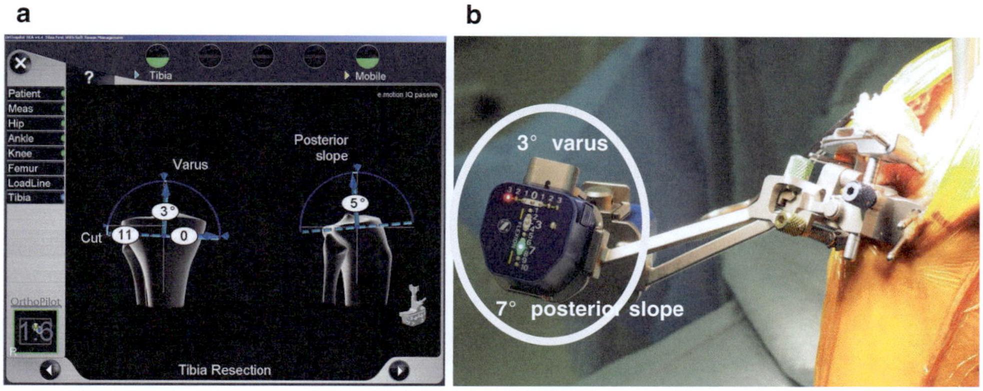

Fig. 58.2 Proximal tibial osteotomy. (**a**) When using the navigation system (Orthopilot 4.2. B. Braun Aesculap). The osteotomy guide is placed, and osteotomy is performed as follows: 3° varus, approximately 7° posterior tilt (anatomical posterior tilt, <10°), and approximately 10 mm lateral level (3° posterior tilt is built into the insert of this model). (**b**) When using portable navigation (iASSIST Zimmer-Biomet), an osteotomy guide is placed, and osteotomy is performed with 3° varus, approximately 7° posterior tilt (anatomic posterior tilt, <10°), and approximately 10 mm lateral level

3. Proximal tibial osteotomy

In all cases, the tibia is osteotomized with 3° varus, approximately 7° posterior tilt (anatomical posterior tilt <10°), and a definite lateral osteotomy of approximately 10 mm. The placement of the osteotomy guide is adjusted using a conventional navigation system (Fig. 58.2a) or a portable navigation system (Fig. 58.2b), and osteotomy is performed along the guide that is placed according to the plan after bone island preservation. Estimating the amount of medial and lateral osteotomy by preoperative simulation and measuring and confirming it intraoperatively using calipers is important. In the accuracy evaluation of 3° varus tibial osteotomy, no significant difference was found between the conventional and portable navigation systems, confirming the usefulness of portable navigation, which is easier to use [16].

4. Meniscectomy

After distal femoral and proximal tibial osteotomies, sufficient space is reserved to resect the medial and lateral menisci in one lump using a spreader in the extended position of the knee joint. After meniscectomy, the final soft tissue balance can be predicted using an offset-type tensor to measure the gap and varus/valgus balance during extension and flexion before posterior condylar osteotomy.

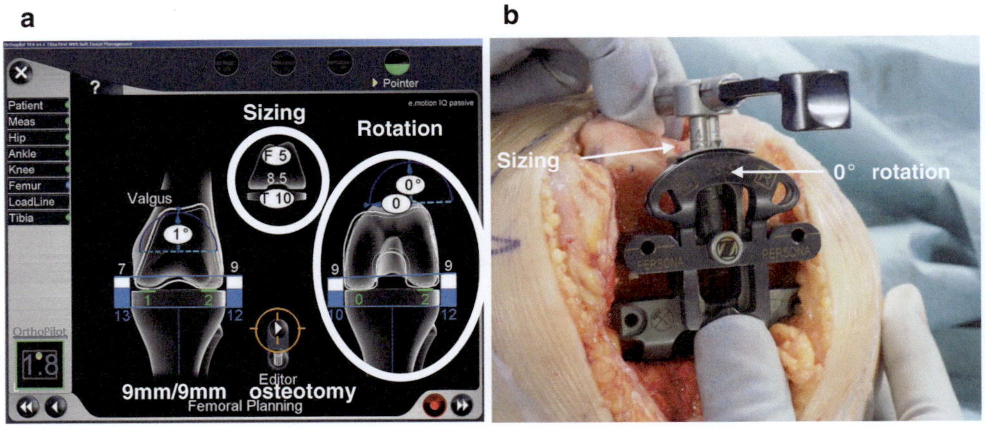

Fig. 58.3 Final osteotomy of the femur. (**a**) When using the navigation system (Orthopilot 4.2. B. Braun Aesculap), select a 9 mm medial and lateral size to the posterior condylar axis on the screen with no anterior notching and perform placement and osteotomy. (**b**) When using the portable navigation system, the osteotomy guide is placed and osteotomized using the posterior reference method, with the size selected such that there is no anterior notching 9 mm from the posterior condylar axis and 0° of internal/external rotation

5. Confirmation of extension gap

After the meniscectomy, a spacer block is inserted into the extended position to ensure that a sufficient extension gap is secured. The alignment of the lower extremities should be checked using an alignment rod or navigation system. Modified KA-TKA should result in osteotomy with 0°–3° valgus of the femur and 3° varus of the tibia, and the lower extremity alignment should be 0°–3° varus.

6. Final femoral osteotomy

Posterior osteotomy of the femur is performed 9 mm lateral and 8–9 mm medial (considering the thickness of cartilage wear) according to the method described by Howell et al. When the conventional navigation system is used, an osteotomy guide is placed at the planned component size and position (Fig. 58.3a). Portable navigation is an osteotomy support device for the distal femur and proximal tibia only and cannot be used for posterior osteotomies. An appropriate component size is selected using the posterior reference method for a posterior condylar osteotomy of 9 mm (parallel to the posterior condylar axis), and a conventional osteotomy guide is placed for the final osteotomy (Fig. 58.3b). After the osteotomy, the medial and lateral osteotomies are measured with a caliper to confirm that the osteotomy is parallel to the posterior condylar axis and that the anterior osteotomy surface is symmetrical (grand piano sign in MA-TKA).

7. Confirmation of rotational tibial placement and evaluation of soft tissue balance

The femoral trial component is inserted, the tibial trial component and insert are placed before the tibial finish, and the position of tibial rotation is confirmed and marked to avoid rotational mismatch using the ROM method. With the femoral trial component still in place, an offset-type tensor should be placed to measure the gap and varus/valgus balance and confirm the final soft tissue balance.

8. Finishing the tibia

The tibial trial component is placed and finished by selecting the appropriate size and position of the tibial rotational placement using the Akagi line and ROM method. The femoral trial component insert is placed again, and the ROM is confirmed. Patellar tracking is checked using the no-thumb method, and if lateral shift or tilt is observed, lateral retinacular release is added, and tracking is adjusted as appropriate. In a comparison of single-radius and multi-radius models in KA-TKA, it was reported that intraoperative lateral retinacular release was more frequently observed in the multi-radius model, although there was no significant difference in the radiographic parameters of the postoperative PFJ [17]. In CR-TKA, good postoperative flexion cannot be achieved if flexion is tightened. With the trial component and insert in place, the PCL's flexion tension is checked with a pull-out lift-off (POLO) test and the amount of femoral rollback was compared with the preoperative level using a rollback gauge. If necessary, pie crusting is added to the PCL, or the PCL is partially detached from the bone island attachment site.

9. Closing

After a thorough cleaning, each implant was cemented, and an appropriate insert is placed. The surgery is completed by suturing the joint capsule to each layer. At this time, ROM and patellar tracking should be reconfirmed when half of the joint capsule is sutured.

4 Clinical Results

There have already been several meta-analyses and systematic reviews on the short-term clinical outcomes of KA-TKA compared to MA-TKA, and many have shown that KA-TKA has significantly better functional scores and knee flexion angles. However, there were no significant differences in postoperative alignment and complication rates between the two groups [18–23]. A meta-analysis conducted in Japan examined five randomized clinical trials (RCTs) and found that the Western Ontario and McMaster Universities Osteoarthritis Index (WOMAC), Oxford Knee Score (OKS), combined Knee Society Score (C-KSS), Knee Function Score (KFS), and KA-TKA was significantly superior to MA-TKA [24]. However, a meta-analysis of four RCTs published by the same group, which were limited to studies conducted 2 years F/U or more postoperatively, reported that there were no significant differences between the two groups in OKS, KSS, KSF, and ROM [25]. The number of RCT reports is limited, and reliable meta-analyses, including mid- and long-term results, are awaited.

Two groups pioneered the use of KA-TKA in Japan and reported the short-term clinical results and related clinical studies. In a comparison between 45 KA-TKAs and 45 MA-TKAs, the functional activity score, especially the advanced activity score in the 2011 KSS, was significantly better in KA-TKA [26]. In a gait analysis of KA-TKA and MA-TKA, the authors reported that KA-TKA enabled a stable gait owing to a decrease in the adduction moment caused by the medial tilting of the joint surface [27]. The same group also reported that in a multivariate analysis of 100 KA-TKAs and 100 MA-TKAs, highly active males showed higher functional scores and were more suitable for KA-TKA [28].

In the comparison between 30 modified KA-TKAs and 30 MA-TKAs in the other group, in MA-TKA, the joint plane was in a laterally inclined position, and the mechanical axis considering the calcaneus, passed slightly lateral to the knee joint center. In comparison, in modified KA-TKA, the joint plane is parallel to the ground, the calcaneus-considered mechanical axis (ground mechanical axis) passed through the center of the knee joint, and the functional activity score and postoperative flexion angle in the 2011 KSS are significantly better [14]. The group also performed soft tissue balance (measured by an offset-type tensor) and kinematics analyses (measured by navigation) compared to MA-TKA in another series. They reported that the modified KA-TKA demonstrated medial stability throughout the ROM and maintenance of tibial internal rotation motion during flexion, contributing to better clinical outcomes, including acquiring a better flexion angle and patient satisfaction [29]. Furthermore, in plantar pressure measurements in modified KA-TKA, MA-TKA, and healthy knees, the hindfoot pressure during gait in modified KA-TKA was more evenly distributed medially and laterally, similar to healthy knees, than in MA-TKA [30].

5 Future Prospects

This chapter introduces the concept, surgical techniques, and clinical outcomes of KA-TKA. MA-TKA remains the gold standard; however, reports suggest that KA-TKA may improve knee joint function and patient satisfaction by reproducing the physiological knee joint, which cannot be achieved with MA-TKA. However, there are many different surgical techniques for KA-TKA, and surgeons must be familiar with each concept and technique. KA-TKA is attracting attention as an alternative to MA-TKA for improving patient satisfaction; [31] however, there are no reports on long-term results or implant survival rates, and careful follow-up is necessary.

References

1. Hungerford DS, Krackow KA. Total joint arthroplasty of the knee. Clin Orthop Relat Res. 1985;192:23–33.
2. Bellemans J, et al. The Chitranjan Ranawat award: is neutral mechanical alignment normal for all patients? The concept of constitutional varus. Orthop. 2012;470:45–53.
3. Parratte S, et al. Effect of postoperative mechanical axis alignment on the fifteen-year survival of modern, cemented total knee replacements. J Bone Joint Surg Am. 2010;92:2143–9.
4. Vanlommel L, et al. Slight undercorrection following total knee arthroplasty results in superior clinical outcomes in varus knees. Knee Surg Sports Traumatol Arthrosc. 2013;21:2325–30.
5. Howell SM, et al. Does a kinematically aligned total knee arthroplasty restore function without failure regardless of alignment category? Orthop Relat Res. 2013;471:1000–7.
6. Eckhoff DG, et al. Three-dimensional mechanics, kinematics, and morphology of the knee viewed in virtual reality. J Bone Joint Surg Am. 2005;87(Suppl 2):71–80.
7. Nedopil AJ, et al. Does calipered kinematically aligned TKA restore native left to right symmetry of the lower limb and improve function? J Arthroplast. 2018;33:398–406.
8. Howell SM, et al. Implant survival and function ten years after kinematically aligned total knee arthroplasty. J Arthroplast. 2018;33:3678–84.
9. Howell SM, et al. Accurate alignment and high function after kinematically aligned TKA performed with generic instruments. Knee Surg Sports Traumatol Arthrosc. 2013;21:2271–80.
10. Almaawi AM, et al. The impact of mechanical and restricted kinematic alignment on knee anatomy in total knee arthroplasty. J Arthroplast. 2017;32:2133–40.
11. Matsumoto T, et al. A radiographic analysis of the alignment of the lower extremities–initiation and progression of varus-type knee osteoarthritis. Osteoarthr Cartil. 2015;23:217–23.
12. Nishida K, et al. Remaining mild varus limb alignment leads to better clinical outcome in total knee arthroplasty for varus osteoarthritis. Knee Surg Sports Traumatol Arthrosc. 2017;25:3488–94.
13. Kuroda Y, et al. Varus deformity in the proximal tibia and immediate postoperative varus alignment result in varus progression in

limb alignment in the long term after total knee arthroplasty. Knee Surg Sports Traumatol Arthrosc. 2017;25:3488–94.
14. Matsumoto T, et al. Radiological and clinical comparison of modified kinematically versus mechanically aligned total knee arthroplasty. Bone Joint J. 2017;99-B:640–6.
15. Oussedik S, et al. Alignment in total knee arthroplasty. Bone Joint J. 2020;102:276–9.
16. Tsubosaka M, et al. Accelerometer-based portable navigation system is useful for tibial bone cutting in modified kinematically aligned total knee. J Knee Surg. 2021;34:870–6.
17. Matsumoto T, et al. Comparison between single- and multiradius prostheses used in modified kinematically aligned cruciate-retaining total knee. J Knee Surg. 2020; https://doi.org/10.1055/s--0040-1721413. Online ahead of print
18. Courtney PM, et al. Early outcomes of kinematic alignment in primary total knee arthroplasty: a meta-analysis of the literature. J Arthroplast. 2017;32:2028–32.
19. Li Y, et al. Does kinematic alignment improve short-term functional outcomes after total knee arthroplasty compared with mechanical alignment? A systematic review and meta-analysis. J Knee Surg. 2018;31:78–86.
20. Yoon JR, et al. Comparison of kinematic and mechanical alignment techniques in primary total knee arthroplasty: a meta-analysis. Medicine. 2017;96:e8157.
21. Xu J, et al. Kinematic versus mechanical alignment for primary total knee replacement: a systematic review and meta-analysis. J Orthop. 2019;16:151–7.
22. Luo Z, et al. Similar results with kinematic and mechanical alignment applied in total knee arthroplasty. Knee Surg sports Traumatol Arthrosc 2020. Knee Surg Sports Traumatol Arthrosc. 2020;28:1720–35.
23. Gao ZX, et al. Comparison of kinematic alignment and mechanical alignment in total knee arthroplasty: a meta-analysis of randomized controlled. Orthop Surg. 2020;12:1567–78.
24. Takahashi T, et al. Kinematically aligned total knee arthroplasty or mechanically aligned total knee arthroplasty. J Knee Surg. 2018;31:999–1006.
25. Hiyama S, et al. Kinematically aligned total knee arthroplasty did not show superior patient-reported outcome measures: an updated meta-analysis of. J Knee Surg. 2022;35:634–9.
26. Niki Y, et al. Joint line modification in kinematically aligned total knee arthroplasty improves functional activity but not patient satisfaction. J Arthroplast. 2018;33:2125–30.
27. Niki Y, et al. Kinematically aligned total knee arthroplasty reduces knee adduction moment more than mechanically aligned total knee arthroplasty. Knee Surg Sports Traumatol Arthrosc. 2017;26:1629–35.
28. Niki Y, et al. Who will benefit from kinematically aligned total knee arthroplasty? Perspectives on patient-reported outcome measures. J Arthroplast. 2020;35:438–42.
29. Matsumoto T, et al. Intraoperative soft tissue balance/kinematics and clinical evaluation of modified kinematically versus mechanically aligned. J Knee Surg. 2020;33:777–84.
30. Kamenaga T, et al. Comparison of plantar pressure distribution during walking and lower limb alignment between modified kinematically. J Biomech. 2021;120:110379.
31. Matsumoto T, et al. Kinematically aligned total knee arthroplasty: an alternative standardized technique? Ann Transl Med. 2017;5(Suppl 1):S10.

59 Patellar Resurfacing

Shu Kobayashi

Consensus is lacking about whether to resurface the patella during total knee arthroplasty (TKA). It is necessary to understand the advantages and disadvantages of each option and carefully consider the implant choices before choosing a product. Here we compare the clinical results, procedural knacks, pitfalls of patellar resurfacing, and recent findings.

1 Current Status of Patellar Resurfacing in Various Countries

The use of patellar resurfacing in TKA varies among countries. The percentage of patellar resurfacing in TKA in 2004–2014 was reportedly >80% in the United States, 70–80% in Denmark, 40–60% in Australia, and < 10% in Norway and Sweden [1] (Fig. 59.1). According to a joint reg-

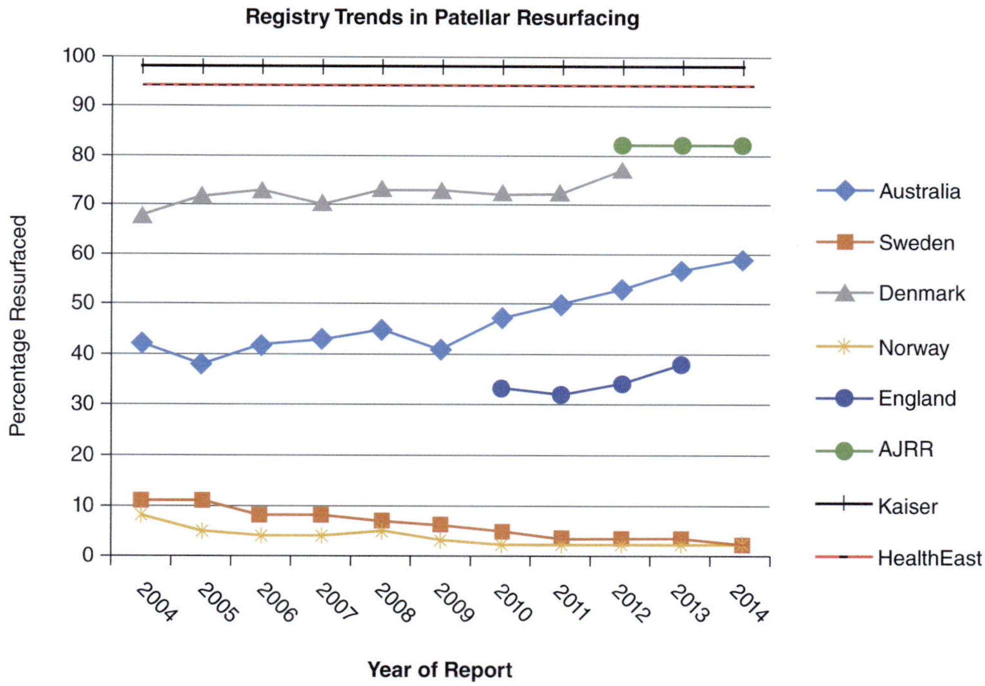

Fig. 59.1 Percentage of patellar resurfacing in total knee arthroplasty by country. (Cited from Ref. [1])

S. Kobayashi (✉)
Department of Orthopedic Surgery, Keio University School of Medicine, Shinjuku-ku, Tokyo, Japan
e-mail: kobashu@keio.jp

S. Matsuda (ed.), *Advances in Total Knee Arthroplasty*, https://doi.org/10.1007/978-981-97-4920-1_59

istry survey conducted by the Japanese Society for Replacement Arthroplasty, the percentage of patellar resurfacing in Japan was approximately half each. Thus, consensus is currently lacking regarding patellar resurfacing or nonresurfacing.

2 Respective Advantages of Patellar Resurfacing and Nonresurfacing

The advantages of patellar nonresurfacing include a shorter operative time, cost savings, and no risk of patellar fracture or polyethylene wear of the patellar component. Moreover, recent improvements in implant geometry have made the design of the femoral groove and patellar component more anatomical; therefore, patellofemoral (PF) joint complaints are becoming less common, even with patellar nonresurfacing. However, the advantage of patellar resurfacing is the lack of concern about future wear of the patellar cartilage. Fluoroscopic studies demonstrated that patellar replacement has little effect on tibiofemoral joint kinematics [2].

3 Types of Patellar Resurfacing

There are two patellar resurfacing methods: the onlay method, in which the entire patellar articular surface is osteotomized and the component is placed on the osteotomy surface, and the inlay method, in which reaming is performed only at the placement site and the surrounding articular cartilage is preserved. The advantages of the inlay method include ease of replacement, reduced bone removal, increased resistance to shear forces, and increased fixation force [3, 4]. The advantages of the onlay technique include its technical simplicity and low revision rate [5]. Using the onlay technique increases the probability of postoperative patellar crepitus in posterior-stabilized (PS) type TKA [6].

The patellar component designs include domed, anatomical (asymmetrical), and cylindrical. The domed design provides excellent PF conformity, but the contact pressure tends to increase with increasing flexion angle [7]. The anatomical design provides the most appropriate contact pressure and reduces patellar tilt [8], although misalignment can cause instability, increased wear, increased shear forces, and anterior knee pain [9]. The cylindrical design reduces the contact pressure but increases the risks of fracture and binding at the peg [7].

Although sometimes used in the 1980s, the metal-back type is no longer used because of polyethylene insert breakage, patellar dislocation, component fracture, and metallosis [10].

There have also been several reports on the number of pegs used for patellar component fixation. In a study comparing fixation with a single large peg in the center to fixation with multiple smaller pegs around the center, the fracture risk was higher with the former, whereas the risk of laxity was higher with the latter [11]. Another study reported that the fracture rate was lower with three 6.4-mm-diameter pegs than with one 10-mm-diameter peg [12].

4 Cases in Which Patellar Resurfacing Is Recommended

Patellar resurfacing is desirable in cases of inflammatory diseases such as rheumatoid arthritis (RA) or severe deformity of the PF joint and is reportedly associated with less pain during standing and stair climbing and less PF joint tenderness in a 6-year or longer follow-up of TKA for RA [13]. Patellar resurfacing should also be considered depending on the implant model used and the femoral groove shape. In PS-type TKA, a box is created on the femoral side that may adversely affect the PF joint, and patellar resurfacing is generally preferred. Comfort et al. reported that patellar nonresurfacing increases the risk of revision by 115% at 1 year and 50% at 8 years postoperative compared with patellar resurfacing when the PS type is selected [14]. They also reported that patellar resurfacing reduced the frequency of postoperative patellar crepitus in PS-type TKA [15]. A study using PF joint pressure sensors for different implant models reported that PF pressure increased significantly in cruciate retaining and cruciate stabilized types, and the distance of PF offset decreased with flexion in normal knees; however, the distance remained the same for all ranges of motion (ROMs) in knees after TKA [16]. The location of stress concentration in the PF joint after TKA without patellar resurfacing also reportedly differs among implant models [17]; moreover, since the PF contact pressure changes depending on the shape of the patellar component and flexion angle [18], it is necessary to pay attention to the implant model to be used. From the patient's perspective, it is desirable to resurface the patella with a steep facet angle of ≤126° because the patella is translucent on postoperative radiography and causes postoperative anterior knee pain [19, 20]. In addition, patellar resurfacing is desirable because the patella may impinge on the tibial component during postoperative deep flexion [21].

5 Results of Resurfacing Versus Nonresurfacing

Several meta-analyses recently reported results of patellar resurfacing versus nonresurfacing. In a meta-analysis of 7075 patients, Pavlou et al. reported no significant intergroup difference in the frequency of complaints of anterior knee pain among patients with versus without patellar resurfacing and that the frequency of revision surgery was higher in the nonresurfacing group [22]. In a meta-analysis of 3465 patients, Pilling et al. also reported no significant intergroup differences in anterior knee pain or postoperative satisfaction with a higher revision surgery rate in the nonresufacing group [23]. In a recent meta-analysis, Kai et al. reported no differences in ROM, Oxford score, Knee injury and Osteoarthritis Outcome Score, visual analog scale score, Feller score, patellar tilt, and patient satisfaction in 6887 patients; however, the resurfacing group performed better in terms of frequency of revision, Knee Society Score, function score, and PF crepitus [24]. Moreover, the difference in the frequency of revision surgery was due to bias in that patients in the nonresurfacing group were more likely to choose revision surgery because they had the option of patellar resurfacing. Another study reported that no cases of TKA without patellar resurfacing using the Miller-Galante model required revision after 12 years of follow-up [25], and consensus is lacking.

6 Knacks and Pitfalls of Patellar Resurfacing

A thicker patella after versus before patellar resurfacing increases the contact pressure on the PF joint, causing component wear and the risk of a reduced ROM. A study of pressure sensors implanted in the PF joint reported a 20% increase or decrease in PF pressure when the patellar thickness increased or decreased by 2 mm, respectively [26]. Therefore, the author's goal was to remove the same amount as the thickness of the component plus 0.5–1 mm. However, an excessive osteotomy increases the risk of patellar fracture; therefore, we try to leave a patellar thickness of 11–12 mm after the osteotomy. Inlay-type patellar replacements recently became available that enable the replacement of even a thin patella. There are two osteotomy methods, freehand and guided. The knack of freehand osteotomy involves fixing the patella with a sharp cloth or Kocher forceps around the patella (Fig. 59.2), touching the anterior surface of the patella with a finger, and cutting it parallel to it (Fig. 59.3). After a patellar osteotomy, the component is placed on the medial side (Fig. 59.4), and the lateral excess bone is resected with a bone saw or similar tool (Fig. 59.5). Reportedly, dropping the outer excess bone improves the fit of the PF and reduces postoperative anterior pain [27].

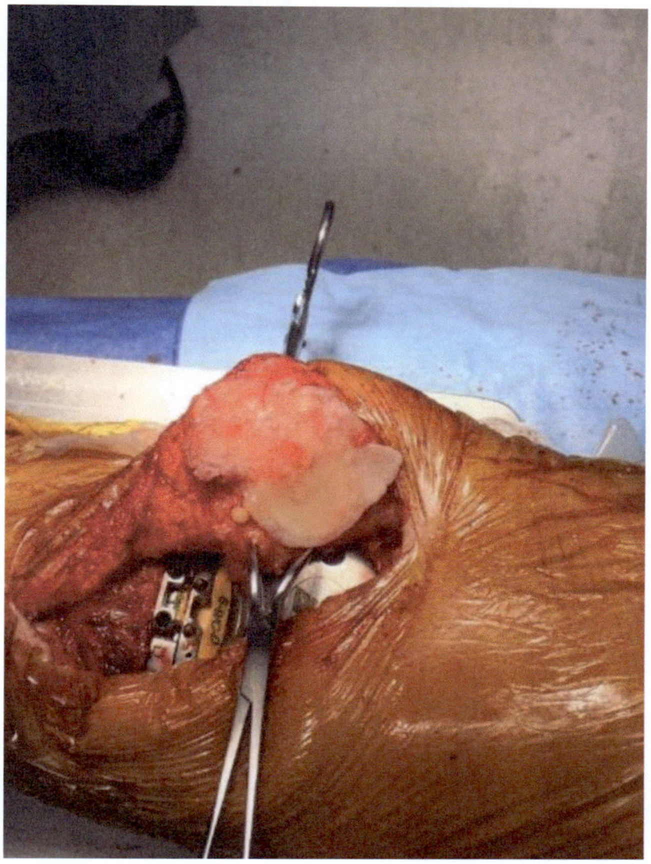

Fig. 59.2 Fixation of the patella during patellar osteotomy

After resurfacing the patellar component trial, tracking of the PF joint must be confirmed, and an additional lateral release should be considered if there is a tendency toward lateral subluxation. The osteotomy angle during patellar replacement may affect postoperative PF joint alignment, and an excessive medial patellar cut can reportedly lead to increased patellar tilt and require lateral release [28]. A study using a PF pressure sensor reported that the PF pressure at 60° and 140° of flexion correlated with patient satisfaction, Forgotten Joint Score-12, and Patella Score after TKA [29]. Further research and clarification, including ROM and pressure measurements, are warranted.

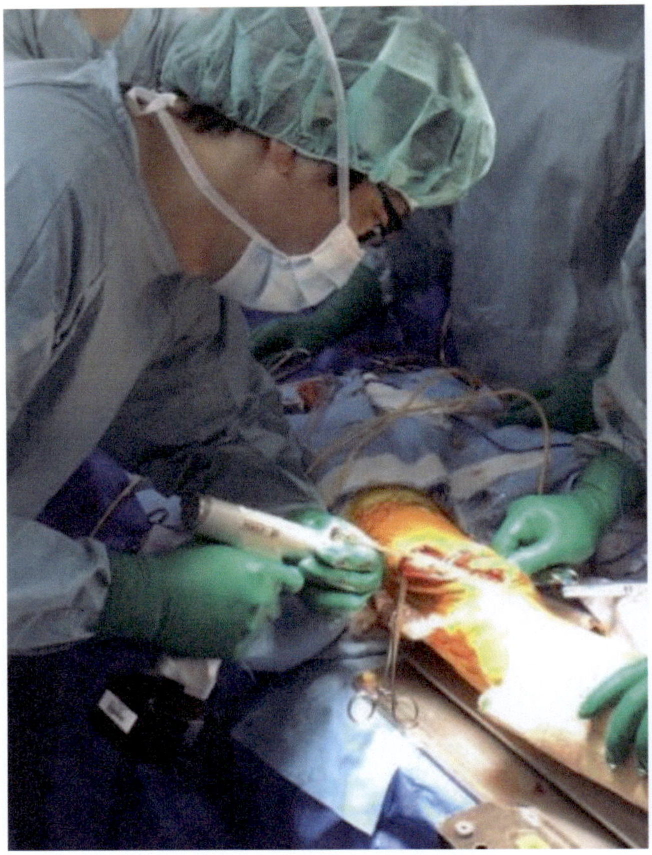

Fig. 59.3 Patellar osteotomy

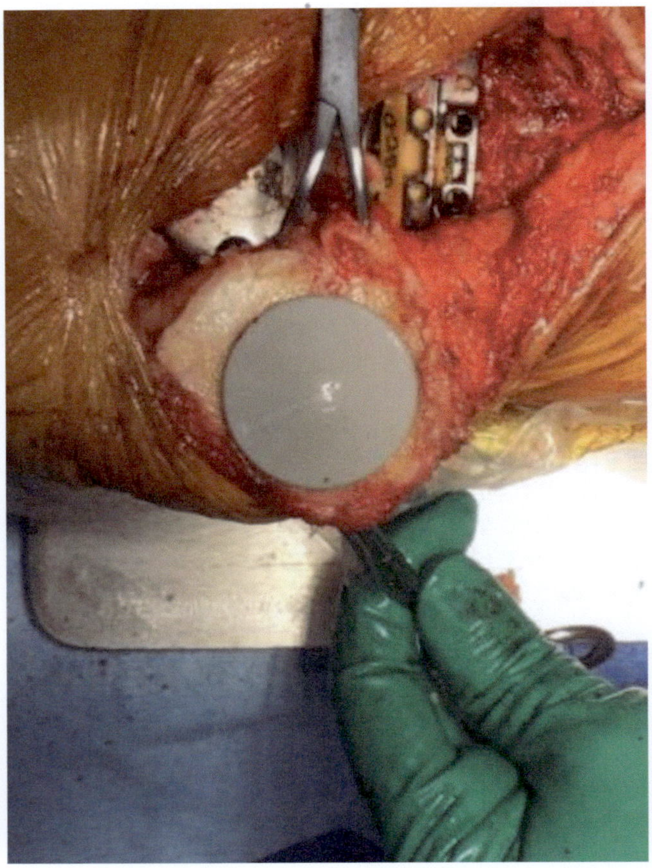

Fig. 59.4 Implantation of the patellar component. The lower and upper sides of the figure show the medial and lateral sides, respectively

Fig. 59.5 Resection of the lateral excess bone (**a**) Osteotomy of lateral excess bone after patellar component placement. (**b**) Preoperative axial radiograph (red area: excess bone resection site). (**c**) Postoperative axial X-ray

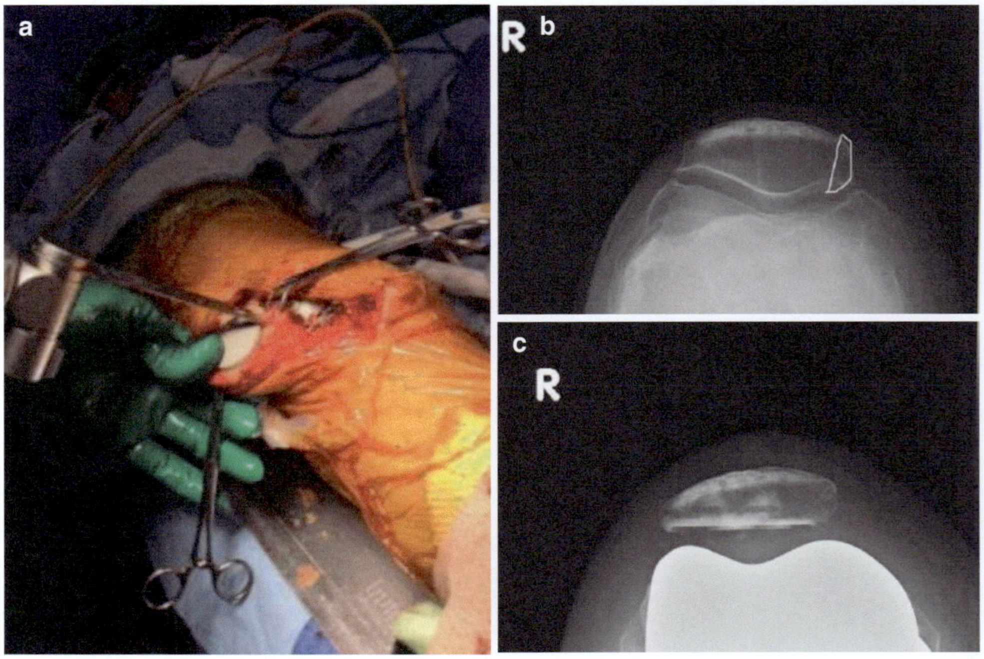

References

1. Fraser JF, et al. International rates of patellar resurfacing in primary total knee arthroplasty, 2004-2014. J Arthroplast. 2017;32:83–6.
2. Kono K, et al. Patellar resurfacing has minimal impact on in vitro tibiofemoral kinematics during deep knee flexion in total knee arthroplasty. Knee. 2021;30:163–9.
3. Lachiewicz PF. Implant design and techniques for patellar resurfacing in total knee arthroplasty. Instr Course Lect. 2004;53:187–91.
4. Rosenstein AD, et al. Fixation strength comparison of onlay and inset patellar implants. Knee. 2007;14:194–7.
5. Coory JA, et al. The outcome of total knee arthroplasty with and without patellar resurfacing up to 17 years: a report from the Australian orthopaedic. J Arthroplast. 2020;35:132–8.
6. Thiengwittayaporn S, et al. Onlay patellar resurfacing in a posterior-stabilized total knee arthroplasty increases patellar crepitus complication: a randomized, controlled trial. J Arthroplast. 2021;36:3443–50.
7. Schindler OS. The controversy of patellar resurfacing in total knee arthroplasty: ibisne in medio tutissimus? Knee Surg Sports Traumatol Arthrosc. 2012;20:1227–44.
8. Mochizuki T, et al. Effect on patellar kinematics of different patellar component designs in total knee arthroplasty: intraoperative measurement of dome type versus anatomic type. Eur J Orthop Surg Traumatol. 2020;30:419–24.
9. Schindier OS. Basic kinematics and biomechanics of the patellofemoral joint. Part 2: the patella in total knee arthroplasty. Acta Orthop Belg. 2012;78:11.
10. Lombardi AV Jr, et al. Fracture/dissociation of polyethylene in metal-backed patellar components in total knee arthroplasty. J Bone Joint Surg Am. 1988;70:675–9.
11. Brick GW, Scott RD. The patellofemoral component of total knee arthroplasty. Clin Orthop. 1988;231:163–78.
12. Larson CM, et al. One-peg versus three-peg patella component fixation in total knee arthroplasty. Clin Orthop Relat Res. 2001;392:94–100.
13. Kajino A, et al. Comparison of the results of bilateral total knee arthroplasty with and without patellar replacement for rheumatoid arthritis. Joint Surg Am. 1997;79:570–4.
14. Comfort T, et al. International comparative evaluation of fixed-bearing non-posterior-stabilized and posterior-stabilized total knee replacements. J Bone Joint Surg Am. 2014;96(Suppl 1):65–72.
15. Ogawa H, et al. Effect of patellar resurfacing on patellofemoral crepitus in posterior-stabilized total knee arthroplasty. J Arthroplast. 2016;31:1792–6.
16. Tanikawa H, et al. Influence of total knee arthroplasty on patellar kinematics and patellofemoral pressure. J Arthroplast. 2017;32:280–5.
17. Takahashi A, et al. Patellar morphology and femoral component geometry influence patellofemoral contact stress in total knee arthroplasty without patellar resurfacing. Knee Surg Sports Traumatol Arthrosc. 2012;20:1787–95.
18. Matsuda S, et al. Patellofemoral joint after total knee arthroplasty: effect on contact area and contact stress. J Arthroplast. 1997;12:790–7.
19. Takahashi A, et al. Radiolucent zone of the patella following total knee arthroplasty without patellar resurfacing. J Orthop Sci. 2014;19:558–63.
20. Inoue A, et al. Differences in patellofemoral alignment as a result of patellar shape in cruciate-retaining total knee arthroplasty without patellar resurfacing at a minimum three-year follow-up. Knee. 2017;24:1448–53.
21. Maeno S, et al. Patellar impingement against the tibial component after Total knee arthroplasty. Clin Orthop Relat Res. 2006;452:265–9.
22. Pavlou G, et al. Patellar resurfacing in total knee arthroplasty: does design matter? A meta-analysis of 7075 cases. J Bone Joint Surg Am. 2011;93:1301–9.
23. Pilling RW, et al. Patellar resurfacing in primary total knee replacement: a meta-analysis. J Bone Joint Surg Am. 2012;94:2270–8.
24. Kai C, et al. Patellar resurfacing versus nonresurfacing in total knee arthroplasty: an updated meta-analysis of randomized controlled trials. J Orthop Surg Res. 2021;16:83.
25. Hasegawa M, et al. Long-term clinical results and radiographic changes in the nonresurfaced patella after total knee arthroplasty. Scandinavica. 2002;73:539–45.
26. Tanikawa H, et al. Influence of patella thickness on patellofemoral pressure in total knee arthroplasty. BMC Musculoskelet Disord. 2021;22:298.
27. Nagashima M, et al. Lateral patellar facetectomy reduces the incidence of anterior knee pain in patellar resurfaced total knee arthroplasty. Surg. 2022;36:773–8.
28. Fukagawa S, et al. Changes in patellar alignment after total knee arthroplasty. Knee Surg Sports Traumatol Arthrosc. 2011;19:99–104.
29. Kaneko T, et al. The influence of compressive forces across the patellofemoral joint on patient-reported outcomes after bi-cruciate stabilized total knee arthroplasty. Bone Joint J. 2018;100-B:1585–15.

Part X

Computer Assisted Surgery

60 Navigation/Portable Navigation

Kazunari Ishida

1 History of Computer-Assisted Surgery (CAS)

Although total knee arthroplasty (TKA) is a successful surgical technique with stable postoperative outcomes, it can lead to loosening in cases of poor alignment [1]. Therefore, there is a need for more precise surgery to improve outcomes [1]. Computer-assisted surgery (CAS) for knee joint replacement has been developed to address this need, and studies verifying its accuracy have been published since the early 2000s [2]. In Japan, the use of navigation in TKA began in 2002, and its validation has been reported since 2004 [3, 4].

Initially, two types of navigation were used in knee arthroplasty: imageless navigation and image-based (CT-based) navigation. Currently, the available navigation systems in Japan are Exactech GPS (Exactech), OrthoPilot Knee Suite (B Braun Aesculap), Knee3 (BrainLab), and OrthoMap Express/Precision Knee (Stryker). These systems do not require preoperative imaging (Fig. 60.1).

Accurate alignment and implant placement are expected to improve the long-term outcomes of TKA [5]. However, the disadvantages of navigation systems include longer operation time and increased medical costs [6]. To address these issues, portable navigation systems have emerged, offering accurate osteotomy angles in the coronal and sagittal planes easily and inexpensively without the need for large main-body equipment. An accelerometer-based imageless navigation system has been used in Japan since 2014, and the currently available systems are Knee Align2 and iASSIST from Zimmer Biomet.

The use of navigation and robotics in TKA is increasing worldwide, with the percentage rising from 1.2% in 2005 to 6.3% in 2014 in the United States [6]. In Australia, navigation was utilized in 31.3% of all first-time TKA procedures performed in 2020, while the proportion of procedures performed without the use of technology, including robotics, decreased to 40% [7]. The future adoption of CAS, including navigation, in knee arthroplasty will depend on its cost-effectiveness in patient-oriented clinical outcomes and its ability to accurately reproduce ideal alignment and soft tissue balance for each patient, thus improving long-term durability.

K. Ishida (✉)
Department of Orthopaedic Surgery, Kobe Kaisei Hospital, Nada, Kobe, Japan

S. Matsuda (ed.), *Advances in Total Knee Arthroplasty*, https://doi.org/10.1007/978-981-97-4920-1_60

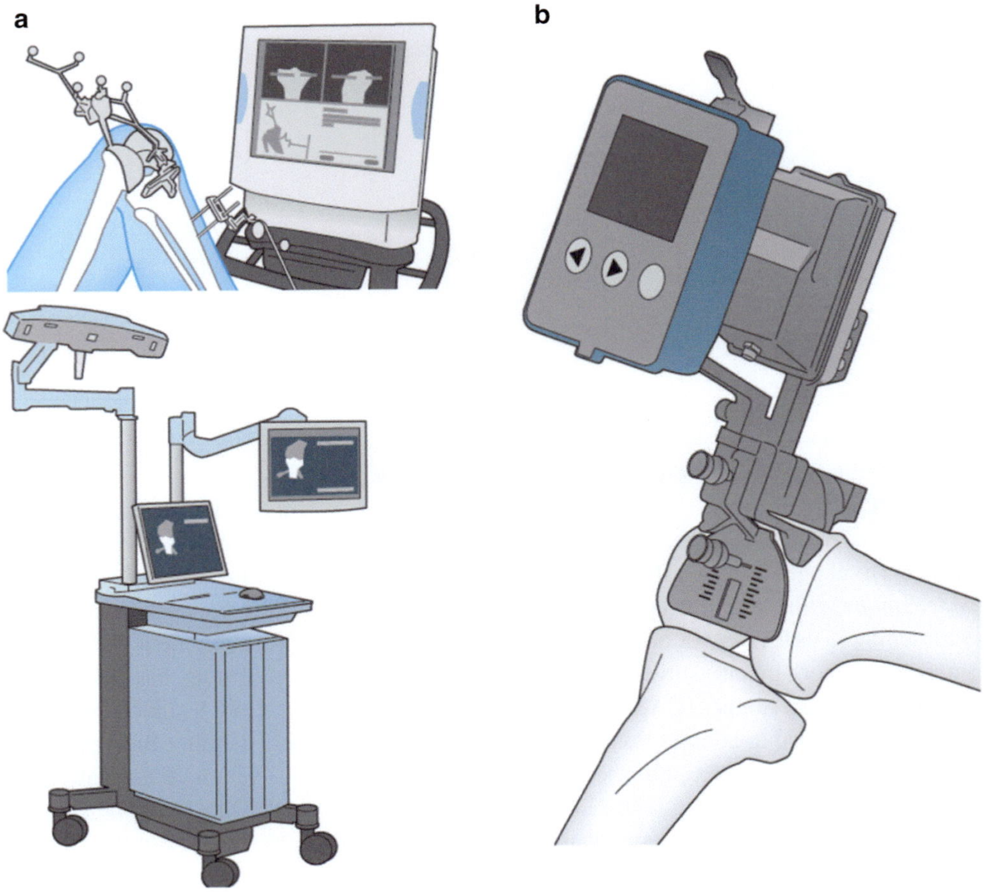

Fig. 60.1 Navigation (**a**) and Portable navigation (**b**). Data such as osteotomy angles are displayed on a monitor outside the surgical field for navigation and on a monitor inside the surgical field for portable navigation

2 Image-Less Navigation

Image-less navigation uses an algorithm to determine the center of joint motion and lower limb alignment based on anatomical landmark registration and kinematic data acquisition. The system is also superior in not requiring image cost or preoperative planning.

The intraoperative position inputs were transmitted to the navigation system using trackers attached to the femur and tibia. The proximal femoral center is usually calculated and identified by moving the hip joint conically, and the ankle joint center is determined by each navigation algorithm, such as by calculating the ratio between the two points between the medial and lateral malleoli or calculating the kinematic center by plantar dorsiflexion motion of the ankle joint.

The intraarticular landmarks of the distal femur and proximal tibia were determined by direct pointing. In the femur, the distal center of the mechanical axis, medial and lateral epicondyles, most distal points of the medial and lateral condyles, and most posterior portions of the medial and lateral posterior condyles are usually observed. The anterior cortex of the femur memorizes the anterior shape. This step is important for determining the size of the femur and evaluating it without causing notch formation. The Whiteside line [8] was recorded to determine the axis of rotation. On the tibial side, the proximal center of the mechanical axis, articular surfaces of the medial and lateral condyles (usually around the posterior 1/3), and the axis of rotation were recorded. If cartilage remains when pointing to the articular surface, care should be taken not to puncture it.

After registration, the bone-cutting guide was fixed based on the position information displayed on the monitor, and the surgeon performed bone resection. Navigation can validate the angle and bone resection thickness after bone resection and during cementing [9, 10], which is a function that is not available with portable navigation.

In addition to obtaining accurate bone sections, navigation provides intraoperative information useful for surgical maneuvers. Depending on the type of navigation, the virtual component gap can be evaluated after preoperative planning, verification, and intraoperative kinematic evaluation, such as the internal/external rotation and internal/external rotation angle (Fig. 60.2). It is important to accurately understand the measurement principles of the presented numerical values and incorporate data that are considered useful in surgical operations in a moderate manner. The intraoperative data changed with each manual surgical step. Attempting to make everything as close to perfect as possible by adhering to the displayed data is unlikely to produce good results, even though it increases the operation time.

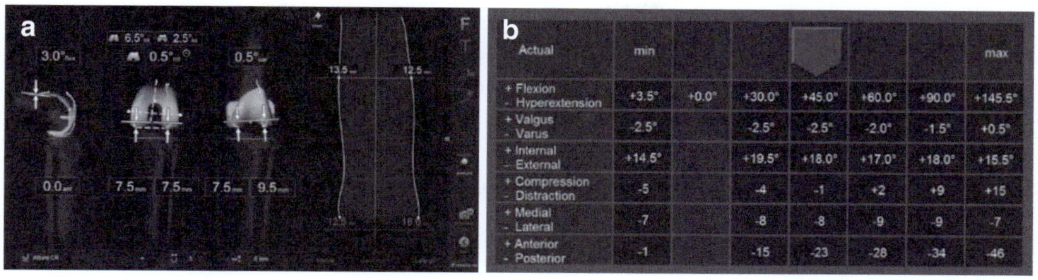

Fig. 60.2 Example of intraoperative data displayed in image-less navigation. (**a**) surgical planning and component gap in extension and flexion (Knee 3, Brain Lab). (**b**) intraoperative kinematics (OrthoMap Precision Knee, Stryker)

Fractures at the pin insertion site have been reported as a complication. It is essential to insert a pin into the center of the bone [11].

2.1 Navigation Accuracy

(i) Accuracy in the coronal plane

It has been reported that navigation provides a better implant placement position than manual group [12–15]. In comparing CT-based navigation TKA and jig-based manual TKA, significantly better coronal, sagittal, and transverse alignment and placement accuracies were obtained with CT-based navigation TKA [13, 14]. In a comparison of post-operative clinical results between CT-free navigation TKA and manual TKA at a minimum of 5 years postoperatively, outliers of 3°in lower limb alignment were significantly fewer in the navigation group (18.5%) than in the manual group (33.3%). The femoral external rotation was significantly higher in the navigation group [15]. It was also reported that the degree of preoperative varus deformity did not affect navigation accuracy of navigation [12]. Although there is a difference in the algorithm used to define the angle of accuracy in the coronal plane, a comparison of the three models, including image-free and CT-based models [16], suggests no significant difference in placement accuracy among the three models. In contrast, in minimally invasive (MIS) TKA, the navigation group showed significantly fewer outliers in the overall lower extremity alignment; however, there were no significant differences in implant positioning in the femur and tibia, and there were no significant differences in clinical outcomes at six months after surgery [17].

To evaluate implant placement accuracy, 3D CT seems to be preferable to 2D CT using X-rays [18, 19]. In the case of TKA using navigation, 44.4% of those judged to be outliers of >2 °on the femoral side and 100% of those on the tibial side were not outliers, suggesting that outliers may not be as common as those reported by 2D evaluation [18].

(ii) Accuracy in the sagittal plane

It has been noted that the accuracy of navigating TKA in the sagittal plane is inferior to that in the coronal plane. It has been noted that the target angles of manual TKA and various types of navigation TKA are different, even if the same values are obtained, because of the different distal reference points [20–22]. It has been reported that the sagittal plane angle and selected implant size may differ depending on the navigation technique used [22] and that the femoral-tibial implant may be placed in a more extended position [20, 21].

A cautionary note regarding the use of navigation systems to achieve osteotomy perpendicular to the mechanical axis has been reported in the sagittal plane of the femur. It has been reported that there is a risk of anterior notch formation if an osteotomy perpendicular to the mechanical axis is attempted because the line of the femoral anterior cortex is in extension 40–85% of men and 65–100% of older women in relation to the mechanical axis [23]. For the same reason, it has also been reported that if osteotomy is performed perpendicular to the mechanical axis, the anterior-posterior width of the femur must be increased compared to that of manual TKA to avoid anterior notch formation, which may result in oversizing of the femur [4].

(iii) Accuracy in the transverse plane

In imageless navigation, the accuracy in the transverse plane or rotational axis is inferior to that in the coronal or sagittal plane [24]. In imageless navigation, each axis is defined by pointing to anatomical landmarks. However, the condyles required to determine the rotational axis are located in a position that is difficult to point to posteriorly, and the distance between the two points is short; therefore, small errors are reflected in the accuracy of the navigation. A change of 2 mm was considered to cause an error of 1° [24]. In some navigation systems, the rotational axis is defined as the average of two reference lines (the transepicondylar axis and the line perpendicular to the whiteside line) to minimize variation. Inui et al. reported a method for improving the accuracy of the rotational axis by adding another reference line evaluated from preoperative radiographic images (the condylar twisting axis) [25]. Some studies have also reported that the combination of image-less and CT-based navigation can improve the accuracy of the rotational placement of tibial implants [26].

Good accuracy can be achieved using image-based navigation by controlling the angle under navigation until the implant is cemented [27]. Even if the implant is placed under image-based navigation, there is a risk that the rotational position may change during final cementation. The accuracy of implant placement is influenced by several factors.

2.2 Clinical Results of Navigational TKA

A report from Japan showed no significant difference in clinical outcomes between navigation TKA and manual TKA at an average of 7.7 years postoperatively [28], and a report from the same institution showed that navigation TKA was significantly better than manual TKA at a minimum of 5 years postoperatively in terms of ROM and KSS [15], there was no significant difference in KSS 2011 at a mean of 10 years postoperatively [29]. A recent meta-analysis [30] found that it can reduce the outliers of hip-knee-ankle (HKA) angles of 3° or greater by approximately 10%. However, no other effects were found, and we do not recommend the use of navigation in TKA, considering its impact on operative time and medical costs. The American Academy of Orthopaedic Surgeons (AAOS) 2015 guidelines also recommend against the use of intraoperative navigation as strong evidence [31]. The guidelines state that there is no difference in clinical outcomes in terms of patient quality of life, knee function, and pain at 90 days postoperatively compared to the non-use group, that there is no significant difference in the risk of infection, and that non-use is recommended in terms of operative time. In addition, the authors stated that although there is a theoretical advantage of navigation for knees after malunion of lower limb fractures and after insertion of intramedullary nails, the current evidence is inconclusive. In cases of strong femoral bowing, outliers in the coronal plane are common; however, navigation is more appropriately placed [32]. However, the answer to whether using navigation has a merit to long-term survivorship is not yet available. A subgroup analysis of Australian registry data limited to patients aged 65 years showed a lower rate of non-infectious revision in the navigation TKA group than in the manual TKA group (6.2% vs. 6.8%) [7] at a follow-up >9 years after surgery. Revisions due to malalignment are known to increase by 10 years postoperatively [33, 34]. Although the optimal implant position has not yet been determined, navigated TKA may be superior in terms of long-term survivorship owing to accurate implant placement [31], and further reports are awaited [5].

2.3 Other Navigation Functions

Navigation has the advantage of real-time intraoperative evaluation of parameters (flexion-extension, varus/valgus, and internal/external rotation angles). This intraoperative kinematic evaluation is reproducible regardless of the evaluator [35], and this function allows evaluation of the influence of surgical procedures on the parameters [36–38] and the relationship between the parameters and postoperative function and results [39–42]. In Japan, this method is widely used for evaluation and research.

3 Portable Navigation

3.1 Portable Navigation Accuracy

In TKA, it has been reported that the use of portable navigation improves the placement angle [43–49]. In a multicenter prospective study, accurate implant placement and lower extremity alignment were achieved with accelerometer-based portable navigation without increasing the operative time or complications compared to the conventional technique [46]. In a prospective randomized trial, portable navigation significantly reduced the number of outliers and errors from the target placement angle compared to the conventional technique and significantly reduced postoperative blood loss [47]. However, some reports concluded that portable navigation had no clear advantage because it could reduce outliers in the lower extremity functional axis and coronal femoral plane but not on the tibial side [48].

The usefulness of portable navigation for kinematic alignment (KA) TKA has been previously reported. In KA TKA with a fixed tibial osteotomy of 3° varus and 7° posterior slope, there was no significant difference in osteotomy accuracy, flexion angle at 1 year postoperatively or KSS 2011 between the use of navigation and portable navigation [49]. In KA-TKA, which requires accurate implant placement, CAS technology should be used to ensure postoperative safety. The use of portable navigation may be a solution to ensure that KA-TKA is widely and safely performed.

The usefulness of portable navigation for UKA has also been reported. The postoperative implant position was evaluated using X-rays and compared with the intraoperative target osteotomy angle of the tibia; when an outlier was defined as 3°, there were no outliers in the coronal or sagittal plane with portable navigation. In the absolute value evaluation, the portable navigation showed significantly less error than the manual navigation compared to the target osteotomy angle than manual navigation [50].

The following points should be noted during the registration of portable navigation systems. During hip movement to determine the proximal femoral center, fluoroscopic confirmation of the movement of the femoral head revealed that the femoral head's greatest movement was seen during the femur's adduction. Therefore, avoiding excessive adduction of the femur can improve the accuracy of femoral placement [51].

3.2 Clinical Results of Portable Navigation System

There are few reports on the clinical outcomes of TKA performed using portable navigation. One of the few studies from Japan reported good results at an average follow-up of 14 months after surgery on the JOA, JKOM, and KSS [52]. Further studies on the clinical usefulness of portable navigation systems are warranted.

4 Future of Navigation

Navigation using augmented reality technology has been developed to enable accurate bone resection [53, 54]. The CAS is expected to bifurcate in the future. One is the trend toward simplification, represented by portable navigation, and the other is the trend toward the advancement of surgical technology systems, including big data and artificial intelligence represented by robotics.

Surgery that automatically derives and implements an individualized preoperative plan for each patient can clearly improve clinical outcomes. The key to maximizing the benefits of CAS technology remains in surgeons' hands; therefore, surgeons should continue improving their proficiency in knee arthroplasty.

References

1. Jeffery RS, et al. Coronal alignment after total knee replacement. J Bone Joint Surg Br. 1991;73:709–14.
2. Mielke RK, et al. Navigation in knee endoprosthesis implantation–preliminary experiences and prospective comparative study with conventional implantation technique. Z Orthop Ihre Grenzgeb. 2001;139:109–16.
3. Nabeyama R, et al. The accuracy of image-guided knee replacement based on computed tomography. J Bone Joint Surg Br. 2004;86:366–71.
4. Matsumoto T, et al. Prosthetic alignment and sizing in computer-assisted total knee arthroplasty. Int Orthop. 2004;28:282–5.
5. Jones CW, et al. Current role of computer navigation in total knee arthroplasty. J Arthroplast. 2018;33:1989–93.
6. Antonios JK, et al. Trends in computer navigation and robotic assistance for total knee arthroplasty in the United States: an analysis of patient and hospital factors. Arthroplast Today. 2019;5:88–95.
7. Australian Orthopedic Association National Joint Replacement Registry (AOANJRR). Hip, knee & shoulder Arthroplasty 2021 Annual Report. Adelaide: AOA; 2021. pp. 1–432.
8. Whiteside LA, et al. The anteroposterior axis for femoral rotational alignment in valgus total knee arthroplasty. Clin Orthop Relat Res. 1995;321:168–72.
9. Hasegawa M, et al. Cutting and implanting errors in minimally invasive total knee arthroplasty using a navigation system. Int Orthop. 2013;37:27–30.
10. Nakahara H, et al. Cutting error of the distal femur in total knee arthroplasty by use of a navigation system. J Arthroplast. 2012;27:1119–22.
11. Jung HJ, et al. Fractures associated with computer-navigated total knee arthroplasty: a report of two cases. J Bone Joint Surg Am. 2007;89:2280–4.
12. Maniwa K, et al. Accuracy of image-free computer navigated total knee arthroplasty is not compromised in severely deformed varus knees. J Arthroplast. 2013;28:802–6.
13. Mizu-uchi H, et al. The evaluation of postoperative alignment in total knee replacement using a CT-based navigation system. J Bone Joint Surg Br. 2008;90:1025–31.
14. Miyasaka T, et al. Accuracy of computed tomography-based navigation-assisted total knee arthroplasty: outlier analysis. J Arthroplast. 2017;32:47-. 52.
15. Ishida K, et al. Mid-term outcomes of computer-assisted total knee arthroplasty. Knee Surg Sports Traumatol Arthrosc. 2011;19:1107–12.
16. Sasaki H, et al. Comparison of coronal prosthetic alignment after total knee arthroplasty using 3 computer-assisted navigation systems. Orthopedics. 2018;41:e621–8.
17. Hasegawa M, et al. Minimally invasive total knee arthroplasty: comparison of jig-based technique versus computer navigation for clinical and alignment outcome. Knee Surg Sports Traumatol Arthrosc. 2011;19:904–10.
18. Tsukeoka T, et al. False malalignment after computer-navigated total knee arthroplasty. Knee. 2021;30:100–5.
19. Mizu-Uchi H, et al. Three-dimensional analysis of computed tomography-based navigation system for total knee arthroplasty: the accuracy of computed. J Arthroplast. 2009;24:1103–10.
20. Minoda Y, et al. TKA sagittal alignment with navigation systems and conventional techniques vary only a few degrees. Clin Orthop Relat Res. 2009;467(4):1000–6.
21. Sugama R, et al. Conventional or navigated total knee arthroplasty affects sagittal component alignment. Knee Surg Sports Traumatol Arthrosc. 2012;20:2454–9.
22. Matsumoto T, et al. Differing prosthetic alignment and femoral component sizing between two computer-assisted CT-free navigation systems in TKA. Orthopedics. 2011;34:e860–5.
23. Minoda Y, et al. The risk of notching the anterior femoral cortex with the use of navigation systems in total knee arthroplasty. Knee Surg Sports Traumatol Arthrosc. 2010;18:718–22.
24. Amanatullah DF, et al. Identification of the landmark registration safe zones during total knee arthroplasty using an imageless navigation system. Arthroplasty. 2013;28:938–42.
25. Inui H, et al. An additional reference axis improves femoral rotation alignment in image-free computer navigation assisted total knee arthroplasty. J Arthroplast. 2013;28:766–71.
26. Mitsuhashi S, et al. Combined CT-based and image-free navigation systems in TKA reduce postoperative outliers of rotational alignment of the tibial component. Arch Orthop Trauma Surg. 2018;138:259–66.
27. Kuriyama S, et al. Tibial rotational alignment was significantly improved by the use of a CT-navigated control device in total knee arthroplasty. J Arthroplast. 2014;29:2352–6.
28. Hasegawa M, et al. Comparison of midterm outcomes of minimally invasive computer-assisted vs minimally invasive jig-based total knee arthroplasty. J Arthroplast. 2017;32:43–6.
29. Nakano N, et al. Long-term subjective outcomes of computer-assisted total knee arthroplasty. Int Orthop. 2013;37:1911–5.
30. Bouche PA, et al. Are there differences in accuracy or outcomes scores among navigated, robotic, patient-specific instruments or standard cutting a network meta-analysis. Clin Orthop Relat Res. 2020;478:2105–16.
31. McGrory BJ, et al. Surgical management of osteoarthritis of the knee: evidence-based guideline. J Am Acad Orthop Surg. 2016;24:e87–93.

32. Kobayashi H, et al. Influence of coronal bowing on the lower alignment and positioning of components in navigation and conventional total knee. Orthop Traumatol Surg Res. 2017;103:251–6.
33. Ritter MA, et al. Postoperative alignment of total knee replacement: its effect on survival. Clin Orthop Relat Res. 1994;299:153–6.
34. Berend ME, et al. Tibial component failure mechanisms in total knee arthroplasty. Clin Orthop Relat Res. 2004;428:26–34.
35. Wada K, et al. Measurement of rotational and coronal alignment in total knee arthroplasty using a navigation system is reproducible. Arch Orthop Trauma Surg. 2016;136:271–6.
36. Takagi H, et al. Case series report of navigation-based in vivo knee kinematics in total knee arthroplasty with a gradually reducing femoral radius design. Ann Med Surg (Lond). 2017;17:33–7.
37. Toyooka S, et al. Assessing the role of minimal medial tissue release during navigation-assisted varus total knee arthroplasty based on the degree of preoperative varus deformity. J Knee Surg. 2022;35(11):1236–41.
38. Sasaki E, et al. Posterior clearance increases the knee extension angle in cruciate retaining type total knee arthroplasty: intraoperative. J Orthop Sci. 2020;25:861–7.
39. Hasegawa M, et al. Prediction of postoperative range of motion using intraoperative soft tissue balance in total knee arthroplasty with navigation. Comput Aided Surg. 2015;20:47–51.
40. Ishida K, et al. Navigation-based femorotibial rotation pattern correlated with flexion angle after total knee arthroplasty. Knee Surg Sports Traumatol Arthrosc. 2016;24:89–95.
41. Matsuzaki T, et al. Tibial internal rotation is affected by lateral laxity in cruciate-retaining total knee arthroplasty: an intraoperative kinematic study using a navigation system and offset-type tensor. Knee Surg Sports Traumatol Arthrosc. 2014;22(3):615–20.
42. Hino K, et al. Mid-flexion laxity is greater after posterior-stabilized total knee replacement than with cruciate-retaining procedures: a computer navigation study: a computer navigation study. Bone Joint J. 2013;95-B(4):493–7.
43. Matsumoto K, et al. Comparative study for alignment of extramedullary guides versus portable, accelerometer-based navigation in total knee. J Knee Surg. 2018;31:92–8.
44. Ueyama H, et al. An accelerometer-based portable navigation system improved prosthetic alignment after total knee arthroplasty in 3D measurements. Knee Surg Sports Traumatol Arthrosc. 2019;27:1580–6.
45. Ueyama H, et al. Using accelerometer-based portable navigation to perform accurate total knee arthroplasty bone resection in Asian patients. Orthopedics. 2017;40:e465–72.
46. Minoda Y, et al. Usefulness of an accelerometer-based portable navigation system for total knee arthroplasty: a multicenter prospective randomized. J Bone Joint Surg Am. 2020;102:1993–2000.
47. Ikawa T, et al. Usefulness of an accelerometer-based portable navigation system in total knee arthroplasty. Bone Joint J. 2017;99-B:1047–52.
48. Shigemura T, et al. No definite advantage of a portable accelerometer-based navigation system over conventional techniques in total knee. Orthop Traumatol Surg Res. 2019;105:619–26.
49. Tsubosaka M, et al. Accelerometer-based portable navigation system is useful for tibial bone cutting in modified kinematically aligned total knee. J Knee Surg. 2021;34:870–6.
50. Suda Y, et al. Improved implant alignment accuracy with an accelerometer-based portable navigation system in medial unicompartmental knee. Knee Surg Sports Traumatol Arthrosc. 2020;28:2917–23.
51. Fujimoto E, et al. Technical considerations and accuracy improvement of accelerometer-based portable computer navigation for performing distal. J Arthroplast. 2017;32:53–60.
52. Shoji H, et al. Radiographic assessment and clinical outcomes after total knee arthroplasty using an accelerometer-based portable navigation device. Arthroplast Today. 2018;4:319–22.
53. Tsukada S, et al. Augmented reality-based navigation system applied to tibial bone resection in total knee arthroplasty. J Exp Orthop. 2019;6:44.
54. Tsukada S, et al. Augmented reality-assisted femoral bone resection in total knee arthroplasty. JB JS Open Access. 2021;6(3):e21.00001.

Patient-Specific Instrumentation (PSI)

61

Kazumasa Yamamura

1 The Birth of PSI and Recent Trends

PSI was developed as a new concept for providing patient-specific surgical instruments through computerized preoperative planning [1]. The process begins with preoperative CT or MRI image data being used to construct 3D images, which are then superimposed on a computer with the 3D image data of the TKA implant to be used. A virtual surgery called "surgical planning" is performed to confirm implant size, alignment, and position, as well as osteotomy volume. To this point, the process is similar to preoperative planning using 3D templates; however, PSI is an extension of this process.

Based on the created surgical plan, a physical guide (containing cutting slots and/or pin insertion points) is created using 3D printing from CAD data (Fig. 61.1). This guide makes it possible to replicate the preoperative plan intraoperatively. In other words, the preoperative plan, which was initially displayed on a 2D monitor, can now be visualized in a 3D world.

According to overseas registries, the number of TKAs using this revolutionary technology is increasing every year [3].

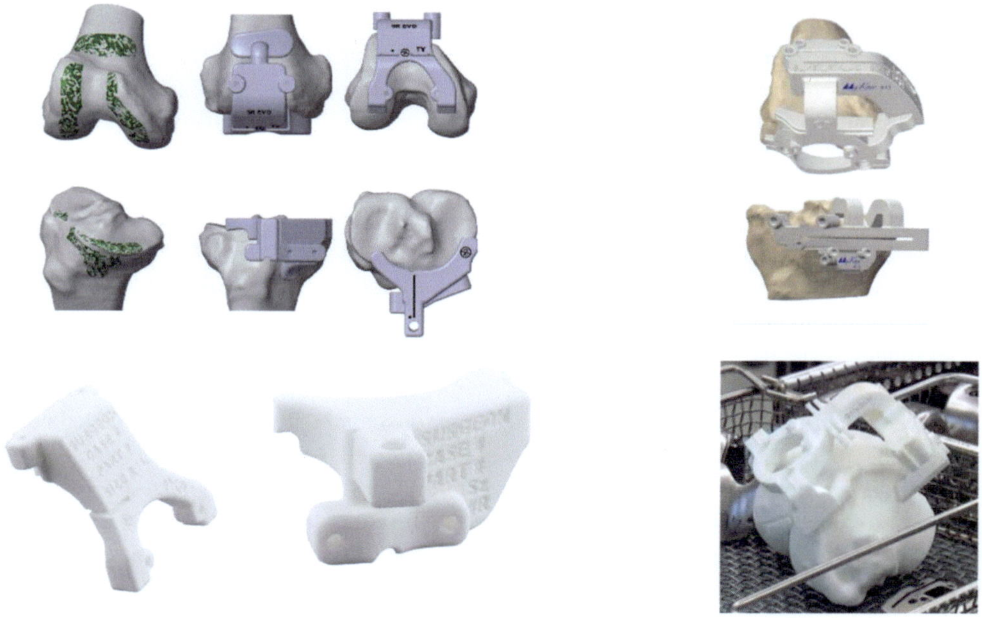

Fig. 61.1 3D templating and 3D printed pin guides or cutting blocks. (Reprinted with permission from Springer Nature [2])

K. Yamamura (✉)
Department of Orthopaedic Surgery, Osaka General Hospital of West Japan Railway Company, Osaka, Japan

S. Matsuda (ed.), *Advances in Total Knee Arthroplasty*, https://doi.org/10.1007/978-981-97-4920-1_61

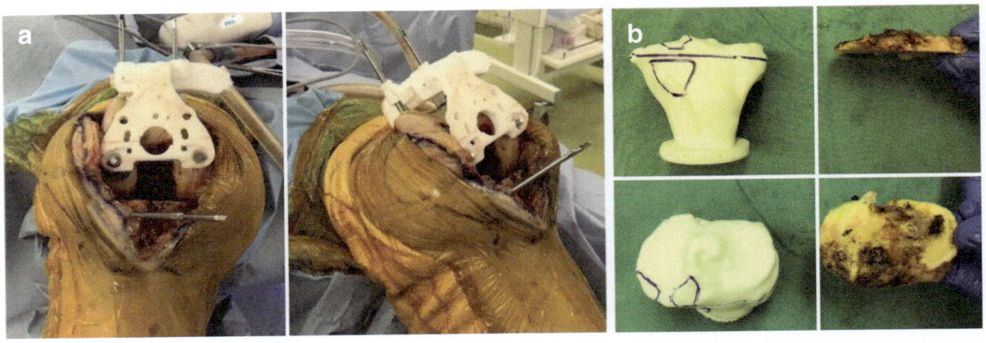

Fig. 61.2 Surgical procedure using PSI. (**a**) PSI is placed on the femur. (**b**) Comparison of the bone model of the tibial side PSI with the actual resected bone. The area where the PSI will be placed in the bone should be drawn in advance using a marker. When using a CT-based PSI, the soft tissue and residual articular cartilage at the placement site should be adequately removed. (Reprinted with permission from Springer Nature [8])

2 Actual TKA Using PSI

2.1 CT-Based or MRI-Based?

Large-scale comparative studies are limited in this field, and the reported results vary [4, 5]. This is a field in which surgeons' preferences are often reflected.

The advantages of the CT-based PSI are as follows: [6].

No contraindications to imaging.

- Short imaging time
- Lesser halation than MRI in the presence of metal implants in the surrounding tissue
- Easy image editing using computer software

The advantages of the MRI-based PSI are as follows:

- No radiation exposure
- Because PSI can be fabricated by considering articular cartilage components, they are relatively easy to place on expanded bone [4].

However, identification of articular cartilage components during the preoperative planning stage is a complicated manual procedure and requires accuracy. In addition, PSI cannot be placed properly when large osteophytes are involved [7], and MRI-based PSI may be suitable for surgeons with a certain level of expertise in the technique.

2.2 Surgical Technique

Since PSI is often designed to use a specific osteophyte as a steppingstone, it is necessary to confirm in advance that the necessary osteophyte will not be resected during deep deployment. Placement can be performed on either the femoral or tibial side.

When using a CT-based PSI, soft tissues and residual articular cartilage at the corresponding locations are not reflected in the preoperative plan; therefore, they must be adequately removed during surgery. Specifically on the tibial side, inadequate soft tissue removal and large osteophytes have been reported to interfere with PSI fitting [7]; thus, we recommend careful resection when the surgeon is inexperienced. Similarly, osteophytes serve as indicators for PSI placement on the femoral side, although placement is relatively easy because of the presence of many characteristic structures that can be used as a foothold. After placement, osteotomy can be directly performed with a bone saw through the attached cutting slot. Alternatively, the PSI can be changed to a metal-cutting guide, while leaving only the pin in place (Fig. 61.2a).

2.3 Amount of Bone Resected

A study was conducted to compare the accuracy of TKA osteotomies using PSI between skilled and inexperienced surgeons using a simulated bone model. According to this study, the accuracy was not related to the surgeon's experience or skill, and the results were comparable between the two surgeons [9]. However, in actual surgery, the accuracy of resected bone volume varies due to the differences in PSI models, leading to the differences in the congruency of resected areas [8, 10, 11]. In addition, intraoperative correction or fine adjustment of PSI placement is often necessary [12]. These findings suggest that the accuracy of bone resection may not solely depend on PSI, but could be influenced by the differences in the basic surgical techniques, such as soft tissue handling, surgical field development, and securing the field of view, which are not accounted for in simulated bone studies (Fig. 61.2b).

2.4 Clinical Scores

Compared with conventional TKA, no significant differences were observed in the KSS, KOOS, Oxford knee score (OKS), WOMAC, 5-dimensions of European quality of life using the visual analog scale (EQ- 5D VAS), and 12-Item Short Form Survey (SF-12) in the early evaluation from 3 months to <1 year after surgery [13, 14]. However, when the follow-up period was 1 year or longer, significant differences were observed in the KSS Functional score, KOOS pain, KOOS ADL, and KOOS QoL in favor of PSI-based TKA [14, 15]. The direct cause of this significant difference in scores is still unclear and remains a subject of debate.

2.5 Radiographic Outcomes (2D and 3D)

Radiography, CT, and MRI are mainly used for postoperative image evaluation (2D), whereas CT and MRI can be evaluated after 3D reconstruction, and their accuracy is superior to that of the 2D reconstructions [16, 17]. In recent years, the use of intraoperative or preoperative and postoperative 3D reconstructed data has become more prevalent in evaluating the accuracy of PSI [2, 18–20] (Fig. 61.3).

1. Mechanical Axis (MA)

TKA using PSI has postoperative MA values similar to that of conventional TKA, but it generates fewer outliers. Certain PSI models may have higher frequency of outliers, suggesting differences in accuracy among different PSI models [13, 15].

2. Component alignment

In coronal alignment, PSI significantly improves the component placement accuracy on the femoral side than conventional TKA does. On the tibial side, the placement angle is similar to that in conventional TKA; however, the risk of outlier occurrence is reported to be higher [13, 15].

In sagittal alignment, the accuracy of PSI is equivalent to that of conventional TKA on the femoral side, and the use of PSI may increase accuracy. However, the use of PSIs on the tibial side appears to increase the risk of inaccurate placement and outlier occurrence [13, 15].

Axial alignment requires postoperative CT or MRI for measurement, and only few reports are available on this technique. Among them, the effectiveness of PSI for both the femoral and tibial aspects is controversial [13, 22–25]; however, no studies have indicated that PSI is inferior to conventional TKA in terms of effectiveness.

2.6 Cost

When discussing the cost of PSI-based TKA, the costs of imaging tests, PSI fabrication, and instrument sterilization are often taken into consideration. Outside Japan, the use of PSI has reportedly reduced the number of surgical instruments and sterilized trays required [26, 27], shortened the sterilization process time, and reduced costs [28]. However, some studies have raised the issue of the cost of preoperative imaging tests [29]. The cost-effectiveness of PSI can be considered relatively high when considered as a single product, and somewhat inexpensive when considered in terms of total cost-effectiveness.

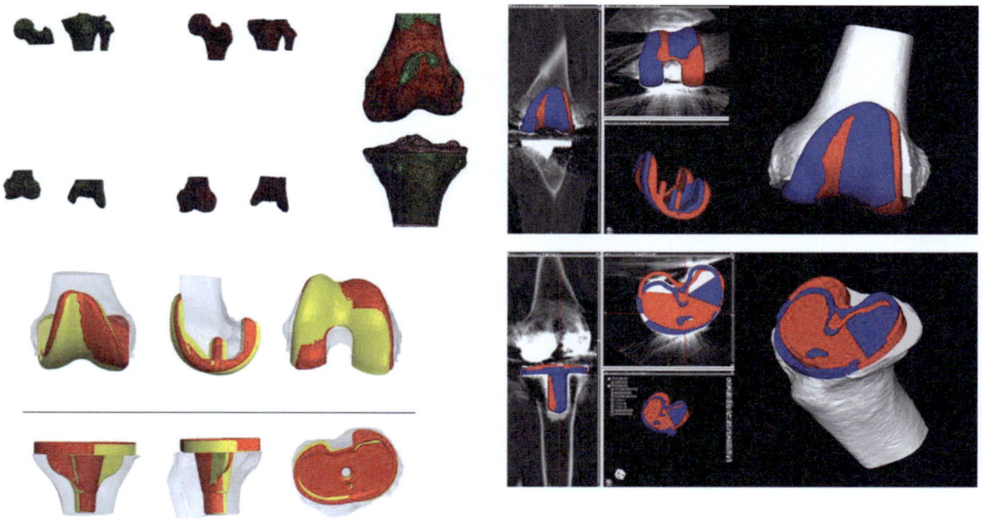

Fig. 61.3 Accuracy verification of PSI using postoperative images and preoperative planning. Top left: Bone models reconstructed from preoperative and postoperative CT are superimposed on a computer. Bottom left: The resulting alignment difference between the preoperative plan and postoperative implant is visualized. Right: Similar procedures and measurements can be performed using a commercially available 3D templating software. (Reprinted with permission from Springer Nature 20 and Georg Thieme Verlag KG [21])

2.7 Surgical Duration and Blood Loss

PSI-based TKA has a reduced operating time by 3–18 min than that for conventional TKA [14, 30]. However, a meta-analysis did not demonstrate a significant difference in the tourniquet use time [15]. PSI-based TKA has shown a learning curve for [6], with varying opinions among surgeons having different proficiency levels.

TKA using PSI has demonstrated a certain amount of reduction in blood loss than conventional TKA [14]. In addition, it has demonstrated a lower percentage of patients (1.1%) requiring blood transfusions than that of conventional TKA (7.7%), indicating a level of effectiveness [30].

2.8 Survival Rates Based on Revision

The revision rates of TKA using PSI and conventional TKA are comparable from early to mid-term [3].

However, according to the 2020 annual report of the Australian Orthopaedic Association National Joint Replacement Registry, TKA using custom-made pin guides or cutting TKA with custom-made pin guides or cutting blocks is associated with a higher revision rate than that of TKA without custom-made pin guides or cutting blocks. However, such an increase in revision rate is specific to certain models and cannot be attributed to the PSI per se. In patients aged <65 years, no increase in the revision rate was noted with the use of PSI.

2.9 Other Complications

A meta-analysis reported no significant differences in the incidences of DVT, PE, or surgical site infection between conventional TKA and TKA using PSI [14].

3 TKA Using PSI: Problems with and Usefulness of 3D Image Evaluation

As mentioned above, reports on the accuracy evaluation of PSI are divided, with most of them based on the evaluation of 2D images using radiography, CT, and MRI. Compared with 3D images, the accuracy of 2D image evaluation is low [16], and variations exist among evaluators [17]. Variations in the measured values depending on the limb position at the time of imaging [31] and the slice direction of the image [32] are thought to be one of the reasons for the differences in the opinions [21]. PSI produced using 3D images should also be evaluated using 3D images, and an increasing number of reports have utilized intraoperative or preoperative 3D reconstructed CT data for its evaluation [2, 18–20] (Fig. 61.3). In addition, multiple PSI models have been evaluated using the same method, suggesting that the PSI accuracy may vary depending on the model [21].

4 Future Prospects

In brief, TKA using PSI is "non-inferior" to conventional TKA in terms of radiographic outcomes and clinical scores. Although some advantages in terms of postoperative blood loss and operation time have been reported, blood transfusions are rarely used because the use of tranexamic acid in recent years has greatly reduced the amount of blood loss. In addition, the effect of the learning curve cannot be ignored in terms of the reduction in operation time. The cost is difficult to evaluate depending on the extent to which it is defined. Even if the hospital stay is shortened, it is uncommon in Japan for patients to be discharged within a few days of surgery, unlike in other countries.

This is not a positive statement for PSI. However, this developing field still has some unresolved issues, which are discussed below; therefore, further research is required to address these concerns.

4.1 Image Evaluation Method

Compared to 3D images, the accuracy of 2D image evaluation is low [16], and variations exist among evaluators [17]. One reason for this may be that the measured values tend to vary depending on the limb position at the time of imaging [31] and slice direction of the image [32]. Recent reports on 3D image evaluation have demonstrated that PSIs accurately reproduce preoperative planning [2, 18–20]; however, 3D image evaluation is still not widely used because of its complexity. The evaluation of PSI may change with advances in accurate evaluation methods.

4.2 PSI Design

PSIs are not uniform and vary among different companies, each with its own characteristics, and postoperative alignment accuracy of some models is low. However, several reports in recent years [33, 34] have shown significant improvements in accuracy by improving the PSI design and the placement method. Further design improvements may lead to a more convenient and accurate device.

References

1. Hafez MA, et al. Computer-assisted total knee arthroplasty using patient-specific templating. Clin Orthop Relat Res. 2006;444:184–92.
2. Yamamura K, et al. Novel alignment measurement technique for total knee arthroplasty using patient specific instrumentation. Arch Orthop Trauma Surg. 2017;137:401–7.
3. McAuliffe MJ, et al. Impact of image-derived instrumentation on total knee arthroplasty revision rates: an analysis of 83,823 procedures from the Australian orthopedic association National Joint Replacement Registry. J Bone Joint Surg Am. 2019;101:580–8.
4. Ensini A, et al. Intra- and post-operative accuracy assessments of two different patient-specific instrumentation systems for total knee replacement. Knee Surg Sports Traumatol Arthrosc. 2014;22:621–9.
5. Asada S, et al. Comparison of MRI- and CT-based patient-specific guides for total knee arthroplasty. Knee. 2014;21:1238–43.
6. Hafez MA, Moholkar K. Patient-specific instruments: advantages and pitfalls. Sicot J. 2017;3:66.
7. Seon JK, et al. Assessing the accuracy of patient-specific guides for total knee arthroplasty. Knee Surg Sports Traumatol Arthrosc. 2016;24:3678–83.
8. Yamamura K, et al. Evaluation of the accuracy of resected bone thickness based on patient-specific instrumentation during total knee arthroplasty. Arch Orthop Trauma Surg. 2021;141:1583–90.
9. Antoniadis A, et al. Accuracy of tibial cuts with patient-specific instrumentation is not influenced by the surgeon's level of experience. Knee Surg Sports Traumatol Arthrosc. 2019;27:1535–43.
10. Levy YD, et al. The accuracy of bony resection from patient-specific guides during total knee arthroplasty. Knee Surg Sports Traumatol Arthrosc. 2017;25:1678–85.
11. Zambianchi F, et al. Assessment of patient-specific instrumentation precision through bone resection measurements. Knee Surg Sports Traumatol Arthrosc. 2017;25:2841–8.
12. Okada Y, et al. Preoperative corrections are required for planning of patient-specific instrumentation in total knee arthroplasty. Knee. 2017;24:1492–7.
13. Huijbregts HJ, et al. Patient-specific instrumentation does not improve radiographic alignment or clinical outcomes after total knee arthroplasty. Acta Orthop. 2016;87:386–94.
14. Kizaki K, et al. Total knee arthroplasty using patient-specific instrumentation for osteoarthritis of the knee: a meta-analysis. BMC Musculoskelet Disord. 2019;20:561.
15. Thienpont E, et al. Efficacy of patient-specific instruments in total knee arthroplasty: a systematic review and meta-analysis. J Bone Joint Surg Am. 2017;99:521–30.
16. Ueyama H, et al. Two-dimensional measurement misidentifies alignment outliers in total knee arthroplasty: a comparison of two- and three-dimensional measurements. Knee Surg Sports Traumatol Arthrosc. 2019;27:1497–503.
17. Wai Hung CL, et al. Interobserver and intraobserver errors in distal femur transepicondylar axis measurement with computed tomography. J Arthroplast. 2009;24:96–100.
18. De Vloo R, et al. Three-dimensional analysis of accuracy of component positioning in total knee arthroplasty with patient specific and conventional instruments: a randomized controlled trial. Knee. 2017;24:1469–77.
19. Kuwashima U, et al. Three-dimensional analysis of accuracy of patient-matched instrumentation in total knee arthroplasty: evaluation of intraoperative techniques and postoperative alignment. J Orthop Sci. 2017;22:1077–83.
20. Kobashi S, et al. Image-based evaluation of patient specific instrument attachment in TKA. Procedia Computer Sci. 2013;22:976–82.
21. Yamamura K, et al. Prosthetic accuracy depends on the Design of Patient-Specific Instrumentation: results of a retrospective study using three-dimensional imaging. J Knee Surg. 2020; https://doi.org/10.1055/s-0040-1721127. Online ahead of print
22. Victor J, et al. Patient-specific guides do not improve accuracy in total knee arthroplasty: a prospective randomized controlled trial. Clin Orthop Relat Res. 2014;472:263–71.
23. Heyse TJ, Tibesku CO. Improved tibial component rotation in TKA using patient-specific instrumentation. Arch Orthop Trauma Surg. 2015;135:697–701.
24. Pauzenberger L, et al. Patient-specific instrumentation improved three-dimensional accuracy in total knee arthroplasty: a comparative radiographic analysis of 1257 total knee arthroplasties. J Orthop Surg Res. 2019;14:437.
25. Heyse TJ, Tibesku CO. Improved femoral component rotation in TKA using patient-specific instrumentation. Knee. 2014;21:268–71.
26. Noble JW Jr, et al. The value of patient-matched instrumentation in total knee arthroplasty. J Arthroplast. 2012;27:153–5.
27. Hamilton WG, et al. Patient-specific instrumentation does not shorten surgical time: a prospective, randomized trial. J Arthroplast. 2013;28:96–100.
28. Barrack RL, et al. Patient specific cutting blocks are currently of no proven value. J Bone Joint Surg Br. 2012;94:95–9.
29. Goyal N, et al. Does implant design influence the accuracy of patient specific instrumentation in total knee arthroplasty? J Arthroplast. 2015;30:1526–30.
30. Vide J, et al. Patient-specific instrumentation in total knee arthroplasty: simpler, faster and more accurate than standard instrumentation-a randomized controlled trial. Knee Surg Sports Traumatol Arthrosc. 2017;25:2616–21.
31. Lee YS, et al. Effect of foot rotation on the mechanical axis and correlation between knee and whole leg radiographs. Knee Surg Sports Traumatol Arthrosc. 2013;21:2542–7.
32. Okamoto S, et al. Two-dimensional planning can result in internal rotation of the femoral component in total knee arthroplasty. Knee Surg Sports Traumatol Arthrosc. 2016;24:229–35.
33. Yamamura K, et al. Design improvement in patient-specific instrumentation for total knee arthroplasty improved the accuracy of the tibial prosthetic alignment in the coronal and axial planes. Knee Surg Sports Traumatol Arthrosc. 2020;28:1560–7.
34. Kwon OR, et al. Patient-specific instrumentation development in TKA: 1st and 2nd generation designs in comparison with conventional instrumentation. Arch Orthop Trauma Surg. 2017;137:111–8.

Robotic Assisted TKA

62

Daisuke Hamada

Although TKA has been shown to achieve pain relief and excellent long-term survival, it has also been reported that 15–20% of patients still have some dissatisfaction with their knees after surgery, and various attempts are being made to further improve the surgical results. The success of TKA requires implant placement based on accurate osteotomy and proper soft tissue balance. Although the introduction of navigation systems, 3D preoperative planning using CT and MRI, and patient-matched instruments have improved the accuracy of implant placement, soft tissue balance is still dependent on the skill of the surgeon and does not always lead to a significant improvement in postoperative outcomes [1, 2]. Although mechanical alignment has been the gold standard for postoperative alignment of the lower limb, the concepts of constitutional varus and kinematic alignment have become increasingly popular in recent years, and the debate on these concepts has become more active. Robotic-assisted surgery is expected to be a tool to achieve alignment and soft tissue balance with high accuracy and reproducibility, which can make the difference between success and failure in surgery.

The development of ROBODOC for THA began in the late 1980s, and it was clinically applied in the U.S. in 1992, and clinical trials were conducted in Japan in 2000. ROBODOC can also be used for TKA, and a study on CT-base placement planning using ORTHODOC, a workstation for preoperative planning of the same system, has been reported [3]. In the early 2000s, Nakashima Propeller Co., Ltd. developed a TKA surgical support robot in Japan, but it has not been used for practical applications [4]. Since then, new platforms have been developed and are now in clinical use in Japan.

1 Surgical Robots Available in Japan

As of this writing (November 2021), MAKO (Stryker), NAVIO and its successors CORI (Smith & Nephew) and ROSA (Zimmer Biomet) are available in Japan.

1.1 MAKO (Fig. 62.1)

The RIO Robotic Arm Interactive Orthopedic System, developed by Mako Surgical Corporation and applied to UKA, TKA, and THA in the United States since 2006. It was acquired by Stryker in 2013 and marketed as MAKO and it is currently being used worldwide. In Japan, the robot has been approved for THA, TKA, and UKA since November 2021. Surgical planning performed with CT-base, and imaging information is sent to Stryker before surgery to create a surgical plan under the surgeon's prescribed conditions. Based on this information, intraoperative landmark points are used for registration, and the surgical plan is verified to create a CT based 3D model. Bone resection is performed with high accuracy using a blade fixed to the robotic arm. It is a semi-active type robotic arm that can perform osteotomy on both the femoral and tibial sides, and only the area defined by the computer is osteotomized by the surgeon's manipulation. The robot is controlled to stop the blade if it attempts to go outside the defined area, and the area to be osteotomized is strictly defined by the blade to minimize invasion to the soft tissue [5]. Postoperative soft-tissue balance can be simulated by applying varus and valgus stress to the knee joint in extension and flexion after registration, and the measured values can be used to fine-tune the placement plan before osteotomy. The use of implants is limited to those manufactured by Stryker.

D. Hamada (✉)
Department of Orthopedics, Institute for Biomedical Sciences, Tokushima University Graduate School, Tokushima, Japan
e-mail: daisuke.hamada@tokushima-u.ac.jp

S. Matsuda (ed.), *Advances in Total Knee Arthroplasty*, https://doi.org/10.1007/978-981-97-4920-1_62

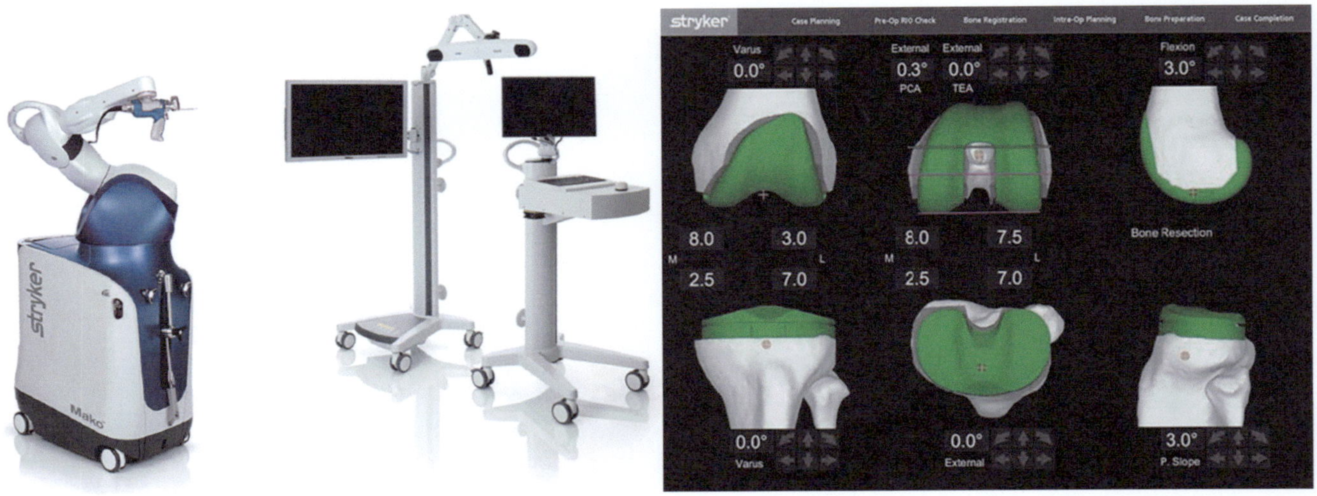

Fig. 62.1 MAKO. (Provided by Stryker)

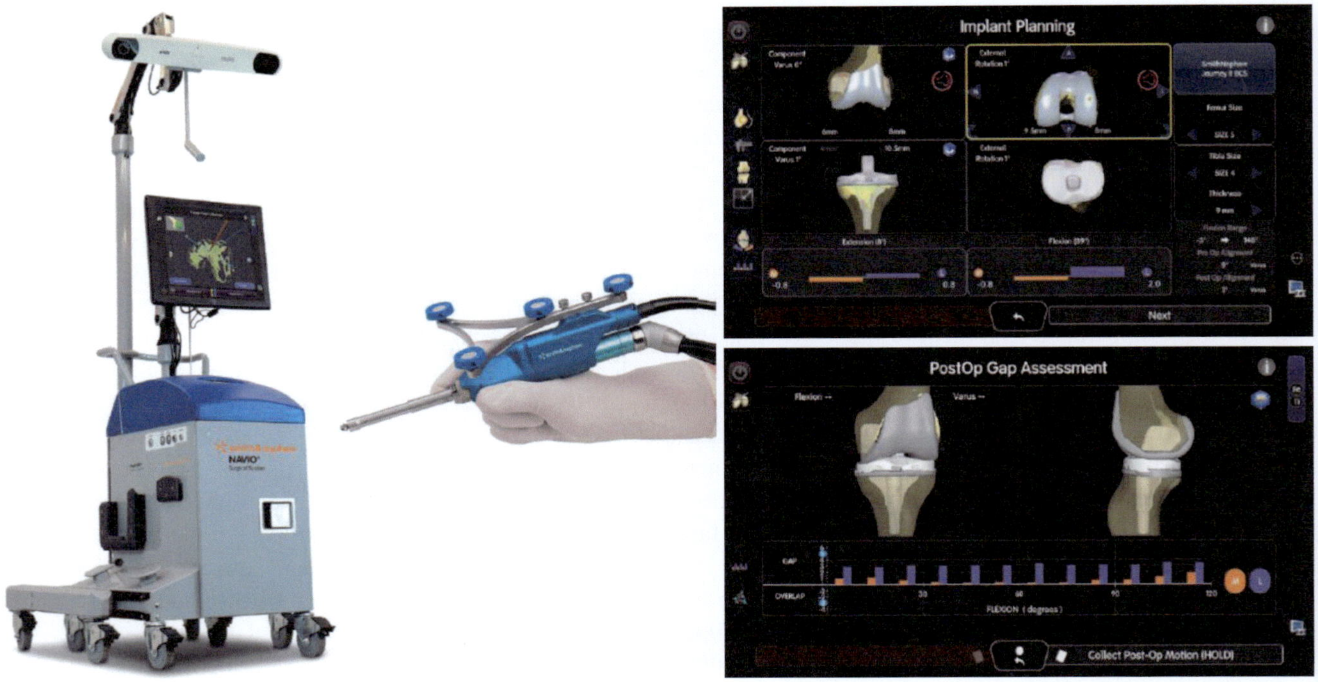

Fig. 62.2 NAVIO. (Provided by Smith & Nephew)

1.2 NAVIO/CORI (Fig. 62.2)

NAVIO, manufactured by Blue Belt Technologies, received FDA approval in 2011 as a surgical robot for UKA. Navio was acquired by Smith & Nephew in 2014 and received US FDA approval for TKA in 2016. In Japan, clinical use began after regulatory approval in 2018, and the successor model CORI has been in use in Japan since 2021. The system is image-free and does not necessarily require preoperative imaging information. The system performs intraoperative surface mapping, constructs a 3D model of the knee joint, and allows the surgeon to adjust the implant position on the screen. The surface mapping is characterized by the inclusion of cartilage information and the ability to confirm the location of implants in the coronal, sagittal, and transverse planes using the cross section view, making it easy to grasp the shape of the articular surface. Ligament balance can be evaluated by applying varus/valgus stress to the entire ROM, and the surgical plan can be fine-tuned according to the results of the gap simulation. Bone resection is performed using a milling device controlled by the robot, but a cutting device can also be used in combination. The robot is a semi-active type robot that performs osteotomy using a hand-held burr under robot control according to a surgical

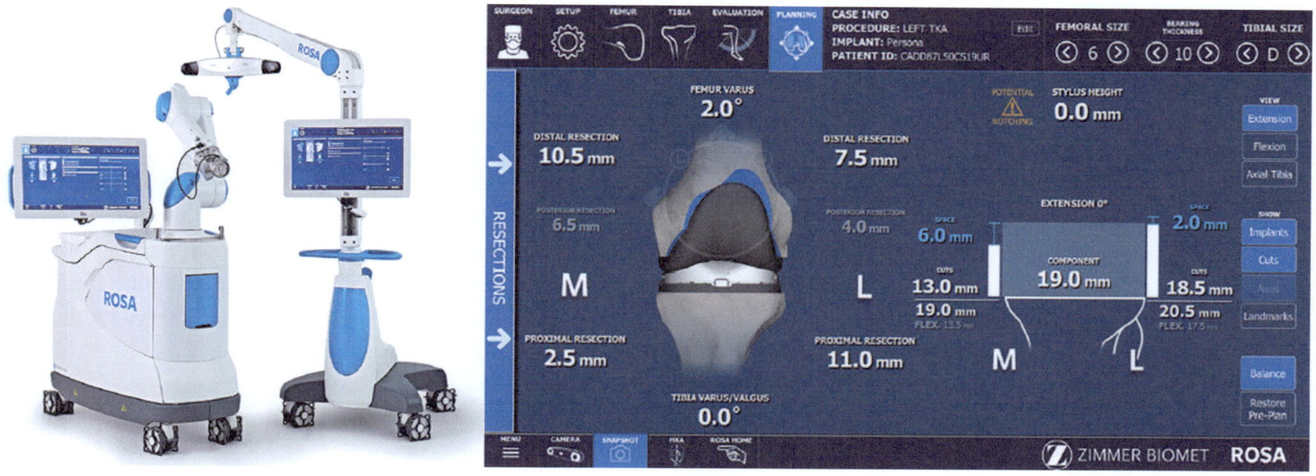

Fig. 62.3 ROSA. (Provided by Zimmer Biomet)

plan set by the surgeon. Burring is controlled by the robot to avoid excessive bone removal by retracting or stopping the drill into the sleeve when it is about to leave the defined area. The robot can be used only with implants manufactured by Smith & Nephew, but it is also compatible with BCR-TKA.

1.3 ROSA (Fig. 62.3)

The ROSA Recon robotic system is based on Medtech's ROSA One surgical robot for neurosurgery and spine surgery, developed by Zimmer Biomet, which acquired Medtech in 2016 and received FDA approval for TKA in January 2019 and UKA in April 2019. In Japan, regulatory approval for the use on TKA and UKA was obtained in July 2020 and September 2021, respectively, and clinical use for only TKA has begun since November 2021. The product is compatible with both image-based and image-free methods. Image-based surgical planning is mainly performed with X-Atlas using X-rays, and is also possible with MR. The 3D model in X-Atlas is selected from more than 100 existing models to be used for intraoperative planning, while no 3D model is created for image-free planning. The ligament balance can be evaluated in extension and flexion, and the surgical plan can be fine-tuned based on this information. The robotic arm guides the cutting guide according to the intraoperative plan, and after pin fixation, the surgeon performs osteotomies of the distal femur and proximal tibia using a bone saw. Femoral rotation and size can also be determined using the robotic arm. The use of implants is limited to those manufactured by Zimmer Biomet.

2 Advantages of Robotic Surgery

Although there are some differences among the models, the following features are provided: (1) placement planning based on a 3D model constructed for each case, (2) ligament balance evaluation, (3) intraoperative adjustment of the surgical plan, (4) accurate robot-controlled bone resection, and (5) feedback from surgical data recording and postoperative results. Although the methods are different from CT-based and image-free techniques, the detailed planning of placement angles and positions is possible through 3D modeling of the patient's individual bone configuration, and the planning can be fine-tuned intraoperatively according to the results of ligament balance evaluation, which is different from existing techniques. Especially in the case of UKA and BCR-TKA, where the original joint surface of the patient should be reproduced as much as possible, surgical planning can be performed while overlaying the implant on the virtual 3D model, thus reducing the stress on the surgeon. Robot-controlled osteotomy has been reported to be more accurate than conventional techniques using a conventional guide and bone saw. Although alignment and ligament balance have various targets, they are currently the most useful tools to achieve the surgeon's intended concept more accurately and reproducibly. In addition, the intraoperative decision-making process regarding alignment, bone resection, and ligament balance can be recorded in detail, and the relationship with postoperative outcomes can be investigated to improve surgical techniques [6].

3 Clinical Results

3.1 Robotic-Assisted (RA)-UKA

Bell et al. reported that the MAKO and Batailler et al. reported that NAVIO was more accurate and had fewer outliers than the conventional method, and other studies have generally shown improved accuracy of MAKO and NAVIO [7, 8]. While there are also reports [9–11] that the complication and revision rates are lower than those of the conventional method, there are also reports that patient reported outcome measures (PROMs) show no difference from the conventional group at 2 or 5 years. Therefore the advantage of robotics on clinical outcomes is still controversial [11, 12]. A study comparing the results of MAKO and NAVIO reported no difference in accuracy of placement, results, or frequency of complications, with MAKO having a slightly shorter operative time [13]. The survival rate of implants is reported to be 5 years with MAKO and approximately 2 years with NAVIO, and the revision rate is reported to be lower than those of conventional methods [8, 14].

3.2 RA-TKA

There are reports that MAKO improves placement accuracy [15] and reproduces posterior condylar offset [16], while TKA using NAVIO and ROSA has also been able to achieve accurate placement [17–19]. Similarly, many studies have also reported improvement to placement accuracy in case of UKA. Although, there are some reports that PROMs and revision rates are better than those of conventional methods in the short term, long-term results with a large number of patients have not been reported for any of the available models in Japan at this time.

The results of both RA-UKA and RA-TKA have been accumulated, and their usefulness and problems are expected to be clarified in the future. At least the accuracy of osteotomy and postoperative alignment of the lower limb are considered to be more accurate and reproducible than those of conventional techniques; however, the short-term clinical results are controversial, some claiming that they are better than those of conventional methods while others' state that they are the same. However, at present, there is a lack of studies that discuss the results in terms of both quality and quantity, and we are awaiting the results of future large-scale studies with long-term follow-up.

4 Future Prospects

A paper reported that there is no reason to introduce new technology unless postoperative function and long-term survival rates are improved because there are disadvantages, such as the cost of introduction and prolonged operation time [20]. On the other hand, it is clear that new technology has advantages over existing methods in terms of improving placement accuracy and achieving the desired alignment. The key to improving results is how the robot is being used rather than the use of the robot. New concepts such as "functional alignment" have been proposed for robotic surgery [21], and the future results are expected. It may be that it is not the robot itself that improves results but rather the newly developed concept and implant design generated by the data obtained from RA-UKA and RA-TKA [6].

References

1. Lee DY, et al. No differences in mid- to long-term outcomes of computer-assisted navigation versus conventional total knee arthroplasty. Knee Surg Sports Traumatol Arthrosc. 2020;28:3183–92.
2. Kim YH, et al. The clinical outcome of computer-navigated compared with conventional knee arthroplasty in the same patients: a prospective, randomized, double-blind, long-term study. J Bone Joint Surg Am. 2017;99:989–96.
3. Hananouchi T, et al. CT-based planning of a single-radius femoral component in total knee arthroplasty using the ROBODOC system. Comput Aided Surg. 2008;13:23–9.
4. Sugita N, et al. Development of a novel robot-assisted orthopedic system designed for total knee arthroplasty. Lect Notes Comput Sci. 2004;3217(1 PART 2):153–60.
5. Kayani B, et al. Iatrogenic bone and soft tissue trauma in robotic-arm assisted total knee arthroplasty compared with conventional jig-based total knee arthroplasty: a prospective cohort study and validation of a new classification system. J Arthroplast. 2018;33:2496–501.
6. Elliott J, et al. Robotic-assisted knee arthroplasty: an evolution in progress: a concise review of the available systems and the data supporting them. Arch Orthop Trauma Surg. 2021;141:2099–117.
7. Bell SW, et al. Improved accuracy of component positioning with roboticassisted unicompartmental knee arthroplasty: data from a prospective, a randomized controlled study. J Bone Joint Surg Am. 2016;98:627–35.
8. Batailler C, et al. Improved implant position and lower revision rate with robotic-assisted unicompartmental knee arthroplasty. Knee Surg Sports Traumatol Arthrosc. 2019;27:1232–40.
9. Kunze KN, et al. Comparing clinical and radiographic outcomes of robotic-assisted, computer-navigated and conventional unicompartmental knee arthroplasty: a network meta-analysis of randomized controlled trials. J Orthop. 2021;25:212–9.
10. Mergenthaler G, et al. Is robotic-assisted unicompartmental knee arthroplasty a safe procedure? A case control study. Knee Surg Sports Traumatol Arthrosc. 2021;29:931–8.
11. Banger M, et al. Robotic arm-assisted versus conventional medial unicompartmental knee arthroplasty: five-year clinical outcomes of a randomized. Bone Joint J. 2021;103-B:1088–95.
12. Gilmour A, et al. Robotic-arm-assisted versus conventional unicompartmental knee arthroplasty: the 2-year clinical outcomes of a randomized controlled trial. J Arthroplast. 2018;33(7S):s109–15.
13. Leelasestaporn C, et al. Comparison of 1-year outcomes between MAKO versus NAVIO robot-assisted medial UKA: a nonrandomized, prospective, comparative. Knee Surg Relat Res. 2020;32:13.
14. Burger JA, et al. Mid-term survivorship and patient-reported outcomes of robotic-arm assisted partial knee arthroplasty. Bone Joint J. 2020;102-B:108–16.
15. Kayani B, et al. Robotic-arm assisted total knee arthroplasty has a learning curve of seven cases for integration into the surgical work-

flow but no learning curve effect for accuracy of implant positioning. Knee Surg Sports Traumatol Arthrosc. 2019;27:1132–41.
16. Sultan AA, et al. Robotic-arm assisted total knee arthroplasty more accurately restored the posterior condylar offset ratio and the Insall–Salvati index compared to the manual technique: a cohort-matched study. Surg Technol Int. 2019;34:409–13.
17. Kaneko T, et al. Robotic-assisted total knee arthroplasty improves the outlier of rotational alignment of the tibial prosthesis using 3DCT. Knee. 2021;31:64–76.
18. Casper M, et al. Accuracy assessment of a novel image-free handheld robot for total knee arthroplasty in a cadaveric study. Comput Assist Surg (Abingdon). 2018;23:14–20.
19. Seidenstein A, et al. Better accuracy and reproducibility of a new robotically assisted system for total knee arthroplasty compared to conventional. Knee Surg Sports Traumatol Arthrosc. 2021;29:859–66.
20. Bouché PA, et al. Are there differences in accuracy or outcomes scores among navigated, robotic, patient-specific instruments or standard cutting. Clin Orthop Relat Res. 2020;478:2105–16.
21. Kayani B, et al. A prospective double-blinded randomized controlled trial comparing robotic arm-assisted functionally aligned total knee arthroplasty. Trials. 2020;21:194.

Sensor-Assisted TKA

63

Hiroshi Mikami

1 Significance of Intraoperative Sensors

Before the introduction and widespread use of the knee navigation system (navigation), the final intraoperative balance of TKA was left to the subjective judgment of the operating surgeon. However, with the introduction and widespread use of navigation, objective evaluation of implant balance and kinematic analysis through the full range of motion (ROM) is now possible. While navigation is not always an absolute answer, it is difficult to visually understand the presence or absence of contact between implants and the degree of compressive force through a full ROM with a closed reduction of the patellar. Therefore, the evaluation of "stability and mobility" must be subjective. This is the " blind spot of navigation". However, the edge localization of compressive force, lift-off of the femoral and tibial joint surfaces, and localized axial compression due to medial pivot motion are important factors affecting clinical outcomes, including long-term durability and postoperative ROM, such as premature wear and tear of inserts. Hence, intraoperative pressure sensors can compensate for this "blind spot of navigation" and achieve the appropriate balance between "stability and mobility." The intraoperative use of pressure sensors is significant because it provides objective evaluation and scientific evidence of the distribution of compressive force on the tibial joint surface, which is reflected in the "establishment of objective surgical techniques," the "improvement of long-term postoperative outcomes," and the "education of future generations.

2 Pressure Sensor Types and Issues

There are two types of pressure sensors: (1) a sheet-type pressure sensor that directly measures compressive force by fixing a sheet-type device consisting of a grid of tiny conductive chips covered with a highly flexible material to the voluntary joint surface [1, 2] and (2) a load cell-type pressure sensor that indirectly measures compressive force through the joint surface by inserting a pressure receptor (load cell) inside a tibial trial component [3]. The sheet type has many measurement points and a wide measurement range, making it clinically practical and versatile. However, its reliability and reproducibility are a concern due to problems with sheet fixation, deflection and strain of the flexible material. In contrast, the load cell type is physically limited in the number of units that can be installed, which reduces the number of measurement points and measurement area, and its technical application has been hindered by economic problems, such as the development of materials for joint surfaces. Recent advances in medical engineering and innovations have resulted in the development of unique load cell pressure sensors with high reliability and reproducibility and improved clinical outcomes [3–5].

In 2009, we developed the first load cell-type pressure sensor in Japan. Herein, we discuss the advantages and disadvantages of the pressure sensor with a load cell.

H. Mikami (✉)
TAKAMATSU Knee Arthropathy Clinic, 1-9-22 Hananomiya-cho, Takamatu-shi, Kagawa, Japan
e-mail: mikami@shikoku-ms.com

S. Matsuda (ed.), *Advances in Total Knee Arthroplasty*, https://doi.org/10.1007/978-981-97-4920-1_63

3 Prior Development of Independent Pressure Centers for Medial and Lateral Joint Surfaces

Our basic design consisted of three concepts: first, we placed three load cells each under the medial and lateral joint surfaces of a tibial trial component copied from the joint surface of an original insert; second, we cut a triangular joint surface connecting the center points of each load cell out of a tibial trial body, thus freeing the triangular joint surface, and third, we supported the quarried triangular joint surface by three load cells directly. The feature of this device is that not only implant balance but also the point and pressure of center of gravity are calculated from pressure values of the medial and lateral three load cells, and the line connecting the center points of gravity on the medial and lateral joint surfaces is illustrated as the load mapping, which allows the trajectory of the center of gravity and relative relationship between medial and lateral centers of gravity during full ROM to be applied for kinematic analysis. Thus, cruciate-retaining (CR) and posterior-stabilizing (PS) types of TKAs were developed. Clinical reports [4, 5] comparing the load mapping with the two CR-types of different TKAs using our device suggest the possibility of different design concepts.

4 New Development of the 4-Point Pressure Sensor

In 2016, we developed an improved pressure sensor with four load cells, following the change to the Persona™ TKA CR model (Zimmer Biomet. Warsaw. Indiana. USA). The 4-point pressure sensor is expected to improve the accuracy of kinematic analysis by load mapping because of its significantly improved measurement area of pressure distribution and joint surface coverage compared to the 3-point pressure sensor.

Due to the above reasons, first, Fig. 63.1 shows the pressure value of each load cell and the medial/lateral pressure value of the center of gravity through intraoperative full ROMs in the 24 cases. In extension, approximately 10 to 20 pounds were applied to medial joint surfaces (3) and (8), which were < 6 pounds in the middle ROM, and 12–18 pounds were applied to medial joint surfaces (1), (3), and (6) in deep flexion. These results suggest that in maximal extension and deep flexion, the innermost aspect of the medial joint surface is highly loaded. However, contrary to expectations, the medial pivot motion is induced even without high axial pressure. The pressure value of the center of gravity was consistently higher on the medial than on the lateral joint surface, indicating tension in the medial cruciate ligament.

Next, load mapping showed that the medial pivot was induced at the most medial-middle position in the early phase of flexion. It then rolled back approximately 8 mm to induce a second medial pivot in the late phase of flexion, suggesting the possibility of future detailed kinematic analysis (Fig. 63.2).

5 VERASENSE (3-Point Pressure Sensor) Commercially Available

In 2019, VERASENSE™ (V-type: OrthoSensor, Dania, Florida, USA), a load cell-type pressure sensor dedicated to Persona™ TKA, became commercially available in Japan,

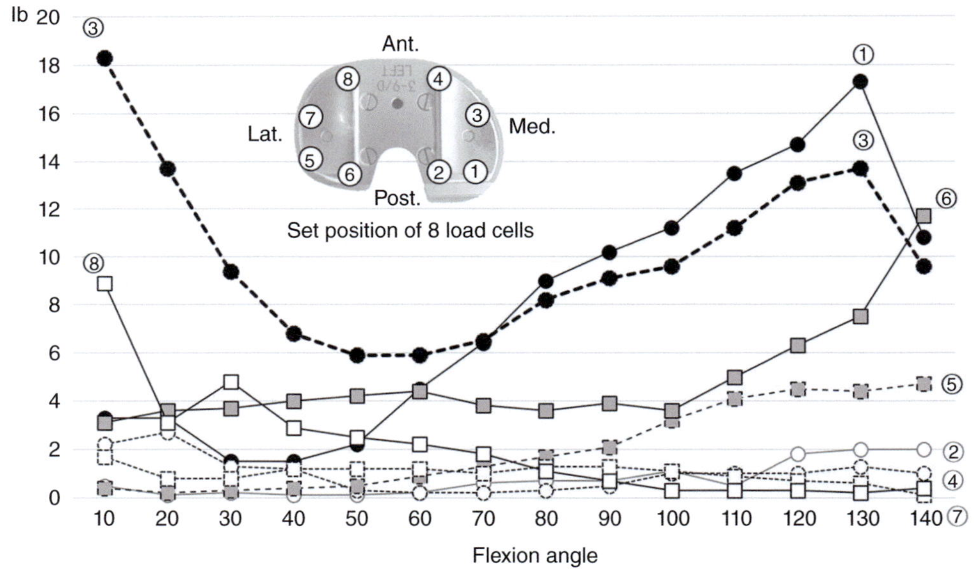

Medial/ lateral pressure value of the center of gravity

Flex. angle	Medial	Lateral
10	24.4	14.1
20	19.2	7.7
30	12.5	9.6
40	9.6	4.5
50	8.5	5.9
60	10.9	8.7
70	13.7	6.9
80	18.2	7.7
90	20.5	8.0
100	21.9	8.2
110	25.3	5.4
120	29.6	11.8
130	32.3	12.7
140	20.4	17.0

Fig. 63.1 Each load cell value and the medial/lateral center-of-gravity pressure value through the full (ROMs)

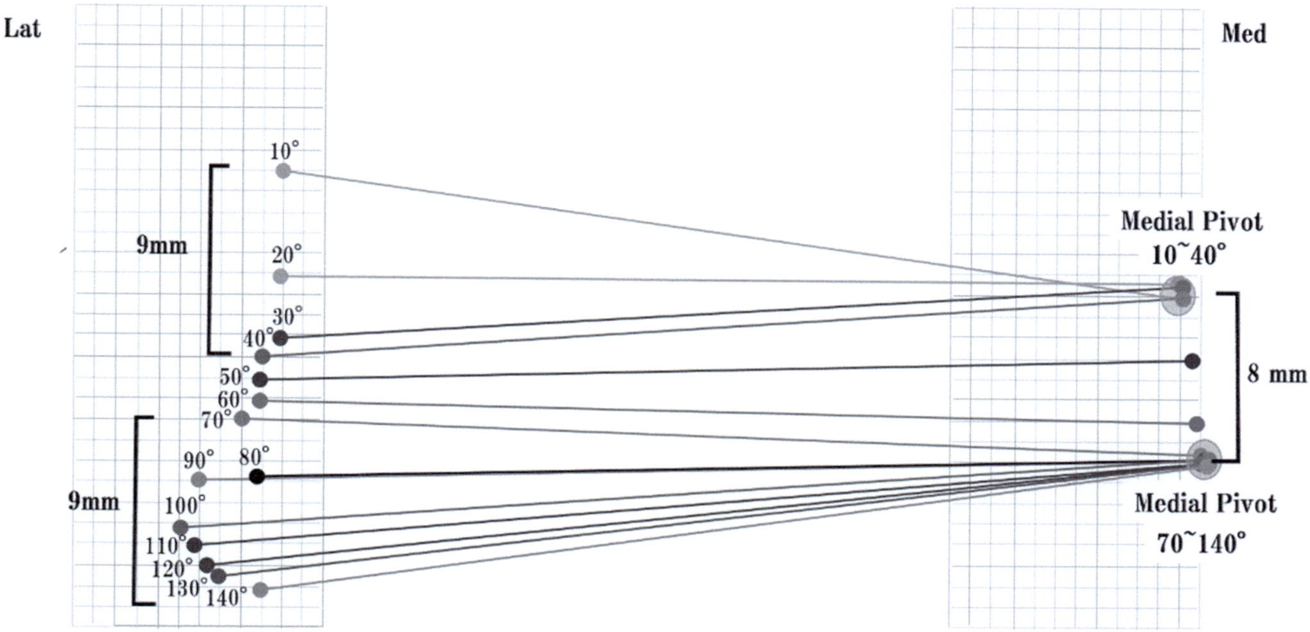

Fig. 63.2 Kinematic analysis by load mapping through the full ROMs

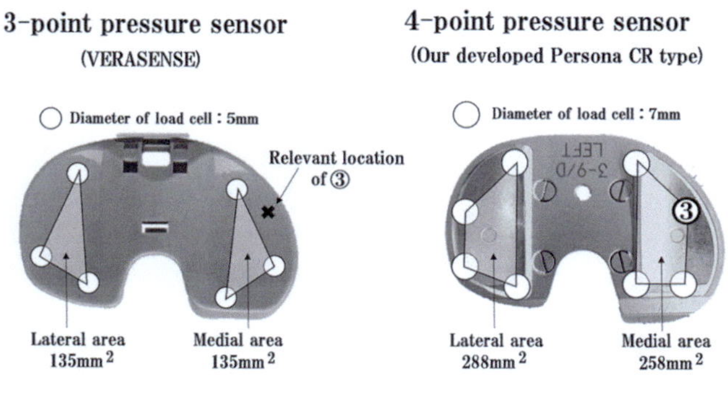

Fig. 63.3 Difference in measurement range area between 3-point and 4-point pressure sensors

and it is expected that new findings will be reported from many facilities as its use grows. Concerns exist regarding the 3-point load cell specification. This is because both the medial and lateral triangular-shaped coverage area of the V-type is approximately 135 mm^2. In comparison, the trapezoid-shaped coverage area of the 4-point pressure sensor is approximately 258 mm^2 on the medial and 288 mm^2 on the lateral, approximately twice the coverage area of the V-type (Fig. 63.3). Furthermore, based on the findings of the previous 4-point pressure sensor, high contact pressure was applied to the load cell (Figs. 63.1, 63.2 and 63.3) at the medial anterior position of the medial joint surface through full ROMs, considered to be an important measurement site. However, V-type (3-point pressure sensor) may exclude the third areas and be concerned with measurement precision. Future studies are needed to compare the V-type pressure sensor and the 4-point pressure sensor in the same case.

However, the commercial diffusion of the V-type pressure sensor is a significant first step in determining the usefulness and importance of pressure sensors, and we are pleased that the 4-point pressure sensor we developed previously is a prototype of the V-type pressure sensor. We are confident that the development of a 4-point pressure sensor based on our accumulated knowledge and a novel kinematic analysis method will improve the intraoperative evaluation method and clinical outcomes of TKA.

References

1. Som MHM, et al. A development of force distribution measurement system with high resolution for total knee arthroplasty. J Adv Comput Intell Intell Inform. 2014;18-2:213–20.
2. Tanabe F, et al. Flexible sensor for real time pressure monitoring in artificial knee joint during total knee arthroplasty. Proceedings of the Annual International Conference of the IEEE Engineering in Medicine and Biology Society. EMBS 2017; 8E+06: pp. 1591–4.
3. Kaneko T, et al. The influence of tibiofemoral joint forces on patient- reported outcome measurements after bicruciate stabilized total knee arthroplasty. J Orthop Surg (Hong Kong). 2020;28:2309499020915106.
4. Mikami H, et al. Intra-operative findings of compressive forces on the tibial joint surface in two different cruciate-retaining total knee arthroplasties. JPN J Joint Dis. 2014;33:41–8.
5. Wada K, et al. Significance of tibiofemoral compression force evaluation in total knee arthroplasty using a digital force sensor. Orthop Sur. 2019;75:206–9.

Part XI

Range of Motion

64 Surgical Techniques to Achieve Full Extension

Ken Okazaki

1 Impact of Residual Extension Limitation

Residual extension limitation after TKA is considered as a contributing factor to poor clinical outcomes [1]. Biomechanical studies using cadaveric knees have shown that the quadriceps muscle required 22% more strength to stabilize the knee joint against loading stresses in 15° flexion than in full extension [2]. A gait analysis study of normal knees with orthotic extension restriction showed that the leg length difference and the left-right difference in stride length applied unnatural stress to both knees, increasing the extension moment and adduction moment by up to 15% in the knee with 15° extension restriction [3]. Extension restriction also affected the trunk alignments; the coronal plane showed that the trunk tilted toward the extension-limited side during standing and walking, while the sagittal plane showed increased pelvic posterior tilt during standing and anterior tilt during walking [4]. In a study of postoperative TKA patients with residual limitation of extension, it was shown that the load during gait was increased on the nonoperated knee [5]. In a clinical study of more than 5000 patients who were evaluated over a minimum of 3 years, patients with residual limitation of 10° or more showed significantly lower Knee Society Scores for pain and function items than those of patients who achieved full extension [6]. Another study reported significantly lower Oxford Score and satisfaction in the patients with extension limitation [7]. The postoperative extension limitation tends to improve between 3 months and 2 years [8, 9], however, it was reported that if extension limitation remained greater than 15° at 3 months postoperatively, the risk of residual limitation greater than 5°was high at 2 years postoperatively [10]. Although an immediate postoperative extension limitation of approximately 10° may be acceptable, the goal should be to achieve a fully extendable knee after the surgery.

K. Okazaki (✉)
Department of Orthopaedic Surgery, Tokyo Women's Medical University, Shinjuku-ku, Tokyo, Japan
e-mail: okazaki.ken@twmu.ac.jp

2 Risk Factors for Residual Extension Limitation

Several studies have reported on the risk factors for residual postoperative extension limitation; for example, a retrospective study of 811 patients showed that the risk factors were male, older age, and preoperative flexion contracture [7]. Intraoperative factors include the importance of obtaining the necessary extension gap for implant placement. A multiple regression analysis of the relationship between preoperative patient factors and extension gap values measured intraoperatively with a tensor showed that the degree of preoperative varus deformity was significantly associated with extension gap values in 59 patients [11]. A study of implant alignments showed that among 120 CR TKA patients, the sagittal alignment of the femoral component in 33 patients with a residual extension limitation of 10° or more at 2 years postoperatively was a mean of 7.3° in flexion, which was significantly more flexed than the mean of 4.2°in the group that achieved full extension [12].

3 Points to Consider in Intraoperative Extension Gap Evaluation

Intraoperative gap assessment is often performed with a spacer block or a tensor with a constant pull-off tension. In both methods, the presence or absence of a posterior femoral condyle influences the assessment of the extension gap. After the osteotomy for posterior femoral condyle, the extension gap without the femoral component (interosteotomy gap) is overestimated, and if this is used as a reference, there is a risk of limitation of extension due to an insufficient extension gap. The extension gap should be evaluated with the femoral component in place.

S. Matsuda (ed.), *Advances in Total Knee Arthroplasty*, https://doi.org/10.1007/978-981-97-4920-1_64

In 42 cases of CR-type TKA for varus knees with the extension-gap-first technique, when the extension gap was remeasured after the resection of posterior femoral condyle, the gap was significantly increased compared with the extension gap that has been evaluated before the posterior osteotomy [13]. The extension gap decreased by approximately 1 mm after the femoral trial component was placed, and the increase in tension of the posterior capsule due to the posterior femoral condyle was considered to be a factor [14]. In fact, in 40% of cases where the osteotomy of the posterior condyle was thinner than the thickness of the posterior condyle of the femoral component and the posterior condylar offset (PCO) of the posterior femoral condyle increased, the extension gap narrowed by more than 2 mm. The use of a custom-made trial component that increased the posterior femoral condyle by 4 mm narrowed the extension gap by 2 to 3 mm, indicating that the PCO has a significant influence on the extension gap. In vitro studies have also shown that the PCO is larger than that of the normal knee in most commercially available TKA models [15]. An investigation of 73 PS-type TKA cases that were evaluated by the relationship between the angle of extension at 4 weeks postoperatively and three values measured intraoperatively (inter-osteotomy extension gap with no femoral trial component, the extension component gap measured with the trial component in place, and the joint laxity with the selected insert thickness subtracted from the extension gap) was reported. The results indicated that there was no correlation with the inter-osteotomy gap, but correlation with the component gap and laxity did exist [16]. The final extension gap was evaluated with the trial component in place, and the difference between that gap and the thickness of the selected tibial insert proved important.

4 Intraoperative Extension Gap Required to Achieve Full Extension

Some reports suggest that mild extension limitation in the immediate postoperative period may improve over time [17]. However, as mentioned above, an extension limitation of 15° or greater is a risk for residual extension limitation [10]. It has also been reported that the extension gap that is measured intraoperatively with a tensor correlates with the extension angle at 1 year postoperatively [18]. Several studies have investigated the extension gap that is required to avoid residual extension limitation. In 127 cases of TKA, the relationship between the extension angle at 2 years postoperatively and the extension gap that was calculated by subtracting the implant thickness of the selected insert from the extension gap as was measured by an intraoperative tensor was examined and revealed that residual limitation of extension existed if the extension gap was less than 2 mm [9]. In a similar study of 75 patients, 20% of the patients with an extension gap of 1 mm or less still had an extension limitation of 5° or more at 1 year postoperatively. The extension gap with the femoral component in place should be greater than the thickness of the tibial component to achieve full postoperative extension.

5 Techniques to Enlarge the Extension Gap

Several methods of magnification have been shown to be effective in cases of an insufficient intraoperative extension gap. Intraoperative extension manipulation was reported as ineffective in 61 of 136 patients, and neither the intraoperative extension gap measurement nor the extension angle that was measured immediately after surgery at 3 months and at 6 months were improved [19]. Resection of osteophytes that are posterior to the femur is effective in widening the extension gap. In 27 CR-type TKAs, changes in the extension gap and ROM were measured by navigation before and after resection of the osteophytes at the posterior femur, and increases in the medial extension gap, extension angle, and flexion angle were reported as 0.7 mm, 4.9°, and 6.5°, respectively [20]. The magnification of the extension angle correlated with the size of the resected osteophyte. The most common method of increasing the extension gap is additional resection of the distal femur, but it should be noted that the amount of additional resection of the distal femur does not equally correspond to the amount of the extension gap expansion. In a study using custom-made femoral trial components with the same shape but that were 4 mm thinner distally to reproduce the same condition as a 4 mm additional distal femoral resection, the increase in the extension gap was only 2 mm compared with that which was measured with the normal trial component in place [21]. It has been argued that the proximal placement of the femoral component may cause the posterior condyle to tense the posterior capsule, which in turn, reduces the extension gap. Thus, tension of the posterior capsule has a significant effect on the extension gap, and many reports suggest that dissection of the posterior capsule is effective in widening the extension gap. In 54 cases of PS-type TKA, 30 patients with an intraoperative extension limitation of 5° or greater underwent dissection of the posterior capsule by approximately 10 mm in the proximal direction, and an approximately 2 mm increase in the extension gap was obtained [22]. A technique to longitudinally dissect the posterior medial capsule is also effective. In 37 cases of PS-type TKA, the extension gap was measured by navigation before and after longitudinal dissection of the posterior medial capsule, and it was reported that the extension gap was increased by approximately 2.7 mm, while the flexion gap was unchanged [23]. Thus, it is necessary to secure an appropriate extension gap to obtain full extension after surgery while using techniques such as complete removal of the osteophytes at the posterior femur, additional distal femoral osteotomy, dissection of the posterior capsule, and longitudinal dissection of the posterior medial capsule.

References

1. Su EP. Fixed flexion deformity and total knee arthroplasty. J Bone Joint Surg Br. 2012;94(11 Suppl A):112–5.
2. Perry J, et al. Analysis of knee-joint forces during flexed-knee stance. J Bone Joint Surg. 1975;57:961–7.
3. Harato K, et al. Knee flexion contracture will lead to mechanical overload in both limbs: a simulation study using gait analysis. Knee. 2008;15:467–72.
4. Harato K, et al. A gait analysis of simulated knee flexion contracture to elucidate knee-spine syndrome. Gait Posture. 2008;28:687–92.
5. Harato K, et al. Extension limitation in standing affects weight-bearing asymmetry after unilateral total knee arthroplasty. J Arthroplast. 2010;25:225–9.
6. Ritter MA, et al. The role of flexion contracture on outcomes in primary total knee arthroplasty. J Arthroplast. 2007;22:1092–6.
7. Goudie ST, et al. Flexion contracture following primary total knee arthroplasty: risk factors and outcomes. Orthopedics. 2011;34:e855–9.
8. Quah C, et al. Fixed flexion deformity following total knee arthroplasty: a prospective study of the natural history. Knee. 2012;19:519–21.
9. Kishimura Y, et al. Changes in postoperative extension angle after total knee arthroplasty: effect of polyethylene insert thickness. J Orthop Sci. 2019;24:674–9.
10. Mitsuyasu H, et al. Flexion contracture persists if the contracture is more than 15° at 3 months after total knee arthroplasty. J Arthroplast. 2011;26:639–43.
11. Wakama H, et al. Preoperative factors associated with extension gap in cruciate-retaining total knee arthroplasty: a retrospective study on. J Orthop Sci. 2020;25:1035–9.
12. Okamoto Y, et al. Sagittal alignment of the femoral component and patient height are associated with persisting flexion contracture after primary. J Arthroplast. 2019;34:1476–82.
13. Minoda Y, et al. Flexion gap preparation opens the extension gap in posterior cruciate ligament-retaining TKA. Knee Surg Sports Traumatol Arthrosc. 2007;15:1321–5.
14. Mitsuyasu H, et al. Enlarged postoperative posterior condyle tightens extension gap in total knee arthroplasty. J Bone Joint Surg Br. 2011;93:1210–6.
15. Onodera T, et al. Posterior femoral condylar offset after total knee replacement in the risk of knee flexion contracture. J Arthroplast. 2013;28:1112–6.
16. Nagai K, et al. Influence of intraoperative soft tissue balance on postoperative active knee extension in posterior-stabilized total knee. J Arthroplast. 2015;30:1155–9.
17. Tsukeoka T, Tsuneizumi Y. Residual medial tightness in extension is corrected spontaneously after total knee arthroplasty in varus knees. Knee Surg Sports Traumatol Arthrosc. 2019;27:692–7.
18. Asano H, et al. Soft tissue tension in extension in total knee arthroplasty affects postoperative knee extension and stability. Knee Surg Sports Traumatol Arthrosc. 2008;16:999–1003.
19. Matsui Y, et al. Intraoperative manipulation for flexion contracture during total knee arthroplasty. J Orthop Surg (Hong Kong). 2019;27:2309499019825574.
20. Sasaki E, et al. Posterior clearance increases the knee extension angle in cruciate retaining type total knee arthroplasty: intraoperative. J Orthop Sci. 2020;25:861–7.
21. Minoda Y, et al. Four-millimeter additional bone resection in the distal femur does not result in an equivalent increase in the extension joint gap. J Arthroplast. 2021;36:958–62.
22. Okamoto Y, et al. Capsular release around the intercondylar notch increases the extension gap in posterior-stabilized rotating-platform total knee arthroplasty. Knee. 2016;23:730–5.
23. Masuda S, et al. Posteromedial vertical capsulotomy selectively increases the extension gap in posterior stabilized total knee arthroplasty. Knee Surg Sports Traumatol Arthrosc. 2020;28:1419–24.

Factors Affecting Postoperative Flexion Angle

65

Yasuo Niki

Range of motion (ROM), particularly knee flexion angle after total knee arthroplasty (TKA), undoubtedly influences clinical scores and patient-reported outcomes in patients engaging in high-flexion daily activities. While the preoperative knee flexion angle is the most critical determinant of postoperative knee flexion angle, several factors are reported to be prerequisites for achieving optimal postoperative flexion.

In addition to non-surgical factors such as obesity, postoperative pain control [1], and patient motivation [2], surgeons should also consider various surgical factors. The following factors are likely to play a role: medial-lateral gap balance [3, 4], flexion-extension gap balance [5, 6], degree of deep knee flexion gap [6–8], quadriceps strain [9], PCL tightness [10–12], posterior tibial slope [13–15], posterior condylar offset (PCO) [15–18], joint line height [19, 20], patella thickness [21], and anterior condyle height (ACH) [22].

Table 65.1 summarizes the contributions of these factors, which may vary depending on the type of prosthesis used (posterior cruciate ligament retaining (CR) or posterior stabilized (PS)-TKA). For instance, the postoperative flexion angle has been reported to improve by 1.5 degrees with a 1 mm increase in PCO and by 0.7 degrees with a 1-degree increase in posterior tibial slope. Conversely, it may worsen by 1.3 degrees with a 1 mm increase in patella thickness [21, 23] and by 1.7 degrees with a 1 mm increase in ACH [22]. The following factors are considered important:

Table 65.1 Surgical factors affecting range of motion after TKA

Factor	References
Medial-lateral gap balance	[3, 4]
Flexion-extension gap balance	[5, 6]
Deep flexion gap	[6–8]
Quadriceps excursion	[9]
PCL tension	[10–12]
Posterior tibial slope	[13–15]
Posterior condylar offset (PCO)	[15–18]
Joint line height	[19, 20]
Patella thickness	[21, 23]
Anterior condyle height (ACH)	[22]

1 PCO

Bellmans et al. pointed out that maintaining a large PCO delays the impingement of the posterior aspect of the femoral bone and posterior margin of the tibial insert during deep knee flexion, resulting in an increased flexion angle [16]. In CR-TKA, rollback is not induced by the post-cam mechanism as in PS-TKA, so in full squat and other deep flexion activities, the impingement of the posterior aspect of the femoral bone and posterior edge of the tibial insert easily occur. Therefore, it becomes crucial to replicate at least the preoperative PCO postoperatively.

Using a posterior reference guide makes it relatively easy to control the amount of osteotomy to maintain the medial PCO. However, in mechanically aligned (MA) TKA, both the measured resection technique and the gap balancing technique can lead to a change in lateral PCO. On the other hand, kinematically aligned (KA) TKA does not cause any PCO change. Generally, the medial side exhibits less rollback and is more likely to cause impingement of the posterior aspect of the femoral bone and tibial inserts, while the lateral side has a larger amount of rollback and is less likely to cause impingement. As a result, a change in lateral PCO has not been considered a drawback of MA-TKA. In CR-TKA, it is essential to maintain the preoperative medial PCO, while appropriately managing the medial flexion gap and posterior cruciate ligament (PCL) tension.

Y. Niki (✉)
Center for Musculoskeletal Surgery and Research, Fujita Medical Innovation Center Tokyo, Haneda, Tokyo, Japan
e-mail: yasuo.niki@fujita-hu.ac.jp

S. Matsuda (ed.), *Advances in Total Knee Arthroplasty*, https://doi.org/10.1007/978-981-97-4920-1_65

2 Posterior Tibial Slope

Increasing the posterior tibial slope improves flexion in CR-TKA by reducing impingement of the posterior margin of the tibial insert and by reducing tension in the PCL. However, excessive posterior slope can cause the femur to be positioned posteriorly when the knee is in extension, resulting in paradoxical knee motion, which negatively affects flexion. Conversely, excessive posterior slope in posterior-stabilized total knee arthroplasty (PS-TKA) increases the risk of a large flexion gap, delays the timing of post-cam contact, and worsens flexion. It has been reported that a 5° increase in posterior slope delays post-cam contact by 18° in flexion, thus emphasizing the importance of avoiding excessive posterior slope in PS-TKA.

3 ACH

Overstuffing of the patellofemoral joint has long been associated with various complications, including increased shearing stress, anterior knee pain, polyethylene wear of the patellar component, and impaired flexion. Both patellar thickness and ACH play pivotal roles in this, with an increased medial ACH particularly exacerbating the flexion. As previously mentioned, a 1 mm increase in patellar thickness has been reported to worsen flexion by approximately 1 to 1.3 degrees, while a 1 mm increase in ACH is associated with a flexion deterioration of 1.7 degrees. Additionally, computer simulations have revealed that an increase in ACH leads to elevated patellofemoral contact stress [22]. By being mindful of these factors, surgeons can better manage the patellofemoral joint and avoid potential complications related to overstuffing. Careful consideration of patella thickness and ACH during TKA is crucial to achieve optimal postoperative flexion and overall patient reported outcomes.

4 Deep Flexion Gap

Traditionally, many Japanese elderly patients prefer seiza-sitting due to the floor-sitting lifestyle, and achieving deep flexion motion greatly impacts patient satisfaction. Seiza-sitting is a highly stressful position in which the feet are turned sole upwards underneath the buttocks with the tibia internally rotated. Although the number of patients performing seiza-sitting has recently been decreasing, it is crucial to maintain at least the preoperative flexion angle postoperatively. Data from Japanese patients who underwent PS-TKA reveal that approximately 7% can perform seiza-sitting after TKA [24].

Two primary factors influencing the ability to perform seiza-sitting are the preservation of the deep flexion gap and the appropriate rotational alignment of the tibial component [8]. If the tibial component is excessively internally rotated, and the tibia is externally rotated in relation to the femur, the patella tends to shift laterally with increasing knee flexion, resulting in elevated patellofemoral contact pressure.

Furthermore, cases with limited quadriceps excursion exhibit a smaller deep flexion gap and reduced flexion angle. In CR-TKA, the deep flexion gap is related to PCL tension. However, in PS-TKA, where the PCL does not affect the flexion gap, quadriceps excursion becomes more critical. Therefore, when the quadriceps muscle is contracted or the patellar tendon length is shortened in PS-knees, the deep flexion gap becomes smaller, and flexion is limited. To address this, a rectus snip or V-Y quadricepsplasty is performed to enhance flexion.

References

1. Fu P, et al. Efficacy of intra-articular cocktail analgesic injection in total knee arthroplasty – a randomized controlled trial. Knee. 2009;16:280–4.
2. Ayers DC, et al. Total knee replacement outcome and coexisting physical and emotional illness. Clin Orthop Relat Res. 2005;440:157–61.
3. Matsuda Y, et al. Varus-valgus balance and range of movement after total knee arthroplasty. J Bone Joint Surg Br. 2005;87:804–8.
4. Unitt L, et al. Short-term outcome in total knee replacement after soft-tissue release and balancing. J Bone Joint Surg Br. 2008;90:159–65.
5. Matsumoto T, et al. Influence of intra-operative joint gap on post-operative flexion angle in osteoarthritis patients undergoing posterior-stabilized total knee arthroplasty. Knee Surg Sports Traumatol Arthrosc. 2007;15:1013–8.
6. Higuchi H, et al. Relationship between joint gap difference and range of motion in total knee arthroplasty: a prospective randomised study between. Int Orthop. 2009;33:997–1000.
7. Watanabe T, et al. Intraoperative joint gaps affect postoperative range of motion in TKAs with posterior-stabilized prostheses. Clin Orthop Relat Res. 2013;471:1326–33.
8. Niki Y, et al. Factors affecting the achievement of Japanese-style deep knee flexion after total knee arthroplasty using posterior-stabilized. J Orthop Sci. 2015;20:1012–8.
9. Gejo R, et al. Joint gap changes with patellar tendon strain and patellar position during TKA. Clin Orthop Relat Res. 2008;466:946–51.
10. Dennis DA, et al. In vivo anteroposterior femorotibial translation of total knee arthroplasty: a multicenter analysis. Clin Orthop Relat Res. 1998;356(356):47–57.
11. Dennis DA, et al. Multicenter determination of in vivo kinematics after total knee arthroplasty. Clin Orthop Relat Res. 2003;416:37–57.
12. Wyss T, et al. Tension controlled ligament balanced total knee arthroplasty: 5-year results of a soft tissue orientated surgical technique. Arch Orthop Trauma Surg. 2008;128:129–35.
13. Fujimoto E, et al. Significant effect of the posterior tibial slope and medial/lateral ligament balance on knee flexion in total knee arthroplasty. Knee Surg Sports Traumatol Arthrosc. 2013;21:2704–12.
14. Shi X, et al. The effect of posterior tibial slope on knee flexion in posterior-stabilized total knee arthroplasty. Knee Surg Sports Traumatol Arthrosc. 2013;21:2696–703.
15. Kim JH. Effect of posterior femoral condylar offset and posterior tibial slope on maximal flexion angle of the knee in the posterior cruciate ligament. Knee Surg Relat Res. 2013;25:54–9.

16. Bellemans J, et al. Fluoroscopic analysis of the kinematics of deep flexion in total knee arthroplasty. Influence of posterior condylar offset. Joint Surg Br. 2002;84:50–3.
17. Massin P, et al. Optimization of the posterior condylar offset, tibial slope, and condylar roll-back in total knee arthroplasty. J Arthroplast. 2006;21:889–96.
18. Hanratty BM, et al. The influence of posterior condylar offset on knee flexion after total knee replacement using a cruciate-sacrificing mobile-bearing implant. J Bone Joint Surg Br. 2007;89:915–8.
19. Figgie HE 3rd, et al. The influence of tibial-patellofemoral location on the function of the knee in patients with a posterior stabilized condylar knee. J Bone Joint Surg Am. 1986;68:1035–40.
20. Meneghini RM, et al. The effect of the Insall-Salvati ratio on outcome after total knee arthroplasty. J Arthroplast. 2006;21(6 Suppl 2):116–20.
21. Abolghasemian M, et al. Effect of patellar thickness on knee flexion in total knee arthroplasty:a biomechanical and experimental study. J Arthroplast. 2014;29:80–4.
22. Nishitani K, et al. A greater reduction in the distal femoral anterior condyle improves flexion after total knee arthroplasty in patients with. Knee. 2019;26:1364–71.
23. Kim AD, et al. The effect of patellar thickness on intraoperative knee flexion and patellar tracking in patients with arthrofibrosis undergoing total. J Arthroplast. 2016;31:1011–5.
24. Niki Y, et al. Is greater than 145 degrees of deep knee flexion under weight-bearing conditions safe after total knee arthroplasty? Bone Joint J. 2013;95:782–7.

66 Surgical Techniques to Achieve Deep Flexion

Yoshinori Kadoya

In Japan, a wide range of knee ROM is required in daily life due to the unique lifestyle practice of sitting on the floor, and there is a strong potential demand for deep flexion even after TKA. However, most conventional knee joint motion analyses have focused on flexion angles up to 130°, and little analysis has been conducted on flexion angles that are greater than that measurement. Given this limitation, interest in normal knee joints and deep flexion after TKA has increased since the 1990s and has been the subject of focus at academic conferences and symposia; moreover, several TKA designs for deep flexion have been introduced recently.

In recent years, with the general increase of ROM due to improved surgical techniques and the further westernization of lifestyles, the widespread interest in deep flexion has decreased considerably. However, important findings regarding deep flexion have accumulated in Japan, and a proper understanding of these findings will provide valuable guidelines for safely increasing ROM in the future. In this chapter, we summarize the findings on deep flexion in the normal knee and those after TKA and discuss surgical techniques and implant designs that are used to achieve deep flexion, including a review of literatures.

1 What Is Deep Bending?

The maximum active flexion angle of the normal knee joint (the flexion angle obtained by contraction of the knee flexor muscles) is approximately 130°, and any flexion beyond this angle is achieved only when an external force is applied. The ultimate state, sitting upright in a kneeling position, is "maximum flexion by body weight," and the flexion angle exceeds 160°. The key point of this extreme deep flexion position is that it is a "forced position by external force" that cannot be achieved by the patient's own muscle contraction. In fact, various kinematic abnormalities are observed at flexion angles of 130° or greater, showing a different aspect from the previous flexion range [1, 2]. It is therefore appropriate to define flexion of 130° or greater as deep flexion in the normal knee and to distinguish it from flexion that occurs at an angle of less than 130°. Furthermore, at the "Challenges to Deep Flexion after TKA" symposium that was held at the 33rd Annual Meeting of the Japanese Society for Artificial Knee Arthroplasty in 2003, it was agreed that deep flexion after TKA should be defined as 130° or greater.

Y. Kadoya (✉)
Hanwa Joint Reconstruction Center, Osaka, Japan
e-mail: kadoya.yoshinori@kinshukai.or.jp

2 Knee Kinematics in Deep Flexion

2.1 Kinematics of the Tibiofemoral Joint [1] (Fig. 66.1)

The posterior movement of the medial femoral condyle on the tibiofemoral joint surface during flexion is clearly smaller than that of the lateral condyle (Fig. 66.1c), and medial pivot movement centered on the medial joint surface is seen overall. However, from 120° to deep flexion, the posterior movement of the medial femoral condyle also increases, resulting in what is called bicondylar rollback [3]. Notable here is the marked posterior shift of the lateral femoral condyle and subluxation of the tibiofemoral joint that are accompanied by marked dislocation of the meniscus (Fig. 66.1f).The medial condyle is lifted from the articular surface, with impingement of the proximal part of the medial femoral condyle and the posterior tibial border (Fig. 66.1c).

2.2 Kinematics of the PF Joint [2]

The contact surface at 90° is continuous medially and laterally proximal to the patella (Fig. 66.2a). At maximum active flexion, the medial patellar facet loses contact, but the odd facet comes into contact, and the total contact area remains almost unchanged (Fig. 66.2b). However, the patellar contact

S. Matsuda (ed.), *Advances in Total Knee Arthroplasty*, https://doi.org/10.1007/978-981-97-4920-1_66

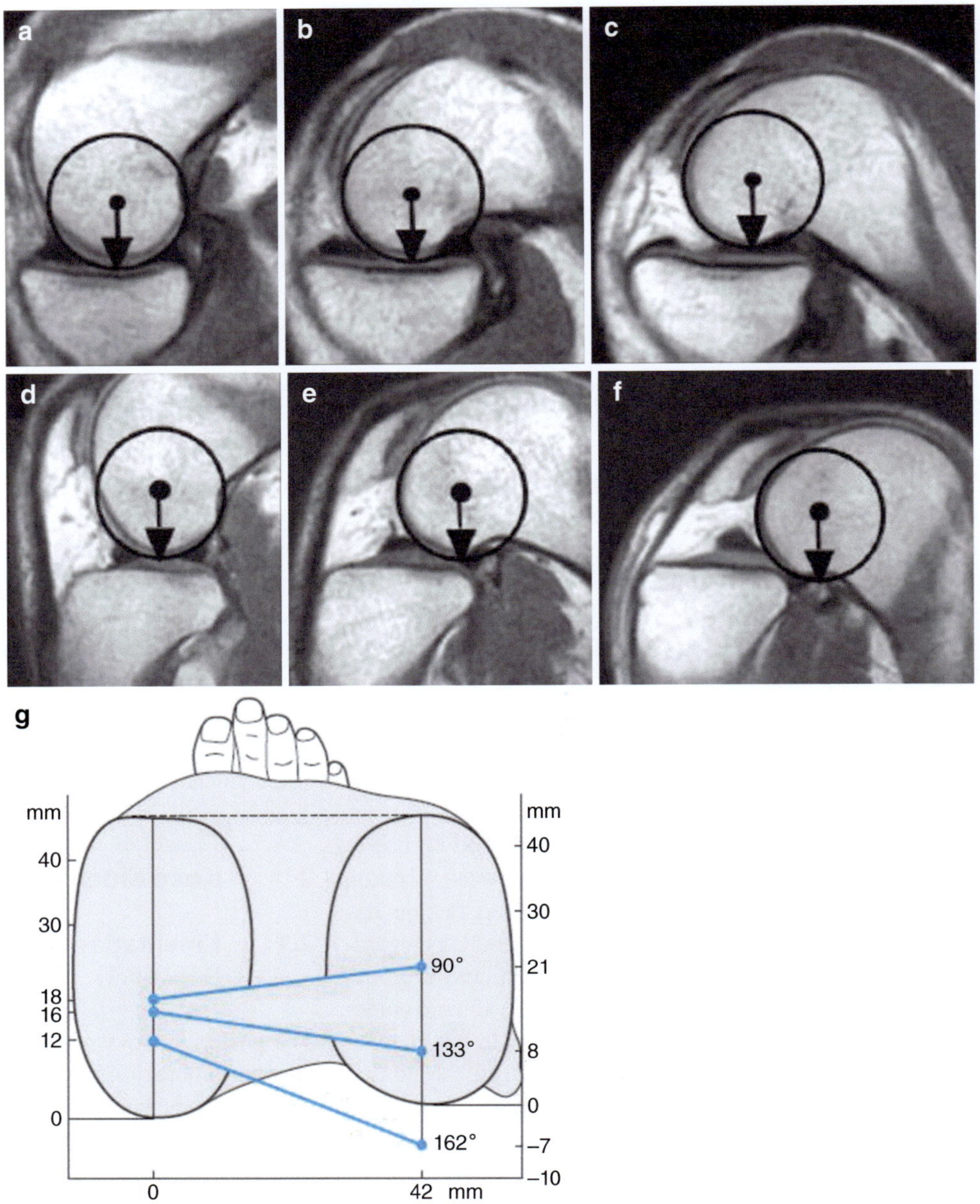

Fig. 66.1 Tibiofemoral joint kinematics. (**a**) Medial side:90° flexion. (**b**) Medial side: Active maximum flexion. (**c**) Medial side: Passive maximum flexion. (**d**) Lateral side: 90° flexion. (**e**) Lateral side: Active maximum flexion. (**f**) Lateral side: Passive maximum flexion. (**g**) Active maximum flexion m = 133° ± 9°, Passive maximum flexion = 162° ± 9°

area decreases when the patella reaches passive maximum flexion with external force, which is due to the loss of contact of the superior aspect of the lateral facet (Fig. 66.2c). The patellar odd facet begins to make contact with the femoral intercondylar area when the knee is flexed more than 130°, and as the flexion progresses further, the odd facet becomes stuck in the intercondylar area of the corresponding distal femoral medial condyle, In deep flexion, the patellar contact area decreases, suggesting that various decompression mechanisms work in the normal knee to prevent an excessive increase in contact pressure. These mechanisms include the anteromedial shift of the tibial rough surface due to internal rotation of the tibia, contact of the quadriceps tendon with the condylar area, and hydrostatic pressure from the fat pad.

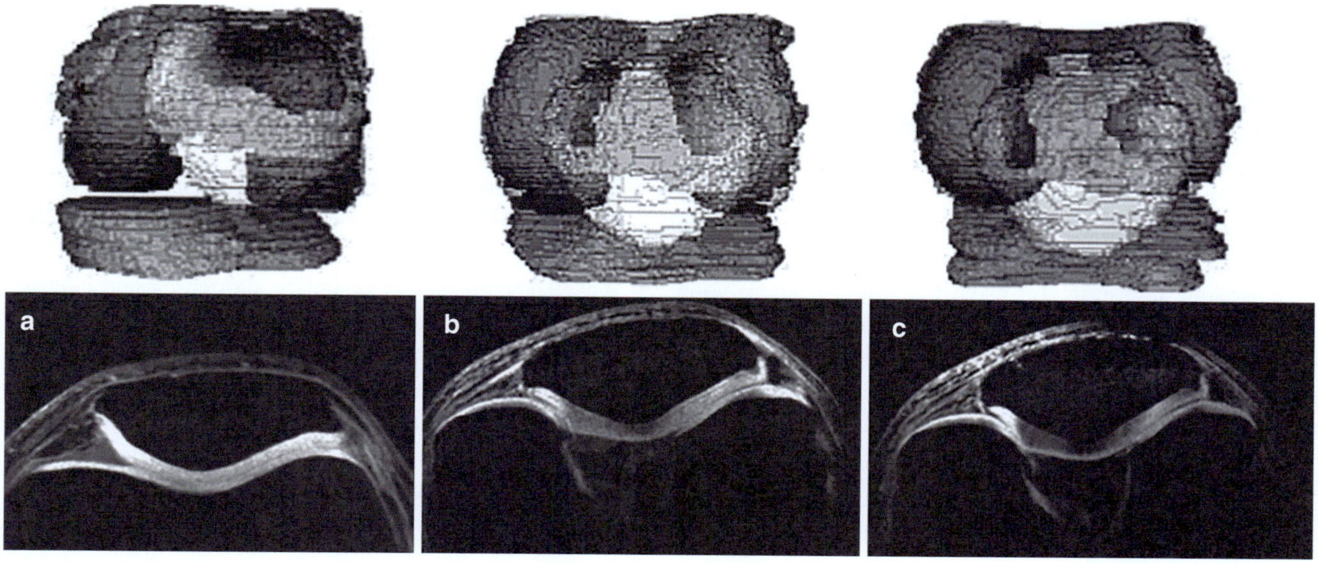

Fig. 66.2 Pattelofemoral joint kinematics. (**a**) 90° flexion. The contact surface is proximal to the patella and continuous with the internal and external articular surfaces (arrow). (**b**) Active maximum flexion. The contact surface is divided into an odd facet and a lateral facet; the medial facet is no longer in contact. The contact area is almost unchanged. (**c**): Passive maximum flexion. The distal portion of the odd facet begins to come into contact. The overall contact area is significantly reduced

3 Deep Flexion After TKA and Its Complications

As shown in the analysis of the normal knee, deep flexion, especially in the upright sitting position, is a "forced limb position by external force" and is considered as a special position that is similar to subluxation. It is not difficult to imagine that the knee structures (ligaments, meniscus, and cartilage) are subjected to a large mechanical load in this position, although various cushioning mechanisms are thought to be in operation in the normal knee. In deep flexion after TKA, the joint surface is replaced with metal and polyethylene, and the normal motion patterns and cushioning mechanisms are lost, so the risk of wear and tear is high.

It has been reported that deep flexion of 130° or more is possible in approximately half of patients after TKA, depending on the patient's background and the skill of the surgeon [4]. Although the majority of patients have good results and no significant complications have been reported [5], complications have been reported in Japan in patients who routinely sit upright or who are highly active [6]. The complications include dislocation, breakage of the polyethylene, breakage of the PS-type post, lifting of the mobile mechanism, inflammation and pain of the infra-patellar fat pad, and so on. Lift-off motion and pressure on the fat pad are factors that contribute to complications and require sufficient attention owing to the relatively early onset of disease [6]. On the other hand, in a dynamic analysis of deep flexion after TKA, Kanekasu et al. reported that lateral lift-off was observed in approximately half of patients with flexion of 130° or greater, highlighting the risk of insert injury in deep flexion [7].

The most important goal after TKA is to maintain a pain-free and supportive joint for a long period of time; the goal is never to achieve deep flexion. Considering that an increase in ROM beyond a certain level, especially in the ability to sit upright, does not directly lead to increased patient satisfaction, it is essential to teach patients to avoid activities that involve deep flexion and sitting upright.

4 Surgical Technique to Obtain Deep Flexion

Unfortunately, there is no special trick to achieve deep flexion, and the only way to improve ROM is to understand the various techniques that are involved and to perform each of them carefully and confidently. Moreover, it is important to select an implant design that is safe even in deep bending.

Postoperative ROM is influenced by various factors, but the most important factor is the extensibility of the extensor mechanism (degree of contraction). Therefore, it is well known that preoperative ROM is the factor that has the greatest influence on postoperative ROM. Other factors that can be controlled by the surgeon include the following:

- PCL
- Posterior Clearance

(a) Maintenance of the posterior condylar offset(PCO) and prevention of posterior impingement
  - Extensor mechanism (quadriceps and patella)

4.1 PCL

In general, it is often reported that the PS type is superior to the CR type in terms of ROM. The reasons for this are that the preserved PCL does not always function and rollback is not reproduced, and if the PCL tension is not adequately adjusted, ROM restriction may occur due to overstretching in the flexed position. Therefore, it is essential to adjust the PCL tension appropriately in the CR type to obtain good ROM. On the other hand, the posterior impingement that is described below is less likely to occur in the PS type because the post-cam mechanism induces rollback forcibly and mechanically, which is considered advantageous for ROM.

4.2 Posterior Clearance

Theoretically, if posterior clearance is secured, flexion can be maximized to the extent that the extensor mechanism permits. Thus, it is essential to prevent impingement of the posterior component, and for this purpose, it is important to perform surgery while keeping the rollback amount and PCO in mind.

Rollback theoretically prevents posterior impingement; therefore, the PS type is advantageous (Fig. 66.3). Another factor affecting posterior clearance is the presence of osteophytes on the posterior surface of the femoral condyle, which must be removed to avoid postoperative impingement. In this case, resection of the PCL facilitates access to the posterior femur and removal of the osteophytes (Fig. 66.4).

PCO is defined as the distance between the posterior surface of the femoral cortex and the posterior femoral condyle [8]. As the PCO decreases, impingement that is associated with bending is more likely to occur (Fig. 66.5). Moreover, when the PCO is decreased, the decrease in ROM is more pronounced in the CR type, while the PS type is relatively unaffected.

4.3 Extensor Mechanisms (Quadriceps and Patella)

It is well known that preoperative shortening and contracture of the quadriceps muscle has a significant influence on postoperative ROM. Although there are reports that dissection of the quadriceps tendon from the anterior aspect of the femur contributes to achieving deep flexion [9], we have no experience with this. We believe that the indications for invasion of the extensor mechanism are very limited. Considering that the most important goals after TKA are pain relief and improvement of ADLs and that achieving deep flexion does not directly lead to improved patient satisfaction, we believe that the indications for invasion of the extension mechanism are very limited. Although deepening the patella groove of the femoral component by design can reduce contact pressure on the PF joint and increase ROM, this approach also has its limitations because it is associated with a reduction in the quadriceps lever arm and decreased muscle efficiency.

The most important thing is to ensure that the basic techniques are performed to obtain good PF tracking. Specifically, the rotational position of the femoral and tibial components must be carefully positioned to avoid internal rotation. Moreover, it is important to subsequently place the patellar component medially and in line with the superior patellar border. In addition, peripatellar synovectomy can prevent patella clunk in PS. Anterior knee pain during deep flexion is a relatively common complication that occurs when the

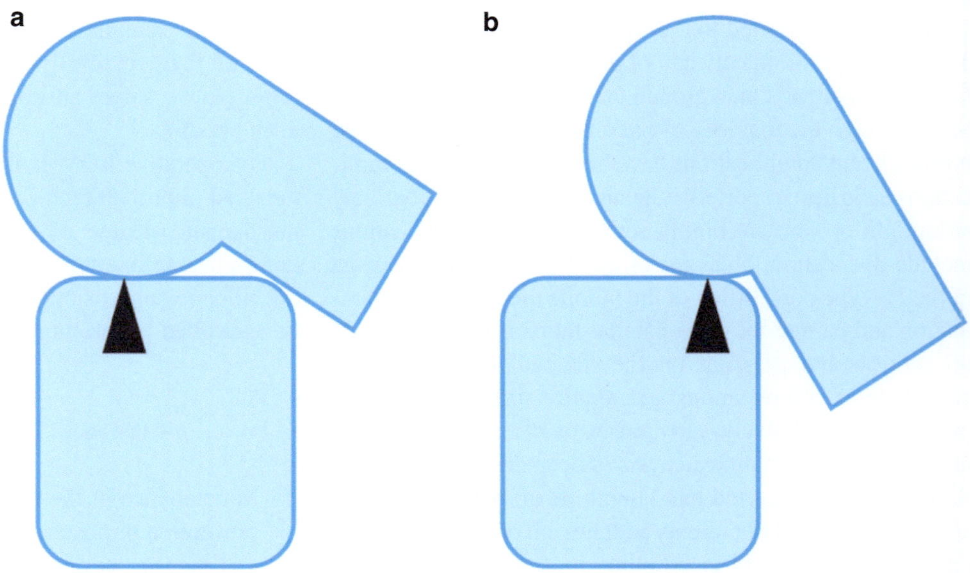

Fig. 66.3 Rollback and backward impingement. (**a**) If rollback does not occur, backward impingement occurs, and ROM is limited. (**b**) When rollback occurs, backward impingement is avoided, and ROM increases

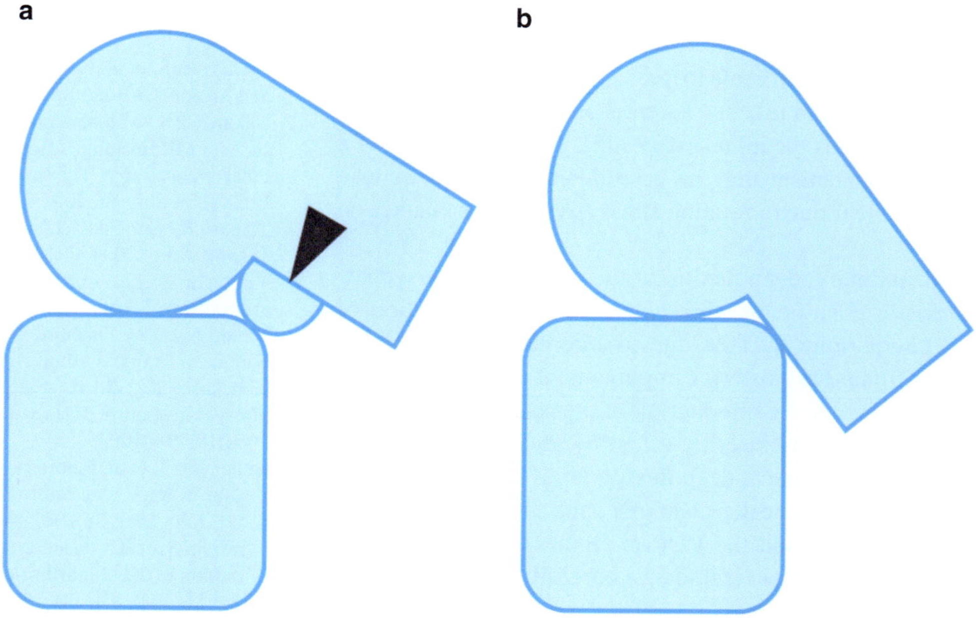

Fig. 66.4 Removal of osteophytes from the posterior femur. (**a**) The presence of osteophytes causes posterior impingement and limits ROM. (**b**) Osteophyte removal avoids posterior impingement and increases ROM

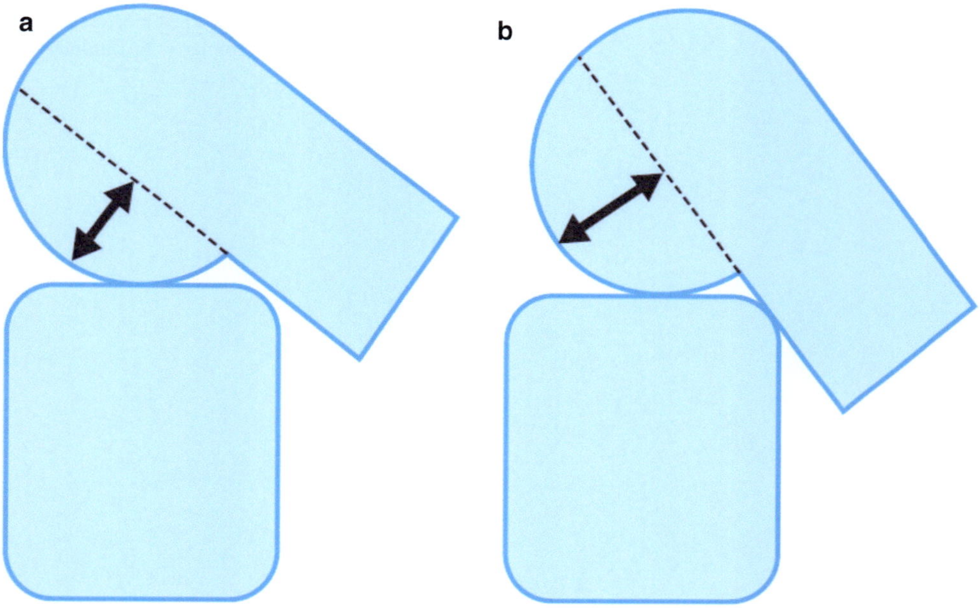

Fig. 66.5 Posterior condylar offset (PCO) and posterior impingement. (**a**) Small PCO causes posterior impingement and limits ROM. (**b**) A large PCO avoids posterior impingement and increases ROM

infrapatellar fat pad is repeatedly trapped between the patellar component and the polyethylene insert. Resection of the fat pad (approximately 70–80% of the total volume) is recommended for prevention of this complication [6].

5 Implant Design for Deep Flexion

In knees with good ROM, impingement between the extensor mechanism and the anterior portion of the polyethylene insert is likely to occur at maximum flexion. For this reason, many implants are designed to prevent impingement by reducing the anterior part of the polyethylene insert.

A major issue in achieving deep flexion after TKA has been to address the internal rotation of the tibia, as seen in the normal knee. However, even in patients who have achieved deep flexion after TKA, this angle of rotation is smaller than that of the normal knee [7], and intraoperative measurements show that the direction and magnitude of rotation in the maximal flexion position with other forces are not constant. The bisurface (KU) TKA that was developed in Japan is characterized by a ball and socket-shaped secondary joint set posterior to the joint [10]. This structure is designed to achieve both reproducible rollback in the PS position and rotational motion in the flexed position. The ball and socket shape is an excellent design for deep flexion in terms of stability and rotational motion tolerance [11]. Even if deep flexion can be achieved after TKA, the rotation position is difficult to predict, and it is necessary to address wear and tear due to stress on the post-cam mechanism as well as the joint surfaces [5]. The introduction of a mobile mechanism may be a promising option to reduce wear and tear due to rotation stress (see also the chapter on PSRP).

In summary, deep flexion of the normal knee is a characteristic position in which (1) posterior subluxation of the lateral joint surface due to internal rotation of the tibia, (2) medial-posterior impingement and lift-off of the joint surface, and (3) reduction of the patellofemoral joint contact area and buffering effect by the fat pad, etc., are observed. To achieve deep flexion after TKA, it is important to secure posterior clearance, and the only way to do so is to understand the various factors that are involved and to perform basic techniques carefully and confidently. This understanding is also key to preventing complications.

References

1. Nakagawa S, et al. Knee kinematics in deep flexion: analysis with MR imaging. J Bone Joint Surg Br. 2000;82:1201–3.
2. Nakagawa S, et al. Kinematics of the patella in deep flexion: analysis with MR imaging. J Bone Joint Surg Am. 2003;85:1238–42.
3. Pinskerova V, et al. The knee in full flexion: an anatomical study. J Bone Joint Surg Br. 2009;91:830–4.
4. Tomari K. Surgical techniques to achieve deep flexion. Orthop Disaster Surg. 2004;47:137–44.
5. Tarabichi S, et al. Achieving deep flexion after primary total knee arthroplasty. J Arthroplast. 2010;25:219–24.
6. Kondo M, et al. Outcome and complications of knee arthroplasty with deep flexion. Orthop Disaster Sur. 2004;47:171–80.
7. Kanekasu K, et al. Fluoroscopic analysis of knee arthroplasty kinematics during deep flexion kneeling. J Arthroplast. 2004;19:998–1003.
8. Bellemans J, et al. Fluoroscopic analysis of the kinematics of deep flexion in total knee arthroplasty: influence of posterior condylar offset. Joint Surg Br. 2002;84:50–3.
9. Tarabichi S, et al. Can anterior quadriceps release improve the range of motion in the stiff arthritic knee? J Arthroplast. 2010;25:571–5.
10. Akagi M, et al. The bisurface TKA: a unique design for flexion. J Bone Joint Surg Am. 2000;82:1626–33.
11. Nakamura S, et al. Fluoroscopic and computed tomographic analysis of knee kinematics during very deep flexion after total knee arthroplasty. J Arthroplast. 2010;25:486–91.

Part XII

Postoperative Rehabilitation

67 Range of Motion (ROM) and Gait Function

Eiichi Tsuda

1 Range of Motion (ROM)

1.1 ROM in Normal Knee Joints and Osteoarthritis

The primary motions of the normal knee joint are extension and flexion, and the passive ROM is approximately 0–160°. Active extension ROM is 0°. In contrast, the flexion ROM is affected by the hip joint position and is 140° in hip flexion and 120° in hip extension because the medial and lateral hamstrings, which are the agonist muscles of knee flexion, are biarticular muscles. Osteoarthritis of the knee causes various degrees of ROM limitation depending on the degree of progression, which leads to difficulties in carrying out daily activities (Table 67.1) [1]. Factors that cause ROM limitation include pain, knee joint effusion, periarticular muscle tightness, osteophyte formation, and ligament and joint capsule degeneration.

As advanced-stage osteoarthritis of the knee, for which TKA is indicated, presents with various clinical symptoms and complications, patients have a wide range of expectations for TKA [2]. Table 67.2 shows the expectations of the patients who underwent TKA based on preoperative interviews. Many of these expectations include activities that require a certain degree of knee joint flexion ROM, such as stair climbing, sitting and squatting. In other words, improvement in ROM after TKA is essential for meeting preoperative expectations and increasing postoperative satisfaction.

Table 67.1 Knee joint flexion-extension angles in activities of daily living

	Healthy knees	Pre-TKA	Post-TKA[a]
Active flexion	135.5 ± 5.6	104.3 ± 13.7	96.9 ± 13.2
Level walking	67.4 ± 6.3	50.0 ± 14.0	59.1 ± 10.0
Slope ascent[b]	64.5 ± 7.7	48.2 ± 14.5	57.7 ± 9.6
Slope descent[b]	72.1 ± 8.2	53.5 ± 14.6	61.9 ± 9.9
Stairs ascent[c]	98.5 ± 10.3	68.1 ± 20.1	78.3 ± 12.1
Stair descent[c]	97.3 ± 8.7	67.1 ± 23.9	75.7 ± 16.1
Into low chair[d]	101.7 ± 11.9	77.4 ± 17.8	77.9 ± 16.5
Out of low chair[d]	104.9 ± 9.5	80.1 ± 17.9	79.4 ± 14.2
Into standard chair[e]	99.0 ± 11.4	77.3 ± 16.0	78.9 ± 14.0
Out of standard chair[e]	99.3 ± 10.1	79.6 ± 14.9	81.3 ± 12.6
Into bath[f]	131.0 ± 13.4	76.5 ± 23.1	78.0 ± 14.8
Out of bath[f]	138.1 ± 13.6	77.1 ± 22.1	77.5 ± 17.7

Adapted from Ref. [1]
[a]18–24 months postoperatively
[b]inclination 5°
[c]kick height 16.5 cm, tread 28.0 cm
[d]seat height 38.0 cm
[e]seat height 46.0 cm
[f]straddle height 59.0 cm

1.2 ROM Exercise with Preoperative Rehabilitation Therapy

When surgery is decided upon, the patient must visit the rehabilitation department for preoperative evaluation. The knee ROM is measured during both passive and active movements. If possible, basic daily activities are recorded on video, and the maximum extension and flexion angles are also measured during several activities. As preoperative ROM strongly affects postoperative ROM, any existing limitations should be corrected as much as possible before surgery [3]. Pain medication should be continued, and joint mobilization exercises should be performed after puncture for extreme joint effusion that contributes to ROM limitations. Home-based rehabilitation programs include ROM exercises using the active and active-assisted exercises described below and stretching of the periarticular muscles.

E. Tsuda (✉)
Department of Rehabilitation Medicine, Hirosaki University Graduate School of Medicine, Hirosaki, Aomori, Japan
e-mail: eiichi@hirosaki-u.ac.jp

S. Matsuda (ed.), *Advances in Total Knee Arthroplasty*, https://doi.org/10.1007/978-981-97-4920-1_67

Table 67.2 Expectations for TKA

Order	Expectations
1	Improved ability to walk
2	Relieve daytime pain in joint
3	Improve ability to go down stairs
4	Improve ability to climb stairs
5	Improve ability to get in or out of a bed, chair, car, or bus
6	Improve ability to perform daily activities around the home
7	Relieve pain in the joint that interferes with sleep
8	Improve ability to perform daily activities away from the home
9	Improve ability to kneel
10	Improve psychological well-being
11	Make knee or leg straight
12	Remove the need for a stick or other assistive device
13	Improve ability to squat
14	Improve ability to exercise or take part in recreational sports
15	Improve ability to participate in recreational or social activities
16	Sexual activity
17	Be employed for monetary reimbursement

Adapted from Ref. [2]

1.3 ROM Exercise with Postoperative Rehabilitation Therapy

Information regarding the surgery (approach, TKA implants used, bone grafting, intraoperative ROM, intraoperative complications, and pain control procedures) is obtained from the surgical records, and ROM training is planned in the postoperative rehabilitation program. Postoperative lateral radiographs are used to evaluate femoral component alignment, posterior condylar offset, and posterior tibial component tilt, which affect extension and flexion ROM.

The following three modalities of ROM exercises are usually used as needed, depending on the involvement of the periarticular muscles and the presence or absence of external force assistance.

1. Active-mode ROM exercise

A training style in which the target joint is moved by voluntary contraction of the periarticular muscles. Depending on the posture, the weight of the body part distal to the knee joint may assist; however, in principle, no external force is used to assist. It is often used for ROM exercise when the occurrence or exacerbation of pain is a concern.

2. Active-assisted-mode ROM exercise

A training style in which the knee joint is moved with the aid of an external force, in addition to voluntary muscle contraction. The source of the external force is the therapist or the patient's own limb, other than the involved limb.

3. Passive-mode ROM exercise

A form of exercise in which the joint is moved by an external force without contraction of the muscles acting on the knee joint. The source of the external force is the therapist, the patient's own limb other than the involved limb, or specialized equipment, such as continuous passive motion (CPM) devices (Fig. 67.1).

ROM training should be started the day after surgery, and although the effectiveness of the CPM device in improving ROM remains controversial, it is considered an effective method because it can be performed at the bedside in the early postoperative period when the patient has a strong fear of experiencing pain [4]. If possible, the patient is transferred to the rehabilitation room in a wheelchair for early release from bed, and the ROM exercise is performed under the guidance of a therapist. The ROM exercises mainly consist of manual extension training. The knee joint is gradually flexed to prevent defensive muscle contraction, with the quadriceps muscle relaxing in the hip-flexion position (Fig. 67.2a). In flexion ROM exercises beyond 90° flexion, the posterior aspect of the lower leg is supported to prevent it from falling backward and inducing rollback of the femoral condyle (Fig. 67.2b). When advancing to a deeper flexion position, the thigh on the operative side is immobilized, and flexion is deepened while palpating the quadriceps muscle tension (Fig. 67.3). During the extension ROM exercise, force is applied in the direction of knee extension while pulling the foot in the distal direction (Fig. 67.4). The first goal of ROM is to achieve 0° extension and 90° flexion within 1 week after surgery.

Once the pain resolves and the periarticular muscles are able to contract sufficiently, active-assisted and active ROM exercises are included. In active-assisted ROM exercises, the knee joint is flexed by lifting the distal femur (Fig. 67.5a) or

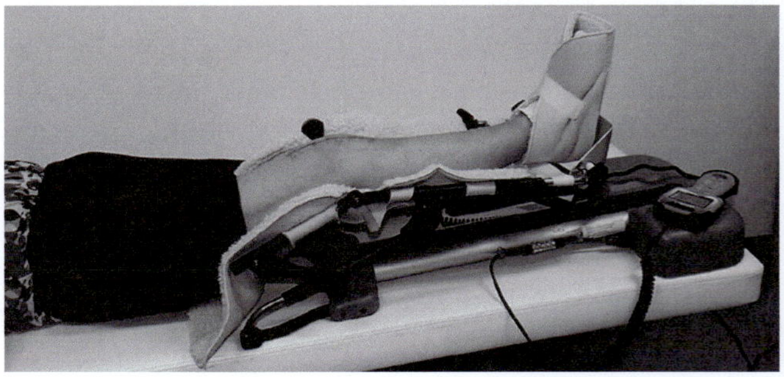

Fig. 67.1 Range of motion (ROM) exercise with a continuous passive motion (CPM) device. The ROM is 0–120° and can be set within a speed range of 40–220°/min, depending on the device. The knee joint is set with the center of rotation aligned with the hinge of the frame. The lower extremity on the operated side is loosely fixed to the device; therefore, there is a difference between the displayed and actual joint angles

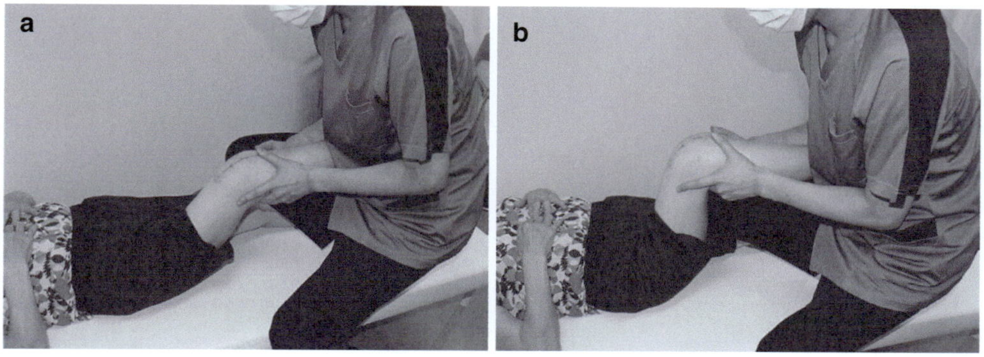

Fig. 67.2 Exercise of passive flexion range of motion (ROM) with the manual stretching technique. (**a**) With the patient in the supine position on the examination table, the knee joint is flexed while releasing the quadriceps muscle tension in the hip flexion position. The patient's facial expression is checked and the flexion angle is increased slowly to avoid defensive muscle contraction due to pain. (**b**) In the deep flexion exercise, the posterior aspect of the lower leg is supported such that it does not drop backward and induce rollback of the femoral condyle

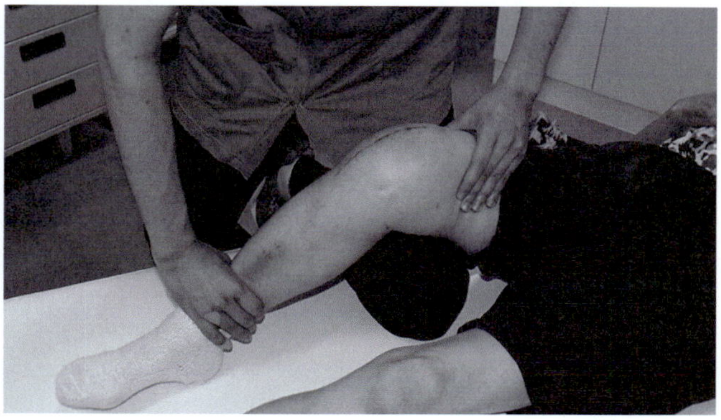

Fig. 67.3 Exercise of passive flexion range of motion (ROM) in the deep flexion position. The operative side is placed on the therapist's thigh and immobilized, and knee flexion is increased while palpating the quadriceps muscle tension

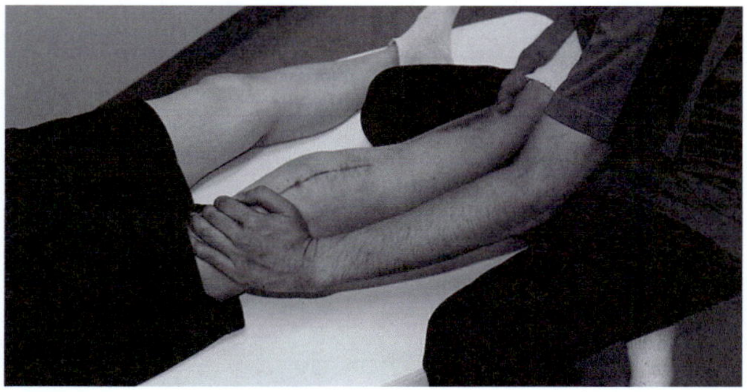

Fig. 67.4 Exercise of passive extension range of motion (ROM) with the manual stretching technique. The foot is grasped with one hand such that the knee joint is lifted off the examination table, and force is applied using the other hand in the direction of knee extension while pulling the foot distally

pulling a towel over the foot (Fig. 67.5b). In the sitting posture, the lower leg is lowered naturally, and the knee joint is passively flexed with assistance from the uninvolved lower limb (Fig. 67.6). Once the patient is able to maintain a stable standing position, the knee joint is flexed while weight is applied to the lower leg on a low table (Fig. 67.7). Recently, robotic rehabilitation has been introduced for the treatment of musculoskeletal disorders, and its effectiveness has been reported. It has been shown that rehabilitation therapy using the Hybrid Assistive Limb single-joint type (HAL-SJ) results in early improvement in active extension ROM compared to nonuse (Fig. 67.8) [5]. When patellofemoral joint mobilization is incorporated into the rehabilitation program, wound healing progresses and pain is no longer elicited. Joint mobilization refers to a physical therapy in which the joint is moved repeatedly in various directions at a low velocity and with various amplitudes to improve joint laxity. Patellofemoral joint mobilization is performed mainly in the medial and distal directions to improve patellar mobility (Figs. 67.9 and 67.10). The tibiofemoral joint is mobilized in the same manner (Fig. 67.11).

1.4 ROM Exercise in Life Stage Rehabilitation

Because of the efforts in medical institutions to shorten the length of hospital stay, patients are often discharged as soon as they achieve stable walking with a T-cane. The patients rarely visit for rehabilitation therapy after discharge. Although at the time of discharge, ROM exercises as home-based rehabilitation programs are taught, it is difficult to determine their effectiveness in terms of adherence. Recently, telerehabilitation in an internet environment has been reported to be useful for the recovery of knee joint function as a possible approach of home rehabilitation [6]. Telerehabilitation is expected to be an option for rehabilitation treatment during the daily life phase.

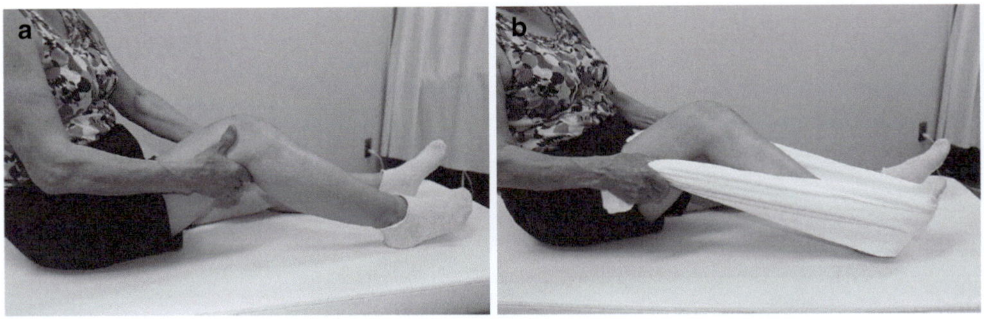

Fig. 67.5 Exercise of active-assisted flexion range of motion (ROM) in bed. (**a**) Active flexion of the knee joint while assisting in lifting the distal femur. (**b**) The knee joint is actively flexed while assisting with pulling a towel placed over the foot in the proximal direction

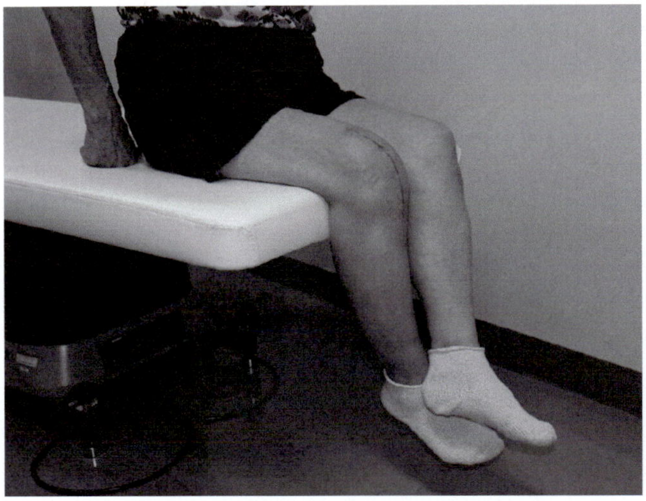

Fig. 67.6 Exercise of active-assisted flexion range of motion (ROM) in the sitting-on position. The lower leg is spontaneously lowered to promote relaxation of the quadriceps muscle and active flexion of the operated knee joint, with assistance from the non-operated lower limb

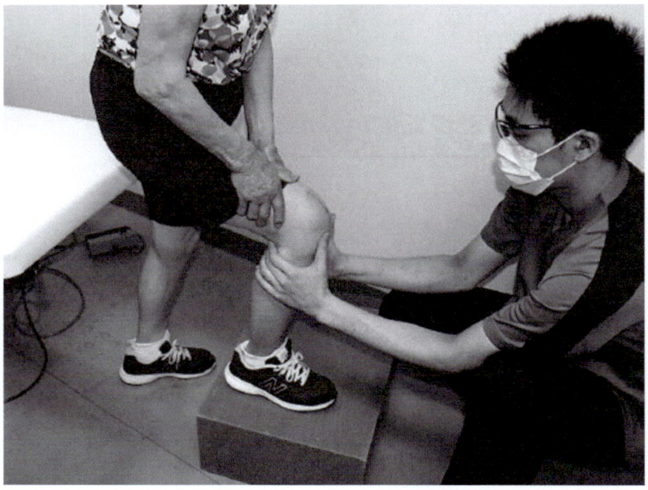

Fig. 67.7 Exercise of active-assisted flexion range of motion (ROM) in the standing position. Once the patient is able to maintain a stable standing position, the knee joint is flexed while the patient's lower extremity is placed on a low table and weight is gradually applied to the knee joint. The therapist assists the patient at the beginning of the procedure to ensure safety

2 Walking Ability

2.1 Gait in the Normal Knee Joint and after TKA Surgery

The roles of the knee joint in walking are: (1) to absorb the impact of ground contact by bending during the loading response phase, (2) to reduce the excessive upward shift of the center of gravity by bending during the stance phase, and (3) to secure toe clearance by bending during the swing phase [7]. To fulfill these roles, the knee joint is flexed by approximately 20° during the loading response phase and approximately 70° during the swing phase, resulting in a double knee action during a single gait cycle.

Improvement in walking ability is the most expected outcome of TKA (Table 67.1). Even after TKA, the knee joint extension and flexion pattern during gait shows the double knee action similar to that of the normal knee. However, owing to the dysfunction of the remaining quadriceps muscle and compensatory overactivity of the hamstring as a dynamic stabilizing mechanism during loading, the extension and flexion patterns are affected by the stabilizing strategy with joint immobilization. This results in a decrease in the knee joint flexion angle during the loading response phase, which reduces the shock absorption capacity and negatively affects forward propulsion [8].

2.2 Gait Training with Postoperative Rehabilitation Therapy

If the patients' general condition is satisfactory, they should begin standing training at the bedside the day after surgery, except in special cases, such as extensive bone grafting. Vital signs are monitored to ensure that there are no symptoms, such as dizziness or discomfort, and the patient is progressively moved from a supine position to a bed-up sitting position, and then to a sitting-on position. The patient is placed in maximum extension of the knee joint on the operative side, and standing training is performed with the therapist's assistance or under supervision. These exercises utilize the

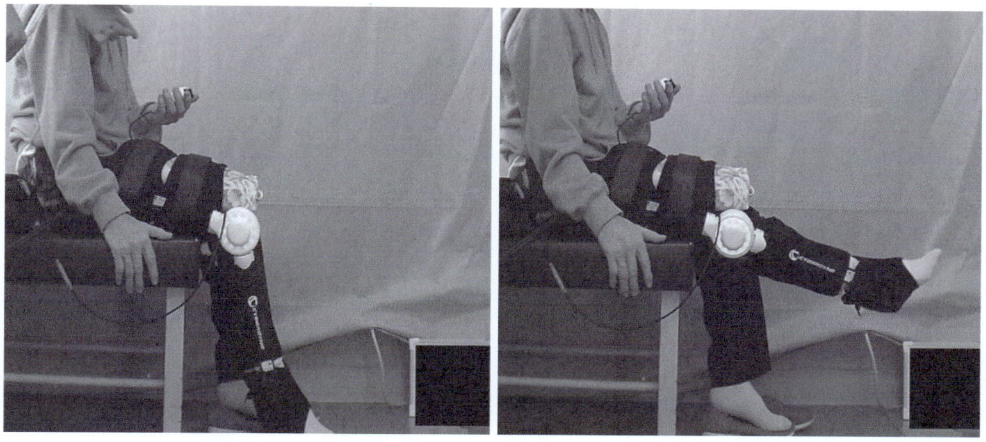

Fig. 67.8 Rehabilitation treatment using Hybrid Assistive Limb single joint type (HAL-SJ). HAL-SJ assists in the active extension and flexion of the knee joint with an optimal amount of assistance according to the bioelectrical signals derived from the quadriceps and hamstring muscles. (Provided by Dr. Tomoyuki Sasaki, Department of Orthopedic Surgery, Hirosaki Memorial Hospital)

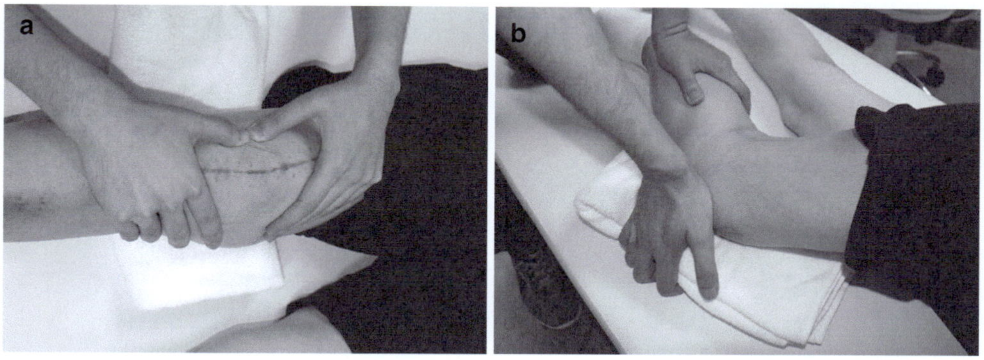

Fig. 67.9 Patellofemoral joint mobilization in the medial direction. The knee joint is placed in shallow flexion in the supine position (**a**) or in the lateral recumbent position with the operated side up (**b**), and joint mobilization is performed by applying medial pressure to the lateral edge of the patella

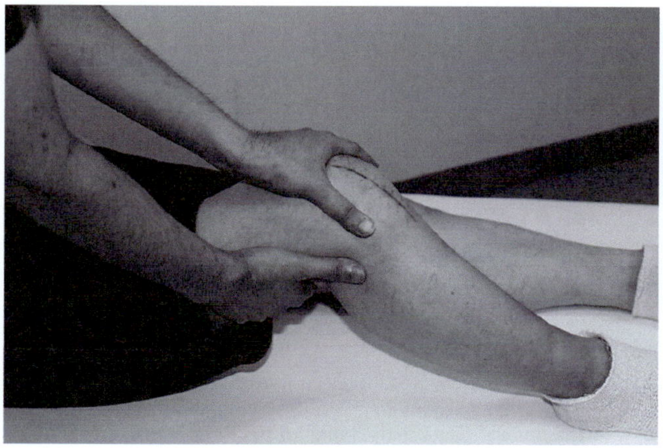

Fig. 67.10 Patellofemoral joint mobilization in the distal direction. The knee joint is placed in a slightly shallower flexion position than the maximum flexion angle, and joint mobilization is performed by applying a distal compression force to the superior pole of the patella

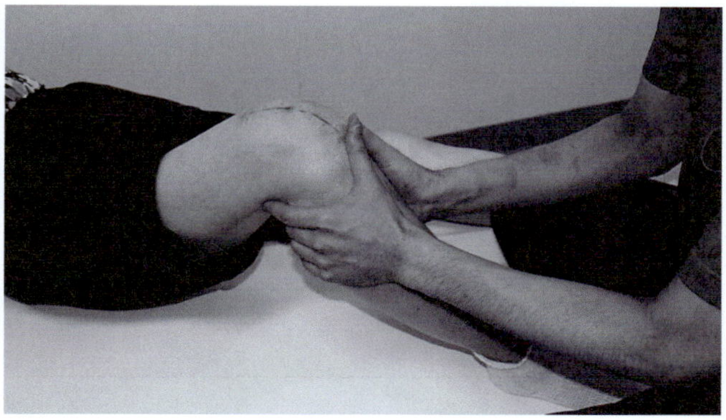

Fig. 67.11 Mobilization of the tibiofemoral joint. With the knee joint in flexion, joint mobilization is performed by sliding the proximal end of the tibia anteriorly and posteriorly against the femoral joint surface

anterior tilt of the trunk and involve both upper limbs to push up (Fig. 67.12). In the standing posture, voluntary contraction of the quadriceps muscle on the operative side should be achieved, and the knee should not bend under a load.

Three to four days after surgery, the patients begin gait training using a parallel bar. The first step is to start with a forward-aligned gait in which the patient steps forward with the surgical leg and aligns it with the non-surgical leg (Fig. 67.13). Once the patient is able to perform a stable forward-aligned gait, he/she switches to a normal gait. The patient then moves to a T-cane gait, starting with a three-action gait (cane → surgical leg → nonsurgical leg) (Fig. 67.14) and progressing to a two-action gait (cane and surgical leg simultaneously → nonsurgical leg). Treadmill gait training allows various tasks to be performed and provides the training necessary for real-life situations (Fig. 67.15). Patients who need to use stairs after discharge

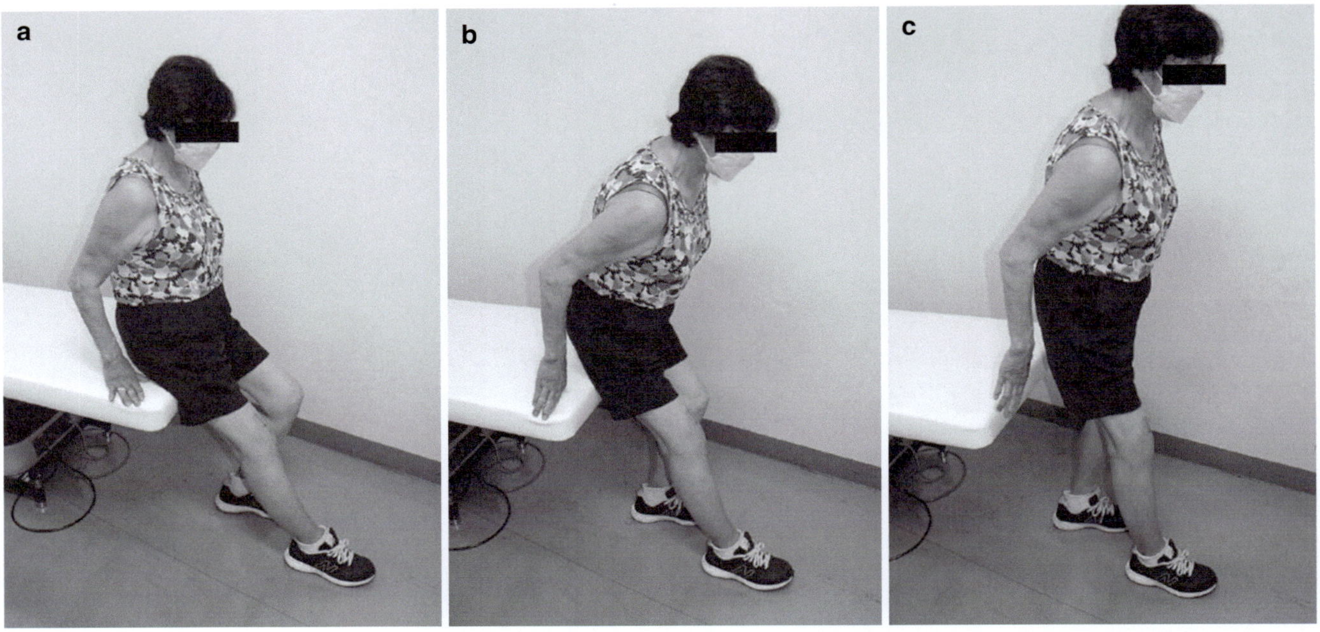

Fig. 67.12 Bedside standing training. The patient is placed in maximum knee extension with the operated lower limb thrown forward and the trunk tilted forward. The patient stands by pushing on the edge of the bed using both upper limbs in a push-up manner

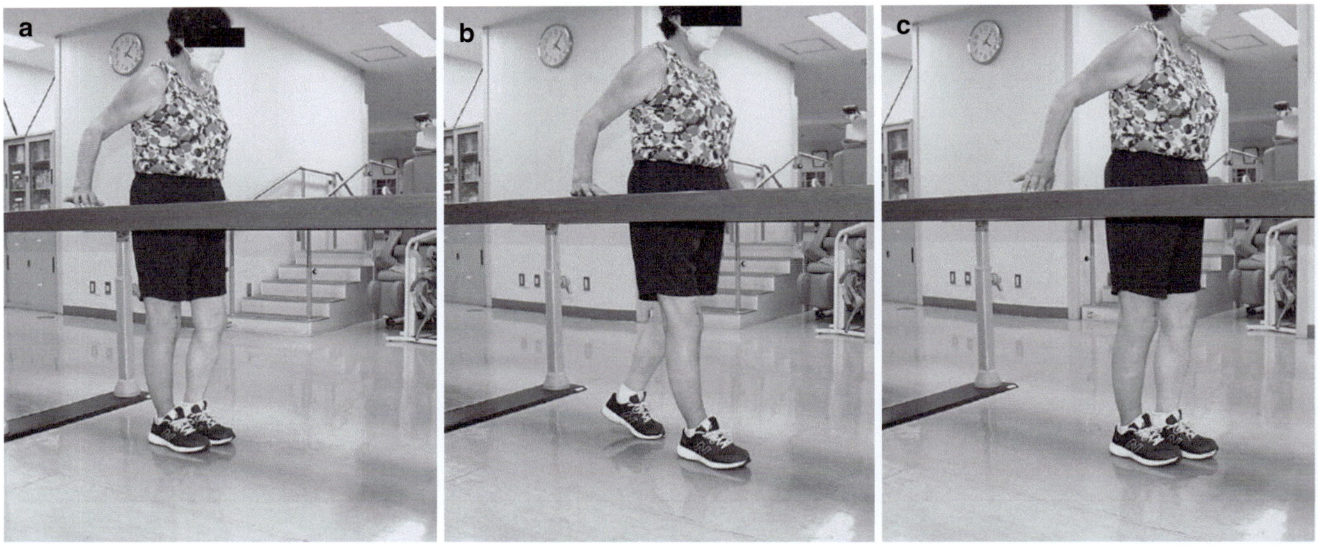

Fig. 67.13 Training a forward-aligned gait within a parallel platform. With both lower limbs aligned (**a**), the operated lower limb is first stepped forward (**b**), and then the non-operated lower limb is aligned (**c**). The time required to stand on one leg on the operated side is shortened, allowing the patient to perform the procedure even during periods of residual loading pain

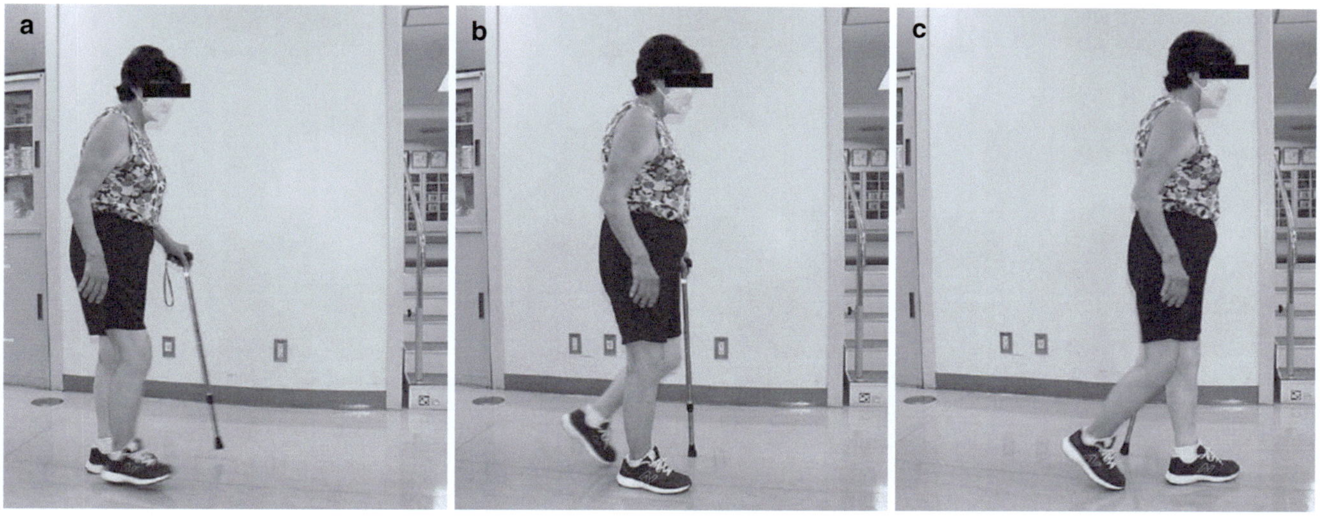

Fig. 67.14 Three-action gait with T-cane. First, the T-cane is thrusted forward (**a**), followed by the operative leg (**b**), and then the nonoperative leg (**c**)

Fig. 67.15 Gait training using a treadmill. Using a variety of task settings, such as straddling and avoiding obstacles, it is possible to train the walking skills necessary for real-life applications. Therefore, suspension devices and harnesses are necessary to prevent falls

should learn, during hospitalization, to ascend and descend stairs in two steps for each stair -raiser (Fig. 67.16).

2.3 Rehabilitation Treatment to Improve Gait Pattern

To acquire an efficient and stable gait, it is necessary to learn a gait pattern that closely resembles the normal gait. To evaluate gait patterns in detail, it is ideal to measure joint kinematics and kinetics by three-dimensional motion analysis. However, gait evaluation via visual observation is the most common method used in daily outpatient rehabilitation. To eliminate the subjectivity of the examiner as much as possible and to extract improvement points more objectively, it is effective to use a gait analysis sheet. Figure 67.17 shows the gait analysis sheet of the Observation Gait Instruction Group [7]. This sheet contains check items for each phase of the gait cycle, allowing the items necessary to improve the gait pattern to be organized and easily reflected in subsequent rehabilitation treatment. The following section describes the items often observed after TKA related to the knee joint.

1. Decreased knee joint flexion during the loading response phase

Knee flexion during the loading response phase is controlled by the eccentric contraction of the quadriceps muscle; thus, residual quadriceps muscle weakness causes a decrease in the flexion angle. Training for eccentric contraction of the quadriceps muscle using closed kinetic chains, such as squatting, is performed.

2. Poor knee joint extension in mid-to-terminal stance

Knee joint extension during this phase is caused by the reaction of the contralateral lower limb swinging outward. Insufficient knee joint extension results in the weakening of the forward propulsive force generated by the contraction of the ankle plantar flexor muscles at the end of the stance phase.

3. Decreased knee joint flexion during the swing phase

Knee joint flexion during this period is caused by the inertial force of the lower leg associated with the swing of the thigh. A decrease in flexion angle makes it difficult to maintain suf-

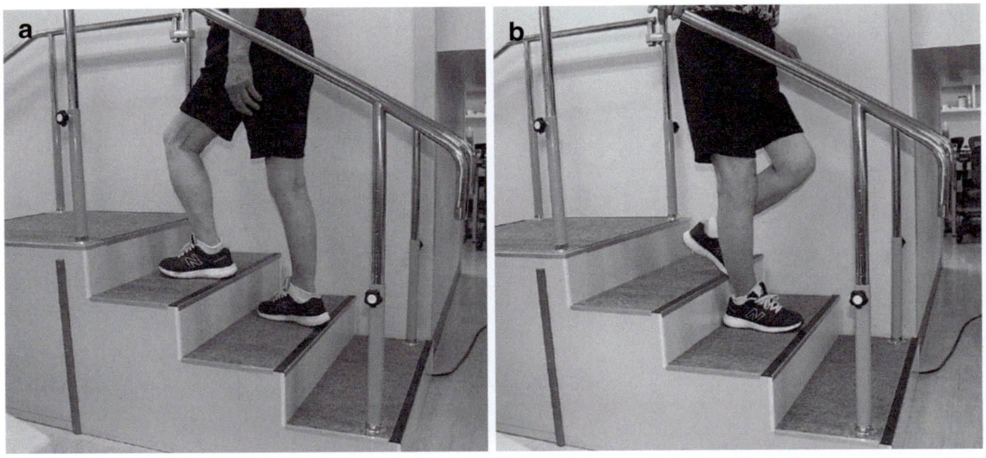

Fig. 67.16 Stair climbing training with two steps for each stair-raiser. The ascending step is taken from the non-operated leg, and the descending step is taken from the operated leg

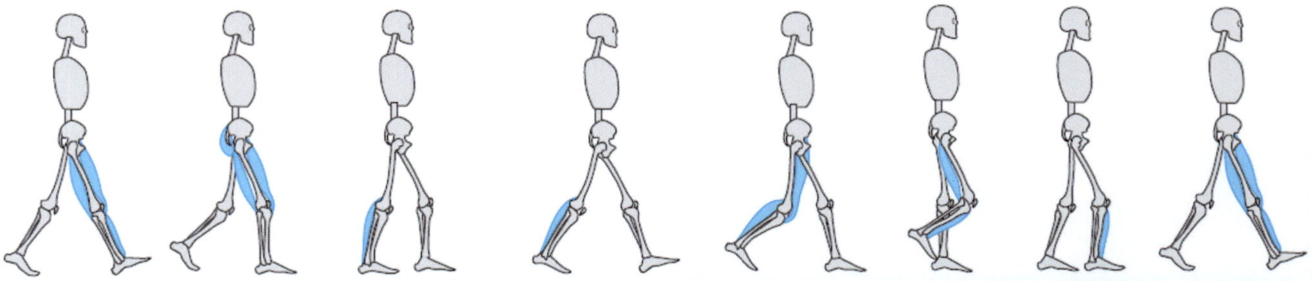

<table>
<tr><th colspan="2">Weight Acceptance</th><th colspan="2">Single Limb Stance</th><th colspan="4">Swing Limb Advancement</th></tr>
<tr><th>IC</th><th>LR</th><th>MSt</th><th>Tst</th><th>PSw</th><th>ISw</th><th>MSw</th><th>TSw</th></tr>
<tr><td colspan="2">Heel Strike
○ Yes
○ No</td><td colspan="2">Ankle Dorsiflexion
○ Adequate
○ Inadequate</td><td></td><td colspan="3">Ankle Dorsiflexion
○ Adequate
○ Inadequate</td></tr>
<tr><td></td><td>Ankle Plantar Flexion
○ Adequate
○ Inadequate</td><td colspan="3">Heel Off Timing
○ Early ○ Adequate ○ Late
Pelvis
○ Stable
○ Unstable</td><td colspan="3"></td></tr>
<tr><td colspan="2">Knee Flexion
○ Adequate
○ Inadequate</td><td colspan="2">Knee Extension
○ Adequate
○ Inadequate</td><td colspan="2">Knee Flexion
○ Adequate
○ Inadequate</td><td colspan="2">Knee Extension
○ Adequate
○ Inadequate</td></tr>
<tr><td colspan="2"></td><td colspan="2">Hip Extension
○ Adequate
○ Inadequate</td><td colspan="4">Hip Flexion
○ Adequate
○ Inadequate</td></tr>
<tr><td colspan="2">Heel Rocker
○ Insufficient
○ Excessive
○ Normal</td><td>Ankle Rocker
○ Insufficient
○ Excessive
○ Normal</td><td>Forefoot Rocker
○ Insufficient
○ Excessive
○ Normal</td><td colspan="4">Foot Clearance
○ Adequate
○ Inadequate</td></tr>
</table>

Fig. 67.17 Gait analysis sheet by Observation Gait Instruction Group. By establishing check items for each phase of the gait cycle, it is easy to organize the essential components needed to improve the gait pattern and incorporate them during rehabilitation treatment

ficient toe clearance, which can lead to stumbling. The patient is trained to smoothly swing the lower limb by quickly contracting the iliopsoas and ankle plantar flexors from a fully stretched state, replicating the end of the stance phase.

2.4 Fall Prevention in Life Stage Rehabilitation

The risk of falling is reported to be 1.39 times higher in patients with knee osteoarthritis than in healthy individuals, which is one of the main causes of locomotive syndrome [9]. TKA has been reported to reduce the risk of falls, and half of the patients with a history of falls before surgery were able to avoid falls after surgery [10]. Falls after TKA can cause periprosthetic fractures that are often difficult to treat and require osteosynthesis or revision arthroplasty. When patients are discharged home, they should be advised to modify their lifestyle according to their living environment and mobility, and active approaches to fall prevention using "locomotion training" and other methods are necessary in rehabilitation therapy during the daily life phase [11].

References

1. Myles CM, et al. Knee joint functional range of movement prior to and following total knee arthroplasty measured using flexible electrogoniometry. Gait Posture. 2002;16:46–54.
2. Scott CEH, et al. Patient expectations of arthroplasty of the hip and knee. J Bone Joint Surg Br. 2012;94:974–81.
3. Kotani A, et al. Factors influencing range of motion after contemporary total knee arthroplasty. Arthroplast. 2005;20:850–6.
4. Yang X, et al. Continuous passive motion after total knee arthroplasty: a systematic review and meta-analysis of associated effects on clinical outcomes. Arch Phys Med Rehabil. 2019;100:1763–78.
5. Yoshioka T, et al. Feasibility and efficacy of knee extension training using a single-joint hybrid assistive limb, versus conventional. J Rural Med. 2021;16:22–8.
6. Jiang S, et al. The comparison of telerehabilitation and face-to-face rehabilitation after total knee arthroplasty: a systematic review and meta- analysis. J Telemed Telecare. 2018;24:257–62.
7. Neumann KG, Gehen verstehen. Ganganalyse in der Physiotherapie. 4th ed. Stuttgart: Thieme Verlag; 2016.
8. Milner CE. Is gait normal after total knee arthroplasty? Systematic review of the literature. J Orthop Sci. 2009;14:114–20.
9. Doré AL, et al. Lower limb osteoarthritis and the risk of falls in a community-based longitudinal study of adults with and without osteoarthritis. Arthritis Care Res (Hoboken). 2015;67:633–9.
10. Moutzouri M, et al. The effect of total knee arthroplasty on patients' balance and incidence of falls: a systematic review. Knee Surg Sports Traumatol Arthrosc. 2017;25:3439–51.
11. Nakamura K, et al. Locomotive syndrome: definition and management. Clin Rev Bone Miner Metab. 2016;14:56–67.

68 Muscle Strengthening

Hideki Ueyama

1 What Is Muscle Strength?

The most important basic movements of the knee joint are flexion and extension, and the quadriceps and hamstrings (biceps femoris, semimembranosus, and semitendinosus) are the primary muscles involved in this motion. When the term "muscle strength" is used to describe muscular force, it is important to note whether the author intends to convey the concept of "muscle strength", which refers to the maximum contractile force of a muscle, or "muscular endurance", which refers to the ability to sustain a workload. This section focuses on "muscle strength", referring to the maximum contractile force of a muscle.

2 Muscle Strength and Muscle Mass

Muscle strength and muscle mass are similar but different concepts. As described above, muscle strength refers to the amount of work exerted, whereas muscle mass is a quantitative measure of area, weight or volume. More specifically, muscle mass may refer to the total skeletal muscle mass of the limbs (in this case, mainly volume), as used in the diagnostic criteria for sarcopenia [1], or to the cross-sectional area of a muscle as measured using magnetic resonance imaging or echocardiography. As such, it is necessary to exercise caution in what is defined as muscle mass when reading the literature. Among elderly Japanese patients (≥ 75 years of age), there is a generally positive correlation between lower limb muscle strength and muscle mass, and it is often appropriate to consider them to be similar. However, it should be noted that muscle strength and mass are not necessarily correlated in middle-age females (≤ 74 years of age); therefore, they cannot be considered the same for all individuals [2]. Nevertheless, there is little doubt that muscle strength and muscle mass are closely related, and it is reasonable, therefore, to set them as goals of rehabilitation therapy. Muscle mass appears to be more suitable for quantitative evaluation because it is a visible entity.

3 Muscle Strength Measurement Methods

Muscle strength is defined as an absolute quantity expressed for Newton (N) in physics; the physical quantity required to accelerate a body and change its state of motion is known as "force". Muscular strength around the knee is often expressed in terms of torque (Newton•meter [Nm]), which takes into account the moment of rotation around the knee and has a different meaning (Fig. 68.1). The vector quantity at the action point (measurement point) is the muscle strength (N). The moment of force calculated as a vector product multiplied by the length of the lower leg (m) is expressed as torque (Nm). The actual workload is calculated by multiplying torque by the distance of movement; therefore, torque is a useful and practical parameter for evaluating the activities of daily living (ADL). Some modified comparisons have been made in clinical studies by dividing muscle strength by body weight to obtain normality (N/kg or Nm/kg). This method of analysis considers differences in body size. There is no acceptable or unacceptable method depending on what is used as an index; however, it is important to note that values expressed using different units should not be compared, and there is a risk for misunderstanding if these concepts are not understood to some extent when reviewing studies and/or data.

The standard metric for strength during knee flexion and extension is isokinetic muscle strength [3] (Fig. 68.2). A large amount of information, such as torque, range of motion, and power (i.e., workload), can be obtained during exercise at an arbitrarily set angular velocity, and the measured values are highly reliable. Disadvantages of measuring isokinetic muscle strength are that it requires bulky and expensive equipment, is a time-consuming procedure and imposes a

H. Ueyama (✉)
Osaka Rosai Hospital, Sakaishi, Osaka, Japan

S. Matsuda (ed.), *Advances in Total Knee Arthroplasty*, https://doi.org/10.1007/978-981-97-4920-1_68

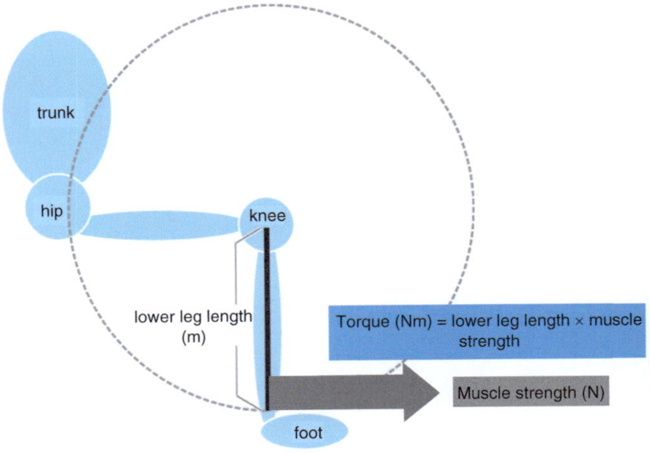

Fig. 68.1 Conceptual diagram illustrating muscle strength. In this chapter, muscle strength is described as the amount of absolute force (N) exerted at the action point (measurement point), and torque (Nm) is shown as a vector product obtained by multiplying the distance from the center of rotation (knee center in this figure) by the action point (approximated by the lower leg length)

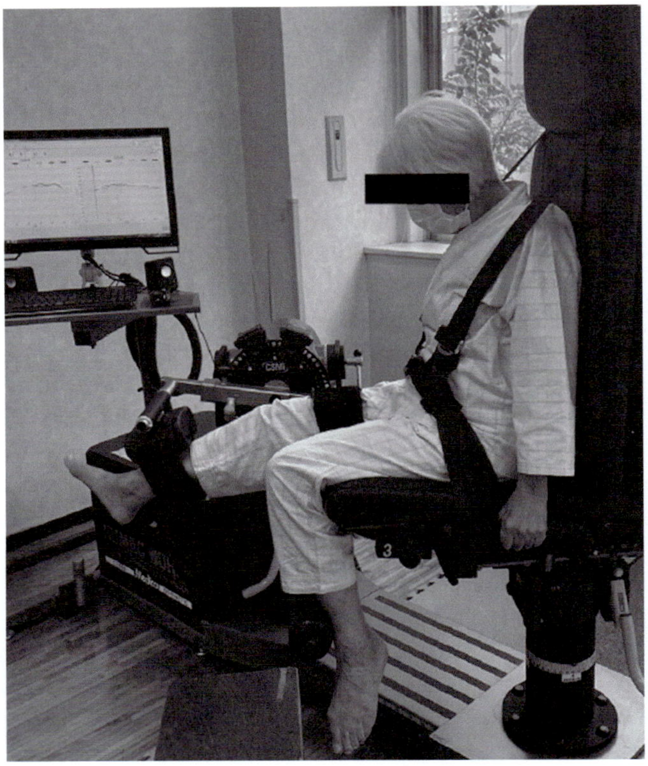

Fig. 68.2 Isokinetic measurement of muscle strength. This method of muscle strength measurement is considered to be the most reliable because it can calculate the average and peak torque values by measuring muscle strength multiple times at a set angular velocity. However, it has some disadvantages, such as large equipment, and load on the knee joint. In the photograph, muscle strength in the right knee is measured

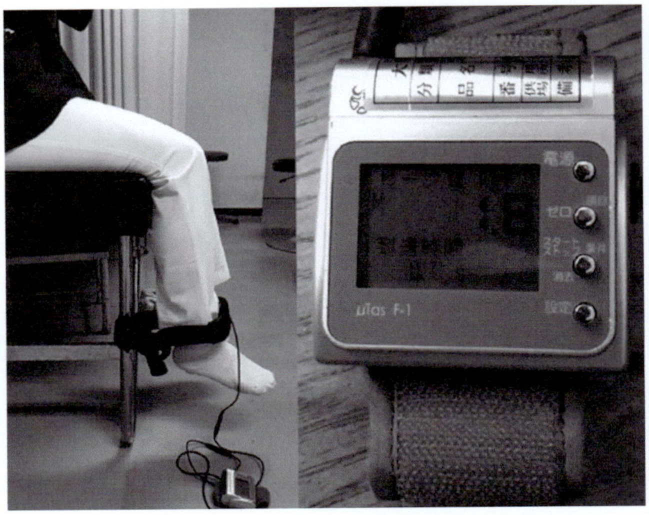

Fig. 68.3 Isometric measurement of muscle strength. Isometric measurement of muscle strength is method of assessing exerted muscle strength with the joint immobilized. Although this metric can be easily measured and does not require load on the knee joint, it has the disadvantage of comparatively lower reliability than isokinetic muscle strength. The use of a manual pullout muscle tester, as shown in the photograph, improves the reliability of measurement

heavy motion load on the knee. This may make it difficult to use in elderly patients undergoing total knee arthroplasty (TKA) [4].

Other muscle strength metrics include isometric muscle strength, typified by manual strength testing. Because this strength measurement can be evaluated without joint movement, it is easy to use in patients with knee pain caused by movement. Other advantages of measuring isometric muscle strength include relatively low cost, portability, and ease of use in the clinical field. However, the reliability of the measurement is lower than that of isokinetic measurement, which is a disadvantage [5] (Fig. 68.3). In clinical practice, isometric measurements using manual muscle testers are simpler and more practical.

4 TKA and Muscle Strength

4.1 Importance of Muscle Strength Around the Knee During TKA

Muscle strength around the knee is not only extremely important for the ADL in humans who typically use both legs to walk but also for maintaining quality of life, including cultural events such as sporting activities. In elderly patients, who commonly undergo TKA, there is a correlation between quadriceps muscle weakness and the risk for falls, and muscle weakness around the knee cannot be overlooked from a safety standpoint [6]. It has been shown that muscle strength around the knee is decreased in elderly patients with disease(s) affecting the knee [7]. In addition to the increased risk for falls mentioned above, increased nursing care costs [8] and rehospitalization rates [9] have been reported to be associated with lower-limb muscle weakness after TKA. It has been reported that preventing quadriceps muscle weakness after TKA not only overcomes these problems, but also improves postoperative knee joint awareness (Forgotten Joint Score-12) [10], as well as patient satisfaction [11]. In terms of improving patient-reported outcome measures, which has been recognized as an important outcome related to TKA in recent years, it is reasonable to improve postoperative muscle strength around the knee to improve clinical results after TKA.

4.2 Muscle Strength After TKA

In a study evaluating quadriceps muscle strength after TKA in patients with an average age of 71 years by isometric muscle strength measurement (knee flexion 90°), there was a 52% change at 3 weeks, 81% at 3 months, 96% at 6 months, and 112% at 1 year after surgery when preoperative muscle strength was defined as 100% [12]. The hamstrings exhibited a similar trend: 42% at 3 weeks, 77% at 3 months, 85% at 6 months, and 108% at 1 year after surgery [12]. The graphs in Fig. 68.4 illustrate the data obtained by our research group, in which quadriceps strength was measured prospectively for 2 years after TKA using isometric measurements (knee flexion, 90°) in patients 75 years of age. Generally, muscle strength after TKA reached the preoperative level in the first 6 months after surgery, surpassed the preoperative level, and further increased in the second year after surgery (Fig. 68.4).

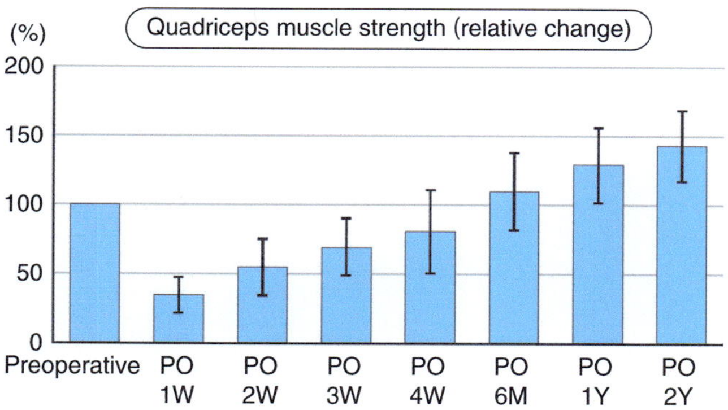

Fig. 68.4 Quadriceps muscle strength after total knee arthroplasty. The relative change in muscle strength was observed compared to the preoperative condition of 100%. Muscle strength generally improves to the preoperative level in approximately 6 months and continues to improve with time, surpassing preoperative strength in many patients 1–2 years postoperatively. *PO* postoperative, *W* weeks, *M* months, *Y* years

Although no studies have examined long-term trends in muscle strength around the knee after TKA, it has been reported that quadriceps muscle strength at 10 years postoperatively in TKA patients, measured isometrically (knee flexion, 20°), was approximately 74% of that of healthy subjects of the same age [13]. This suggests that, among elderly patients with diseases affecting the knee undergoing TKA, muscle strength around the knee after TKA improved over time compared to preoperative levels but did not reach the level of healthy individuals. Therefore, it is important to use medical interventions to improve muscle strength during the perioperative period to avoid muscle weakness, as described in the previous section.

5 Findings for Muscle Recovery

5.1 Factors Affecting Muscle Strength After TKA

1. Age

Age is reported to affect muscle strength, and it has been shown that strength of the muscles around the knee is low in elderly individuals with disease(s) affecting the knee [7].

2. Diseases

A study comparing postoperative muscle strength improvement between patients with osteoarthritis (OA) and rheumatoid arthritis after TKA reported that the latter group recovered muscle strength significantly faster at 1 month postoperatively than the former, and that the effect of rehabilitation was significant [4]. This is considered to be because patients with OA have already used their reserve function to the limit, but it is possible that changes in postoperative muscle strength may differ depending on the characteristics of each disease.

3. Prostheses of TKA

The effect of different prostheses on muscle strength has been evaluated by comparing changes in postoperative quadriceps and hamstring muscle strength between bicruciate substituting (BCS) TKA and posterior stabilized (PS) TKA [14], and by comparing postoperative quadriceps and hamstring muscle activity using electromyography between BCS and cruciate-retaining TKAs [15]. No differences were found between the prosthetic models. Although further studies are desired, even models that can reproduce the physiological kinematics of the normal knee may not be affected enough by the model design to cause differences in postoperative muscle strength around the knee.

4. Preoperative muscle strength and mass

Preoperative quadriceps muscle strength and quadriceps muscle cross-sectional area were predictors of muscle recovery 6 months postoperatively [16]. It has been shown that postoperative recovery tends to be delayed in patients with preoperative weakness of the quadriceps muscle and in those with atrophied quadriceps cross-sectional area.

5. Nutritional status

Patients with preoperative malnutrition (albumin, ≤ 3.4 g/dL) tended to experience worse clinical outcomes and a longer hospital stay after TKA than those without malnutrition [17]. To our knowledge, there have been no reports describing the direct effects on lower limb muscle strength; nevertheless, this finding is important and is presented herein.

5.2 Interventions to Improve Muscle Strength After TKA

1. Rehabilitation

Rehabilitation is useful in practice because it is a non-invasive intervention that can be introduced immediately. Sufficient preoperative rehabilitation has been shown to have a favorable effect on muscle strength improvement after TKA [18]. Starting exercises focused on quadriceps muscle strength in the early postoperative period (days 2–14) have been reported to be helpful in improving pain, gait, and other clinical outcomes [19]. In addition to self-exercise, electrical stimulation, which induces action potentials by electrically stimulating muscles, and hybrid training, which combines electric stimulation and self-exercise, have been reported to be effective in maintaining quadricep muscle strength after TKA [20]. Perioperative rehabilitation is recommended because it is essentially non-invasive and provides definite benefits when performed safely.

2. Nutrition

Because clinical outcomes after TKA are worse in under- or malnourished patients [17], it is reasonable to attempt to address these problems with nutritional supplementation. One study reported that oral administration of 2400 mg/day beta-hydroxy beta-methylbutyrate (i.e., "HMB"), 14,000 mg/day glutamine, and 14,000 mg/day arginine, a metabolite of leucine, from 5 days before surgery to 4 weeks after surgery, improved quadriceps muscle strength at 2 weeks postoperatively [11]. In another investigation, oral administration of ES-polytamine (9 g/day essential amino acid preparation), from 1 week before to 2 weeks after surgery according to the dosage and administration, suppressed atrophy of the quadriceps muscle cross-sectional area at 3 and 4 weeks after surgery, contributing to early postoperative improvement in lower limb function [21] (Fig. 68.5). A comparison of a large group of undernourished patients, who were preoperatively educated on a high-protein diet to improve their nutritional status, and a group that was not educated preoperatively showed that the postoperative outcome of TKA was significantly improved in the group that received nutritional guidance [17]. Catabolism is considered to be a physiological response to surgical invasion. The concept of preoperative nutritional supplementation to alleviate the progressive degradation of muscle tissue caused by catabolism has long attracted attention in other fields. The importance of perioperative nutritional management in the field of orthopedic surgery has been slowly—but surely—recognized.

3. Surgical invasiveness

A study comparing the postoperative cross-sectional area of the quadriceps muscle with and without a tourniquet in TKA reported that muscle atrophy was significantly greater when the tourniquet was used at 1–3 months postoperatively [22]. To the best of our knowledge, no other study has demonstrated a clear relationship between surgical invasiveness and the postoperative status of limb muscles. Considering the

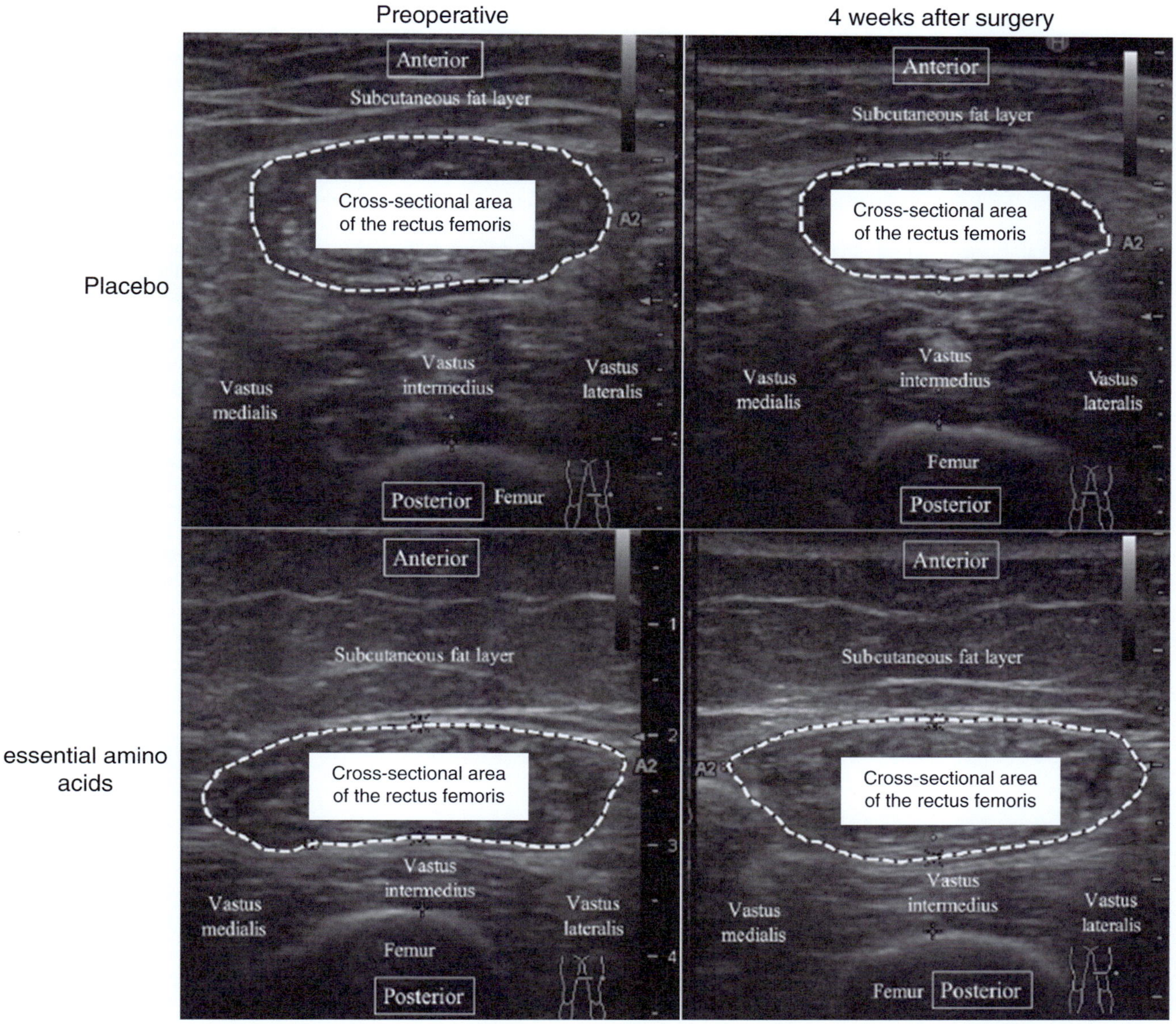

Fig. 68.5 Cross-sectional area of the rectus femoris muscle before and after total knee arthroplasty. The quadriceps muscle mass was atrophied, and strength was decreased postoperatively. However, perioperative supplementation with essential amino acids suppressed this atrophy in the early phase after total knee arthroplasty (lower panel). *CSA* Cross-sectional area

physiological response of increased catabolism to stress, surgical invasion may affect postoperative muscle atrophy and strength, which warrants future research in this area.

References

1. Chen LK, et al. Sarcopenia in Asia: consensus report of the Asian working Group for Sarcopenia. J Am Med Dir Assoc. 2014;15:95–101.
2. Hayashida I, et al. Correlation between muscle strength and muscle mass, and their association with walking speed, in community-dwelling elderly. PLoS One. 2014;9:e111810.
3. Schroer WC, et al. Isokinetic strength testing of minimally invasive total knee arthroplasty recovery. J Arthroplast. 2010;25:274–9.
4. Sasaki K, et al. Changes in lower limb muscle strength and walking speed before and after knee joint replacement surgery: comparison of rheumatoid arthritis and knee osteoarthritis patients. Physiotherapy. Science. 2002;17:259–64. (Japanese)
5. Lesnak J, et al. Validity of hand-held dynamometry in measuring quadriceps strength and rate of torque development. Int J Sports Phys Ther. 2019;14:180–7.
6. Moreland JD, et al. Muscle weakness and falls in older adults: a systematic review and meta-analysis. J Am Geriatr Soc. 2004;52:1121–9.
7. Conroy MB, et al. Muscle strength, mass, and quality in older men and women with knee osteoarthritis. Arthritis Care Res (Hoboken). 2012;64:15–21.
8. Ferrucci L, et al. Hospital diagnoses, medicare charges, and nursing home admissions in the year when older persons become severely disabled. JAMA. 1997;277:728–34.

9. Klingenstein GG, et al. Rapid discharge to home after total knee arthroplasty is safe in eligible medicare patients. J Arthroplast. 2017;32:3308–13.
10. Furu M, et al. Quadriceps strength affects patient satisfaction after total knee arthroplasty. J Orthop Sci. 2016;21:38–43.
11. Nishizaki K, et al. Effects of supplementation with a combination of β-hydroxy-β-methyl butyrate, L-arginine, and L-glutamine on postoperative recovery of quadriceps muscle strength after total knee arthroplasty. Asia Pac J Clin Nutr. 2015;24:412–20.
12. Sanada Y, et al. Isometric muscle strength of quadriceps and hamstrings by knee joint angle in patients after total knee arthroplasty. Physiotherapy. Science. 2017;32:553–7. (Japanese)
13. Ishii Y, et al. Quadriceps strength impairment in the mid- to long-term follow-up period after total knee arthroplasty. Knee Surg Sports Traumatol Arthrosc. 2017;25:3372–7.
14. Takubo A, et al. Comparison of muscle recovery following bicruciate substituting versus posterior stabilized total knee arthroplasty in an Asian population. J Knee Surg. 2017;30:725–9.
15. Hyodo K, et al. Gait analysis comparing kinematic, kinetic, and muscle activation data of modern and conventional total knee arthroplasty. Arthroplast Today. 2020;6:338–42.
16. Furumoto T, et al. Preoperative factors predicting cases of muscle weakness 6 months after total knee arthroplasty. Physiotherapy Science. 2020;35:629–33. (Japanese)
17. Schroer WC, et al. Chitranjan S. Ranawat award: elective joint arthroplasty outcomes improved in malnourished patients with nutritional intervention: a prospective population analysis demonstrates a modifiable risk factor. Bone Joint J. 2019;101-B(7 Suppl C):17–21.
18. Calatayud J, et al. High-intensity preoperative training improves physical and functional recovery in the early postoperative periods after total knee arthroplasty: a randomized controlled trial. Knee Surg Sports Traumatol Arthrosc. 2017;25:2864–72.
19. Kondo Y, et al. Short-term effects of isometric quadriceps muscle exercise with auditory and visual feedback on pain, physical function, and performance after total knee arthroplasty: a randomized controlled trial. J Knee Surg. 2022;35:922–31.
20. Tsukada Y, et al. Combined application of electrically stimulated antagonist muscle contraction and volitional muscle contraction prevents muscle strength weakness and promotes physical function recovery after total knee arthroplasty: a randomized controlled trial. Kurume Med J. 2020;65:145–54.
21. Ueyama H, et al. Chitranjan S. Ranawat award: perioperative essential amino acid supplementation suppresses rectus femoris muscle atrophy and accelerates early functional recovery following total knee arthroplasty. Bone Joint J. 2020;102-B(6 Suppl A):10–8.
22. Guler O, et al. Comparison of quadriceps muscle volume after unilateral total knee arthroplasty with and without tourniquet use. Knee Surg Sports Traumatol Arthrosc. 2016;24:2595–605.

Part XIII

Imaging and Functional Assessment

Postoperative Evaluation (E.g., PROMs)

69

Masafumi Itoh

1 Significance and Necessity of Patient-Reported Outcome Measures (PROMs)

Previously, clinical outcomes of total knee arthroplasty (TKA) were objectively evaluated based on indicators considered important by surgeons, such as prosthesis durability and complications [1, 2]. Recent advances in knee-prosthesis materials and designs are overcoming the issue of durability [3]. Moreover, patient expectations for TKA include pain relief and improved knee function, which do not necessarily correspond to surgeon expectations for objective indices such as durability. Patients expect pain relief and many also expect to be able to perform activities of daily living (ADLs) without restrictions. However, 10–25% of patients are dissatisfied with their knee at 1–3 years postoperatively. The main causes of dissatisfaction are residual symptoms and failure to meet preoperative patient expectations [4, 5].

With advances in surgical techniques and prosthesis durability, the number of young patients undergoing TKA has increased [6]. More recently, patient expectations have gone beyond simple improvements in range of motion (ROM) and ADL to an ability to engage in sports and other recreational activities [7].

When considering the value/importance of TKA, objective quantification of the balance between the cost to the healthcare system and the effectiveness of the procedure is required; that is, the cost-effectiveness of TKA 1, [8–10]. Therefore, the "success of surgery" should be evaluated not only from the surgeon's point of view based on objective measures but also from the patient's point of view based on subjective measures; that is, patient-reported outcome measures (PROMs) [11]. In other words, the merits of TKA can be objectively demonstrated by evaluating the surgical "success" in terms of the relief of patient symptoms, recovery of physical function, and realization of preoperative and postoperative wishes [12, 13]. In addition, PROMs are expected to become increasingly important indicators for clinical evaluation as standards of quality of care and patient satisfaction play increasingly important roles in determining the quality of medical care [14]. Therefore, PROMs play a major role in monitoring the quality of care at medical institutions and play an essential role in evaluating the outcomes of clinical trials [2].

2 Use of PROMs That Are Appropriate for the Purpose

Health-related quality of life (HRQOL) [15] is a comprehensive concept that encompasses many areas, including psychological and social well-being, pain, and function. Therefore, improvement in function after TKA is thought to have a positive impact on HRQOL; however, this may not always be the case, depending on the influence of factors other than the affected knee (e.g., psychological status, musculoskeletal disease, and other medical conditions). For example, a study assessing HRQOL using the Short Form-36 (SF-36) at 6 months after TKA, found that patient psychological distress and the presence of comorbidities explained 23% of the variance in SF-36 scores [16, 17] and that economic deprivation was also related to poor outcome on the Short Form-12 (SF-12) [18, 19]. Therefore, PROMs such as the EuroQOL questionnaire (EQ-5D/EQ-VAS) [20] and SF-12/36, which are primarily intended to measure HRQOL, have been applied to various diseases such as cancer and neurological diseases. Using these generic PROMs alone as proxies for specific domains, such as function, pain, and satisfaction after TKA, may lead to incorrect estimates of the treatment effects. Moreover, although the SF-12/36 and EQ-5D/EQ-VAS are widely used PROMs in HRQOL assessment after TKA [21], they have demonstrated low responsiveness to changes in addition to ceiling effects [22, 23].

M. Itoh (✉)
Department of Orthopedic Surgery, Tokyo Women's Medical University, Shinjuku, Tokyo, Japan
e-mail: itou.masafumi@twmu.ac.jp

S. Matsuda (ed.), *Advances in Total Knee Arthroplasty*, https://doi.org/10.1007/978-981-97-4920-1_69

3 PROMs Used to Evaluate TKA

The purpose of TKA is to relieve pain and improve knee joint function. Since sports and other recreational activities have also been required as goals in recent years, it is essential to incorporate these important aspects into the functional assessment. As mentioned previously, the PROMs used for the generic assessment of HRQOL do not fully meet these requirements; therefore, PROMs appropriate for TKA assessment are required. PROMs comprise many aspects, including the balance between subjective and objective scores, patient satisfaction, and factors critical to patients and surgeons; thus, the combination is vast. The PROMs for knee arthroplasty are typically designed to quantify changes in functional assessments. Previously, terms such as "excellent," "good," and "fair" were used as crude indicators of performance in evaluation methods from the patient's perspective [24, 25]. Currently, PROMs are subdivided into several indicators, and many types of PROMs exist; therefore, the selection of the appropriate PROM is important. Appropriate PROMs are those that measure clinical outcomes appropriately and are valid, reliable, responsive, and easy to implement (1); however, it can be difficult to select the appropriate PROM from various PROMs [1]. The indiscriminate use of various PROMs makes it difficult to standardize clinical outcomes and compare them across studies. Furthermore, as the era of big data and large joint registries matures, the identification and implementation of optimal PROMs will become more important [14]. Although a clear definition of the outcome of "successful surgery" has not yet been determined for the current PROMs, narrowing down the choices of useful PROMs may facilitate their analysis [26].

Previous studies attempted to narrow down the PROMs suitable for evaluating TKA. In a systematic review of 38 studies involving 85,541 patients at a mean of 26.3 ± 30.8 months after TKA surgery, Ramkumar et al. [26] reported that the most studied PROMs were the Oxford Knee Score (OKS), Knee Society Score (KSS)-2011, Knee Injury and Osteoarthritis Outcome Score (KOOS), and WOMAC [27–30]. Siljander et al. [14] reviewed abstracts in four major journals; namely, *The Journal of Bone and Joint Surgery-American*, *The Bone and Joint Journal*, *Clinical Orthopaedics and Related Research*, and *The Journal of Arthroplasty* in 2004, 2009, 2014, and 2016 for TKA, unicompartmental knee arthroplasty (UKA), and total hip arthroplasty. They reported that 42 different PROMs were used 1073 times in 644 studies and that the number of publications describing the use of PROMs in these four journals increased from 97 in 2004 to 228 in 2016, with the former KSS [31] being the most commonly used in TKA studies and OKS in UKA.

These reports indicate that the WOMAC, KOOS, OKS, and KSS (−2011) are widely used PROMs; therefore, this section discusses these four instruments as representative PROMs for evaluating TKA performance, in addition to the Forgot Joint Score-12 (FJS) [32], which assesses joint awareness. The reliability and validity of the Japanese versions of the above five instruments have been proven [33–36]. In addition, although Siljander et al. reported that the old KSS is the most frequently used PROM for evaluating TKA performance, it was not included in this section because of its higher ceiling effect and lower responsiveness than the KSS-2011 [37]. Therefore, it was not considered to be the PROM currently in general use.

3.1 WOMAC

The WOMAC consists of 24 questions in total, divided into three subscales (pain subscale: 5 questions, stiffness subscale: 2 questions, physical function subscale: 17 questions). Its reliability and validity were reported by Bellamy et al. in 1988 [30]. Each question is rated on a five-point Likert scale. The total score ranges from 0 to 96, with a lower score indicating a better outcome. The WOMAC has been translated into more than 100 languages and is widely used in the evaluation of knee and hip replacement surgery [36].

3.2 KOOS

The KOOS consists of five subscales and was first developed by Roos et al. in 1998 [29]. It has been universally used in knee osteotomies and anterior cruciate ligament reconstructions. The subscales assess pain (nine questions), symptoms (seven questions), activities of daily living (17 questions), sports and recreation function (five questions), and knee-related quality of life (four questions), for a total of 42 questions to expand on those in the WOMAC. Each question is rated on a five-point Likert scale. The scores range from 0 to 100, with a higher score indicating a better outcome for each subscale. This is the only PROM discussed in this section that asks about quality of life. Its validity is reportedly superior and it is equally responsive to the WOMAC for postoperative TKA patients [38].

3.3 OKS

The OKS is a PROM used to evaluate knee pain and function, and was first described by Dawson et al. in 1998 [27]. The difficulties in performing daily life activities, such as "knee pain", "downstairs", "getting in and out of a car", and "squatting" caused by knee disorders, and 12 items including "limping" and "giving way" caused by knee symptoms, are evaluated by five-point Likert scales (1–5 points). The total score ranges from 12 to 60 points, with lower scores indicating less difficulty. This reverse-scoring notation is based on an earlier system proposed by a group at Oxford University. To convert the reverse scoring notation to the score notation commonly used in modern literature, the following formula is used: (60 − score in reverse scoring notation) = score in modern notation. The resulting OKS is generally indicated by a score between 0 and 48, with a score of 40 or higher considered a good result. The internal consistency, test reliability, and construct validity of the OKS have been demonstrated, as well as an appropriate level of responsiveness to clinical changes [39, 40]. A report [41] suggested that OKS was highly associated with pain and less associated with postoperative function in the postoperative evaluation of mobile-bearing-type UKA.

3.4 KSS-2011

The KSS-2011, reported by Scuderi et al. [28], is a revised version of the KSS [31] first reported in 1989. The previous KSS had three questions for both surgeon-reported (ROM, instability, and alignment) and patient-reported (pain and function in level/stair walking) components, whereas the KSS-2011 has a PROM component consisting of 30 questions. The PROM component consists of I. Symptoms: Current knee symptoms (three questions; 25-point scale), II. Patient satisfaction (five questions; 40-point scale); III. Patient expectations (three questions, 15-point scale); and IV. Functional activities: Four parts to assess current activities (19 questions; 100-point scale) with a total score of 180 points. Functional activities are further divided into walking and standing (five questions; 30-point scale), standard activities (six questions; 30-point scale), advanced activities (five questions; 25-point scale), and knee activities that the patients chose at their own discretion (three questions; 15-point scale). The discretionary knee activity section reflects lifestyle changes in individual patients. The KSS-2011 is the only one of the five PROMs discussed in this section to include a surgeon-reported component; however, this component is not used to calculate the numerical score. The fulfillment of patient expectations after TKA is also considered a predictive criterion for postoperative satisfaction [42]. Therefore, assessing the achievement of expectations after TKA could complement HRQOL assessment. The KSS-2011 includes an evaluation of HRQOL because it also includes an evaluation of patient expectations. However, questions regarding patient expectations differ between the pre-and postoperative periods; therefore, care should be taken when comparing pre- and postoperative outcomes. For example, if a pre-TKA patient "a lot" expects to be able to engage in recreational and sports activities after surgery, the preoperative score would be 5. Even if the patient can perform recreational and sports activities "as expected" after TKA, the score would still be 3, which is lower than the preoperative score.

3.5 FJS-12

The FJS-12 is a PROM developed by Behrend et al. [32] to assess joint awareness. Joint awareness is defined as the degree to which a person becomes aware of a joint when there is a slight abnormality. Five-point Likert scales are used to assess the degree of awareness of the knee joint in 12 situations in daily life, ranging from low-activity situations such as "in bed at night" and "sitting in a chair for more than one hour" to relatively high-activity situations such as "standing up from a low sitting position" and "doing a favorite sport." The FJS-12 is evaluated on a scale of 0–100 points, with higher scores indicating lower joint awareness and superior outcome. Knees without joint awareness are considered the ultimate goal of knee replacement surgery because this means that "the patients are not at all concerned about their knee." In addition, as TKA performance has improved in recent years, ceiling effects have been reported in the performance evaluation with existing PROMs; however, the FJS is less likely to produce ceiling effects and is highly responsive [32]. As it is simple, with only 12 questions, it may be useful in combination with the above PROMs. In recent years, the FJS-12 has been used not only for TKA but also for the evaluation of high tibial osteotomy [43].

4 Issues Regarding PROMs

Self-reported function may not adequately reflect actual function as PROMs are essentially subjective assessment methods; thus, they measure the patient's perception of function rather than the actual ability [2]. For example, an analysis of data from >150,000 patients assessed by OKS reported that self-reports of pain and function are not distinguished from each other [44], and patients assessed by WOMAC showed the significant influence of the pain subscale on the physical function subscale [45]. Additionally, as increasingly younger patients undergo TKA, a ceiling effect may be observed regarding the evaluation of postoperative measures of function [46]. Moreover, the coexistence of pain from other causes, such as low back pain, influences the PROMs in evaluating TKA performance [47]. Therefore, TKA performance can be evaluated in more detail by using not only PROMs specialized for the evaluation of knee joint replacement performance but also a combination of surgeon-reported objective evaluation and general-purpose PROMs used to evaluate HRQOL.

References

1. Collins NJ, Roos EM. Patient-reported outcomes for total hip and knee arthroplasty: commonly used instruments and attributes of a "good" measure. Clin Geriatr Med. 2012;28:367–94.
2. Hossain FS, et al. The assessment of outcome after total knee arthroplasty: are we there yet? Bone Joint J. 2015;97-B:3–9.
3. Parratte S, et al. Effect of postoperative mechanical axis alignment on the fifteen-year survival of modern, cemented total knee replacements. J Bone Joint Surg Am. 2010;92:2143–9.
4. Bourne RB. Measuring tools for functional outcomes in total knee arthroplasty. Clin Orthop Relat Res. 2008;466:2634–8.
5. Bourne RB, et al. Patient satisfaction after total knee arthroplasty: who is satisfied and who is not? Clin Orthop Relat Res. 2010;468:57–63.
6. Kurtz SM, et al. Future young patient demand for primary and revision joint replacement: national projections from 2010 to 2030. Clin Orthop Relat Res. 2009;467:2606–12.
7. Nilsdotter AK, Toksvig-Larsen S, Roos EM. Knee arthroplasty: are patients' expectations fulfilled? A prospective study of pain and function in 102 patients with 5-year follow-up. Acta Orthop. 2009;80:55–61.
8. Noble PC, et al. Does total knee replacement restore normal knee function? Clin Orthop Relat Res. 2005;431:157–65.
9. Gandhi R, Davey JR, Mahomed N. Patient expectations predict greater pain relief with joint arthroplasty. J Arthroplast. 2009;24:716–21.
10. Bourne RB, Maloney WJ, Wright JG. An AOA critical issue. The outcome of the outcomes movement. J Bone Joint Surg Am. 2004;86:633–40.
11. Ghomrawi HM, et al. How often are patient and surgeon recovery expectations for total joint arthroplasty aligned? Results of a pilot study. HSS J. 2011;7:229–34.
12. Noble PC, et al. Development of a new knee society scoring system. Clin Orthop Relat Res. 2012;470:20–32.
13. Alviar MJ, et al. Do patient-reported outcome measures used in assessing outcomes in rehabilitation after hip and knee arthroplasty capture issues relevant to patients? Results of a systematic review and ICF linking process. J Rehabil Med. 2011;43:374–81.
14. Siljander MP, et al. Current trends in patient-reported outcome measures in Total joint arthroplasty: a study of 4 major Orthopaedic journals. J Arthroplast. 2018;33:3416–21.
15. Parsons JT, Snyder AR. Health-related quality of life as a primary clinical outcome in sport rehabilitation. J Sport Rehabil. 2011;20:17–36.
16. Ware JE, Sherbourne CD. The MOS 36-item short-form health survey (SF-36). I. Conceptual framework and item selection. Med Care. 1992;30:473–83.
17. Desmeules F, et al. Determinants of pain, functional limitations and health-related quality of life six months after total knee arthroplasty: results from a prospective cohort study. BMC Sports Sci Med Rehabil. 2013;5:2.
18. Ware J Jr, Kosinski M, Keller SD. A 12-item short-form health survey: construction of scales and preliminary tests of reliability and validity. Med Care. 1996;34:220–33.
19. Clement ND, et al. Socioeconomic status affects the Oxford knee score and short-form 12 score following total knee replacement. Bone Joint J. 2013;95-B:52–8.
20. Rabin R, de Charro F. EQ-5D: a measure of health status from the EuroQol group. Ann Med. 2001;33:337–43.
21. Ethgen O, et al. Health-related quality of life in total hip and total knee arthroplasty. A qualitative and systematic review of the literature. J Bone Joint Surg Am. 2004;86:963–74.
22. Giesinger K, et al. Comparative responsiveness of outcome measures for total knee arthroplasty. Osteoarthr Cartil. 2014;22:184–9.
23. Busija L, et al. Magnitude and meaningfulness of change in SF-36 scores in four types of orthopedic surgery. Health Qual Life Outcomes. 2008;6:55.
24. Wordsworth BP, Shakespeare DT, Mowat AG. MacIntosh arthroplasty for the rheumatoid knee: a 10-year follow up. Ann Rheum Dis. 1985;44:738–41.
25. Shaw NE, Chatterjee RK. Manchester knee arthroplasty. J Bone Joint Surg Br. 1978;60-B:310–4.
26. Ramkumar PN, Harris JD, Noble PC. Patient-reported outcome measures after total knee arthroplasty: a systematic review. Bone Joint Res. 2015;4:120–7.
27. Dawson J, et al. Questionnaire on the perceptions of patients about total knee replacement. J Bone Joint Surg Br. 1998;80:63–9.
28. Scuderi GR, et al. The new knee society knee scoring system. Clin Orthop Relat Res. 2012;470:3–19.
29. Roos EM, et al. Knee injury and osteoarthritis outcome score (KOOS)–development of a self-administered outcome measure. J Orthop Sports Phys Ther. 1998;28:88–96.
30. Bellamy N, et al. Validation study of WOMAC: a health status instrument for measuring clinically important patient relevant outcomes to antirheumatic drug therapy in patients with osteoarthritis of the hip or knee. J Rheumatol. 1988;15:1833–40.
31. Insall JN, et al. Rationale of the knee society clinical rating system. Clin Orthop Relat Res. 1989;248:13–4.
32. Behrend H, et al. The "forgotten joint" as the ultimate goal in joint arthroplasty: validation of a new patient-reported outcome measure. J Arthroplast. 2012;27:430–436.e1.
33. Takeuchi R, et al. Cross-cultural adaptation and validation of the Oxford 12-item knee score in Japanese. Arch Orthop Trauma Surg. 2011;131:247–54.
34. Nakamura N, et al. Cross-cultural adaptation and validation of the Japanese knee injury and osteoarthritis outcome score (KOOS). J Orthop Sci. 2011;16:516–23.
35. Hamamoto Y, et al. Cross-cultural adaptation and validation of the Japanese version of the new knee society scoring system for osteoarthritic knee with total knee arthroplasty. J Orthop Sci. 2015;20:849–53.
36. Moskowitz RW. The burden of osteoarthritis: clinical and quality-of-life issues. Am J Manag Care. 2009;15((8 (Suppl)):S223–9.
37. Maniar RN, et al. What is the responsiveness and respondent burden of the new knee society score? Clin Orthop Relat Res. 2017;475:2218–27.
38. Roos EM, Toksvig-Larsen S. Knee injury and osteoarthritis outcome score (KOOS)—validation and comparison to the WOMAC in total knee replacement. Health Qual Life Outcomes. 2003;1:17.
39. Collins NJ, et al. Measures of knee function: international knee documentation committee (IKDC) subjective knee evaluation form, knee injury and osteoarthritis outcome score (KOOS), knee injury and osteoarthritis outcome score physical function short form (KOOS-PS), knee outcome survey activities of daily living scale (KOS-ADL), Lysholm knee scoring scale, Oxford knee score (OKS), Western Ontario and McMaster universities osteoarthritis index (WOMAC), activity rating scale (ARS), and Tegner activity score (TAS). Arthritis Care Res, Hoboken. 2011;63(Suppl 11):S208–28.
40. Ko Y, et al. Comparison of the responsiveness of the SF-36, the Oxford knee score, and the knee society clinical rating system in patients undergoing total knee replacement. Qual Life Res. 2013;22:2455–9.
41. van Hove RP, et al. High correlation of the Oxford knee score with postoperative pain, but not with performance-based functioning. Knee Surg Sports Traumatol Arthrosc. 2016;24:3369–75.
42. Dunbar MJ, et al. I can't get no satisfaction after my total knee replacement: rhymes and reasons. Bone Joint J. 2013;95-B(11 Suppl. A):148–52.

43. Itoh M, et al. Good validity and high internal consistency of the forgotten joint Score-12 in patients after medial opening wedge high Tibial osteotomy in. J Arthroplast. 2021;36:2691–7.
44. Harris K, et al. Can pain and function be distinguished in the Oxford knee score in a meaningful way? An exploratory and confirmatory factor analysis. Qual Life Res. 2013;22:2561–8.
45. Stratford PW, Kennedy DM. Performance measures were necessary to obtain a complete picture of osteoarthritic patients. J Clin Epidemiol. 2006;59:160–7.
46. Talbot S, et al. Use of a new high-activity arthroplasty score to assess function of young patients with total hip or knee arthroplasty. J Arthroplast. 2010;25:268–73.
47. Clement ND, et al. Total knee replacement in patients with concomitant back pain results in a worse functional outcome and a lower rate of satisfaction. Bone Joint J. 2013;95-B:1632–9.

Imaging

70

Kazunori Hino

The residual pain related factors after total knee arthroplasty (TKA) vary and include loosening, instability, and malalignment. Moreover, multiple factors occasionally overlap, and extraarticular factors may exist. To identify these factors efficiently, imaging examinations, such as X-rays with specific positions, stress views of X-rays, and 3D-imaging tools such as computed tomography (CT) and magnetic resonance imaging (MRI), are also effective.

1 Radiographs

Clinicians require proficient interpretation of radiographic images as X-ray examinations provide a large amount of information. The Kellgren and Lawrence classification is the most common imaging modality for classifying the severity of preoperative knee osteoarthritis (OA) [1]. The X-ray incident angle should be aligned with the posterior tibial slope when evaluating the joint space of the tibiofemoral joint in the frontal view while in a standing position [2, 3]. On the immediate postoperative TKA radiograph, the overall alignment, implant placement angle, and placement position are first evaluated. The Knee Society Radiographic Evaluation System and Methodology is often used to measure the angle of implant placement [4]. In addition, the evaluation before and after TKA should include examining changes in the joint line, removal of osteophytes, presence of anterior femoral notching, identification of residual cement, detection of free bodies and fractures, and assessment of patellar position on the skyline view. Moreover, postoperative follow-up radiographs should be used to identify radiolucent lines and bone resorption. The full-length standing AP view or the Rosenberg view, imaged from behind while standing at 45° knee flexion, or other specialized imaging techniques can be utilized as needed to obtain additional information [2]. The Rosenberg view is particularly useful in cases of severe pain despite limited joint space narrowing during extension and in the evaluation of lateral posterior condylar OA with valgus alignment.

2 Stress Radiographs

Varus and valgus stress radiographs in the AP view are commonly conducted to evaluate the medial and lateral joint gaps. Generally, lateral laxity is greater than medial laxity in a normal knee joint [5].

3 Femoral Epicondylar Imaging (Kneeling View, Epicondylar View, and Stress Epicondylar View)

In the kneeling view, patients assume a kneeling position with their knee placed on a radiographic table with 70–80° knee flexion, and the X-ray incident angle should be aligned from behind (Fig. 70.1) [6]. In the epicondylar view, the patient is seated on an X-ray table with their lower leg hanging down, and weights are placed on the ankle joint. The X-ray incident angle is adjusted upward by approximately 10° from behind, and the beam center is aligned with the center of the patella (Fig. 70.2).

The stress epicondylar view is a quantitative evaluation method for varus and valgus laxity during knee flexion. To evaluate valgus laxity, the patient actively rotates their hip joint externally and swings their lower leg inward while the examiner applies an antagonistic force (valgus force) of approximately 10 kg from the inside of the ankle joint. A total varus and valgus stress angle of ≤10° is considered a good indicator of varus–valgus stability. This method is expected to reduce patient pain compared to conventional flexion stress imaging as a result of the examinee's active hip motion [7].

K. Hino (✉)
Department of Joint Reconstruction, Ehime University School of Medicine, Ehime, Japan

S. Matsuda (ed.), *Advances in Total Knee Arthroplasty*, https://doi.org/10.1007/978-981-97-4920-1_70

Fig. 70.1 Kneeling view

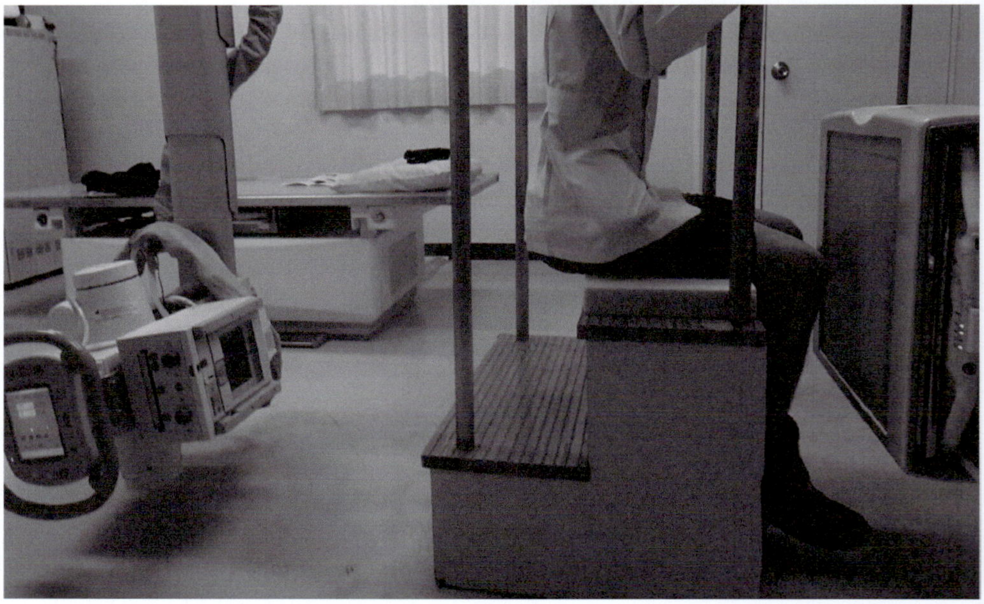

Fig. 70.2 Epicondylar view

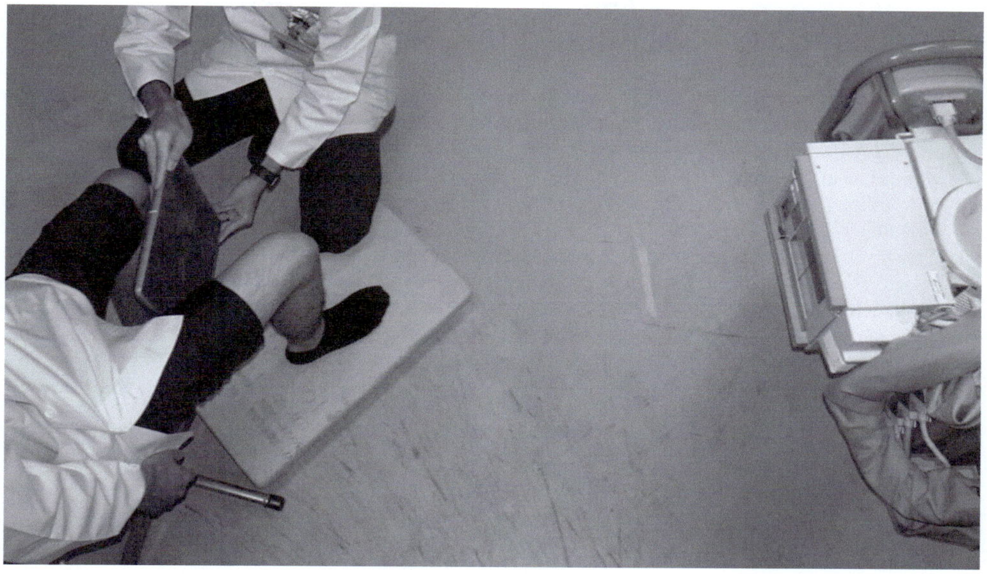

Fig. 70.3 Oblique posterior condylar view

4 Oblique Posterior Condylar View (OPC View)

Although, osteolysis commonly occurs in the posterior part of the femoral condyles, this is an area which is often difficult to evaluate on a true-lateral radiograph; an abnormality in one condyle may be obscured by the other. The oblique posterior condylar (OPC) view, which was developed to evaluate bone resorption in the posterior condyle of the femur, captures images from both oblique angles, with the knee joint flexed at approximately 90°. Notably, the use of the OPC view has improved the detection rate of osteolysis in the posterior condyle compared to conventional LM views (Fig. 70.3) [8].

5 MRI

MRI is a noninvasive and useful method for evaluating soft tissues, as these are difficult to assess through radiographic examinations. Whole-Organ Magnetic Resonance Imaging Scores (WORKS), osteoarthritis knee scores (MOAKS), and Boston Leeds Osteoarthritis Knee Scores have been used to classify OA severity classification using MRI [9, 10]. However, issues such as a long inspection time, high cost, and lack of image accuracy may occur due to the halation caused by metal components. Recently, the development of sequencing techniques has rendered MRI valuable even after TKA.

Updated MRI methods enable the quantitative evaluation of changes in cartilage composition and structure and are applied in clinical practice. T2 mapping is a useful imaging method that evaluates collagen arrangement and water content in cartilage and is useful for detecting early OA changes and quantitatively evaluating cartilage degeneration over follow-up [11, 12].

6 CT

CT is the most reliable method for evaluating bone structures, with excellent spatial resolution. Three-dimensional (3D)-CT is widely used for preoperative planning and postoperative evaluation of TKA, providing valuable insights into the rotational position of components. Compared to radiographs or 2D-CT, 3D-CT offers superior accuracy [13]. Recently, the effectiveness of bone single photon emission computed tomography (SPECT/CT) has been reported in detecting postoperative TKA complications such as loosening and anterior knee pain [14]. While CT has several advantages over MRI, including a lower cost and shorter inspection time, it should be performed after clarifying the purpose and indications for the examination and recognizing that radiation exposure is an issue.

References

1. Kellgren JH, Lawrence JS. Radiological assessment of osteoarthrosis. Ann Rheum Dis. 1957;16:494–502.
2. Rosenberg TD, et al. The forty-five-degree posteroanterior flexion weight-bearing radiograph of the knee. J Bone Joint Surg Am. 1988;70:1479–83.
3. Davies AP, et al. Plain radiography in the degenerate knee: a case for change. J Bone Joint Surg Br. 1999;81:632–5.
4. Ewald FC. The knee society total knee arthroplasty roentgenographic evaluation and scoring system. Clin Orthop Relat Res. 1989;248:9–12.
5. Okazaki K, et al. Asymmetry of mediolateral laxity of the normal knee. J Orthop Sci. 2006;11:264–6.

6. Takai S, et al. Kneeling view: a new roentgenographic technique to assess rotational deformity and alignment of the distal femur. J Arthroplast. 2003;18:478–83.
7. Stahelin T, et al. Fluoroscopically assisted stress radiography for varus-valgus stability assessment in flexion after total knee arthroplasty. J Arthroplast. 2003;18:513–5.
8. Miura H, et al. The oblique posterior femoral condylar radiographic view following total knee arthroplasty. J Bone Joint Surg Am. 2004;86:47–50.
9. Peterfy CG, et al. Whole-organ magnetic resonance imaging score (WORMS) of the knee in osteoarthritis. Osteoarthr Cartil. 2004;12:177–90.
10. Hunter DJ, et al. The reliability of a new scoring system for knee osteoarthritis MRI and the validity of bone marrow lesion assessment: BLOKS (Boston Leeds osteoarthritis knee score). Ann Rheum Dis. 2008;67:206–11.
11. Eckstein F, et al. Recent advances in osteoarthritis imaging – the osteoarthritis initiative. Nat Rev Rheumatol. 2012;8:622–30.
12. Kumahashi N, et al. A longitudinal study of the quantitative evaluation of patella cartilage after total knee replacement by delayed gadolinium. Osteoarthr Cartil. 2013;21:126–35.
13. Watanabe S, et al. Intra- and interobserver reliability of implant positioning evaluation on a CT-based three-dimensional postoperative. BMC Musculoskelet Disord. 2021;22:363.
14. Murer AM, et al. Bone SPECT/CT has excellent sensitivity and specificity for diagnosis of loosening and patellofemoral problems after total knee. Knee Surg Sports Traumatol Arthrosc. 2020;28:1029–35.

71 Postoperative Sports Activities

Takashi Nakamura

Currently, several countries worldwide have an aging population [1], and the rate of OA is increasing accordingly. Epidemiological studies of knee and hip osteoarthritis in 195 countries worldwide have shown that the prevalence and annual incidence rates of knee and hip osteoarthritis have increased by 9.3% and 8.2%, respectively, from 1990 to 2017, with an estimated 303.1 million people suffering from knee and hip osteoarthritis in 2017 [2]. Although various treatment methods are available for OA, total knee arthroplasty (TKA) is a well-established treatment for end-stage OA owing to its superior pain relief and improvement in knee joint function. As the number of patients with OA is expected to increase in the future, the number of TKA cases is also expected to increase.

With the gradual increase in the average life expectancy [1], people's awareness of healthy life expectancy is increasing annually, and there is a growing expectation that people will continue to engage in recreational activities, including sports, after TKA. Exercise improves not only physical functions, such as respiratory and circulatory function, muscle strength, balance, and obesity prevention, but also psychological effects, such as depression and anxiety [3]. While sports activities after TKA provide these benefits, there are concerns about the increased risk of injury as well as increased component loosening and wear of polyethylene after TKA. This section outlines the current status and benefits of sports activities involving TKA, as well as the problems associated with these activities.

1 What Are the Recommended Sports Activities After TKA?

In 1999, the Knee Society Survey provided the first guidelines for sports activities following TKA [4]. The survey classified sports into four categories: allowed, allowed with experience, no consensus, and not recommended. It also allowed low-contact and low-impact sports, such as bowling, golf, walking, and swimming, but not high-contact and high-impact sports, such as soccer, volleyball, basketball, and jogging. Subsequently, a similar survey was conducted in 2005, and the number of events classified as not recommended decreased from 12 to 5, whereas the number of events that were allowed or allowed with experience increased from 18 to 19 (Table 71.1) [5]. The reasons for this are stated to be the excellent clinical results of TKA, improved surgical techniques, and improved implants; however, as they are not based on sufficient evidence, participation in competitions should be done with caution.

Some authors argue that not only sport but also its frequency should be considered when deciding whether a sports activity can be recommended after TKA. Mont et al. [6] defined the sports activity score as the score obtained by multiplying the impact point (recommended: 1 point; if experienced: 2 points; not recommended: 3 points) by the type of sport and the number of times the sport was performed per week. According to this definition, they classified sports activity scores of 0–8 as low activity and scores of 9 or more

T. Nakamura (✉)
Department of Orthopaedics Surgery, School of Medicine, Toho University, Tokyo, Japan
e-mail: takashin@med.toho-u.ac.jp

S. Matsuda (ed.), *Advances in Total Knee Arthroplasty*, https://doi.org/10.1007/978-981-97-4920-1_71

Table 71.1 Classification of activities after TKA

Allowed			Allowed with experience			No consensus			Not recommended		
	1999	2005		1999	2005		1999	2005		1999	2005
Stationary cycling	✓	✓	Bowling	✓		Square dancing	✓		Baseball	✓	
Ballroom dancing	✓	✓	Canoeing	✓		Fencing	✓	✓	Basketball	✓	✓
Golf	✓	✓	Road cycling	✓		Rowing	✓		Football	✓	✓
Shuffleboard	✓	✓	Hiking	✓		Ice skating	✓		Gymnastics	✓	
Swimming	✓	✓	Horseback riding	✓	✓	Roller skating	✓		Handball	✓	
Doubles tennis	✓		Cross-country skiing	✓	✓	Downhill skiing	✓		Hockey	✓	
Normal walking	✓	✓	Rowing		✓	Stationary skiing	✓		Jogging	✓	✓
Bowling		✓	Ice skating		✓	Speed walking	✓		Rock climbing	✓	
Canoeing		✓	Roller skating		✓	Weight lifting	✓		Soccer	✓	✓
Road cycling		✓	Downhill skiing		✓	Weight machine	✓		Squash/racquetball	✓	
Square dancing		✓	Stationary skiing		✓	Baseball		✓	Singles tennis	✓	
Hiking		✓	Doubles tennis		✓	Gymnastics		✓	Volleyball	✓	
Speed walking		✓	Weight lifting		✓	Handball		✓			
			Weight machine		✓	Hockey		✓			
						Rock climbing		✓			
						Squash/racquetball		✓			
						Singles tennis		✓			
						Volleyball		✓			

as high activity. For example, if a person cycles three times and swims four times per week, the sports activity score is (2 × 3) + (1 × 4) = 10. They reported no significant differences in clinical outcomes or complications between the low- and high-activity groups at an average follow-up of 7 years postoperatively. However, they do not recommend high-impact sports because the long-term outcomes are unknown.

Regarding sports activities after TKA, the degree of impact force applied to the knee joint was classified as high or low impact, and the degree of recommendation was evaluated. Currently, there is no clear conclusion regarding the types of sports that can be recommended after TKA; however, low-contact sports, low-impact sports, and sports in which athletes have experience prior to surgery are considered acceptable.

2 Current Status of Sports Participation Following Total Knee Arthroplasty

Wylde et al. [7] conducted a survey on the postoperative return to sports in 2085 patients who underwent arthroplasty of the lower extremities, such as THA, TKA, and UKA. They reported that of the 726 patients who were active in sports preoperatively, 446 (61.4%) were able to return to sports activities postoperatively, whereas 192 (26.4%) were unable to return, with particularly low rates of return to high-impact sports, such as badminton, tennis, and dance. In addition, they reported that the only factor involved in returning to sports was sex (male), with no correlation with age or surgical technique.

In recent years, an increasing number of reports on sports activities after TKA, as well as systematic reviews and meta-analyses have been published. Witjes et al. [8] reported that the return rate to sports activities after TKA ranged from 36% to 89%, and that the average time to return to sports was 13 weeks postoperatively. They also reported that after surgery, 80% of patients played low-impact sports, 13% played intermediate-impact sports, 7% played high-impact sports, 87% played low-impact sports, 9% played intermediate-impact sports, and 4% played high-impact sports. A systematic review/meta-analysis of 25 references on sports activity after TKA conducted by Hanreich et al. [9] revealed that the rate of return to sports activity after TKA was 71.2%, with a decrease in the proportion of intermediate- or high-impact sports and an increase in the proportion of low-impact sports, such as walking, cycling, and swimming. Matsushita et al. [10] also reported that low-impact sports, such as stretching, gardening, walking, swimming, and cycling, were the main sports activities emphasized by patients after TKA. The cur-

Table 71.2 UCLA activity score

Level description	
1.	Wholly inactive, dependent on others, and cannot leave residence
2.	Mostly inactive or restricted to minimum activities of daily living
3.	Occasionally participates in mild activities, such as walking, limited housework, and limited shopping
4.	Regularly participates in mild activities
5.	Occasionally participates in moderate activities, such as swimming, or could do unlimited housework or shopping
6.	Regularly participates in moderate activities
7.	Regularly participates in active events, such as cycling
8.	Regularly participates in active events, such as golf or bowling
9.	Occasionally participates in impact sports, such as jogging, tennis, skiing, acrobatics, ballet, heavy labor, or backpacking
10.	Regularly participates in impact sports

rent return to sports after TKA is approximately 60–70%; furthermore, the number of recommended high-impact sports is decreasing, while the number of recommended low-impact sports is increasing due to concerns about implant durability.

On the other hand, in a study of high-impact sports after TKA, Mayr et al. [11] examined 81 knees of patients aged 65 years or older and reported that approximately one-fourth of the patients who performed high-impact sports, such as alpine skiing, rock climbing, dancing, and tennis, did not show any wear or loosening 6 years postoperatively. Meftah et al. [12] reported a good survival rate of 98% at an average of 12.3 years after surgery in 392 knees of patients that were younger than 60 years, which included patients with a high UCLA activity score [13] (Table 71.2) of 5 or higher. In addition, Crawford et al. [14] investigated implant survival rates in 2038 knees classified into two groups: a low-activity group with a UCLA activity score of 5 or less and a high-activity group with a UCLA activity score of 6 or more. The survival rate of the high-activity group was higher than that of the low-activity group at 11.4 years postoperatively. Patients who can return to high-impact sports are considered pain-free and have excellent ROM and stability. If a patient has previous experience in high-impact sports and a high clinical score after TKA, lowering the patient's postoperative activity levels may be unnecessary. However, these few reports did not lead to recommendations for high-impact sports.

3 Impact of Sports Activities After TKA

With the aging population in many countries worldwide, the number of TKA cases and the target age for TKA are expected to increase in the future. A survey [15] conducted by the National Health Service in the United Kingdom on age at surgery reported that the percentage of patients aged 75 years or older increased from 14.4% in 1999 to 22.9% in 2015. The prevalence of chronic diseases increases with advancing age, particularly lipid metabolic disorders, vascular diseases, and ischemic diseases [16]. Owing to the older target age for TKA, there are concerns that the comorbidity rates of chronic diseases in patients undergoing TKA will increase. Exercise improves respiratory and circulatory functions, muscle strength, balance, and physical functions, such as obesity prevention [3]. Hence, moderate sports activities, such as low-impact sports after TKA, are expected to play an important role in the treatment and prevention of chronic diseases.

The effects of exercise on the body are diverse; however, it is also known to increase bone density. In a clinical study that examined the effects of exercise interventions on bone density, Bonaiuti et al. [17] found that aerobic loading exercises, such as walking, increased lumbar spine bone density by 1.31% and proximal femur bone density by 0.92%. Howe et al. [18] also reported that lower limb muscle strength training increased proximal femur bone density by 1.03%. Kuster [19] reported that moderate sports activity after TKA may improve bone quality and implant fixation by stimulating loading stress. In addition, Valle et al. [20] found that in a follow-up study of 130 knees at an average of 12 years after TKA, the revision rate was 15.2% in the group of patients who were continuously active in sports after surgery and 23.8% in the group of patients who were not active in sports. The authors concluded that moderate sports activity had no adverse effects on knee prostheses, and may lead to improved implant fixation. In other words, moderate sports activities, such as low-impact sports, may increase bone strength and further improve long-term clinical outcomes after TKA.

In recent years, patient-reported outcome measures have been mainly used to assess the clinical outcomes of TKA, reflecting patient-derived functional assessments and sense of health. It has been shown that expected postoperative and preoperative activities have a significant impact on patient satisfaction [21, 22]. Mayr et al. [11] reported that, in a survey of patients aged 65 years or older who underwent TKA, the WOMAC total score and Knee injury and Osteoarthritis Outcome Score (KOOS)-QOL score were significantly higher in those patients who engaged in high-level sports activities. Plassard et al. [23] also investigated the return to sports, activity, and satisfaction after TKA in 443 knees and found that the return to sports rate was as high as 85% and the UCLA activity score increased from 4.45 preoperatively to 5.92 postoperatively, which were considered the reasons for extremely high satisfaction (93% satisfied or extremely satisfied). To achieve high patient satisfaction after TKA, it is necessary to improve patients' quality of life. Participation in or return to sports activities appropriate to the patient's activity level may improve patient satisfaction with TKA.

However, because after TKA, the knee joint becomes an artificial joint, there are concerns that increasing the load stress on the knee joint due to exercise increases the risk of component loosening and polyethylene wear. Moreover, some reports suggest that high activity levels after TKA increase insert wear [24, 25]. Ponzio et al. [26] reported a revision rate of 3.2% in a group of highly active patients 5–10 years after TKA, compared with a revision rate of 1.6% in a group of less active patients. The authors reported that high activity levels increased the risk of revision. The stress applied to the knee joint is 4–6 times the body weight when climbing stairs, 8 times the body weight when going downhill, and as much as 14 times the body weight during jogging or running [19]. Even if sports activities after TKA have a low impact, careful postoperative follow-up is necessary.

4 Issues Related to the Resumption of Sports Activities

The rate of return to sports activities after TKA is not high' moreover, there are various reasons for not returning to sports activities after TKA [22, 27], and a lack of explanation by the surgeon is considered one of the factors [7]. Previously, surgeons did not actively recommend postoperative sports activities from the viewpoint of component loosening and polyethylene wear. With remarkable progress in surgical techniques and technology, we believe that moderate sports activities after TKA should be encouraged. To increase the rate of return to sports activities, it is important for the surgeon to understand the patient's physical abilities and expectations and provide sufficient explanations regarding sports activities after TKA. It has also been reported that the level of activity immediately before TKA has a significant impact on the return rate to sports activities [28]. It is important to plan TKA when the patient's ADL is relatively well maintained and his/her motivation to engage in sports activities is preserved.

The issue of sports activity after TKA remains inadequately explored, and there is a dearth of established evidence in this area. The number of TKA procedures has continued to increase. Although the most important factor is the aging society, the stable long-term clinical results of TKA may also be one of the reasons the indications for TKA are expanding to younger patients. Young patients are more active than elderly patients and are more willing to return to sports and other recreational activities after surgery [22]. With the current growing interest in sports activities after TKA, we should endeavor to engage in comprehensive research on sports activities after TKA to foster evidence-based practices.

References

1. United Nations, Department of Economic and Social Affairs, Population Division. 2019.
2. Safiri S, et al. Global, regional and national burden of osteoarthritis 1990-2017: a systematic analysis of the global burden of disease study 2017. Rheum Dis. 2020;76:819–28.
3. Paffenbarger RS Jr, et al. Physical activity, all-cause mortality, and longevity of college alumni. N Engl J Med. 1986;314:605–13.
4. Healy WL, et al. Athletic activity after total knee arthroplasty. Clin Orthop Relat Res. 2000;380:65–71.
5. Healy WL, et al. Athletic activity after total joint arthroplasty. J Bone Joint Surg Am. 2008;90:2245–52.
6. Mont MA, et al. Knee arthroplasties have similar results in high- and low-activity patients. Clin Orthop Relat Res. 2007;460:165–73.
7. Wylde V, et al. Return to sport after joint replacement. J Bone Joint Surg Br. 2008;90:920–3.
8. Witjes S, et al. Return to sports and physical activity after total and unicondylar knee arthroplasty; a systematic review and meta-analysis. Sports Med. 2016;46:269–92.
9. Hanreich C, et al. Sport and physical activity following primary total knee arthroplasty: a systematic review and meta-analysis. J Arthroplast. 2020;35:2274–85.
10. Matsushita Y, et al. Recreational sports, workout and gym activities after total knee arthroplasty: an Asian cohort study. J Orthop. 2019;16:41–4.
11. Mayr HO, et al. Sports activity following total knee arthroplasty in patients older than 60 years. J Arthroplast. 2015;30:46–9.
12. Maftah M, et al. Long-term results of total knee arthroplasty in young and active patients with posterior stabilized design. Knee. 2016;23:318–21.
13. Amstutz HC, et al. Treatment of primary osteoarthritis of the hip: a comparison of total joint and surface replacement arthroplasty. J Bone Joint Surg Am. 1984;66:228–41.
14. Crawford DA, et al. Higher activity level following total knee arthroplasty is not deleterious to mid-term implant survivorship. J Arthroplast. 2020;35:116–20.
15. Fowler AJ, et al. Age of patients undergoing surgery. Br J Surg. 2019;106:1012–8.
16. Atella V, et al. Trends in age-related disease burden and healthcare utilization. Aging Cell. 2019;18:1–8.
17. Bonaiuti D, et al. Exercise for preventing and treating osteoporosis in postmenopausal women. Cochrane Database Syst Rev. 2002;3:CD000333.
18. Howe TE, et al. Exercise for preventing and treating osteoporosis in postmenopausal women. Cochrane Database Syst Rev. 2011;7:CD000333.
19. Kuster MS. Exercise recommendations after total joint replacement: a review of the current literature and proposal of scientifically based guidelines. Sports Med. 2002;32:433–45.
20. Valle C, et al. Dose sports activity influence total knee arthroplasty durability? Analysis with a follow-up of 12 years. Sportverlrtz Sportschaden. 2017;31:111–5.
21. Noble PC, et al. The John Insall award; patient expectations affect satisfaction with total knee arthroplasty. Clin Orthop Relat Res. 2006;452:35–43.
22. Bonnin M, et al. Can patients truly do sport after TKA? Knee Surg Sports Traumatol Arthrosc. 2010;18:853–62.
23. Plassard J, et al. Factors leading to return to sports and recreational activity after total knee replacement: a retrospective study. SICOT-J. 2020;6:1–6.

24. Lavemia CJ, et al. Activity level and wear in total knee arthroplasty: a study of autopsy retrieved specimens. J Arthroplast. 2001;16:446–53.
25. Kuster MS, et al. Factors affecting polyethylene wear in total knee arthroplasty. Orthopedics. 2002;25:235–42.
26. Ponzio D, et al. An analysis of the influence of physical activity level on total knee arthroplasty expectations, satisfaction, and outcomes. J Bone Joint Surg Am. 2018;100:1539–48.
27. Chang MJ, et al. Why patients do not participate in sports activities after total knee arthroplasty. Orthop J Sports Med. 2015;3:1–8.
28. Konings MJ, et al. Effect of knee arthroplasty on sports participation and activity levels: a systematic review and meta-analysis. BMJ Open Sports Exerc Med. 2020;6:e000729.

Improving Patient Satisfaction

72

Ryota Yamagami

1 Patient Satisfaction After TKA

TKA is a well-established surgical treatment for advanced knee OA and RA, providing pain relief and functional recovery to patients. It is expected to reduce complications and further improve the long-term implant survival rate by enhancing surgical techniques and implant designs. However, approximately 20% of patients who underwent TKA are not satisfied with the surgical outcomes [1, 2], and patient satisfaction is lower than that of total hip arthroplasty [3]. Conventionally, outcomes are objectively evaluated by surgeons. Given the discrepancy between surgeon and patient evaluations [4], subjective evaluations, such as patient-reported outcomes and patient satisfaction, have become widely used among surgeons.

The Knee Society developed the 2011 Knee Society Score (KSS) as a new knee joint assessment system to better evaluate the expectations and satisfaction levels of patients, in addition to their symptoms and physical activities. The Japanese version of the KSS was subsequently validated [5]. The questionnaire is divided into four sections with a total of 180 points: (1) knee symptoms, (2) satisfaction, (3) expectation, and (4) activity. The satisfaction section adopts five rating scales to evaluate knee pain and joint function in daily life and recreational activities: "very satisfied" (8 points), "satisfied" (6 points), "normal" (4 points), "dissatisfied" (2 points), and "very dissatisfied" (0 points).

With increasing interest in patient satisfaction in recent years, many studies have been conducted to elucidate the intraoperative and postoperative factors that affect patient satisfaction. This section outlines the factors that influence postoperative patient satisfaction, focusing on studies conducted in Japan.

R. Yamagami (✉)
Department of Orthopaedic Surgery, Faculty of Medicine, The University of Tokyo, Bunkyo-ku, Tokyo, Japan
e-mail: yamagamir-ort@h.u-tokyo.ac.jp

2 Factors Affecting Postoperative Patient Satisfaction

2.1 Ligament Balance

In recent years, a technique that emphasizes medial stability in accordance with physiological knee joint balance has been proposed [6]. Joint instability during the mid-flexion range has been reported to affect postoperative joint function [7]. Thus, an increasing number of surgeons have focused on achieving proper knee–joint ligament balance during TKA.

An observational study [8] examining the relationship between the intraoperative component gap and 2011 KSS patient satisfaction 1 year postoperatively in BCS-TKA found that an increase in the medial gap at 30° knee flexion was associated with postoperative dissatisfaction. In another study [9] evaluating the relationship between the medial joint gap opening under varus/valgus stress and 2011 KSS patient satisfaction 1 year postoperatively in CR-TKA, a greater medial joint opening was associated with lower patient satisfaction in both knee extension and flexion. Both studies suggested the importance of postoperative medial stability. However, the appropriate ligament balance may vary with the implant design, and surgeons should understand the characteristics of each implant before surgery.

2.2 Discomfort and Noise

An ultimate goal of TKA is to create a natural and comfortable joint. The Forgotten Joint Score-12 (FJS-12) [10], developed in 2012, is used to assess the degree of discomfort in the knee joint during daily life and sports activities. In a report [11] that investigated the relationship between the FJS-12 and other patient-reported outcome measures for 295 knees after TKA with various implant designs, the FJS-12 showed a moderate positive correlation not only with the 2011 KSS total score, but also with the 2011 KSS patient satisfaction. This finding suggests that a knee without dis-

S. Matsuda (ed.), *Advances in Total Knee Arthroplasty*, https://doi.org/10.1007/978-981-97-4920-1_72

comfort leads to patient satisfaction. The sacrifice of the cruciate ligament is assumed to be one of the causes of postoperative discomfort after TKA. Hence, BCR-TKA, which preserves both cruciate ligaments, is expected to result in a knee with less discomfort by achieving physiological knee kinematics postoperatively, but this has not yet been reported.

Noise, which is sometimes observed after TKA, is a possible factor that can affect satisfaction; however, none of the reports from Japan showed any significant effect on postoperative patient satisfaction [11, 12]. This is presumably because noise itself is more likely to be found in patients with good postoperative ROM, and patients who have achieved such sufficient functional recovery are less affected by the discomfort caused by noise [11, 12].

2.3 Lower Extremity Alignments

Conventionally, mechanical alignment of the lower limb after TKA is the gold standard, which means that the aim of coronal alignment is to achieve a straight lower limb with a hip-knee-ankle angle (HKA) of 0°. In Japan, a previous retrospective survey of 500 patients who underwent TKA for more than 2 years showed that postoperative varus limb alignment was associated with patient dissatisfaction [4]. However, a new concept, "kinematic alignment," has recently been proposed to achieve the patient's original (i.e., pre-osteoarthritis) lower limb alignment. As the concept of surgery changes in the future, what to aim for as optimal lower limb alignment may shift to a more personalized alignment for each patient.

2.4 ROM

Several reports from the United States and Europe have shown that postoperative knee joint ROM is not necessarily associated with joint function or patient satisfaction. By contrast, in Japan, postoperative joint ROM and the 2011 KSS activity score are significantly associated [4]. Moreover, better KOOS in postoperative TKA patients with a flexion of 120° or more and high patient satisfaction when the postoperative joint ROM improved by 5° or more have been reported [13]. Therefore, postoperative ROM is an important factor that affects patient satisfaction in the Japanese population. Further improvement in the postoperative joint ROM is expected to be achieved by implant design enhancement, appropriate postoperative ligament balance, and acquisition of more physiological kinematics.

2.5 Physical Activity and Lower Limb Muscle Strength

The 2011 KSS patient satisfaction positively correlates with the 2011 KSS activity score [14]. Among the physical activities, basic activities of daily living, such as ascending and descending stairs, getting in and out of a car, moving sideways, walking time, and standing time, are strongly associated with patient satisfaction [15]. Therefore, the ability to perform them after surgery is essential to improve patient satisfaction. In terms of clinical physical performance assessment items, the 10-m walk test (measuring the time required to walk 10 m as fast as possible) and the time up-and-go test (measuring the time required to get up from a chair without a handrail, walk 3 m and back, and then sit down again) are associated with the 2011 KSS patient satisfaction at 1 year postoperatively [16], which may serve as indicators of patient satisfaction in postoperative rehabilitation interventions.

Quadriceps muscle strength is considered to play a key role in all the above movements. However, the strength of the postoperative lower limb muscle improves by only approximately 70% compared with that of the unaffected knee 1 year after surgery, and postoperative rehabilitation intervention is essential for improvement. A study [16] reported that knee extensor strength was associated not only with the 2011 KSS patient satisfaction at 1 year postoperatively but also with the 2011 KSS symptom and activity scores. Another study [17] found that improvement in knee extensor strength up to 1 year postoperatively was also associated with improvement in the FJS-12. These findings suggest that improvement in knee extensor strength is a useful indicator that should be achieved after TKA [17].

2.6 Psychological/Psychological Factors

Preoperative depression, anxiety disorders, and catastrophic thinking have been reported to exacerbate postoperative pain [18, 19]. A systematic review of 181 studies [20] found that mental health problems were the most frequently reported factors that affect patient satisfaction after TKA. The same review stated that psychological factors, such as achieving preoperatively-expected results postoperatively, also influenced patient satisfaction. Psychological factors cannot be excluded from the consideration of patient satisfaction. Providing consultation, medication, and counseling is necessary to patients before surgery for problems that can be addressed and to dispel unrealistic expectations.

2.7 Others

PFJ is another factor that affects patient satisfaction. A study [21] that measured PFJ pressure intraoperatively with a sensor device shaped like a component in BCS-TKA found no relationship between PFJ pressure and postoperative ROM, but reported a negative correlation between PFJ pressure at 140° flexion and the 2011 KSS patient satisfaction at 2 years postoperatively. On the other hand, a systematic review also reported that patellar nonreplacement was associated with patient dissatisfaction. Therefore, patellar replacement should be carefully performed to avoid overstuffing [20].

An underlying disease prior to TKA is also associated with patient satisfaction. A study comparing patients with RA and those with OA in the 2011 KSS at 2 years after TKA surgery [22] reported that although the 2011 KSS activity score was higher in patients with OA, the diagnosis of RA itself was a factor that improved patient satisfaction in the 2011 KSS compared with OA diagnosis. Although the conflicting results between activity scores and satisfaction should be interpreted with caution, this information is useful for predicting postoperative satisfaction with TKA in patients with RA.

3 Can Postoperative Patient Satisfaction Be Predicted Preoperatively?

In recent years, with increasing interest in patient satisfaction, attempts have been made to develop models to preoperatively predict patient satisfaction after surgery. Such models have recently been reported in Europe and other countries [23]. A model reported in Belgium in 2016 predicted postoperative patient satisfaction based on 10 items. These items include basic patient information, such as sex and age, and items extracted from the KOOS, 2011 KSS, EurQol 5 Dimension, and Pain Catastrophizing Scale. However, a validation study conducted in Japan [24] concluded that the model was not useful for Japanese patients and that further improvement was necessary.

As previously mentioned, numerous factors are involved in patient satisfaction. In addition to those mentioned above, patient age [4], preoperative deformity, type of implant used, and presence or absence of back pain [20] have been identified as factors involved in patient satisfaction. Further research is currently ongoing to analyze postoperative joint kinematics using fluoroscopic imaging, gait analysis using motion capture systems, and prediction of postoperative knee joint dynamics using musculoskeletal models. More information is expected to become available in the future. Presently, predicting patient satisfaction after TKA based on this vast amount of information is difficult. In my opinion, AI is expected to be utilized in the future to predict postoperative patient satisfaction from this detailed information and to construct a model that incorporates the essence of the surgical technique to achieve improvement in patient satisfaction.

4 Points to Improve Patient Satisfaction

Factors influencing patient satisfaction after TKA were reviewed, with a focus on Japanese studies. Among the factors listed in this section, ligament balance, noise, lower limb alignment, and lower limb muscle strength can be improved by surgeons through surgical techniques, implant selection, and postoperative rehabilitation interventions. With advances in technology and computer-assisted surgery, including robotic arthroplasty and navigation systems, more accurate implant placement, lower extremity alignment, and more detailed intraoperative ligament balance can be obtained, making these issues easier to address for many orthopedic surgeons. However, patient satisfaction is notably influenced by numerous factors, including those that are difficult for surgeons to intervene in, such as the patient's mental state and age. Obtaining a comprehensive picture of the patient as much as possible and planning the surgery with appropriate preoperative explanations are essential for surgeons.

References

1. Bourne RB, et al. Patient satisfaction after total knee arthroplasty: who is satisfied and who is not? Clin Orthop Relat Res. 2010;468:57–63.
2. Becher R, et al. Expectation, satisfaction and clinical outcome of patients after total knee arthroplasty. Knee Surg Sports Traumatol Arthrosc. 2011;19:1433–41.
3. Bourne RB, et al. Comparing patient outcomes after THA and TKA: is there a difference? Clin Orthop Relat Res. 2010;468:542–6.
4. Matsuda S, et al. Postoperative alignment and ROM affect patient satisfaction after TKA. Clin Orthop Relat Res. 2013;471:127–33.
5. Hamamoto Y, et al. Cross-cultural adaptation and validation of the Japanese version of the new Knee Society Scoring System for osteoarthritic knee with total knee arthroplasty. J Orthop Sci. 2015;20:849–53.
6. Tsubosaka M, et al. Comparison of intraoperative soft tissue balance between cruciate-retaining and posterior-stabilized total knee arthroplasty performed by a newly developed medial preserving gap technique. J Arthroplast. 2018;33:729–34.
7. Mochizuki T, et al. Association between anteroposterior laxity in mid-range flexion and subjective healing of instability after total knee arthroplasty. Knee Surg Sports Traumatol Arthrosc. 2017;25:3543–8.
8. Inui H, et al. Influence of surgical factors on patient satisfaction after bi-cruciate stabilized total knee arthroplasty: retrospective examination using multiple regression analysis. BMC Musculoskelet Disord. 2021;22:215.

9. Kamenaga T, et al. The influence of postoperative knee stability on patient satisfaction in cruciate-retaining total knee arthroplasty. J Arthroplast. 2018;33:2475–9.
10. Behrend H, et al. The "forgotten joint" as the ultimate goal in joint arthroplasty: validation of a new patient-reported. J Arthroplast. 2012;27:430–6.
11. Taniguchi H, et al. Noise after total knee arthroplasty has limited effect on joint awareness and patient-reported clinical outcomes: retrospective study. BMC Musculoskelet Disord. 2020;21:115.
12. Kuriyama S, et al. Noise generation with good range of motion but without femorotibial instability has small effect on patient satisfaction after total knee arthroplasty. J Arthroplast. 2017;32:407–12.
13. Kubo M, et al. Good postoperative flexion angle improves knee function and improvement of flexion angle increases patient satisfaction after total knee arthroplasty. J Arthroplast. 2021;36:3137–40.
14. Hasegawa M, et al. Factors contributing to patient satisfaction and expectations following computer-assisted total knee arthroplasty. J Knee Surg. 2018;31:448–52.
15. Nakahara H, et al. Correlations between patient satisfaction and ability to perform daily activities after total knee arthroplasty: why aren't patients satisfied? J Orthop Sci. 2015;20:87–92.
16. Furu M, et al. Quadriceps strength affects patient satisfaction after total knee arthroplasty. J Orthop Sci. 2016;21:38–43.
17. Hiyama Y, et al. Joint awareness after total knee arthroplasty is affected by pain and quadriceps strength. Orthop Traumatol Surg Res. 2016;102:435–9.
18. Brander V, et al. Pain and depression influence outcome 5 years after knee replacement surgery. Clin Orthop Relat Res. 2007;464:21–6.
19. Bonnin MP, et al. What are the factors of residual pain after uncomplicated TKA? Knee Surg Sports Traumatol Arthrosc. 2011;19:1411–7.
20. Nakano N, et al. Why are patients dissatisfied following a total knee replacement? A systematic review. Int Orhop. 2020;44:1971–2007.
21. Kaneko T, et al. The influence of compressive forces across the patellofemoral joint on patient-reported outcome after bi-cruciate stabilized total knee arthroplasty. Bone Joint J. 2018;100-B:1585–91.
22. Kobayashi S, et al. Rheumatoid arthritis patients achieve better satisfaction but lower functional activities as compared to osteoarthritis patients after total knee arthroplasty. J Arthroplast. 2019;34:478–82.
23. Van Onsem S, et al. A new prediction model for patient satisfaction after total knee arthroplasty. J Arthroplast. 2016;31:2660–7.
24. Itou J, et al. Assessing the validity of a new prediction model for patient satisfaction after total knee arthroplasty: a retrospective cross-sectional study. Orthop Res Rev. 2020;12:133–7.

Part XIV

Difficult Cases

Severe Varus Deformity

73

Yasushi Oshima and Tokifumi Majima

Preoperative evaluation and surgical planning are important in the treatment of severe varus knee deformity, which is associated with medial soft tissue contracture, lateral laxity, limited ROM, and bone defect. This section outlines TKA for severe varus deformity.

1 Severe Varus Deformity

The mechanical axis (Mikulicz line) is the line connecting the center of the femoral head to the center of the ankle joint, which represents the lower limb weight bearing line, and normally passes through the center of the knee joint in the loading position. Similarly, the angle formed by the line from the center of the femoral head to the center of knee joint and the line from the center of knee joint to the center of the ankle talus is called the hip-knee-ankle (HKA) angle, and the HKA is almost 0° in normal knee, which is called neutral alignment. In contrast, in medial OA, the HKA angle is less than 0° (negative value means varus knee) due to varus deformity. Although the severe varus knee deformity has not been clearly defined, when the varus deformation exceeded 20°, it was called severe varus deformity [1]. Another classification defined the deformation from 0° to 3° as well-aligned, 4° to 10° as common, 11° to 20° as substantial, 21° to 30° as important, and >30° as extreme, depending on the degree of deformations [2].

On the other hand, in a simple standing anteroposterior X-ray image of the full length of the lower extremity, the normal lower limb alignment is a lateral femorotibial angle (FTA) of approximately 176°, the FTA of more than 180° was called varus deformity [3], and the FTA of more than 195° was called severe varus deformity [4].

Some articles have defined both HKA > 20° and FTA > 15° as severe varus deformity, however, it should be carefully considered that the relationship between HKA and FTA may change depending on the femoral valgus angle and the bowing deformity of the bone axis [5]. Therefore, we defined severe varus deformity as HKA < −20° or FTA > 195°.

2 Causes of Severe Varus Deformity

Causes include intra-articular and extra-articular factors. Intra-articular factors include articular cartilage and bone defect due to end-stage knee OA, fragility fractures, post-traumatic nonunion and malunion.

As for extra-articular factors, there are postfracture pseudoarthrosis, fragility fractures, and previous osteotomy around the knee, as well as bone dysplasia and bone metabolism disorders such as bone bowing or thinning deformity due to osteogenesis imperfecta, varus deformity of the proximal tibia and distal femur due to Blount disease, worsened bone quality and a variety of deformity due to hereditary hypophosphatemia or hyperthyroidism, and bone tumors and Paget disease [1, 6].

3 Preoperative Evaluation

Knee ROM, as well as existence of extension and flexion contractures, anterior-posterior and medial-lateral instability, and hyperextension should be examined.

Simple radiographs are taken in three directions of both knees and the full-length of the standing anteroposterior view of the lower extremities to evaluate osteophytes of the medial and posterior femoral and medial tibial condyles, lower extremity alignment, medial tilt of the joint line, bone defects, and patella baja. Since the femoral osteotomy is generally performed with the intramedullary rod, the insertion point of the intramedullary rod and the valgus osteotomy angle of the

Y. Oshima (✉) · T. Majima
Department of Orthopaedic Surgery, Nippon Medical School, Tokyo, Japan
e-mail: y-oshim@nms.ac.jp

S. Matsuda (ed.), *Advances in Total Knee Arthroplasty*, https://doi.org/10.1007/978-981-97-4920-1_73

femur are measured, and the size and the position of the implants are planned. The degrees of extra-articular and lateral/anterior bowing deformities were also evaluated. Because tibial extra-articular deformity is often not included in the normal-sized knee joint image, it may be failed to notice, and the full-length lower extremity image should also be evaluated carefully. In addition, stress test is performed to confirm the medial and lateral instability, hyperextension, and the possibility of manually correction of lower extremity alignment. Simple CT images are used to measure the angle of femoral rotation, hypoplasia of the lateral condyle, and amount of bone defect. Based on the data obtained, we consider the implant to be used, and the need for metal augmentations and stem extensions, and the constrained designed implants.

4 Implant Selection

The CR implants may be used if there is only a severe varus deformity. However, in cases with flexion contracture, at least the PCL is likely to be degenerated, and the PS implants have been recommended to reduce the postoperative limitation of the knee extension, not to decrease the implant durability, and not to increase the revision rates [3].

The constrained type implants are rarely applied for severe varus deformities alone. However, in knees, which manual correction of the varus deformity is difficult and the PS implant cannot provide the stability after the excessive soft tissue release, after multiligament knee injury, and with severe flexion contracture, the constrained type implants may be necessary to be applied. Moreover, the hinge-type implants are preferred for the knees, in which the knee stability is not obtained, and with the recurvatum pattern. Therefore, the careful preoperative evaluations and the implant preparations are necessary to be performed.

5 Extra-articular Deformity

It is essential to consider whether extra-articular deformities can be corrected by TKA alone or whether corrective osteotomy is necessary in combination with TKA.

To evaluate the level of the deformity, the center of rotation of angulation (CORA) is firstly identified, and its distance from the knee joint line is measured. In general, the further the CORA is from the knee joint line, the less the effect of the CORA on alignment. If the CORA is further from the joint line than the metaphyseal region and the femoral deformity is less than 20° or the tibial deformity is less than 30°, the correction of the deformity can be achieved with TKA alone. Conversely, if the CORA is closer to the knee joint line than the metaphyseal region or if the angle of deformity is greater, corrective osteotomy is required before TKA [6].

Another evaluation method is to draw a distal femoral osteotomy line perpendicular to the functional axis drawn from the center of the femoral head to the center of the femoral intercondylar fossa. If this line is distal to the attachment of the collateral ligament (usually 2.5 mm from the joint line), it is considered to be correctable by TKA alone. Similarly, on the tibial side, if the extension of the line drawn from the center of the talus to the tibial shaft, which is distal to the CORA, passes through the tibial articular surface, TKA alone is considered sufficient for correction [6, 7].

In the sagittal plane, if the deformity is smaller than 10° of flexion or 20° of extension, TKA alone can be sufficient. In the horizontal plane, on the other hand, it is difficult to correct a large malrotational alignment with TKA. Therefore, in the malrotational deformities, corrective osteotomy is recommended to be performed first, followed by TKA, hence in two phases.

6 Surgical Approach

A median longitudinal incision of approximately 12 cm is made from 3 cm proximal to the patella to the medial aspect of the tibial tuberosity. If the knee exposure is difficult or in cases with bone defect, the incision should be extended as needed so that the skin incision does not extend distally to the tibial tuberosity. The medial parapatellar approach is applied, and the medial capsule is incised 2 cm distally from the articular surface under the tibial periosteum and is released posteriorly to the anterior margin of the dMCL, however, the dMCL itself is preserved. The patella is shifted laterally as much as possible, but is not everted, because forceful maneuver of the patella can damage the patellar tendon. The significant osteophytes around the patella prevents this process, the osteophytes are removed as possible. The infrapatellar fat pad and suprapatellar synovium are partially resected, and the knee joint is gradually flexed. At this stage, the knee is placed in full extension, and varus and valgus stress tests are applied manually to check the soft tissue balance of the knee, but in many cases valgus stress cannot correct the knee to neutral alignment.

Next, the MCL of the medial femoral epicondyle and the dMCL of the tibial insertion are protected with an elevatorium, while the femoral and tibial medial osteophytes are resected using bone rongeur and bone chisel. The posteromedial tibia is exposed using the Ransall maneuver with knee flexion, foot external rotation, and anterior drawer of the tibia, and the posteromedial osteophyte is resected while protecting the posterior oblique ligament (POL) and semimembranosus tendon [3]. At the same time, the joint capsule release from the posteromedial side of the tibia to the PCL attachment area under the periosteum is useful for posterior knee exposure. Since the MCL is tense with the bow-string effect caused by the overhang of the osteophytes, osteophyte resection relaxes the MCL and allows manual correction to neutral alignment in many cases of varus deformity. The angle of alignment corrected by osteophyte resection has been reported to be 5.3° in males and 9.1° in females [8]. The relationship between the width of the resected osteophyte and the angle of correction of lower limb alignment has been reported to be 0.4° per 1 mm of osteophyte [9].

7 Femoral and Tibial Osteotomies

Femoral osteotomy is performed using the conventional method, except in cases which the femoral intramedullary rod cannot be inserted due to significant bowing deformity or other reasons.

In the case of severe varus deformity, the amount of bone resection should be 8 mm, which is less than the thickness of the implant, with reference to the apex of the intercondylar fossa, and additional osteotomy should be made as necessary. The femoral valgus osteotomy angle is usually 5–7°, but in severe varus deformity, it varies from 2° to 12° depending on the progression and the shape of the deformity [10]. When osteophytes are evident, it is difficult to palpate the medial condyle depression and identify the SEA, which is the center of rotation of the knee. Therefore, the authors used the CT template method, in which the SEA was identified from multiple slices of preoperative CT images, and was copied onto a CT image at the distal femoral osteotomy level and projected onto the osteotomy surface of the femur in the operative field to accurately determine the SEA [11]. The osteophyte of the posterior condyle of the femur is resected sufficiently because it causes limitation of extension, and if necessary, the adhesion of the posterior joint capsule is released under the periosteum with a cob retractor.

After the four-sided femoral osteotomy, a femoral trial is inserted, and a spacer block is used to confirm the level of the tibial osteotomy and to determine whether there is any residual bone defect. Tibial osteotomy is performed using an extramedullary rod in which the lateral articular surface of the tibia is referred to an indicator because it is often preserved in varus knee deformity. The amount of the resection should be 8 mm, which is less than the thickness of the implant. However, in cases of severe varus deformity with the joint space wider due to the lateral soft tissue laxity, the amount of resection should be diminished to be 5–7 mm, and additional osteotomy should be made as necessary. This technique may obviate the need for a thick tibial insert.

After tibial osteotomy, a spacer block is inserted to check the balance in extension and flexion. It has been known that the gap of the normal knee is slightly tighter in extension than in flexion and medial than lateral [12]. Therefore, we allow a trapezoidal gap with a narrow medial side of up to 3 mm instead of preserving the dMCL to ensure medial stability [13–15]. Furthermore, it has been known that resection of the PCL significantly increased the flexion gap compared to the extension gap [16]. However, with the MCL preserved, which provides the first restraint against valgus stress load, the operators should understand that the extension and flexion gaps do not increase significantly after the PCL resection [17].

If manual correction of alignment is not obtained at this stage, soft tissue release is initiated.

8 Soft Tissue Balancing

Excessive soft tissue release cannot be reversed and causes joint instability, so it is important to proceed with careful step-by-step release.

Clayton et al. recommended the proceeding with the soft tissue release of the proximal tibia in four stages after osteotomy [18].

Stage I: The proximal tibia is exposed 1 cm below the periosteum to the tibial attachment of the semimembranosus tendon. Next, the marginal osteophyte of the medial the tibia is removed, and the joint capsule and dMCL are released.

Stage II: The pes anserinus is preserved, and the sMCL, which is located beneath the pes anserinus is released under the periosteum 5–6 cm distal to the articular surface. In addition, the medial capsule, including the tibial attachment of the semimembranosus tendon, is released posteriorly subperiosteally.

Stage III: The tibial insertion of the pes anserinus is released subperiosteally. The medial capsule is further released to the attachment of the PCL.

Stage IV: The adductor magnus is preserved, and the MCL is released subperiosteally from the femur.

The dMCL and the medial joint capsule are generally released during the joint exposure phase. The tibial attachment area of the dMCL is 18.0 mm in anteroposterior diameter and 5.1 mm in length, centered 6.5 mm distal to the tibial articular surface [19]. Therefore, in cases of medial tibial defect larger than 10 mm, the dMCL may have degenerated or disappeared. However, since the release of the dMCL and POL causes rotational instability, it is desirable to preserve them as much as possible to reproduce more physiological knee kinematics after TKA [20].

Even when significant osteophytes are present, the dMCL often remains distal to the osteophytes because the osteophytes form proximally to the tibial attachment of the dMCL. Therefore, we carefully exposed and resected the osteophyte, preserved the dMCL, and only if it was still difficult to correct the alignment, the tibial attachment of the dMCL was released posteromedial to the dMCL.

If medial tightness remains at this stage, tibial implant size down and medial tibial reduction osteotomy should be considered. The reduction osteotomy provides 1° of the correction of the alignment with a 2 mm osteotomy of the tibia [21].

Because transverse dissection of the sMCL causes significant medial instability in the range of mid-flexion and deep-flexion that is difficult to recover, the semimembranosus tendon is released first. The release of anterior tissues, such as the sMCL, causes the flexion gap to be larger than the extension gap, while the release of posterior tissue, such as the POL and semimembranosus tendon, causes the extension gap to be larger than the flexion gap. In addition, the release of the semimembranosus tendon decreases the internal rotation of the tibia during flexion, which may result in a decrease in the postoperative knee flexion angle [22].

The sMCL is carefully released from anterior to posterior at the distal tibial attachment or pie crust incision is performed at the distal attachment of the sMCL [23]. After 2–3 pie crust incisions, a spacer block is inserted, and valgus stress is applied to correct the alignment of the lower extremity. This process is repeated carefully. The pes anserinus is released from a proximal to distal direction. If the flexion gap is too wide open, some procedures such as suturing the pes anserinus in 90° knee flexion should be considered [3]. We avoided dissection of the sMCL as much as possible and applied reduction osteotomy of the medial tibia and pie crust incision of the sMCL.

In most cases, well-balanced soft tissues can be obtained with the PS inserts, but if varus or valgus instability is observed, the use of the constrained type inserts should be considered. In such cases, a metal stem extension should be attached to the tibia to prevent the early loosening of the implants [3].

In significant deformities where an intramedullary femoral rod cannot be used, a computer navigation system or PSI osteotomy may be useful. However, these systems involves some problems in terms of installation cost and the time required to fabricate osteotomy guides. In contrast, the recently developed portable navigation system is a useful option because it is inexpensive, can be used with implants from multiple companies and is as accurate in the coronal and sagittal planes as conventional navigation systems [24].

9 How to Manage Bone Defects

In severe varus deformities, bone defects are often present in the medial tibia. However, the amount of bone defect is not considered in proximal tibial osteotomy, and the tibial osteotomy with a fewer thickness of the tibial implant is performed, in which the less degenerated lateral part of the tibia is used as an indicator. Then, the remaining depth of the bone defect of the medial tibia is evaluated. The bone defect is usually an asymmetric peripheral type, although a contained type surrounded by bone cortex can also occur. The bone strength of the proximal tibia decreases beyond 20 mm from the articular surface, so care should be taken to increase the amount of bone resection.

Bone defects are classified according to the depth: <5 mm, 5–10 mm, and > 10 mm. If the bone defect in the medial tibia is small, increasing the amount of osteotomy should be considered. However, if the tibial osteotomy line is lowered, Gerdy's tubercle should be identified to avoid dysfunction of the iliotibial band. For this reason, tibial osteotomies are limited to a maximum of 12–14 mm. It is also necessary to remind that the tibial implant decreases the size as the osteotomy amount increases.

If the bone defect is less than 5 mm, cementation alone can be used. If the bone defect is a contained type, autogenous bone grafting using the resected bone fragments after osteotomy is also available. However, for a bone defect larger than 5 mm, bone grafting with screw fixation, or metal augmentation should be used. There are block and wedge types of metal augmentations, and these are selected according to the shape of the bone defect. The block type is often selected because it is less susceptible to shearing forces. In recent years, there has been wide variations of shapes and thicknesses of metal augmentations, making it easier to use metal augmentations than to harvest autologous bone from other parts of the body.

In cases with bone defects, the tibial stem extension is recommended to be used. However, if the bone quality is sufficient and the bone defect area does not extend into the keel of the tibial implant, a short stem of 25–30 mm will provide sufficient stability for the implant.

Even in cases where there is no bone defect after tibial osteotomy, there is often a residual osteosclerotic area at the medial condyle of the tibia. In this case, several bone perforations are made in the osteosclerotic area using a 2.4-mm-diameter Kirschner wire or a stopper pin used to create an entry hole for a four-sided femoral osteotomy guide to facilitate the penetration of bone cement.

10 MCL Advancement

This technique is performed in place of the release of the pes anserinus and sMCL, when the aforementioned staged soft tissue release does not correct the varus deformity and is effective for both flexion and extension contractures.

After the osteotomy of the sMCL insertion of the medial femoral condyle, the extension gap is increased when the bone fragment is moved distally, and the flexion gap is affected when the bone fragment is moved proximally. The bone fragment is fixed with screws after confirming the optimal position, but it is rarely performed for varus deformities.

11 Procedures at Our Hospital

In the secondary OA with the extra-articular deformity after the lower extremity trauma, the alignment should be confirmed over the entire length of the lower extremity. Even in the severe varus deformity, as the CORA was 50 mm from the tibial articular surface, and the deformity was 15°, this case could be treated with TKA alone (Fig. 73.1). However, because the bone defect in the medial part of the tibia was 18 mm, the metal augmentations, the tibial stem extensions, and thicker variations of the tibial inserts were included in the preparation.

If the knee joint stress test shows contracture on the medial side and laxity on the lateral side of the affected limb compared to the healthy side, it is safer to prepare constrained type implants as a backup, but if there is no excessive instability or hyperextension, it is considered unnecessary to use hinge type implants.

Although the dMCL may degenerate or disappear in some cases depending on the size of the bone defect, some dMCL fibers often remain at the inferior margin of the osteophyte, along with scar tissue and the joint capsule, and should be preserved as much as possible. After resection of the medial femoral and medial to posteromedial tibial osteophytes, the valgus stress test is performed, and if the alignment can be corrected to neutral, no further soft-tissue release is performed.

After the osteotomy of the distal femur and proximal tibia at 8 mm, confirm that a 9 mm (or the thinnest) spacer block can be inserted, and make additional osteotomies as needed. The bone defect in the medial proximal tibia is evaluated, and an additional osteotomy is performed while considering the thickness of the metal augmentation. The remaining bone defect of less than 5 mm on the medial margin of the tibia is drilled with a 2.4-mm k-wire to create 3 holes and filled with bone cement (Figs. 73.2 and 73.3). TKA can be performed without additional soft tissue release and with soft tissue

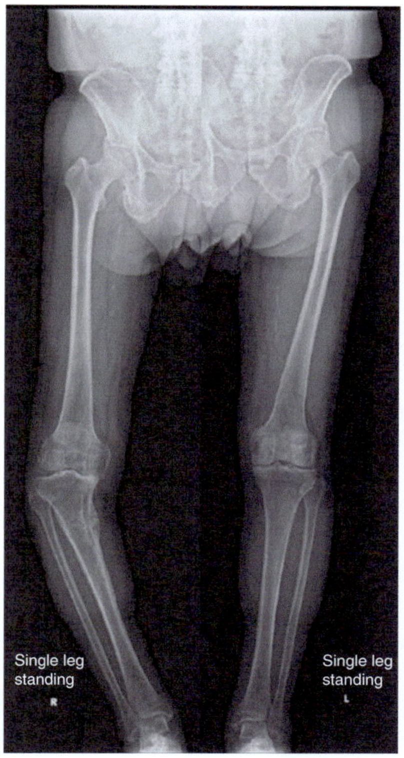

Fig. 73.1 Severe varus knee deformity due to extra-articular deformity after tibial fracture surgery. The anteroposterior view of a standing full-length simple radiograph of the lower extremity showed HKA −29° and FTA 203°

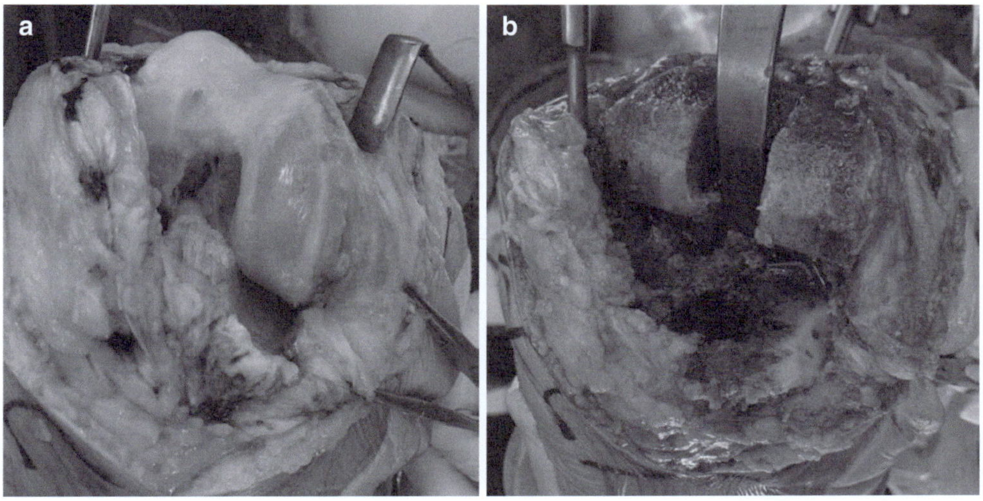

Fig. 73.2 Intraoperative findings. (**a**) On knee exposure, there showed eburnation of the medial femoral condyle articular surface, osteophyte formation of the medial femur and tibia, and significant bone defect in the medial tibia. (**b**) After the tibial osteotomy, three drilling holes were created for osteosclerotic area of the remaining tibial bone defect of 5 mm or less

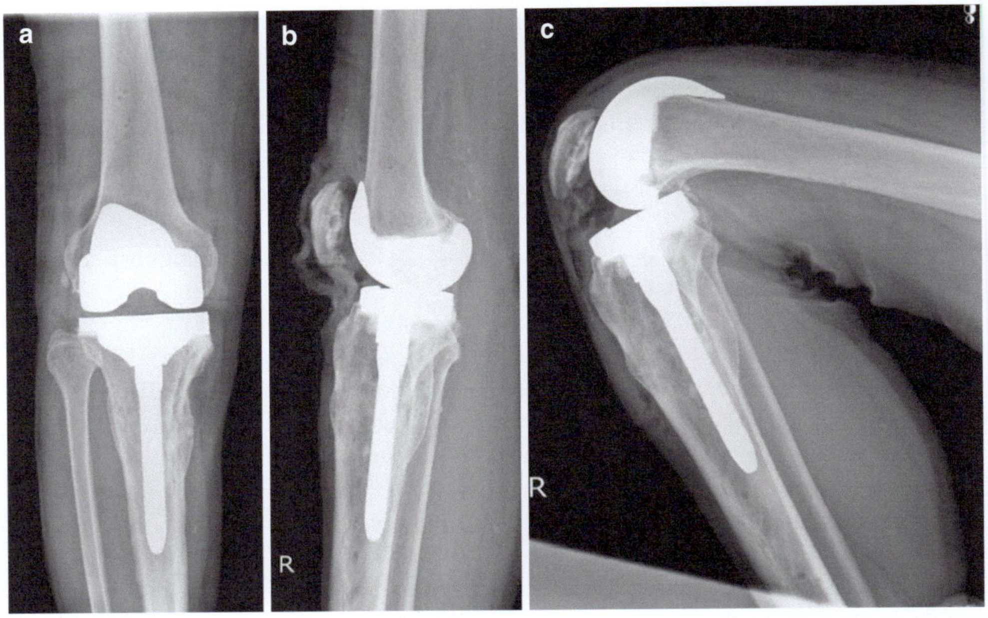

Fig. 73.3 Postoperative simple radiograph, anteroposterior, lateral maximal extension and lateral maximal flexion, the FTA improved to 177°, gait balance was regained, and the knee ROM was 0° in extension and 120° in flexion. (**a**) AP view. (**b**) Lateral view (maximum knee extension). (**c**) Lateral view (maximum knee flexion)

preservation using the usual technique, even with severe varus deformity, as shown in the figure, if the critical instability cases are excluded (Fig. 73.3).

12 Future Tasks

In many cases of severe varus deformity, when the tibial osteophyte is resected posteromedially, the lower limb alignment could be corrected, as severe osteophyte formation is generally evident. If the implants are inserted in this condition, medial soft tissue can be stabilized, and the correction to neutral alignment can be achieved, however, the lateral laxity remains. At this point, there is some debate as to what degree of lateral laxity is acceptable. However, since the normal knee has pseudo laxity, with the lateral is wider than the medial, and the flexion gap is wider than the extension gap [25], we believe that a trapezoidal gap with a slightly wider lateral is more likely to reproduce physiological knee kinematics than a rectangular gap with an equal medial and lateral, which has been conventionally recommended in the past.

Additional osteotomy guides are used to resect the bone defect sites, however, accurate bone resections are difficult to achieve, and final adjustments are often made freehand with a bone saw. Recently, robotic technology has been introduced to perform the bone resections as precise as 0.1° or 0.5 mm. Although it has not been established at this stage, it is expected to be a useful tool for additional bone resection and for adjusting osteotomies for bone defects.

References

1. Mullaji A, et al. Correcting varus deformity. J Arthroplast. 2007;22(4 Suppl 1):15–9.
2. De Muylder J, et al. Total knee arthroplasty in patients with substantial deformities using primary knee components. Knee Surg Sports Traumatol Arthrosc. 2015;23:3653–9.
3. Rossi R, et al. Total knee arthroplasty in the varus knee: tips and tricks. Int Orthop. 2019;43:151–8.
4. Niki Y, et al. Accuracy of implant positioning for minimally invasive total knee arthroplasty in patients with severe varus deformity. J Arthroplasty. 2010;25:381–6.
5. Kim MS, et al. Correcting severe varus deformity using trial components during total knee arthroplasty. J Arthroplast. 2017;32:1488–95.
6. Sculco PK, et al. Management of extra-articular deformity in the setting of total knee arthroplasty. J Am Acad Orthop Surg. 2019;27:e819–30.
7. Mullaji A, et al. Quantification of the effect of sequential posteromedial release on flexion and extension gaps: a computer-assisted study in cadaveric. J Arthroplast. 2009;24:795–805.
8. Toyooka S, et al. Assessing the role of minimal medial tissue release during navigation-assisted varus total knee arthroplasty based on the degree of. J Knee Surg. 2021 Jan 20; https://doi.org/10.1055/s-0040-1722658.
9. Iizawa N, et al. Effect of medial osteophyte removal on correction of varus deformity in total knee arthroplasty. J Nippon Med Sch. 2020;87:215–9.
10. Mullaji AB, Shetty GM. Correcting deformity in total knee arthroplasty: techniques to avoid the release of collateral ligaments in severely deformed. Bone Joint J. 2016;98-B(1 Suppl A):101–4.
11. Oshima Y, et al. A computed-tomography-scan-based template to place the femoral component in accurate rotation with respect to the surgical epicondylar axis in total knee arthroplasty. Knee. 2018;25:195–202.

12. Okazaki K, et al. Asymmetry of mediolateral laxity of the normal knee. J Orthop Sci. 2006;11(3):264–6.
13. Tanaka Y, et al. Intraoperative physiological lateral laxity in extension and flexion for varus knees did not affect short-term clinical outcomes. Knee Surg Sports Traumatol Arthrosc. 2020;28:3888–98.
14. Aunan E, et al. A new method to measure ligament balancing in total knee arthroplasty: laxity measurements in 100 knees. Arch Orthop Trauma Surg. 2012;132:1173–81.
15. In Y, et al. Agreements between different methods of gap balance estimation in cruciate-retaining total knee arthroplasty. Knee Surg Sports Traumatol Arthrosc. 2009;17:60–4.
16. Kadoya Y, et al. Effects of posterior cruciate ligament resection on the tibiofemoral joint gap. Clin Orthop Relat Res. 2001;391:210–7.
17. Oshima Y, et al. The influence of posterior cruciate ligament resection on tibiofemoral joint gap in varus osteoarthritic knees. J Knee Surg. 2022;35:323–30.
18. Clayton ML, et al. Correction of alignment deformities during total knee arthroplasties: staged soft-tissue releases. Clin Orthop Relat Res. 1986;202:117–24.
19. Liu F, et al. Morphology of the medial collateral ligament of the knee. J Orthop Surg Res. 2010;5:69.
20. Iizawa N, et al. Influence of the medial knee structures on valgus and rotatory stability in total knee arthroplasty. J Arthroplast. 2016;31:688–93.
21. Mullaji AB, Shetty GM. Correction of varus deformity during TKA with reduction osteotomy. Clin Orthop Relat Res. 2014;472:126–32.
22. Matsumoto T, et al. Semimembranosus release reduces tibial internal rotation and flexion angle in cruciate-retaining total knee arthroplasty. J Arthroplast. 2015;30:1537–41.
23. Bellemans J, et al. A new surgical technique for medial collateral ligament balancing: multiple needle puncturing. J Arthroplast. 2010;25:1151–6.
24. Nam D, et al. Extramedullary guides versus portable, accelerometer-based navigation for tibial alignment in total knee arthroplasty: a randomized, controlled trial. J Arthroplast. 2014;29:288–94.
25. Risitano S, Indelli PF. Is "symmetric" gap balancing still the gold standard in primary total knee arthroplasty? Ann Transl Med. 2017;5:325.

74 Severe Valgus Deformity

Takashi Nakamura

Valgus knee, defined as a knee with a preoperative valgus angle greater than 10° [1–3], can be caused by rheumatoid arthritis, posttraumatic arthritis, osteoarthritis, arthropathic changes associated with hip joint disease (coxitis knee), or metabolic bone disease. The frequency of its occurrence is approximately 10–15% in cases requiring total knee arthroplasty (TKA) [4, 5]. Severe valgus knee, defined as a knee with a preoperative valgus angle greater than 20° [5, 6], comprises approximately 5% of all cases of valgus knee [7].

The anatomical characteristics of the valgus knee includes a shortened iliotibial band (ITB), hypoplasia of the lateral femoral condyle, bone defects on the lateral tibial plateau, and laxity of the medial collateral ligament (MCL). These anatomical characteristics must be fully understood prior to performing TKA on knees with severe valgus. In particular, it is necessary to appropriately address lateral bone defects, external tibial rotation, and MCL laxity to ensure excellent clinical outcomes of severe valgus knee. It is also important to consider that soft-tissue treatment differs in extension and flexion [8].

TKA of the severe valgus knee is among the most challenging procedures, requiring detailed preoperative planning and appropriate surgical techniques. The lateral approach of TKA to for treating valgus knee has several advantages, including the rationality of direct lesion exposure and the advantage of patellar tracking owing to simultaneous lateral release and joint exposure. In this section, we review the surgical techniques for TKA of severe valgus knee using the lateral approach.

1 Preoperative Planning

Because valgus knee is often associated with hypoplasia of the lateral femoral condyle and bone defects on the lateral tibial plateau, performing an accurate osteotomy when relying solely on the information obtained in the operative field is difficult. To achieve accurate lower-extremity alignment, understanding the unique bony morphology of the valgus knee in each case and adequate preoperative planning are of paramount importance.

1.1 Bone Morphology

In the valgus knee, the hip joint may be disabled; thus, plain radiography including the hip joint should always be performed. The degree of hypoplasia of the lateral femoral condyle and bone defects on the lateral tibial plateau should be evaluated, while the degree of autogenous bone grafting and need for augmentation should be confirmed. Lateral patellar subluxation may also occur in patients with severe valgus knee. In addition, hypoplasia of the lateral femoral condyle affects the rotational alignment of the femoral component; therefore, evaluating bone morphology in detail using computed tomography and magnetic resonance imaging is important.

1.2 Prediction of Soft-Tissue Dissociation

There are two types of valgus knee: non-fixed, in which the valgus deformity can be corrected by the application of stress; and fixed, in which the valgus deformity cannot be corrected by the application of stress [5, 7]. Preoperative stress testing is recommended to predict the extent of the lateral soft-tissue release. It is also important to evaluate MCL function, as its support greatly influences postoperative clinical outcomes. On valgus stress imaging, if the medial joint space is open by more than 5°, the MCL is considered dysfunctional and should be treated with a constrainted implant.

Preoperative planning should be comprehensive, based on both image inspection and physical findings such as gait, range of motion (ROM), degree of contracture, and patellar tracking. Total hip arthroplasty is preferred when the hip

T. Nakamura (✉)
Department of Orthopaedics Surgery, School of Medicine, Toho University, Tokyo, Japan
e-mail: takashin@med.toho-u.ac.jp

S. Matsuda (ed.), *Advances in Total Knee Arthroplasty*, https://doi.org/10.1007/978-981-97-4920-1_74

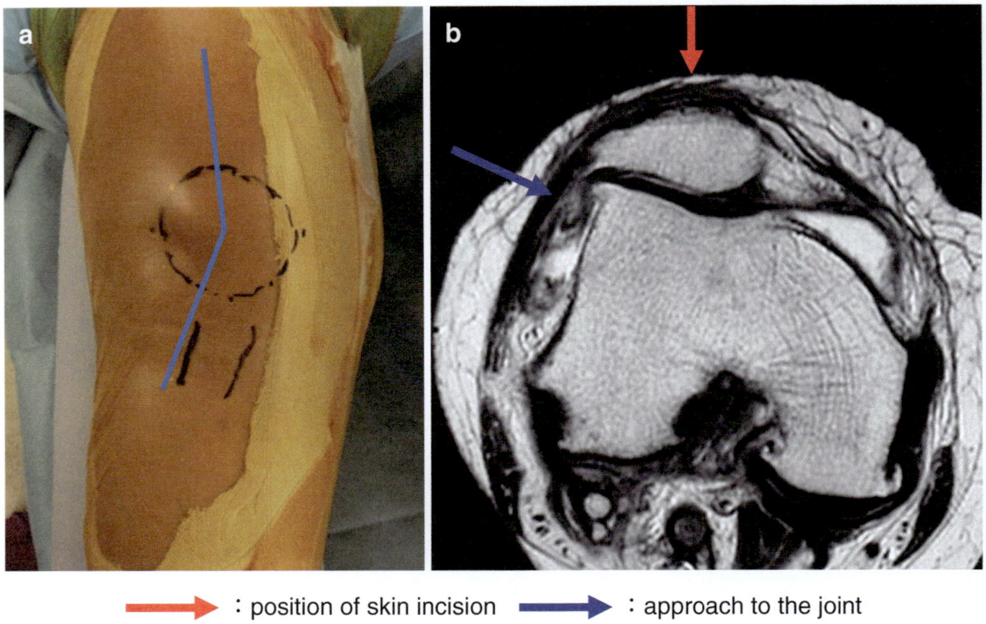

Fig. 74.1 Skin incision. (**a**) Position of skin incision. (**b**) Position of skin incision and approach to the joint. Using a median longitudinal incision (blue line), even if the suture of the lateral patellar retinaculum is insufficient, the skin incision is not positioned directly above the lateral patellar retinaculum in an effort to prevent a postoperative infection

joint is severely impaired with a poor ROM, such as during adduction contracture.

2 Surgical Techniques

2.1 Joint Exposure

A median longitudinal skin incision is made along the Q-angle (Fig. 74.1a). It is important to avoid lateral placement of the incision. Suturing of the lateral patellar retinaculum may be inadequate due to correction of the valgus deformity. (See below for the key points of wound closure.) To prevent delayed wound healing and postoperative infection, it is important to intentionally shift the position of the incision toward the joint and skin (Fig. 74.1b). A lateral parapatellar approach is used to expose the joint. Once an incision is made from the lateral border of the rectus femoris tendon along the lateral border of the patella and patellar tendon, the ITB is released from Gerdy's tubercle. The key to this release is to extend it slightly distally and maintain continuity with the fascia of the tibialis anterior muscle. Release of the ITB is first performed up to the anterior one-half of Gerdy's tubercle; if joint exposure is difficult, the next anterior two-thirds is added. It is important to treat the infrapatellar fat pad as a single mass while keeping it attached to the patellar tendon (Fig. 74.2). This covers the joint with the infrapatellar fat pad when the suture of the lateral patellar retinaculum is inadequate at the time of closure, the points of which are described below.

Next, the patella is everted inward. First, the lower leg is fully internally rotated with knee extension and the patella is everted inward. In cases where everting the patella is difficult, the process can be facilitated by sufficiently removing the osteophytes of the patellofemoral joint and expanding the exposure in the proximal direction. If the patella can be everted inward during knee extension, the proximal side of the rectus femoris is pulled inward with a retractor and the knee joint flexed slowly while the lower leg is rotated internally (Fig. 74.3).

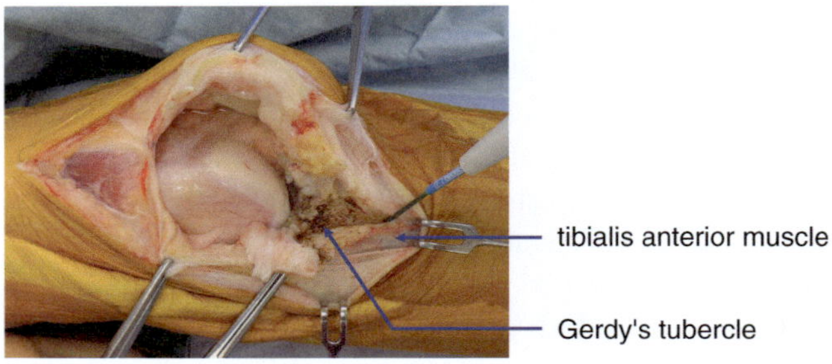

Fig. 74.2 Iliotibial band release. If the iliotibial band is released from Gerdy's tubercle while maintaining continuity with the fascia of the tibialis anterior muscle, it can be reliably dissected. It is important to treat the infrapatellar fat pad as a single mass while keeping it attached to the patellar tendon

2.2 Bone Cuts

The principle of the osteotomy procedure is the same for valgus and varus knee because mechanical alignment is the common goal. However, some care is required in valgus knee because the bone morphology differs from that of varus knee.

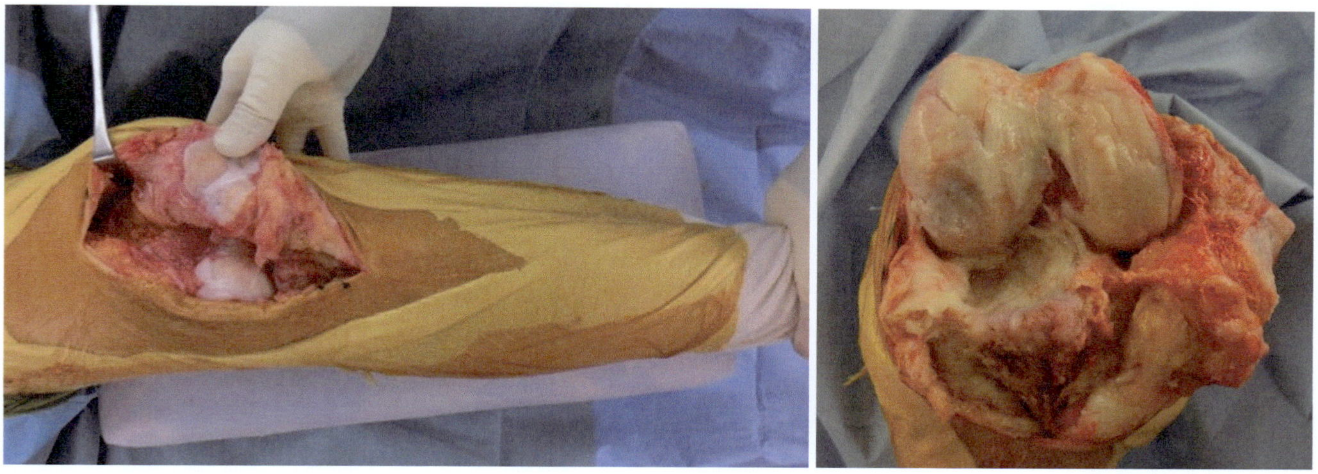

Fig. 74.3 Eversion of the patella. Patellar eversion was performed by internal rotation of the lower leg and pulling of the proximal rectus femoris inward. When it is difficult to evert the patella, the joint exposure expands proximally, and the patellar tendon is released to one-third of the tibial tuberosity attachment

1. Femur

Keblish [9] reported that the angle of the intramedullary femoral rod was 3°. However, the actual intramedullary femoral rod angle should be planned for each case using the mechanical alignment method, and the intramedullary rod insertion point should be shifted slightly medial to the center of the femoral groove to avoid undercorrection. If the lateral femoral condyle is hypoplastic, its resection becomes almost impossible (Fig. 74.4). If femoral resection is judged as insufficient during surgery and the osteotomy line is elevated proximally, distal medial condyle thickness will increase. Rossi et al. [10] reported that distal medial condyle thickness should not exceed 10 mm. An increased distal medial resection may lead to medial knee joint instability. The osteotomy line is carefully determined by measuring the degree of osteotomy in the medial condyle using caliper.

Careful attention should also be paid to rotational alignment. In some cases, the condylar twist angle, or the angle between the clinical epicondylar axis (CEA) and the posterior condylar axis (PCA), exceeds 10° due to hypoplasia or severe erosion of the lateral femoral condyle in severe valgus knee. In such cases, even if the femoral component is placed in an external rotation position of 3°–5° relative to the PCA, as in the case of varus knee, the femoral component is placed in internal rotation relative to the CEA and surgical epicondylar axis (Fig. 74.5). Internal rotation of the femoral component causes patellar tracking failure [11–13] and should be avoided.

Preoperative computed tomography images should be used to accurately determine the condylar twist angle and other bone morphologies, and intraoperative judgments should be made comprehensively, including the transepicondylar axis, Whiteside axis, and shape of the anterior femoral incision.

2. Tibia

Because the lateral tibial side of valgus knee shows severe erosion and bone defects, the medial articular surface of the tibia is the reference point for determining the joint line. In valgus knee, the focus tends to be on the bone morphology of the femoral side, such as the hypoplasia of the lateral femoral condyle. However, the medial proximal tibial angle (MPTA) [14] tends to be larger in valgus knees than in normal knees. Approximately 23.6% of valgus knees have femur-based deformities, and 41% have tibia-based deformities [15]. The bone morphology on the tibial side should also be considered. To maintain the joint line based on the medial articular surface, cutting the tibia 6–8 mm below its medial articular surface is recommended (Fig. 74.6). However, because the lateral articular surface of the tibia is often affected by bone defects, bone resection on the lateral tibial condyle may not be possible. Bone defects should be treated with autologous bone grafting or augmentation, and care should be taken to avoid judging bone resection as inadequate.

2.3 Soft-Tissue Balance

The soft-tissue processing of valgus knees differs in extension and flexion: balance is first achieved in extension and then adjusted in flexion. In severe valgus knee, ITB contracture occurs; therefore, ITB release is extremely important for achieving balance during extension. In severe varus knee, posterior cruciate ligament (PCL) removal effectively corrects the varus deformity; however, in valgus knee, PCL removal is not as effective [16], and reliable ITB release is necessary. The lateral approach is advantageous for achieving balance because the ITB is released as the joint is

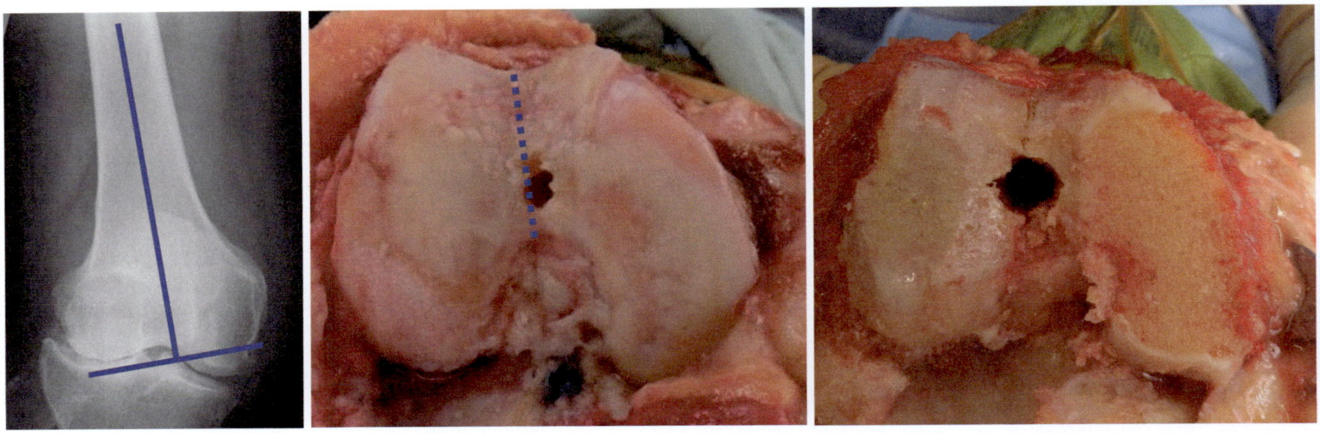

Fig. 74.4 Cutting of the femur. Undercorrection is less likely if the intramedullary rod insertion point is slightly shifted medially from the center of the femoral groove. In some cases, the lateral tibial condyle cannot be resected. Thus, the shape of the distal femoral cut surface can be imaged during preoperative planning

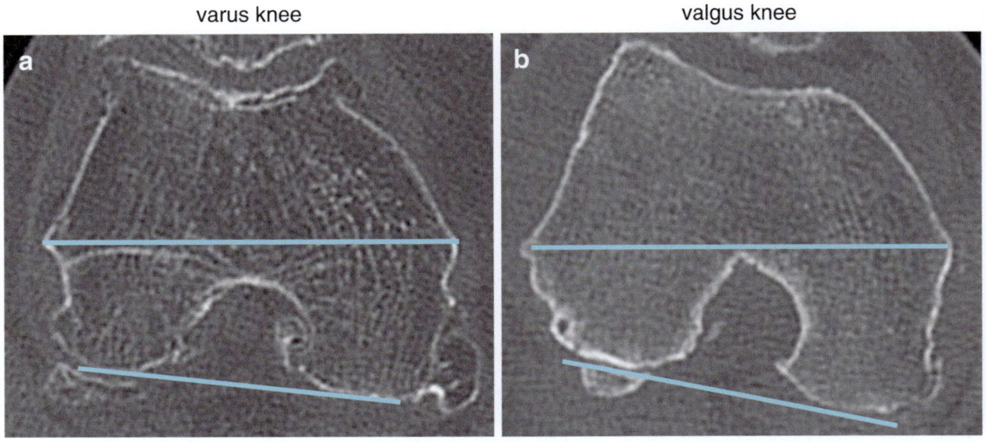

Fig. 74.5 Rotational alignment of the femur. (**a**) Varus knee: condylar twist angle of 5°. (**b**) Valgus knee: condylar twist angle of 13°. The posterior condylar axis was not helpful in determining the rotational alignment of the femur. The condylar twist angle was greater than that for varus knee. A thorough understanding is required of the bone morphology in each case on preoperative computed tomography images

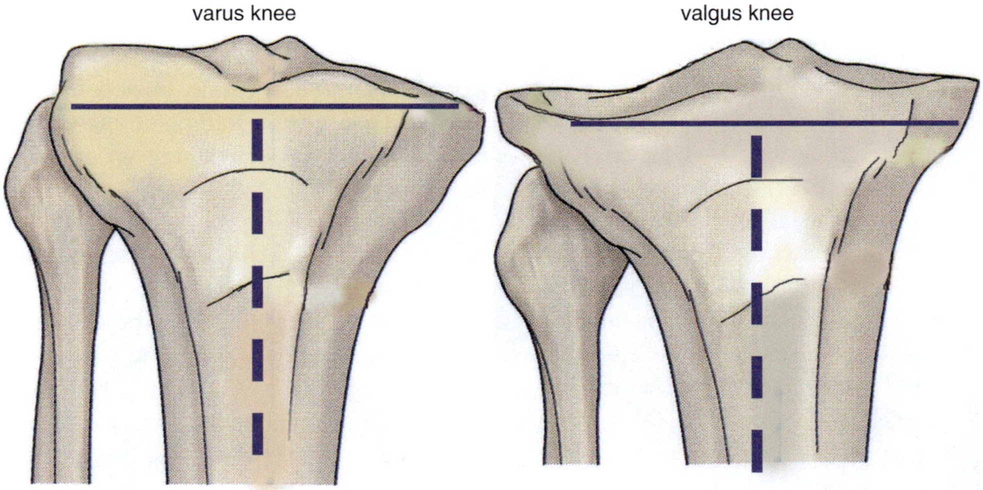

Fig. 74.6 Cutting of the tibia. (**a**) Varus knee. (**b**) Valgus knee. The lateral articular surface cannot be used as a landmark for the tibial resection. If the tibia is cut based on the lateral tibial articular surface, the joint line is excessively lowered

exposed. ITB release is already performed up to the anterior one-half to two-thirds of Gerdy's tubercle during joint exposure, and balance in extension has not been achieved. The ITB is completely released from Gerdy's tubercle, and if further adjustment is needed, the lateral posterior capsule is released. In some cases, complete ITB release is not suffi-

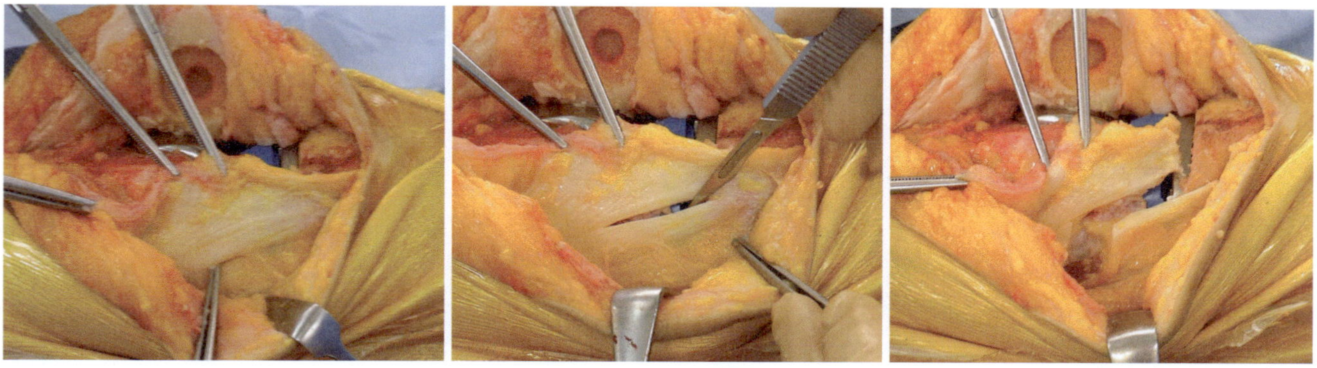

Fig. 74.7 Iliotibial band Z-plasty lengthening. A distal transverse incision of the iliotibial band was made at the level of the tibial cut surface so that its distal side can be covered with an infrapatellar fat pad

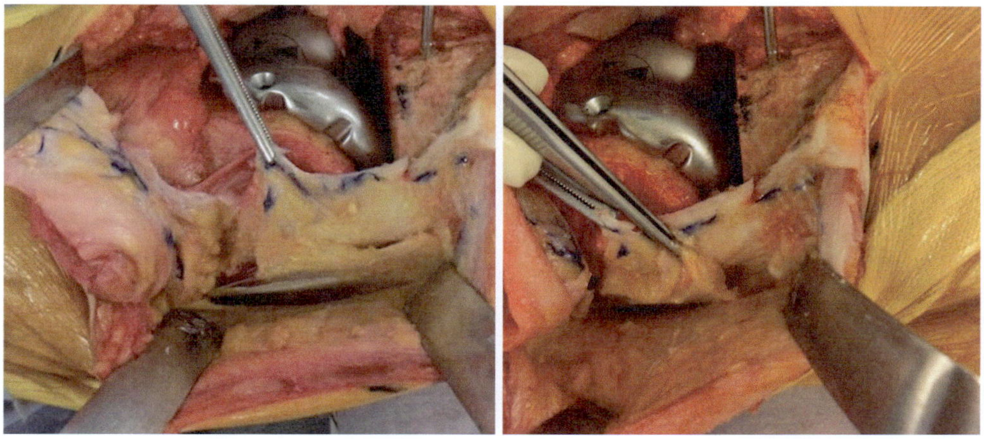

Fig. 74.8 Transection of the biceps femoris tendon. The biceps femoris is the only external rotator of the knee joint. Indications for biceps femoris transection are limited to cases in which excessive tibial external rotation or flexion contractures persist

cient to achieve balance in a severe valgus knee, in which case Z-plasty [17] or transverse cutting [18] (Fig. 74.7) of the ITB becomes necessary. However, if further balance adjustment is needed, the biceps femoris tendon should be transected from the fibular insertion (Fig. 74.8). The biceps femoris is a component of the hamstrings involved in hip extension, knee flexion, and external rotation of the lower leg. The biceps femoris tendon is transected if the valgus deformity correction in extension is insufficient, even after the ITB is cut transversely or excessive tibial external rotation or flexion contracture remains. Biceps femoris tendon dissection reportedly increases the medial extension gap by 1.1 mm and the lateral extension gap by 3.1 mm [19], and it is recommended as a method of adjusting soft-tissue balance in extension for severe valgus knee.

To adjust the balance in flexion, the lateral cruciate ligament (LCL) and popliteus tendon (PT) are released. First, the LCL is released from the lateral femoral epicondyle; if further adjustment is necessary, the PT is released from the lateral femoral epicondyle (Fig. 74.9). In most cases, a constrained implant is selected to accommodate changes in the joint line due to correction of the valgus deformity and support MCL laxity in severe valgus knee. As the PCL is removed, the adjustment in flexion is often not as difficult as that in extension. In fact, in a report in which PS TKA was performed using a lateral approach for a fixed valgus knee,

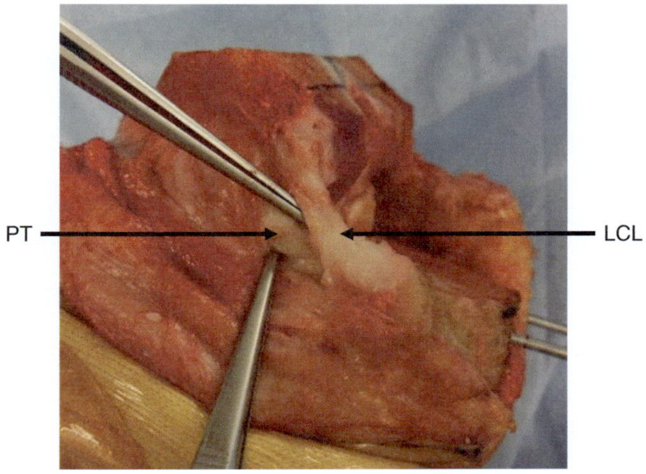

Fig. 74.9 Release of the lateral cruciate ligament and popliteus tendon. Although the attachments of the lateral cruciate ligament (LCL) and popliteus tendon (PT) were close to each other, it was easy to distinguish between them. In extension, the PT is posterior to the LCL; in flexion, it resides just anterior to the LCL

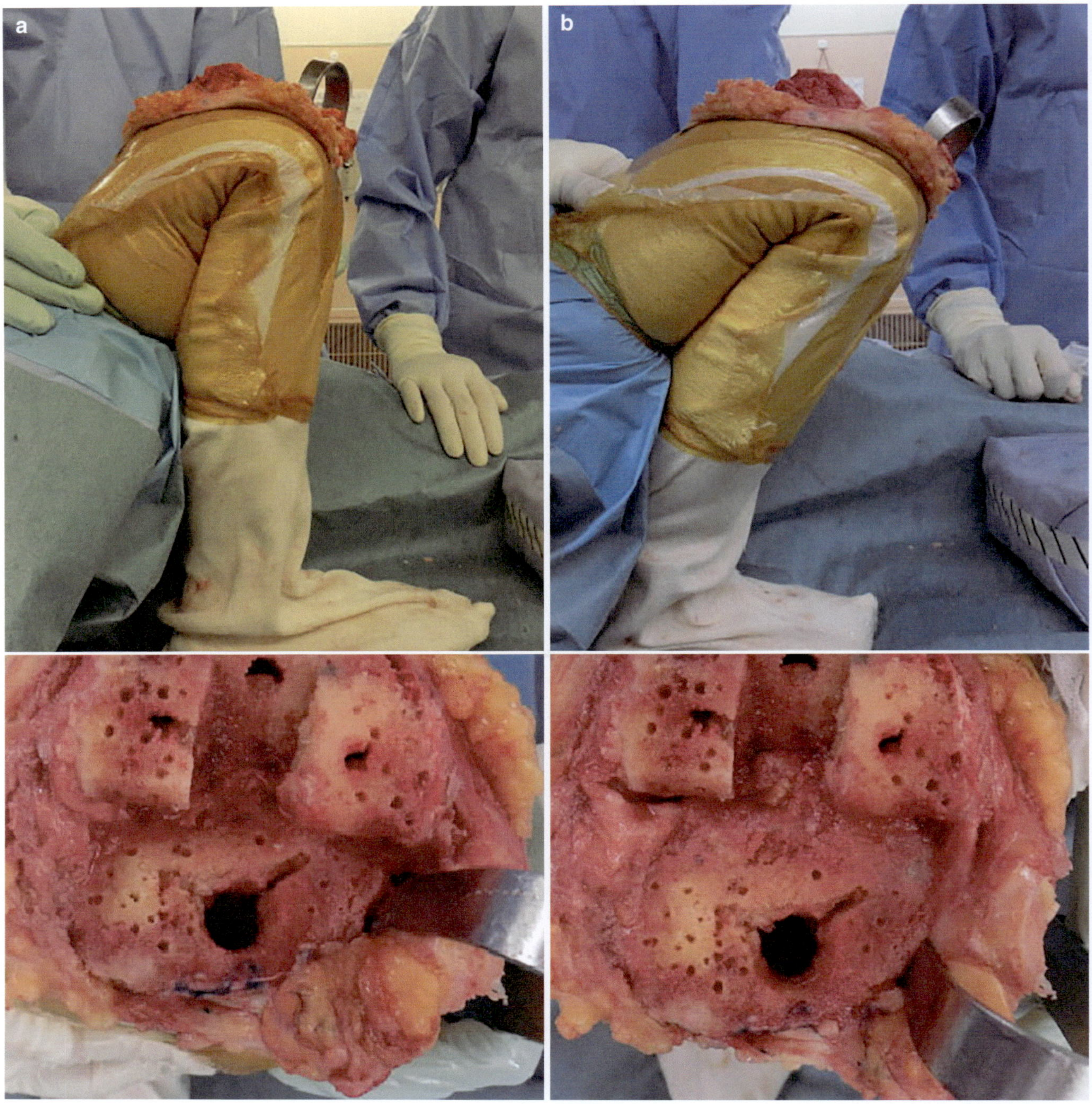

Fig. 74.10 Leg position. (**a**): If the leg is not pulled forward sufficiently, the medial femoral condyle may cover the cut surface of the medial tibia. (**b**): The tibial implant can be smoothly placed by pulling the lower leg forward with the knee in maximum flexion

only 7.9% of patients required LCL and PT release [20]. The PT acts as an internal rotator of the lower leg. External rotation of the lower leg is generally present in valgus knee; therefore, PT release is usually not necessary. The LCL and PT were cut as necessary when balance in flexion cannot be achieved or correction of the external rotation of the lower leg is insufficient.

The basis for obtaining ligament balance is to release the ligament sequentially while checking soft-tissue balance during extension and flexion. These procedures enable the achievement of good soft-tissue balance.

2.4 Implant Placement and Closure

In the lateral approach, the medial femoral condyle may cover the cut medial tibial surface and interfere with tibial implant placement. The tibial implant can be placed smoothly by pulling the lower leg forward with the knee in maximum flexion (Fig. 74.10). If the implant is to be fixed with cement, the bone cement should be prepared once possible placement of the actual implant is confirmed, not just during the trial.

In the case of severe valgus knee, suturing of the lateral patellar retinaculum may become insufficient due to the correction of the valgus deformity at the time of closure. Covering the lateral patellar retinaculum with an infrapatellar fat pad is a known solution to this problem [9, 21, 22] (Fig. 74.11). The infrapatellar fat pad flap is then raised from the medial border of the patellar ligament and moved to the lateral side to achieve relatively wide coverage. However, the proximal side of the lateral patellar retinaculum may not be adequately covered by the infrapatellar fat pad alone in patients with severe valgus knee. In such cases, it is necessary to use a deep fascia flap [23] or extend the lateral patellofemoral ligament (LPFL) to cover the lateral patellar retinaculum. When using the LPFL to extend and cover the lateral patellar retinaculum, the two layers of the LPFL and lateral patellar retinaculum should be separated at the time of joint exposure and the height of each incision changed. The lateral patellar retinaculum and LPFL can be sutured as a Z-plasty to cover the proximal part of the lateral patellar retinaculum (Fig. 74.12). Adequate joint coverage is critical to preventing subcutaneous hematoma, delayed wound healing, and postoperative infection.

2.5 How to Obtain Stable Clinical Results

The main complications of TKA for valgus knee include instability, recurrent external deformity, poor ROM, delayed wound healing, poor patellar tracking, and peroneal nerve palsy [24, 25]. Nikolopoulos et al. [26] studied recurrence of valgus deformities treated with lateral and medial approaches. They reported a 9% relapse rate using the lateral approach and a 32% relapse rate using the medial approach. The lateral approach should be used for severe valgus knee because of its superior soft-tissue balancing ability. The lateral approach does not require special treatment for poor patellar tracking. In TKA for severe valgus knee, it is important to ensure that the femoral component does not become internally rotated due to hypoplasia of the lateral femoral condyle. If the lateral patellar retinaculum cannot be sutured because of correction of the valgus deformity, it is important to ensure that the joint is properly covered by the infrapatellar fat pad or the extension of the lateral patellar retinaculum described above.

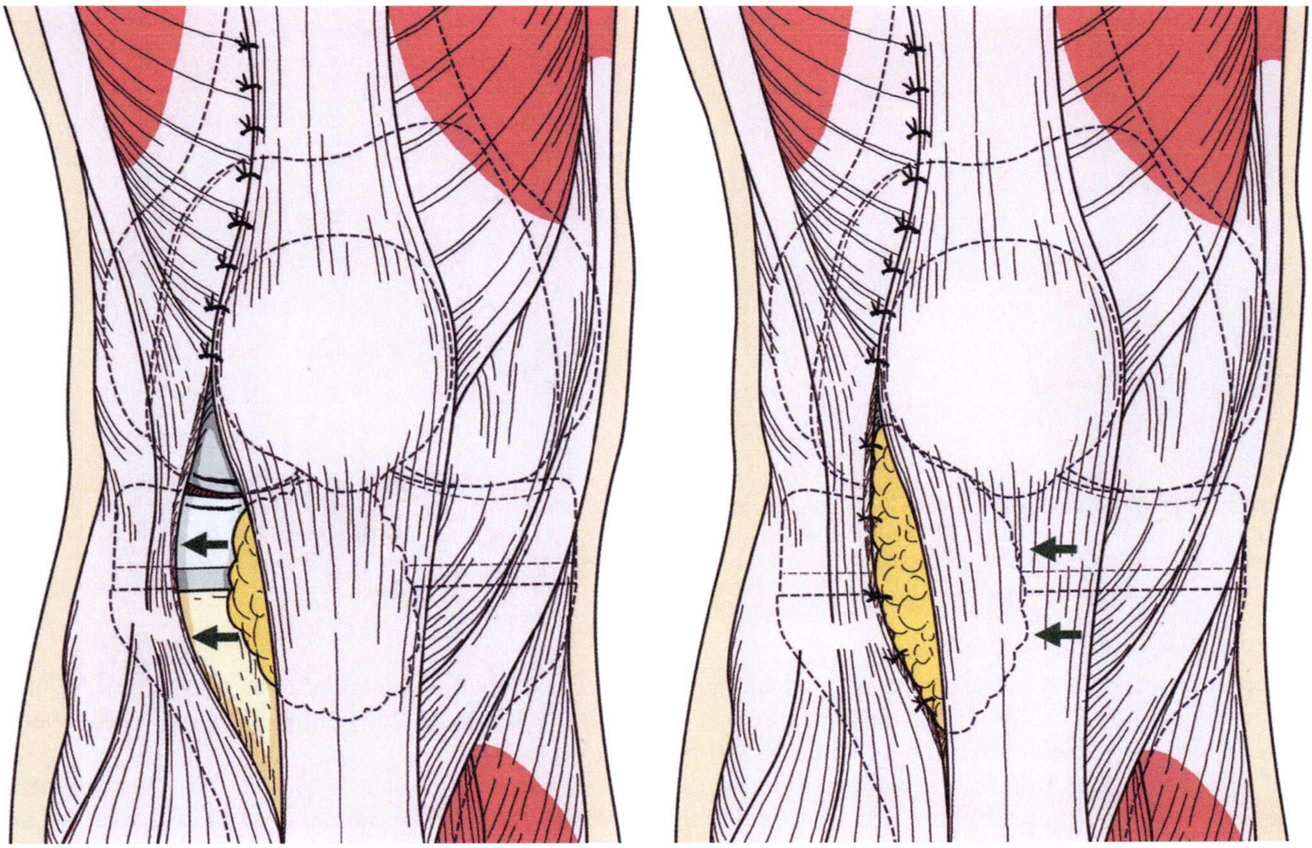

Fig. 74.11 Joint coverage with infrapatellar fat pad. If the suture defect in the lateral patellar retinaculum is small, it can be covered directly with an infrapatellar fat pad. If the defect is large, the infrapatellar fat pad flap is raised from the medial border of the patellar ligament and moved laterally

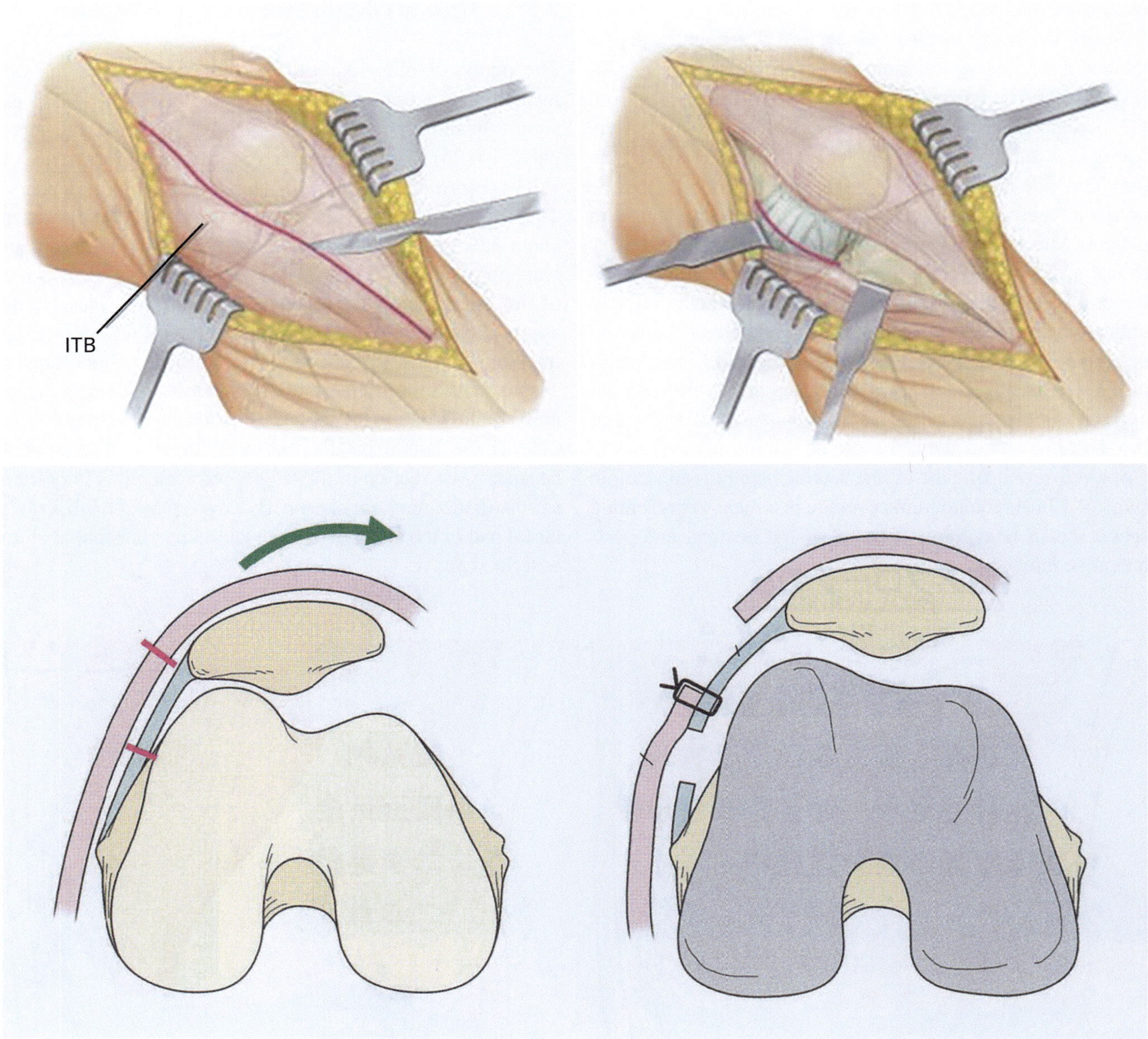

Fig. 74.12 Tips for joint coverage. The lateral patellar retinaculum and lateral patellofemoral ligament (LPFL) should not be incised using a single incision; rather, the two layers should be separated and the height of each changed. The LPFL and lateral patellar retinaculum are sutured to allow joint coverage and preserve the effect of the lateral release

Peroneal nerve palsy occurs when the axon is prolonged by 4–11% [27, 28], with a frequency of occurrence of 0.5–4% [2, 9, 29]. Careful attention should be paid when severe valgus knee is accompanied by severe flexion contracture. However, most peroneal nerve palsies that occur after surgery for severe valgus knee are incomplete and usually improve within 6 months postoperative. Although peroneal nerve release is reportedly effective [30], it is important to start extension training gradually while maintaining mild flexion to avoid peroneal nerve palsy.

It is important to achieve dynamic soft-tissue balance in extension-flexion, correction of tibial external rotation, and MCL support in severe valgus knee. Detailed preoperative planning and appropriate surgical techniques are essential for achieving excellent clinical outcomes.

References

1. Miyasaka KC, et al. 10- to 20-year followup of total knee arthroplasty for valgus deformities. Clin Orthop Relat Res. 1997;345:29–37.
2. Stern SH, et al. Total knee arthroplasty in valgus knees. Clin Orthop Relat Res. 1991;273:5–8.
3. Whiteside LA. Correction of ligament and bone defects in total arthroplasty of the severely valgus knee. Clin Orthop Relat Res. 1993;288:234–45.
4. Fiddian NJ, et al. Replacement arthroplasty of the valgus knee: a modified lateral capsular approach with repositioning of the valgus lateralis. J Bone Joint Surg Br. 1998;80:859–61.
5. Ranawat AS, et al. Total knee arthroplasty for severe valgus deformity. J Bone Joint Surg Am. 2005;87(Suppl):271–84.
6. Beverland D. Management of the severe varus and valgus knee using a low contact stress rotating platform. Orthopedics. 2006;29:S60–3.
7. Lombardi AV, et al. An algorithmic approach to total knee arthroplasty in the valgus knee. J Bone Joint Surg Am. 2004;86(Suppl):62–71.
8. Whiteside LA. Selective ligament release in total knee arthroplasty of the valgus. Clin Orthop Relat Res. 1999;367:130–40.
9. Keblish PA. The lateral approach to the valgus knee: surgical technique and analysis of 53 cases with over two-year follow-up evaluation. Relat Res. 1991;271:52–62.
10. Rossi R, et al. Total knee arthroplasty in the valgus knee. Int Orthop. 2014;38:273–83.
11. Matsuda S, et al. Effect of femoral and tibial component position on patellar tracking following total knee arthroplasty: 10-year follow-up of. Am J Knee Surg. 2001;14:152–6.
12. Berger RA, et al. Malrotation causing patellofemoral complications after total knee arthroplasty. Clin Orthop Relat Res. 1998;356:144–53.
13. Akagi M, et al. Effect of rotational alignment on patellar tracking in total knee arthroplasty. Clin Orthop Relat Res. 1999;366:155–63.
14. Paley D. Principles of deformity correction. New York: Springer; 2002. p. 1–97.
15. Eberbach H, et al. Geometry of the valgus knee: contradicting the dogma of a femoral-based deformity. Am J Sports Med. 2017;45:909–14.
16. Kayani B, et al. Posterior cruciate ligament resection in total knee arthroplasty. The effect on flexion-extension gaps, mediolateral laxity, and fixed flexion deformity. Bone Joint J. 2019;101-B:1230–7.
17. Stehlík J, et al. Z-plasty for valgus deformity in total knee arthroplasty. Acta Chir Orthop Traumatol Cechoslov. 2006;73:169–75.
18. Whiteside LA. Soft tissue balancing: the knee. J Arthroplast. 2002;17:23–7.
19. Peters CL, et al. Primary total knee arthroplasty in the valgus knee: creating a balanced soft tissue envelope. J Arthroplast. 2001;16:721–9.
20. Boyer P, et al. Total knee replacement in the fixed valgus deformity using a lateral approach: role of the automatic iliotibial band release for a successful balancing. Int Ortop. 2009;33:1577–83.
21. Buechel FF. A sequential three-step lateral release for correcting fixed valgus knee deformities during total knee arthroplasty. Clin Orthop Relat Res. 1990;260:170–5.
22. Bassaine M, et al. The composite meniscal-capsular-fat pad flap in a lateral approach to the fixed valgus knee: an anatomical study. J Arthroplast. 2007;22:601–4.
23. Jun J, et al. A lateral approach defect closure technique with deep fascia flap for valgus knee TKA. J Orthop Surg Res. 2015;10:173.
24. Favorito PJ, et al. Total knee arthroplasty in the valgus knee. J Am Acad Orthop Surg. 2002;10:16–24.
25. Clarke HD, et al. Anatomic risk of peroneal nerve injury with the "pie crust" technique for valgus release in TKA. J Arthroplast. 2004;19:40–4.
26. NIkolopoulos DD, et al. Total knee arthroplasty in severe valgus knee deformity: comparison of a standard medial parapatellar approach combined with tibial tubercle osteotomy. Knee Surg Sports Traumatol Arthrosc. 2011;19:1834–42.
27. Lundborg G, et al. Effects of stretching the tibial nerve of the rabbit: a preliminary study of the intraneural circulation and the barrier function of the perineurium. J Bone Joint Surg Br. 1973;55:390–401.
28. Rydevik BL, et al. An in vitro mechanical and histological study of acute stretching on the rabbit tibial nerve. J Orthop Res. 1990;8:694–701.
29. Krackow KA, et al. Primary total knee arthroplasty in patients with fixed valgus deformity. Clin Orthop Relat Res. 1991;273:9–18.
30. Xu J, et al. Common peroneal nerve 'prerelease' in total knee arthroplasty for severe valgus deformities. Knee. 2020;27:980–6.

75 Extra-articular Deformities

Hiroshi Inui

1 TKA for Patients with Extra-articular Deformities

There are two types of lower limb deformities: intra-articular deformities caused by bone loss in the joint or deformation of the joint surface and extra-articular deformities caused by deformation of the femur or tibia. The causes of extra-articular deformity include deformity healing after fracture (including epiphyseal injury), osteotomy, congenital and metabolic diseases. This section describes TKA in patients with extra-articular deformities.

2 Merits and Demerits of Intra- and Extra-articular Deformity Correction for "Extra-articular Deformity"

The advantages of intra-articular deformity correction include (1) no invasion of areas other than the knee joint, and (2) only a single operation is needed, while the disadvantages include (1) a large amount of bone resection is required in many cases, and (2) difficulty in adjusting the soft tissue balance.

The advantages of extra-articular deformity correction include (1) the ability to correct severe deformities, (2) normal amount of bone resection in the joint, and (3) less difficulty in adjusting the soft tissue balance, while the disadvantages include (1) invasion of areas other than the knee joint, (2) possibility of nonunion or delayed union at the osteotomy site, (3) effect of internal fixation material at the osteotomy site, and (4) possibility of multiple operations.

3 Whether to Choose "Intra-articular Correction" or "Extra-articular Correction"?

3.1 Deformity on the Femoral Side

If the femoral condyle osteotomy line drawn perpendicular to the line drawn from the center of the femoral head to the joint center (mechanical axis) is proximal to the MCL or LCL attachment site, an extra-articular corrective osteotomy is necessary, and if it is more distal, it can be corrected in the joint [1]. It has been reported that intra-articular correction is possible for deformities of up to 20° in the coronal plane and up to 15° in both flexion and extension in the sagittal plane [1, 2].

3.2 Deformity on the Tibial Side

If the line drawn on the distal axis of the deformed tibia is within the articular plane of the tibia, intra-articular correction is possible, whereas if it lies outside the articular plane, extra-articular corrective osteotomy should be used. Specifically, intra-articular correction is considered possible for deformities from 15–30° in the coronal plane (Fig. 75.1) [3].

H. Inui (✉)
Department of Orthopaedic Surgery, Saitama Medical Center, Saitama Medical University, Saitama, Japan

S. Matsuda (ed.), *Advances in Total Knee Arthroplasty*, https://doi.org/10.1007/978-981-97-4920-1_75

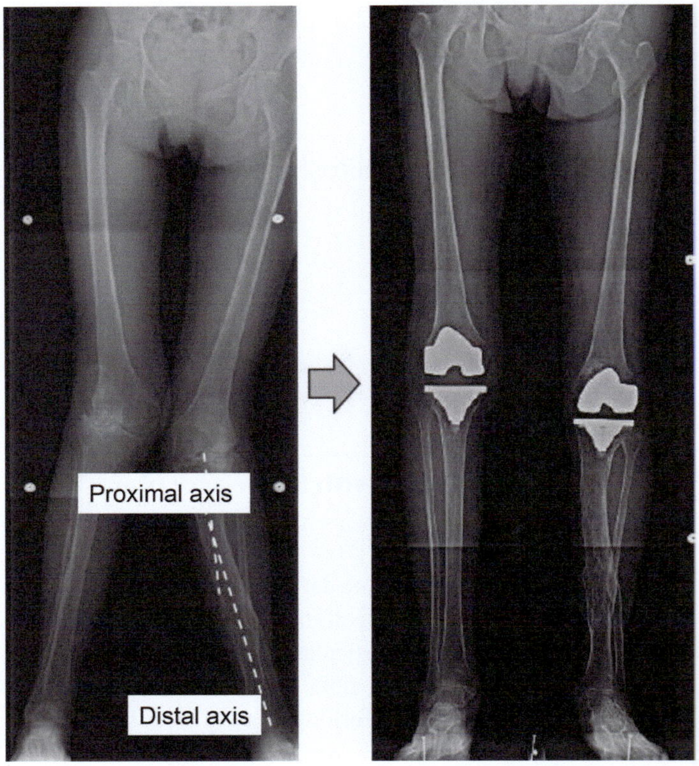

Fig. 75.1 Intra-articular correction of extra-articular deformity (coronal plane 15° after left tibial fracture)

4 Surgical Procedure

4.1 Deformity Correction Within a Joint

In most cases, the parapatellar approach can be used. If an intramedullary guide is used as usual on the femoral side, careful preoperative planning is necessary to determine its insertion site and external rotation angle. In cases where intramedullary rod insertion is difficult, an extramedullary guide may be used. In recent years, there have been many reports of achieving good alignment by using navigation during TKA for patients with extra-articular deformities [4, 5].

In a report from Japan comparing TKA patients with extra-articular deformity with and without image-free navigation, the hip-knee-ankle (HKA) angle was within a safe range of 180 ± 3° in all cases when image-free navigation was used, and the clinical results were significantly better in the group using navigation than in the group not using navigation. In addition, the clinical outcome of the navigation group was significantly better than that of the non-navigation group in terms of the Knee Society function score [4]. It has been shown that navigation allows accurate implant placement even in patients with extra-articular deformity, and that navigation can be used to preserve the internal fixation material in OA cases after internal fixation of a periprosthetic knee fracture, even when there is a risk of refracture due to removal of the internal fixation material [5].

Along with console type navigation systems, good postoperative alignment in patients with extra-articular deformity has also been reported using portable navigation systems and PSI [6]. It is desirable to use navigation and PSI to obtain accurate alignment in TKA for patients with extra-articular deformities.

However, even when the osteotomies are performed accurately, there are limitations to soft tissue balancing in the operative field. There is a risk of excessive soft tissue dissection to correct severe medial or lateral instability after osteotomy or elevation of the joint line due to the use of thick tibial inserts. Therefore, when there is a deformity on the femoral side, it is necessary to reconstruct the appropriate joint line using a bone graft or augmentation block (Fig. 75.2).

4.2 Deformity Correction Outside a Joint

As mentioned above, corrective osteotomy should be considered in cases of extra-articular deformity of 20° or more in the coronal plane or 15° or more in the sagittal plane on the femoral side and 15–30° or more in the coronal plane on the tibial side [1–3].

There are two types of corrective osteotomies: one-stage TKA and two-stage TKA after confirming bone fusion after corrective osteotomy. One-stage surgery has advantages such as a shorter treatment time and early return to society,

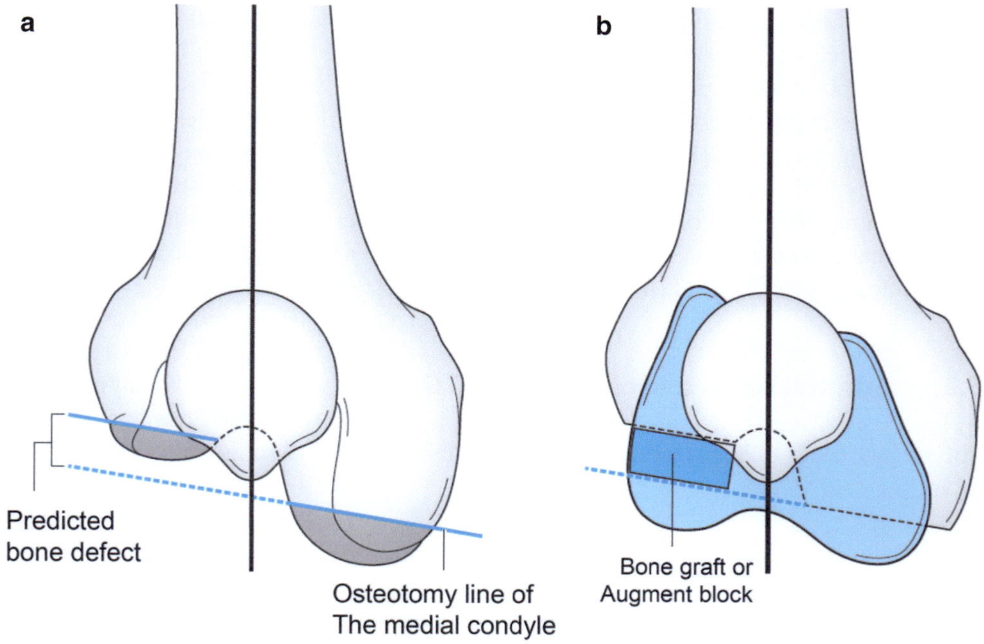

Fig. 75.2 Use of bone graft or augmentation block. (**a**) Osteotomy line of the distal femur. (**b**) Augmentation block. If the amount of medial osteotomy is adjusted to preserve the function of the MCL, extensive bone loss at the lateral condyle is expected. This defect should be replaced with a bone graft or augmentation block. (From Toru and Sakae [7])

but it also has disadvantages such as a large surgical invasion, complicated surgical procedures, and the possibility of pseudoarthrosis at the osteotomy site. Two-stage surgery can be a standard TKA procedure if the osteotomy is properly performed during the initial surgery. However, there may be problems such as joint contracture after the initial surgery and prolonged treatment period; therefore, each case should be individually considered for indications.

In a report of a one-stage surgery in Japan in a 71-year-old woman with OA of the knee after deformity healing of a proximal tibia fracture (deformity of 9°in varus in the coronal plane and 27° in flexion in the sagittal plane), the patient underwent a three-dimensional (3D) closed osteotomy to correct varus and flexion deformity, and a full-scale 3D bone model was used for preoperative simulation [8]. Postoperative joint ROM training was conducted in the early postoperative period, and full-load walking was performed 6 weeks after surgery with good results. In contrast, in a 65-year-old woman with end-stage OA of the knee after femur fracture deformity fusion (20° in the coronal plane), a two-stage surgery instead of one-stage surgery was planned, taking into consideration the age, correction site, and correction angle. The patient underwent corrective osteotomy using a locking plate and temporary Ilizarov external fixator at the initial surgery, and the internal fixator was removed 5 months after surgery [9].

In recent years, there have been increasing reports of one-stage surgery using locking plates, retrograde intramedullary nails, components with stems, etc. Although not a corrective osteotomy, one-stage TKA surgery with internal fixation has also been used for supracondylar or condylar femur fractures with knee OA [10].

However, as mentioned above, one-stage surgery can potentially be complicated and invasive, and there have been reports of complications in one-stage surgery in Japan. Specifically, a 78-year-old man with end-stage OA after deformity healing of a proximal tibia fracture underwent a one-stage corrective osteotomy and TKA using an extension stem on the tibial side; however, thermal necrosis occurred on the tibial side, possibly due to the reaming operation, and the patient experienced difficulty in subsequent treatment. The necrosis was caused by the use of an extension stem on the distal tibial side of the osteotomy [11]. The authors considered the excessive reaming that was performed to fix the tibia distal to the osteotomy using only a cementless extension stem as the potential cause. Although implant replacement components such as internal fixation materials and stem extensions have been steadily evolving in recent years, careful surgical planning and techniques are still needed.

5 After HTO

HTO is an effective surgery for medial OA. However, 10 years or more after surgery, symptoms may reoccur, and TKA may be necessary. In most cases, surgery is indicated after the progression of the valgus deformity.

5.1 Problems with Post-HTO TKA

Problems with TKA after HTO include (1) low patella and shortened patellar tendon adhesions, which make deployment difficult; (2) imbalance between lateral soft tissue contracture and medial soft tissue laxity in knees with excessive external correction; (3) possibility of excessive osteotomy of the medial tibial condyle due to severe external tibial deformity; (4) possibility of excessive osteotomy of the medial tibial condyle due to a high degree of external tibial deformity; (5) possibility of lateral placement of the tibial component and tibial stem due to external soft tissue contracture and external deformity; and (6) presence of a false joint at the osteotomy site of HTO [12].

5.2 Surgical Technique of TKA after HTO

The surgery needs to be performed with the above issues in mind.

1. Difficult deployment due to low patella and shortened patellar tendon adhesions

   If deployment is difficult with the usual approach, a quadriceps snip or V-Y plasty of the quadriceps may be necessary.
2. Imbalance due to contraction of the lateral soft tissues and relaxation of the medial soft tissues due to excessive correction of the valgus knee

   Soft tissue balancing is similar to the surgical technique for the valgus knee, but the joint line may be elevated when using a thick tibial insert along with excessive lateral dissection for balancing, and it is advisable to prepare an implant that is compatible with (semi)constrained inserts.
3. Possibility of excessive osteotomy of the medial tibial condyle due to severe valgus deformity of the tibia

   In cases of excessive tibial valgus deformity due to HTO, if the amount of osteotomy on the medial condyle is increased by performing osteotomy on the tibial articular surface, the lax MCL must be further released, resulting in increased instability of the medial compartment. In such cases, it is desirable to avoid excessive medial osteotomy while considering the balance of the ligaments, and to use a bone graft if a bone defect occurs on the lateral side (Fig. 75.3). Fine adjustment is also possible if the amount of osteotomy can be confirmed by using navigation. Good surgical results have been reported using the reversed tibial flip autograft technique as a reconstruction method for bone defects in the posterior tibial articular surface after HTO [13].
4. Possible lateral deviation of the proximal tibia after HTO, resulting in lateral placement of the tibial component and tibial stem

   After HTO, the mechanical and anatomical axes are likely to be deviated [14]. It has also been reported that the proximal tibia is more laterally deviated after closed wedge HTO compared to opening wedge HTO [15]. Careful preoperative planning is needed, as the tibial stem may interfere with the lateral bony cortex during TKA surgery. Therefore, careful preoperative planning is needed, and the use of an offset stem should be considered in some cases (Fig. 75.4).
5. Patellar dislocation and subluxation due to soft tissue contracture of the lateral compartment and valgus deformity

   If a lateral approach is used, there is no problem; however, if a medial approach is used, sufficient dissection of

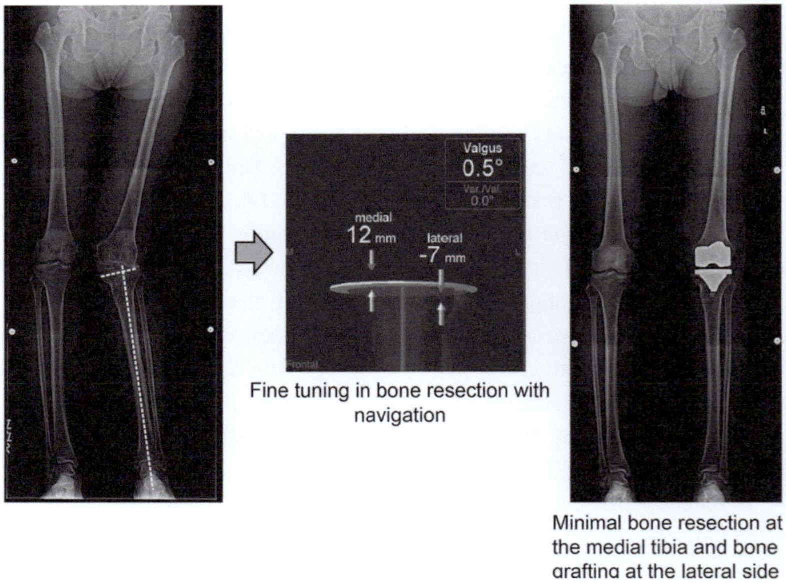

Fig. 75.3. A case after HTO. Excessive amount of osteotomy performed on the medial condyle is deeply concerning

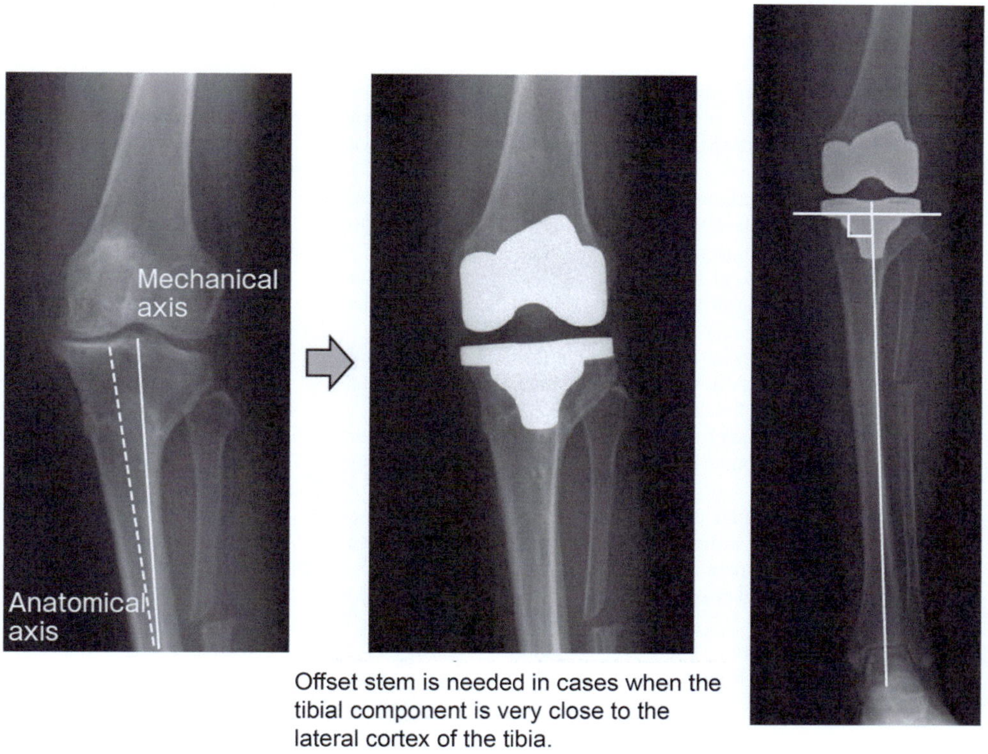

Fig. 75.4 Lateral deviation of the proximal tibia

the lateral soft tissue and lateral release of the lateral patellar band should be added as needed.

6. Nonunion at the osteotomy site of HTO

If there is nonunion after HTO, surgical manipulation to obtain bone fusion at the pseudoarthrosis is necessary. Although two-stage surgery is sometimes undertaken to ensure bone fusion, there are reports that one-stage surgery can be performed by correcting the deformity at the false joint, performing sufficient bone grafting, and using a long stem [16]. Therefore, one-stage surgery may be considered under careful surgical planning.

5.3 Surgical Outcomes of TKA After HTO

Although TKA after HTO can generally achieve good results, it has been reported that the surgical outcome is slightly inferior to that of primary TKA, with more complications such as infection, higher revision rates, and poor joint ROM [17, 18] compared to primary TKA. In a study of surgical results in Japan, Noda et al. [19] reported that the mean 2-year postoperative results of 28 patients who underwent TKA after HTO were inferior to those with primary TKA at the same institution, amounting to excellent in 14%, good in 47%, fair in 21%, and poor in 18% patients, with a mean joint ROM of 90.3%. Patellar subluxation, excessive residual instability (requiring bracing), and avulsion fracture of the tibial tuberosity in patients who underwent osteotomy of the tibial tuberosity at the time of the surgery were reported as complications. In a comparison of surgical results (average 9.6 years postoperatively) between the CR and PS types of TKA after HTO, the PS type showed significantly higher postoperative KSS, with particularly good stability and ROM score; accordingly, the PS type of TKA was recommended for TKA after HTO [20]. TKA after HTO requires various special surgical manipulations compared to normal TKA in different cases. Shibano et al. [21] developed a PSI guide based on 3D simulation and used intraoperative CT-based navigation to achieve appropriate corrective osteotomy for significant tibial deformity after HTO. They reportedly used intraoperative CT-based navigation to perform a one-stage operation. We hope that appropriate surgical planning and careful surgical techniques for TKA after HTO will lead to surgical outcomes comparable to those of primary TKA.

References

1. Wang JW, et al. Total knee arthroplasty for arthritis of the knee with extra-articular deformoity. J Bone Joint Surg Am. 2002;84:1769–74.
2. Wang JW, et al. Total knee replacement with intra-articular resection of bone after malunion of a femoral fracture: can sagittal angulation be. J Bone Joint Surg Br. 2010;92:1392–6.
3. Radke S, et al. Total knee arthroplasty in combination with a one-stage tibial osteotomy: a technique for the correction of gonarthrosis with a severe. J Arthroplast. 2002;17:533–7.
4. Tani I, et al. Navigated total knee arthroplasty for osteoarthritis with extra-articular deformity. Acta Orthop Bras. 2018;26:170–4.

5. Hamada D, et al. Navigation-assisted total knee arthroplasty for osteoarthritis with extra-articular femoral deformity and/or retained hardware. Case Rep Orthop. 2013;2013:174384.
6. Matassi F, et al. Total knee arthroplaty in patients with extra-articular demormity: restoration of mechanical alignment using an accelerometer-based navigation system. J Arthroplast. 2019;34:676–81.
7. Toru K, Sakae T, editors. All about total knee arthroplasty [TKA]. 2nd ed. Medical View; 2-30 Ichigaya Motommura Cho, Shinjuku-Ku, Tokyo, Japan, 2017.
8. Hosokawa T, et al. Total knee arthroplasty with corrective osteotomy for knee osteoarthritis associated with malunion after tibial plateau fracture a case report. BMC Res Notes. 2017;10:223.
9. Yagi K, et al. Treatment of knee osteoarthritis associated with extraarticular varus deformity of the femur: staged total knee arthroplasty. J Orthop Sci. 2006;11:386–9.
10. Yoshino N, et al. Primary total knee arthroplasty for supracondylar/condylar femoral fracture in osteoarthritic knees. J Arthroplasty. 2001;16:471–5.
11. Ishida K, et al. Thermal necrosis after simultaneous tibial osteotomy and total knee arthroplasty using press-fit extension stem. Knee Surg Sports Traumatol Arthrosc. 2011;19:112–4.
12. Matsuno M, et al. Total knee arthroplasty-basic and clinical. Tokyo: Bunkodo; 2005.
13. Nagumo A, et al. A reversed tibial flip autograft for correcting overvalgus knee after high tibial closing-wedge osteotomy in total knee. J Arthroplast. 2006;21:771–4.
14. Kawano T, et al. Alignment in total knee arthroplasty following failed high tibial osteotomy. J Knee Surg. 2003;16:168–72.
15. Kuwashima U, et al. Comparison of the impact of closing wedge versus opening wedge high tibial osteotomy on proximal tibial deformity and subsequent. Knee Surg Sports Traumatol Arthrosc. 2017;25:869–75.
16. Yoshino N, et al. Total knee arthroplasty with long stem for treatment of nonunion after high tibial osteotomy. J Arthroplast. 2004;19:528–31.
17. Chen X, et al. Higher risk of revision in total knee arthroplasty after high tibial osteotomy: a systematic review and updated meta-analysis. BMC Musculoskelet Disord. 2020;21:153.
18. Sun X, et al. A meta-analysis of total knee arthroplasty following high tibial osteotomy versus primary total knee arthroplasty. Arch Orthop Trauma Surg. 2020;140:527–35.
19. Noda T, et al. Clinico-radiological study of total knee arthroplasty after high titial osteotomy. J Orthop Sci. 2000;5:25–36.
20. Akasaki Y, et al. Total knee arthroplasty following failed high tibial osteotomy: mid-term comparison of posterior cruciate-retaining versus posterior stabilized prosthesis. Knee Surg Sports Traumatol Arthrosc. 2009;17:795–8799.
21. Shibano K, et al. Total knee arthroplasty with concomitant corrective tibial osteotomy using patient specific instrumentation and computed. Arthroplast Today. 2020;6:742–6.

Flexion Contracture and Ankylosis of the Knee

76

Ryuji Nagamine

1 Causes of Contracted Knee

In general, articular cartilage wears away in the flexed position as OA progresses. If patients are not properly guided and treated during this period, they will continue to hold the knee in a flexed position to avoid pain, resulting in posterior osteophyte formation, contracture of the posterior capsule, shortening of the hamstrings, contracture of the collateral ligaments, and contracture of the PCL, which will complete flexion contracture. In RA, flexion contracture may occur rapidly due to the dissolution of articular cartilage with worsening arthritis.

2 Problems of Flexion Contracture and Countermeasures Before TKA

Flexion contracture of the knee causes an increased quadriceps muscle load and abnormal gait [1]. Preoperatively clarifying the cause of contracture and developing countermeasures is necessary to safely improve the range of motion after TKA.

The degree of flexion contracture was evaluated preoperatively. Generally, flexion contractures are classified as mild, moderate, or severe; mild contractures are defined as contractures of less than 10°, moderate contractures from 10° to 30°, and severe contractures greater than 30° [2]. If the contracture is greater than 30°, bed traction therapy and ROM training are recommended before surgery [3]. The next step is to determine whether surgery is feasible in one stage, as previous reports have suggested that contractures between 45° and 60° can be treated in a one-stage operation. The usefulness of posterior soft tissue release in the first stage of a two-stage surgery has also been reported. However, as described below, the degree of correction varies depending on the cause of contracture; hence, the angles indicated in the literature should only be used as guides. It is important to consider the cause of contracture in each case and make decisions based on the skill of the surgeon. The aim is to achieve an extension limit of less than 6° after TKA [4], but this is often difficult in patients with strong flexion contractures.

3 Causes of Contractures and Preoperative Planning

Two main factors affect the flexion contracture. The first is soft tissue contracture with limited extension, which is common in RA with minimal bone destruction. The second is bony contracture, in which the femoral condyle becomes stuck in the tibial condyle due to progressive joint destruction. In the case of soft tissue contracture only, TKA can be performed without any problems if the soft tissue in the posterior joint is released, and the postoperative lower extremity alignment can be adjusted appropriately. Therefore, preoperative planning and bone cuts were the same as those in conventional TKA.

Careful preoperative planning is required in the presence of a bony contracture. Three difficult cases are presented. In the first case of OA shown in Fig. 76.1, the preoperative ROM was −20° to 110°. During TKA, the knee joint was exposed using the usual medial parapatellar approach. When manual force was applied to the ankle joint to expose the tibial articular surface during flexion, a longitudinal fracture occurred in the medial femoral condyle. It was thought that the fracture was caused by the absence of a joint gap necessary for deep flexion due to osteophytes in the posterior part of the knee joint and that the medial soft tissue was overstrained during the procedure to expose the tibial articular surface. In the present case, all bone cuts were performed without fracture treatment. After all bone-cutting procedures of bone cuts, the medial condyle was fixed with two screws before component insertion. Figure 76.2 shows the second patient with long-term RA. The preoperative ROM was −25°

R. Nagamine (✉)
Nagamine Rheumatology & Orthopedics Clinic, Fukuoka, Japan
e-mail: takaakane@do6.enjoy.ne.jp

S. Matsuda (ed.), *Advances in Total Knee Arthroplasty*, https://doi.org/10.1007/978-981-97-4920-1_76

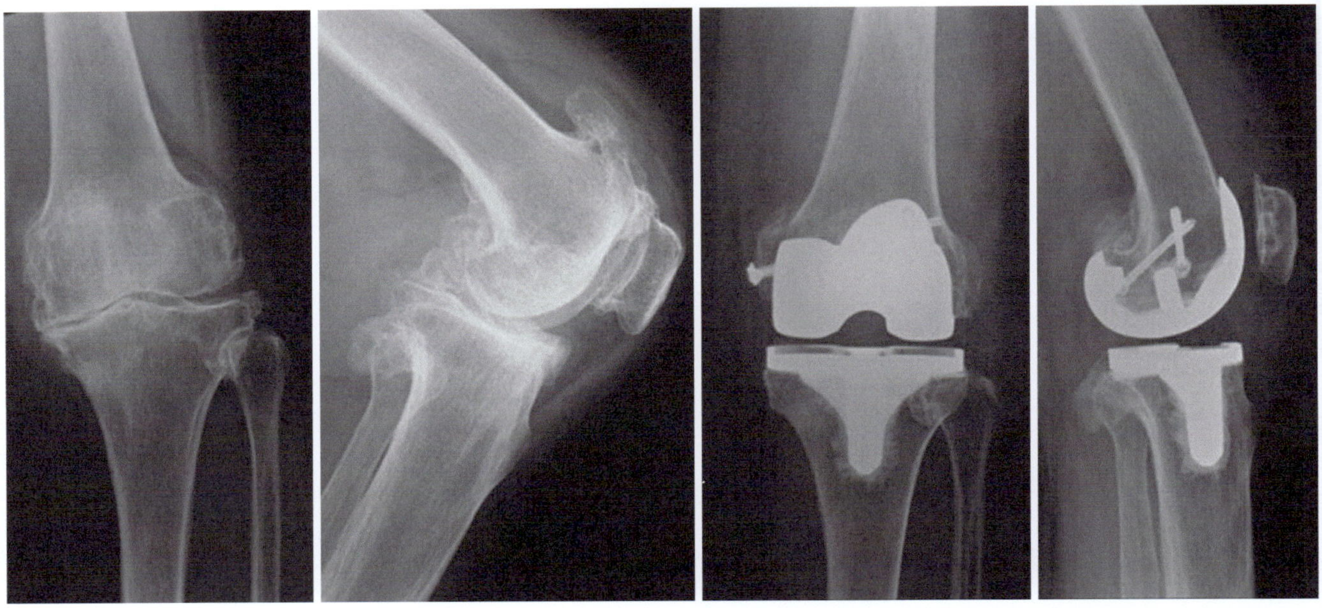

Fig. 76.1 A case of fracture of the medial femoral condyle during joint exposure. The joint space in deep flexion was narrowed by posterior osteophytes, and the medial soft tissues became overstressed during deep flexion, resulting in a fracture of the medial femoral condyle. After all osteotomies were made, fracture was fixed with two screws and components were inserted

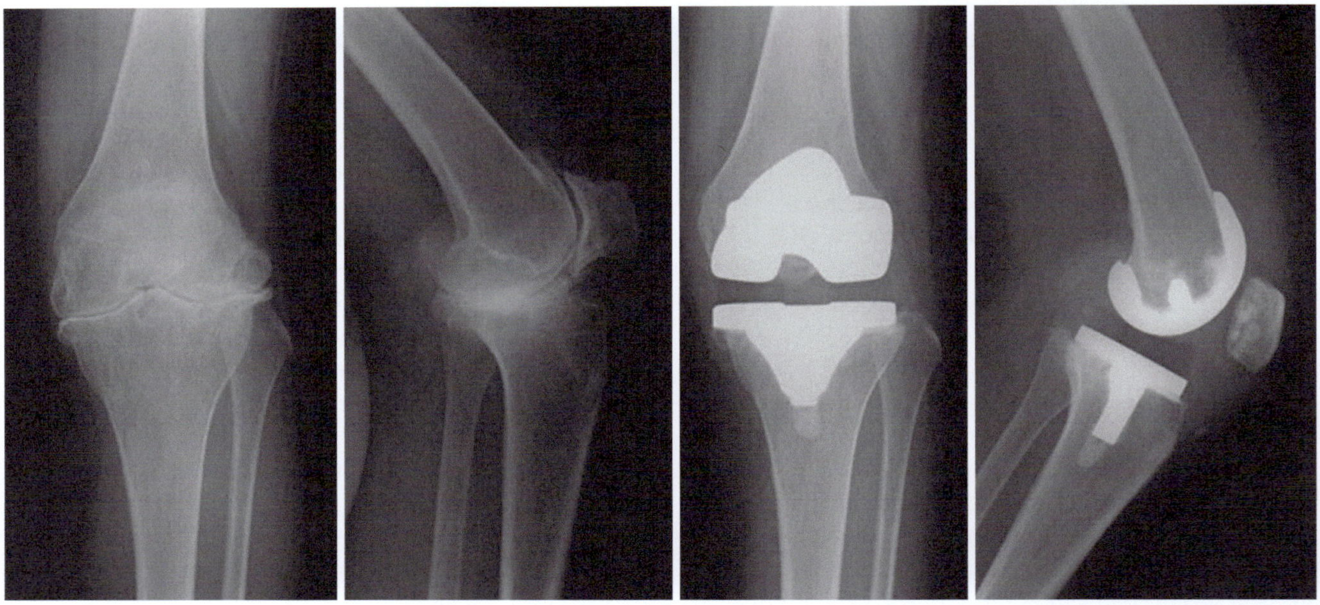

Fig. 76.2 Preoperative images of a long-term RA patient with bony contractures. TKA was performed using a measured resection technique. External rotation of the femoral component was set at 4°, but soft tissue balance could not be achieved in flexion

to 120°. The knee joint was exposed with a rectus snip, and bone cuts were performed with 7° valgus and 4° external rotation of the femoral component. The soft tissue balance at the time of femoral trial component insertion was 0° in extension and 9° valgus in flexion. Soft tissue balance could not be achieved during flexion. We considered using a highly constrained system; however, because of her young age (48 years), we performed TKA using the PS system. After TKA, RA caused foot deformity and the lower leg became more externally rotated over time, eventually resulting in marked instability, even in extension and patellar dislocation (Fig. 76.3). In previous reports on flexion-contracted knees in patients with RA, dislocation of the tibiofemoral joint after TKA has also been reported [3]. Prolonged flexion contractures and ankylosis indicated soft tissue degeneration. Preoperative planning should consider the difficulty of achieving soft tissue balance in RA; furthermore, backup of a highly constrained system is necessary [3]. The third case,

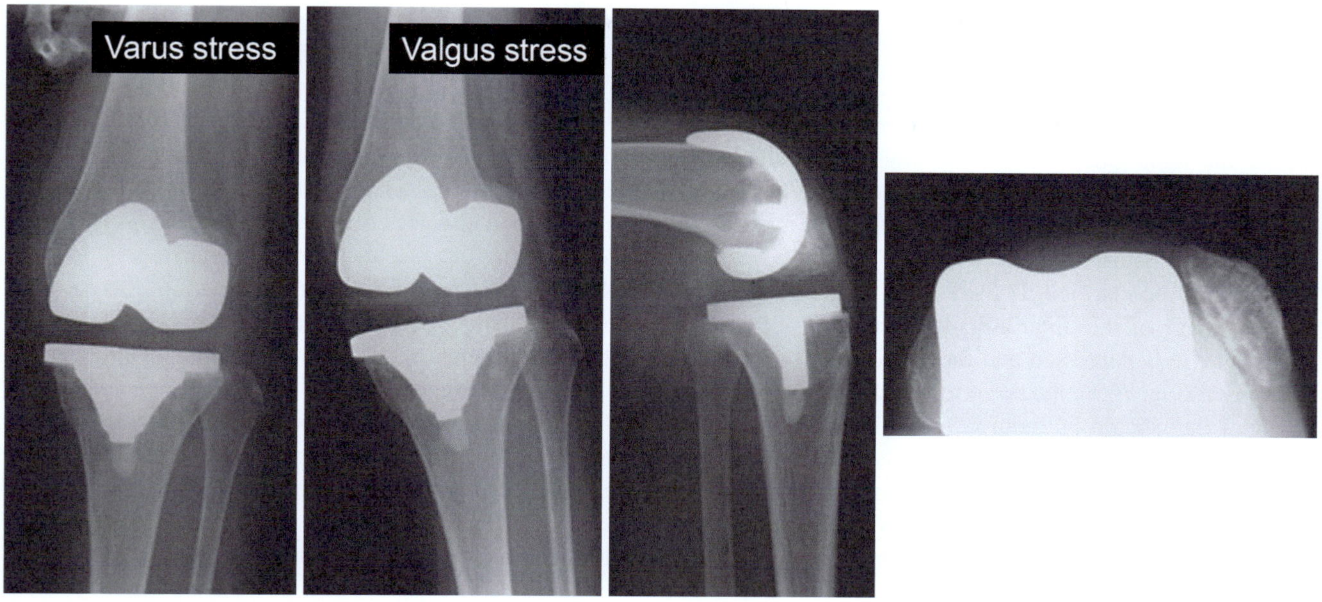

Fig. 76.3 Postoperative images of a long-term RA patient with bony contractures

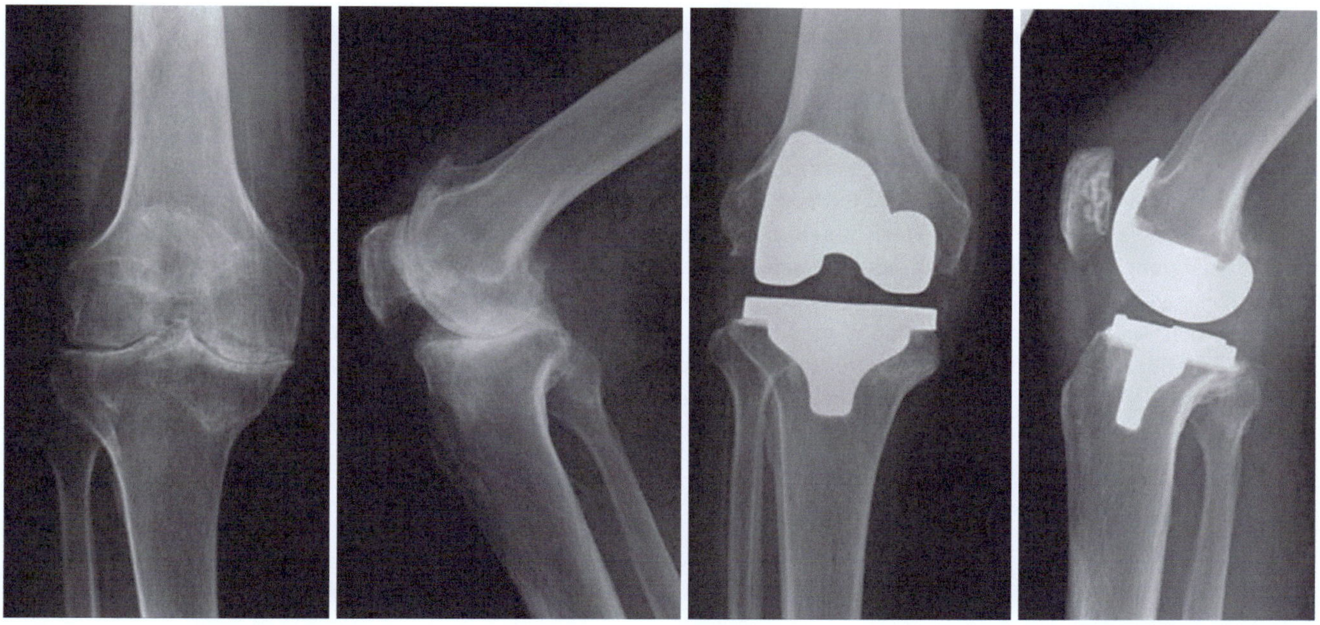

Fig. 76.4 Flexion contracture by RA in the knee with large posterior rotation of the tibial condyle. In addition to preoperative flexion contracture, knee full extension is difficult to be achieved in the knee with large posterior tilt of the tibia

shown in Fig. 76.4, was a patient with long-term RA. The preoperative ROM ranged from −55° to 90°. The patient underwent TKA with a rectus snip, and the external rotation of the femoral component was set perpendicular to the AP axis. At discharge, the patient had −10° extension and 110° flexion. The posterior rotation angle of the tibial condyle relative to the anterior cortex of the tibial diaphysis was 33.3°, which was larger than the usual average of 18.6° [5]. The posterior slope angle of the tibial articular surface was 17.7°, which was larger than the usual average of 7.0°. In this case, the patient had a large posterior rotation of the tibial condyle and a large posterior slope of the articular surface, which, combined with the flexion contracture, made it extremely difficult to achieve full extension. Adjustment of the tibial component placement angle should be considered during preoperative planning.

4 Points to Keep in Mind in TKA

Exposure of the joint space in the flexion-contracted knees is often difficult. A parapatellar approach is preferable to fully expose the joint space. When exposure is difficult, rectus snipping or turndown should be considered [6]. The osteophyte is thoroughly resected, and the deep MCL is released. If the tibial articular surface cannot be fully exposed, the cause should be clarified. If there is an osteophyte on the posterior edge of the tibial articular face, or if the medial tibial condyle is depressed and the articular surface cannot be exposed (Fig. 76.1), the posterior condyle not covered by the tibial tray should be removed using an osteotome. The posterior bone fragments are temporarily left in place. The bony factor can be eliminated using this procedure. Additionally, reduction osteotomy is useful for exposing the tibial articular surface. The tibial tray is placed with respect to the central line of the tibial diaphysis, and a smaller tibial component is used. The bone of the medial condyle that is not covered by the tray is resected. With these procedures, fractures of the medial femoral condyle can be prevented.

In flexion contracture and ankylosis, the most important issue reported thus far has been the difficulty in equalizing the joint gap during extension and flexion [2, 3, 6–9]. This is due to the combination of degeneration of all soft tissues, including posterior osteophytes, and bone deformities. If all femoral bone cuts are easily performed, the gap between extension and flexion may be completely different, which may cause significant extension impairment. Therefore, even in the measured resection technique, the gap should be measured in extension and flexion. If the gap is small in extension, an additional bone cut of the distal femur should be performed [2, 8]. With PCL resection, a clear view of the posterior part of the knee joint can be obtained, and it is much easier to remove the osteophyte and release the posterior capsule. However, PCL resection may widen the joint gap not in extension but in flexion [10], and it may be more difficult to equalize the gap during extension and flexion. Therefore, even when a PS system is used, adjusting the levels of the distal and posterior femoral bone cuts is necessary to equalize the gaps in extension and flexion.

Another important issue is the treatment of the posterior osteophytes. The Leie classification is useful for evaluating posterior osteophytes [7]. Grade 0, no osteophytes; grade 1, up to 10 mm in length; grade 2, 11–15 mm in length; grade 3, 16 mm or longer. In grades 2 and higher, osteophyte removal improves the angle of extension. A safe and precise osteophyte excision using an osteotome (posterior clearance) is required (Figs. 76.1 and 76.4). If osteophyte removal is not performed properly, the posterior lip of the tibial insert may impinge on the osteophyte in a deep flexion position, causing wear debris. In PS, the posterior lip of the tibial insert is difficult to identify radiographically; however, fluoroscopic images show that the post of the tibial insert and the cam of the femoral component are almost completely dislocated in flexion (Fig. 76.5). Posterior clearance is indispensable for knees with flexion contracture. It is recommended not to remove osteophytes in grade 1 in the PCL-sparing TKA [7].

If the contracture is mild and less than 10°, it can be treated with the usual technique after confirming that it is a soft-tissue contracture; however, close attention should be paid to obtain the optimal gap in extension and flexion.

If the contracture is moderate, posterior capsule release from 10° to less than 30° is required. It is important to measure the length of the posterior condyle of the femoral component, resect the posterior femoral condyle and osteophytes

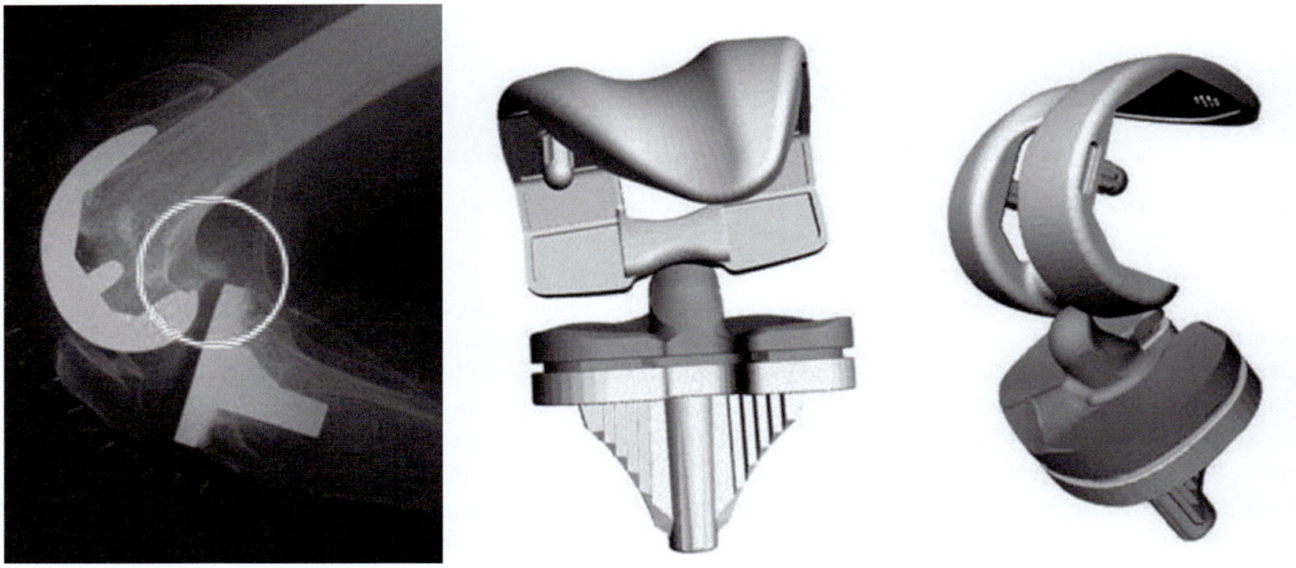

Fig. 76.5 Computer simulation in the knee with unresected osteophytes. In deep flexion, posterior osteophyte and posterior insert are impinged, anterior joint is open, and post-cam mechanism is subluxated. Plain X-rays cannot confirm this condition

not covered by the component, and release the posterior capsule using an osteotome. If release is performed under the periosteum in the posterior femoral condyle, there are no issues, such as concerns about damage to the popliteal artery. It is necessary to know that the capsule posterior to the intercondylar notch cannot be released if the PCL is present. Some techniques aim to detach the posterior soft tissues by manually holding the extension position intraoperatively; however, their effectiveness is limited [9]. When synovitis is evident in RA, the PCL is highly likely to be damaged. Therefore, it is recommended that the PS type be selected.

If the contracture is severe and >30°, it is often due to bony factors, requiring further posterior capsule release and treatment of osteophytes. In many cases, medial soft tissue contracture is also present, and if medial soft tissue release is performed, the PCL will shorten relatively; therefore, the PS type is recommended. In some cases, the posterior capsule contractures and becomes attached to the MCL/LCL. In these cases, transverse or longitudinal dissection of the posterior capsule to dissect the internal and external posterior capsules from the collateral ligament has been reported [2]. Some cases also require dissection of the medial gastrocnemius and lateral biceps femoris muscle [6, 11]. It is necessary to determine the goal after TKA by considering the preoperative condition of each case and making concessions during the operation. It is recommended to back up a highly constrained system.

References

1. Su EP. Fixed flexion deformity and total knee arthroplasty. J Bone Joint Surg Br. 2012;94B:112–5.
2. Scuderi GR, et al. Management of flexion contracture in total knee arthroplasty. J Arthroplast. 2007;22(4 Suppl 1):20–4.
3. Abe S, et al. Total knee arthroplasty for rheumatoid knee with bilateral, severe flexion contracture: report of three cases. Mod Rheumatol. 2008;18:499–506.
4. Ritter MA, et al. The role of flexion contracture on outcomes in primary total knee arthroplasty. J Arthroplast. 2007;22:1092–6.
5. Nagamine R, et al. The posterior tibial slope is mainly created by the posterior rotation of the tibial condyles. J Orthop Surg (Hong Kong). 2020;28:2309499020975580.
6. Ojima T, et al. Bilateral total knee arthroplasty after spontaneous osseous ankylosis in rheumatoid arthritis. Mod Rheumatol. 2005;15:139–43.
7. Leie MA, et al. Large osteophyte removal from the posterior femoral condyle significantly improves extension at the time of surgery in a total knee. J Orthop. 2019;19:76–83.
8. Liu DW, et al. The effect of distal femoral resection on fixed flexion deformity in total knee arthroplasty. J Arthroplast. 2016;31:98–102.
9. Matsui Y, et al. Intraoperative manipulation for flexion contracture during total knee arthroplasty. J Orthop Surg (Hong Kong). 2019;27:2309499019825574.
10. Ma Y, et al. Comparative evaluation of posterior cruciate ligament in total knee arthroplasty. J Orthop Surg (Hong Kong). 2017;25:2309499017690976.
11. Rocco J, et al. The effect of gastrocnemius resection on knee flexion in a total knee arthroplasty model. Arch Orthop Trauma Surg. 2022 Oct;142(10):2503–11. https://doi.org/10.1007/s00402-020-03695-x.

Bone Defects and Osteoporosis

77

Mitsuru Saito

1 Bone Defects

Osteoarthritis (OA) patients are increasing with the ageing population. It is not unusual to perform bilateral total knee replacements (TKA) in patients over 90 due to advances in preoperative and perioperative medical care. Therefore, there are needs for treating severe deformities, contractures and defects in these patients. Furthermore, treatment of osteoporosis reduces the rate of revision arthroplasties, and reduces the risk of peri-prosthetic fractures. This is an immediate problem that we face in Japan, where lifeexpectancy is approaching 100.

2 Evaluation of Bone Defects

Preoperative plain X-rays and computed tomography (CT) are performed to evaluate bone defects. First, the presence or absence of a cortical wall should be confirmed [1, 2]. MRI is also performed when joint instability is evident on examination, since cases of severe bone loss are often accompanied by ligament instability. Especially in cases of severe valgus knee, evaluation of the MCL is important in selecting the type of implant.

- Cortical wall intact: constrained type
- Cortical wall defects: unconstrained type
- Bone defect size: less than 5 mm, 5–10 mm, more than 10 mm

To evaluate osteoporosis, preoperative bone mineral density of the lumbar spine and femur, bone metabolic markers, and history of osteoporotic fractures are assessed.

M. Saito (✉)
Department of Orthopaedic Surgery, The Jikei University School of Medicine, Tokyo, Japan

3 How to Address Bone Defects

3.1 Methods for Filling of Bone Defects

Bone cement, autologous bone graft [3], modular metal augmentation + stem extension [4]. The autologous bone used for grafting is the surplus bone obtained during osteotomy. In general, the bone is filled in block form, but it is also possible to use morselized bone [5, 6]. In the case of total hip arthroplasty (THA), morselized bone can be easily obtained by reaming the acetabulum, whereas in TKA, morselized bone has to be made from excised bone, which is a time-consuming process. If the autologous bone is insufficient to cover the bone defect, artificial bone (beta-tricalcium phosphate: β-TCP) is used to supplement the area which comes into contact with the implant.

3.2 Dealing with Bone Defects

1. Amount of osteotomy

For medial tibial bone defects (<10 mm) seen in OA, metal augmentation may be unnecessary by adjusting the osteotomy level by +2 to +4 mm distally. If the osteotomy is set distally and the bone defect is reduced to approximately 1/4 of the tibial osteotomy surface, a conventional implant can be placed using bone grafting or cement filling, without the need for metal augmentation and stem extension. If the remaining defect is more than 1/4, a stem should be used to distribute the shear and compressive stress to the bone defect. In the case of femoral bone defects, the osteotomy should be limited to 2 mm because the resulting joint line rises if the osteotomy plane is moved proximally.

2. Bone cement filling

Bone defects of 5 mm or less should be filled with bone cement. However, bone cement is not glue. It penetrates the

S. Matsuda (ed.), *Advances in Total Knee Arthroplasty*, https://doi.org/10.1007/978-981-97-4920-1_77

bone microstructure and attaches to it. In addition, since bone cement is weak against shear force, wedge-shaped bone defects may be created in a step-like fashion. Screws may also be inserted into the bone for support during the cementation process.

3. Autologous bone graft

After osteotomy, the bone is thoroughly examined, and in the case of a contained type where there is a strong cortical wall, the cortex is used as a "shell" and filled with crushed cancellous bone. In the uncontained type, a bone block is used to recreate the wall, but depression due to osteonecrosis may occur. For this reason, repair using morselized bone (Fig. 77.1) or metal augmentation + stem is sometimes selected. If the bone defect is more than 5 mm and a bone graft is used, the bone fragment is fixed with a screw or a bone peg (Fig. 77.2). In the case of simultaneous bilateral TKA, this is not a problem, but when TKAs are performed in series, i.e., one knee after the other, the side with less defect should be performed first to obtain more excess bone from osteotomy. In cases where bone volume is insufficient for contained type bone defects or cysts, β-TCP is used to fill the void, and then covered with autologous bone before fixation with the cemented implant.

4. Metal augmentation

If the bone defect is greater than 5 mm, a wedge or block of metal is used to repair the defect. Depending on the implant, additional osteotomy may require the removal of a significant amount of healthy bone. Based on the preoperative planning and intraoperative findings, the choice is made between bone-sparing repair using only autologous bone or bone resection with metal augmentation. As many elderly Japanese women are small, care should be taken to avoid

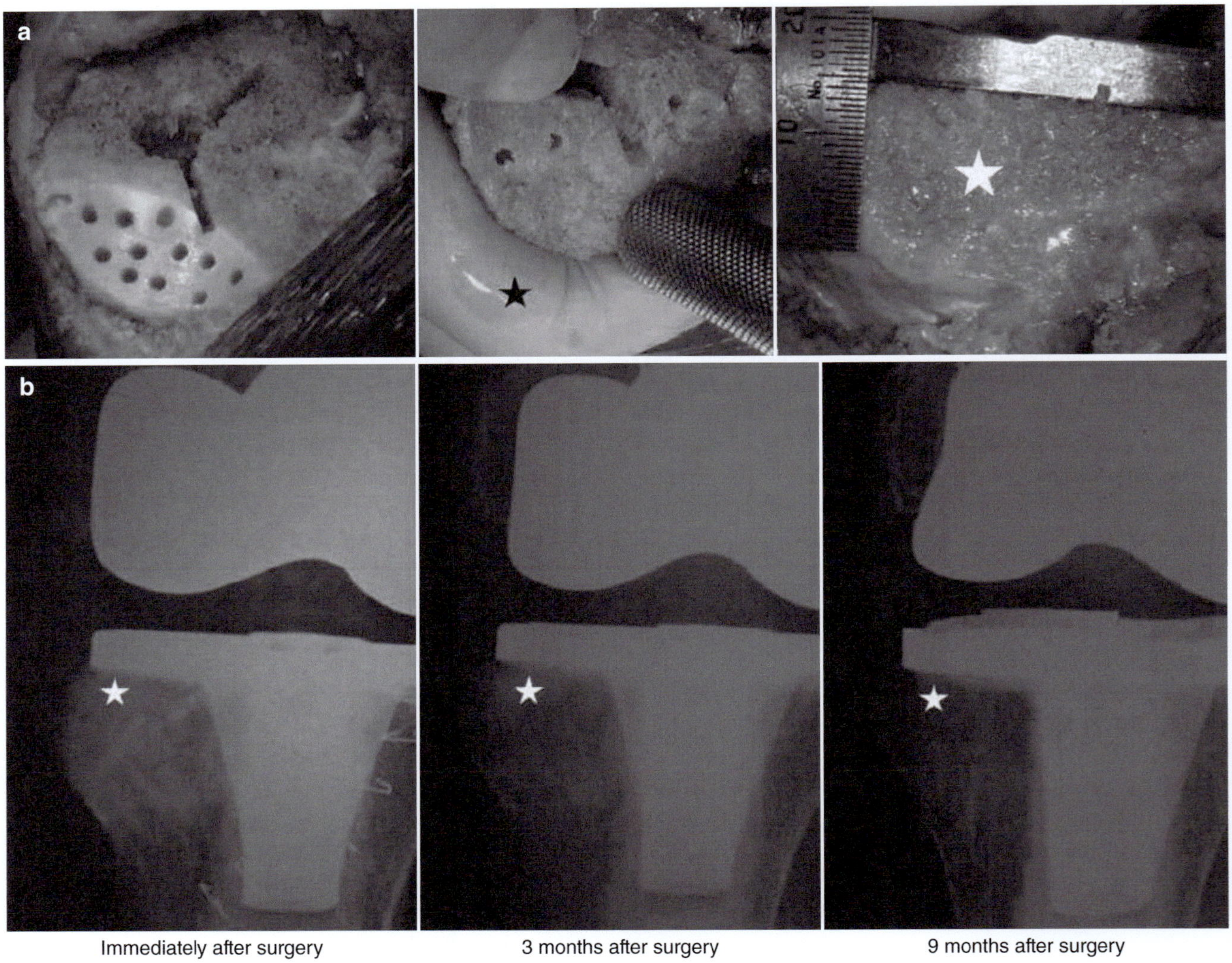

Fig. 77.1 Repair of a bone defect in the proximal tibia using morselized bone (autologous bone). (**a**) Drilling a hole in the wedge-shaped defect and compressing morselized bone made from autologous cancellous bone. The bone is formed using a hammer and the index finger. The implant is then cemented. (**b**) Changes over time at the transplant site. (Ref. 7)

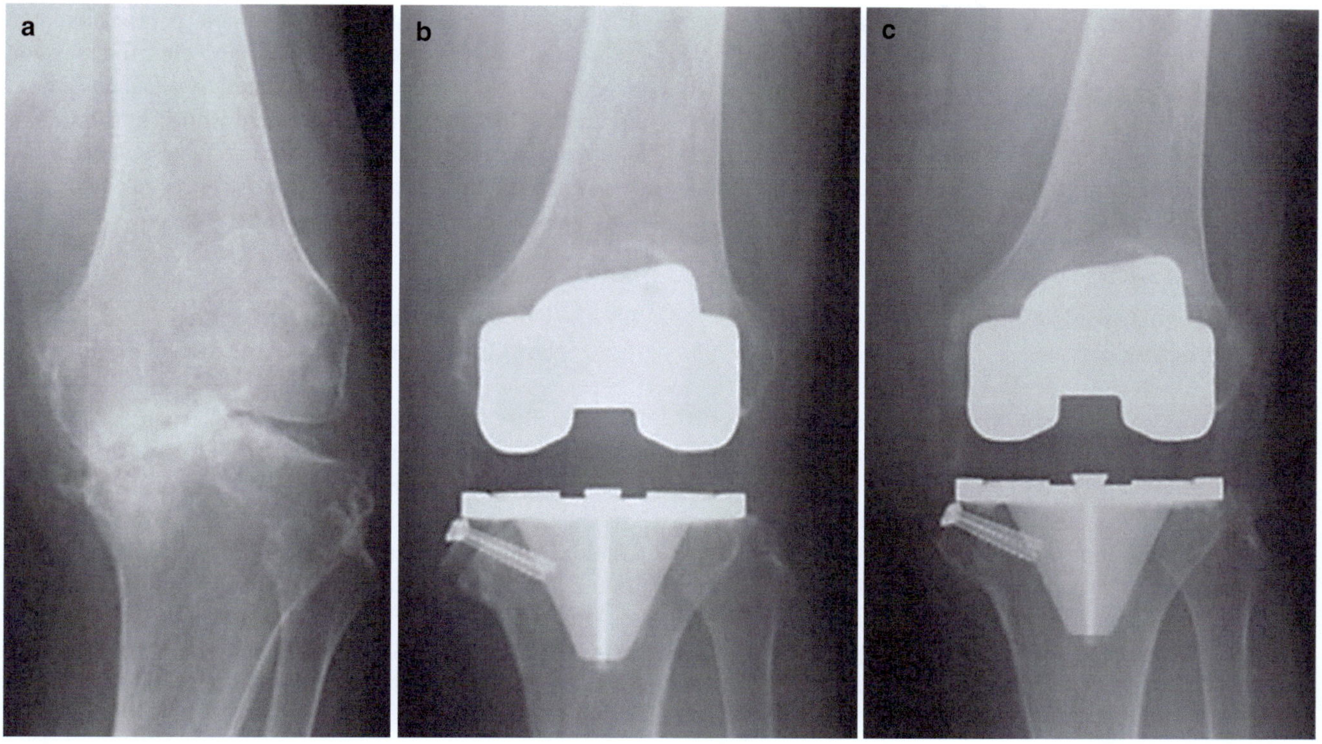

Fig. 77.2 A technique for stabilizing bone grafts for bone defects in the proximal tibia using screws. (**a**) Preoperative, (**b**) 1 week after surgery, (**c**) 5.5 years after surgery

fracture of the femoral condyle during the box cut or additional cuts. Therefore, positioning of the femoral condyle box cut and stem insertion must be precisely determined preoperatively. If the offset stems are selected, care should be taken in determining the insertion position to minimize the amount of offset, since it requires additional bone resection at the femoral condyle. Metal augmentations can distribute the load to the osteotomy surface by using a stem, and in cases of augmentation greater than 10 mm, the stem should be press-fit and canal-filled so that it is in contact with the cortex unless the tibia has thin cortical bone, in which case an offset stem should be considered. It is better to aim for a fit and fill even if it requires the use of an offset stem. Unlike the femur, osteotomy for tibial offset does not cause fracture. For long stems (>10 cm), cement should be applied to the tibial tray and stem junction but not to the distal portion of the stem. Constrained implants should also be made available because the bone defect may extend to the attachment of the collateral ligament.

4 Perioperative Osteoporosis Intervention

Postmenopausal women or men over 60 years of age should be evaluated for osteoporosis as follows: If there is low bone density or an existing fracture, osteoporosis treatment should be started before surgery. However, there is no need to delay the surgery for therapeutic effects. Bilateral TKA may performed on the same day for bilateral cases regardless of the presence or absence of osteoporosis [1].

4.1 Essential Examination and Important Imaging Findings

In the preoperative examination, osteoporosis should be evaluated by bone mineral density and the presence of previous fractures. According to the Japanese diagnostic criteria for osteoporosis, osteoporosis is diagnosed even in patients with high bone density if there is a previous vertebral fracture or a proximal femur fracture [8]. The Rotterdam cohort reported that OA patients are at high risk of vertebral and nonvertebral fractures despite high bone density not only around the knee joint but also in the proximal femur and vertebral body (Fig. 77.3) [9]. This means that OA patients are at high risk of vertebral and nonvertebral fractures. It is important to keep in mind that knee osteoarthritis patients have "osteoporosis with poor bone quality (hard and brittle bones)" that can cause fragility fractures despite high bone density.

4.2 Osteoporosis Cases: Intraoperative Precautions

Osteosclerosis on the osteotomy surface, represents the stiff and brittle bone, which implicates "poor bone quality" osteoporosis. Caution should also be exercised when there is no osteosclerosis on the osteotomy surface and there is lack of trabecular number and continuity. In this case, sufficient bone cement should be pressed into the keel area to adequately fill the bone defect. In many cases, a lucent line appears on the medial side of the tibial implant due to postoperative changes in lower limb alignment, so cementation of the keel is essential to prevent bone depression and debonding. Also extra care should be used when the assisting surgeon uses a retractor for intraoperative maneuvers, as unintended "defects" may occur unknowingly. In addition, if high-pressure gas by CO_2 bone preparation system is used to clean the bone prior to cementation, it is necessary to be cautious so as not cause microcracks in osteoporotic cases.

4.3 Postoperative Care: The Need for Osteoporosis Treatment

Bone fragility increases with age in both men and women due to increased bone remodeling with a predominance of bone resorption. Therefore the older age group undergoing TKA should be considered to be at a greater risk. Bone around implants continues to undergo remodeling throughout life. It is necessary to take measures against osteoporosis not only preoperatively but also postoperatively, bearing in mind that decreased bone metabolism around implants can cause loosening and peri-prosthetic fractures.

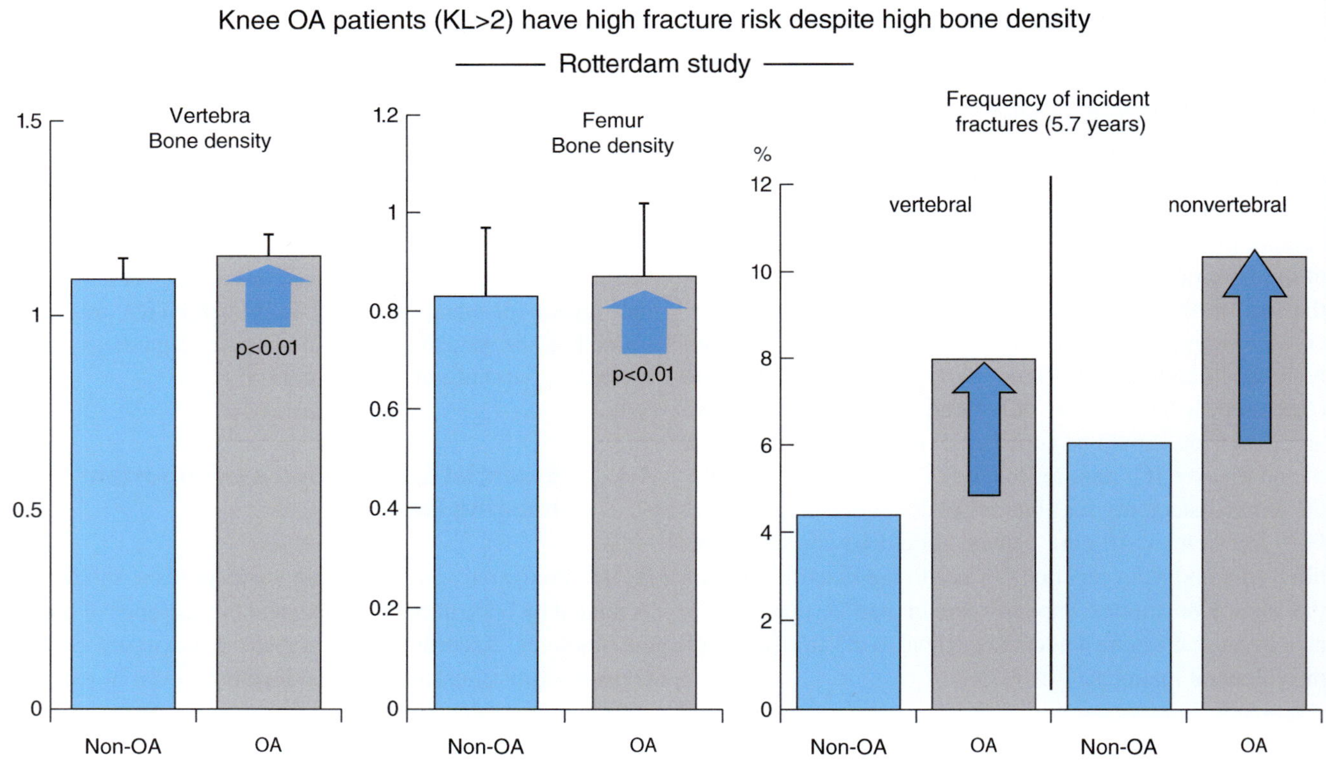

Fig. 77.3 OA patients have high fracture risk despite high bone density. OA patients are at high risk of vertebral and nonvertebral fractures despite high bone density not only around the knee joint but also in the proximal femur and vertebral bodies. OA patients have osteoporosis with "poor bone quality". (Ref. 9, 20)

4.4 Peri-prosthetic Fractures and Osteoporosis

Peri-prosthetic fractures in TKA often occur proximal to the femoral component, whereas in unicompartmental knee arthroplasty (UKA), fractures are more common just below the tibial component. Factors leading to peri-prosthetic fractures include preoperative bone fragility due to osteoporosis, changes in the mechanical environment due to decreased walking immediately after surgery followed by subsequent improvement in walking, and a decrease in bone density around the implant over time. Factors that may contribute to the loss of bone density around implants over time include implant design, cementation, and changes in alignment. In TKA, a decrease in bone mineral density around the femoral component has been reported in the anterior and central regions of the femur regardless of implant design or cement use [10–12]. Shimada et al. analyzed bone mineral density changes after TKA and reported a 6–10% decrease at 6 months postoperatively compared to 1 month postoperatively in the femur, and this did not recover even at 24 months postoperatively [13]. This trend is also observed in the tibia. In the same study, Shimada et al. reported that medial bone mineral density loss was significant in the tibia, with a residual loss of approximately 30–40% at 24 months postoperatively compared with 1 month postoperatively. The main reason for the medial bone mineral density loss is thought to be the reduced stress on the medial aspect of the tibia due to improved lower limb alignment due to correction of the varus knee. Further study is hence needed in the future.

The decrease in bone mineral density after TKA is most pronounced approximately 6 months postoperatively, and these changes are also associated with changes in bone resorption markers. The authors measured serum type 1 collagen cross-linked N-terminal telopeptide (NTx), a bone resorption marker, at 1 day, 3 days, 1, 2, 3, 4, 5, 6, 8, and 24 weeks after surgery in 22 women aged 70 to 84 (mean 74 years) who underwent unilateral TKA (cementless) [14]. The results showed that the postoperative serum NTX increased above the minimum significant change (MSC) value of 14.2% [14]. An increase of 14.2% or more was observed in 15 patients on the third postoperative day (68%) and the peak occurred between 3 and 4 weeks postoperatively. The increase ranged from 11% to 168%, with a mean of 62%. At 6 months postoperatively, 16 patients (73%) had decreased to the preoperative level of +14.2%, while the remaining 6 patients showed a decreasing trend but still had high values. However, bone resorption markers normalized at 6 months postoperatively, but bone scintigraphy showed increased bone resorption around the implants in these patients. Bone metabolism markers are merely averages of systemic metabolism, and therefore if bone resorption is increased only around implants, the metabolic markers do not necessarily show an increase (Fig. 77.4). Based on these facts, it may be necessary to use bone resorption inhibitors for longer than 6 months after surgery to normalize bone metabolism around the implants to achieve implant fixation. Population-based studies in Europe and the United States have already shown that the revision rate was significantly reduced in patients who underwent arthroplasty with oral bisphosphonates [15, 16]. In a report analyzing the association between oral bisphosphonate use and revision TKA in 18,726 patients who underwent TKA between 1986 and 2006 in the United Kingdom [15], the revision rate due to aseptic loosening at 5 years after surgery was 0.93% when oral bisphosphonates were used for at least 6 months after surgery, which was half that of the 1.96% in the untreated group. The long-term revision rate was similarly significantly lower in the bisphosphonate-treated patients (hazard risk 0.54; 95% confidence interval 0.29–0.99). This trend was also true for hip replacements performed at the same time in the same study. Similar results were obtained in a Danish population-based study [16]. The revision rate was significantly lower in patients treated with oral bisphosphonates than in those not treated with oral bisphosphonates (hazard risk 0.41; 95% confidence interval 0.27 to 0.61). The study concluded that bisphosphonate therapy for the prevention of complications after TKA should be started as early as possible after surgery, continued for at least 1 year, and with an adherence rate of at least 80%.

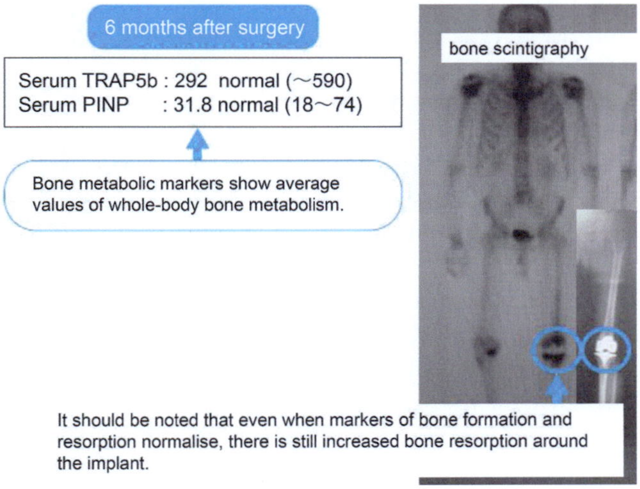

Fig. 77.4 Increased bone resorption after TKA surgery persists for more than 6 months. Bone resorption markers are elevated after TKA but normalize by 6 months. It is important to note that bone resorption and bone formation markers are averaged over the entire body's bone metabolism. In this case, bone formation and resorption markers normalized at 6 months postoperatively, but bone scintigraphy showed vigorous uptake around the implant. This means that even if bone metabolic markers normalize, abnormal bone metabolism continues around the implant. (Bone scintigraphy was performed for follow-up of a separate urological problem)

However, there was a concern that early postoperative use of bisphosphonates may inhibit bone ingrowth/ongrowth around the implants because bone formation is inhibited as well as bone resorption. However, Mochida et al. inserted a cementless prosthesis into a dog and administered alendronate, a bisphosphonate, immediately after surgery and found that no change was observed with regards to bone formation on the implant surface [17]. In other words, early postoperative use of bone resorption inhibitors, mainly bisphosphonates, may not only reduce systemic osteoporotic fractures but also inhibit bone resorption around implants, while maintaining bone density, as well as preventing peri-prosthetic fractures and aseptic loosening caused by bone fragility. In contrast, teriparatide, an osteogenesis-promoting agent, has been reported to promote fracture healing in peri-prosthetic fractures where adequate bone healing was not observed with low intensity pulsed ultrasound (LIPUS) alone [18]. In one case of bilateral TKA, simultaneous bilateral patellar fractures occurred during standard training [19]. This patient had diabetes mellitus, which causes bone quality deterioration despite a normal bone density [20]. Teriparatide was started every day from the day after surgery, as the intraoperative osteotomy revealed fragile bones. However, on postoperative day 21, she experienced severe pain in both knees while climbing stairs, resulting in bilateral patellar fractures. Conservative treatment was chosen, and LIPUS [21] was commenced in addition to the daily teriparatide, to accelerate fracture healing and improve bone quality. The combination of LIPUS and teriparatide was found to synergistically promote bone healing in a rat femur fracture model by Warden et al. [22]. Therefore the combination of LIPUS and teriparatide may be effective for patients with delayed healing.

In another case of loosening due to infection, teriparatide was commenced in advance to increase trabecular bone which achieve good bone fixation with the implant in revision TKA [23].

In conclusion, perioperative osteoporosis evaluation and drug treatment in TKA are necessary to improve long-term postoperative outcomes especially in Japan, where patients aim to live for 100 years.

References

1. Engh GA, et al. Bone defect classification. In: Engh GA, Rorabeck CH, editors. Revision total knee arthroplasty. Boltimore: Williams and Wilkins; 1997. p. 63–120.
2. Dorr LD, et al. Bone graft for tibial defect in total knee arthroplasty. Clin Orthop. 1986;205:153–65.
3. Watanabe W, et al. Autologous bone grafting without screw fixation for tibial defects in total knee arthroplasty. J Orthop Sci. 2001;6:481–6.
4. Takeda S, et al. Metal block augmentation for bone defects of the medial tibia during primary total knee arthroplasty. J Orthop Surg Res. 2013;208:36.
5. Sugita T, et al. Preliminary results of managing large medial tibial defects in primary total knee arthroplasty: autogenous morcellised bone graft. Int Orthop. 2017;41:931–7.
6. Sugita T, et al. Autologous morselised bone grafting for medial tibial defects in total knee arthroplasty. J Orthop Surg (Hong Kong). 2015;23:185–9.
7. Hosaka K, et al. Union knee alignment and clinical outcomes of patients treated with autologous bone grafting for medial tibial defects in primary total. Orthopedics. 2017;140:e604–8.
8. Saito M, Marumo K, et al. Future surgical techniques – bilateral same-day TKA. In: Ishibashi Y, editor. Perfect knee replacement. Kyoto: Kinpo-Do; 2016. p. 218–22. [Japanese].
9. Bergink AP, et al. Osteoarthritis of the knee is associated with vertebral and nonvertebral fractures in elderly individuals: the Rotterdam study. Arthritis Rheum. 2003;15(49):648–57.
10. Spittlehouse AJ, et al. Measurement of bone mineral density by dual-energy X-ray absorptiometry around an uncemented knee prosthesis. J Arthroplast. 1999;14:957–63.
11. Petersen MM, et al. Changes in bone mineral density of the distal femur following uncemented total knee arthroplast. J Arthroplast. 1995;10:7–11.
12. Soininvaara TA, et al. Periprosthetic femoral bone loss after total knee arthroplasty: a 1-year follow-up study of 69 patients. Knee. 2004;11:297–302.
13. Shimada K. Secular changes in bone mineral density around artificial joints. Jpn J Arthroplasty. 2006;36:210–1.
14. Tanaka T, et al. Serum type I collagen cross-linked N-telopeptide (NTX) levels in osteoporosis patients – effects of fracture. Medecine. 2005;59:604–7.
15. Prieto-Alhambra D, et al. Association between bisphosphonate use and implant survival after primary total arthroplasty of the knee or hip. BMJ. 2011;343:d7222.
16. Prieto-Alhambra D, et al. Oral bisphosphonate use and total knee/hip implant survival: validation of results in an external population-based cohort. Arthritis Rheumatol. 2014;66:3233–40.
17. Mochida Y, et al. Alendronate does not inhibit early bone apposition to hydroxyapatite-coated total joint implants: a preliminary study. Joint Surg Am. 2002;84:226–35.
18. Ochi K, et al. Administration of teriparatide treatment for a challenging case of nonunion of periprosthetic fracture after total knee arthroplasty. Arch Osteoporos. 2013;8:159.
19. Yonemoto K, et al. A case of bilateral patellar fracture due to external medial force after bilateral knee arthroplasty. Bone Joint Nerve. 2005;6:219–22.
20. Saito M, et al. Collagen cross-links as a determinant of bone quality: a possible explanation for bone fragility in aging, osteoporosis, and diabetes. Osteoporos Int (Review). 2010;21:195–214.
21. Saito M, et al. Intensity-related differences in collagen posttranslational modification in MC3T3-E1 osteoblasts after exposure to low and high. Bone. 2003;35:644–55.
22. Warden SJ, et al. Recombinant human parathyroid hormone (PTH 1-34) and low-intensity pulsed ultrasound have contrasting additive effects during. Bone. 2009;44:485–94.
23. Nishikawa M, et al. Bone stock reconstruction for huge bone loss using allograft-bones bone marrow and teriparatide in an infected total knee arthroplasty. J Clin Orthop Trauma. 2019;10:329–33.

78 Severe Instability

Ryuichi Gejo

1 Managing Instability

In primary total knee arthroplasty (TKA), instability can be addressed in most cases with cruciate-retaining (CR) and posterior-stabilized (PS) standard implants by achieving good alignment and proper soft tissue balancing. However, a more rigid restraining implant may be necessary in cases of severe instability or hyperextension with significant collateral ligament dysfunction or when soft tissue balance cannot be properly achieved intraoperatively.

2 Implant Restraints

The movement of the tibiofemoral joint has six degrees of freedom, which are represented by translation in the anteroposterior, medial/lateral, and distal directions, and rotation around the axes, i.e., extension-flexion, varus-valgus, and internal-external rotation. In addition to the morphology of the bone itself and the meniscus, the tissues that control the degrees of freedom of movement are mainly ligaments and joint capsules, and their failure causes knee joint instability.

If the cause of joint instability in TKA is a bone defect, it can be corrected by replacing the defect with a bone graft or by augmentation (see the previous section on bone defects). However, joint instability should be controlled by the restraining qualities of the appropriate implants if it is brought on by bone loss that extends to the ligamentous attachments, extensive soft tissue release caused by severe deformity or contracture, ligament dysfunction, or joint capsule laxity.

Implants have various degrees of restraint, ranging from no restraint to firm restraint, depending on the shape of the joint surface [1]. The various degrees of restraint are described in the following examples: Partial knee arthroplasty, CR, and bicruciate-retaining, in which the implants themselves have almost no restraint; cruciate-substituting, in which the surface shape provides anterior-posterior restraint; and PS and bicruciate-substituting, in which the post-cam mechanism provides anterior-posterior restraint. These models do not provide restraint for varus/valgus and internal/external rotation; therefore, they cannot be used for knees with medial or lateral instability, and the following models are required:

Semiconstrained PS-type implants with a large tibial post have recently emerged as implants that can accommodate moderate collateral ligament insufficiency with a certain degree of varus/valgus and internal/external rotation constraint. Good mid-term results without stem extension into the medulla [2] and reproduction of natural kinematics in deep bending [3] have been reported. There is also a constrained condylar knee that strongly constrains varus/valgus and internal/external rotation and a rotating hinge that constrains extension, used in cases of high instability in multiple directions (Table 78.1).

R. Gejo (✉)
Department of Orthopaedic Surgery, Faculty of Medicine, University of Toyama, Toyama, Japan
e-mail: rgejo@med.u-toyama.ac.jp

S. Matsuda (ed.), *Advances in Total Knee Arthroplasty*, https://doi.org/10.1007/978-981-97-4920-1_78

Table 78.1 Constraints and indications for each implant

	Constraint				Ligament insufficiency			
Implant	AP	VV	IR	Extension	PCL	MCL	LCL	Recruvatum
CS, PS, BCS	+				+			
Semi-constrained PS	+	+	+		+	±		
							±	
Constrained condylar knee	+	+	+		+	+		
							+	
Rotating hinge	+	+		+	+	+	+	+

AP anterior-posterior, *VV* varus-valgus, *IR* internal-external rotation

3 Constrained Condylar Knee (CCK)

3.1 CCK Characteristics

The CCK implant provides knee joint stability through the post-cam mechanism of the polyethylene liner. The tibial post, which is higher and wider than the PS, fits snugly between the condyles of the femoral component, constraining the posterior and lateral translations, varus and valgus, and internal and external rotations (Fig. 78.1). First-generation CCK implants are nonmodular and have a high rate of aseptic loosening [4] and problems with patella-related complications such as poor tracking, fracture, and osteonecrosis [5]. Second-generation CCK implants improved the patellofemoral joint surface shape, introduced a modular-type stem for the tibia and femur, and introduced a new locking mechanism for polyethylene liners. Satisfactory clinical results and radiographic findings have been reported [6].

3.2 CCK Indications

Indications for CCK include bone defects extending to the medial collateral ligament (MCL)/lateral collateral ligament (LCL) attachments, MCL or LCL insufficiency, and valgus deformities with medial laxity. CCK is also considered when extensive release is necessary because of a significant varus or valgus deformity or contracture resulting in the loss of soft tissue balance. Residual medial instability has a negative impact on postoperative activities of daily living and patient satisfaction [7]. Therefore, careful attention should be paid to the acquisition of medial stability. Although there is no consensus on what degree of instability is indicated for CCK, it has been reported that a restrictive design should be used when intraoperative medial or lateral ligament balance cannot be achieved and laxity exceeds 7–10 mm [8].

3.3 Precautions During Surgery

Implant loosening is the greatest concern associated with the use of CCK. As the restraint of the articular surface increases, the stress on the implant-bone interface theoretically increases, which may lead to implant loosening in the long term. Therefore, using an intramedullary extension stem is recommended to distribute stress on the implant-bone interface and strengthen the fixation of the implant [9]. Moussa et al. reported that the revision rate due to implant loosening tends to be higher for CCKs without stems than those with stems, even in the short term of 2 years [10].

3.4 Case Presentation

A 73-year-old male had secondary traumatic osteoarthritis (Fig. 78.2). The preoperative joint range of motion (ROM) was 5–120°. Although the details of the injury are unknown, a physical examination revealed posterolateral and anteroposterior instability. Intraoperatively, the anterior cruciate ligament disappeared, the posterior cruciate ligament was scarred and dysfunctional, and the MCL was intact. However, the LCL and posterolateral structures were avulsed with small bony fragments (Fig. 78.3), with lateral capsule laxity. Component gap measurements at 30 pounds using a seesaw-type tensor showed a center gap of 12 mm with a 5° varus gap tilt angle at 0° knee extension. A center gap of 20 mm with an 11° varus gap tilt angle at 90° knee flexion indicated an imbalance in extension-flexion and medial-lateral directions, with a significantly increased flexion-lateral gap (Fig. 78.4). Therefore, a CCK-type component was selected, and a tibial stem was used (Fig. 78.5). The postoperative joint ROM was satisfactory, ranging from 0° to 135°, with no posterior or lateral instability in the total ROM.

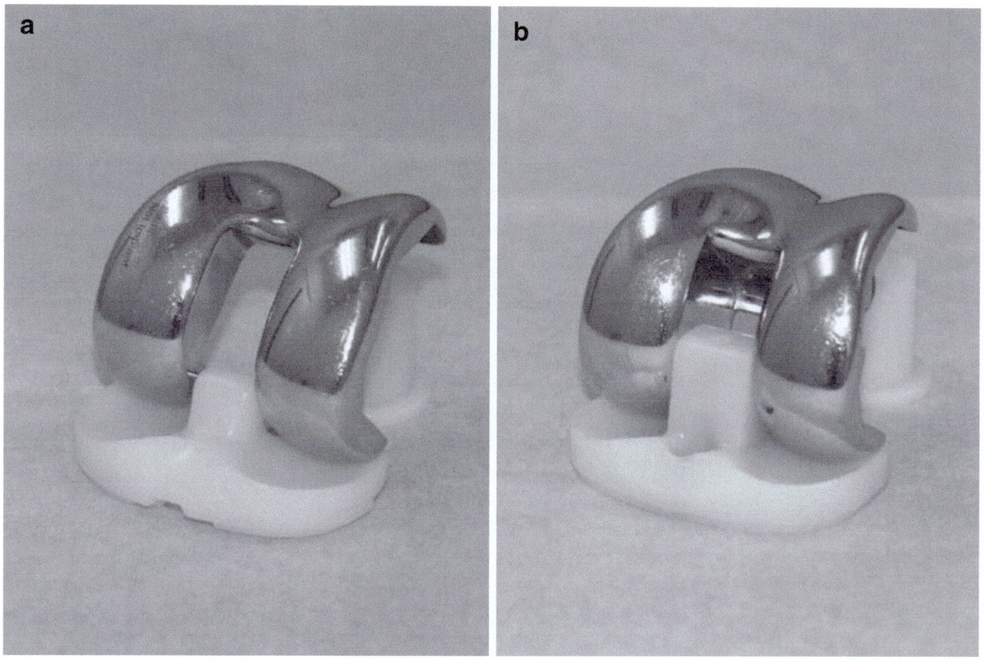

Fig. 78.1 CCK (right) has a higher and wider tibial post than PS (left)

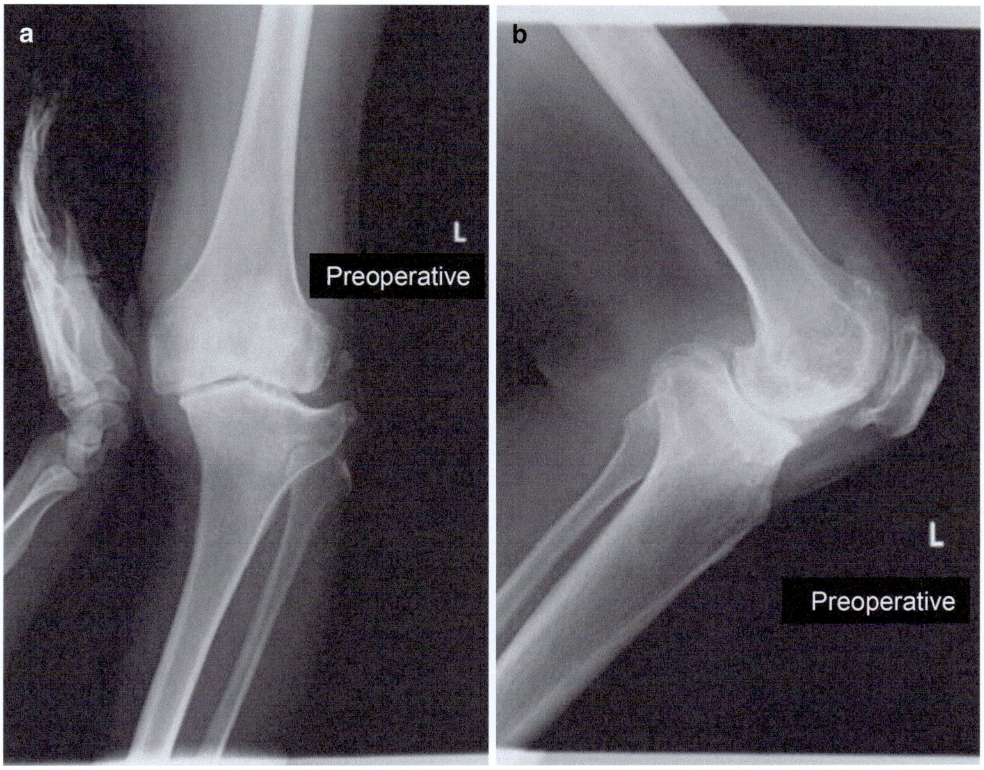

Fig. 78.2 Preoperative radiograph. Lateral and posterior instability is observed

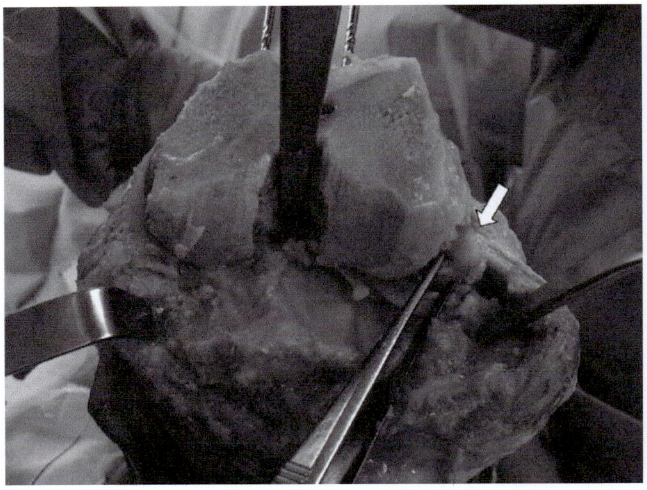

Fig. 78.3 LCL and posterolateral structures detached from condyle with small bone fragments (arrows)

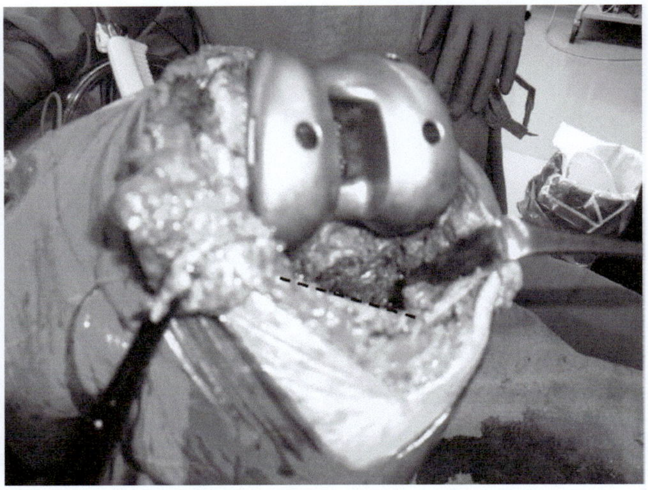

Fig. 78.4 The flexion lateral gap was markedly opened (dotted line is the tibial osteotomy plane)

Fig. 78.5 Postoperative radiograph. Arthroplasty was performed using CCK with the long tibial stem

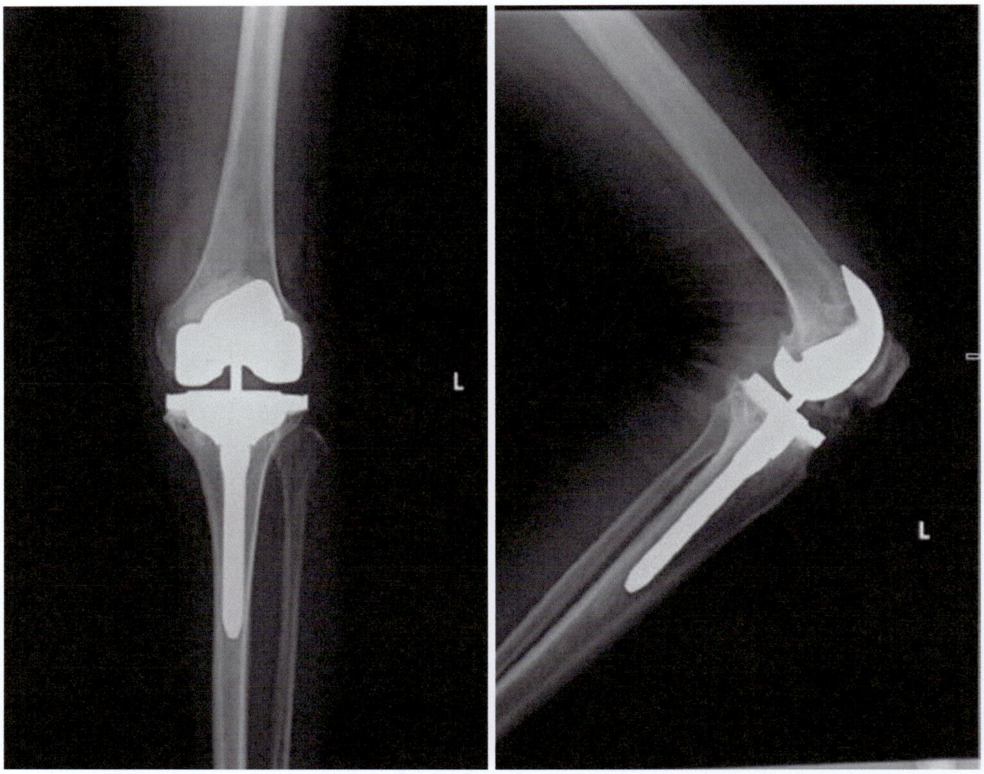

3.5 Postoperative Results

The short- to mid-term postoperative results of CCK are excellent, with clinical success reported at an average of 49.4 months after revision TKA [11]. The implant survival rate is 91.7% at an average of 7 years after primary TKA for rheumatoid arthritis with severe deformity and joint instability [12]. The 10-year survival rate is 90.0%, and the 20-year survival rate is 72.8% [13], but there are still concerns regarding long-term durability.

4 Rotating Hinge

4.1 Characteristics of Hinge Design Implants

In the 1950s, uniaxial hinges were developed, in which everything but flexion was constrained, and the prosthesis provided stability [14, 15]. However, simple hinged implants cannot reproduce complex three-dimensional knee joint motion, and increased loading on the implant-bone interface is inevitable, resulting in a high rate of loosening and other problems. Recently, a rotating hinge (Fig. 78.6), a biaxial hinge that allows movement in the rotation and pull-off directions, was developed and used. The load is applied by polyethylene on the tibial articular surface, as in the case of conventional implants. The mobile mechanism allows internal and external rotation, thus avoiding torsional loading on the implant-bone interface and soft tissues. Hinge-design implants also allow hyperextension to be controlled by restrained extension.

4.2 Rotating-Hinge Indications

The indications include bone defects extending to the MCL and LCL attachments, MCL and LCL insufficiency, significant joint laxity, and joint laxity due to neuromuscular disorders, such as poliomyelitis, Charcot's joint [16], and ankylosing knees [17]. For genu recurvatum, if osteoarthritis is the primary pathology, satisfactory results without recurrence have been reported using a conventional implant with a thin distal femoral osteotomy and a thick insert [18]. In contrast, patients with neuromuscular disorders and other conditions that cause weak quadriceps muscle strength always walk with the knee joint hyperextended to compensate for the lack of extensor muscle strength, resulting in progressive laxity of the posterior joint capsule and genu recurvatum. Such cases are difficult to treat using models that cannot control extension, and a rotating hinge is indicated.

4.3 Precautions During Surgery

Rotating hinges are highly constrained within the joint, resulting in high implant-bone interface stress. Therefore, stem extension into the medulla is essential for stress dispersion and implant fixation. Finite element analyses of stem fixation methods have shown that cementation results in less implant micromotion and better stress dispersion over the entire stem than press-fitting [19], and cementation of a long stem is recommended for use with a rotating hinge.

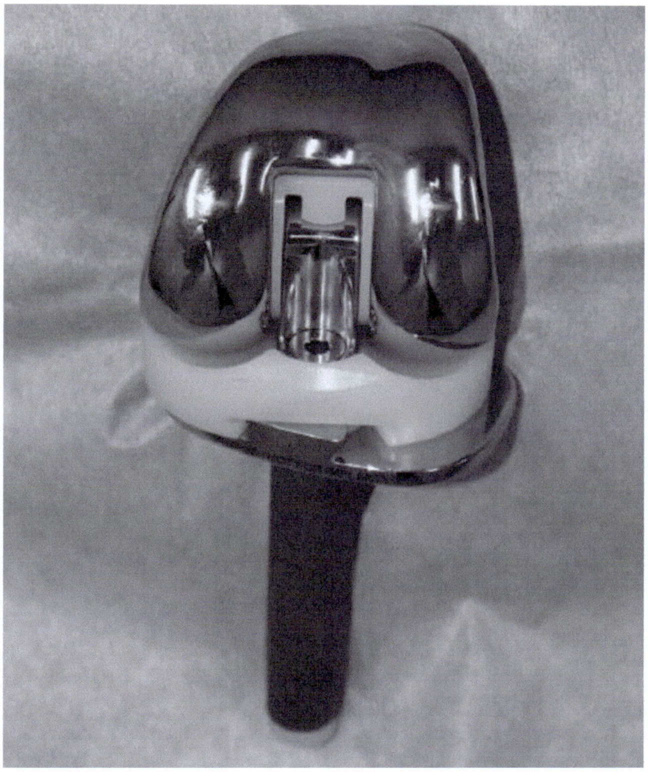

Fig. 78.6 Rotating hinge. The load is applied with polyethylene and shows rotational tolerance because of the mobile mechanism

4.4 Postoperative Results

Although the cumulative survival rate of implants at an average of 13 years post-surgery is relatively satisfactory (90%) [20], long-term results are not guaranteed, as reported to be 74.6% at 10 years and 40.3% at 20 years postoperatively [13]. The frequency of postoperative complications is also higher than that of conventional implants [21]. Therefore, using rotating hinges requires careful long-term follow-up and should be considered carefully, except in unavoidable cases.

References

1. Lombardi AV Jr, et al. Posterior-stabilized constrained total knee arthroplasty for complex primary cases. J Bone Joint Surg Am. 2007;89:90–102.
2. Crawford DA, et al. Midlevel constraint without stem extensions in primary total knee arthroplasty provides stability without compromising fixation. J Arthroplast. 2018;33:2800–3.
3. Sumino T, et al. Semiconstrained posterior stabilized total knee arthroplasty reproduces natural deep knee bending kinematics. BMC Musculoskelet Disord. 2020;21:107.
4. Donaldson WF 3rd, et al. Total condylar III knee prosthesis: a long-term follow-up study. Clin Orthop Relat Res. 1988;226:21–8.
5. Lachiewicz PF, et al. Ten-year survival and clinical results of constrained components in primary total knee arthroplasty. J Arthroplast. 2006;21:803–8.

6. Lachiewicz PF, et al. The results of a second-generation constrained condylar prosthesis in primary total knee arthroplasty. J Arthroplast. 2011;26:1228–31.
7. Tsukiyama H, et al. Medial rather than lateral knee instability correlates with inferior patient satisfaction and knee function after total knee. Knee. 2017;24:1478–84.
8. Sculco TP. The role of constraint in total knee arthroplasty. J Arthroplast. 2006;21:54–6.
9. Rawlinson JJ, et al. Cancellous bone strains indicate efficacy of stem augmentation in constrained condylar knees. Clin Orthop Relat Res. 2005;440:107–16.
10. Moussa ME, et al. Clinical outcomes following the use of constrained condylar knees in primary total knee arthroplasty. J Arthroplast. 2017;32:1869–73.
11. Nakano N, et al. Revision total knee arthroplasty using a modern constrained condylar knee prosthesis. Acta Ortop Bras. 2016;24:304–8.
12. Fujiwara T, et al. Mid-term clinical outcome of constrained condylar knee prosthesis for patients with rheumatoid arthritis. Mod Rheumatol. 2019;29:596–601.
13. Martin JR, et al. Complex primary total knee arthroplasty: long-term outcomes. J Bone Joint Surg Am. 2016;98:1459–70.
14. Shiers LG. Arthroplasty of the knee: preliminary report of a new method. J Bone Joint Surg Br. 1954;36:553–60.
15. Walldius B. Arthroplasty of the knee with an endoprosthesis. Acta Chir Scand. 1957;113:445–6.
16. Yoshikawa R, et al. Staged bilateral total knee arthroplasty in a patient with charcot knees: a case report. J Orthop Case Rep. 2020;10:9–12.
17. Nishitani K, et al. Rotating-hinge total knee arthroplasty in a patient with genu recurvatum after osteomyelitis of the distal femur. J Arthroplast. 2007;22:630–3.
18. Seo SS, et al. Outcomes of total knee arthroplasty in degenerative osteoarthritic knee with genu recurvatum. Knee. 2018;25:167–76.
19. El-Zayat BF, et al. Fixation techniques and stem dimensions in hinged total knee arthroplasty: a finite element study. Arch Orthop Trauma Surg. 2016;136:1741–52.
20. Gehrke T, et al. The role of hinges in primary total knee replacement. Bone Joint J. 2014;96:93–5.
21. Guenoun B, et al. Complications following rotating hinge Endo-Model (Link) knee arthroplasty. Orthop Traumatol Surg Res. 2009;95:529–36.

79 Patellar Dislocation/Subluxation

Yasuyuki Ishibashi

1 Pathophysiology of Patellar Dislocation and Subluxation

Malalignments of the patellofemoral (PF) joint, such as recurrent patellar dislocation/subluxation, habitual patellar dislocation, and permanent patellar dislocation, cause uneven and excessive loading on the PFJ and result in PF osteoarthritis (OA) and potentially lead to femorotibial (FT) OA. Such patients often exhibit anatomical variations, such as trochlea dysplasia and valgus knee, which are predisposing factors for dislocation (Table 79.1). The anatomy of patients with PF-OA differs significantly from that of patients with medial OA, which is more common in Japan. Therefore, special caution should be taken during their surgeries.

Postoperative patellar dislocation or subluxation after total knee arthroplasty (TKA) are relatively frequent complications [1, 2]. These complications can adversely affect the postoperative outcome of TKA, necessitating revision surgery. The causes of postoperative patellar dislocation include the patients' anatomical anomalies, component design, and errors in surgical technique (Table 79.2).

Table 79.1 Predisposing factors for PFJ alignment abnormalities

Bone	Soft tissue
Increased femoral anteversion	Soft tissue imbalance (tightness of the lateral retinaculum, etc.)
Valgus knee	Length of the patellar tendon (patella alta)
Hypoplasia of the trochlear groove	General joint laxity
Lateralized tibial tuberosity	Recurvatum knee
External rotation of the proximal tibia	
Hypoplasia of the patella	

Table 79.2 Factors contributing to PFJ dislocation after TKA

Factors contributing to PFJ dislocation
Anatomical factors (Table 79.1)
Implant design
Surgical approach (medial retinaculum resection)
Medialized femoral component
Internal rotated femoral component
Internal rotated tibial component
Lateralized patellar component
Inadequate patellar preparation (amount/angle of bone resection)

2 Treatment of Patellofemoral Osteoarthritis with Patellar Subluxation: PFJ Replacement

Patellofemoral arthroplasty (PFA) is mainly indicated for isolated PF-OA, with favorable long-term results [3, 4]. Although PFA is contraindicated in patients with severe PF instability, maltracking, or malalignment, positive results have been reported by combining medial patellofemoral ligament (MPFL) reconstruction with PFA [3–5].

For PFA, both the medial and lateral parapatellar approaches can be used. However, the lateral parapatellar approach, which allows simultaneous lateral retinacular release (LRR), is more suitable for patients with patellar dislocation. A previous study comparing the medial and lateral approaches for PFA in PF-OA showed no difference in clinical outcomes, but the lateral approach had better postoperative alignment [6]. The shape of PFA helps reproduce the normal patellar groove (metallic trochleoplasty), improves congruity, and occasionally enhances PF tracking. However, if the tendency to dislocate remains after PFA, MPFL reconstruction or tibial tubercle transfer should be considered (Figs. 79.1 and 79.2) [5, 7]. In cases of PF-OA caused by patellar dislocation, identifying the Whiteside line can be challenging due to the presence of trochlea dysplasia and significant PF-OA. It is important to determine the position and direction of the femoral component by comprehensively evaluating the epicondylar and femoral axes.

Y. Ishibashi (✉)
Department of Orthopedic Surgery, Hirosaki University Graduate School of Medicine, Hirosaki, Japan
e-mail: yasuyuki@hirosaki-u.ac.jp

S. Matsuda (ed.), *Advances in Total Knee Arthroplasty*, https://doi.org/10.1007/978-981-97-4920-1_79

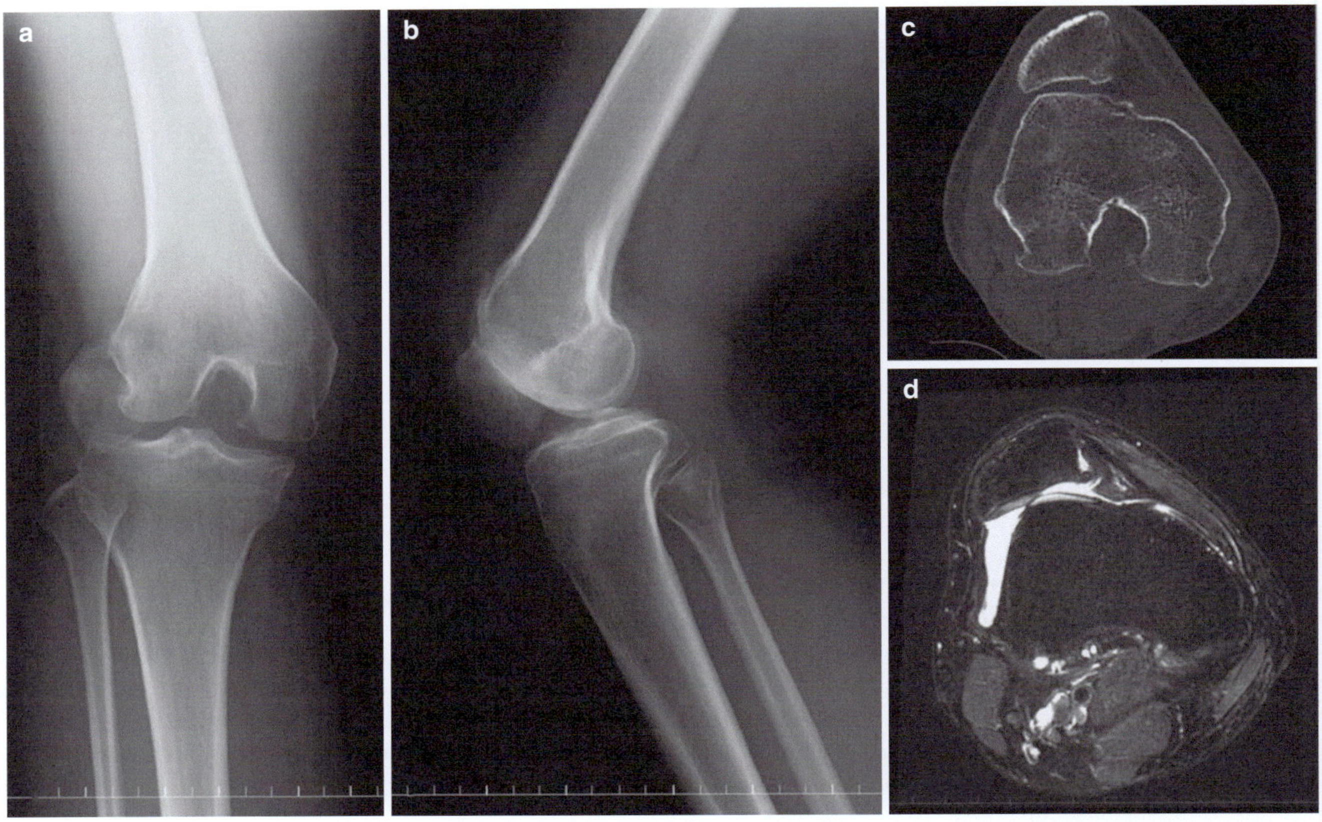

Fig. 79.1 PFJ-OA due to habitual patellar dislocation (46-year-old woman). In the first year of junior high school, the right patella was dislocated for the first time, followed by repetitive patellar dislocation. At 40 years old, the patella was easily dislocated laterally in flexion, which caused pain. (**a**) Patella is dislocated laterally at intercondylar fossa imaging. (**b**): Lateral view showing thick patella and pulley dysplasia. (**c**) CT (in extension) shows the patella in a repositioned position but with an outward tilt. (**d**) MRI shows cartilage defects on the lateral patellar articular surface and lateral side of the glenoid

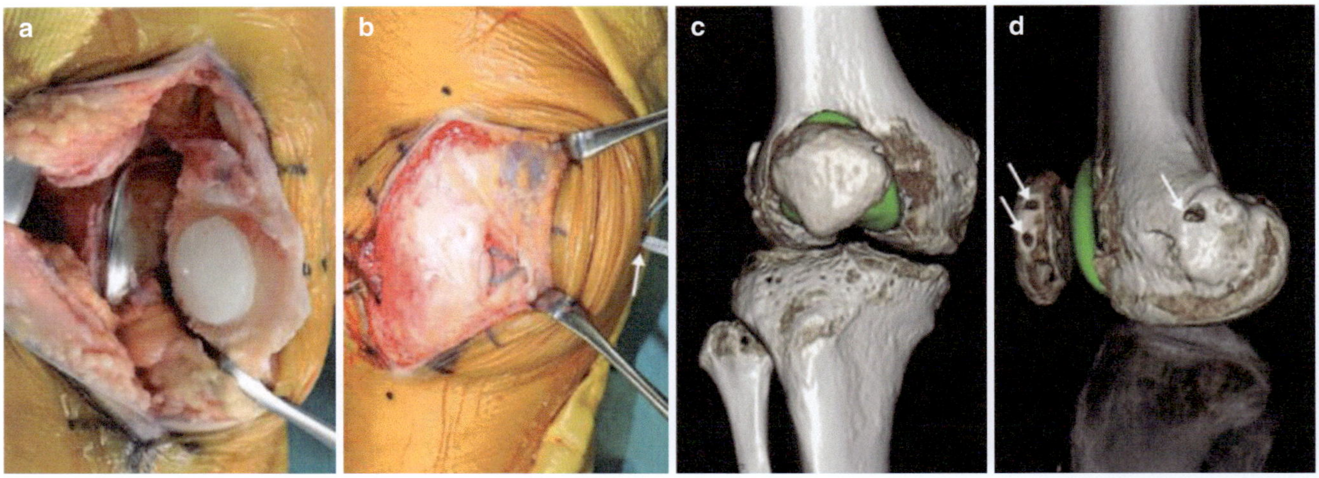

Fig. 79.2 PFJ replacement for a case of habitual patellar dislocation (the same case as Fig. 79.1). (**a**) The skin incision was made through a median longitudinal incision and developed through a lateral parapatellar approach, and PFA was performed. (**b**) Medial augmentation using FiberTape and SwiveLock according to the MPFL reconstruction method for repetitive patellar dislocation (Ishibashi). Arrowheads indicate small skin incisions to fix the FiberTape at the Schöttle point. (**c, d**) Postoperative X-ray images

3 Treatment of Knee OA with Patellar Dislocation: Total Knee Arthroplasty

Elderly patients with FT-OA and those with limited joint ROM are good candidates for TKA (Fig. 79.3). Knee OA associated with permanent patellar dislocation is also a good indication for TKA [8–11]. Although some previous reports suggested leaving the dislocated patella during TKA, correcting patella tracking is essential to improve knee extension function [9]. Preoperative assessment of various predisposing risk factors for patellar dislocation is crucial to prevent postoperative complications.

In cases of patellar dislocation, regardless of whether it is associated with medial or lateral FT-OA, a lateral approach is recommended due to the overconstraint of the lateral patellar retinaculum [10] (Fig. 79.4). Although V-Y quadricepsplasty and tibial tubercle osteotomy have been reported for difficult cases, these procedures should be avoided if possible due to

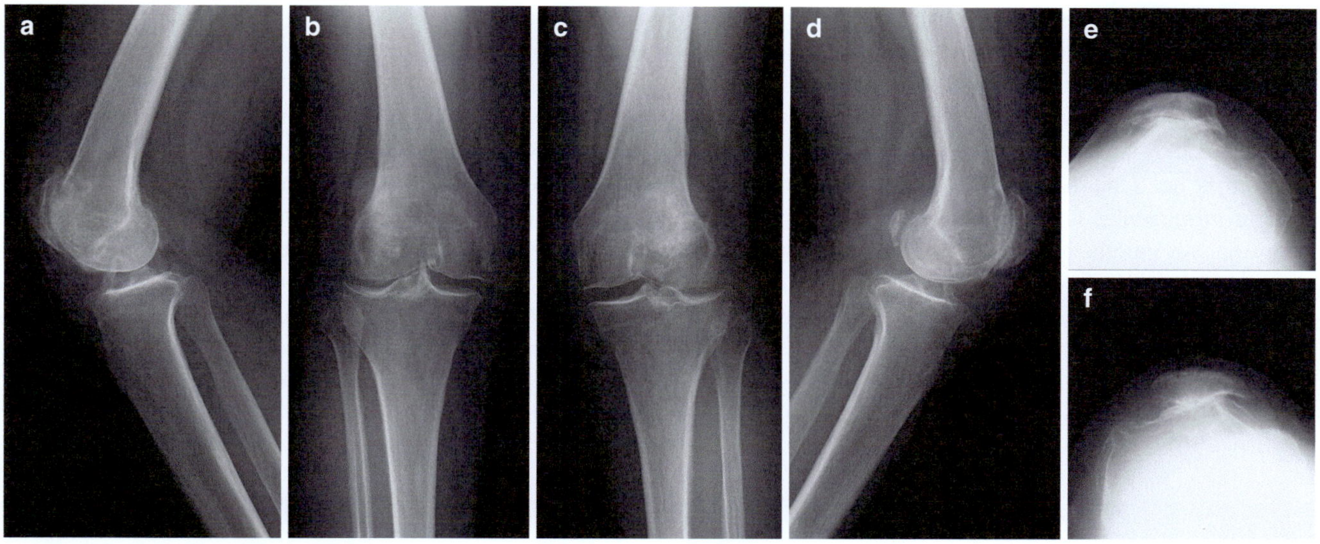

Fig. 79.3 OA due to habitual patellar dislocation (59-year-old woman). In her 50s, she could not sit upright and had difficulty extending her knees. Knee ROM was −30°/120° on the right and −25°/95° on the left. Preoperative plain radiographs of the right knee (**a**, **b**, **e**) and left knee (**c**, **d**, **f**)

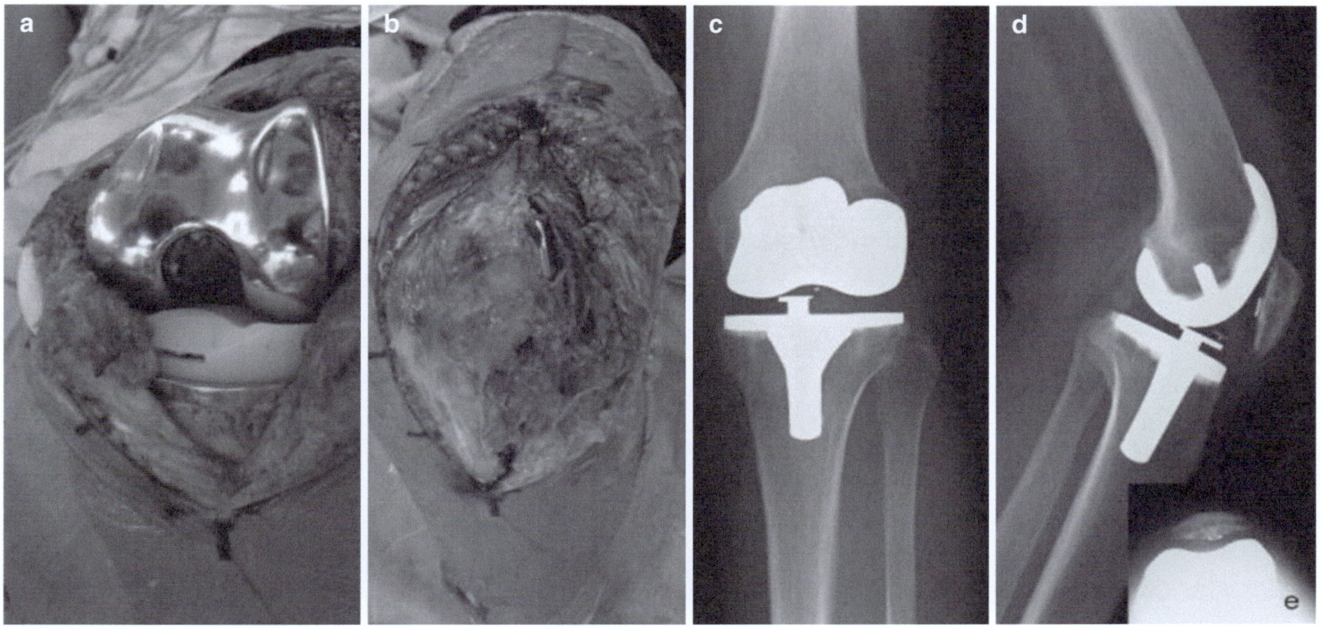

Fig. 79.4 TKA for habitual patellar dislocation (same case as Fig. 79.3). (**a**) TKA was performed on both knees via a lateral approach (left knee shown). (**b**) The incised lateral tibial band was not sutured, and the TKA was covered with a subpatellar fat body. Intraoperative evaluation showed good patellar tracking, so MPFL reconstruction was not performed. (**c**–**e**) Simple radiographs 3 years after surgery

their frequent complications. If necessary, a lateral snip of the quadriceps tendon should be performed.

The usual surgical approach can be used for medial FT-OA. However, special care must be taken to address risk factors and prevent postoperative dislocation, especially in lateral FT-OA cases with valgus deformities. In the coronal plane, valgus alignment should be avoided, and the rotation of the femoral component should be carefully decided because of the hypoplasia of the lateral femoral posterior condyle [9] (for more details, refer to the section on highly externally displaced knees). Special attention should be paid to component placement, avoiding medial placement of the femoral component, and internal rotation of the tibial component as these can increase the postoperative Q angle (Table 79.2).

After placing all components, the separated lateral retinaculum is not repaired, and the components are covered with an infrapatellar fat pad. Patellar tracking is evaluated using the 'no thumb' technique, and if necessary, realignment procedures such as medial retinacular reefing, MPFL reconstruction [8, 12], and tibial tubercle transfer are performed. However, the tibial tubercle transfer technique should be performed with great caution in elderly osteoporotic patients because of the risk of prolonged fusion and fracture. For such cases, MPFL reconstruction using an artificial ligament seems an easier and more reliable alternative (Fig. 79.4).

4 Treatment of Patellar Luxation After TKA

If patellar dislocation occurs after TKA, a plain X-ray or CT scan should be taken to examine for malalignment or component placement problems that may cause dislocation (Fig. 79.5). Salvage surgeries, such as revision or corrective osteotomy, are necessary in cases of malalignment or procedural problems [13, 14]. However, revising a femoral component placed in internal rotation poses technical difficulties. In such cases, tibial tubercle transfer could be considered. When there is no malalignment or component malalignment requiring revision, additional procedures like tibial tubercle transfer [15] or MPFL reconstruction using autologous tendons [1, 16, 17] or artificial ligaments [8] may be useful in addition to LRR. Considering the invasiveness of surgery, MPFL reconstruction is considered technically easier (Fig. 79.6).

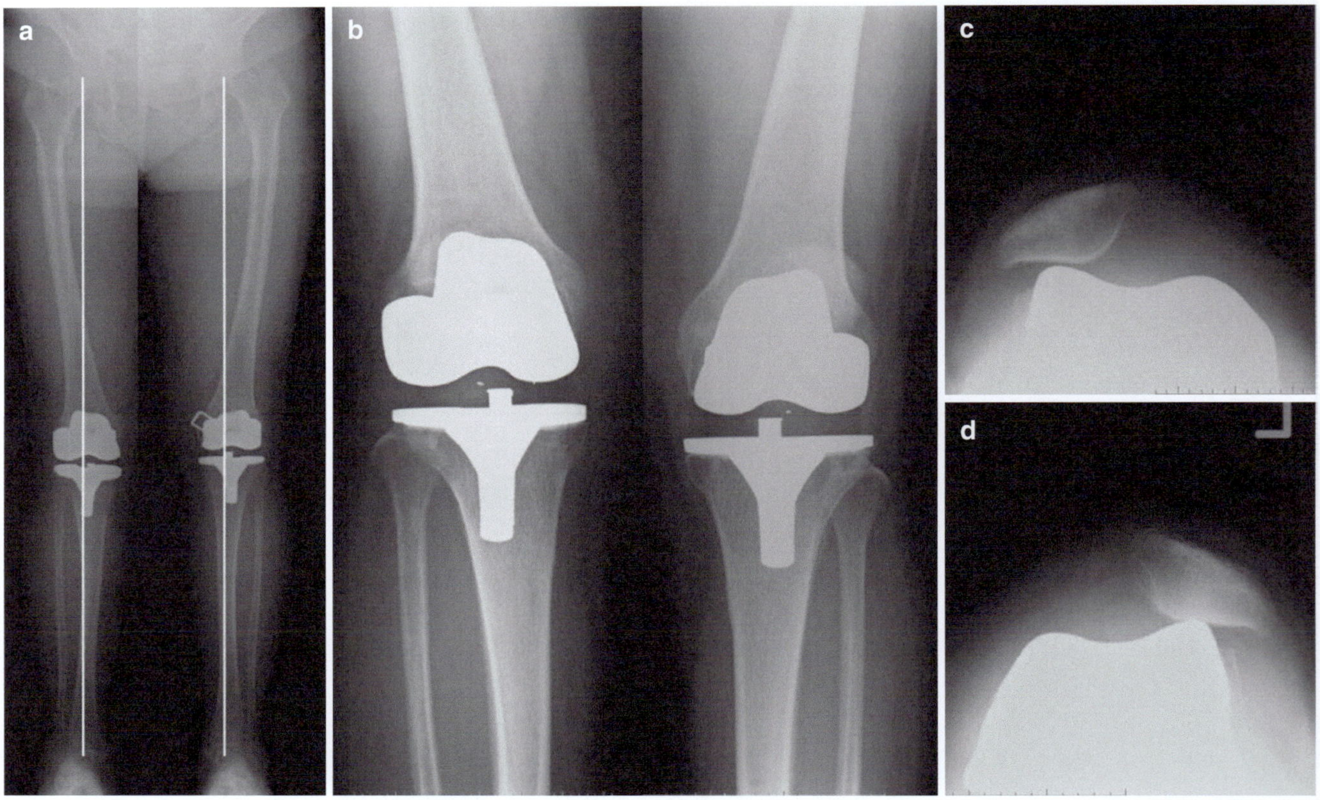

Fig. 79.5 Patellar subluxation after TKA (67-year-old woman). A patient with knee OA with femoral kyphosis and lower leg kyphosis underwent TKA via a medial parapatellar approach using navigation. The Mikulicz line runs almost through the center of the knee (**a**), but the local FTA is externally inclined (**b**). Both knees had residual postoperative anterior knee pain, and axial projection images showed subluxation of the patella (**c**, **d**)

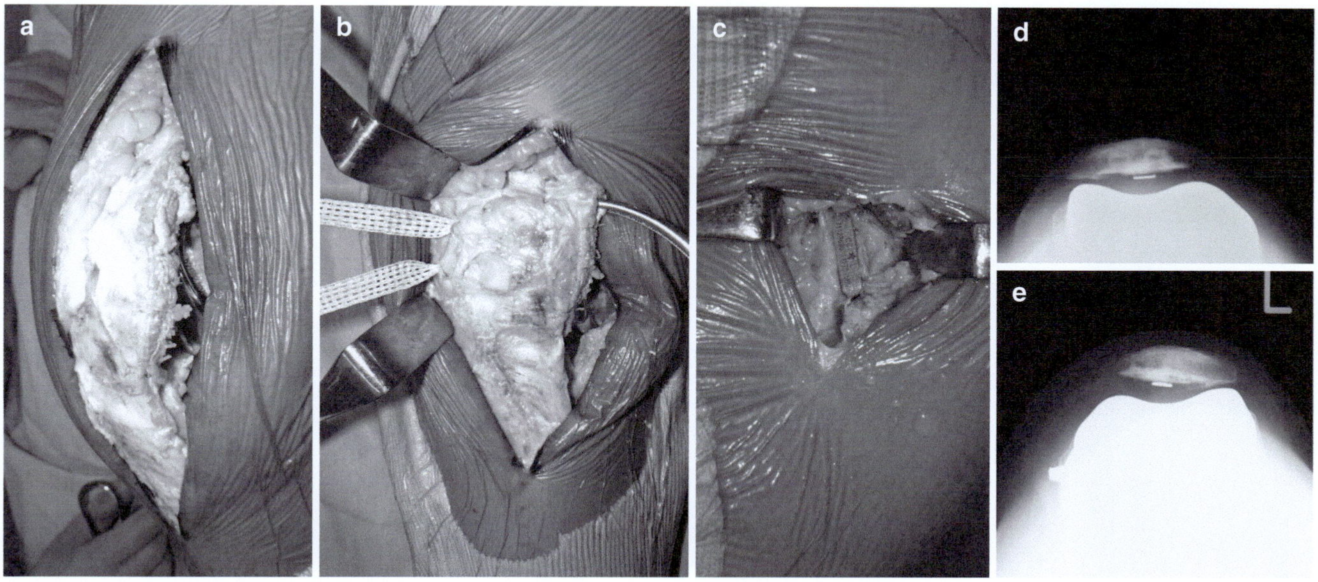

Fig. 79.6 MPFL reconstruction for patellar subluxation after TKA (same case as Fig. 79.5). A patellar replacement was performed through the same incision (**a**) with a lateral patellar band transection (left knee). The left knee was reconstructed with a Leeds-Keio artificial ligament for MPFL reconstruction (**b**) and stapled to the medial femoral epicondyle (**c**). Postoperatively, patellar tracking improved, and anterior knee pain disappeared (**d**, **e**). The MPFL reconstruction of the right knee was fixed with SwiveLock using FiberTape, the same as shown in Fig. 79.2

References

1. Lamotte A, et al. Medial patellofemoral ligament reconstruction for patellar instability following total knee arthroplasty: a review of 6 cases. Orthop Traumatol Surg Res. 2016;102:607–10.
2. Putman S, et al. Patellar complications after total knee arthroplasty. Orthop Traumatol Surg Res. 2019;105(1S):S43–51.
3. Strickland SM, et al. Advances in patellofemoral arthroplasty. Curr Rev Musculoskelet Med. 2018;11:221–30.
4. van Jonbergen HP, et al. Long-term outcomes of patellofemoral arthroplasty. J Arthroplast. 2010;25:1066–71.
5. Yamagami R, et al. Combined patellofemoral arthroplasty and medial patellofemoral ligament reconstruction for chronic patellar instability with trochlear dysplasia: a report of two cases. Mod Rheumatol Case Rep. 2020;4:135–40.
6. Jeong SH, et al. Patellofemoral arthroplasty surgical technique: lateral or medial parapatellar approach. J Arthroplast. 2020;35:2429–34.
7. Ishibashi Y, et al. Medial patellofemoral ligament reconstruction using FiberTape and Knotless SwiveLock anchors. Arthrosc Tech. 2020;9:e1197–202.
8. Sato H, et al. Total knee arthroplasty for gonarthrosis with patellar dislocation. J Orthop Sci. 2005;10:656–60.
9. Kumagi M, et al. Total knee replacement for osteoarthritis of the knee with congenital dislocation of the patella. J Bone Joint Surg Br. 2007;89:1522–4.
10. Fujiwara S, et al. Osteoarthritis with permanent dislocation of the patella treated by total knee arthroplasty through a lateral approach: a report. Int J Surg Case Rep. 2017;38:73–7.
11. Kumagai K, et al. Total knee arthroplasty for treatment of osteoarthritis with prolonged patellar dislocation. Arthroplast Today. 2017;3:25–8.
12. Matsushita T, et al. Total knee arthroplasty combined with medial patellofemoral ligament reconstruction for osteoarthritic knee with preoperative valgus deformity and chronic patellar dislocation. J Arthroplast. 2011;26(505):e17–20.
13. Yoshino N, et al. Permanent patellar dislocation due to malunion of supracondylar fracture after total knee arthroplasty. J Orthop Sci. 2006;11:521–3.
14. Saito H, et al. Successful treatment of a habitual patellar dislocation after a total knee arthroplasty with a closing-wedge distal femoral varus osteotomy and medial patella-femoral ligament reconstruction. J Exp Orthop. 2020;7:63.
15. Nakajima A, et al. The Elmslie–Trillat procedure for recurrent patellar subluxation after total knee arthroplasty. J Arthroplast. 2010;25(7):1170.e1–5.
16. Asada S, et al. Medial patellofemoral ligament reconstruction for recurrent patellar dislocation after total knee arthroplasty. J Orthop Sci. 2008;13:255–8.
17. Goto T, et al. Medial patellofemoral ligament reconstruction for patellar dislocation due to rupture of the medial structures after total knee arthroplasty: a case report and review of the literature. J Med Investig. 2014;61:409–12.

Rheumatoid Arthritis (RA)

80

Shuji Asai

1 Joint Destruction in RA

In joints affected by RA, inflammatory cytokines produced by activated T cells and macrophages increase the receptor activator of nuclear factor kappa-B ligand (RANKL) expression in synovial fibroblasts and induce osteoclasts [1], which results in bone destruction, such as bone erosion and juxtaarticular osteoporosis. Inflammatory cytokines also induce the production of extracellular matrix proteases (MMP-3 and ADAMTS-4). MMP-3 degrades proteoglycans and collagen, whereas ADAMTS-4 degrades aggrecan, resulting in cartilage destruction. Typical radiographic findings in knees with RA include bone erosion and atrophy, joint surface destruction, and joint space narrowing (Fig. 80.1a, b). In knees with

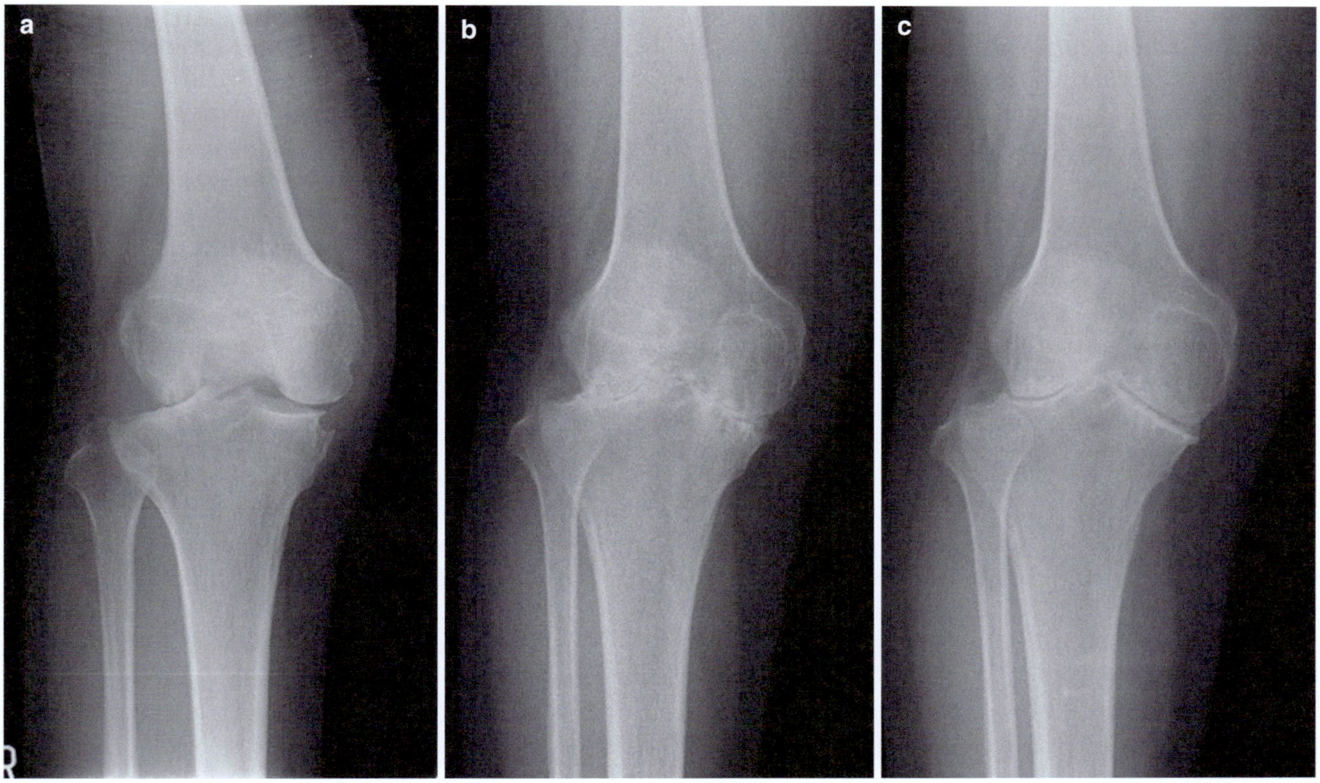

Fig. 80.1 Plain X-ray images of the knee joint in the same RA patient. (**a**) Year X. Joint space narrowing and periarticular bone erosion of the proximal tibia. (**b**) X + 5 years. Disappearance of the joint space and joint surface destruction of the femur and tibia. (**c**) X + 10 years. Secondary OA changes: Subchondral bone sclerosis of the femur and tibia. *RA* rheumatoid arthritis, *OA* osteoarthritis

S. Asai (✉)
Department of Orthopedic Surgery and Rheumatology, Nagoya University Graduate School of Medicine, Nagoya, Aichi, Japan
e-mail: asai.syuji.i8@f.mail.nagoya-u.ac.jp

S. Matsuda (ed.), *Advances in Total Knee Arthroplasty*, https://doi.org/10.1007/978-981-97-4920-1_80

RA having joint destruction, secondary osteoarthritis (OA) changes, such as periarticular osteophyte formation and subchondral bone sclerosis, may be observed when inflammation is adequately controlled [2] (Fig. 80.1c).

2 Drug Therapy and Surgery

In recent years, newer medications, including methotrexate and biologics, and more aggressive treatment strategies, such as treat-to-target (T2T), have enabled more patients with RA to achieve good disease control [3]. Tight control with T2T has been shown to suppress not only the destruction of small joints but also that of large joints requiring joint replacement [4, 5]. Many reports have indicated that the number of total knee arthroplasties (TKAs) in patients with RA has been decreasing in Japan in recent years owing to advances in drug therapy [6–10]. The decrease is more obvious when compared to the increase in the number of TKAs in patients with OA due to an aging society [6]. Additionally, in patients with RA undergoing TKA, more knee joints show secondary OA changes as inflammation is controlled at the time of surgery and less typical RA changes in the knees are observed [6]. Since the age of patients with RA undergoing TKA is increasing, but is still lower than that of OA patients [6], longer term stability is required for TKA in patients with RA than in those with OA.

3 Key Points of Surgery

3.1 Indications for Cruciate-Retaining (CR)-TKA

CR-TKA may be an option in knees with RA having a robust posterior cruciate ligament (PCL).

Persistent joint synovitis destroys bone, cartilage, and ligaments in RA. The anterior cruciate ligament (ACL) and PCL, which are intraarticular ligaments, are particularly susceptible to synovitis. CR-TKA is not indicated for knee joints with impaired PCL function, where posterior-stabilized (PS)-TKA should be used.

In some knees with RA having bone and cartilage destruction requiring TKA, PCL does not undergo complete destruction, as synovitis is quiescent prior to surgery. CR-TKA may be an option for knees with RA if PCL function is preserved. A study using the KT-2000 arthrometer in 65 knees of 48 patients with RA who underwent CR-TKA in Japan showed that PCL function was maintained with no obvious antero-posterior instability for an average of 7.5 years postoperatively [11]. For long-term results, a study in the USA reported an implant survival rate of 93% at 20 years after surgery, with posterior instability at the end point [12]. Although studies since 2000 have reported good postoperative results, many studies in the 1990s reported posterior instability after CR-TKA in RA knees [13–15]. For example, the incidence of posterior instability 6 years after surgery was 50% for CR-TKA and 1% for PS-TKA [15]. Furthermore, patients who underwent revision for instability had a higher incidence of recurrent synovitis, and PCL was absent during revision. Suppression of synovitis with advanced drug therapy could potentially improve the postoperative outcomes of CR-TKA. In other words, continuous control of disease activity is important for achieving good long-term results with CR-TKA in RA.

The extent to which residual PCL function is required for CR-TKA in knees with RA needs further investigation. If intraoperative findings indicate residual synovitis or inadequate control of disease activity, it would be safer to use PS-TKA because PCL function may be impaired over time.

3.2 Synovectomy

Joint synovium is resected as much as possible in knees with RA.

A previous study examined the pathology of the synovium in knee joints that had undergone revision arthroplasty. Synovitis findings, such as rheumatoid necrosis and proliferation of synovial stromal cells, were observed in 26% of knees with RA that underwent revision for aseptic loosening [16]. In contrast, no evidence of synovitis was observed in 64 knees with OA that underwent revision for aseptic loosening and in 18 knees with RA that underwent revision for causes other than loosening. Persistent postoperative synovitis can cause aseptic loosening, impaired PCL function, and instability in knees that have undergone CR-TKA. Intraoperative resection of the synovium and appropriate systemic drug therapy may be effective in controlling postoperative synovitis.

A randomized control trial conducted in Japan showed that synovectomy reduces anterior pain after TKA in patients with RA. The proportion of patients with postoperative anterior pain was significantly lower among those who underwent infrapatellar synovectomy than among those who did not (27% vs. 4%, $p < 0.01$) [17]. However, 43% of the patients who underwent infrapatellar synovectomy, including infrapatellar fat plate resection, experienced anterior pain after surgery. Therefore, the extent of resection should be carefully considered when resecting synovial joints.

3.3 Bone Fragility

Implant fixation with cement is a measure of bone fragility in knees with RA (Fig. 80.2).

A recent study investigated the relationship between tibial component migration and preoperative bone mineral density (BMD) using dual-energy X-ray absorptiometry in 92 patients with OA who underwent cementless CR-TKA. Lower preoperative BMD in the proximal tibia was shown to be related to high migration of the tibial components after 3–24 months [18]. Another study showed a significant relationship between preoperative BMD in the proximal tibia and migration of the tibial component 2 years postoperatively in cementless TKA; however, there was no significant association between the two in cemented TKA [19].

In a study of nine knees with RA having cemented all-polyethylene tibial components in Japan, no tibial component loosening was observed at an average of 12.8 years after surgery, even when severe osteoporosis was observed in the bone tissue proximal to the tibia that was collected intraoperatively for histological evaluation [20].

These results suggest that cemented tibial components should be recommended for knees with RA having bone fragility [21]. However, there are no clear criteria for the degree of bone fragility that requires cementation. Nevertheless, the component-fixation method should be chosen considering the possibility that the bone quality may change during the course of RA, even after surgery.

3.4 Patella

Patellar replacement is recommended in RA.

In a randomized study of 26 patients with RA who underwent simultaneous bilateral TKA with patellar replacement on one side and without patellar replacement on the contralateral side, there were no significant differences in the Hospital for Special Surgery Knee Score (overall pain, function, ROM, muscle strength, flexion contracture, and instability) 6 years after the surgery between the knees with and without patellar replacement. However, bone erosion and joint space narrowing were observed over time in some knees that did not undergo patellar replacement. Pain on standing and ascending or descending stairs and tenderness of the patellofemoral joint were only noted in knees without patellar replacement [22].

In a comparison of 17 knees with OA and 43 with RA that underwent TKA without patellar replacement, 11.8% of patients with OA and 27.9% of patients with RA had postoperative peripatellar pain. The patellar thickness significantly decreased over time in patients with RA. Approximately 70% of patients with RA whose patellar thickness decreased to ≤80% complained of peripatellar pain [23].

Although studies on the necessity of patellar replacement in RA are limited, there are no data showing the superiority of TKA without patellar replacement. Therefore, patellar replacement is recommended in patients with RA. However, patellar fracture, a complication unique to patellar replace-

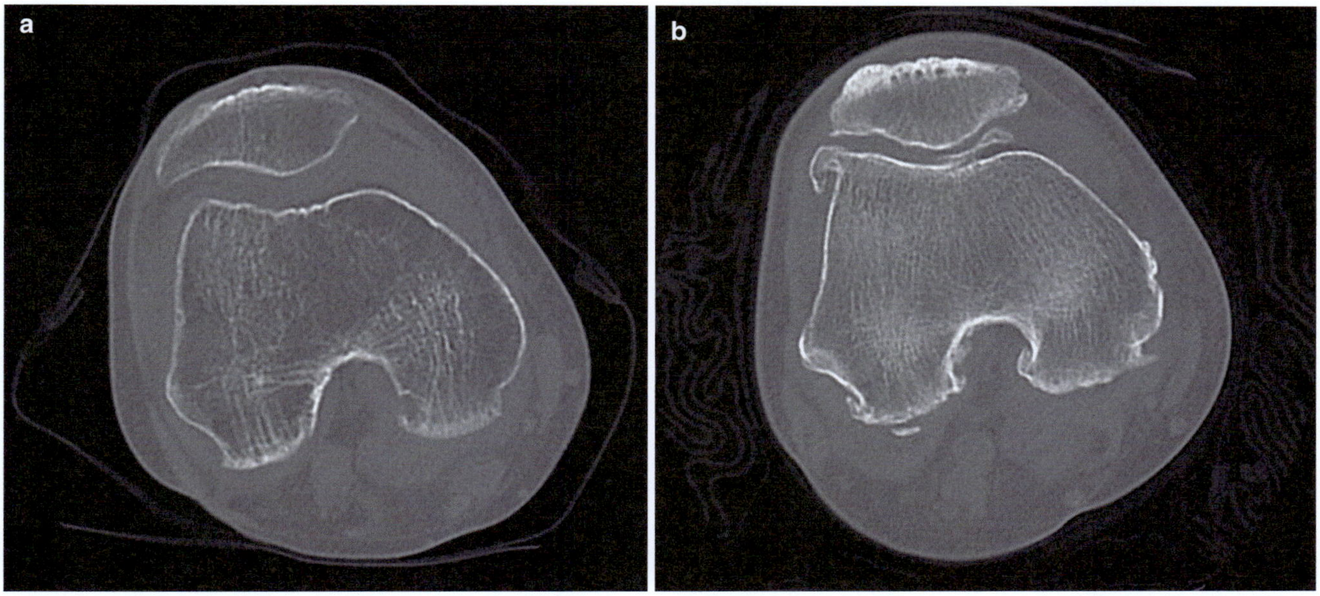

Fig. 80.2 CT images of the femoral condyle. RA knees with bone fragility (**a**) show a lower trabecular bone density than OA knees (**b**). *CT* computed tomography, *OA* osteoarthritis

ment, should be noted. Patellar fracture was observed in 1.51% of RA knees with patellar replacement, and a thin patella and PS-TKA were reported as the risk factors [24].

4 Clinical Results

4.1 Clinical Manifestations

According to a large registry study in Europe, the implant survival rate for RA knees with revision surgery as the endpoint was 95–96% at 10 years [25, 26]. Despite a small number of cases, satisfactory implant survival rates and clinical outcomes have been reported in Japan. In a study of 32 knees with RA that underwent cementless CR-TKA, the implant survival rate was 96.9% after 12 years [27]. In a study of 36 knees in 25 patients with RA who underwent cemented CR-TKA, the implant survival rate was 93.7% at 15 years, and 77.7% patients had a Hospital for Special Surgery Knee Score of good or excellent [28]. Another study compared the Knee Society Scores of 75 knees of 70 patients with RA and 534 knees of 501 patients with OA who underwent TKA. Compared to patients with OA, those with RA had significantly lower activity scores 2 years postoperatively but significantly higher range-of-motion and satisfaction scores [29]. Furthermore, patients with RA who underwent TKA with synovectomy showed a significant reduction in disease activity 6 months postoperatively compared to preoperatively, and this reduction was sustained for 2–3 years after the surgery [30, 31]. TKA for RA demonstrates stable long-term results and can improve not only knee joint function, but also the disease activity. Therefore, TKA for the treatment of RA is expected to have multifaceted effects.

4.2 Postoperative Complications

1. Infection

Although some studies have suggested that RA is not associated with a higher risk of surgical site infection (SSI) than OA, several studies have shown that RA is a risk factor for SSI. In a Danish cohort study comparing 3913 patients with RA and 120,499 with OA who underwent TKA or total hip arthroplasty, the risk of developing prosthetic joint infection was significantly higher in RA than in OA (adjusted hazard ratio [95% confidence interval] 1.46 [1.13–1.88]) [32]. In this study, the use of biologics was not a significant risk factor for SSI. In contrast, a retrospective study in Japan showed that the use of biologics was an independent risk factor for early SSI in RA patients (odds ratio [95% confidence interval] 5.69 [2.07–15.61]) [33]. Thus, there is no consensus on whether the use of biologics increases the incidence of SSI. In the perioperative management of RA, it is worth noting that steroid use and diabetes mellitus are risk factors for SSI, in addition to RA itself.

2. Venous Thromboembolism (VTE)

In a study comparing 238 knees of 199 patients with RA and 169 knees of 156 patients with OA who underwent TKA, the proportion of patients with deep vein thrombosis on Doppler ultrasonography 7 days postoperatively was significantly higher among patients with OA than those with RA (43.2% vs. 20.6%, $p < 0.001$) [34]. However, there was no significant difference between the two groups after adjusting for age and NSAID use. In another study comparing 204 patients with RA and 1084 with OA who underwent TKA, there was no significant difference in the incidence of VTE between the two groups (24.0% vs. 24.3%) [35]. In this study, female sex, use of tourniquets, and nonuse of fondaparinux were the independent risk factors for VTE. Furthermore, in a large Swedish cohort study of 46,316 patients with RA, the incidence of VTE was significantly higher in the high disease activity group than in the remission group (1.08% vs. 0.52%, relative risk ratio [95% confidence interval] 2.03 [1.73–2.38]) [36].

TKA in patients with RA does not require special prophylaxis against VTE and can be performed in the same manner as in patients with OA. Nevertheless, patients with RA and high disease activity require careful postoperative follow-up.

References

1. Tateiwa D, et al. Cartilage and bone destruction in arthritis: pathogenesis and treatment strategy: a literature review. Cells. 2019;8:818.
2. Kaneko A, et al. Development and validation of a new radiographic scoring system to evaluate bone and cartilage destruction and healing of large joints with rheumatoid arthritis: ARASHI (Assessment of rheumatoid arthritis by scoring of large joint destruction and healing in radiographic imaging) study. Mod Rheumatol. 2013;23:1053–62.
3. Yamanaka H, et al. A large observational cohort study of rheumatoid arthritis, IORRA: providing context for today's treatment options. Mod Rheumatol. 2020;30:1–6.
4. Asai S, et al. Concomitant methotrexate protects against total knee arthroplasty in patients with rheumatoid arthritis treated with tumor necrosis. J Rheumatol. 2015;42:2255–60.
5. Asai S, et al. Effects of concomitant methotrexate on large joint replacement in patients with rheumatoid arthritis treated with tumor necrosis factor. Arthritis Care Res (Hoboken). 2015;67:1363–70.
6. Asai S, et al. Fifteen-year trends in the number and age of patients with rheumatoid arthritis undergoing total knee and hip arthroplasty: a comparison study with osteoarthritis. Mod Rheumatol. 2021;31:768–71.

7. Asai S, et al. Characteristics of patients with rheumatoid arthritis undergoing primary total joint replacement: a 14-year trend analysis (2004–2017). Mod Rheumatol. 2020;30:657–63.
8. Matsumoto T, et al. Trends in treatment, outcomes, and incidence of orthopedic surgery in patients with rheumatoid arthritis: an observational cohort study using the Japanese National Database of Rheumatic Diseases. J Rheumatol. 2017;44:1575–82.
9. Yasui T, et al. Prevalence of total knee arthroplasty and its predictive factors in Japanese patients with rheumatoid arthritis: analysis using the NinJa cohort. Mod Rheumatol. 2016;26:36–9.
10. Momohara S, et al. Recent trends in orthopedic surgery aiming to improve quality of life for those with rheumatoid arthritis: data from a large observational cohort. J Rheumatol. 2014;41:862–6.
11. Niki Y, et al. Factors affecting anteroposterior instability following cruciate-retaining total knee arthroplasty in patients with rheumatoid arthritis. Knee. 2008;15:26–30.
12. Miller MD, et al. Posterior cruciate ligament-retaining total knee arthroplasty in patients with rheumatoid arthritis: a concise follow-up of a previous report. J Bone Joint Surg Am. 2011;93:131–6.
13. Waslewski GL, et al. Early, incapacitating instability of posterior cruciate ligament-retaining total knee arthroplasty. J Arthroplast. 1998;13:763–7.
14. Pagnano MW, et al. Flexion instability after primary posterior cruciate retaining total knee arthroplasty. Clin Orthop Relat Res. 1998;356:39–46.
15. Laskin RS, et al. The Insall Award. Total knee replacement with posterior cruciate ligament retention in rheumatoid arthritis: problems and complications. Clin Orthop Relat Res. 1997;345:24–8.
16. Fink B, et al. Recurring synovitis as a possible reason for aseptic loosening of knee endoprostheses in patients with rheumatoid arthritis. J Bone Joint Surg Br. 2001;83:604–8.
17. Tanaka N, et al. Influence of infrapatellar fat pad resection in synovectomy during total knee arthroplasty in patients with rheumatoid arthritis. J Arthroplast. 2003;18:897–902.
18. Andersen MR, et al. Low preoperative BMD is related to high migration of tibia components in uncemented TKA-92 patients in a combined DEXA and RSA study with 2-year follow-up. J Arthroplast. 2017;32:2141–6.
19. Li MG, et al. The effect of preoperative bone quality on the fixation of the tibial component in total knee arthroplasty. J Arthroplast. 2000;15:744–53.
20. Sugita T, et al. Influence of tibial bone quality on loosening of the tibial component in total knee arthroplasty for rheumatoid arthritis: long-term results. Orthopedics. 1999;22:213–5.
21. Lee JK, et al. Total knee arthroplasty in rheumatoid arthritis. Knee Surg Relat Res. 2012;24:1–6.
22. Kajino A, et al. Comparison of the results of bilateral total knee arthroplasty with and without patellar replacement for rheumatoid arthritis: a follow-up note. J Bone Joint Surg Am. 1997;79:570–4.
23. Kawakubo M, et al. Radiographic changes in the patella after total knee arthroplasty without resurfacing the patella: comparison of osteoarthrosis and rheumatoid arthritis. Bull Hosp Jt Dis. 1997;56:237–44.
24. Jujo Y, et al. Patellar fracture after total knee arthroplasty for rheumatoid arthritis. J Arthroplast. 2013;28:40–3.
25. Robertsson O, et al. Knee arthroplasty in rheumatoid arthritis: a report from the Swedish Knee Arthroplasty Register on 4,381 primary operations 1985–1995. Acta Orthop Scand. 1997;68:545–53.
26. Himanen AK, et al. Survival of the AGC total knee arthroplasty is similar for arthrosis and rheumatoid arthritis. The Finnish Arthroplasty Register report on 8,467 operations carried out between 1985 and 1999. Acta Orthop. 2005;76:85–8.
27. Yamanaka H, et al. Clinical results of hi-tech knee II total knee arthroplasty in patients with rheumatoid athritis: 5- to 12-year follow-up. J Orthop Surg Res. 2012;7:9.
28. Ito J, et al. 15-year follow-up study of total knee arthroplasty in patients with rheumatoid arthritis. J Arthroplast. 2003;18:984–92.
29. Kobayashi S, et al. Rheumatoid arthritis patients achieve greater satisfaction but lower functional activities as compared to osteoarthritis patients after total knee arthroplasty. J Arthroplast. 2019;34:478–82.
30. Yano K, et al. Effect of total knee arthroplasty on disease activity in patients with established rheumatoid arthritis: 3-year follow-up results of combined medical therapy and surgical intervention. Mod Rheumatol. 2010;20:452–7.
31. Kumagai K, et al. Total knee arthroplasty improves both knee function and disease activity in patients with rheumatoid arthritis. Mod Rheumatol. 2017;27:806–10.
32. Cordtz RL, et al. Predictors of revision, prosthetic joint infection and mortality following total hip or total knee arthroplasty in patients with rheumatoid arthritis: a nationwide cohort study using Danish healthcare registers. Ann Rheum Dis. 2018;77:281–8.
33. Momohara S, et al. Prosthetic joint infection after total hip or knee arthroplasty in rheumatoid arthritis patients treated with nonbiologic and biologic disease-modifying antirheumatic drugs. Mod Rheumatol. 2011;21:469–75.
34. Niki Y, et al. Rheumatoid arthritis: a risk factor for deep venous thrombosis after total knee arthroplasty? Comparative study with osteoarthritis. J Orthop Sci. 2010;15:57–63.
35. Izumi M, et al. Risk of venous thromboembolism after total knee arthroplasty in patients with rheumatoid arthritis. J Rheumatol. 2015;42:928–34.
36. Molander V, et al. Risk of venous thromboembolism in rheumatoid arthritis, and its association with disease activity: a nationwide cohort study from Sweden. Ann Rheum Dis. 2021;80:169–75.

81 Charcot Joints

Shinichi Miyazawa

1 Charcot Joint Disorder

Charcot joint disorder occurs when the pain defense mechanism fails due to an abnormal sensation caused by neuropathy. The joints are easily overloaded beyond their natural pain threshold, resulting in articular cartilage degeneration, marked bone destruction, and irregular reactive bone proliferation. Charcot first described this disease in 1868, reporting the relationship between locomotor ataxia caused by syphilis and neuropathic arthropathy. Later, he reported the relationship between diabetes mellitus and neuropathic arthropathy in Jordan in 1936 [1]. In addition to locus coeruleus and diabetes mellitus, other primary diseases that cause this syndrome include syringomyelia, spinal meningocele, adhesive arachnoiditis, spinal cord injury, polyradiculoneuritis, congenital hyperalgesia, amyloidosis, and Charcot-Marie-Tooth disease.

TKA of the Charcot joint is contraindicated owing to the high incidence of poor outcomes and postoperative complications. Therefore, knee arthrodesis surgery is performed in cases where conservative treatment with braces is ineffective; however, the level of satisfaction is not sufficiently high owing to the functional decline of the knee joint. Since Soudry et al., reported the short-term results of TKA for Charcot joints in 1986, TKA has become a viable option for patients with Charcot joints [2].

2 Tips and Points of Surgery

2.1 Surgical Times

The timing of TKA should be determined based on disease stages based on the Eichenholtz system (Stage I: Development, Stage II: Coalescence, Stage III: Reconstruction).

During development or stage I, radiographs show osteopenia, periarticular tissue, bone fragments, subluxation, and dislocation of the joints. Clinically, joint instability may occur because of local swelling, erythema, and warmth.

In the coalescence stage or stage II, radiographs display absorption of debris, sclerosis, and fusion of larger fragments. Clinically, local heat, swelling, and decreased erythema are seen.

In the last stage of reconstruction, the joint deformity is considerably severe, as visible on the radiograph in addition to prominent osteogenesis. Osteosclerosis is reduced, and the bone fragments became rounded and smooth. Arthropathic changes such as narrowing of the joint space and fibrous or bony ankylosis were observed. Clinically, local heat, swelling, and erythema disappear and the joint becomes stabilized or ankylosed (Table 81.1) [3, 4].

Table 81.1 Eichenholz classification of Charcot arthropathy

Disease stag	Radiographic findings	Clinical findings
I (development)	Osteopenia, joint subluxation, dislocation	Swelling, erythema, warmth, ligamentous laxity
II (coalescence)	Absorption of debris, sclerosis, fusion of larger fragments	Decreased warmth, decreased swelling, decreased erythema
III (reconstruction)	Deformity consolidation, fibrous ankyloses, rounding and smoothing of bone fragments	Fixed deformity, absence of warmth, swelling, erythema

S. Miyazawa (✉)
Department of Orthopedic Surgery, National Hospital Organization Fukuyama Medical Center, Hiroshima, Japan

S. Matsuda (ed.), *Advances in Total Knee Arthroplasty*, https://doi.org/10.1007/978-981-97-4920-1_81

As surgical treatment at the developmental stage is considered to have poor results, surgical intervention is preferred after the coalescence stage [5].

2.2 Implant Selection

The type of implant that may be used for surgery remains debatable. Various case reports are currently available about prostheses with low constraints to those with high constraints [6–11]. Considering the poor bone quality and strong instability seen with Charcot arthropathy, the use of prostheses that are minimally constrained or severely constrained is considered unsuitable; therefore, we believe that prostheses that are condylar constrained and rotating hinges should be used. Chun et al. reported 100% implant survival after an average follow-up period of 10.7 years (8.5–11.4 years) in 20 cases [1]. The rotating-hinge type is suitable in cases of severe deformities, bone loss, or instability. Bae et al. performed TKA with a rotating hinge in 11 patients with severe deformity, bone loss, or instability, and reported that implant survival was 81.8% as seen in a mean follow-up of 12.3 (10–22) years [12]. In all models, surgery and follow-up were performed with caution to avoid complications. The complication rate reached 26.4%, with instability being the most common complication; peri-implant fractures, periprosthetic infections, implant loosening, and ligament injuries have been reported [13].

2.3 Case Presentation

Case 1

A 52-year-old male with neurosyphilis presented with swelling of the left knee joint at the hospital. The patient was referred for a knee joint puncture, and 575 mL of joint fluid was obtained. Subsequently, the patient frequently underwent knee joint punctures at his primary care clinic. The patient did not complain of pain; however, there was evidence of knee joint destruction. The patient was referred to the hospital for further treatment.

On follow-up radiography, small bone fragments were observed around the joint, indicative of the developmental stage (Fig. 81.1a). In the subsequent 3 months, osteosclerosis and fusion of bone fragments were observed; therefore, the knee joint was considered to be in the coalescence stage, and TKA was performed using a rotating hinge (Fig. 81.1b). The bone defect in the proximal medial tibia was grafted using the resected bone (Fig. 81.1c). At the time of the last follow-up, 5 years after the surgery, the grafted bone had united, and no laxity in the TKA was observed (Fig. 81.1d).

Case 2

A 76-year-old man with diabetes mellitus experienced instability of the left knee after a fall and was referred for treatment of a left tibial plateau fracture on radiography

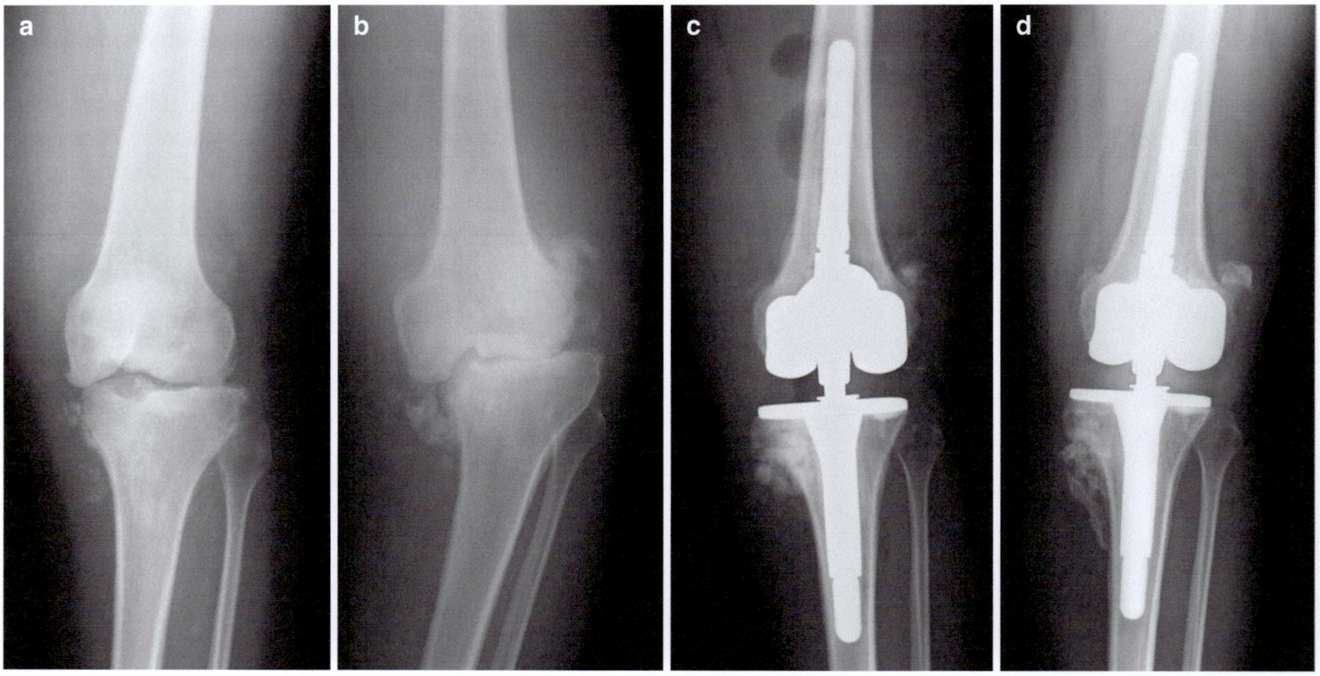

Fig. 81.1 Case 1- radiographs. (**a**) At the time of the return visit to our clinic, (**b**) Preoperative, (**c**) Immediately after surgery, (**d**) 5 years after surgery at the time of the last survey

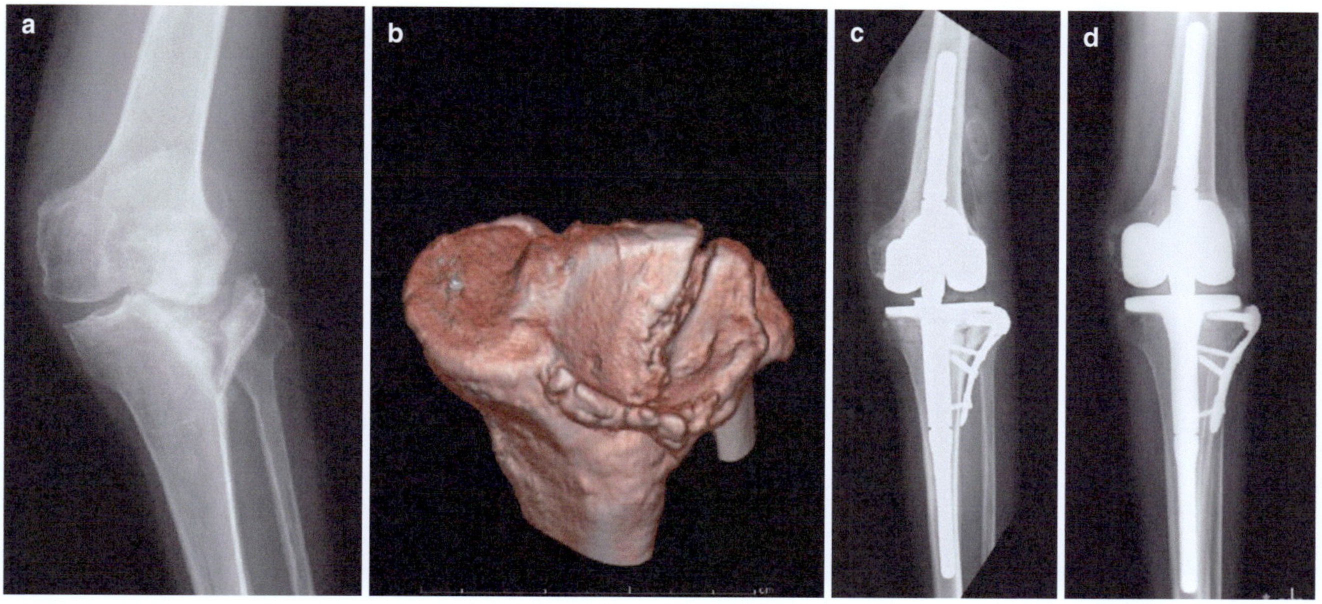

Fig. 81.2 Case 2- radiographs. (**a**) At the initial visit to our clinic, (**b**) 3D CT at the initial visit, (**c**) Immediately after surgery, (**d**) 5 years and 8 months postoperatively at the time of the last survey

(Fig. 81.2a). Preoperative computed tomography (CT) revealed a bone defect and fracture of the proximal lateral tibia (Fig. 81.2b). TKA was performed using a rotating hinge. The bone defect on the proximal lateral side of the tibia was grafted with the resected bone and the fracture was fixed with a plate (Fig. 81.2c). At the last follow-up, 5 years and 8 months after surgery, the fracture site and grafted bone had united, and there was no laxity in the TKA (Fig. 81.2d).

TKA for the Charcot joint is considered to have inferior outcomes compared with conventional TKA. Therefore, when surgery is performed, the timing of the surgery and the choice of TKA prosthesis should be thoroughly considered. Patients should be sufficiently informed regarding complications, outcomes, and the need for revision surgery before surgery, and informed written consent should be obtained. Therefore, careful postoperative follow-ups are necessary to prevent complications.

References

1. Chun KC, et al. The fate of allogenic head bone grafts using varus-valgus constrained total knee arthroplasty in neuropathic joints. J Arthroplast. 2016;31:2778–83.
2. Soundry M, et al. Total knee arthroplasty in Charcot and Charcot-like joints. Clin Orthop Relat Res. 1986;208:199–204.
3. Rosenbaum AJ, et al. Classifications in brief: eichenholz of Charcot arthropathy. Clin Orthop Relat Res. 2015;473:1168–71.
4. Lu V, et al. Charcot knee- presentation, diagnosis, management: a scoping review. Clin Rheumatol. 2021;40:4445–56.
5. Zeng M, et al. Total knee arthroplasty in patients with Charcot joints. Knee Surg Sports Tolaumatol Arthrosc. 2016;24:2672–7.
6. Yoshino S, et al. Total knee arthroplasty in Charcot's joint. J Arthroplast. 1993;8:335–40.
7. Yasuda M, et al. A giant thigh mass in a patient with total knee arthroplasty for Charcot joint. Clin Orthop Relat Res. 1995;317: 159–61.
8. Sugitani K, et al. Total knee arthroplasty for neuropathic joint disease after severe bone destruction eroded the tibial tuberosity. Orthop. 2012;35:e1108–11.
9. Sugita T, et al. Autologous morselised bone grafting for medial tibial defects in total knee arthroplasty. J Orthop Surg. 2015;23: 185–9.
10. Nozaka K, et al. Simultaneous total knee arthroplasty and ankle arthrodesis for Charcot neuroarthropathy. Case Rep Orthop. 2019;2019:6136409.
11. Yoshikawa R, et al. Staged bilateral total knee arthroplasty in a patient with Charcot knees: a case report. J Orthop Case Rep. 2020;10:9–12.
12. Bae DK, et al. Long-term outcome of total knee arthroplasty in the Charcot joint. J Arthroplast. 2009;24:1152–6.
13. Rattanaprichavej P, et al. Mid- to long-term outcomes of contemporary total knee arthroplasty in Charcot neuroarthropathy: a systematic review. EFFORT Open Rev. 2021;6:556–64.

Part XV

Complications

82 Perioperative Pain Management

Sachiyuki Tsukada

Improving pain management after TKA enhances knee joint function and increases patient satisfaction during the early postoperative period [1]. Effective pain reduction not only yields favorable short-term outcomes but also plays a role in ameliorating long-term outcomes by minimizing the likelihood of developing chronic pain [1]. TKA is associated with severe postoperative pain; however, the widespread use of multimodal analgesia and pre-emptive pain management strategies has facilitated successful pain control.

1 Multimodal and Pre-emptive Analgesias

1.1 Multimodal Analgesia

Multimodal analgesia is a comprehensive approach that aims to maximize pain relief while minimizing potential side effects by using multiple analgesic techniques [2]. Multimodal analgesia in TKA involves a combination of various methods such as periarticular multidrug injections, peripheral nerve blocks, and systemic administration of medications.

Epidural anesthesia and peripheral nerve blocks have previously been used alone to treat postoperative pain after TKA. If these methods alone do not offer adequate pain relief, opioids and NSAIDs are introduced as additional measures. However, relying on a single analgesic method poses issues of inadequate pain relief and the potential for excessive opioid doses. Opioid epidemics have emerged as a significant social issue, particularly in Europe and the United States, resulting in a notable increase in deaths related to opioid misuse. The inappropriate administration of opioids during the perioperative period has been identified as a contributing factor. In Japan, numerous multimodal analgesic regimens that do not rely on opioids have been extensively studied [2, 3].

1.2 Pre-emptive Analgesia

Pre-emptive analgesia is an approach aimed at minimizing postoperative pain by administering analgesics before the pain stimulus is introduced, thereby preventing sensitization of the central nervous system [4]. Consequently, epidural anesthesia and peripheral nerve blocks should be administered before making a skin incision. It is crucial to initiate systemic administration of analgesics and periarticular multidrug injections as early as possible. A study comparing patients who began taking oral anti-inflammatory analgesics two hours after TKA with those who started on the second day after TKA revealed that the early administration group experienced superior pain relief [5]. Another study indicated that periarticular multidrug injection performed immediately before arthrotomy (early stage injection) was more effective than administration after prosthesis implantation (late-stage injection) [6].

2 Epidural Anesthesia

Epidural anesthesia is a commonly used technique for perioperative pain management in TKA. It involves the injection of drugs, primarily local anesthetics, into the epidural space. There are two approaches to administer epidural anesthesia: single administration of the drug or continuous administration through a catheter placed in the epidural space. However, epidural anesthesia is used less frequently in the perioperative period of TKA because it blocks not only the sensory nerves but also the motor nerves and can cause serious complications such as epidural hematoma [7].

A study comparing the pain-relieving effects of epidural anesthesia with those of femoral nerve block have concluded

S. Tsukada (✉)
Department of Orthopaedic Surgery, Hokusuikai Kinen Hospital, Ibaraki, Japan
e-mail: s8058@nms.ac.jp

S. Matsuda (ed.), *Advances in Total Knee Arthroplasty*, https://doi.org/10.1007/978-981-97-4920-1_82

that epidural anesthesia is superior for pain relief after TKA [7]; however, another study revealed that continuous epidural anesthesia is inferior for pain relief compared to continuous femoral nerve block [8].

3 Peripheral Nerve Block

Peripheral nerve block is a procedure in which a local anesthetic is injected near a specific peripheral nerve, often accompanied by the simultaneous administration of corticosteroids to extend the duration of the anesthetic effect. Similar to epidural anesthesia, peripheral nerve blocks can be administered as a single-shot injection or continuously using a catheter. While femoral nerve blocks have been employed to manage anterior knee pain and sciatic nerve blocks for posterior knee pain, these conventional approaches have the drawback of blocking the motor nerves, causing motor function impairment [9].

In recent years, motor-sparing peripheral nerve blocks, which avoid blocking the motor nerves while providing effective pain relief, have gained popularity. In TKA, two motor-sparing blocks are commonly used: the adductor canal block, in which the femoral nerve block is targeted at the adductor canal level, and the infiltration between the popliteal artery and capsule of the posterior knee (iPACK), which involves performing a sciatic nerve block at the location where a branch of the tibial nerve enters the posterior knee joint.

3.1 Conventional Peripheral Nerve Block

Femoral nerve block provides analgesia to the anterior aspect of the knee joint. Because a femoral nerve block alone does not relieve pain in the posterior aspect of the knee joint, a sciatic nerve block is often required. A selective tibial nerve block is often performed, in which only the tibial nerve is blocked distal to the level where the sciatic nerve bifurcates into the tibial and common peroneal nerves [8]. In addition, a femoral nerve block is often combined with periarticular multidrug injections into the posterior aspect of the knee joint to provide additional pain relief in the posterior knee joint [10].

As a pain-relieving effect of conventional peripheral nerve blocks, it was reported that the time required to obtain 120-degree flexion of knee ROM was shorter when a one-shot selective tibial nerve block was combined with a continuous femoral nerve block compared with continuous epidural anesthesia [8]. A comparative study of continuous and one-shot sciatic nerve blocks combined with femoral nerve blocks reported that the postoperative opioid consumption was lower in the continuous nerve block group [11].

The conventional femoral nerve block blocks not only the sensory nerves but also the motor nerves, resulting in postoperative muscle weakness, especially in the quadriceps muscles [9]. Local anesthetic concentrations were adjusted to minimize muscle weakness. A study comparing high (1.0 mg/mL) and low (0.5 mg/mL) concentrations of levobupivacaine in continuous femoral nerve block reported no differences in muscle strength up to 48 hours after surgery and in pain scores between the two groups [9].

A femoral nerve block offers the advantage of allowing the placement of a continuous catheter, enabling its use in patient-controlled analgesia. A study compared the effectiveness of a 24-h continuous femoral nerve block followed by a continuous femoral nerve block at a consistent rate of administration with a switch to patient-controlled analgesia and concluded that both approaches provided comparable pain relief [12].

Many studies have compared the effectiveness of a combination of sciatic and femoral nerve blocks to that of periarticular multidrug injection [13–16]. Some previous studies reported that a combination of femoral and sciatic nerve blocks [13] and a femoral block and periarticular multidrug injection [14] are more effective; however, some other studies reported no difference in the pain-relieving effect between the two groups [15, 16].

3.2 Motor-Sparing Blocks

A motor-sparing block is performed to avoid blocking the motor nerve, which is a problem with conventional femoral nerve blocks. Conventional femoral nerve blocks are performed using at the femoral triangle. A technique for performing the block more distally to minimize muscle weakness has been reported [17]. Recently, techniques for blocking the adductor canal have been widely used.

iPACK is a motor-sparing sciatic nerve block used in TKA [18]. The common peroneal nerve branches from the sciatic nerve innervate the posterior knee joint, forming the tibial nerve. iPACK is performed in the proximal part, where the sensory branch of the tibial nerve enters the posterior knee joint.

4 Periarticular Multidrug Injection

Periarticular multidrug injection is a technique for injecting a mixture of drugs with different mechanisms of action into the periarticular tissues (Fig. 82.1). Despite their technical ease, periarticular multidrug injections provide robust pain relief. This aligns well with the concept of multimodal analgesia and has emerged as a central method of pain management after TKA. Instead of injecting medication into the

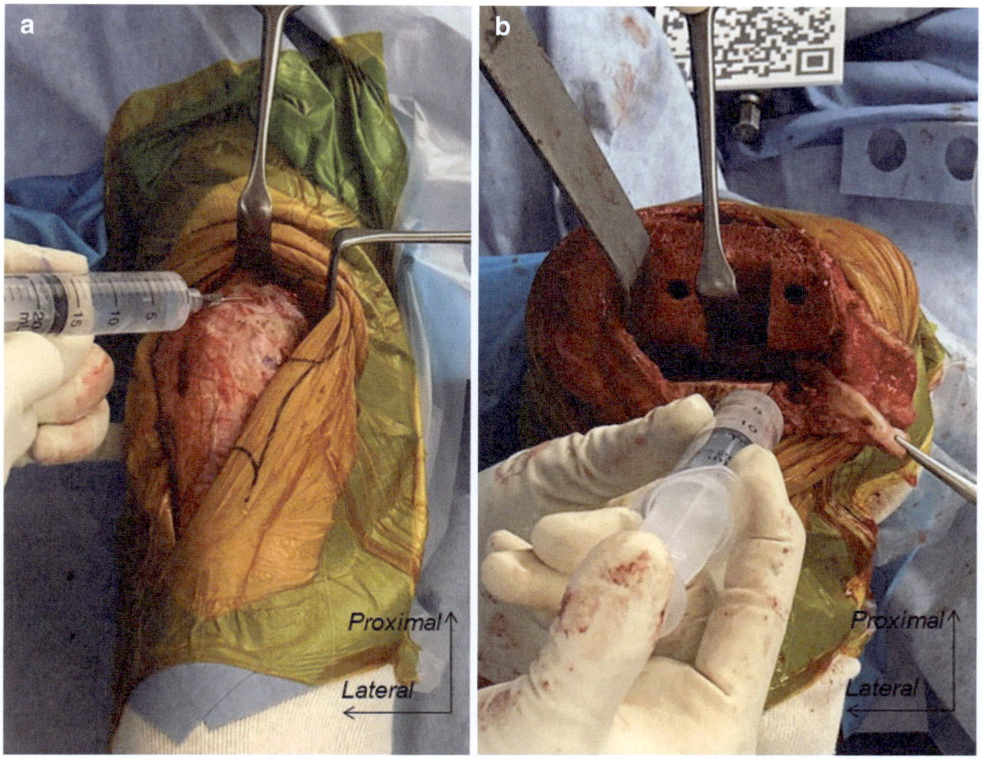

Fig. 82.1 Periarticular multidrug injection technique (right knee). A solution containing local anesthetics, corticosteroids, NSAIDs, opioids, and epinephrine is injected around the joints. (**a**) Injection into the anterior aspect of the knee joint, (**b**) Injection into the posterior aspect of the knee joint. To minimize the risk of common peroneal nerve palsy, a small amount of the solution is administered to the posterolateral area

joint cavity, it should be administered to the tissues surrounding the knee joint. Studies have indicated that periarticular administration is more effective for pain relief than intra-articular administration [19].

4.1 Effectiveness of Periarticular Multidrug Injections

Several studies have compared the pain-relieving effects of periarticular multidrug injections and other techniques. A study comparing continuous epidural anesthesia with periarticular multidrug injection reported equal pain scores between the two techniques [20]. However, some studies concluded that periarticular multidrug injection is superior to continuous epidural anesthesia in terms of pain relief and has fewer complications in both unilateral and simultaneous bilateral TKAs [21, 22].

Several studies reported that periarticular multidrug injections are superior to continuous femoral nerve blocks for postoperative pain relief [23, 24]. This may be because the femoral nerve block alone does not provide pain relief to the posterior knee.

Previous studies have compared combined femoral and sciatic nerve blocks with periarticular multidrug injections. Compared to combined femoral and sciatic nerve blocks, the pain-relieving effect of periarticular multidrug injection was reported to be inferior up to 6 h postoperatively, with no difference up to 18 h thereafter, and superior at 24–30 h [25]. A study using a diagnostic procedure combination, one of the real-world data collected in Japan, reported no difference in postoperative walking ability between patients treated with peripheral nerve block and those treated with periarticular multidrug injection [26].

4.2 Drugs Used for Periarticular Multidrug Injections

Ongoing research is focused on identifying the most effective drug combination for periarticular multidrug injections. Corticosteroids and opioids have received considerable attention in recent years.

Corticosteroids possess strong anti-inflammatory properties and their incorporation into periarticular multidrug injections is believed to alleviate postoperative pain. However, it is imperative to consider that their usage may also increase the risk of complications, such as deep infections. Several reports have shown that postoperative pain scores improve when corticosteroids are added to periarticular multidrug injections [1, 27]. These studies did not report an increase in the incidence of complications. Another report showed that corticosteroid use has decreased the frequency of postoperative nausea and vomiting [28]. Corticosteroids are often administered intravenously during the perioperative period of TKA, and the use of corticosteroids for periarticular multidrug injections has been reported to be superior to the use of intravenous corticosteroids for pain relief [29].

Morphine mixed with periarticular multidrug injections is expected to have systemic as well as local effects on opioid receptors that are expressed by peripheral tissue invasions [30]. However, there have been several reports on the effects of mixing morphine with periarticular multidrug injection [31–34]. It has been reported that including morphine in periarticular multidrug injection did not improve pain scores after TKA [31, 32]. In a comparison of the three groups, one in which morphine (10 mg) was locally administered through a periarticular multidrug injection, one in which morphine 0.1 mg was administered into the epidural space during epidural anesthesia, and another in which morphine was not used in either periarticular multidrug injection or epidural anesthesia, the group in which morphine 0.1 mg was mixed with the epidural anesthetic had the lowest pain score after TKA [33]. Although local administration of morphine is expected to reduce the risk of systemic side effects, the development of respiratory depression due to morphine used in periarticular multidrug injections has been reported [34].

4.3 Rebound Pain After Periarticular Multidrug Injections

Although periarticular multidrug injection demonstrates remarkable pain relief in the immediate postoperative period, pain scores tend to worsen approximately 24 h after surgery [21]. One study indicated that intravenous acetaminophen effectively alleviated rebound pain, particularly when the effect of periarticular multidrug injections began to diminish [2].

Owing to the significant pain-relieving effects of intraoperative periarticular multidrug injections, there have been reports of additional injections being performed after TKA. When a double-lumen catheter was placed around the joint during TKA and local anesthetic and corticosteroids were administered after TKA, the postoperative pain score was lower than that of the group that did not receive anesthetic and corticosteroids [35]. It has been reported that percutaneous injection of local anesthetics, corticosteroids, and epinephrine into the vastus medialis muscle one day after surgery resulted in better pain relief than the group that did not receive the injections [3, 36].

5 Systemic Administration of Opioids

Intravenous and subcutaneous opioids continue to be recognized as essential means of pain relief, although their usage has decreased in the perioperative period of TKA owing to the growing awareness of reducing opioid consumption. Patient-controlled analgesia, which allows patients to self-administer pain medications, offers advantages over continuous opioid administration by allowing better adaptation to pain fluctuations and reducing the risk of opioid overdose. The use of subcutaneous fentanyl for patient-controlled analgesia decreases the frequency of postoperative rescue analgesia [37].

6 Patch

A patch is a medication product designed for application to the skin to deliver active components to the target organ. They can be classified into two types: topical patches, which exert their effects only at the site of application, and systemically acting patches, in which the drug is released into the blood via the skin and acts on the entire body.

NSAIDs are the most commonly used topical patches. When NSAIDs patches were used in the perioperative period of TKA, postoperative pain scores were better than those in patients who did not receive them [38].

Fentanyl is often used as a systemic patch. In TKA, the pain score is better in patients treated with a fentanyl patch than in those treated without it [39].

References

1. Ikeuchi M, et al. Effects of dexamethasone on local infiltration analgesia in total knee arthroplasty: a randomized controlled trial. Knee Surg Sports Traumatol Arthrosc. 2014;22:1638–43.
2. Murata-Ooiwa M, et al. Intravenous acetaminophen in multimodal pain management for patients undergoing total knee arthroplasty: a randomized, double-blind, placebo-controlled trial. J Arthroplast. 2017;32:3024–8.
3. Iseki T, et al. Percutaneous periarticular multi-drug injection at one day after total knee arthroplasty as a component of multimodal pain management: a randomized controlled trial. BMC Musculoskelet Disord. 2019;20:61.
4. Kurosaka K, et al. Comparison of early-stage and late-stage periarticular injection for pain relief after total hip arthroplasty: a double-blind randomized controlled trial. J Arthroplast. 2020;35:1275–80.
5. Mammoto T, et al. Short-term effects of early postoperative celecoxib administration for pain, sleep quality, and range of motion after total knee arthroplasty: a randomized controlled trial. J Arthroplast. 2021;36:526–31.
6. Tsukada S, et al. Early-stage periarticular injection during total knee arthroplasty may provide better postoperative pain relief than late-stage periarticular injection: a randomized-controlled trial. Knee Surg Sports Traumatol Arthrosc. 2019;27:1124–31.
7. Kodaka M, et al. A prospective, randomized, blinded trial to compare continuous epidural block and femoral nerve block for total knee arthroplasty. J Anesth Clin Res. 2014;5:2.
8. Sakai N, et al. Continuous femoral versus epidural block for attainment of 120° knee flexion after total knee arthroplasty: a randomized controlled trial. J Arthroplast. 2013;28:807–14.
9. Sakai N, et al. Equivalence of postoperative quadriceps strength during 1 or 0.5 mg ml (-1) levobupivacaine administration for continuous femoral nerve block following total knee arthroplasty: a double-blinded, randomized controlled trial. Eur J Anesthesiol. 2015;32:658–9.
10. Aso K, et al. Additional benefit of local infiltration of analgesia to femoral nerve block in total knee arthroplasty: double-blind

randomized control study. Knee Surg Sports Traumatol Arthrosc. 2019;27:2368–74.
11. Sato K, et al. Continuous versus single-injection sciatic nerve block added to continuous femoral nerve block for analgesia after total knee arthroplasty: a prospective, randomized, double-blind study. Reg Anesth Pain Med. 2014;39:225–9.
12. Sakai N, et al. Patient-controlled bolus femoral nerve block after knee arthroplasty: quadriceps recovery, analgesia, local anesthetic consumption. Acta Anesthesiol Scand. 2016;60:1461–9.
13. Nagafuchi M, et al. Femoral nerve block-sciatic nerve block vs. femoral nerve block-local infiltration analgesia for total knee arthroplasty: a randomized controlled trial. BMC Anesthesiol. 2015;15:182.
14. Gi E, et al. Effects of local infiltration analgesia for posterior knee pain after total knee arthroplasty: comparison with sciatic nerve block. J Anesth. 2014;28:696–701.
15. Tanikawa H, et al. Comparison of local infiltration of analgesia and sciatic nerve block in addition to femoral nerve block for total knee arthroplasty. J Arthroplast. 2014;29:2462–7.
16. Tanikawa H, et al. Local infiltration of analgesia and sciatic nerve block provide similar pain relief after total knee arthroplasty. J Orthop Surg Res. 2017;12:109.
17. Ishiguro S, et al. Day zero ambulation under modified femoral nerve block after minimally invasive surgery for total knee arthroplasty. J Orthop Surg Res. 2017;12:109.
18. Ohgoshi Y, et al. Use of IPACK block with continuous femoral triangle block for total knee arthroplasty: a clinical experience. J Clin Anesth. 2019;54:52–4.
19. Nakai T, et al. Controlling pain after total knee arthroplasty using a multimodal protocol with local periarticular injections. J Orthop. 2013;10:92–4.
20. Ukai T, et al. Comparison of periarticular multidrug infiltration and epidural catheter use in total knee arthroplasty: a prospective randomized controlled study. J Orthop Surg (Hong Kong). 2020;28:2309499020910663.
21. Tsukada S, et al. Postoperative epidural analgesia compared with intraoperative periarticular injection for pain control following total knee arthroplasty under spinal anesthesia: a randomized controlled trial. J Bone Joint Surg Am. 2014;96:1433–8.
22. Tsukada S, et al. Pain control after simultaneous bilateral total knee arthroplasty: a randomized controlled trial comparing periarticular injection. J Bone Joint Surg Am. 2015;97:367–73.
23. Kurosaka K, et al. Local infiltration analgesia versus continuous femoral nerve block in pain relief after total knee arthroplasty: a randomized controlled trial. J Arthroplast. 2016;31:913–7.
24. Ishida K, et al. Periarticular multimodal drug injection improves postoperative pain and functional recovery after total knee arthroplasty. J Orthop Sci. 2016;21:178–83.
25. Uesugi K, et al. Comparison of peripheral nerve block with periarticular injection analgesia after total knee arthroplasty: a randomized, controlled trial. Knee. 2014;21:848–52.
26. Fukuda T, et al. Comparison of peripheral nerve block with local infiltration analgesia regarding walking ability after total knee replacement: a retrospective, propensity-score matched-pair cohort study. J Orthop Surg (Hong Kong). 2020;28:2309499020931656.
27. Tsukada S, et al. The impact of including corticosteroid in a periarticular injection for pain control after total knee arthroplasty: a double-blind randomised controlled trial. Bone Joint J. 2016;98:194–200.
28. Yano T, et al. Relationship between dexamethasone added to periarticular anesthetic infiltration and postoperative nausea and vomiting following total knee arthroplasty under general anesthesia: a retrospective observational study. JA Clin Rep. 2020;6:66.
29. Hatayama K, et al. Comparison of intravenous and periarticular administration of corticosteroids in total knee arthroplasty: a prospective, randomized controlled study. J Bone Joint Surg Am. 2021;103:319–25.
30. Stein C. The control of pain in peripheral tissues by opioids. N Engl J Med. 1995;332:1685–90.
31. Iwakiri K, et al. Effect of periarticular morphine injection for total knee arthroplasty: a randomized, double-blind trial. J Arthroplast. 2017;32:1839–44.
32. Iwakiri K, et al. Local efficacy of periarticular morphine injection in simultaneous bilateral total knee arthroplasty: a prospective, randomized, double-blind trial. J Arthroplast. 2017;32:3637–42.
33. Miyamoto S, et al. The effect of morphine added to periarticular multimodal drug injection or spinal anesthesia on pain management and functional recovery after total knee arthroplasty. J Orthop Sci. 2018;23:801–6.
34. Yoshida K, et al. Respiratory depression after intraoperative opioid-containing periarticular multimodal drug injection for total hip and total knee arthroplasty. J Clin Anesth. 2019;54:147–8.
35. Ikeuchi M, et al. Local infusion analgesia using intra-articular double lumen catheter after total knee arthroplasty: a double blinded randomized control study. Knee Surg Sports Traumatol Arthrosc. 2013;21:2680–4.
36. Iseki T, et al. Percutaneous periarticular analgesic injection at one day after simultaneous bilateral total knee arthroplasty: an open-label randomized control trial. J Orthop Surg Res. 2021;16:357.
37. Nakagawa Y, et al. Benefit of subcutaneous patient-controlled analgesia after total knee arthroplasty. Asia Pac J Sports Med Arthrosc Rehabil Technol. 2019;18:18–22.
38. Tsuji M, et al. Effect of flurbiprofen and s-flurbiprofen patches on multimodal pain management after total knee arthroplasty: a prospective randomized controlled trial. J Arthroplast. 2020;35:2033–8.
39. Matsumoto S, et al. Transdermal fentanyl patch improves postoperative pain relief and promotes early functional recovery in patients undergoing primary total knee arthroplasty: a prospective, randomized, controlled trial. Arch Orthop Trauma Surg. 2015;135:1291–7.

83 Evaluation and Management of Painful Total Knee Arthroplasty

Masahiko Ikeuchi

Total knee arthroplasty (TKA) is the only treatment that can reconstruct joints severely deformed or destroyed by osteoarthritis (OA) or rheumatoid arthritis, and its effectiveness is undisputed. Although the acute phase of postoperative pain is usually the most intense after TKA, it is almost always relieved within approximately 3 months and decreases over the first year postoperatively [1] (Fig. 83.1). This pain-relieving effect is usually maintained for 20–30 years, but there are cases in which pain persists postoperatively, including implant malpositioning, joint instability, and chronic infection, or new pain arises several years after surgery, including implant loosening, periprosthetic fractures, and delayed infections, that require treatment. These problems require revision surgery, which often improves pain. However, chronic post-surgical pain (CPSP), which does not require reoperation and remains painful, has attracted attention in recent years. In this section, we discuss the pathogenesis of CPSP and the possibility of its prevention and treatment, with a focus on the differential diagnosis of post-TKA pain.

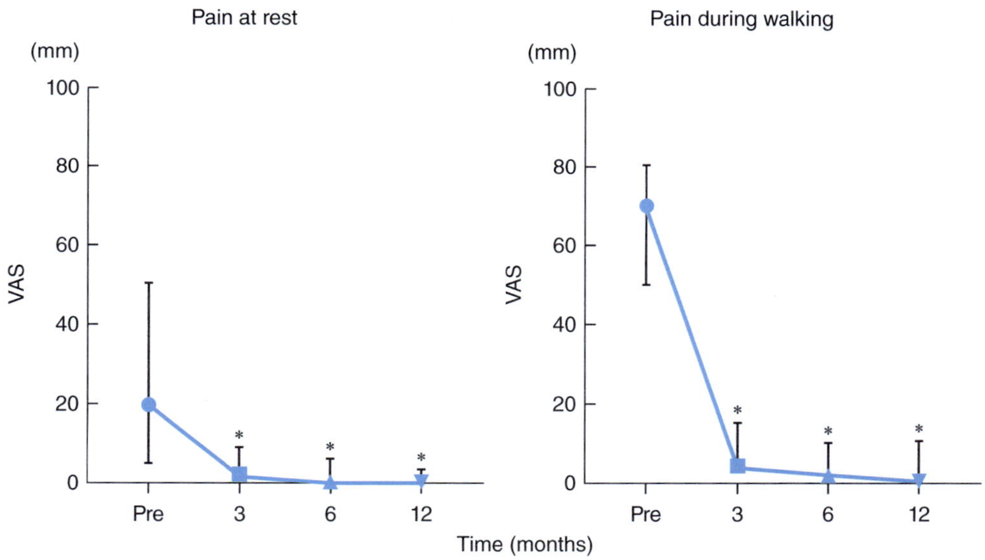

Fig. 83.1 Postoperative pain after TKA

The original version of the chapter has been revised. A correction to this chapter can be found at https://doi.org/10.1007/978-981-97-4920-1_89

M. Ikeuchi (✉)
Department of Orthopadic Surgery, Kochi Medical School, Kochi University, Nankoku, Kochi, Japan
e-mail: ikeuchim@kochi-u.ac.jp

S. Matsuda (ed.), *Advances in Total Knee Arthroplasty*, https://doi.org/10.1007/978-981-97-4920-1_83

1 Differential Diagnosis and Common Causes of Painful TKA

There are several causes of painful TKA [2] (Table 83.1). Its diagnosis is important because appropriate treatment can only be provided once the cause is identified. Medical history, physical examination, and plain radiography can provide pain-related clues. The medical history includes the patient's psychosocial background, history of surgery, pain treatment, and comorbidities. In addition to local conditions (erythema, heat, and presence of fistulas), the physical examination should include patellar tracking during knee flexion and extension, anteroposterior and varus-valgus instability, and medial-lateral thrust during gait. A simple radiographic examination including antero-posterior, lateral, and axial views of the knee joint is required, with oblique views added as necessary. The presence of fracture or osteolysis around the implant, implant malpositioning, loosening, overhang or underhang, and joint lines are checked. The common causes of painful TKA that should be differentiated include the following.

1.1 Surgical Site Infections

When treating patients with painful TKA, it is important to exclude infection. Acute infection with strong inflammatory findings or chronic infections with fistulas and implant loosening (Fig. 83.2) can be easily diagnosed but are often difficult to recognize. If infection is suspected due to the findings of the physical examination, blood tests, or radiographic imaging, joint fluid should be obtained immediately to confirm pus accumulation. The collected joint fluid is subjected to bacterial culture tests and various fluid tests. The results of these tests are compared with the diagnostic criteria for peri-prosthetic infection [3] (Table 83.2) to differentiate between infectious and non-infectious conditions. Once infection is diagnosed, patients should be treated promptly. In cases of acute infection, it is possible to preserve the implants if irrigation and debridement are performed at an early stage. Patients with chronic infection often require implant removal and revision.

Table 83.1 Differential diagnosis of painful TKA (Reprinted with permission. © 2008 The British Editorial Society of Bone & Joint Surgery. All rights reserved. Mandalia V et al: J Bone Joint Surg[Br] 2008; 90-B(3): 265–271.) [2]

Intrinsic factors	
Infection	Early post-operative infection
	Acute haematogeneous infection
	Late (chronic) infection
Instability	Mediolateral
	Anteroposterior
	Flexion/extension
Malalignment	Coronal plane
	Sagittal plane
	Rotational
Soft-tissue impingement	Patellar clunk
	Fabellar impingement
	Popliteus tendon impingement
	Component overhang
Arthrofibrosis	
Wear, osteolysis and aseptic loosening	
Recurrent haemarthrosis	
Extensor mechanism problems	Patellar maltracking
	Extensor mechanism disruption (patellar fracture, patellar tendon rupture, quadricep tendon rupture)
	Unresurfaced patella
	Undersized patella button with lateral facet impingement
	Oversized patella button with overstuffing of patellofemoral joint
	Patella baja + alta

Extrinsic factors	
Hip pathology	
Neurological	Spinal disorders – stenosis/prolapsed disc
	Neuroma
	Complex regional pain syndrome
Vascular	Deep vein thrombosis
	Vascular claudication
Pes anserinus bursitis	
Stress fracture and peri-prosthetic fracture	
Tendinopathy (patella/quadricep)	
Heterotopic ossification	
Psychological disorder	
Others	Paget's disease
	Pigmented villonodular synovitis
	Rheumatoid arthritis
	Foot and ankle pathology

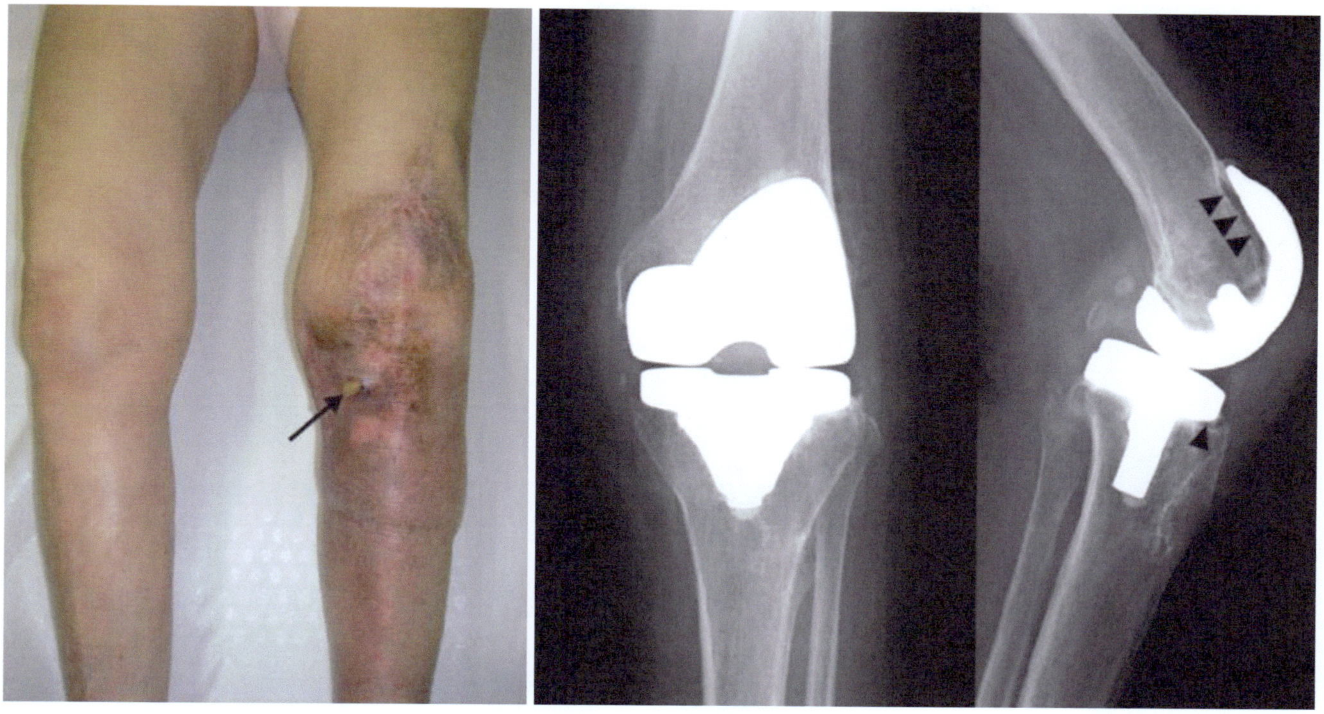

Fig. 83.2 Woman in her 70 s with delayed chronic infection after TKA. Fistula formation (arrow) and bone resorption are visible around the implant (arrowhead)

Table 83.2 Diagnostic criteria for periprosthetic joint infection. (International Consensus Meeting on Musculoskeletal Infection 2018)

Major Criteria(at least one of the following)				Decision
Two positive growths of the same organism using standard culture methods				Infected
Sinus tract with evidence of communication to the joint or visualization of the prosthesis				
Preoperative diagnosis		Minor criteria[a]	Score	Decision
	Serum	Elevated CRP (> 1 mg/dL) or D-dimer (> 860 ng/mL)	2	≧ 6 infected 2~5 inconclusive 0~1 not infected
	Synovia	Elevated ESR (> 30 mm/h)	1	
		Elevated synovial WBC count (> 3000/μL) or leukocyte esterase (++)	3	
		Positive α-defensin	3	
		Elevated synovial PMN (> 80%)	2	
		Elevated synovial CRP (>6.9 mg/L)	1	
Intraoperative diagnosis	Inconclusive pre-op score or dry tap		Score	Decision
	Preoperative score		–	≧6 infected 4~5 inconclusive ≦3 not infected
	Positive histology		3	
	Positive purulence		3	
	Single positive culture		2	

[a]These criteria were never validated on acute infections

1.2 Non-infectious Diseases that Are Indications for Reoperation

The typical conditions requiring revision surgery include joint instability and implant malpositioning and loosening. In addition, periprosthetic fractures and disruption of the knee extensor mechanism due to trauma require revision surgery such as osteosynthesis and tissue repair/reconstruction. In most cases, these diseases can be diagnosed by history, physical examination, and simple radiography; however, additional examinations such as stress radiography, computed tomography (CT), and magnetic resonance imaging are performed as necessary (Fig. 83.3). A diagnosis of loosening is generally made when a radiolucent zone of 2 mm or more is observed around the implant. In addition to the coronal and sagittal views, it is important to evaluate the rotational position of the implant in the transverse plane on CT. Careful observation of the combined plain radiograph and CT images can also reveal the diagnosis of implant overhang or underhang and lateral facet impingement between

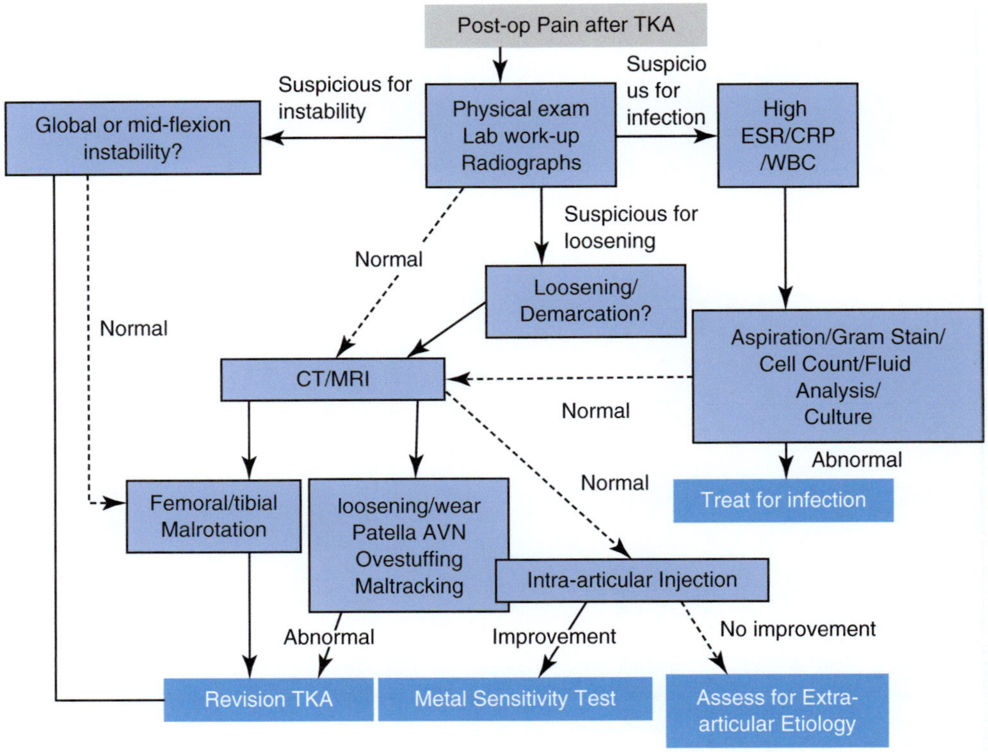

Fig. 83.3 Diagnostic algorithm for painful TKA. (Park CN, et al. Orthopedics 2016; 39: e246-e252)

the lateral patellar border and femoral component. Joint instability is evaluated in the varus-valgus and anteroposterior directions at full extension, mid-flexion, and 90° of flexion. Hyperextended knees can also cause pain after TKA. Patellar clunk syndrome, in which crepitus with pain is observed during knee joint extension, is diagnosed by physical examination and often appears normal on imaging. It is often resolved by the arthroscopic excision of a fibrous nodule proximal to the patella.

1.3 Others

In addition to the diseases and conditions described above, other causes of pain can be identified, such as metal allergy, pain originating from the spine or hip joint, and neuroma of the infrapatellar branch of the saphenous nerve. However, there have been cases of unexplained pain despite the use of various tests. CPSP was first defined by Macrae et al. as a condition that met the following four criteria: (1) occurrence after surgery; (2) persistence for more than 2 months; (3) exclusion of pain due to other causes, such as residual tumor or infection; and (4) exclusion of pain that existed before surgery [4]. In the new International Classification of Diseases, Eleventh Revision, of the World Health Organization, which was revised in 2019, CPSP is classified as "pain that lasts more than 3 months or is recurrent," thereby unifying the above requirement [2] as 3 months or longer.

2 Chronic Post-surgical Pain

2.1 Epidemiology

A systematic review by Beswick et al. found that the incidence of "moderate or high CPSP" at >3 months postoperative was 10–34%, and in a limited number of high-quality studies, the incidence was reportedly 20% [5]. The incidence of "moderate or high CPSP" is reportedly 10–34% at >3 months postoperative and approximately 20% in high-quality studies [5]. Incidentally, postoperative pain after total hip arthroplasty (THA) is less severe than that after TKA, and the incidence of CPSP after THA was reportedly 9% in the same systematic review. According to Aso et al., the incidence of "moderate or worse CPSP" was 10.3% at 6 months and 8.8% at 1 year postoperatively [1]. The incidence of CPSP after TKA in Japan may differ from that in Europe and the United States, as indications for surgery and postoperative pain management differ among countries and insurance systems. Ethnic differences in pain sensitivity have also been reported.

2.2 Pathophysiology

The most common surgical approach for TKA is the medial parapatellar approach, in which the medial femoral cutaneous nerve and infrapatellar branch of the saphenous nerve are severed subcutaneously through a skin incision. The medial retinacular nerve, a branch of the femoral nerve distributed medially to the joint capsule, is severed. Neuropathic pain originating from these peripheral neuropathies accounts for approximately 20% of CPSP cases [6]. Furthermore, chronic pain associated with peripheral neuropathy causes central sensitization, which further prolongs the pain and renders it intractable. Patients undergoing TKA experience chronic knee pain, and central sensitization occurs before surgery. This central sensitization is normalized after TKA [7] and considered maintained by peripheral input. The risk factors for CPSP reportedly include female sex, younger age, severe knee pain, pain in other parts of the body, central sensitization, mental health problems, and catastrophic thinking 1, [8–10].

2.3 Prevention

The most important preoperative step is to accurately assess the indications for surgery. TKA is generally indicated for patients with end-stage OA who have lost their articular cartilage. However, surgery is often indicated for patients with symptoms and functional impairment, even if the disease has not yet reached the end stage. Postoperative pain tends to be prolonged when TKA is performed on knees with mild arthritis [11], and other treatment options must be thoroughly considered before surgery.

Many patients who undergo TKA are elderly and have degeneration in the knee as well as various other musculoskeletal organs, which can be a source of pain. Periprosthetic knee pain associated with lumbar nerve root disease or hip OA can easily be mistaken for knee pain. The lidocaine test is effective for differential diagnosis.

Preoperative risk factors that can be modified by interventions include preoperative knee pain and central sensitization. Mental health and catastrophic thoughts about pain are also considered modifiable through intervention [12]. Pain management in the acute postoperative period after TKA has progressed significantly in recent years. In addition to various analgesics, multifaceted pain management is currently performed, including intravenous patient-controlled analgesia, continuous epidural anesthesia, peripheral nerve blocks, and periarticular cocktail injections. To date, no method has been established for preventing CPSP after TKA. However, CPSP is reportedly reduced in open chest surgery and breast cancer surgery by thorough pain management in the acute phase [13], and future studies are warranted.

2.4 Treatment

Once prolonged, postoperative pain is difficult to treat and should be prevented. Analgesics, including common nonsteroidal anti-inflammatory drugs, are often ineffective for managing CPSP after TKA. Recently, drugs such as pregabalin, opioids, and selective serotonin reuptake inhibitors have become available. However, their efficacy for CPSP after TKA is limited, and no treatment protocols have been established. Other treatment methods that have been used include the intra-articular injection of botulinum toxin [14], radiofrequency thermal coagulation [15], peripheral neurotomy [16], ultrasound-guided steroid injection [17], and spinal cord electrical stimulation [18]. In addition to these treatment methods, exercise therapy as a physical approach and cognitive-behavioral therapy as a psychosocial approach should be provided in a multidisciplinary manner.

References

1. Aso K, et al. Chronic postsurgical pain after total knee arthroplasty: a prospective cohort study in a Japanese population. Mod Rheumatol. 2021;18:1–11.
2. Mandalia V, et al. Evaluation of patients with a painful total knee replacement. J Bone Joint Surg Br. 2008;90:265–71.
3. Parvizi J, et al. The 2018 definition of periprosthetic hip and knee infection: an evidence-based and validated criteria. J Arthroplast. 2018;33(1309–1314):e2.
4. Macrae WA. Chronic pain after surgery. Br J Anaesth. 2001;87:88–98.
5. Beswick AD, et al. What proportion of patients report long-term pain after total hip or knee replacement for osteoarthritis? A systematic review. BMJ Open. 2012;2:e000435.
6. Phillips JR, et al. The natural history of pain and neuropathic pain after knee replacement: a prospective cohort study of the point prevalence of pain and neuropathic pain to a minimum three-year follow-up. Bone Joint J. 2014;96-B:1227–33.
7. Graven-Nielsen T, et al. Normalization of widespread hyperesthesia and facilitated spatial summation of deep-tissue pain in knee osteoarthritis. Arthritis Rheum. 2012;64:2907–16.
8. Lewis GN, et al. Predictors of persistent pain after total knee arthroplasty: a systematic review and meta-analysis. Br J Anaesth. 2015;114:551–61.
9. Singh JA, et al. The impact of gender, age, and preoperative pain severity on pain after TKA. Clin Orthop Relat Res. 2008;466:2717–23.
10. Kim SH, et al. Influence of centrally mediated symptoms on postoperative pain in osteoarthritis patients undergoing total knee arthroplasty: a prospective observational evaluation. Pain Pract. 2015;15:E46–53.
11. Valdes AM, et al. Inverse relationship between preoperative radiographic severity and postoperative pain in patients with osteoarthritis. Semin Arthritis Rheum. 2012;41:568–75.
12. Khatib Y, et al. Do psychological factors predict poor outcomes in patients undergoing TKA? A systematic review. Clin Orthop Relat Res. 2015;473:2630–8.
13. Andreae MH, Andreae DA. Regional anesthesia to prevent chronic pain after surgery: a Cochrane systematic review and meta-analysis. Br J Anaesth. 2013;111:711–20.
14. Singh JA, et al. Intraarticular botulinum toxin a for refractory painful total knee arthroplasty: a randomized controlled trial. J Rheumatol. 2010;37:2377–86.
15. Ikeuchi M, et al. Percutaneous radiofrequency treatment for refractory anteromedial pain of osteoarthritic knees. Pain Med. 2011;12:546–51.

16. Dellon AL, et al. Partial denervation for persistent neuroma pain after total knee arthroplasty. Clin Orthop Relat Res. 1995;316: 145–50.
17. Clendenen S, et al. Infrapatellar saphenous neuralgia after TKA can be improved with ultrasound-guided local treatments. Clin Orthop Relat Res. 2015;473:119–25.
18. Lowry AM, Simopoulos TT. Spinal cord stimulation for treatment of chronic knee pain following total knee replacement. Pain Physician. 2010;13:251–6.

84 Blood Management Strategy

Sachiyuki Tsukada

TKA is associated with considerable perioperative blood loss that sometimes necessitates allogeneic blood transfusions. Allogeneic blood transfusions carry the risk of transfusion-associated viral infections and graft-versus-host disease [1]. Furthermore, allogeneic blood transfusion in the perioperative period after TKA is an independent risk factor for periprosthetic joint infections and DVT [2]. Numerous studies have focused on managing blood loss during TKA to reduce the need for allogeneic blood transfusions.

1 Bleeding Characteristics of TKA

1.1 Use of Bone Cement

When the prostheses are cemented, bleeding from the resected bone surface is suppressed by the cement layer. Therefore, postoperative bleeding is lower with cemented TKA than with cementless TKA [3]. Countermeasures such as treating the osteotomy surface with fibrin glue have been attempted for cementless TKA [4]. There was no difference in postoperative blood loss between cementless and hybrid TKAs, in which only the tibial component was cemented and no cement was used to fix the femoral component [5]. One study showed that hybrid TKA resulted in a similar amount of blood loss as cemented TKA [6].

1.2 Rheumatoid Arthritis

Blood loss is greater in patients diagnosed with RA than in those diagnosed with OA, and allogeneic blood transfusion is more likely to be required in RA [3]. Fibrinogen levels decrease in patients receiving tocilizumab, a biologic agent [7]. One study reported that patients with RA receiving tocilizumab experienced more blood loss in the perioperative period of TKA [7].

2 Pre-deposit Autologous Blood Transfusion

Pre-deposit autologous blood transfusion is a technique in which blood is collected before surgery and transfused into patients after surgery. Theoretically, there is no risk of transfusion-associated viral infection or graft-versus-host disease. In a study of patients with preoperative hemoglobin levels of less than 13.0 g/dL, the number of patients requiring allogeneic blood transfusion was lower among those who received pre-deposit autologous blood transfusion than among those who did not receive such transfusion [1].

However, recent advances in perioperative blood management have re-evaluated the need for pre-deposit autologous blood transfusions. A study comparing patients who underwent pre-deposit autologous blood transfusion with those who did not undergo it but underwent bleeding control measures, including tranexamic acid administration, reported no difference in the number of patients who required allogeneic blood transfusion after unilateral TKA [8]. Another study compared the rate of necessary allogeneic transfusion for patients undergoing simultaneous bilateral TKA and reported that allogeneic transfusion requirements in patients who did not deposit autologous blood in advance were non-inferior to those who pre-deposit blood [9].

3 Tourniquet

Tourniquets are used to apply pressure higher than the arterial pressure to the proximal femur to minimize bleeding during TKA. The use of a tourniquet has been reported to reduce the total amount of perioperative bleeding compared with no tourniquet application [10].

S. Tsukada (✉)
Department of Orthopaedic Surgery, Hokusuikai Kinen Hospital, Mito, Ibaraki, Japan
e-mail: s8058@nms.ac.jp

S. Matsuda (ed.), *Advances in Total Knee Arthroplasty*, https://doi.org/10.1007/978-981-97-4920-1_84

Obstruction of blood flow caused by a tourniquet may lead to complications, such as DVT and delayed wound healing. A randomized controlled trial comparing patients with and without a tourniquet showed that distal DVT was more prevalent in patients in whom a tourniquet was applied [11].

While utilizing a high-pressure tourniquet during TKA is beneficial for reducing blood flow to the knee, excessively high tourniquet pressure may increase the risk of complications due to impaired blood circulation. A study comparing patients with a tourniquet pressure of 350 mmHg and those with a tourniquet pressure of each patient's systolic blood pressure plus 100 mmHg found no significant difference in either bleeding volume and rate of complications [12].

Although the tourniquet is generally deflated after wound closure, it is also common to release the tourniquet immediately before wound closure to identify the bleeding points. A study comparing these two approaches to tourniquet deflation revealed that total blood loss was lower when the tourniquet was deflated after wound closure [13]. A study conducted using transesophageal echocardiography during tourniquet deflation revealed the presence of embolic masses measuring ≥1 cm in diameter in the hearts of all patients [14].

In summary, the use of a tourniquet should be thoroughly assessed in patients who are at high risk of DVT or have a history of cardiac disease.

4 Drain Clamping

Drain clamping is a technique in which saline and epinephrine are administered into the joint cavity immediately after TKA, and the drain is released after clamping for a certain period. This method has been reported to reduce postoperative blood loss compared to that without clamping [15, 16]. The addition of tranexamic acid or carbazochrome to intra-articular drugs has also been reported to be effective [17–20]. A previous study reported fewer complications when a clamped drain was released 1 h after clamping than when it was released 24 h after clamping [21].

However, as described below, there have been an increasing number of negative reports on intra-articular drain placement. There are many reports on the local administration of drugs into the joint cavity without drain [22].

5 Intra-articular Drain

Drains have traditionally been inserted into the joint cavity because they help remove blood from the knee joint and are believed to reduce postoperative swelling and the potential for wound complications. Nevertheless, there have been numerous adverse reports on the clinical effectiveness of intra-articular drains in TKA.

A comparison of patients who underwent bilateral primary TKA with intra-articular drains in both knees, only in one knee, and no intra-articular drains in both knees showed that patients with intra-articular drains experienced more bleeding, and there was no difference in complications among the three groups [23]. The mean length of hospital stay was longer in patients with intra-articular drains than in those without intra-articular drains [24].

6 Tranexamic Acid

Tranexamic acid is a hemostatic agent developed in Japan [25]. The risk of bleeding events must always be considered in relation to the increased risk of thromboembolism after TKA. However, tranexamic acid is theoretically not associated with an increased risk of thromboembolism because it specifically targets the fibrinolytic system without affecting the coagulation system. Therefore, tranexamic acid is considered the ideal drug for perioperative use in TKA.

Tranexamic acid is most commonly administered intravenously. It can also be administered orally and intra-articularly. A new method of administration has been proposed in which tranexamic acid is injected directly into the surgical field. Several studies have investigated the most effective administration routes and doses.

6.1 Intravenous Administration

When a tourniquet is used, intravenous tranexamic acid should be administered immediately before the release. Patients who received 1000 mg intravenous tranexamic acid immediately before deflating the tourniquet had less drainage than those who did not receive tranexamic acid [26].

The effectiveness of postoperative intravenous tranexamic acid administration in addition to intraoperative intravenous tranexamic acid has also been reported [27]. The amount of blood loss was reported to be lower in patients who received an additional dose of intravenous tranexamic acid 3 h after surgery, in addition to the dose administered immediately before releasing the tourniquet, than in patients who received tranexamic acid immediately before releasing the tourniquet [28]. In a study examining the most effective timing of administration of 2000 mg intravenous tranexamic acid, 1000 mg each administered preoperatively and at the time of tourniquet release was reported to cause the least blood loss [29].

Conflicting results have been reported regarding the efficacy of additional postoperative tranexamic acid administration for TKA without a tourniquet, which is thought to cause less postoperative bleeding than TKA with a tourniquet. In a study of patients who received combined intraoperative intravenous and intra-articular tranexamic acid administration, there was no difference in perioperative blood loss between patients who received additional postoperative tranexamic acid administration and those received a placebo [2].

6.2 Combined Intravenous and Intra-articular Administration

In a comparison of patients treated with intravenous tranexamic acid alone and those treated with intravenous plus intra-articular administration, significantly less blood loss was reported in the latter group for both unilateral and bilateral primary TKAs [22, 30].

6.3 Addition of Tranexamic Acid to Periarticular Multidrug Injections

Periarticular multidrug injections, in which multiple drugs with different mechanisms of action are injected around the joint, are commonly used for pain management after TKA [31]. The addition of tranexamic acid to a periarticular multidrug injection solution has been reported to be effective in reducing perioperative blood loss [32, 33].

7 Continuation of Antithrombotic Drugs

With the increasing number of older adults undergoing TKA, there has been a corresponding increase in the number of patients taking antiplatelet and anticoagulant medications. Ensuring safe TKA performance in patients receiving chronic antithrombotic therapy is of paramount importance.

In the past, it was common practice to discontinue the use of antiplatelet agents and anticoagulants for a specific duration prior to surgery, with heparin often administered as an alternative medication. However, recently, there have been many negative reports on the effectiveness of heparin bridging therapy. In a randomized controlled trial including 1884 patients (9.2% of whom underwent orthopedic surgery) who received heparin bridging therapy versus a placebo, there was no difference in the frequency of thrombotic complications between the two groups [34].

There have been reports of attempts to perform TKA in patients who continue to take oral antithrombotic agents without heparin bridging therapy [35, 36]. Although the risk of thrombotic complications is not increased when antithrombotic medication is continued, there is concern regarding the potential increase in perioperative blood loss. One study found no difference in perioperative blood loss in patients continuing antiplatelet agents compared with those who discontinued antiplatelet agents [35]. Another study found no difference in total perioperative blood loss in patients who underwent TKA without withdrawal of antithrombotic drugs, including warfarin and direct-acting oral anticoagulants, compared to patients who did not take antithrombotic drugs [36].

References

1. Minoda Y, et al. Blood management for patients with hemoglobin levels lower than 130 g/l in total knee arthroplasty. Arch Orthop Trauma Surg. 2004;124:317–9.
2. Tsukada S, et al. Intraoperative intravenous and intra-articular plus postoperative intravenous tranexamic acid in total knee arthroplasty: a placebo-controlled randomized controlled trial. J Bone Joint Surg Am. 2020;102:687–92.
3. Fujimoto H, et al. Blood loss in patients for total knee arthroplasty. Knee Surg Sports Traumatol Arthrosc. 2003;11:149–54.
4. Akizuki S, et al. A new method of hemostasis for cementless total knee arthroplasty. Bull Hosp Jt Dis. 1997;56:222–4.
5. Ishii Y, et al. Perioperative blood loss in cementless or hybrid total knee arthroplasty without patellar resurfacing: a prospective, randomized study. J Arthroplast. 2005;20:972–6.
6. Ishii Y, et al. Characteristics of hidden blood loss following hybrid total knee arthroplasty. J Orthop Sci. 2021;26:1064–8.
7. Imamura H, et al. Tocilizumab treatment in patients with rheumatoid arthritis is associated with reduced fibrinogen levels and increased blood loss after total knee arthroplasty. Mod Rheumatol. 2018;28:976–80.
8. Sasanuma H, et al. Efficient strategy for controlling postoperative hemorrhage in total knee arthroplasty. Knee Surg Sports Traumatol Arthrosc. 2011;9:921–5.
9. Tsukada S, et al. Predeposited autologous blood transfusion in single-anesthetic bilateral total knee arthroplasty with modern blood conservation strategy. JBJS Open Access. 2023;8:e22.00125.
10. Nishiguchi M, et al. Estimation of blood loss in total knee arthroplasty with and without tourniquet. Acta Med Nagasaki. 2009;53:105–9.
11. Mori N, et al. Use of a pneumatic tourniquet in total knee arthroplasty increases the risk of distal deep vein thrombosis: a prospective, randomized study. Knee. 2016;23:887–9.
12. Ishii Y, et al. Effect of tourniquet pressure on perioperative blood loss associated with cementless total knee arthroplasty: a prospective, randomized study. J Arthroplast. 2005;20:325–30.
13. Ishii Y, et al. Effect of the timing of tourniquet release on perioperative blood loss associated with cementless total knee arthroplasty. J Arthroplast. 2005;20:977–83.
14. Wauke K, et al. Comparative study between thromboembolism and total knee arthroplasty with or without tourniquet in rheumatoid arthritis patients. Arch Orthop Trauma Surg. 2002;122:442–6.
15. Ryu J, et al. The postoperative drain-clamping method for hemostasis in total knee arthroplasty: reducing postoperative bleeding in total knee arthroplasty. Bull Hosp Jt Dis. 1997;56:251–4.
16. Tsumura N, et al. A prospective comparison of clamping the drain or postoperative salvage of blood in reducing blood loss after total knee arthroplasty. J Bone Joint Surg Br. 2006;88:49–53.
17. Ishida K, et al. Intra-articular injection of tranexamic acid reduces not only blood loss but also knee joint swelling after total knee arthroplasty. Int Orthop. 2011;35:1639–45.

18. Onodera T, et al. Risk of deep venous thrombosis in drain clamping with tranexamic acid and carbazochrome sodium sulfonate hydrate in total knee arthroplasty. J Arthroplast. 2012;27:105–8.
19. Mutsuzaki H, et al. Intra-articular injection of tranexamic acid via a drain plus drain-clamping to reduce blood loss in cementless total knee arthroplasty. J Orthop Surg Res. 2012;7:32.
20. Kamatsuki Y, et al. Intra-articular 1 g tranexamic acid administration during total knee arthroplasty is safe and effective for the reduction of blood loss and blood transfusion. Eur J Orthop Surg Traumatol. 2019;29:1737–41.
21. Yamada K, et al. Comparison between 1-hour and 24-hour drain clamping using diluted epinephrine solution after total knee arthroplasty. J Arthroplast. 2001;16:458–62.
22. Iseki T, et al. Intravenous tranexamic acid only versus combined intravenous and intra-articular tranexamic acid for perioperative blood loss in patients undergoing total knee arthroplasty. Eur J Orthop Surg Traumatol. 2018;28:1397–402.
23. Watanabe T, et al. Closed suction drainage is not necessary for total knee arthroplasty: a prospective study on simultaneous bilateral surgeries of a mean follow-up of 5.5 years. J Arthroplast. 2016;31:641–5.
24. Nishitani K, et al. A multivariate analysis on the effect of no closed suction drain on the length of hospital stay in total knee arthroplasty. Knee Surg Relat Res. 2019;31:25–30.
25. Okamoto S, et al. Amino-methyl-cyclohexane-carboxylic acid: AMCHA a new potent inhibitor of fibrinolysis. Keio J Med. 1962;11:105–15.
26. Ishii Y, et al. Effect of a single injection of tranexamic acid on blood loss after primary hybrid TKA. Knee. 2015;22:197–200.
27. Ido K, et al. Reduction of blood loss using tranexamic acid in total knee and hip arthroplasties. Arch Orthop Trauma Surg. 2000;120:518–20.
28. Iwai T, et al. Repeat-dose intravenous tranexamic acid further decreases blood loss in total knee arthroplasty. Int Orthop. 2013;37:441–5.
29. Tanaka N, et al. Timing of the administration of tranexamic acid for maximum reduction in blood loss in arthroplasty of the knee. J Bone Joint Surg Br. 2001;83:702–5.
30. Tsukada S, et al. Combined intravenous and intra-articular tranexamic acid in simultaneous bilateral total knee arthroplasty without tourniquet use. JBJS Open Access. 2017;2:e0002.
31. Ikeuchi M, et al. Effects of dexamethasone on local infiltration analgesia in total knee arthroplasty: a randomized controlled trial. Knee Surg Sports Traumatol Arthrosc. 2014;22:1638–43.
32. Yozawa S, et al. Periarticular injection of tranexamic acid reduces blood loss and the necessity for allogeneic transfusion after total knee arthroplasty using autologous transfusion: a retrospective observational study. J Arthroplast. 2018;33:86–8.
33. Hirose H, et al. Periarticular injection of tranexamic acid promotes early recovery of the range of knee motion after total knee arthroplasty. J Orthop Surg (Hong Kong). 2019;27:2309499019864693.
34. Douketis JD, et al. Perioperative bridging anticoagulation in patients with atrial fibrillation. N Engl J Med. 2015;373:823–33.
35. Tsukada S, et al. Continuing versus discontinuing antiplatelet drugs, vasodilators, and/or cerebral ameliorators on perioperative total blood loss in total knee arthroplasty without pneumatic tourniquet. Arthroplast Today. 2018;4:89–93.
36. Tsukada S, et al. A strategy of continued antiplatelet agents, vitamin K antagonists, and direct oral anticoagulants throughout the perioperative period of total knee arthroplasty in patients receiving chronic antithrombotic therapy. JBJS Open Access. 2019;4:e0057.1–6.

85 Deep Vein Thrombosis (DVT) Prevention

Kohei Kawaguchi

1 DVT and Pulmonary Embolism (PE) in Total Knee Arthroplasty (TKA)

DVT is a condition in which a thrombus forms in a vein deeper than the fascia. PE is a condition in which a functional pulmonary vessel is occluded by an embolus, especially when the embolus is a thrombus, which is called pulmonary thromboembolism (PTE). The term venous thromboembolism (VTE) is widely used because DVT and PTE are two consecutive conditions. VTE can be classified into asymptomatic and symptomatic VTE, with symptomatic VTE being the most clinically problematic form. However, because of its low incidence, many clinical studies have been conducted on asymptomatic VTE for prevention and screening. In Japan, "Guidelines for the Prevention of Pulmonary Thromboembolism/Deep Vein Thrombosis" [1] was published in 2004, while "Guidelines for the Prevention of Venous Thromboembolism" [2] was published by the Japanese Orthopedic Association in 2008, and prophylactic anticoagulants are now covered by insurance for orthopedic surgery, including TKA. However, as anticoagulants sometimes induce serious bleeding events, the uniform use of anticoagulants in all patients, including asymptomatic patients, to prevent VTE has become controversial [3]. In the USA, the American College of Chest Physicians (ACCP) issued new guidelines in 2012 [4] that differ significantly from previous guidelines in that they define all VTE, symptomatic and fatal PTE, and symptomatic DVT as targets for prophylaxis and proposed a policy for the prophylaxis of all VTE while considering the balance between efficacy and safety. Similarly, in 2017, the Japanese Orthopedic Association published the "Guidelines for the Prevention of Symptomatic Venous Thrombosis 2017" [5], which changed the scope of prophylaxis from all VTE to symptomatic and fatal PTE and symptomatic DVT only and presented a policy for the prevention of VTE in orthopedic surgery. This section describes the current necessity, methods, and problems associated with symptomatic VTE prophylaxis based on these guidelines.

DVT in the inferior vena cava, iliac vein, or femoral vein is called "proximal (central) type DVT," while DVT in the anterior tibial vein, posterior tibial vein, or peroneal vein is called "distal (lower leg) type DVT." Most clinically problematic, fatal, and symptomatic PTEs are thought to be caused by proximal DVT. Virchow proposed the following three DVT etiological factors: (1) blood stagnation, (2) vascular endothelial injury, and (3) increased coagulability. PTE is classified into three types according to the shape of the thromboembolus: extensive PTE, which occludes two or more pulmonary lobar arteries; subextensive PTE, which occludes fewer than two lobar arteries from the regional pulmonary artery; and microscopic PTE, which occludes a small artery. DVTs formed in deep veins are thought to be released from the vessel wall and occlude the pulmonary artery as an embolus, resulting in PTE, suggesting that thrombi in the inferior vena cava may extend to the central side, forming a secondary thrombus that releases and causes PTE [6]. Therefore, the prevention of DVT in the lower extremities is important to prevent the development of symptomatic PTE.

2 DVT in TKA

The 2012 ACCP guidelines [4] and the 2009 guidelines of pulmonary thromboembolism and deep vein thrombosis by the Japanese Circulation Society [7] stated that TKA is associated with a high risk of VTE, and the Guidelines for the Prevention of Symptomatic Venous Thrombosis 2017 [5] also state that TKA, along with total hip arthroplasty (THA), carries a high risk of developing VTE. VTE is one of the most important complications of TKA and its prevention is important.

K. Kawaguchi (✉)
Department of Orthopaedics, The University of Tokyo, Bunkyo, Tokyo, Japan

S. Matsuda (ed.), *Advances in Total Knee Arthroplasty*, https://doi.org/10.1007/978-981-97-4920-1_85

2.1 Occurrence Frequency

The 2012 ACCP guidelines [4] reported that DVT occurs in 41–85% of patients, proximal DVT in 5–22%, and symptomatic PE in 1.5–10% after TKA, THA, and hip fracture surgery when special prophylaxis was not used. In Japan, DVT (distal type, 60.3%; proximal type, 2.2%) and asymptomatic PTE occurred in 62.5% and 1.8% of patients post-TKA, respectively [8], while DVT and symptomatic PTE occurred in 48.6% (distal type, 34.1%; proximal type, 14.5%) and 1.4% [9] of patients post-TKA, respectively. In a Japanese report using the national database, the incidence of PE was 0.55% in 27,542 patients who underwent TKA and THA [10].

2.2 Symptoms

Clinical symptoms of DVT include subjective symptoms such as pain in the affected limb, edema, swelling, and superficial vein distention. Objective findings include the Homans sign (pain in the calf area when the ankle joint is dorsiflexed) and Lowenberg sign (grasping pain in the calf area, calf pain by applying low pressure with a manchette). These symptoms are associated with thrombi of the complete occlusion type and patients may be asymptomatic in cases of incomplete occlusion or when the collateral pathway is fully formed [11]. The symptoms of PTE include decreased oxygen saturation (SpO_2), cold sweats, chest pain, dyspnea, hypotension, shock, cardiac arrest, and syncope. The onset of PTE is often observed after the patient is released from bed rest after surgery (at the start of bed rest or walking) and the onset of PTE should be strongly suspected when the patient presents with the above symptoms after bed rest is terminated.

2.3 Risk Factors

To prevent symptomatic VTE, it is necessary to understand its risk factors. The 2011 AAOS guidelines [12] reported that a history of symptomatic VTE is a risk factor for postoperative symptomatic VTE after TKA. Obesity is also a risk factor for VTE due to exercise limitations and decreased fibrinolytic activity [13], and prolonged bed rest is a risk factor for VTE due to decreased muscle contractions of lower legs, resulting in venous stagnation [14]. Antithrombin deficiency, protein C deficiency, and protein S deficiency, which are present in congenital thrombophilia, are also risk factors [15]. Antiphospholipid antibody syndrome is also a risk factor for VTE because antiphospholipid antibodies bind to the phospholipids necessary for protein C activation, which promotes thrombogenesis. The release of clots from tumor cells and direct damage to venous endothelial cells by malignant cells are believed to cause DVT formation [16]. Female sex, rheumatoid arthritis [17], dyslipidemia [18], and perioperative blood loss [19] have also been reported as risk factors for DVT. Some reports indicate that taniquet use is a risk factor for DVT; however, no consensus has been reached [20, 21].

3 DVT Prevention

Guidelines for the Prevention of Symptomatic Venous Thromboembolism 2017 [5] recommend either physical or pharmacoprophylaxis, or a combination of both, for patients with TKA at a high risk of developing symptomatic VTE. In particular, it is recommended that physical therapy and pharmacoprophylaxis be used in combination in patients with a history of VTE. However, the authors stated that pharmacoprophylaxis should be used with caution in patients with a high risk of bleeding, that is, the choice of prophylaxis should be based on a balance between the risk of VTE and the risk of bleeding from pharmacoprophylaxis. In other words, there is no standard prophylaxis for TKA at this time, and prophylaxis should be determined on a case-by-case basis, considering the risk of VTE and its side effects.

3.1 Physical Prophylaxis

1. Active lower-limb exercise, early weaning, and gait

Ankle plantar dorsiflexion exercises in the bed activate the muscle pump function and are highly effective movements that can be performed even before getting out of bed rest [22]. The 2011 AAOS guidelines for TKA also recommend early getting out of bed and early ambulation as effective methods for preventing VTE with few complications [12]. Early ambulation helps to activate the muscle pumping function of the lower extremities and pushes up the blood stored in the plantar venous plexus, preventing DVT by increasing venous blood flow.

2. Elastic stockings

Elastic stockings prevent DVT by compressing the superficial veins, decreasing the total cross-sectional area of the veins, increasing deep blood flow velocity, preventing deep vein enlargement, and increasing blood flow velocity. However, the preventive effect is limited in high-risk surgeries such as TKA [23], and therefore, it is necessary to consider the need for anticoagulation therapy and intermittent pneumatic compression in each case. In addition, patients with arteriosclerotic obliterans may have exacerbated hemodynamic disturbances [24].

3. Intermittent pneumatic compression (IPC) method

IPC, in which air is intermittently injected into a cuff wrapped around the lower extremity to help blood return, and venous foot pump, in which blood in the plantar venous plexus is returned by rapid plantar foot compression. The 2012 ACCP guidelines [4] recommend the use of IPC in patients at a high risk of bleeding during TKA, and there have been some reports on the use of anticoagulants in combination with IPC in patients undergoing TKA [25]. However, IPC is not generally used in patients with suspected thrombus formation due to the risk of PTE associated with thrombus release. In addition, although rare, compartment syndrome, peroneal nerve palsy, and PTE may occur [26, 27]. Caution should be exercised in patients with arteriosclerosis obliterans because of the possibility of impeding arterial blood flow.

3.2 Drug Prophylaxis

When the 2004 Guidelines for the Prevention of Pulmonary Thromboembolism/Deep Vein Thrombosis [1] were published, only warfarin and unfractionated heparin were recommended for pharmacological prophylaxis in Japan. Subsequently, enoxaparin and fondaparinux were approved as new anticoagulants from 2007 to 2008, and the Japanese Orthopedic Association Guidelines for VTE Prevention [2] were revised in 2008, stating that pharmacological therapy, including new anticoagulants, may be considered for patients at high risk. Subsequently, the oral anticoagulant edoxaban was approved for VTE prophylaxis in 2011 and, the guidelines were revised to include oral anticoagulants. However, the 2017 Japanese Orthopedic Association Guidelines for Symptomatic Venous Thromboembolism [5] state that anticoagulants should be considered for patients at high risk of VTE; nevertheless, the risk of bleeding must be carefully evaluated, and use should be avoided in patients with a high bleeding risk. Edoxaban, enoxaparin, fondaparinux, unfractionated heparin, and warfarin are all indicated for postoperative TKA and should be selected based on a thorough understanding of their characteristics.

1. Edoxaban

Edoxaban is an oral anticoagulant that selectively inhibits activated coagulation factor ten-a (factor Xa) and decreases the incidence of VTE after TKA [28, 29]. In Japan, the incidence of VTE and DVT is lower with edoxaban administration than with other drugs [30, 31]. Nevertheless, the dosage of edoxaban should be adjusted or its use should be refrained from depending on renal function. As with other anticoagulants, bleeding may occur as an adverse reaction to edoxaban, in which case the drug should be discontinued, and hemostatic measures should be taken. Currently, no drug is known to neutralize the anticoagulant effects of edoxaban.

2. Fondaparinux

The pentasaccharide structure of heparin was scientifically synthesized. Intraoperative monitoring is not required because of its high bioavailability, and its biological half-life of 17.2 h means that a once-daily administration is sufficient. In Japan, the incidence of DVT was significantly reduced in a group of patients treated with fondaparinux 2.5 mg once daily after TKA or THA compared to a placebo group [32]. In addition, the use of fondaparinux after THA or TKA significantly reduced PE compared with physical prophylaxis alone [10, 33]. However, the dosage of this drug should be adjusted, or its use should be refrained from depending on renal function. As with other drugs, bleeding side effects may occur; however, no drugs are currently known to neutralize the anticoagulant effect.

3. Low-molecular-weight heparin

Low-molecular-weight heparin has a reduced direct effect on thrombin and a reduced risk of bleeding compared with conventional unfractionated heparin. Enoxaparin is the only low-molecular-weight heparin approved for use in Japan to prevent venous thromboembolism. It is renally excreted, and its use should be limited, or the dosing interval should be prolonged depending on renal function. In Japan, 20 mg enoxaparin twice daily significantly reduced the incidence of DVT after TKA and THA compared with a placebo [34], and a report using a large Japanese database showed a significant reduction in the incidence of PTE after TKA and THA [33]. However, if emergency treatment is needed, such as in the case of bleeding, the medication should be discontinued and neutralized with protamine sulfate.

4. Unfractionated heparin

Low-dose unfractionated heparin and dose-adjusted unfractionated heparin are available; however, low-dose unfractionated heparin is used more often because of its ease of use. If a bleeding event occurs during administration, the effect of heparin is rapidly attenuated if administration is discontinued owing to its short biological half-life of 60 min. In cases of severe bleeding such as cerebral hemorrhage, freshly frozen plasma should be supplemented, and the effect of heparin should be neutralized with protamine sulfate. Complications of unfractionated heparin include heparin-induced thrombocytopenia (HIT) [35] described below. Randomized trials investigating the efficacy and safety of DVT have not been conducted in Japan.

5. Warfarin

Its anticoagulant effect is mediated by the inhibition of vitamin K-dependent coagulation factor protein synthesis. Warfarin usually takes 3–5 days from the start of administration to the onset of effect. Warfarin is an oral drug that is easy to use, but it is necessary to monitor and control prothrombin time and international normal ratio (PT-INR) to the target level. If bleeding occurs while taking warfarin, it should be immediately discontinued. If the PT-INR is prolonged, fresh-frozen plasma should be used to replenish the coagulation factors, and intravenous vitamin K should be administered. However, randomized trials investigating the preventive effects and safety of warfarin against DVT have not been reported in Japan.

3.3 Complications Due to Prophylaxis

Complications can occur with both physical therapy and pharmacoprophylaxis; here, we discuss the most common complications associated with pharmacoprophylaxis.

1. Hemorrhagic complications

Bleeding is the most common complication of VTE pharmacoprophylaxis. Pharmacoprophylaxis should be considered on an individual basis, balancing the risk of symptomatic VTE with that of bleeding. Bleeding risk factors include previous major bleeding, severe renal dysfunction, antiplatelet use, anticoagulant use, thrombocytopenia, cirrhosis, acute liver failure, uncontrolled hypertension, and hereditary bleeding disorders. In the event of severe bleeding complications, anticoagulants should be discontinued immediately and appropriate measures should be taken (e.g., hemostatic agents, blood transfusion, fresh-frozen plasma, surgical hemostasis). If antagonists are available, they should be considered depending on the urgency of the situation.

2. Heparin-induced thrombocytopenia (HIT)

HIT is a complication associated with the use of heparin products and anticoagulants [25]. There are two types of thrombocytopenia: type I, in which transient thrombocytopenia is caused by platelet stimulation with heparin, and type II, in which heparin-dependent autoantibodies (HIT antibodies) activate platelets to form thrombi, resulting in a decrease in platelet count. Thromboembolism is the most serious complication of HIT. Treatment of HIT involves immediate cessation of heparin administration. However, anticoagulation with an antithrombin agent or an Xa inhibitor is necessary to prevent thromboembolism if heparin is discontinued.

4 Treatment for VTE

Orthopedic surgeons should consult expert cardiologists and vascular surgeons to treat symptomatic VTE. However, attending physicians may be responsible for the initial treatment of symptomatic VTE (especially PTE) and should be knowledgeable about the management of symptomatic PTE. If a postoperative patient is suspected of having symptomatic PTE, vital signs should be checked, and oxygenation should be initiated. A venous line should be secured, an SpO_2 monitor and electrocardiogram should be attached, arterial blood gas analysis should be performed, and plain chest radiography should be performed. If the patient's general condition is stable, a contrast-enhanced computed tomography scan is performed for a definitive diagnosis; if the patient's general condition is unstable, lifesaving measures are prioritized. In some cases, patients require highly specialized treatment, such as percutaneous cardiopulmonary support, thrombolytic therapy, thrombectomy, and inferior vena cava filter placement.

References

1. Committee for the prevention of pulmonary thromboembolism/deep vein thrombosis (venous thromboembolism) prevention guidelines. Guidelines for the prevention of pulmonary thromboembolism/deep vein thrombosis (venous thromboembolism). 1st ed. Tokyo: Medical Front International Limited; 2004.
2. Japanese orthopedic Association Committee for Revision of guidelines for prevention of pulmonary thromboembolism/deep vein thrombosis (venous thromboembolism). In: Japanese orthopedic association venous thromboembolism prevention guidelines 2008. Nankodo, Tokyo; 2008. [in Japanese].
3. Brown GA, et al. Venous thromboembolism prophylaxis after major orthopedic surgery: a pooled analysis of randomized controlled trials. J Arthroplast. 2009;24:77–83.
4. Flack-Ytter Y, et al. Prevention of VTE in orthopedic surgery patients: antithrombotic therapy and prevention of thrombosis 9th ed: American college of chest physicians evidence-based clinical practice guidelines. Chest. 2012;141(2 Suppl):e278S–325S.
5. Japanese Orthopedic Association Clinical Practice Guideline Committee. Association guideline for prevention of symptomatic venous thromboembolism 2017. Tokyo: Nankodo; 2017. [in Japanese]
6. Labropoulos N, et al. Early thrombus remodeling of isolated calf deep vein thrombosis. Eur J Vasc Endovasc Surg. 2002;23:344–8.
7. JCS Joint Working Group. Guidelines for the diagnosis, treatment and prevention of pulmonary thromboembolism and deep vein thrombosis (JCS 2009). Circ J. 2011;75:1258–81.
8. Tateiwa T, et al. Clinical course of asymptomatic deep vein thrombosis after total knee arthroplasty in Japanese patients. J Orthop Surg. 2019;27:2309499019848095.
9. Fujita A, et al. Deep venous thrombosis after total hip or total knee arthroplasty in patients in Japan. Clin Orthop Relat Res. 2000;375:168–74.
10. Nagase Y, et al. Risk factors for pulmonary embolism and the effects of foudaparinux after total hip and knee arthroplasty: a retrospective observational study with use of a national database in Japan. J Bone Joint Surg Am. 2011;93:e146.

11. Di Nisio M, et al. Deep vein thrombosis and pulmonary embolism. Lancet. 2016;388:3060–73.
12. The American academy of orthpaedics surgeons board of directors. Preventing venous thromboembolic disease in patients undergoing hip and knee arthroplasty; 2011. pp. 1–861.
13. Darvall KAL, et al. Obesity and thrombosis. Eur J Vasc Endovasc Surg. 2007;33:223–33.
14. Gibbs NM. Venous thrombosis of the lower limbs with particular reference to bedrest. Br J Surg. 1957;45:209–36.
15. Miyata T, et al. Genetic risk factors for deep vein thrombosis among Japanese: importance of protein S K196E mutation. Int J Hematol. 2006;83:217–23.
16. Sallah S, et al. Venous thrombosis in patients with solid tumors: determination of frequency and characteristics. Thromb Haemost. 2002;87:575–9.
17. Wakabayashi H, et al. The risk factor for preoperative deep vein thrombosis in patients undergoing total knee arthroplasty. J Orthop Sci. 2017;22:698–702.
18. Shimoyama Y, et al. Perioperative risk factors for deep vein thrombosis after total hip arthroplasty or total knee arthroplasty. J Clin Anesth. 2012;24:531–6.
19. Miyagi J, et al. Predictive indicators of deep venous thrombosis and pulmonary arterial thromboembolism in 54 subjects after total knee arthroplasty using multislice computed tomography in logistic regression models. Int J Cardiol. 2007;119:90–4.
20. Fukuda A, et al. Effect of tourniquet application on deep vein thrombosis after total knee arthroplasty. Arch Orthop Trauma Surg. 2007;127:671–5.
21. Mori N, et al. Use of a pneumatic tourniquet in total knee arthroplasty increases the risk of distal deep vein thrombosis: a prospective, randomized study. Knee. 2016;23:887–9.
22. Sochart DH, et al. The relationship of foot and ankle movements to venous return in the lower limb. J Bone Joint Surg Br. 1999;81:700–4.
23. Wells PS, et al. Graduated compression stocking in the prevention of postoperative venous thromboembolism: a meta-analysis. Arch Intern Med. 1994;154:67–72.
24. Merrett ND, et al. Ischemic complications of graduated compression stockings in the treatment of deep venous thrombosis. Postgrad Med J. 1993;69:232–4.
25. Sakai T, et al. Effects of a foot pump on the incidence of deep vein thrombosis after total knee arthroplasty in patients given edoxaban: a randomized controlled study. Medicine (Baltimore). 2016;95:e2247.
26. Anand A. Complications associated with intermittent pneumatic compression devices. Anesthesiology. 2000;93:1556–7.
27. Siddiqui AU. Pulmonary embolism as a consequence of applying a sequential compression device on the legs of a patient asymptomatic of deep vein thrombosis. Anesthesiology. 2000;92:880–2.
28. Fuji T, et al. A dose-ranging study evaluating the oral factor Xa inhibitor edoxaban for the prevention of venous thromboembolism in patients undergoing total knee arthroplasty. J Thromb Haemost. 2010;8:2458–68.
29. Ishida K, et al. A prospective randomized comparative study to determine the appropriate edoxaban administration period, to prevent deep vein thromboembolism in patients with total knee arthroplasty. J Orthop Sci. 2018;23:1005–10.
30. Fuji T, et al. Safety and efficacy of edoxaban, an oral factor Xa inhibitor, versus enoxaparin for thromboprophylaxis after total knee arthroplasty: the STARS E-3 trial. Thromb Res. 2014;134:1198–204.
31. Sasaki H, et al. Retrospective comparison of three thromboprophylaxis agents, edoxaban, fondaparinux, and enoxaparin, for preventing venous thromboembolism in total knee arthroplasty. Int Orthop. 2014;38:525–9.
32. Fuji T, et al. Fondaparinux prevents venous thromboembolism after joint replacement surgery in Japanese patients. Int Orthop. 2008;32:443–51.
33. Shoda N, et al. Prophylactic effect of fondaparinux and enoxaparin for preventing pulmonary embolism after total hip or knee arthroplasty: a retrospective observational study using the Japanese diagnosis procedure combination database. Mod Rheumatol. 2015;25:625–9.
34. Fuji T, et al. Prevention of postoperative venous thromboembolism in Japanese patients undergoing total hip or knee replacement: 2 randomized, double-blind, placebo-controlled studies. J Orthop Sci. 2008;3:442–51.
35. Chow VW, et al. Heparin-induced thrombocytopenia after total knee arthroplasty, with subsequent adrenal hemorrhage. J Arthroplast. 2012;27(1413):e15–1413.e18.

Infection Control

86

Susumu Mikami

1 Preventive Measures for SSI in TKA: Preoperative Period (Risk Factors for Infection)

1.1 Diabetes Mellitus

Poor glycemic control is a risk factor for SSI, and a Hemoglobin A1c (HbA1c) level < 7% is generally considered a prerequisite for surgery [1]. However, improvement in HbA1c takes a long time, and in recent years, the optimization of postoperative glycemic control has been emphasized as meaningful. The Japanese Orthopedic Association Clinical Practice Guidelines on the Prevention of Postoperative Infection of Osteoarthrotomy 2015 states that controlling postoperative blood glucose levels to <200 mg/dL decreases the risk of SSI [2].

1.2 History of Infection and Previous Surgery

History of pyogenic arthritis is a risk factor for SSI. Surgery should not be performed for at least 6 months after the resolution of the infection [3], and careful evaluation of a previous infection site is needed. Active infections at other sites are also risk factors. A history of surgery for fractures or residual metal fixtures is also a risk factor for infection [4].

1.3 Methicillin-Resistant *Staphylococcus aureus* (MRSA) Nasal Carriage

SSIs caused by MRSA are particularly intractable with poor outcomes [5]. In one study, in 749 orthopedic surgical cases, the surgical field was cultured immediately after disinfection, before the first saline irrigation, and after the final irrigation. The results showed that bacterial positivity increased after the final irrigation even after repeated irrigation, and most of the bacteria detected were coagulase-negative staphylococci; no MRSA was found [6]. Although it is unclear how MRSA invades the surgical field and causes SSIs, preoperative nasal carriage of MRSA is an independent risk factor for SSIs [7]; moreover, in 85.7% of nasal *S. aureus* carriers who developed deep infection after orthopedic surgery, genetic homology was found between the nasal cavity and the *S. aureus* in the infected area [8]. Endogenous infection may be caused by the patient's own nasal bacteria, and preoperative sterilization of MRSA in the nasal cavity may reduce SSIs [9]. Endogenous infections include the possibility of bacteria reaching the surgical field hematogenously via airway and nasal mucosal injury caused by intubation and extubation during general anesthesia [10], bacteria being forced into the intestinal tract through the upper airway by intubation, and lymphatic or hematogenous invasion through the intestinal mucosa, which may be called bacterial translocation [11, 12].

The risk factors for preoperative MRSA nasal carriage include a history of MRSA infection, prior hospitalization or admission to a nursing home, and diabetes (particularly insulin use, HbA1c ≥ 6.6%) [13, 14]. In Japan, targeted preoperative screening is typically performed. Preoperative measures, such as nasal sterilization (mupirocin nasal ointment), skin sterilization (chlorhexidine shower and wipes), and the prophylactic administration of antibacterial agents (Vancomycin (VCM) with a beta-lactam antibiotic) are recommended for patients carrying MRSA [13].

1.4 Others

Other risk factors for SSIs include obesity, male sex, active inflammation, such as oral dental inflammation, smoking, renal dysfunction, rheumatoid arthritis and use of immunosuppressive drugs or steroids [2, 15]. Preoperative evaluation, countermeasures, and optimization of these factors are important.

S. Mikami (✉)
Koyo Ikiiki Clinic, Tomakomai, Japan

S. Matsuda (ed.), *Advances in Total Knee Arthroplasty*, https://doi.org/10.1007/978-981-97-4920-1_86

2 Preventive Measures for SSI in TKA: Perioperative Period

2.1 Operating Room

Recently, the effectiveness of laminar flow ventilation systems in preventing SSIs in bioclean rooms has been questioned. Bacterial contamination downstream of a surgeon's head under laminar flow is 27 times higher than that under laminar flow only, and caution should be exercised when using a bioclean room [16].

The number of people in the operating room and the number of times the door is opened correlate with the number of suspended particulates in the room and may influence the occurrence of SSIs. Countermeasures, such as limiting the number of people in the operating room, ensuring that doors remain closed, and storing frequently used instruments in the room should be implemented [15].

2.2 Full Body Exhaust Suits and Surgical Helmet

As with bioclean rooms, their effectiveness in preventing SSIs has been questioned in recent years, and many guidelines do not recommend their use. Compared with general surgical gowns, space suits create positive pressure inside the gown because of a fan in the helmet area, posing a risk of contaminated air leaking from between the unsealed gown and gloves [17]. The number of leaking airborne particles is significantly higher near the feet of space suits. The airborne particles spread up toward the height of the surgical field after footsteps [18]. However, in one study, when contamination of the front transparent shield of the surgical helmet was investigated after TKA, all shields of the surgeon and assistants (38 patients in total) immediately after surgery were contaminated with blood, bone fragments, oil droplets, and water droplets. If helmets were not worn, surgeons and assistants would be at risk of blood contamination from patients with infectious diseases, such as hepatitis and syphilis. From the perspective of SSI prevention, a risk of contaminated fluid on their unclean faces, which could fall onto the surgical field and cause SSI, exists. Therefore, the value of surgical helmets cannot be completely denied. When they are used, care should be taken to minimize their disadvantages, such as sealing the space between the surgical gown and gloves with tape, minimizing footsteps, and minimizing the fan volume [19, 20].

2.3 Povidone–Iodine-Containing Drape

Some reports indicate that povidone–iodine-containing drapes reduce bacterial colonization in the surgical field and that their use in joint arthroplasty may decrease SSI [2]. However, the risk of infection increases when povidone–iodine-containing drapes are peeled off or removed before wound closure. If they are peeled off or removed, measures such as disinfection with povidone–iodine, should be taken [21].

2.4 Irrigation Solution

The World Health Organization and Centers for Disease Control and Prevention recommend the use of diluted povidone–iodine for wound irrigation in surgical procedures; however, the optimal dilution concentration and volume of irrigation solution are unknown. Povidone–iodine is used at low concentrations, such as 0.35–0.5%, due to the possibility of cytotoxicity at high concentrations [22, 23]. However, a study on the amount of saline required to remove cement and bone fragments from the surgical wound in TKA revealed 5 L or more as an effective irrigation volume [24]. However, bacterial contamination in the surgical field increases by the end of surgery, even after thorough irrigation [5]. The use of antibiotics in irrigation solutions is not recommended because of the risk of resistance, drug hypersensitivity, and cost [15].

2.5 Preoperative Antibiotic Administration

First- (Cefazolin (CEZ)) and second-generation (Cefotiam (CTM), Cefmetazole(CMZ), and Flomoxef (FMOX)) cephems are recommended as prophylactic antibiotics for TKA [2]. CEZ should be administered intravenously within 1 h prior to surgery (2 h for VCM), and the dose of CEZ is usually 1 g. For patients weighing 80 kg or more, the dose should be increased according to body weight. Postoperative administration of Cefazolin every 6–8 h for 24–48 h is recommended [2]·

Although antibiotic-impregnated cement is not indicated for initial TKA due to problems with drug resistance, cement strength changes, and medical costs. Its selective use in the patients with the risk factor for SSI has been discussed [15]. This is indicated for revision TKA.

The efficacy of the local intraoperative administration of VCM powder in spinal surgery remains controversial. Postoperative wound dehiscence and erosion have been reported following the use of VCM powder in TKA and UKA [25]. The use of VCM powder in TKA and UKA is not covered by Japanese medical insurance.

2.6 Anesthesia and Temperature Maintenance

Compared to that associated with general anesthesia, spinal anesthesia has a lower risk of SSI [15]·Propofol, which is administered intravenously for anesthesia induction, maintenance, and sedation, is a fat emulsion that contains no antiseptic agents and must therefore be handled aseptically because of the risk of bacterial contamination and growth, and consequent infection [26]. Perioperative maintenance of normothermia is considered important for preventing infection by maintaining blood flow to subcutaneous tissues. Intraoperative administration of amino acids or amino acids infused with an appropriate amount of glucose (intraoperative blood glucose controlled below 180 mg/dL) purportedly prevents perioperative hypothermia and postoperative complications [27].

3 Post-TKA SSI/Periprosthetic Joint Infection (PJI)

3.1 Incidence Rates

SSI/PJI is estimated to occur in approximately 0.2–3.8% of cases with initial joint replacement and 0.5–17.3% of cases with revision joint replacement, for all joint arthroplasties, including TKA [2].

3.2 Diagnosis

The revised 2018 International Consensus Meeting (ICM) on Musculoskeletal Infection criteria for the diagnosis of PJI are shown in Fig. 86.1. For a diagnosis of PJI, one of the following two major criteria must be met: two positive growths of the same organism using standard culture methods and sinus tract with evidence of communication to the joint or visualization of the prosthesis. In addition to existing serum markers, joint fluid, and histological examination, new laboratory tests, such as D-dimer and alpha-defensin, as well as positive intraoperative purulence, were added to the minor criteria. Moreover, a new scoring system was introduced for the minor criteria. The combined preoperative and postoperative scores are defined as follows: ≥ 6, infected; 3 to 5, inconclusive; < 3, not infected [15].

Identifying the pathogen is the key to successful treatment of PJI; however, cultures are often negative. To increase the identification rate, we do not hesitate to perform aspiration of joint fluid; collect two, three, or more specimens; submit them for culture, use blood culture bottles for enrichment culture; do not administer antimicrobial agents prior to culture (if already administered, stop for 2 weeks); and submit at least two blood cultures if bacteremia is suspected. Recently, molecular biological diagnostic methods, such as polymerase chain reaction and next-generation sequencing have been reported to be helpful, particularly in culture-negative cases [15, 28].

Fungal infections are difficult to diagnose and identify as a cause of PJI because they are rare and require culture in specialized fungal medium [28]. Crystal-induced arthritis including calcium pyrophosphate deposition are examples of cases that require differentiation from PJI. To this end, crystal components of the joint fluid, cell counts, sugar levels, and culture results are evaluated. However, there are currently no reliable clues to distinguish between crystal-induced arthritis and PJI [29].

Major Criteria (at least one of the following)	Decision
Two positive growths of the same organism using standard culture methods	Infected
Sinus tract with evidence of communication to the joint or visualization of the prosthesis	

Minor Criteria	Threshold		Score	Decision
	Acute ※1	Chronic		
Serum CRP (mg/L) or D-Dimer (ug/L)	100 Unknown	10 860	2	Combined preoperative and postoperative score: ≥6 Infected 3 to 5 Inconclusive ※3 <3 Not Infected
Elevated Serum ESR (mm/hr)	No role	30	1	
Elevated Synovial WBC (cells/μL) or Leukocyte Esterase or Positive Alpha-defensin (signal/cutoff)	10,000 ++ 1.0	3,000 ++ 1.0	3	
Elevated Synovial PMN (%)	90	70	2	
Single Positive Culture			2	
Positive Histology			3	
Positive Intraoperative Purulence ※2			3	

※1 : This criteria were never validated on acute infections. ※2 : No role in suspected adverse local tissue reaction.
※3 : Consider further molecular diagnostics such as next-generation sequencing

Fig. 86.1 Proposed 2018 International Consensus Meeting (ICM) criteria for PJI

3.3 Classification of Post-TKA Infections and Treatment Strategies

The classification by Segawa et al. [30] for post-TKA infections is widely used in Japan, and treatment strategies are typically based on this classification system.

Type 1: Positive intraoperative culture after revision TKA for aseptic loosening. Antibiotic therapy is initiated.

Type 2: Early infection within 4 weeks of primary TKA. The treatment for superficial early infection includes debridement and antibiotic therapy. Deep early infections are treated with debridement, replacement of the polyethylene insert, retention of the prosthesis, and intravenous antibiotic administration.

Type 3: Acute hematogenous infection at other sites by 4 weeks postoperatively. Acute hematogenous infections are treated with debridement, replacement of the polyethylene insert, retention of the prosthesis if it is not loose, or intravenous antibiotic administration. If the prosthesis is loose, a revision is indicated.

Type 4: Late chronic infection is one that developed 4 weeks or more after primary TKA and that has an insidious clinical presentation, and requires treatment with revision surgery [30]; however, the treatment strategy should be thoroughly considered because the background of each case is different [31].

3.4 Treatment

3.4.1 Debridement, Antibiotics, and Implant Retention (DAIR)

DAIR is indicated in cases of early infection within 4 weeks of onset or acute hematogenous infection with no prosthesis loosening. DAIR is not an emergency procedure but should be performed urgently when patients with acute PJI are medically and surgically optimized. Delayed DAIR may result in biofilm formation and reduced therapeutic efficacy. Surgery should not be delayed to identify the pathogen. Moreover, the exchange of all modular components during DAIR reduces the risk of PJI recurrence.

[15]. The conditions for implant preservation include no implant loosening, a short period since infection (within 3–4 weeks), healthy soft tissues, and availability of sensitive antimicrobial agents [32].

Historically, various other treatments have been attempted in Japan to preserve implants. In addition to synovectomy and debridement, these include: closed continuous irrigation with saline, povidone–iodine solution, or ozone nanobubble water [33, 34]; filling of antibiotic-impregnated cement or hydroxyapatite around the bone defect or preserved implant [35]; and intra-articular infusion of high-dose antibiotics once or twice a day by implanting catheters at the time of DAIR [36].

3.4.2 One-Stage Exchange Arthroplasty

In Japan, one-stage exchange arthroplasty for PJI is currently uncommon, and its potential advantages over two-stage revision surgery, include lower complication and mortality rates, earlier functional recovery, and lower medical costs; however, its rate of resolution and certainty are unknown [28].

3.4.3 Two-Stage Exchange Arthroplasty

In the initial surgery, TKA removal, debridement, and irrigation are performed, and antibiotic-impregnated cement spacers are inserted to maintain the gap between the femur and the tibia in preparation for revision surgery. Two types of cement spacers are available. First, static spacers (non-articulating spacers) have the disadvantages of muscle atrophy and joint contracture during the waiting period, resulting in reduced joint function; however, they are indicated for cases with major bone loss or major soft tissue defects that require limited motion to promote wound healing. However, articulating spacers with articulating surfaces have the advantage that ROM exercises can be performed during the waiting period and joint function can be relatively maintained [37]. Although off-the-shelf articulating spacers are available, they have the disadvantage of being too large for Japanese patients. Therefore, the use of custom-made articulating spacers, produced by using dental vinyl silicone impressions to create molds of the same size as the removed implants, have been reported [38]. By using antibiotic-impregnated cement spacers, a sustained localized antimicrobial action may be achieved. Antibiotics are selected based on the susceptibility of the causative pathogen [15]. Broad-spectrum antibiotics are administered if the pathogen is unknown. The antibiotic should be water-soluble, diffuse into the surrounding environment, with sustained release over a certain period of time, cause minimal reduction in cement strength, and should not be deactivated by the heat of cement polymerization. The cement strength decreases with an increasing amount of antibiotics, whereas the antimicrobial effect increases with increasing amount of antibiotics. Generally, antibiotics are loaded at a ratio of 5–10% [38]. Two antibiotics with the same mass have a significantly higher elution rate than does a single antibiotic, have less effect on cement strength, and prevent bacterial replacement [39]. The heat of cement polymerization affects the cement strength. β-lactam antibiotics are susceptible to the heat of cement polymerization, while aminoglycosides, tetracyclines, and glycopeptides are more heat-stable. They are often used to treat methicillin-resistant bacterial infections. When the pathogen is unknown, the possibility of methicillin-resistant bacteria should be suspected, and use of a mixture of a broad-spectrum antibacterial agent and VCM should be considered. Furthermore, instead of using a conventional cement that generates a polymerization heat of approximately 70 °C, a cement with a polymerization heat suppressed to approximately 50 °C can be used to avoid antibiotic inactivation [38].

The optimal timing for reimplantation of a two-stage exchange arthroplasty of the knee has not yet been established. The duration of antimicrobial administration after removal varies from 6 to 12 weeks. According to the 2018ICM, evidence supports pathogen-specific, highly bioavailable oral antibiotic therapy as an appropriate choice after resection arthroplasty in the two-stage treatment of periprosthetic joint infections after an initial intravenous antibiotic treatment period of at least 5–7 days [15]. The evaluation of infection quiescence is based on negative joint fluid cultures, normal joint fluid tests, negative C-reactive protein levels in blood tests, and decreased erythrocyte sedimentation rate. The evaluations are performed in an antimicrobial-free state.

During revision surgery, every inch of the surgical field is checked to ensure that the infection has subsided. The cement spacer is removed and the joint is thoroughly debrided and irrigated before revision with a new implant. An antibiotic-impregnated cement is used to fix the implant in place. At least two cultures of scar tissue and joint fluid should be obtained during revision surgery. If there are signs of intraoperative infection, the scar tissue should be subjected to rapid intraoperative pathology if available, and the results should be used to determine whether revision surgery should proceed. If a rapid pathological examination is not available, the decision to proceed with revision should not be made. The joint is treated with debridement, irrigation, and replacement of antibiotic-impregnated cement spacers.

3.4.4 Antibiotics for PJI

With PJI, biofilms have already formed, and may lead to spreading of the causative pathogen to the bone tissue. CEZ recommended for initial TKA have poor bone marrow penetration and are inappropriate for treating PJI. The intraosteoblastic activity of antibiotics is poor for lactams, such as VCM, Daptomycin (DAP), and CEZ. However, Teicoplanin (TEIC), Fosfomycin (FOM), Linezoid (LZD), and Minocyclin (MINO) have good activity, and that of Rifampicin (RFP), Clindamycin (CLDM), Ofloxacin (OFLX), Ciprofloxacin (CPFX), and Sulfamethoxazole-trimethoprim (ST) mixtures is also very good[[40]. RFP is expected to have an antibiofilm effect. Some reports have stated that implant preservation rates increase when RFP is used in combination with sensitive antibiotics, so that RFP can be used as a treatment [15, 41]. The combination of RFP with LZD or CLDM is not recommended due to the decreased efficacy of these drugs. VCM is poorly transferred to osteoblasts. In PJI caused by MRSA, the combination of LZD, TEIC, and other anti-MRSA drugs with RFP, ST mixture, or MINO, which have good bone marrow penetration, is considered important [40].

References

1. Lamloum SM, et al. Relationship between postoperative infectious complications and glycemic control for diabetic patients in an orthopedic hospital. Med Princ Pract. 2009;18:447–52.
2. Japanese Orthopedic Association, Committee on Clinical Practice Guidelines, Committee on Guidelines for Prevention of Postoperative Infection of Bone and Joint. Clinical Practice Guideline on the Prevention of Postoperative Infection of Osteoarthrotomy 2015 (Revised 2nd ed.), Nankodo, Tokyo, 2015. [Japanese].
3. Aggarwal VK, et al. Mitigation and education. J Arthroplast. 2014;29(Suppl):19–25.
4. Suzuki G, et al. Previous fracture surgery is a major risk factor for infection after total knee arthroplasty. Knee Surg Sports Traumatol Arthrosc. 2011;19:2040–4.
5. Nakano N, et al. Factors influencing the outcome of deep infection following total knee arthroplasty. Knee. 2015;22:328–32.
6. Yamada K, et al. Bacterial contamination increases toward the end of the operation in clean orthopedic surgery. J Jpn Soc Bone Joint Infect. 2010;24:13–8. [Japanese]
7. Kalra L, et al. Risk of methicillin-resistant *Staphylococcus aureus* surgical site infection in patients with nasal MRSA colonization. Am J Infect Control. 2013;41:1253–7.
8. Skråmm I, et al. Surgical site infections in orthopedic surgery demonstrate clones similar to those in orthopedic *Staphylococcus aureus* nasal carriers. J Bone Joint Surg Am. 2014;96:882–8.
9. Schweizer ML, et al. Association of a bundled intervention with surgical site infections among patients undergoing cardiac hip, or knee surgery. JMMA. 2015;313:2162–71.
10. Valdés C, et al. The incidence of bacteraemia associated with tracheal intubation. Anesthesia. 2008;63:588–92.
11. Watanabe H, et al. Enterocolitis caused by methicillin-resistant *Staphylococcus aureus*: molecular characterization of respiratory and digestive tract isolates. Microbiol Immunol. 2001;45:629–34.
12. Torii Y, et al. Infection route of SSI. Orthopedics. 2018;31:1–4. [Japanese]
13. Committee for the Development of Treatment Guidelines for MRSA Infections (ed.). Treatment Guidelines for MRSA Infections, Revised Edition, The Japanese Society for Chemotherapy, The Japanese Society for Infectious Diseases, 2019. [Japanese].
14. Tateiwa T, et al. Risk factors for preoperative nasal carriage of methicillin-resistant bacteria in total hip and knee arthroplasty patients. J Orthop Sci. 2022;27:798–803.
15. MSIS. Second International Consensus Meeting (ICM) on Musculoskeletal Information. 2018.; Available at: https://icmphilly.com/document/
16. Taylor GJ, et al. Infection and interposition between ultraclean air source and wound. J Bone Joint Surg Br. 1993;75:503–4.
17. Simon WY, et al. Intraoperative contamination and space suits: a potential mechanism. Eur J Orthop Surg Traumatol. 2014;24:409–13.
18. Nakajima D, et al. Does modern space suit reduce intraoperative contamination in total joint replacement? An experimental study. Eur J Orthop Surg Traumatol. 2017;27:1139–43.
19. Mikami M, et al. Contamination of the front shield of surgical helmets during knee arthroplasty. J Jpn Soc Artif Joints. 2020;50:595–6. [Japanese]
20. Masaoka T, et al. Prevention of SSI – recent topics. Orthopedics. 2018;31:25–31. [Japanese]
21. Hanada M, et al. Intraoperative bacterial contamination in total hip and knee arthroplasty is associated with operative duration and peeling of the iodine-containing drape from skin. Eur J Orthop Surg Traumatol. 2020;30:917–21.

22. Mori N. Postoperative infection prevention in TKA. Orthopedics. 2015;28:39–48. [Japanese]
23. Abe T, et al. Preventive measures for SSI – wound irrigation with povidone–iodine solution. Orthopedics. 2018;31:21–4. [Japanese]
24. Niki Y, et al. Intraoperative irrigation volume required for total knee arthroplasty. Jpn J Orthop Surg. 2005;79(S61):3. [Japanese]
25. Hanada M, et al. Intrawound vancomycin powder increases postoperative wound complications and does not decrease periprosthetic joint infection in primary total and unicompartmental knee arthroplasties. Knee Surg Sports Traumatol Arthrosc. 2019;27:2322–7.
26. Kishimoto M, et al. Suspected periprosthetic joint infection after total knee arthroplasty under propofol versus sevoflurane anesthesia: a retrospective cohort study. Can J Anesth. 2018;65:893–900.
27. Fujita Y, et al. Effect of amino acid and glucose infusion on perioperative body temperature and postoperative infection in patients undergoing. Jpn J Anesthesiol. 2012;61:68–73.
28. Inaba Y, et al. Diagnosis, treatment and infection control of periprosthetic joint infections. Jpn J Orthop Surg. 2020;94:61–9. [Japanese]
29. Iwasaki K, et al. Calcium pyrophosphate deposition disease after total knee arthroplasty: comparison with periprosthetic joint infection. J Orthop. 2019;16:128–32.
30. Segawa H, et al. Infection after total knee arthroplasty: a retrospective study of the treatment of eighty-one infections. J Bone Joint Surg Am. 1999;81:1434–45.
31. Matsumoto T, et al. Treatment of 50 deep infections after total knee arthroplasty. Orthopedics. 2015;38:e529–35.
32. Mont MA, et al. Multiple irrigation, debridement, and retention of components in infected total knee arthroplasty. J Arthroplast. 1997;12:426–33.
33. Tsumura H, et al. Synovectomy, debridement, and continuous irrigation for infected total knee arthroplasty. Int Orthop. 2005;29:113–6.
34. Kawashima M, et al. Continuous irrigation therapy with ozone nano-bubble water and hyperbaric oxygen therapy for bone and joint infections. Orthopedics. 2018;31:140–8. [Japanese]
35. Hasegawa M, et al. Use of antibiotic-impregnated hydroxyapatite for infection following total knee arthroplasty. Mod Rheumatol. 2021;31:1073–7.
36. Fukagawa S, et al. High-dose antibiotic infusion for infected knee prosthesis without implant removal. J Orthop Sci. 2010;15:470–6.
37. Wyles CC, et al. Point/counterpoint: nonarticulating vs articulating spacers for resection arthroplasty of the knee or hip. J Arthroplast. 2020;35:S40–4.
38. Inoue H, et al. Surgical technique for revision knee arthroplasty: two-stage revision for infection. Orthopedics. 2021;34:52–8. [Japanese]
39. Slane J, et al. Antibiotic elution from acrylic bone cement loaded with high doses of tobramysin and vancomycin. J Orthop Res. 2018;36:1078–85.
40. Matsushita K, et al. Selection of antimicrobial agents in bone infection. Orthopedics. 2018;31:149–53. [Japanese]
41. Zimmerli W, et al. Role of rifampin for the treatment of orthopedic implant-related staphylococcal infections: a randomized controlled trial. JAMA. 1998;279:1537–41.

Metal Allergy

87

Masahiro Hasegawa

1 Metal Allergy

There are two types of metal allergies: contact allergies and systemic metal allergies. Metal contact allergy is a delayed-type (type IV) hypersensitivity reaction in which T cells play an important role. Metal ions become haptens, which bind to proteins in the epidermis and mucosal epithelium to form hapten-protein complexes. Hapten-protein complexes are processed as antigens by Langerhans cells (antigen-presenting cells) in the skin, which migrate to the lymph nodes and are recognized by T cells [1]. Nickel allergy is the most common type of allergy.

In total knee arthroplasty (TKA), femoral components are mostly made of CoCr alloys because titanium alloys are less suitable for wear and CoCr alloys are unstable, with Co leaching into the body and Cr remaining in the synovial cells. Ti, AL, and V have all been found to be stable [2]. Although metal sensitization increases after arthroplasty, there is no evidence that CoCr alloy prostheses cause metal allergies. There is no clear causal relationship between metal allergies and prosthesis loosening [3].

2 Frequency of Metal Allergy

The number of metal allergy-positive cases has increased over time with 16%,15%, 21%, and 29% in 1970–1979, 1980–1989, 1990–1999, and 2000–201, respectively [4]. European surveillance in 2007 and 2008 showed that Ni accounted for 12–27% of cases and Co accounted for 5–14% [5]. The standard allergens in Japan were Ni (12%), Cr (7%), and Co (6%) in 2008, which were the same as those in Europe, and Ni (16%), Co (9%), and Cr (8%) in 2012, indicating an increase in the Ni-positive rate [6].

M. Hasegawa (✉)
Department of Orthopaedic Surgery, Mie University Graduate School of Medicine, Tsu City, Mie, Japan
e-mail: masahase@clin.medic.mie-u.ac.jp

3 Metal Allergy Testing

Patch tests are frequently used to diagnose allergies to metals. Patch testing is a standard method for treating skin metal allergies (contact dermatitis) [7]. A patch test alone is not sufficient for diagnosing systemic metal allergy, and the case of exacerbations should be confirmed using other tests.

The following five criteria are used to diagnose skin rash caused by a metal allergy [8]: (1) skin rash develops after metal insertion; (2) there is no other cause, such as infection; (3) chronic skin rash; (4) skin rash heals within 2 months after metal removal; and (5) confirmation of metal corrosion after removal. According to surveys conducted by the European Contact Dermatitis Society in 2012 and the American Contact Dermatitis Society in 2013 [9], the four major criteria for the diagnosis of metal allergy are (1) the onset of dermatitis within weeks to months after metal insertion, (2) the presence of metal, (3) a positive patch test, and (4) a mild reaction after removal of the metal implant. The patch test is a controversial test for hypersensitivity to implants and is not suitable for the diagnosis of metal allergy to artificial hip joints; however, accurate diagnostic methods for metal allergy after THA or TKA have not been established [3, 4]. The lymphocyte stimulation test is commonly used for metal allergies to artificial joints [10, 11].

Of the 16 cases of revision after metal-on-metal THA, 11 (69%) were patch test-positive, including seven cases of Co and Cr and four cases of Ni. The lymphocyte stimulation test was positive in 10 cases (62%): Co in seven cases, Cr in five cases, and Ni in seven cases [12]. In other words, metal allergies require revision. On the other hand, lymphocyte stimulation tests were performed in cases with and without pseudotumors and in controls without metal implants after metal-on-metal hip resurfacing. The results showed that the Ni-positive rate was 80% in cases with pseudotumors, 45% in cases without pseudotumors, and 13% in controls. The percentage of Ni-positive cases with pseudotumors was significantly higher than that of the controls, but there was no significant difference between cases with and without

S. Matsuda (ed.), *Advances in Total Knee Arthroplasty*, https://doi.org/10.1007/978-981-97-4920-1_87

pseudotumors. No difference in reactivity to Ni, Co, or Cr was observed in the lymphocyte stimulation test with or without pseudotumors, and type IV hypersensitivity caused by T cells was unlikely to be involved [11].

4 Metal Ion Concentration

There are many reports on metal ion concentrations after metal-on-metal THA [13]; however, there are few reports on metal ion concentrations after TKA. The metal ion concentration of the drainage fluid after TKA was measured; Co was detected in one of the seven cases, and Ni was detected in the other cases. The metal ion concentration was likely affected during the osteotomy [14]. CD3 + CD45RO+ T cells are a characteristic feature of post-TKA metal allergy [15].

5 Cause of ARMD (Adverse Reactions to Metal Debris)

In addition to delayed metal allergy by T cells, other possible causes of ARMD include cellular injury caused by metal wear debris [11] and patient susceptibility. Magnetic resonance imaging after metal-on-metal THA has revealed pseudotumors [13, 16]. In a histological and immunohistochemical study of 13 patients with ARMD after metal-on-metal THA, the lymphocytes were predominantly B cells. Four patients with positive lymphocyte stimulation tests had T-cell predominance, suggesting type IV allergy, whereas the other nine did not appear to have type IV allergy [16]. In other words, metal allergy was suggested in only one-third of the cases, and none had a skin rash.

6 Report of Metal Allergy with Skin Rash After Joint Replacement Surgery

Preoperative lymphocyte stimulation tests were performed in 92 patients who underwent TKA, and 24 (26%) tested positive [10]. Five of the positive cases presented a skin rash, and four cases had a skin rash that lasted for more than 1 year. One patient was Ni-positive, and the skin rash worsened; therefore, revision surgery was performed using alumina ceramic implants. One patient was Ni- and co-positive and had a fracture; therefore, revision surgery was performed using CoCr implants. In a case report of generalized skin rash after TKA, skin histology showed lymphocytic infiltration, type IV allergy, and a positive patch test for Cr, which resolved after revision with a zirconium alloy implant [17].

7 How to Deal with Metal Allergies

Routine screening for metal allergies before joint replacement surgery is not recommended. However, some reports recommend preoperative patch test screening in patients with a history of metal allergy [4, 18]. One review reported a 9% preoperative positivity rate, which increased to 14% postoperatively [18].

In severe cases of metal allergy after TKA with skin rash, it has been reported that revision surgery using metal-free implants, i.e., alumina ceramic or zirconium alloy implants, is recommended [10]. Implants with a CoCr alloy surface coated with titanium niobium nitride or a titanium alloy, which have disadvantages in terms of wear resistance, can be used. Implants with a CoCr alloy surface coated with titanium niobium nitride or titanium alloys with a nitride surface hardened to improve wear resistance can also be used, although they have disadvantages in terms of wear resistance [19].

In Europe and the United States, CoCr alloy implants with a seven-layered surface structure and a top layer coated with zirconium nitride are available to prevent metal allergies. There was no difference in the patch test results in a 1-year postoperative report of TKA using these coated implants and conventional CoCr alloy implants; however, blood Cr levels were significantly lower in cases with coated implants. No difference in the short-term clinical results up to 3 years postoperatively has been reported [20].

References

1. Kaplan DH. In vivo function of Langerhans cells and dermal dendritic cells. Trends Immunol. 2010;31:446–51.
2. Ichinose S, et al. The study of metal ion release and cytotoxicity in Co-Cr-Mo and Ti-Al-V alloy in total knee prosthesis – scanning electron. J Mater Sci Mater Med. 2003;14:79–86.
3. Hallab N, et al. Metal sensitivity in patients with orthopaedic implants. J Bone Joint Surg Am. 2001;83:428–36.
4. Granchi D, et al. Metal hypersensitivity testing in patients undergoing joint replacement: a systematic review. J Bone Joint Surg Br. 2012;94:1126–34.
5. Uter W, et al. Current patch test results with the European baseline series and extensions to it from the 'European surveillance system on contact allergy' network, 2007–2008. Contact Derm. 2012;67:9–19.
6. Suzuki K. How to utilize the Japanese standard allergen 2008. MB Derma. 2015;231:1–7. [Japanese]
7. Möller H, et al. Gold allergy and artificial sweat. Contact Derm. 1999;41:57–60.
8. Rostoker G, et al. Dermatitis due to orthopaedic implants: a review of the literature and report of three cases. J Bone Joint Surg Am. 1987;69:1408–12.
9. Schalock P, et al. Patch testers' opinions regarding diagnostic criteria for metal hypersensitivity reactions to metallic. Dermatitis. 2013;24:183–5.

10. Niki Y, et al. Screening for symptomatic metal sensitivity: a prospective study of 92 patients undergoing total knee arthroplasty. Biomaterials. 2005;26:1019–26.
11. Kwon YM, et al. Lymphocyte proliferation responses in patients with pseudotumors following metal-on-metal hip resurfacing arthroplasty. J Orthop Res. 2010;28:444–50.
12. Thomas P, et al. Increased metal allergy in patients with failed metal-on-metal hip arthroplasty and peri-implant T-lymphocytic inflammation. Allergy. 2009;64:1157–65.
13. Hasegawa M, et al. Cobalt and chromium ion release after large-diameter metal-on-metal total hip arthroplasty. J Arthroplast. 2012;27:990–6.
14. Niki Y, et al. Metal ion concentrations in the joint fluid immediately after total knee arthroplasty. Mod Rheumatol. 2001;11:192–6.
15. Niki Y, et al. Five types of inflammatory arthritis following total knee arthroplasty. J Biomed Mater Res A. 2007;81:1005–10.
16. Hasegawa M, et al. Immune response in adverse reactions to metal debris following metal-on-metal total hip arthroplasty. BMC Musculoskelet Disord. 2016;17:221.
17. Gao X, et al. Dermatitis associated with chromium following total knee arthroplasty. J Arthroplast. 2011;26(665):e13–6.
18. Mitchelson AJ, et al. Biomaterial hypersensitivity: is it real? Supportive evidence and approach considerations for metal allergic patients following total knee arthroplasty. Biomed Res Int. 2015;2015:137287.
19. Rossi SMP, et al. Ten-year outcomes of a nitrided Ti-6Al-4 V titanium alloy fixed-bearing total knee replacement with a highly crosslinked polyethylene-bearing in patients with metal allergy. Knee. 2020;27:1519–24.
20. Postler A, et al. Similar outcome during short-term follow-up after coated and uncoated total knee arthroplasty: a randomized controlled study. Knee Surg Sports Traumatol Arthrosc. 2018;26:3459–67.

Part XVI

Registry

88 Data from the Japan Arthroplasty Register and JOANR

Shinichi Kuriyama

1 About Arthroplasty Registry

Joint arthroplasty is an established procedure for treating advanced joint destruction in the upper and lower limbs. The number of arthroplasty surgeries continues to increase in aging societies. However, as the number of primary surgeries increases, so does the number of revision surgeries owing to implant loosening and poor postoperative patient satisfaction. Moreover, revision surgeries are technically more difficult than primary surgeries, resulting in low long-term durability, a greater burden on patients, and the medical economy. Therefore, it is necessary to investigate and compare the prognoses of different implant designs and surgical procedures to identify the reasons for revision surgery. However, the number of cases and observation periods are limited in small-scale studies conducted at each institution. Since it is essential to conduct large, longitudinal, and cross-sectional surveys in each country to obtain accurate and useful information that considers racial differences and healthcare systems, an arthroplasty registry system was established.

2 History of the International Arthroplasty Registry

In 1975, the first arthroplasty registry was established in Sweden [1, 2]. Early detection of failures and prompt feedback through longitudinal surveys significantly reduced the patient and economic burdens of revision surgery in Sweden. Trends in implant and surgical procedure selection in Sweden were also identified. For example, cruciate-retaining total knee arthroplasty (TKA) and patellar resurfacing were selected in approximately 90% and 3% of the cases, respectively. Detailed annual reports are available on the website [3]. The success of this registry led to its establishment in other Nordic and European countries, including Finland in 1980, Norway in 1987, Denmark in 1995, and the United Kingdom in 2003 [4, 5].

In Australia, data collection began somewhat later in 1999; however, their number of registrations is now the largest in the world [6]. Patients aged under 55 years have higher revision surgery rates than those aged over 55. The revision rate without patellar resurfacing was higher than that with patellar resurfacing, whereas the patellar resurfacing rate in primary TKA was over 50%. In addition, polyethylene with cross-links exhibited less wear than non-cross-linked polyethylene. In the United States, where more than 1 million arthroplasties are performed annually, the American Academy of Orthopedic Surgeons initiated a registry in 2010 [7]. Although many arthroplasty registries were established worldwide, no registry existed in Asian countries until the Japan Arthroplasty Register was established.

3 Changes in the Japan Arthroplasty Register

In 2006, the Japanese Orthopedic Association launched a pilot project for the Japanese Arthroplasty Register, and 10 institutions (9 universities and 1 hospital) started the trial. In 2011, the Secretariat was transferred from the Japanese Orthopedic Association to the Japanese Society for Replacement Arthroplasty. First, total hip arthroplasty, TKA, and unicompartmental knee arthroplasty (UKA) excluding the bipolar head arthroplasty were registered. Akiyama et al. published the results of a trial in the Journal of Orthopaedic Science in 2012 [8], including details of the registration, based on data from approximately 30,000 records obtained from 83 hospitals. Based on the trial results, the first Japanese Arthroplasty register in Asia opened to public in 2012. Subsequently, patellofemoral arthroplasty (PFA) and reverse and anatomical shoulder arthroplasty were included in the registry.

S. Kuriyama (✉)
Department of Orthopaedic Surgery, Graduate School of Medicine, Kyoto University, Kyoto, Japan
e-mail: kuriyama@kuhp.kyoto-u.ac.jp

S. Matsuda (ed.), *Advances in Total Knee Arthroplasty*, https://doi.org/10.1007/978-981-97-4920-1_88

For management at each hospital, a consent form for the Japanese Arthroplasty Register was first obtained from the patients. Patient data, excluding personal information, diagnosis, date of surgery, and 10 items related to surgery, such as surgical approach and use of a computer navigation system, were collected after surgery. Finally, a barcode label for each implant used was attached to the following page. The format is so simple that the data can be completed within a few minutes after the surgery, and the data are sent to the Japan Arthroplasty Register. Since 2014, the University Hospital Medical Information Network (UMIN) has enabled the registration of data other than barcode labels on its website, and 424 facilities have participated in the UMIN. Detailed aggregate reports are available on the website http://jsra.info/jar-report.html. The Japanese Arthroplasty Register were transferred to the Japanese Orthopedic Association National Registry (JONAR), which was established in April 2020 with 1823 participating institutions (135 universities and 1688 hospitals). As of April 2020, approximately 160,000 cases had been registered.

4 Aggregate Results

Between April 2017 and March 2018, 34,992 initial TKA/UKA/PFA surgeries and 878 revision surgeries were performed in 7291 men, 28,536 women, and 43 unknown cases. The age distribution of the patients was as follows: the most common age group was the 70 s with approximately 16,000 cases, followed by the 80 s with approximately 10,000 cases and the 60 s with approximately 6000 cases. Approximately 90%, 5%, and 4% of the patients underwent primary surgery for knee osteoarthritis (OA), rheumatoid arthritis (RA), and osteonecrosis, respectively, whereas approximately 77%, 11%, and 3% underwent revision surgery for knee OA, RA, and osteonecrosis, respectively. Patellar resurfacing versus no patellar resurfacing was performed in approximately 51% versus 49% of patients, whereas arthroplasty with versus without a computer navigation system was performed in 17% versus 83% of patients, and the surgical approach with versus without minimally invasive surgery was performed in 25% versus 75% of patients.

The most common causes of revision surgery were infection in 264 cases, followed by tibial component loosening in 239 cases, femoral component loosening in 164 cases, joint instability in 112 cases, tibial polyethylene wear in 73 cases, trauma in 28 cases, and knee range of motion restriction in 20 cases. Cumulative TKA revision from 2013 to 2017 was significantly higher in men than in women, and early component loosening was more common in patients aged 50 years or younger but gradually increased in patients aged 65–75 years. There was no difference in the revision rates between knee OA and RA; however, cemented fixation had significantly higher revision rates than uncemented fixation. In addition, revision surgery was more common with minimally invasive surgery than without one, whereas there was no difference in the revision rate between patients with or without patellar resurfacing or with or without a navigation system.

5 Problems with the Japanese Arthroplasty Register

The main problem is that approximately two-thirds of all arthroplasty cases in Japan remain unregistered. However, the number of registrations has increased since the Japan Medical Specialty Board's training program began. Second, there is a time lag between the surgery and sending of the registration form because of the paper-based registration. In addition, the reporting of aggregate data is delayed by 2–3 years owing to the need for manual registration by the Secretariat. Finally, the data uploaded on the web can be used after obtaining permission from the Secretariat to reproduce; however, detailed data cannot be used in principle.

6 JOANR

On April 1, 2020, JOANR started a web-based registration system (https://www.joanr.org) to build evidence based on big data for the surgical treatment of musculoskeletal disorders. The JOANR is a comprehensive registration system for almost all surgical procedures, and more details must be registered for joint arthroplasty and arthroscopic surgery. The arthroplasty registration form of JONAR follows that of the Japanese Arthroplasty Register, and the only additional item is whether robotics are used for TKA. Currently, 2029 facilities are registered, and it is expected that the problems in the Japanese Arthroplasty Register will be solved. The Japanese Arthroplasty Register Office will continue its work and will input data that have not yet been entered into the registry as well as extract and analyze prosthesis data in collaboration with the JONAR.

References

1. Knutson K, et al. The Swedish knee arthroplasty register: a nationwide study of 30,003 knees 1976-1992. Acta Orthop Scand. 1994;65:375–86.
2. Robertsson O, et al. Validation of the Swedish knee arthroplasty register: a postal survey regarding 30,376 knees operated on between 1975 and 1995. Orthop Scand. 1999;70:467–72.
3. SKAR (Swedish Knee Arthroplasty Register). Annual report 2020.
4. Robertsson O, et al. Knee arthroplasty in Denmark, Norway and Sweden: a pilot study from the Nordic arthroplasty register association. Acta Orthop. 2010;81:82–9.
5. NJR (National Joint Registry). 18th Annual Report. 2021.
6. AOANJRR (Australian Orthopedic Association National Joint Replacement Registry). Annual Report. 2021.
7. Etkin CD, et al. The American joint replacement registry-the first 5 years. Arthroplast Today. 2017;3:67–9.
8. Akiyama H, et al. A pilot project for the Japan arthroplasty register. J Orthop Sci. 2012;17:358–69.

Correction to: Advances in Total Knee Arthroplasty

Shuichi Matsuda

Correction to:
S. Matsuda (ed.), *Advances in Total Knee Arthroplasty*,
https://doi.org/10.1007/978-981-97-4920-1

The original versions of these chapters were published with incorrect legends in Figure 52.6 and Table 83.1. They have been corrected and updated in the respective chapters.

The updated versions of these chapters can be found at
https://doi.org/10.1007/978-981-97-4920-1_52
https://doi.org/10.1007/978-981-97-4920-1_83

S. Matsuda (ed.), *Advances in Total Knee Arthroplasty*, https://doi.org/10.1007/978-981-97-4920-1_89

MIX
Papier aus verantwortungsvollen Quellen
Paper from responsible sources
FSC® C105338

If you have any concerns about our products,
you can contact us on
ProductSafety@springernature.com

In case Publisher is established outside the EU,
the EU authorized representative is:
Springer Nature Customer Service Center GmbH
Europaplatz 3, 69115 Heidelberg, Germany

Printed by Libri Plureos GmbH
in Hamburg, Germany